W9-CSK-365

ORGANIC CHEMISTRY
An Introduction

ORGANIC CHEMISTRY

AN INTRODUCTION

Jack E. Fernandez
Department of Chemistry
University of South Florida
Tampa, Florida

Prentice-Hall, Inc.,
Englewood Cliffs, New Jersey 07632

Library of Congress Cataloging in Publication Data

FERNANDEZ, JACK E (date)
Organic chemistry.

Includes index.
1. Chemistry, Organic. I. Title.
QD251.2.F47 547 80-39715
ISBN 0-13-640417-0

ORGANIC CHEMISTRY: An Introduction
Jack E. Fernandez

Editorial/production supervision: Eleanor Henshaw Hiatt
Interior design: Eleanor Henshaw Hiatt
Chapter openings and cover design: Lee Cohen
Manufacturing buyer: John Hall
Cover photograph: A common polymer fabric under magnification
(Photo courtesy National Bureau of Standards, U.S. Department of Commerce)

Printed in the United States of America
10 9 8 7 6 5 4 3 2 1

Prentice-Hall International, Inc., *London*
Prentice-Hall of Australia Pty. Limited, *Sydney*
Prentice-Hall of Canada, Ltd., *Toronto*
Prentice-Hall of India Private Limited, *New Delhi*
Prentice-Hall of Japan, Inc., *Tokyo*
Prentice-Hall of Southeast Asia Pte. Ltd., *Singapore*
Whitehall Books Limited, *Wellington, New Zealand*

TO SYLVIA

CONTENTS

PREFACE

In this text I have tried to do two things; present organic chemistry briefly, but thoroughly, and in the simplest, clearest possible style and format.

This book is intended for a one-semester or two-quarter course in organic chemistry for nonchemistry majors. Such courses typically serve students in health-related sciences: ecology, biology, agriculture, certain medical sciences, and environmental sciences. These students do not need the long exposure encountered in the typical one-year course for chemistry majors.

There are at least two ways to approach a brief course in organic chemistry: one is to water down a complete, traditional one-year course; another is to emphasize fewer areas and treat them thoroughly. I have pursued the latter course.

Topics of importance to students in health-related sciences are nomenclature, concepts of structure that relate to properties, and chemical and physical behavior of the important functional groups. I have tried to develop these areas thoroughly, and, where appropriate, to demonstrate biologically important applications. I have used reaction mechanisms as a way to understand reactions rather than as ends in themselves and have given little attention to multistep synthesis.

Some of the features designed to help the student are listed below.

PROBLEMS

This textbook has over 600 problems. Nearly every section contains problems that reinforce ideas in that section. This approach has been quite successful in the classroom. Additional problems at the end of each chapter reinforce important concepts. All the problems have been tested in class.

SECTIONS

Most sections are limited to a single concept or idea. Organic chemistry is difficult enough; it should be organized in order to make it as understandable and useful as possible. Language and terminology are important, too, especially in a text for nonchemistry majors. These students will not go on in chemistry, so they do not need all its technical jargon. I have introduced new terms and new usages of common ones with caution and careful definition. I have striven to use the clearest and simplest possible writing style so that the student does not have to fight both chemistry and language.

NEW TERMS

New terms are defined as they arise. In addition, they are collected together at the end of each chapter with definitions and section references.

GENERAL REACTIONS

Important general reactions are set apart so that the student can review them easily and quickly. Rather than collect and summarize general reactions in one place in the chapter, I have chosen to call attention to these general key reactions as they appear in the chapter. The student may thus review them in the context of examples, conditions, and limitations, rather than merely memorize an abstract list of equations.

ELECTIVE TOPICS

Although some interesting topics are only peripherally important to a short organic course, others may be regarded as somewhat difficult. Such topics are identified in this text as "Elective Topics." Each instructor may decide whether or not to assign an Elective Topic. None of the Elective Topics is essential to an understanding of the material in the rest of the book, but are intended to enrich and expand the student's understanding.

The choice of two elective topics needs clarification: Chapter 9 includes a discussion of nucleophilic substitution (S_N1 and S_N2) and elimination (E1 and E2) mechanisms as elective topics. Although these mechanisms are mentioned occasionally in Chapters 10 and 11, they are not essential to an understanding of these chapters. The fundamental features of these reactions are summarized briefly in Section 10.9.

ANSWERS TO PROBLEMS

Answers are given at the end of the text for all the In-Chapter Problems and some of the Additional Problems. Answers are not given for Additional Problems whose number or letter designations are enclosed within brackets. The instructor thus has the option of assigning either answered or unanswered problems. Complete solutions to all problems are given in a separate Solutions Manual.

FORMULAS

Several kinds of formulas help students make the transition to the three-dimensional world of organic molecules. I have used, especially in the early chapters, drawings of ball-and-stick models, circle-and-line (Alexander) formulas, line-dash-wedge formulas, and Newman projection formulas. The aim is to help the student learn to translate two-dimensional formulas into three-dimensional mental images as soon as possible.

EXAMPLES

The text includes worked-out examples of new key concepts such as systematic naming of compounds, drawing of structural formulas, and calculation of formal charge.

My pedagogical plan has been to introduce major ideas early, and then to repeat and expand them progressively with each new functional group. Some of these ideas are described below.

INTRODUCTORY CHAPTERS

The first three chapters review principles normally presented in general chemistry. Chapter 1 reviews principles of bonding, Chapter 2 reviews structural principles, and Chapter 3 reviews reactivity principles including acid-base theory, equilibrium constants, reaction rates, and reactive intermediates. Well-prepared students may delete these chapters or read them independently.

FUNCTIONAL GROUPS

The important functional groups and their structural interrelationships are summarized in Chapter 2. The aim is to make the student aware of the building process that will occur in the chapters that follow.

ISOMERISM

Isomerism is introduced early (Chapter 2) and developed gradually throughout the first seven chapters. In Chapter 8, all these ideas are coordinated into a fully developed discussion of stereochemistry.

ANALYSIS

Detection and analysis of the important functional groups and compound classes occur in a progressive manner. Thus with each new functional group, problems include compounds from all the classes studied to that point. In this way, students can review and synthesize material learned earlier. Analysis is important not only in itself but also as a means of reinforcing the chemistry of the functional groups learned throughout the course.

THEORY OF REACTIVITY

This subject is presented at a simple and intuitive level as an aid to understanding basic principles. Brønsted–Lowry acid-base theory serves as a vehicle for understanding equilibrium processes. Rate-controlled reactions are understood qualitatively through the use of reaction rate diagrams. The differences between rate and equilibrium are best tied to the differences between energy of activation and free energy of reaction.

BIOLOGICAL CHEMISTRY

The application of organic chemistry to biological systems has grown rapidly in recent years. This fact, coupled with the interests of students who will use this book, requires the use of biological examples where possible. I have employed biological examples throughout the text in discussions of functional groups rather than relegating them to a separate chapter at the end of the book.

ACKNOWLEDGMENTS

Many people have helped me during the writing of this book. At the University of South Florida, I am indebted especially to my long-time colleague and friend Bob Whitaker for his perceptive suggestions and moral support, George Wenzinger who read the entire manuscript early in its development, George Jurch and Graham Solomons who read parts of the manuscript, Jan Tsokos who read the entire manuscript and suggested many biological applications, and Jeff Davis who helped this organic chemist through a few physical chemistry problems.

I also thank those patient and perceptive faculty members across the country who reviewed the manuscript in its various stages of preparation: Dr. Paul Jones of the University of New Hampshire, Dr. Walter Trahanovsky of Iowa State University, Dr. James G. Traynham of Louisiana State University, Dr. Henry Zimmerman of New York City Community College, Dr. Charles B. Rose of the University of Nevada, Dr. Kenneth L. Marsi of California State University, and Dr. Irving Lillien of Miami-Dade Community College. Their generous advice and suggestions often put me back on a sane track.

The editorial staff of Prentice-Hall, Inc. have been helpful and supportive throughout the preparation of this book. I especially thank Bill Gibson, Fred Henry, Betsy Perry, and Eleanor Hiatt.

I wish to thank the MacMillan Publishing Company for permission to quote parts of Chapter 7 of my text, *Modern Chemical Science* (New York, 1971).

Finally, I thank the students that I have taught. I hope I have taught them as much as they have taught me.

J. E. F.

SUGGESTIONS TO THE STUDENTS

The study of organic chemistry should present a challenging and interesting examination of nature. To be sure, a knowledge of organic chemistry sets the stage for understanding the life sciences including such areas as nutrition, drugs, medicinals, the environment, and molecular genetics. But organic chemistry also opens another door, one that is often neglected: Organic chemistry involves the use of a kind of logic that is unique among the sciences. This logic is qualitative but allows us to comprehend a vast part of nature with a minimal investment.

I believe that to encounter this kind of logic is to open the door to a new world of order and beauty filled with lovely and complex structures and dynamic molecular movements. There is much in organic chemistry that is pure art. I leave it to you to seek out this dimension; it will brighten your tour.

How should you approach a course in organic chemistry? There is, without doubt, one best way—problem solving. For this reason I have included many problems throughout the text, both within each chapter and at the end of each chapter. I suggest that you read the material assigned and solve as many of the in-chapter problems as you can as you come to them *before you come to class*. After class you should, as soon as possible, go back over the material—especially the problems that gave you trouble. After this, you should work the end-of-chapter problems to reinforce your study and class work. Organic chemistry is cumulative. Unlike some subjects, each chapter relies heavily on knowledge gained in earlier chapters. In other words, if you fail to grasp the material in a given chapter, that deficiency will haunt you through every succeeding chapter. Many students realize this too late and find that they cannot catch up. There is no substitute for daily study.

J. E. F.

PRINCIPLES OF BONDING

1.1 ORGANIC CHEMISTRY

Organic chemistry is the chemistry of carbon compounds. Why does the single element, carbon, deserve such special attention? Why isn't there a separate course for the chemistry of silicon, nitrogen, or oxygen? The answers to these questions lie in the one property of carbon that is unique to carbon: the ability to form strong covalent bonds with other carbon atoms. These strong covalent bonds allow carbon to form molecules that contain any number of carbon atoms, from methane with its single carbon to polyethylene with thousands of carbon atoms per molecule. The number of different possible carbon compounds is staggering!

Compounds of carbon always contain other elements as well. In nearly all its compounds carbon is bonded to hydrogen. In fact, a more accurate definition of organic chemistry would be the chemistry of the hydrocarbons and compounds derived from them. (**Hydrocarbons** are compounds that contain only the elements carbon and hydrogen.) Carbon compounds often contain other elements as well. The most common ones are oxygen, nitrogen, the halogens, sulfur, and phosphorus.

When we realize that there are many, many organic compounds—actually two to three million are known—how can we possibly hope to understand the names, properties, reactions, and methods of synthesis of more than a few of them? The beauty of organic chemistry lies in the answer to this question. Nearly all organic compounds contain carbon and hydrogen—a hydrocarbon part. This part of the molecule behaves pretty much the same no matter how many carbon atoms it might contain. Also, as we shall see, the other atoms are always found in one of only about a dozen combinations. For example, ethyl alcohol and butyl alcohol both have the same two groupings of atoms:

```
   H  H               H  H  H  H
   |  |               |  |  |  |
H—C—C—O—H          H—C—C—C—C—O—H
   |  |               |  |  |  |
   H  H               H  H  H  H

ethyl alcohol       butyl alcohol
```

the hydrocarbon part (shaded area) and the OH part (Figure 1-1). The hydrocarbon parts exhibit nearly the same chemical properties in both compounds. The OH part also exhibits the same chemical properties in both compounds. Therefore, ethyl alcohol and butyl alcohol have similar physical and chemical properties. Their names suggest the similarity.

An atom or grouping of atoms that is found commonly and whose chemical behavior is independent of the rest of the molecule is called a **functional group**. Because there are only about a dozen or so functional groups in organic chemistry we can begin to see light in the forest: By learning the chemistry of a few functional groups, we shall have learned the chemistry of thousands of organic compounds.

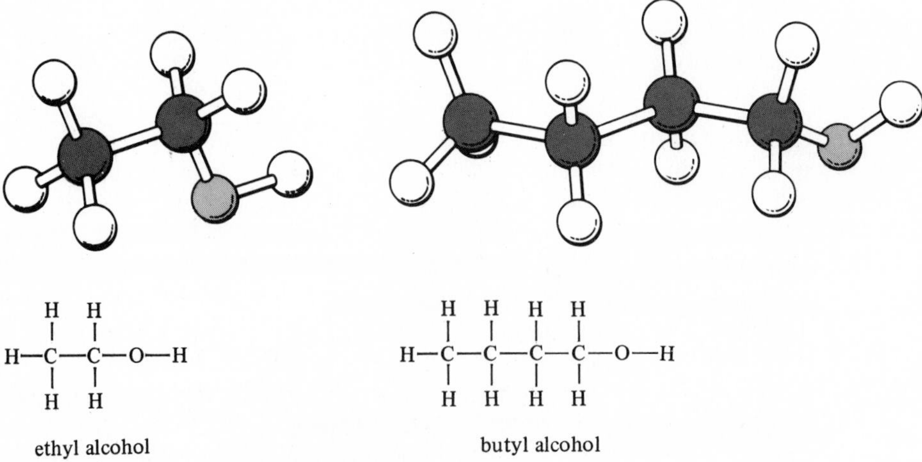

Figure 1.1. Ball-and-stick models of ethyl alcohol and butyl alcohol.

Before we can get into the study of the functional groups, however, we shall have to learn a few principles of structure, and, in the following chapter, some principles of reactivity.

Problem 1.1
What are two characteristics of a functional group?

1.2 IONIC BONDING

Both ionic and covalent bonding were explained in 1916 by W. Kossel and G. N. Lewis, respectively. Both explanations are based on the extreme stability that atoms attain when they have a complete outer shell of eight electrons—the **octet rule**. Hydrogen is an exception to this rule because only two electrons are allowed in its **valence shell**.

In **ionic bonding** atoms acquire a stable octet of valence electrons by the *transfer* of electrons from one atom to another. The formation of sodium chloride is shown as

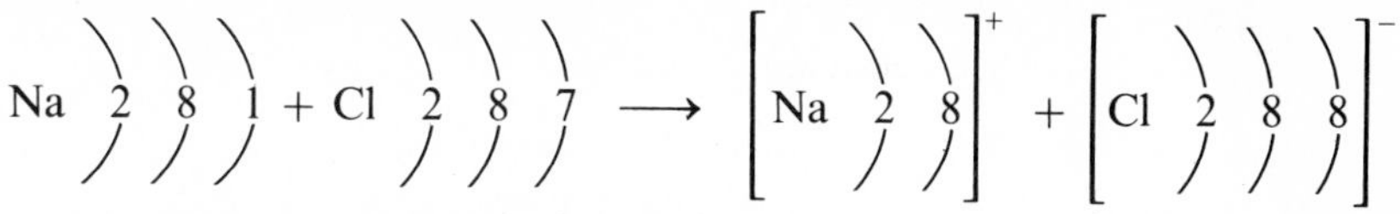

$$\mathrm{Na}\ 2\ 8\ 1 + \mathrm{Cl}\ 2\ 8\ 7 \longrightarrow [\mathrm{Na}\ 2\ 8]^{+} + [\mathrm{Cl}\ 2\ 8\ 8]^{-}$$

where the numbers represent the number of electrons in each shell. We know that sodium and chlorine have one and seven valence shell electrons, respectively, because Na is in group I and Cl is in group VII of the periodic table: The group number of an atom equals its number of valence electrons.

Problem 1.2
Give the number of valence electrons in each of the following atoms.

(a) Li,	(b) B,	(c) C,	(d) O,
(e) N,	(f) Ca,	(g) K,	(h) I

Problem 1.3
Give the electronic configurations of the atoms and ions involved in the following electron transfer reactions.

(a) $\mathrm{Li} + \mathrm{Cl} \longrightarrow$ (b) $\mathrm{Mg} + \mathrm{Br} \longrightarrow$

1.3 ELECTRONEGATIVITY

Ionic bonds are produced when the transfer of one, two, or at most three electrons results in complete octets for both atoms. But such transfer can occur only if the two atoms differ greatly in **electronegativity** (the power to attract electrons). The alkali metals have the lowest electronegativities, and the halogens have the highest. Electronegativity generally increases on moving from left to right in any row of the periodic table and on moving up in any vertical group. Cesium is therefore the least electronegative element (francium has not been studied extensively), and fluorine is the most electronegative.

Problem 1.4

Consult a periodic table to predict which member of each of the following pairs of elements has the higher electronegativity.

(a) Na and K, (b) C and N, (c) N and P, (d) F and Br,
(e) S and Cl, (f) H and Na

1.4 COVALENT BONDING

There is no great tendency for electrons to transfer between atoms that have similar electronegativities. These elements, therefore, do not form ionic bonds. Ionic bonding is also unlikely with atoms whose valence shells contain far from a complete octet. Carbon, for example, with four valence electrons, would either have to gain four electrons to form a C^{4-} ion or lose four electrons to form a C^{4+} ion. Both of these ions are too highly charged to be stable. Sharing electrons between the two atoms solves the problem of completing the octet without producing highly charged and thus unstable ions. The chlorine molecule is an example of bonding between atoms that have identical electronegativities:

$$:\underset{..}{\ddot{Cl}}\cdot + \cdot\underset{..}{\ddot{Cl}}: \longrightarrow :\underset{..}{\ddot{Cl}}:\underset{..}{\ddot{Cl}}:$$

Each chlorine atom completes its octet by sharing one electron with another chlorine atom. The bond formed in this way is called a **covalent bond**. The notation used above to illustrate it was introduced by G. N. Lewis and is known as the **Lewis dot structure**. In the Lewis notation, the symbol of the element, Cl, stands for the nucleus and all electrons except the valence (outer shell) electrons. The dots represent valence shell electrons.

Of more interest to us in organic chemistry is the bonding of carbon atoms to each other and to hydrogen. Examples of these bonds are shown by the Lewis dot structures for the formation of methane (CH_4) and ethane (C_2H_6):

$$\cdot\underset{\cdot}{\dot{C}}\cdot + 4H\cdot \longrightarrow \begin{array}{c} H \\ H:\underset{..}{\ddot{C}}:H \\ H \end{array}$$

$$\cdot\underset{\cdot}{\dot{C}}\cdot + \cdot\underset{\cdot}{\dot{C}}\cdot + 6H\cdot \longrightarrow \begin{array}{c} H\;H \\ H:\underset{..}{\ddot{C}}:\underset{..}{\ddot{C}}:H \\ H\;H \end{array}$$

This simple theory of covalent bonding not only explains the stability of the bonds but also allows us to determine how many bonds will be formed; that is, we construct enough electron pairs to give each atom (except hydrogen) a complete octet. The reaction of one oxygen atom and two hydrogen atoms thus results in H_2O:

$$2H\cdot + \cdot\underset{..}{\ddot{O}}\cdot \longrightarrow H:\underset{..}{\ddot{O}}:H$$

One less hydrogen atom would leave oxygen with only seven valence electrons; one more hydrogen atom would not find an unpaired electron on oxygen with which to form an electron pair.

Problem 1.5
Of the following pairs of elements, which will form covalent bonds?

(a) C and C, (b) Na and F, (c) Cl and F, (d) C and O,
(e) C and N

Problem 1.6
Draw the Lewis dot structures of these molecules.

(a) HCl, (b) H_2, (c) H_2S, (d) F_2

1.5 DOUBLE AND TRIPLE COVALENT BONDS

More than one pair of electrons can be shared between two atoms. The formula of carbon dioxide, CO_2, for example, requires that each atom furnish two electrons to form a **double bond**:

$$:\ddot{\underset{\cdot}{O}}\cdot + \cdot\dot{\underset{\cdot}{C}}\cdot + \cdot\ddot{\underset{\cdot}{O}}: \longrightarrow :\ddot{O}::C::\ddot{O}:$$

Triple bonds exist between atoms that share three pairs of electrons. Hydrogen cyanide, HCN, for example, has the Lewis dot structure

$$H\cdot + \cdot\dot{\underset{\cdot}{C}}\cdot + \cdot\dot{\underset{\cdot}{N}}: \longrightarrow H{:}C\vdots\vdots N{:}$$

Problem 1.7
Draw the Lewis dot structures of these molecules.

(a) NO (with a double bond), (b) HCCH

1.6 COORDINATE COVALENT BONDS

A question immediately arises: If a hydrogen ion, H^+, were available, could it react with a water molecule? After all, the hydrogen ion has no electrons to contribute, but it does have an affinity for electrons, and oxygen has electron pairs that are not used in bonding. In other words, could a covalent bond be produced if the oxygen atom of a water molecule furnished *both* electrons and H^+ furnished none?

$$H^+ + H{:}\ddot{\underset{\cdot\cdot}{O}}{:}H \longrightarrow \left[\begin{matrix} H \\ H{:}\ddot{\underset{\cdot\cdot}{O}}{:}H \end{matrix}\right]^+$$

The answer is yes. Such a bond satisfies all the requirements for a covalent bond. To distinguish it from the ordinary covalent bond, we call it a **coordinate covalent bond.** We must bear in mind, however, that coordinate covalent bonds are actually indistinguishable from ordinary covalent bonds. The case just described for water is very common.

In fact, the hydronium ion, H_3O^+, is always produced when an acid dissolves in water.

$$\mathrm{HCl + H_2O \longrightarrow H_3O^+ + Cl^-}$$

Another common example of a coordinate covalent bond occurs in the ammonium ion, $NH_4{}^+$, which is formed in the reaction of ammonia with acids. We construct the Lewis dot structure of ammonia as follows:

$$3\mathrm{H}\cdot + \cdot\underset{\cdot\cdot}{\dot{\mathrm{N}}}\cdot \longrightarrow \mathrm{H{:}\underset{\cdot\cdot}{\overset{\mathrm{H}}{\overset{\cdot\cdot}{N}}}{:}H}$$

The reaction of ammonia with hydrogen ions is given by the equation

$$\mathrm{H^+ + H{:}\underset{\cdot\cdot}{\overset{\mathrm{H}}{\overset{\cdot\cdot}{N}}}{:}H} \longrightarrow \left[\mathrm{H{:}\underset{\mathrm{H}}{\underset{\cdot\cdot}{\overset{\mathrm{H}}{\overset{\cdot\cdot}{N}}}}{:}H}\right]^+$$

Problem 1.8

Draw the Lewis dot structure of the product of the reaction

$$\mathrm{H^+ + H{:}\underset{\mathrm{H}}{\underset{\cdot\cdot}{\overset{\mathrm{H}}{\overset{\cdot\cdot}{C}}}}{:}\underset{\cdot\cdot}{\overset{\cdot\cdot}{O}}{:}\underset{\mathrm{H}}{\underset{\cdot\cdot}{\overset{\mathrm{H}}{\overset{\cdot\cdot}{C}}}}{:}H} \longrightarrow$$

Is the new bond an ordinary covalent bond or a coordinate covalent bond?

1.7 POLAR COVALENT BONDS

Covalent bonds are usually formed between atoms of similar electronegativity. But hydrogen and oxygen differ considerably in electronegativity. What effect does the rather large electronegativity difference have on the covalent bond between oxygen and hydrogen? The answer to this question is that such bonds are polarized; that is, the electron pair lies, on the average, closer to the oxygen atom (the more electronegative element) than to the hydrogen atom. We show it as

$$\begin{array}{ccc} & \overset{\delta-}{\underset{\cdot\cdot}{\overset{\cdot\cdot}{\mathrm{O}}}} & \\ \overset{\delta+}{\mathrm{H}} & & \overset{\delta+}{\mathrm{H}} \end{array}$$

where $\delta+$ above the H means that there is a fractional positive charge on the hydrogen atom. The $\delta-$ above the O means that the oxygen has a greater than equal share of the bonding electrons, and therefore bears a fractional negative charge. Such bonds are called **polar covalent bonds**. Polar covalent bonds cause molecules to attract each other (Section 1.8).

Problem 1.9

Draw the Lewis dot structure of each of the following molecules and designate the polar covalent bonds by placing a $\delta+$ or a $\delta-$ over the appropriate atoms.

(a) HF, (b) HCl, (c) FOF, (d) HOOH
(e) BrI

1.8 HYDROGEN BONDING

In water, the positively charged hydrogen atoms are attracted to the oxygen atoms of other water molecules. Such attractions are often fairly strong and are called **hydrogen bonds** (shown by the dotted line):

$$\begin{array}{cc} :\ddot{\text{O}}:\text{H}\cdots & :\ddot{\text{O}}:\text{H} \\ \ddot{\text{H}} & \ddot{\text{H}} \end{array}$$

Hydrogen bonding is the most important type of intermolecular attraction. It is important, however, only when hydrogen is bonded to the very electronegative elements O, F, or N. These are the only elements electronegative enough to form polar bonds that leave a positive charge on hydrogen. We could ask why only hydrogen forms hydrogen bonds. Why not lithium, or some of the other elements that are even less electronegative than hydrogen? The answer is that hydrogen is the only element that has no inner shell electrons. All its electrons are involved in bond formation. When that bond is highly polarized, the relatively unprotected hydrogen nucleus can become embedded in the electron shell of another electronegative atom (O, N, or F). All other atoms besides hydrogen have inner shell electrons that screen the positive nucleus.

Hydrogen bonding, then, occurs between a hydrogen atom and an electronegative atom (especially O, N, or F) when the hydrogen is covalently bonded to another such atom (O, N, or F). Thus hydrogen bonds are important when they are of the type

$$\text{X—H}\cdots\text{Y}$$

where both X and Y are highly electronegative, and especially when they are O, N, or F.

Hydrogen bonding is a strong intermolecular attractive force between molecules, equal to about 10% of the strength of an ordinary covalent bond. Hydrogen bonding therefore affects many physical properties. The boiling point, density, and surface tension of a hydrogen-bonded substance such as water are higher than they would be if there were no hydrogen bonding. Compounds such as the hydrocarbons (Sec. 2.6), which cannot form hydrogen bonds, exhibit very little intermolecular attraction. Because of this low intermolecular attraction in hydrocarbon molecules, these substances have relatively low boiling points and densities.

The boiling point depends on the energy required to cause molecules to "fly away from each other" into the vapor phase. In the absence of hydrogen bonding (for example, with hydrocarbons) the boiling point depends primarily on molecular weight: the higher the molecular weight, the higher the boiling point. In a hydrogen-bonded substance, energy must be supplied to break the hydrogen bonds and cause the molecules to boil away. Thus hydrogen bonding results in higher boiling points.

Hydrogen bonding also explains why compounds that contain oxygen or nitrogen atoms are more water-soluble than other compounds of similar molecular weight. An R:Ö:R molecule has no O:H hydrogens through which to form intermolecular hydrogen bonds; it therefore has a low boiling point. We expect this compound to be soluble in water, however, because it can form hydrogen bonds with water molecules through the water hydrogens:

```
                              H
  ..      ..                  ..      ..
R:O:···H:O:     and       R:N:···H:O:
  ..      ..                  ..      ..
  R       H                   H       H
```

Problem 1.10

Explain the large difference in boiling point between ethyl alcohol (boiling point 78°C) and dimethyl ether (boiling point −24°C) even though they have the same molecular formula, C_2H_6O, and thus the same molecular weight.

```
  H H             H   H
  .. ..           ..   ..
H:C:C:O:H       H:C:O:C:H
  .. ..           ..   ..
  H H             H   H
ethyl alcohol   dimethyl ether
```

Problem 1.11

Which compound in each pair exhibits the stronger hydrogen bonding?

(a) H_2O and H_2S (b) HF and HCl, (c) H_2O and HF

1.9 FORMAL CHARGE

Let us return now to the ions H_3O^+ and NH_4^+. We know that both of these ions possess a resultant positive charge. To which atom can we assign the positive charge? Recall that once H^+ has been added to a water or an ammonia molecule, all the bonds are identical.

A useful bookkeeping device for electrons and charges is called **formal charge.** The formal charge of an atom tells us whether that atom has more or fewer electrons when bonded than it does when it is in the hypothetical free state.

We can easily calculate the formal charge of an atom with the formula

$$\text{Formal charge} = G - N - \tfrac{1}{2}B$$

where G = the group number of the atom in the periodic table (this equals the number of valence electrons in the free atom), N = the number of nonbonded electrons on the atom in the molecule, and B = the total number of bonding electrons on the atom.

The formal charges on nitrogen in NH_3 and $NH_4{}^+$ are thus

$$\mathrm{H}:\ddot{\mathrm{N}}:\mathrm{H} \quad (\text{with } \ddot{\mathrm{H}} \text{ below}) \qquad \text{Formal charge on nitrogen} = 5 - 2 - \tfrac{1}{2}(6) = 0$$

$$\left[\begin{array}{c}\mathrm{H}\\ \mathrm{H}:\ddot{\mathrm{N}}:\mathrm{H}\\ \ddot{\mathrm{H}}\end{array}\right]^{+} \qquad \text{Formal charge on nitrogen} = 5 - 0 - \tfrac{1}{2}(8) = +1$$

We see that nitrogen in NH_3 has eight electrons in its valence shell. To determine formal charge we arbitrarily assign half the bonding electrons to the nitrogen and half to the hydrogen. If we now count the electrons assigned to nitrogen, we count two nonbonding electrons plus one electron in each bond. The total is five. Free nitrogen atoms contain five valence electrons. (The number of valence electrons in the free atom is equal to the group number in the periodic table.) We see by this method that nitrogen has neither lost nor gained electrons, and its formal charge is zero. The formal charge of hydrogen is also zero because free hydrogen atoms contain one electron, and bonded hydrogen atoms possess half an electron pair, or one electron. In $NH_4{}^+$, nitrogen also has eight electrons in its valence shell, but they are all used in bonding. We therefore assign only four electrons to nitrogen. Free nitrogen atoms have five electrons; nitrogen in $NH_4{}^+$ therefore has one electron less than it has in the free state, and its formal charge is $+1$.

Problem 1.12

Calculate the formal charges on the atoms in each of the following structures.

(a) $\mathrm{H}:\ddot{\underset{..}{\mathrm{O}}}:\ddot{\mathrm{N}}::\ddot{\mathrm{O}}:$, (b) $\mathrm{H}:\ddot{\underset{..}{\mathrm{O}}}:\underset{:\underset{..}{\ddot{\mathrm{O}}}:}{\mathrm{N}}::\ddot{\mathrm{O}}:$, (c) $:\mathrm{C}\vdots\vdots\mathrm{O}:$,

(d) $:\dot{\mathrm{N}}::\ddot{\mathrm{O}}:$, (e) $:\dot{\mathrm{N}}:\ddot{\underset{..}{\mathrm{O}}}:$

Note that (d) and (e) are examples of molecules that have more than one reasonable structure. We shall deal later with the problem of the structure of a molecule for which several structures can be written.

1.10 GEOMETRY OF THE COVALENT BOND

By the late nineteenth century, but before the discovery of electrons, organic chemists knew that atoms in molecules are arranged in definite geometric configurations. They knew that the hydrogens in methane lie at the corners of a regular tetrahedron that has carbon at its center. A **tetrahedron** is defined by the four alternating corners of a cube (Fig. 1.2). At that time, chemists did not know what holds these atoms together. They represented the covalent bond simply by a line between the atoms.

The discovery of the electron in 1897 was followed by Lewis's representation of bonds as pairs of electrons in 1916. His dot structures accounted for the tetrahedral geometry of carbon by the repulsion of the electron pairs: The greatest separation of four pairs of electrons occurs when they are directed toward the corners of a regular tetrahedron.

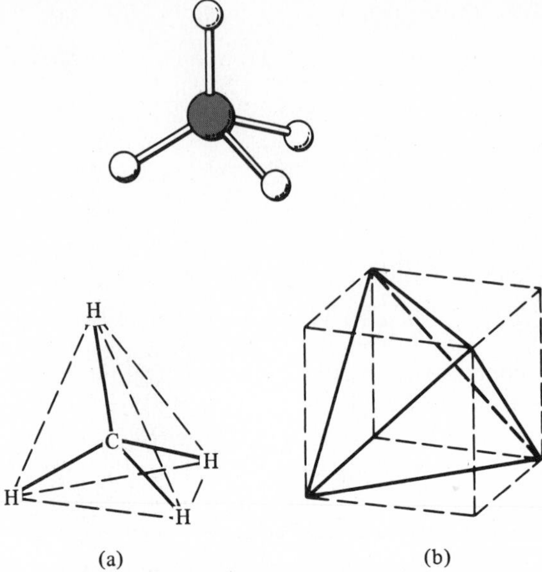

Figure 1.2. The hydrogen atoms in methane are at the corners of a regular tetrahedron (a). A tetrahedron is defined by the four alternating corners of a cube (b).

In 1926, Erwin Schrödinger developed the now accepted quantum mechanical theory of atoms. His theory treats electrons not as particles, but as standing waves. He thus described atomic energy levels as regions in the atom in which an electron can exist as a standing wave. In 1931, Linus Pauling showed how quantum mechanics could be used to explain chemical bonding.

Quantum mechanical theory predicts that an electron in a given energy level, say the valence shell, can occupy one of several **orbitals** *s, p, d*, ... Orbitals are regions in the atom that can contain a maximum of two electrons. Bonding occurs when an orbital of one atom overlaps with an orbital of another atom, and each orbital supplies one electron. The *s* and *p* orbitals for a carbon atom are shown in Figure 1.3. *p* Orbitals are at right angles to one another, and *s* orbitals are spherical.

We can represent the energy levels of orbitals in a free carbon atom as in Figure 1.4(a). According to this arrangement, carbon should be able to form only two covalent bonds because only two orbitals contain one electron each. However, carbon forms CH_4 and not CH_2. We might be tempted to account for the additional bonds by raising one of the 2*s* electrons to the vacant 2*p* orbital as shown in Figure 1.4(b). The energy cost to raise an electron from a 2*s* orbital to a 2*p* orbital would be more than repaid by the formation of two additional bonds. (Recall that bond formation lowers the energy of atoms.) The result is indeed CH_4. Because the *p* orbitals are at right angles to one another, however, the CH_4 molecules constructed from a 2*s* and three 2*p* orbitals should have three equivalent C—H bonds at right angles to one another and a fourth C—H bond equidistant from the other three. These orbitals are obviously not consistent with the observed tetrahedral geometry of carbon compounds.

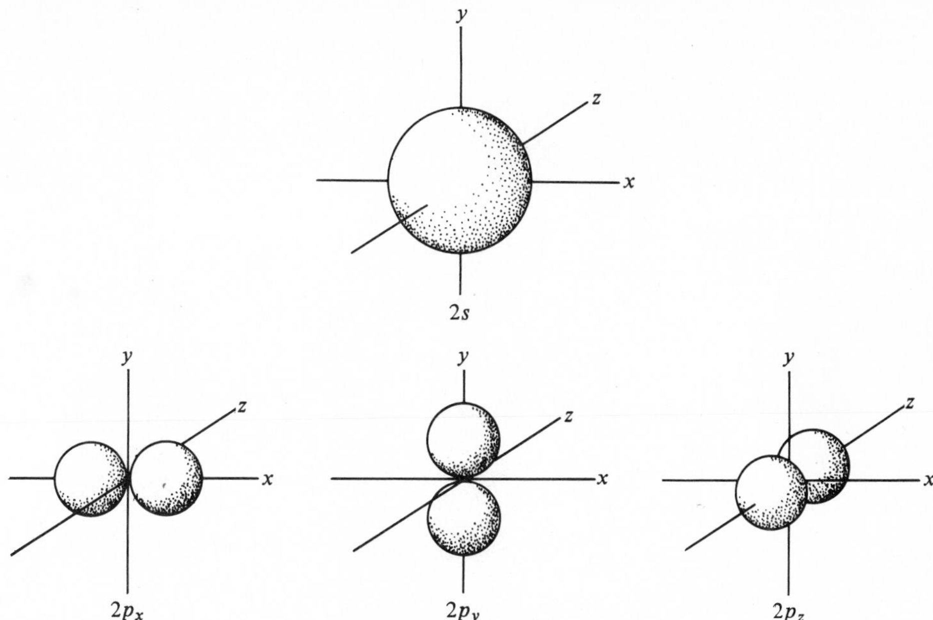

Figure 1.3. The $2s$ and $2p$ orbitals of the valence shell of carbon. The nucleus and inner ($1s$) electrons are at the origin of the axes.

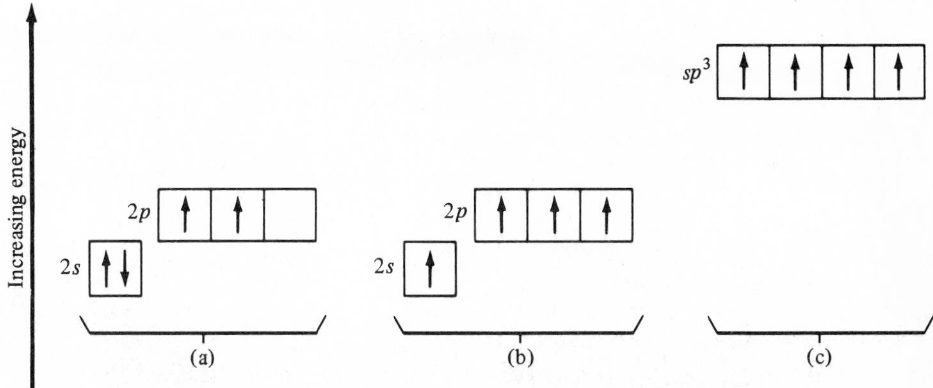

Figure 1.4. Electron energy levels in (a) a free carbon atom, (b) a carbon atom in which one electron has been raised from a $2s$ to a $2p$ orbital, and (c) a carbon atom in which the $2s$ and $2p$ orbitals have been combined (hybridized) to produce four new sp^3 orbitals. Electrons are shown as arrows. An electron pair in one orbital is designated by two arrows ↑↓.

1.11 HYBRID BOND ORBITALS

Pauling solved the problem by showing how the $2s$ and the three $2p$ orbitals could be combined mathematically to arrive at four new orbitals that are larger than s or p orbitals and are directed toward the corners of a regular tetrahedron (Fig. 1.5). He

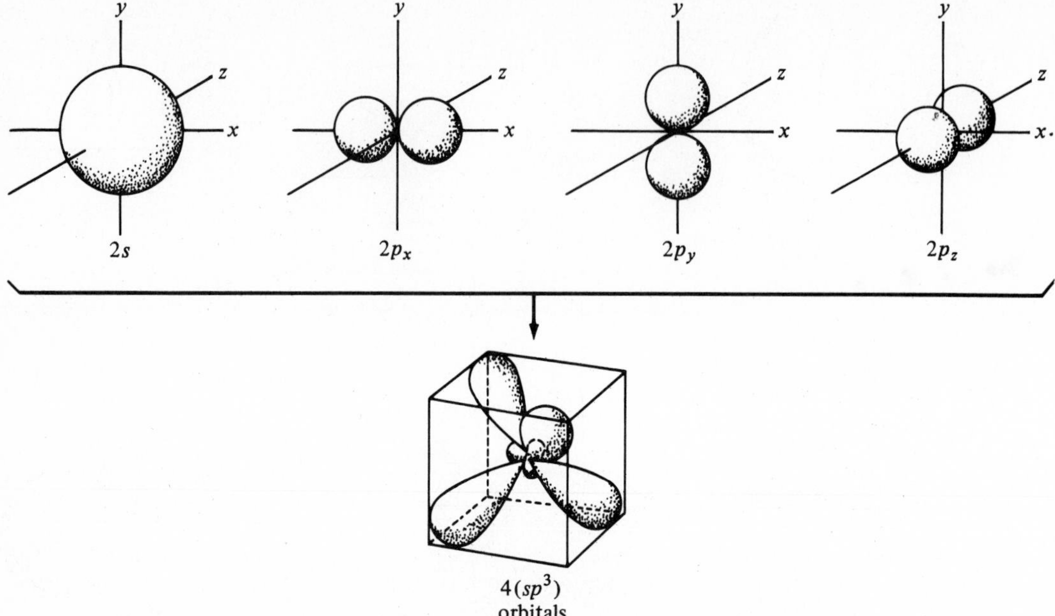

Figure 1.5. The mathematical combination of a 2*s* orbital with three 2*p* orbitals results in four sp^3 orbitals, which are tetrahedrally oriented.

called these new orbitals **hybrid orbitals** and designated them sp^3 to show that they consist of one-fourth *s* character and three-fourths *p* character. Note that *when we hybridize a given number of orbitals we must obtain the same number of hybrid orbitals.*

Problem 1.13

How many hybrid orbitals are formed by the hybridization of the following atomic orbitals? Give their letter designation.

(a) one *s* and three *p* orbitals, (b) one *s* and two *p* orbitals,

(c) one *s* and one *p* orbital

1.12 BOND STRENGTH AND HYBRIDIZATION

Pauling showed that a stable bond forms when an orbital of one atom overlaps with an orbital of another atom (a) if each orbital contains one electron, or (b) if one orbital contains two electrons and the other is empty. The first case, (a), represents an ordinary covalent bond, and the second, (b), represents a coordinate covalent bond (Fig. 1.6).

Pauling stated further that bond strength depends primarily on the degree of overlap. The greater the amount of overlap, the stronger the bond. Because sp^3 hybrid

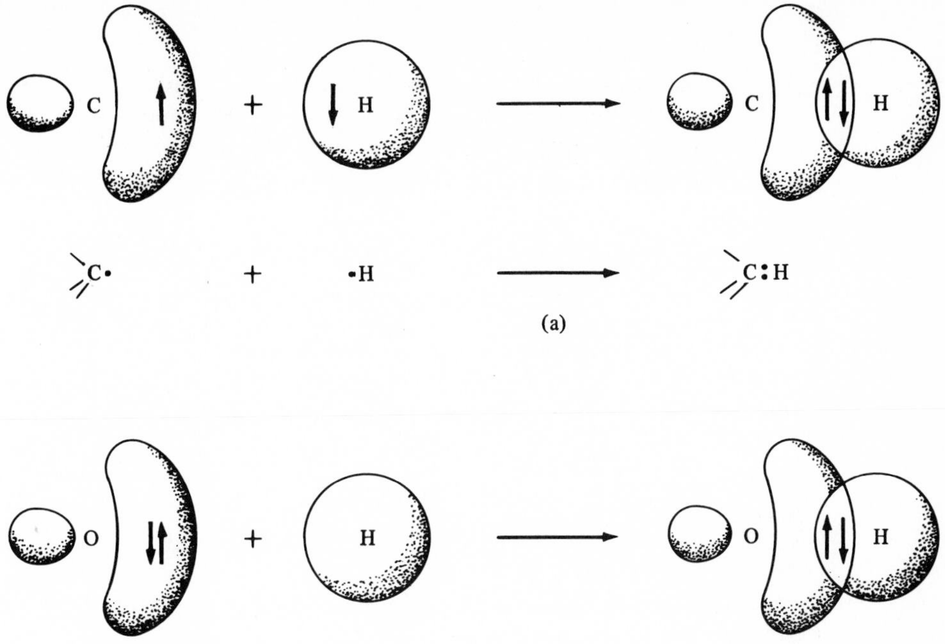

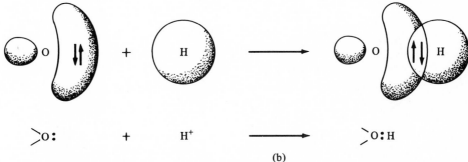

Figure 1.6. (a) Orbital overlap to form a covalent bond. (b) Orbital overlap to form a coordinate covalent bond.

orbitals are larger than *s* or *p* orbitals, they are higher in energy [Fig. 1.4(c)]. The justification for creating higher energy orbitals is that they are able to overlap to a greater extent, and thus to form stronger bonds than pure *s* or *p* orbitals. The overlap of two sp^3 orbitals on different carbons leads to the quite stable carbon-carbon bond (Fig. 1.7).

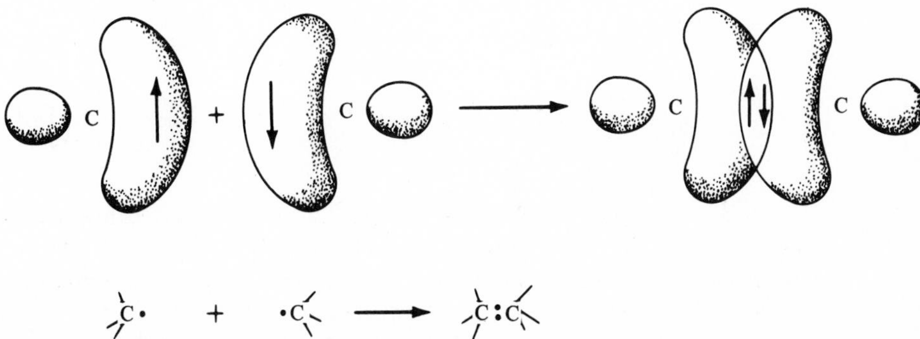

Figure 1.7. Carbon-carbon bond formation by overlap of two sp^3 orbitals.

Other hybridizations are also possible: Three sp^2 hybrid orbitals are formed by combining an s orbital with two p orbitals. Because we use only two p orbitals, say p_x and p_y, our hybrid orbitals all lie in the x–y plane and at angles of 120° to one another (Fig. 1.8). In the same way, we can combine an s orbital with one p orbital to obtain two

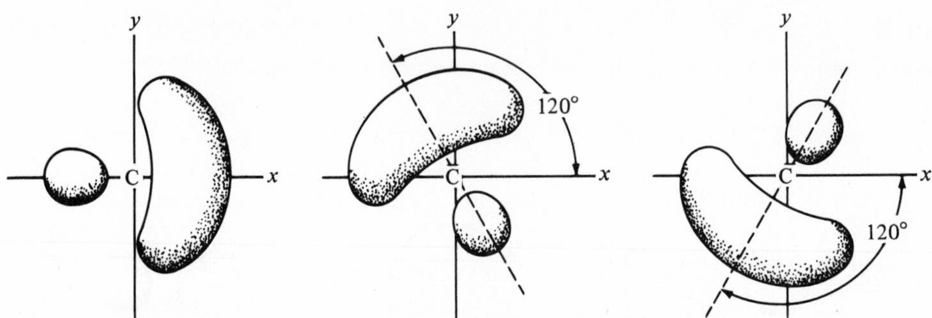

Figure 1.8. Orientation of the three sp^2 hybrid orbitals constructed from an s, a p_x, and a p_y orbital.

sp hybrid orbitals. Here, if we use a p_x orbital, both sp hybrid orbitals will lie along the x axis. They are thus oriented at 180° to one another (Fig. 1.9). We shall have occasion to use these hybrid orbitals later.

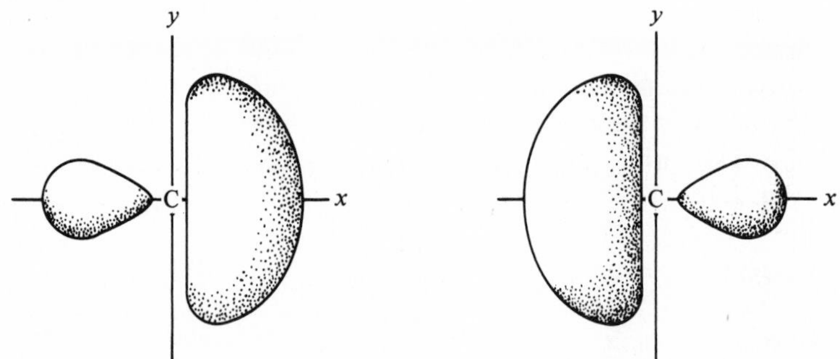

Figure 1.9. Orientation of sp hybrid orbitals constructed from an s and a p_x orbital.

Problem 1.14

Give the geometric orientation of the hybrid orbitals of Problem 1.13 (that is, describe how they are oriented in space).

1.14 PREDICTION OF ORBITAL HYBRIDIZATION

How can we predict which kind of orbital hybridization will occur in a given molecule? This problem is quite simple for compounds that contain only single bonds: All the orbitals that contain valence electrons in an atom are hybridized regardless of whether they are involved in bonding. For example, in water, the oxygen atom has eight electrons:

$$\mathrm{H}:\ddot{\underset{\cdot\cdot}{\mathrm{O}}}: \\ \mathrm{H}$$

All four oxygen orbitals therefore contain electrons, and all four orbitals—$2s$, $2p_x$, $2p_y$, $2p_z$—are hybridized to form four sp^3 orbitals. Two sp^3 orbitals are used in bonding to hydrogen, and each of the other two contains an unshared pair of electrons.

Boron trifluoride, $\mathrm{F}:\overset{\mathrm{F}}{\ddot{\mathrm{B}}}:\mathrm{F}$, uses only three orbitals because the boron atom has no unshared electrons. Since only three orbitals contain electrons—$2s$, $2p_x$, $2p_y$—they hybridize to form three sp^2 orbitals. These orbitals are planar and at 120° to one another.

Similarly, we can predict the situation in beryllium chloride, $BeCl_2$. Beryllium is in group II of the periodic table; thus it has two valence electrons. By sharing each with a chlorine atom (chlorine is in group VII and has seven valence electrons), two bonds are produced: Cl:Be:Cl. Only two beryllium orbitals contain electrons. Therefore a $2s$ and a $2p_x$ orbital are hybridized. The result is two sp hybrid orbitals. Since these orbitals lie along the x axis, they are oriented 180° from one another. The above rules for determining hybridization in singly bonded atoms are summarized in Table 1.1.

Problem 1.15

Make drawings to show why an sp^3–sp^3 bond should be stronger than a bond formed by the overlap of two s orbitals (Remember to keep the nuclei at the same distance in both drawings.)

Table 1.1 Hybridization in Singly Bonded Molecules

Number of Orbitals Involved	Number of Unshared Electron Pairs	Hybridization of Central Atom	Examples
4	0	sp^3	CH_4, $NH_4{}^+$, $BF_4{}^-$
4	1	sp^3	$:NH_3$, $H_3O:^+$
4	2	sp^3	$H_2\ddot{O}:$, $:\ddot{N}H_2{}^-$
3	0	sp^2	BF_3
3	1	sp^2	$:BF_2{}^-$
2	0	sp	$BeCl_2$

Problem 1.16

Draw the Lewis dot structure of each of the following species and predict the hybridization and geometry of the central atom of each. (a) CH_3^+ (a molecule of CH_4 with $:H^-$ removed), (b) CH_3^- (a molecule of CH_4 with H^+ removed).

Problem 1.17

Oxygen has the electronic structure shown. Predict the

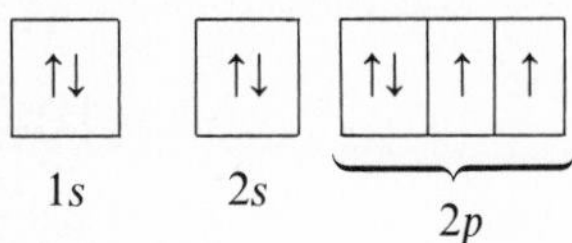

H—O—H bond angle assuming that no hybridization occurs. The actual bond angle (105°) is close to the tetrahedral value of 109.5°. Explain.

NEW TERMS

Coordinate covalent bond A covalent bond in which one atom furnishes both bonding electrons (1.6).

Covalent bond A bond that results when two atoms share a pair of electrons (1.4).

Double bond A covalent bond that consist of two pairs of electrons (1.5).

Electronegativity The power of an atom to attract electrons to itself (1.3).

Formal charge Equals the periodic group number (G) of the atom minus the sum of the number of nonbonded electrons (N) and one-half of the bonding electrons (B) (1.9):

$$\text{Formal charge} = G - (N + \tfrac{1}{2}B)$$

Functional group An atom or grouping of atoms that occurs commonly and confers characteristic chemical properties upon the molecules in which it occurs (1.1).

Hybrid orbitals Orbitals obtained by combination of *s* and *p* atomic orbitals of the same atom. The common hybrid orbitals are sp^3, sp^2, and sp (1.11–1.13).

Hydrocarbon A compound that contains only the elements carbon and hydrogen (1.1).

Hydrogen bond The attractive force between a highly electronegative atom (O, N, or F) and a hydrogen that is bonded to another highly electronegative atom (O, N, or F) (1.8).

Ionic bonding Electrostatic bonding between ions that are formed by the transfer of electrons from one atom to another (1.2).

Lewis dot structure A structural formula in which valence shell electrons including bonding electrons are represented by dots and in which the symbol of the element (C, for example) stands for the nucleus and all electrons except the valence shell electrons (1.4).

Octet rule Atoms of the first period of the periodic table (Li through Ne) can have no more than eight electrons in their valence shells (1.2).

Orbital A region in the atom that can obtain a maximum of two electrons. Orbitals are designated *s*, *p*, and *d* according to their shape (1.10).

Polar covalent bond A covalent bond in which the electron pair lies, on the average, closer to one atom than to the other (1.7).

Tetrahedron A geometric figure defined by the four alternating corners of a cube (Fig. 1.12) (1.10).

Triple bond A covalent bond that consists of three pairs of electrons (1.5).

Valence shell The outer shell of an atom, containing electrons that participate in bonding (1.2).

ADDITIONAL PROBLEMS

1.18 Classify the following compounds as ionic or covalent. Use position in the periodic table as your criterion.

[a]* LiCl	[b] S_2Cl_2	[c] HF	[d] Na_2O
(e) NaBr	(f) SO_2	(g) HCl	(h) K_2O

[1.19] Write the Lewis dot structure for each compound in Problem 1.18.

1.20 Predict whether each of the following pairs of elements will react to form a covalent or an ionic compound. Show the Lewis dot structures of the reactant atoms and product molecules or ions.

[a] $K + I \longrightarrow$	[b] $C + S \longrightarrow$	[c] $O + Cl \longrightarrow$
[d] $Br + Cl \longrightarrow$	[e] $Mg + F \longrightarrow$	[f] $Rb + I \longrightarrow$
[g] $H + Br \longrightarrow$	(h) $H + S \longrightarrow$	(i) $Na + Cl \longrightarrow$
(j) $H + I \longrightarrow$	(k) $Ca + Cl \longrightarrow$	(l) $H + O \longrightarrow$
(m) $S + Cl \longrightarrow$	(n) $Cl + I \longrightarrow$	(o) $Li + O \longrightarrow$

1.21 Which member of each of the following pairs of compounds has the higher boiling point? Explain.

[a] CH_4 and C_6H_{14}

[b]
```
    ..
  H:O:               H H
    ..  ..           .. .. ..
H:C:C:H   and  H:C:C:O:H
  ..             .. .. ..
  H              H H
```

[c]
```
  H H H                H   H H
  .. .. .. ..          ..  .. .. ..
H:C:C:C:O:H  and  H:C:O:C:C:H
  .. .. .. ..          ..  .. .. ..
  H H H                H   H H
```

[d] C_4H_{10} and C_3H_7OH

(e) C_4H_{10} and C_2H_6

(f)
```
    ..
  H:O H               H H H
    ..  ..  ..        .. .. .. ..
H:C:C:C:H  and  H:C:C:C:O:H
  ..    ..            .. .. .. ..
  H     H             H H H
```

(g)
```
   ..
  :O:           H
   ..           ..  ..
H:C:H  and  H:C:O:H
                ..  ..
                H
```

(h)
```
  H H           H
  .. ..         ..  ..
H:C:C:H  and  H:C:O:H
  .. ..         ..  ..
  H H           H
```

[1.22] Which member of each pair in Problem 1.21 is more soluble in water? Explain.

* Answers to problems whose number or letter designation are enclosed within brackets are not given. Answers to all other problems are given in the Selected Answers section at the end of the book.

1.23 Calculate the formal charges on the atoms in the following formulas.

[a] :Ö: / :Ö::S:Ö: [b] :Ö::S:Ö: [c] :S::C::S:

[d] [:Ö: / :Ö::N:Ö:]$^{-}$ (e) [:Ö: / :Ö::S::Ö: / :Ö:]$^{2-}$ (f) [:Ö: / :Ö:S:Ö:]$^{2-}$

(g) :Ö:C::C:C::O: (h) [:Ö:N::Ö:]$^{-}$

1.24 Draw Lewis dot structures and identify the polar bonds in each of the following compounds. Show the direction of polarization by labeling the appropriate atoms $\delta+$ and $\delta-$.

[a] H_2O [b] HBr [c] CCl_4 (d) H_2S
(e) HI (f) CF_2Cl_2

1.25 In which of the following pure compounds would you expect hydrogen bonding to be important?

[a] :O: / H:C:H [b] SO_2 [c] :Ö / H:Ö:S:Ö:H / :Ö

[d] CH_2Cl_2 [e] H_2Se [f] LiH

(g) :Ö / CH_3C:H (h) SO_3 (i) :Ö / H:Ö:S:Ö:H

(j) CH_3F (k) PH_3 (l) NaH

[1.26] Describe the following terms.

(a) an atomic orbital (b) a hybrid orbital
(c) an unshared electron pair

1.27 Predict the hybridization and orbital geometry of the underlined atom in each of the following compounds.

[a] $\underline{C}F_4$ [b] $H_2\underline{S}$ [c] $\underline{P}H_3$ [d] $\underline{S}Cl_2$
[e] $\underline{B}Br_3$ (f) $\underline{C}Br_4$ (g) $\underline{N}F_3$ (h) $H_2\underline{Se}$
(i) $\underline{O}F_2$ (j) $\underline{B}Cl_3$

[1.28] Give the molecular shape of each of the molecules in Problem 1.27. *Note*: Molecular shape is defined here by the centers of the atoms in the molecules (that is, ignore the unshared electron pairs).

[1.29] How did Pauling describe the covalent bond? On what does bond strength depend?

[1.30] Explain in terms of the available atomic orbitals why sp^3 hybrid orbitals are three-dimensional (tetrahedral), why sp^2 hybrid orbitals are planar, and why sp hybrid orbitals are linear.

PRINCIPLES OF STRUCTURE

2.1 STRUCTURAL FORMULAS

Lewis dot structures are the basis of nearly all the structural formulas used by organic chemists. Dot structures are cumbersome to write, however, and the simpler **graphic formulas** are often used instead. We have already used graphic formulas in some of the figures. In a graphic formula we represent the electron pair bond by a line joining the two atoms:

$$\mathrm{H}:\ddot{\underset{\cdot\cdot}{\mathrm{Cl}}}: \text{ is shown as } \mathrm{H}{-}\ddot{\underset{\cdot\cdot}{\mathrm{Cl}}}: \text{ or simply as } \mathrm{H}{-}\mathrm{Cl}.$$

Unshared electrons may be shown as dots if necessary, or they may be omitted. Double and triple bonds are represented by using one line for each pair of electrons in the bond:

$$:\ddot{\mathrm{O}}::\mathrm{C}::\ddot{\mathrm{O}}: \text{ is shown as } :\ddot{\mathrm{O}}{=}\mathrm{C}{=}\ddot{\mathrm{O}}: \text{ or simply } \mathrm{O}{=}\mathrm{C}{=}\mathrm{O};$$

$$\mathrm{H}:\mathrm{C}\vdots\vdots\mathrm{N}: \text{ is shown as } \mathrm{H}{-}\mathrm{C}{\equiv}\mathrm{N}: \text{ or } \mathrm{H}{-}\mathrm{C}{\equiv}\mathrm{N}.$$

So far we have paid no attention to geometry in representing molecular formulas. We know that carbon is tetrahedral, yet we write methane as

```
      H
      |
   H—C—H
      |
      H
```

which implies a square planar structure. The problem, of course, is in attempting to draw a three-dimensional structure on two-dimensional paper. A useful notation for showing three-dimensional perspective formulas on paper is the **line-dash-wedge formula.** The line-dash-wedge formula of methane is

```
     H
     |
     C---H
    /  ◥
   H    H
```

where the solid lines (—) represent bonds in the plane of the page, the dashed line (---) extends away from us behind the page, and the wedge-shaped bond (◄) extends out in front of the page toward us (Figure 2.1). We shall use such perspective formulas only to emphasize the three-dimensional nature of a molecule.

Even graphic formulas become cumbersome at times. For example, the graphic formula of decane (Figure 2.2).

```
    H  H  H  H  H  H  H  H  H  H
    |  |  |  |  |  |  |  |  |  |
 H—C—C—C—C—C—C—C—C—C—C—H
    |  |  |  |  |  |  |  |  |  |
    H  H  H  H  H  H  H  H  H  H
```

can be shortened to $CH_3CH_2CH_2CH_2CH_2CH_2CH_2CH_2CH_2CH_3$. This notation is a **condensed formula.** In it we understand that each hydrogen is attached to the carbon at

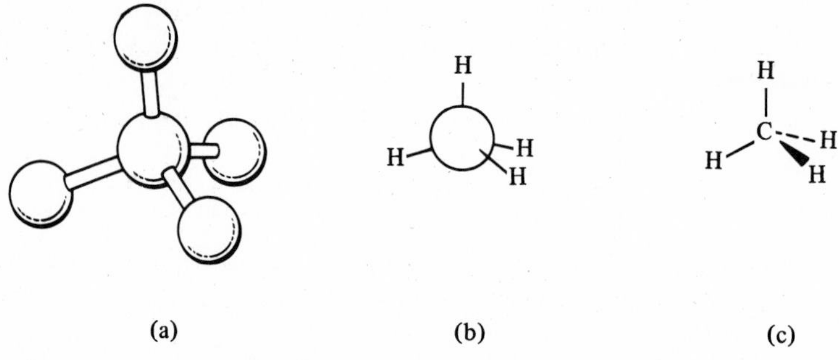

Figure 2.1. Representations of methane: (a) ball-and-stick model; (b) line- and circle; (c) line-dash-wedge formula.

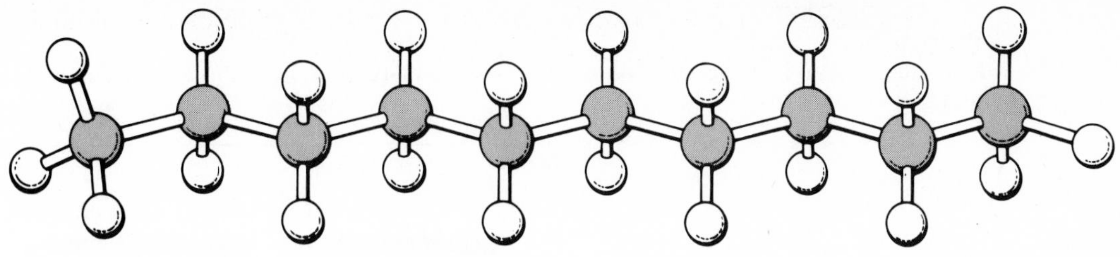

$CH_3CH_2CH_2CH_2CH_2CH_2CH_2CH_2CH_2CH_3$

Figure 2.2. Decane.

its immediate left. The formula $CH_3CHOHCH_3$ similarly implies that the OH is bonded to the nearest carbon to its left, so we know that the graphic formula is

```
   H  O—H  H
   |  |    |
H—C—C——————C—H
   |  |    |
   H  H    H
```

The condensed and graphic formulas of isobutane are

$$CH_3CH(CH_3)CH_3$$

or

```
CH3CHCH3
    |
   CH3
```

and

```
   H      H      H
   |      |      |
H—C——————C——————C—H
   |      |      |
   H   H—C—H    H
          |
          H
```

In many cases we can shorten a condensed formula even further. For example, the decane molecule drawn above can be written as $CH_3(CH_2)_8CH_3$.

Problem 2.1
Draw the graphic and the line-dash-wedge formulas of (a) CH_4, (b) CH_3Cl, (c) isobutane, (d) CH_2Cl_2, (e) $CHCl_3$

Organic chemists usually lump all the above types of formulas together and call them structural formulas. By the term **structural formula** we mean the formula—graphic, condensed, line-dash-wedge or other perspective formula, or combination of these—that shows the structure of the molecule in whatever detail desired. Although the term *structural formula* is somewhat ambiguous, we shall use it in the general sense just described.

2.2 MOLECULAR MODELS

Many different types of molecular models are available commercially. Three different types are shown in Figure 2.3. Probably the oldest and most common are the ball-and-stick models. In these models, carbon atoms are represented by wooden balls that have holes drilled into them at the tetrahedral angles of 109.5°. The balls are connected with wooden dowels that fit into the holes. Other atoms are similarly represented.

A more realistic type of model is the space-filling model. These scale models are designed to represent actual atomic dimensions.

A third type is the framework molecular model. The atoms are represented by the intersection of plastic "bonds." These models emphasize bond angles and bond lengths. They make no attempt to describe the atom except to show the center of its nucleus.

The purpose of these different types of molecular models is to allow us to emphasize different aspects of molecules—atomic sizes, overall molecular shapes, or bond angles and bond lengths. We must always bear in mind that all of them are merely models and not exact representations of the molecules themselves. They provide only an approximation of molecular properties and behavior.

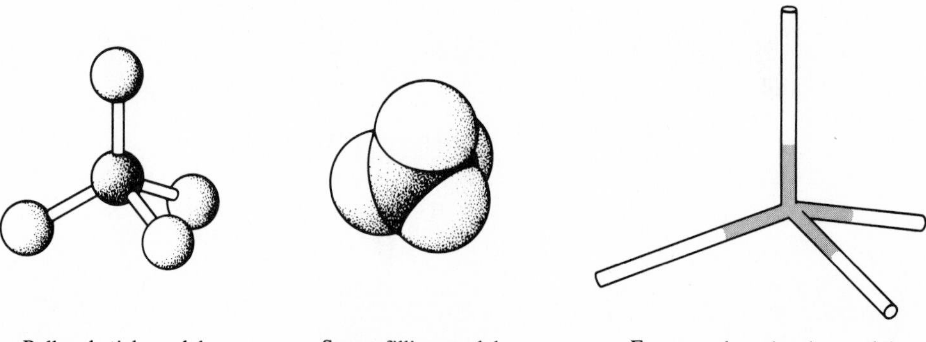

Figure 2.3. Molecular models of methane.

Problem 2.2

Make models of the following molecules.

(a) CH_4 (b) CH_3CH_3 (c) $CH_3CH_2CH_3$

(d) $CH_3CH(CH_3)CH_3$ (e) $CH_3C(CH_3)_2CH_3$ (f) $CH_3CH_2CH_2CH_3$

2.3 ISOMERISM

We know of two compounds that have the molecular formula C_4H_{10}. They are called butane and isobutane. We know that they are actually different compounds because they have different properties (Table 2.1).

Butane and isobutane are examples of **isomers**. *Two substances are isomers of each other if they consist of different molecules that have the same molecular formula.* We know that two substances are different if they have different properties. Molecules that have the same **molecular formula** (in this case, C_4H_{10}) can be different only if they differ somehow in the way their atoms are bonded together. The structures of the two C_4H_{10} isomers are shown in Figure 2.4.

```
   H  H  H  H
   |  |  |  |
H—C—C—C—C—H
   |  |  |  |
   H  H  H  H
```

butane

and

```
   H      H      H
   |      |      |
H—C——————C——————C—H
   |      |      |
   H   H—C—H    H
          |
          H
```

isobutane

Table 2.1 Properties of Butane and Isobutane

	Boiling Point, °C	Melting Point, °C	Density, g/cm³ at −20°C
Butane	0	−138	0.622
Isobutane	−12	−159	0.604

(a) 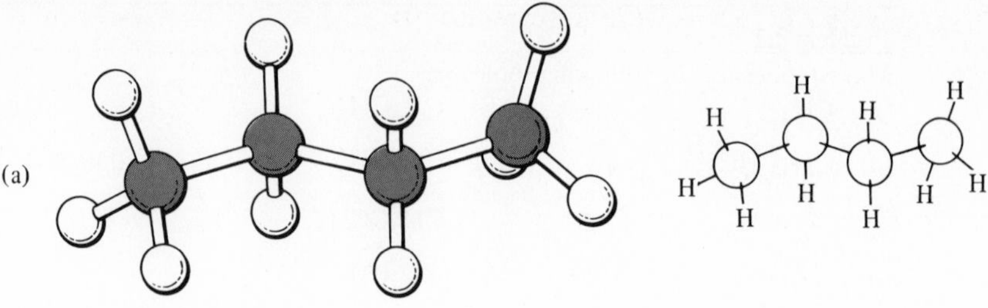

(b) 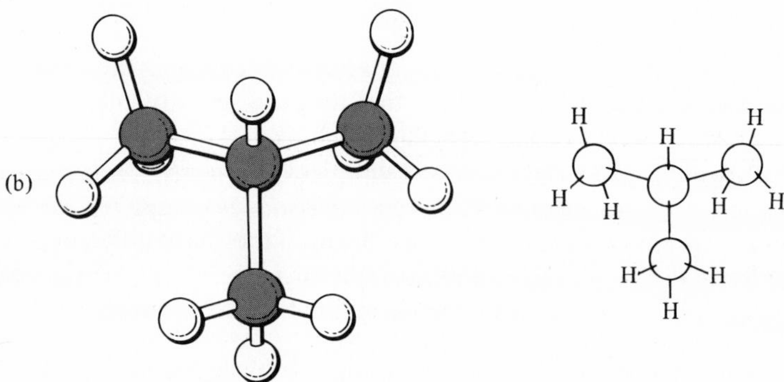

Figure 2.4. (a) Butane and (b) isobutane.

Carbon–carbon covalent bonds are very strong, and they are not readily broken. Thus rearrangement of the atoms does not occur, and butane molecules do not become isobutane molecules or vice versa.

2.4 CONSTITUTIONAL ISOMERS

Isomers like the butanes that differ in the sequential order in which their atoms are bonded together are called **constitutional isomers.** (Some textbooks use the term **structural isomer** instead of constitutional isomer.) Other examples of constitutional isomers are shown in Table 2.2.

Let us examine one such pair of isomers—ethyl alcohol and dimethyl ether. We know that these two substances are indeed different because they have different boiling points:

ethyl alcohol (boiling point 78°C), CH_3CH_2OH
dimethyl ether (boiling point −23.7°C), $CH_3—O—CH_3$

How do we assign the correct structural formula to these two isomers?

Table 2.2 Some Constitutional Isomers

Structural Formulas			Molecular Formulas
CH_3CH_2OH	CH_3OCH_3		C_2H_6O
$CH_3CH_2\overset{O}{\overset{\Vert}{C}}H$	$CH_3\overset{O}{\overset{\Vert}{C}}CH_3$		C_3H_6O
$CH_3CH{=}CHCH_3$	$CH_2{=}CHCH_2CH_3$	$CH_3\overset{CH_3}{\overset{\vert}{C}}{=}CH_2$	C_4H_8
$CH_2{=}CHCH{=}CH_2$	$CH_2{=}C{=}CHCH_3$	$CH_3C{\equiv}CCH_3$	C_4H_6
CH_3CHCl_2	CH_2ClCH_2Cl		$C_2H_4Cl_2$

Ethyl alcohol reacts with metallic sodium to produce one molecule of hydrogen gas per two molecules of ethyl alcohol reacted:

$$2C_2H_6O + 2Na \longrightarrow 2C_2H_5ONa + H_2\uparrow$$

Dimethyl ether does not undergo this reaction. This reaction tells us that ethyl alcohol has one hydrogen that is reactive toward sodium and five hydrogens that are unreactive. In dimethyl ether, all the hydrogens are unreactive. We can easily see that in the first structure above, one hydrogen is different from the other five. The structure CH_3CH_2OH therefore belongs to ethyl alcohol; the structure of dimethyl ether is CH_3OCH_3.

Problem 2.3

Make a model of $H{-}\overset{Cl}{\underset{H}{\overset{\vert}{\underset{\vert}{C}}}}{-}Cl$ and $Cl{-}\overset{H}{\underset{H}{\overset{\vert}{\underset{\vert}{C}}}}{-}Cl$. Are they isomers? Explain.

Problem 2.4

Draw the graphic and condensed formulas for all the isomers of each of the following.

(a) C_5H_{12} (b) C_3H_7Cl (c) $C_3H_6Cl_2$

Problem 2.5

Make a model of each isomer of the compounds in Problem 2.4.

2.5 STEREOISOMERS

Table 2.2 shows three isomers of C_4H_8. There are actually six; two of them do not have a double bond but exist instead as the cyclic structures (Sec. 2.6)

$$\begin{array}{c} H_2C{-}CH_2 \\ \vert \quad\ \ \vert \\ H_2C{-}CH_2 \end{array} \quad \text{and} \quad \begin{array}{l} H_2C \diagdown \\ \ \ \vert \qquad CHCH_3 \\ H_2C \diagup \end{array}$$

Three of the remaining four isomers have the butane skeleton, C—C—C—C, and one has the isobutane skeleton, $\mathrm{C{-}\underset{\underset{C}{|}}{C}{-}C}$. We know this from their reactions with hydrogen.

Hydrogen reacts with a double bond to produce a single bond without altering the carbon skeleton.

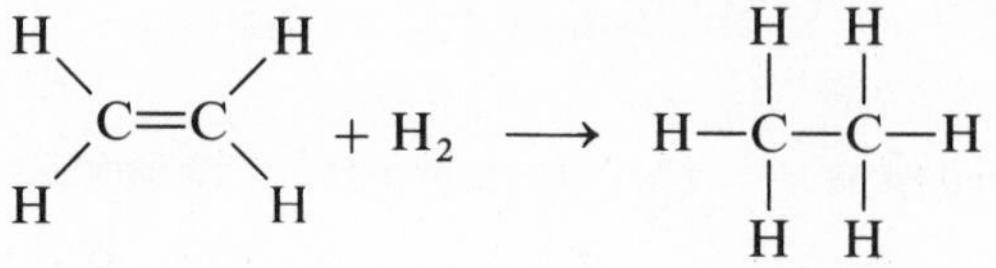

$$\left.\begin{array}{l}\text{Isomer 1}\\ \text{Isomer 2}\\ \text{Isomer 3}\end{array}\right\} + H_2 \longrightarrow CH_3CH_2CH_2CH_3$$

$$\text{Isomer 4} + H_2 \longrightarrow CH_3{-}\underset{\underset{CH_3}{|}}{CH}{-}CH_3$$

Only two isomers in Table 2.2 have the butane skeleton, $CH_3CH{=}CHCH_3$ and $CH_2{=}CHCH_2CH_3$. The only way to account for three isomers with the butane skeleton is to assume that two isomers exist with the double bond between the two middle carbons:

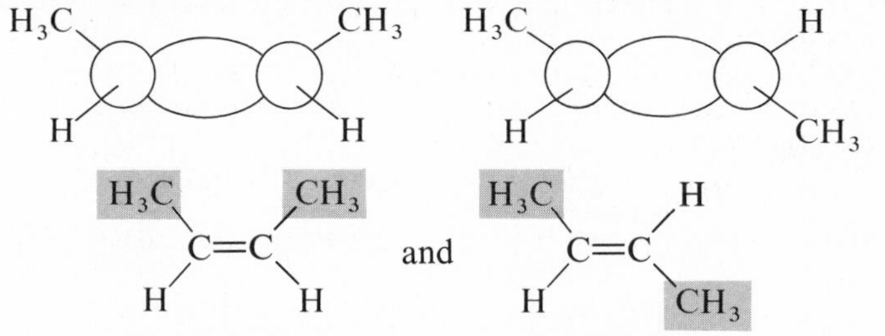

cis-isomer *trans*-isomer

These isomers are called *cis* and *trans* isomers.

We can now summarize the reactions of all four isomers with hydrogen as follows:

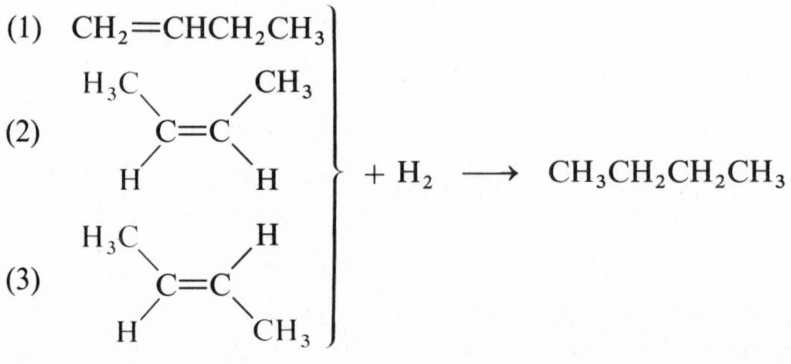

(4) $CH_3C(CH_3){=}CH_2 + H_2 \longrightarrow CH_3CH(CH_3)CH_3$

$$\begin{array}{lcl} CH_3\underset{\displaystyle CH_3}{\underset{|}{C}}{=}CH_2 + H_2 & \longrightarrow & CH_3\underset{\displaystyle CH_3}{\underset{|}{C}}HCH_3 \end{array}$$

The existence of *cis–trans* isomers implies that rotation about the double bond does not occur. Otherwise, *cis* and *trans* isomers would interconvert, and they do not. We shall have more to say about *cis–trans* isomers in Chapter 5. For the moment, we shall merely define what they are and when they can exist. *Cis–trans isomers are possible only when each carbon that forms the double bond is attached to two different groups.* The general formulas

$$\begin{array}{ccc} \text{A} \quad\quad \text{X} & & \text{A} \quad\quad \text{Y} \\ \text{C=C} & \text{and} & \text{C=C} \\ \text{B} \quad\quad \text{Y} & & \text{B} \quad\quad \text{X} \end{array}$$

are *cis–trans* isomers only if **A** ≠ **B** and **X** ≠ **Y**. **A** can be the same as **X** or **Y**, and **B** can be the same as **X** or **Y**.

Some examples will illustrate these conditions for the existence of *cis–trans* isomers.

Example 1: $CH_2{=}CHCH_3$ cannot exist as *cis–trans* isomers because

$$\begin{array}{ccc} \text{H} \quad\quad \text{H} & & \text{H} \quad\quad \text{CH}_3 \\ \text{C=C} & \text{and} & \text{C=C} \\ \text{H} \quad\quad \text{CH}_3 & & \text{H} \quad\quad \text{H} \end{array}$$

are the same; one carbon has two identical groups: **A** = **B** = H.

Example 2:

$$\begin{array}{ccc} \text{H} \quad\quad \text{Cl} & & \text{H} \quad\quad \text{CH}_2\text{CH}_3 \\ \text{C=C} & \text{and} & \text{C=C} \\ \text{H}_3\text{C} \quad\quad \text{CH}_2\text{CH}_3 & & \text{H}_3\text{C} \quad\quad \text{Cl} \end{array}$$

have different functional groups on each carbon; therefore they are *cis–trans* isomers: $H \neq CH_3$ and $Cl \neq CH_2CH_3$.

Cis–trans isomers have the same sequential arrangement of atoms and thus they are *not* constitutional isomers. They are called **stereoisomers** because their atoms are bonded in the same sequence, and *they differ only in the orientation of their atoms in space.*

The relation between these two kinds of isomers is shown in Figure 2.5.

Problem 2.6

Draw all the *cis-trans* isomers (if any) of each of the following compounds.

(a) $CH_2{=}CHCH_2CH_2CH_3$
(b) $ClCH{=}CClCH_2CH_3$
(c) $CClBr{=}CClBr$
(d) $CClBr{=}CHCH_3$

Confirm your answer by making a model of each isomer.

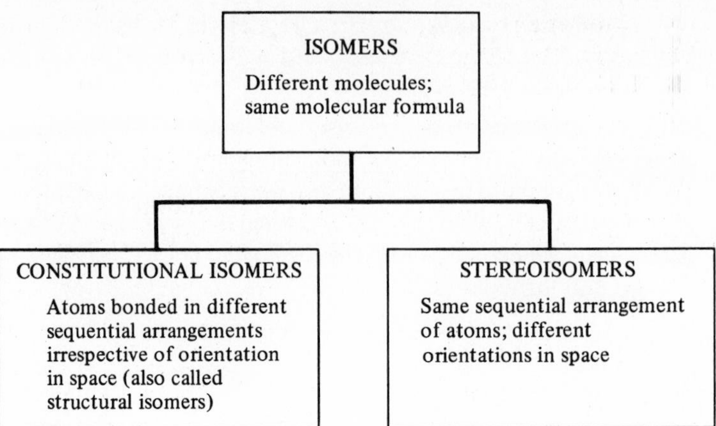

ISOMERS
Different molecules; same molecular formula
CONSTITUTIONAL ISOMERS
Atoms bonded in different sequential arrangements irrespective of orientation in space (also called structural isomers)
STEREOISOMERS
Same sequential arrangement of atoms; different orientations in space

Figure 2.5. Relation between constitutional isomers and stereoisomers.

2.6 HYDROCARBONS

Hydrocarbons are compounds that contain only the elements carbon and hydrogen. We have already introduced the hydrocarbons methane (CH_4), butane (C_4H_{10}), and decane ($C_{10}H_{22}$). The compounds just mentioned all have the general formula C_nH_{2n+2}. This formula applies to all noncyclic hydrocarbons that have only single bonds:

$$H-\left[\begin{array}{c} H \\ | \\ C \\ | \\ H \end{array}\right]_n-H, \text{ where } n = 1, 2, 3, 4, \ldots, n$$

This class of hydrocarbons, the **alkanes**, is referred to as **saturated** to indicate that every carbon has four single bonds to other carbons or hydrogens (see Table 4.1).

Unsaturated hydrocarbons contain fewer hydrogens than are required by the general formula C_nH_{2n+2}. The **alkenes** have the general formula C_nH_{2n}, and the **alkynes** have the general formula C_nH_{2n-2}. Alkenes have a double bond, and alkynes have a triple bond. To form a double bond in a molecule we must remove two hydrogens. A triple bond requires the removal of four hydrogens.

Some hydrocarbons have the formula C_nH_{2n} and do not have a double bond. These are the cyclic hydrocarbons, called **cycloalkanes**. Some examples include

```
 H       H            H  H             H  H
 |       |            |  |              \/
H—C——————C—H        H—C——C—H            C
   \    /             |  |          H—C'   'C—H
     C              H—C——C—H        H'  \   /  'H
    / \               |  |           H—C——C—H
   H   H              H  H            H'    'H
```

cyclopropane — cyclobutane — cyclopentane

Aromatic hydrocarbons are compounds that have a high degree of unsaturation but behave chemically like saturated hydrocarbons. Examples are benzene, C_6H_6, and naphthalene, $C_{10}H_8$.

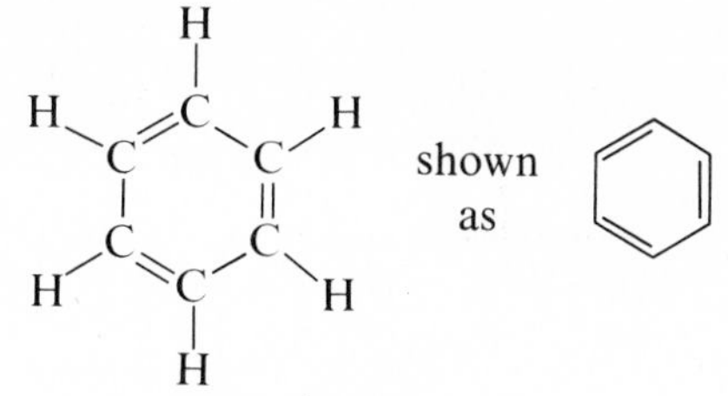

shown as

benzene

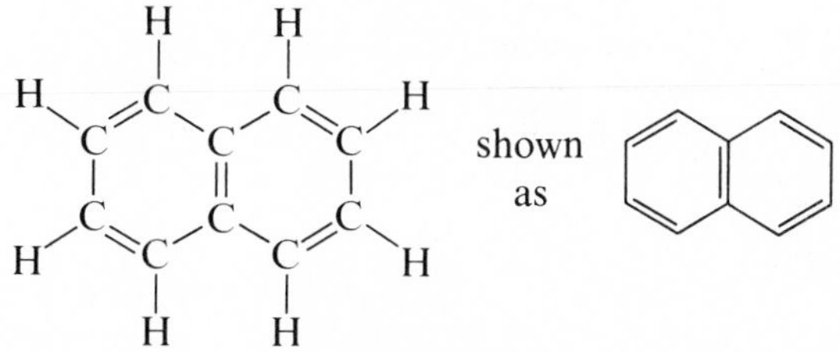

shown as

naphthalene

The unusual properties of aromatic hydrocarbons arise because of the special bonding that exists in certain ring systems in which single and double bonds alternate. We shall examine the aromatic hydrocarbons in Chapter 6.

When a molecule contains a hydrocarbon part and another functional group, as in the case of ethyl alcohol, CH_3CH_2OH, the general name of the hydrocarbon part is **alkyl**. The alkyl group in ethyl alcohol, $CH_3CH_2—$, is called *ethyl* because it derives from ethane. The corresponding general name for an aromatic fragment is *aryl.* Nonaromatic hydrocarbons are collectively called **aliphatic** hydrocarbons.

Problem 2.7

Give an example of (a) an alkane, (b) an alkene, (c) an alkyne, (d) an alkyl chloride, (e) an aryl bromide. Do not repeat any that are given in the text.

2.7 FUNCTIONAL GROUPS

In Section 1.1 we defined a functional group as an atom or grouping of atoms that occurs commonly and whose behavior is independent of the rest of the molecule. We have already seen some examples—the carbon-carbon double and triple bonds.

The major part of this book is devoted to the properties of the various functional groups. It will help us at this point in our study of organic chemistry to become familiar with the major functional groups. Thanks to their small number, the study of organic chemistry is possible; otherwise, we would have to memorize the chemistry of each individual molecule.

Table 2.3 lists the most important simple functional groups and their relationships.

Table 2.3 Important Functional Groups

Functional Group Name	Example with Functional Group Shaded	Occurs in
Alkyl group[a]	$CH_3CH_2CH_2{-}OH$	Most organic compounds
Double bond	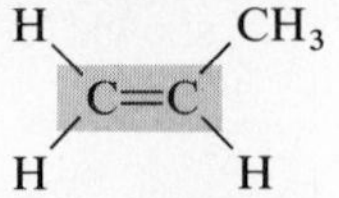 $(H)(H)C{=}C(CH_3)(H)$	Alkenes
Triple bond	$CH_3{-}C{\equiv}C{-}CH_2CH_3$	Alkynes
Aryl group	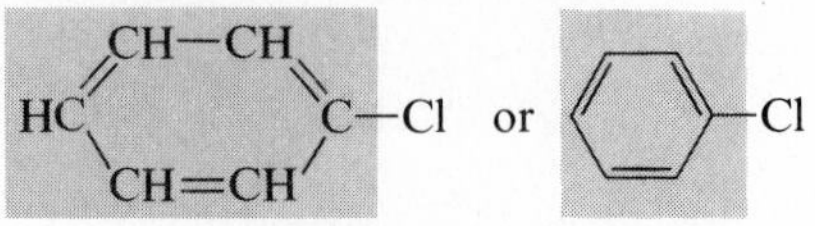 $HC\langle CH{-}CH,\ CH{=}CH\rangle C{-}Cl$ or $C_6H_5{-}Cl$	Aromatic compounds
Hydroxyl group	$CH_3CH_2{-}O{-}H$	Alcohols and phenols
Alkoxy group	$CH_3CH_2{-}O{-}CH_2CH_3$	Ethers
Amino group	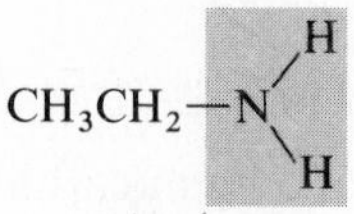 $CH_3CH_2{-}NH_2$	Amines
Aldehyde group	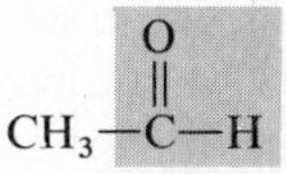 $CH_3{-}C({=}O){-}H$	Aldehydes
Keto group	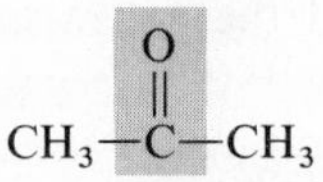 $CH_3{-}C({=}O){-}CH_3$	Ketones
Carboxyl group	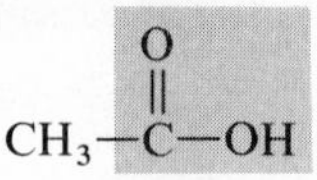 $CH_3{-}C({=}O){-}OH$	Carboxylic acids
Ester group	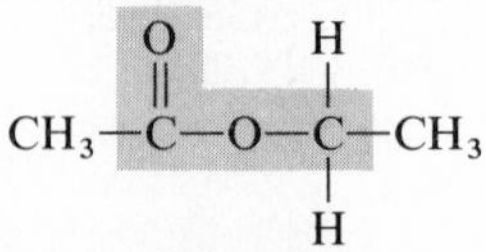 $CH_3{-}C({=}O){-}O{-}CH_2{-}CH_3$	Esters
Amide group	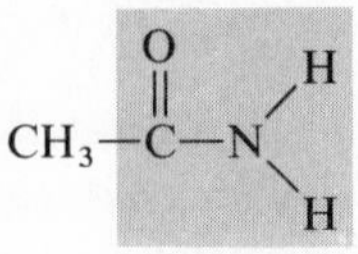 $CH_3{-}C({=}O){-}NH_2$	Amides
Nitro group	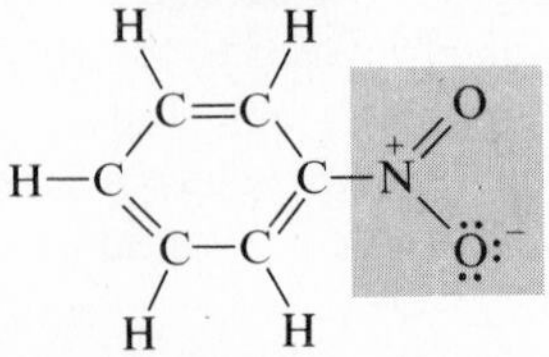 $C_6H_5{-}\overset{+}{N}({=}O){-}\ddot{\underset{\cdot\cdot}{O}}{:}^-$	Nitro compounds
Cyano group	$CH_3{-}C{\equiv}N$	Nitriles

[a] The simple saturated alkyl groups are quite unreactive and are therefore usually not considered as a functional group.

Problem 2.8
Circle and name the functional groups in the following compounds. Do not include saturated alkyl groups.

(a) 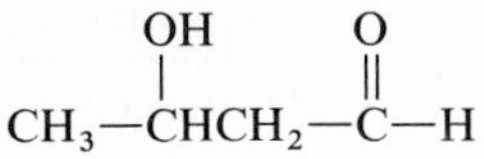 $CH_3-\underset{|}{\overset{OH}{}}CHCH_2-\overset{O}{\overset{\|}{C}}-H$

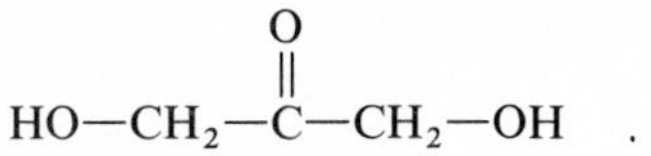

(b) $HO-CH_2-\overset{O}{\overset{\|}{C}}-CH_2-OH$.

(c) $C_6H_5-O-\overset{O}{\overset{\|}{C}}-CH_3$ (benzene ring drawn with H on each ring carbon)

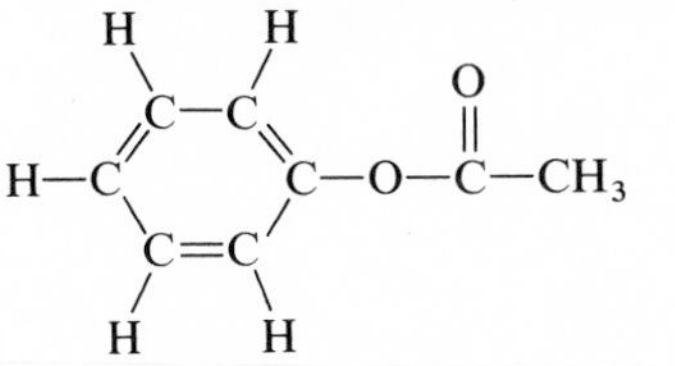

NEW TERMS

Aliphatic hydrocarbon Any nonaromatic hydrocarbon including alkanes, alkenes, alkynes, and cycloalkanes (2.6).

Alkane Noncyclic hydrocarbon containing only single bonds. Alkanes have the general formula C_nH_{2n+2} (2.6).

Alkene Hydrocarbon containing a double bond. Alkenes have the general formula C_nH_{2n} (2.6).

Alkyl group When a molecule contains an aliphatic hydrocarbon part and a functional group, the aliphatic hydrocarbon part is called an alkyl group (2.6).

Alkyne Hydrocarbon containing a triple bond. Alkynes have the general formula C_nH_{2n-2} (2.6).

Aromatic hydrocarbon Compound that has a high degree of unsaturation but behaves chemically like a saturated hydrocarbon (2.6).

Condensed formula Notation in which atoms are given in one line, for example as $CH_3CH_2CH_3$ (2.1).

Constitutional isomers Isomers that differ in the sequential order in which their atoms are bonded together (also called structural isomers) (2.4).

Cycloalkane Saturated hydrocarbon whose carbon atoms are bonded together to form a ring (2.6).

Graphic formula Formula in which an electron pair (covalent) bond is represented by a line joining the two atoms (2.1).

Isomers Identifiably different compounds that have the same molecular formula (2.3).

Line-dash-wedge formula Notation for representing three-dimensional structures. A line (—) represents a bond in the plane of the page, a dashed line (---) represents a bond that extends away from us behind the plane of the page, and a wedge (◄) represents a bond that extends out in front of the page (2.1).

Molecular formula Formula that gives only the number of each kind of atom present in the compound (2.3).

Saturated compound A compound that has only single bonds (2.6).

Stereoisomers Isomers that differ only in the orientation of their atoms in space. Stereoisomers have the same sequential ordering of atoms and are therefore not constitutional isomers (2.5).

Structural formula Any graphic, condensed, line-dash-wedge or other formula that shows the structure of the molecule in whatever detail desired (2.1).

Unsaturated compound A compound that has at least one double or triple bond (2.6).

ADDITIONAL PROBLEMS

[2.9] Draw graphic formulas and line-dash-wedge formulas for the compounds given in Problem 2.2. Use the line-dash-wedge notation only for the central carbon atom.

2.10 Classify each of the following pairs of compounds as constitutional isomers, stereoisomers, not isomers, or identical.

[a] $CH_3CH_2CH_2CH_3$ and $CH_3CH(CH_3)CH_3$

[b]
$$\begin{array}{c}CH_3\\ |\\ CH_3CHCH_3\end{array} \quad \text{and} \quad \begin{array}{c}CH_3CHCH_3\\ |\\ CH_3\end{array}$$

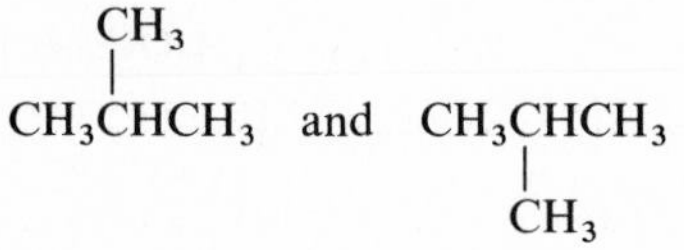

[c]
$$\begin{array}{ccc}H & & H\\ & C{=}C & \\ H & & CH_3\end{array} \quad \text{and} \quad \begin{array}{ccc}H & & CH_3\\ & C{=}C & \\ H & & H\end{array}$$

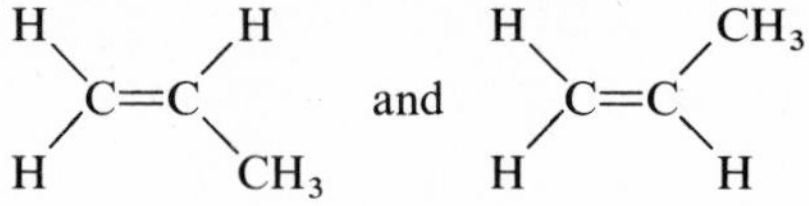

[d]
$$\begin{array}{ccc}H & & H\\ & C{=}C & \\ Cl & & Cl\end{array} \quad \text{and} \quad \begin{array}{ccc}H & & Br\\ & C{=}C & \\ Cl & & H\end{array}$$

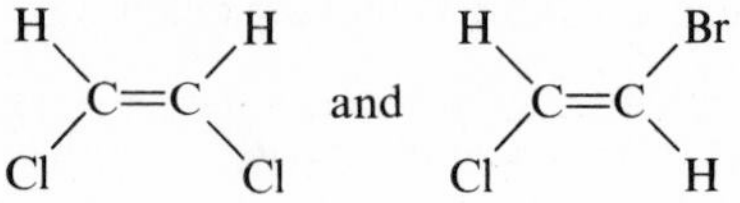

(e) 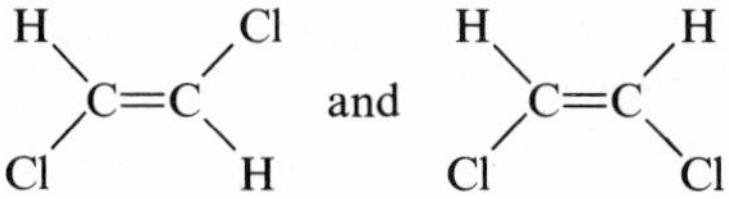
$$\begin{array}{ccc}H & & Cl\\ & C{=}C & \\ Cl & & H\end{array} \quad \text{and} \quad \begin{array}{ccc}H & & H\\ & C{=}C & \\ Cl & & Cl\end{array}$$

(f)
$$CH_3CH{=}CHCH_3 \quad \text{and} \quad \begin{array}{c}H_2C{-}CH_2\\ |\quad\ \ |\\ H_2C{-}CH_2\end{array}$$

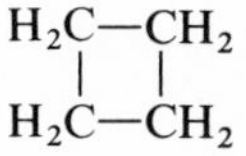

(g)
$$CH_3CH{=}CHCH_2CH{=}CH_2 \quad \text{and} \quad \begin{array}{l}H_2C{-}CH_2\\ |\quad\ \ |\\ H_2C{-}CH{-}CH{=}CH_2\end{array}$$

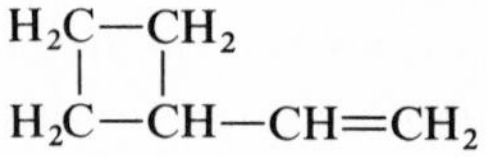

(h)
$$CH_3CH_2CH_2CH_3 \quad \text{and} \quad \begin{array}{c}CH_3\\ |\\ H_3C{-}C{-}CH_3\\ |\\ CH_3\end{array}$$

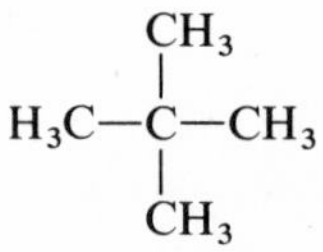

2.11 Draw all the isomers of (a) $C_4H_{10}O$, (b) C_5H_{10}, (c) C_4H_{10}, [d] $C_3H_3Cl_3$, [e] C_4H_6

2.12 Circle and name the functional groups in each of the following compounds. Do not include saturated alkyl groups.

[a] CH_3CH_2OH

[b]
$$\begin{array}{c}CH_3C{-}OH\\ \|\\ O\end{array}$$

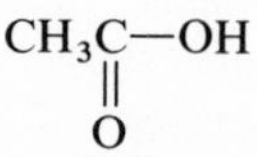

[c] $HC(=O)-O-CH_3$

[d] $C_6H_5-C(=O)-CH_2CH_3$

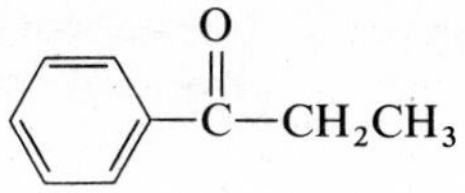

[e] CH_3-OH

(f) 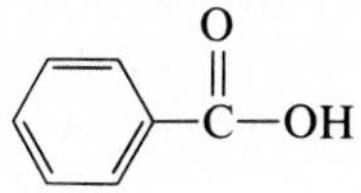 $C_6H_5-C(=O)-OH$

(g) $CH_3CHOHCH_2CH_3$

(h) 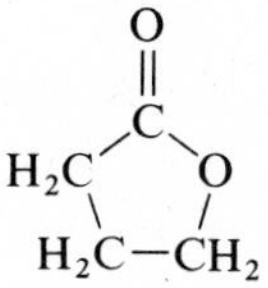 (five-membered ring: $H_2C-C(=O)-O-CH_2-CH_2$)

(i) $HC(=O)-CH(CH_3)CH_3$

(j) $HO-CH_2CH_2-C(=O)-CH_2CH_2-C(=O)-OH$

2.13 Draw structural formulas of two isomers of $C_4H_8O_2$ that have each of the following functional groups.

(a) a carboxyl group,
(b) an aldehyde,
[c] a keto group,
[d] a hydroxyl group,
[e] an alkoxy group

[2.14] Draw the Lewis dot structures of the functional groups shown in Table 2.3. Show all formal charges.

[2.15] Explain how isomers are alike and how they are different.

2.16 Tell whether each of the following hydrocarbons is an alkane, an alkene, an alkyne, or an aromatic hydrocarbon. Which are aliphatic?

[a] $CH_3CH{=}CH_2$
[b] $CH_3CH{=}C(CH_3)CH_3$
[c] $HC{\equiv}CCH_3$
[d] (anthracene structure)
(e) $CH_3CH{=}CHCH_3$
(f) 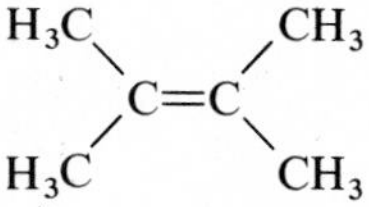 $(H_3C)_2C{=}C(CH_3)_2$
(g) $CH_3C{\equiv}CCH_3$
(h) 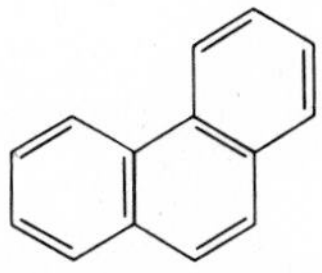 (phenanthrene structure)

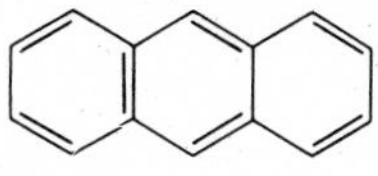

2.17 Identify where possible each of the following compounds as an alkane, an alkene, an alkyne, an aromatic hydrocarbon, or a cycloalkane.

[a] C_6H_{14}
[b] C_2H_4
[c] C_3H_6
[d] C_3H_4
[e] C_6H_6
(f) C_4H_{10}
(g) C_2H_2
(h) C_4H_8
(i) C_4H_6
(j) $C_{10}H_8$

3 PRINCIPLES OF REACTIVITY

3.1 BRØNSTED–LOWRY ACID-BASE THEORY

The best approach to the study of organic reactions is through the study of acid-base reactions and acid-base equilibria. For our purposes the Brønsted–Lowry acid-base theory is the most useful. Brønsted and Lowry based their theory on the following definitions:

1. An **acid** is any substance that is capable of releasing a hydrogen ion, H^+ (called a **proton**).
2. A **base** is any substance that is capable of accepting (reacting with) a proton.
3. Every acid has its conjugate base. The **conjugate base** is the fragment that remains after the acid has lost a proton.

4. Every base has its conjugate acid. The **conjugate acid** is the species that is formed when the base accepts a proton.

Table 3.1 lists some common acids and their conjugate bases. Notice that each acid is the conjugate acid of the base.

A result of these definitions is that every reaction between an acid and a base produces another acid and base—the conjugates of their reactants. Another result is that when we use a water solution of an acid, such as aqueous HCl, we are no longer dealing with HCl. Instead we are dealing with a different acid, the **hydronium ion** (H_3O^+), because water is a base and it reacts completely with HCl:

$$\underset{(\text{acid}_1)}{HCl} + \underset{(\text{base}_2)}{H_2O} \longrightarrow \underset{(\text{acid}_2)}{H_3O^+} + \underset{(\text{base}_1)}{Cl^-}$$

In this scheme acid_1–base_1 and acid_2–base_2 are conjugate acid–base pairs. Every acid–base reaction is a *proton transfer* reaction.

Problem 3.1
Which substances in Table 3.1 are able to act as both acids and bases?

Problem 3.2
What is the conjugate base of (a) NH_3, (b) OH^-, (c) CH_4?

Problem 3.3
What is the conjugate acid of (a) H_2SO_4, (b) HNO_3?

Table 3.1 Some Common Conjugate Acid-Base Pairs

Acid	Base
HCl	Cl^-
$CH_3C(=O)$—OH	$CH_3C(=O)$—O^-
H_2SO_4	HSO_4^-
H_3O^+	H_2O
H_2O	OH^-
NH_4^+	NH_3
HSO_4^-	SO_4^{-2}
C_6H_5—OH	C_6H_5—O^-

3.2 ACID STRENGTH

The strength of an acid is determined by its ability to donate a proton to some standard base. Water is usually chosen as the standard base because it is the most common solvent. The **strength** of an acid is therefore the extent to which the equilibrium

$$HX + H_2O \rightleftharpoons H_3O^+ + X^-$$

favors the products, $H_3O^+ + X^-$. If the reaction occurs completely so that there is no HX present at equilibrium, the acid HX is said to be a **strong acid**. If the reaction occurs to only a slight extent so that there is only a small amount of H_3O^+ and X^- at equilibrium, then the acid HX is said to be a **weak acid**.

Reactants and products both contain an acid and a base. At equilibrium, the weaker acid and base will be present in greater concentrations than the stronger acid and base. In other words, a strong acid will react to form a weaker acid, and a strong base will react to form a weaker base. The reason for this is as follows. The two bases, H_2O and X^-, compete for the proton. The stronger base will acquire the greater proportion of the available protons. This is just another way of saying that reaction favors production of the weaker base. The equilibrium also favors the weaker acid. Both acids, HX and H_3O^+, can donate protons. The stronger one will be more effective in donating protons; therefore it will be present in smaller amounts.

Problem 3.4

Identify the acids and bases and designate the stronger acid and base in each of the equilibria below. The relative sizes of the arrows indicate the direction in which the equilibrium lies.

(a) $HCl + H_2O \longrightarrow H_3O^+ + Cl^-$

(b) $CH_3CO_2H + H_2O \rightleftharpoons CH_3CO_2^- + H_3O^+$

(c) $NH_3 + H_2O \rightleftharpoons NH_4^+ + OH^-$

3.3 ACID AND BASE DISSOCIATION CONSTANTS, K_a AND K_b

We can use acid and base dissociation constants, K_a and K_b, to make quantitative comparisons of acids and bases. We define the acid dissociation constant, K_a, in terms of the reaction of the acid with water:

$$\text{Acid} + H_2O \rightleftharpoons H_3O^+ + \text{conjugate base}$$

The acid dissociation constant is given by

$$K_a = \frac{(H_3O^+)(\text{conjugate base})}{(\text{acid})}$$

The concentration of water does not appear in the equation because water is the solvent and its concentration does not change. The true equilibrium constant K_{eq} is

$$K_{eq} = \frac{(H_3O^+)(\text{conjugate base})}{(H_2O)(\text{acid})}$$

so that K_a is the product of K_{eq} and the constant term (H_2O):

$$K_a = K_{eq}(H_2O) = \frac{(H_3O^+)(\text{conjugate base})}{(\text{acid})}$$

Similarly, we define the base dissociation constant in terms of the reaction of the base with water:

$$\text{Base} + H_2O \rightleftarrows \text{conjugate acid} + OH^-$$

The base dissociation constant is therefore

$$K_b = \frac{(\text{conjugate acid})(OH^-)}{(\text{base})}$$

Table 3.2 lists the K_a values for some common acids. Values of K_b are not given, but we can calculate any K_b easily if we know the K_a of the conjugate acid and if we realize that K_a and K_b are related:

$$K_a \times K_b = K_w$$

where K_w is the **water constant** and has the value 1×10^{-14} at 25°C. Water dissociates slightly:

$$H_2O + H_2O \rightleftarrows H_3O^+ + OH^-$$

The dissociation constant, K_w, for this equilibrium is

$$K_w = (H_3O^+)(OH^-)$$

We can now see that the product of K_a and K_b equals K_w:

$$\frac{(H_3O^+)(\text{conjugate base})}{(\text{acid})} \times \frac{(\text{conjugate acid})(OH^-)}{(\text{base})} = (H_3O^+)(OH^-)$$

if the base is the conjugate base of the acid and the acid is the conjugate acid of the base.

Let us take as an example the acid NH_4^+, and its conjugate base NH_3. The reaction of the acid with water is

$$NH_4^+ + H_2O \rightleftarrows H_3O^+ + NH_3$$

and

$$K_a = \frac{(H_3O^+)(NH_3)}{(NH_4^+)}$$

Table 3.2 K_as for Some Common Acids

Acid	Conjugate Base	K_a (25°C)
HI	I^-	∞
HCl	Cl^-	∞
H_2SO_4	HSO_4^-	∞
H_3O^+	H_2O	55.5
HO—C(=O)—C(=O)—OH	HO—C(=O)—C(=O)—O^-	5.6×10^{-2}
H_3PO_4	$H_2PO_4^-$	5.9×10^{-3}
HF	F^-	6.7×10^{-4}
CH_3C(=O)—OH	CH_3C(=O)—O^-	1.8×10^{-5}
HO—C(=O)—OH	HO—C(=O)—O^-	4.5×10^{-7}
HCN	CN^-	7.2×10^{-10}
C_6H_5—OH	C_6H_5—O^-	1.3×10^{-10}
NH_4^+	NH_3	5.6×10^{-10}
HO—C(=O)—O^-	$^-$O—C(=O)—O^- (CO_3^{-2})	4.7×10^{-11}
H_2O	OH^-	1.8×10^{-16}
OH^-	O^{-2}	less than 10^{-33}

(Arrows in table: Increasing acid strength — pointing up along the acid column; Increasing base strength — pointing down along the conjugate base column.)

The reaction of the conjugate base with water is

$$NH_3 + H_2O \rightleftarrows NH_4^+ + OH^-$$

and

$$K_b = \frac{(NH_4^+)(OH^-)}{(NH_3)}$$

Multiplying K_a and K_b we find

$$K_a \times K_b = \frac{(H_3O^+)(\cancel{NH_3})}{(\cancel{NH_4^+})} \times \frac{(\cancel{NH_4^+})(OH^-)}{(\cancel{NH_3})} = (H_3O^+)(OH^-) = K_w$$

Problem 3.5
What is the hydronium ion concentration, $[H_3O^+]$, in a 0.1 mole/liter aqueous solution of (a) HCl, (b) HF?

Problem 3.6
Refer to Table 3.2 and calculate the K_b of (a) $H_2PO_4^-$, (b) CN^-, (c) HCO_3^-, (d) CO_3^{-2}

3.4 EQUILIBRIUM AND REACTION RATE: MEASURES OF REACTIVITY

So far all our discussions of acids and bases have dealt with equilibrium conditions. In a reaction at equilibrium there is no change in the concentrations of reactants or products. We know that forward and reverse reactions occur continuously, however, because we can upset the equilibrium by altering the conditions under which the reaction takes place. At equilibrium, the rates of the forward and reverse reactions are equal.

We can understand the relationship between equilibrium and rate if we consider the free energy difference between reactants and products. We can define **free energy**, $\Delta G°$, as that type of energy that can do work at a constant temperature and pressure. [The superscript, °, means that all substances are present at a concentration of 1.0 mole per liter, a pressure of 1.0 atmosphere, and a temperature of 25°C (298°K).] Because free energy is easily measured, it is useful in predicting whether an equilibrium will favor reactants or products. $\Delta G°$ is related to the equilibrium constant, K, by the equation

$$\Delta G° = -2.3RT \log_{10} K$$

where R is the gas constant (1.987 cal mole^{-1} degree^{-1}) and T is the absolute temperature (°C + 273). A negative value of $\Delta G°$, therefore, means that the equilibrium favors products.

Figure 3.1 shows the free energy difference between reactants and products for the reaction of acetic acid and water. In this equilibrium $\Delta G°$ is positive; that is, energy must be added to form products, and the equilibrium favors reactants. The free energy change for the reaction of hydrogen gas with oxygen gas is negative, and the equilibrium strongly favors products. However, these reactions are quite different. When we mix acetic acid and water, equilibrium is established very rapidly. The reaction of hydrogen with oxygen is very slow. In fact, a mixture of H_2 and O_2 can be kept for years without any appreciable reaction taking place. Obviously, the free energies tell us nothing about the rates of these reactions.

The difference between the rates of these two reactions is in the **activation energy**, E_a. Molecules of reactant must acquire the necessary energy to surmount the energy barrier and thus become activated; otherwise, the molecules merely bounce apart when they collide. In other words, a collision between molecules is effective only if it occurs with enough energy (violence) to disrupt the molecules and form products. This amount of energy is called the activation energy. The configuration of reacting molecules that have acquired the activation energy is represented by the top of the energy

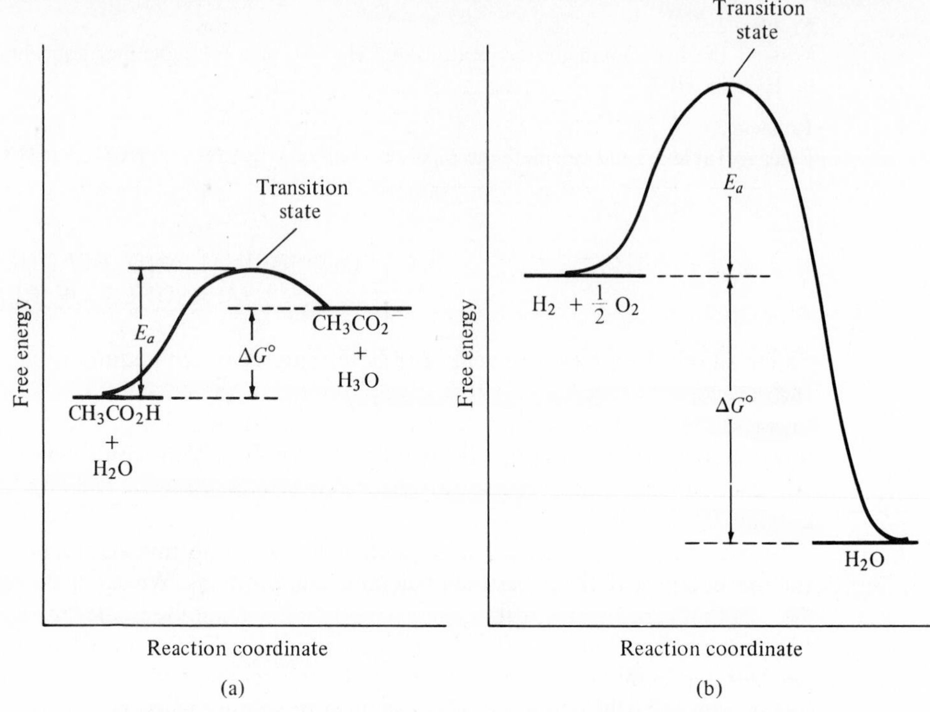

Figure 3.1. Free energy versus reaction coordinate for (a) the reaction of acetic acid with water, and (b) the reaction of hydrogen with oxygen.

barrier and is called the **activated complex** or **transition state**. The reaction of hydrogen and oxygen occurs with explosive violence if a spark or flame is introduced. Once the reaction begins, the released energy is more than enough to supply the necessary activation energy.

The rate of a reaction, therefore, does not depend on the free energy change or on the equilibrium constant K. It depends solely on the activation energy. The higher the activation energy, the slower the reaction rate will be.

Problem 3.7

Draw the reaction diagram (ΔG° versus reaction coordinate) for

(a) a slow reaction whose equilibrium constant is large
(b) a slow reaction whose equilibrium constant is small ($\ll 1$)
(c) a fast reaction whose equilibrium constant is large
(d) a fast reaction whose equilibrium constant is small ($\ll 1$)

3.5 CATALYSTS

We frequently use catalysts to speed up chemical reactions. A **catalyst** is a substance that increases the rate of a chemical reaction without being consumed itself in the reaction. A catalyst, however, does not affect the equilibrium position of a reaction or

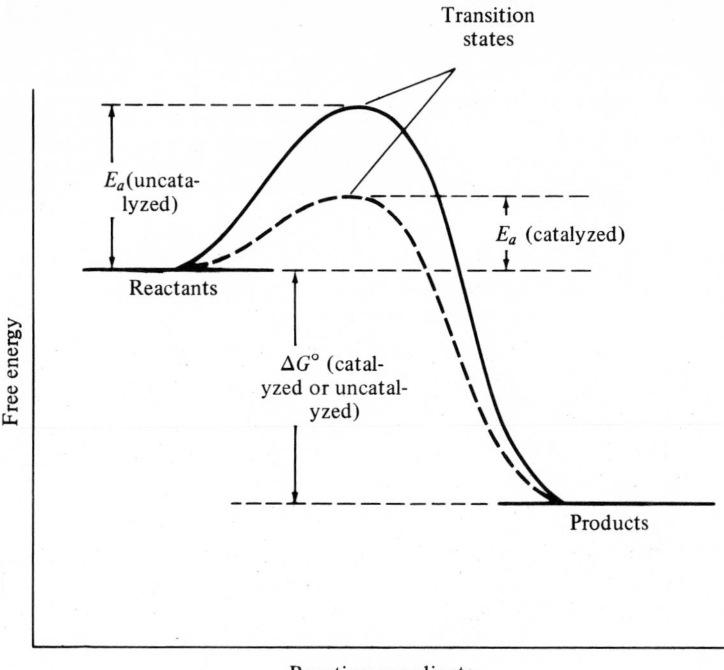

Figure 3.2. Free energy versus reaction coordinate for a hypothetical reaction without catalyst (solid curve) and with catalyst (dashed curve).

its equilibrium constant (and therefore its $\Delta G°$). The only effect of a catalyst is to decrease the activation energy, E_a. The catalyst enters into the reaction in such a way as to lower the energy of the transition state. Figure 3.2 illustrates the energy relationships for a reaction with and without a catalyst.

Problem 3.8
Metallic nickel catalyzes the reaction of H_2 with O_2. Modify the reaction diagram in Figure 3.1(b) to show how nickel affects the reaction.

3.6 REACTIVE INTERMEDIATES IN ORGANIC CHEMISTRY

In our discussions of reactions in later chapters we shall find that many reactions occur through highly unstable, very reactive intermediate molecular fragments (Fig. 3.3). These reactive intermediates are usually ions or free radicals. Note that an intermediate occurs at an energy minimum, and therefore is not a transition state. (A transition state occurs at an energy maximum.)

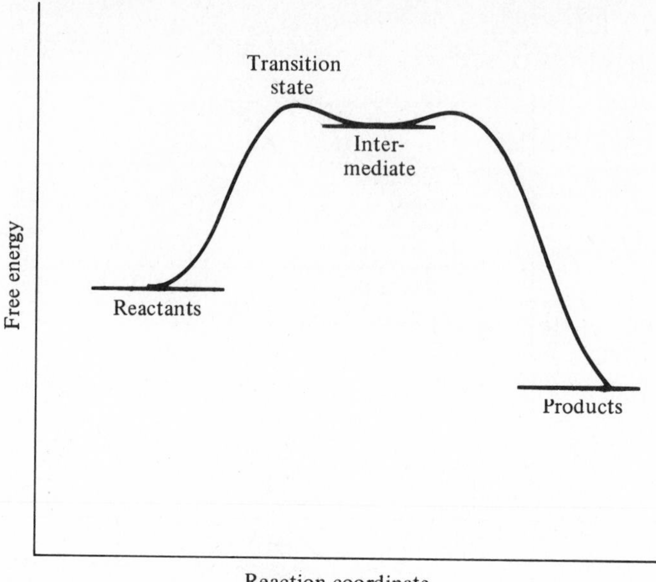

Figure 3.3. Free energy versus reaction coordinate for a reaction that occurs through a reactive intermediate.

These intermediates are encountered so frequently that we shall summarize their nature and properties here. We shall be concerned with three types of reactive intermediates:

Free radicals
Carbocations (also called carbonium ions)
Carbanions

3.7 FREE RADICALS

Free radicals are produced when an electron pair covalent bond breaks homolytically: that is, when the bond breaks in such a way that one of the electrons of the bond ends up on each atom. For example, the **homolytic cleavage** of a carbon–carbon bond would yield two free radicals, as shown:

$$\begin{array}{ccccc} & H & & H & \\ & | & & | & \\ H- & C & : & C & -H \\ & | & & | & \\ & H & & H & \end{array} \longrightarrow \begin{array}{ccc} & H & \\ & | & \\ H- & C\cdot & \\ & | & \\ & H & \end{array} + \begin{array}{ccc} & H & \\ & | & \\ \cdot & C & -H \\ & | & \\ & H & \end{array}$$

A **free radical** is thus an atom or group that bears an unpaired electron (Fig. 3.4). Free radicals are very reactive, and the reaction shown above does not occur spontaneously under ordinary conditions. In fact, free radicals are so reactive that their

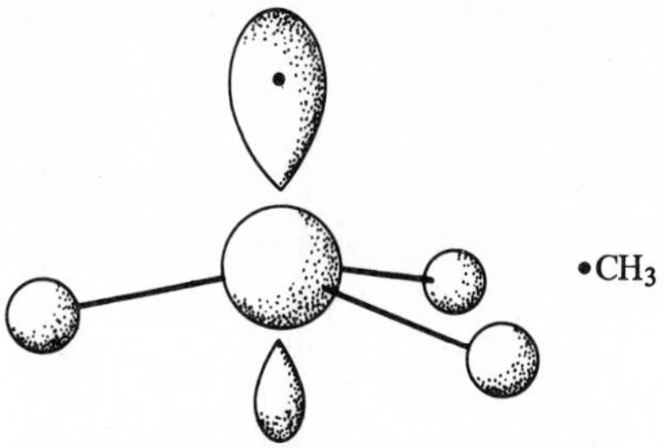

Figure 3.4. The methyl free radical.

recombination has an activation energy equal to zero (Fig. 3.5). The reaction of a free radical with a molecule that is not a free radical always produces another free radical:

$$R\cdot + \underset{\text{ethylene}}{\begin{array}{c} H\ \ H \\ |\ \ \ \ | \\ C{::}C \\ |\ \ \ \ | \\ H\ \ H \end{array}}{}^{*} \longrightarrow R-\begin{array}{c} H\ \ \ \ H \\ |\ \ \ \ \ \ | \\ C-C\cdot \\ |\ \ \ \ \ \ | \\ H\ \ \ \ H \end{array}$$

For this reason, reactions of free radicals are often **chain reactions**—reactions whose products are themselves reactants.

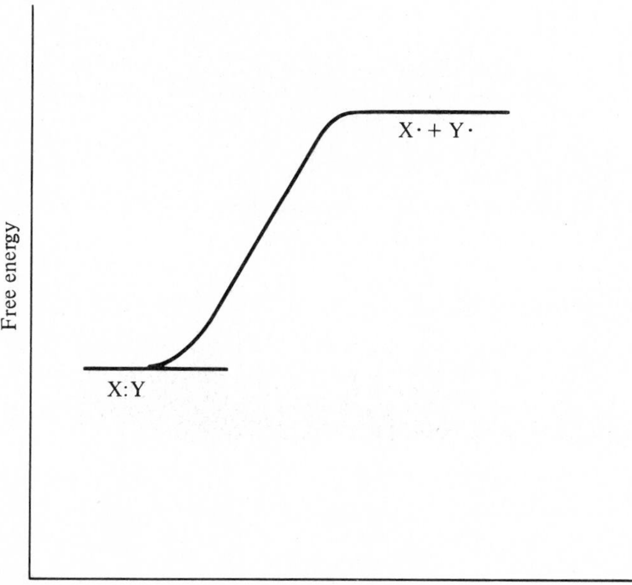

Figure 3.5. Free energy versus reaction coordinate for the homolytic breakage of the covalent bond between the atoms X and Y to form the free radicals X· and Y·. Note that the reverse reaction, X· + Y· → X:Y, has $E_a = 0$.

* The curved, single-barbed arrow indicates the movement of a single electron.

Heterolytic cleavage of an electron pair covalent bond produces ions. In **heterolytic** cleavage both electrons of the bond remain on one atom, the more electronegative one. Heterolytic bond breakage occurs only when the two atoms have very different electronegativities. An example is HCl, which dissociates into H^+ and Cl^- ions. Heterolytic cleavage of a C—X bond can occur if X is an electronegative element such as O, Cl, Br, or I. Heterolytic cleavage of C—X yields a carbocation.* A **carbocation** is an ion in which a carbon atom has only six electrons in its valence shell (Fig. 3.6). The formal charge on a carbocation is +1:

$$R_3C{:}X \xrightarrow[\text{cleavage}]{\text{heterolytic}} R_3C^+ + {:}X^-$$

a carbocation

Carbocations are so reactive that the above reaction does not normally occur spontaneously to yield appreciable amounts of carbocations. For example, an alcohol does not spontaneously dissociate to any measurable extent to produce carbocations:

$$R_3C\text{—}OH \not\rightarrow R_3C^+ + OH^- \quad (K \text{ is very small})$$

The reason for this is that the OH^- ion is a very strong base and the carbocation is a very strong acid (Fig. 3.7). We shall see in Chapter 10 that the reaction can be made to occur by the use of an acid catalyst.

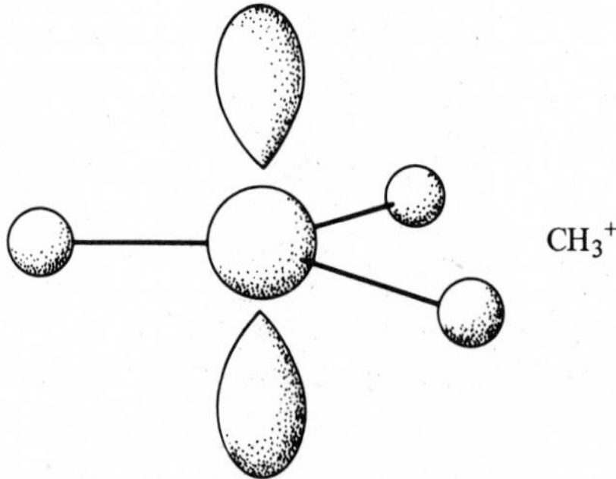

Figure 3.6. The methyl cation.

* The curved, double-barbed arrow indicates the movement of a pair of electrons.

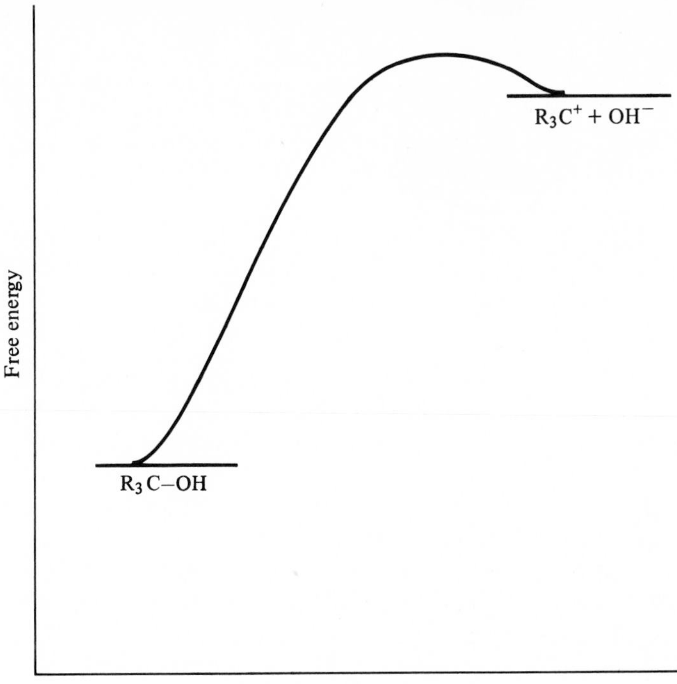

Figure 3.7. Free energy versus reaction coordinate for the dissociation of an alcohol, $R_3C—OH$ to form the carbocation R_3C +. This reaction does not occur to any appreciable extent.

3.9 CARBANIONS

Carbanions are ions in which a carbon atom bears an unshared pair of electrons and has a formal charge of -1 (Fig. 3.8). Such ions are more stable when carbon is bonded to one or more electronegative atoms. When such a carbon atom is bonded to hydrogen, the molecule is an acid.

$$R_3C{:}H \rightleftharpoons R_3C{:}^- + H^+$$

Hydrocarbons are not normally acidic. The above reaction occurs even to a slight extent only when the R groups attached to the carbon atom are very strongly electron-withdrawing. Only then is the negative charge stabilized enough for the carbanion to exist. A well-known example is the cyanide ion, $:N{:::}C{:}^-$.

In most cases, carbanions are so reactive (such strong bases) that they must be prepared by irreversible reactions; that is, reactions that never establish equilibrium:

$$R—H + Na^+H^- \longrightarrow R{:}^-Na^+ + H_2\uparrow$$

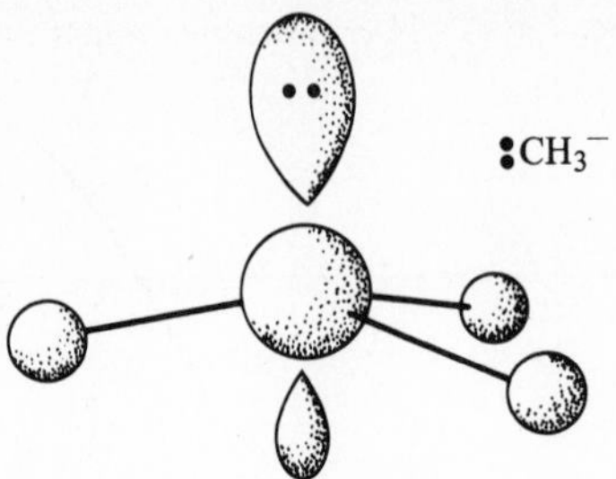

Figure 3.8. The methyl carbanion.

In this case, the basic hydride ion, H:$^-$, accepts the proton from the weak acid, R—H. The loss of gaseous H_2 to the surroundings causes the equilibrium to be displaced completely to the right. Large amounts of very reactive carbanions can be prepared in this way. Of course, the reaction must be carried out in a solvent that is an even weaker acid than R—H; otherwise it will react with the carbanion, R:$^-$, to produce the conjugate base of the solvent:

$$\text{R:}^- + \text{S—H} \rightleftarrows \text{R—H} + \text{S:}^-$$

where S—H is the solvent.

Problem 3.9
Draw the formulas of the free radicals produced by homolytic cleavage of the indicated bonds in each of the following molecules.

(a) H—H, (b) :C̤̈l—C̤̈l:, (c) C_6H_5—CH_3 (ring—CH_3 bond)

Problem 3.10
Draw the formulas of the species produced by heterolytic cleavage of the bonds indicated in Problem 3.9.

Problem 3.11
Draw the structure of the missing species in each of the following equations.

(a) $(CH_3)_3C$—C̤̈l: $\rightleftarrows$:C̤̈l:$^-$ + ________

(b) $(CH_3)_3C$—C̤̈l: $\rightleftarrows$:C̤̈l· + ________

(c) $CH_3CH_2CH_2CH_2Li \rightleftarrows Li^+ +$ ________

[In (d) and (f) only the C—H bond shown undergoes reaction.]

(d)

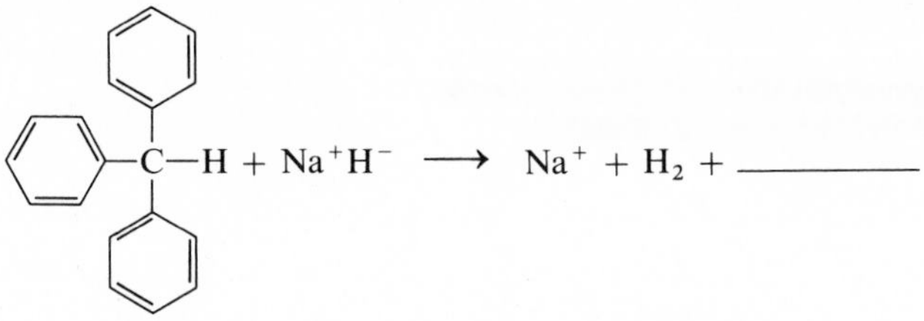

$$(C_6H_5)_3C\text{—}H + Na^+H^- \longrightarrow Na^+ + H_2 + \_\_\_\_\_\_$$

(e)

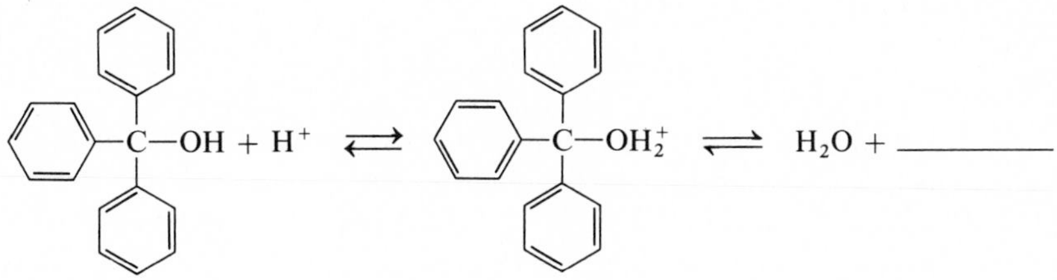

$$(C_6H_5)_3C\text{—}OH + H^+ \rightleftarrows (C_6H_5)_3C\text{—}OH_2^+ \rightleftharpoons H_2O + \_\_\_\_\_\_$$

(f)

$$(C_6H_5)_3C\text{—}H + :\ddot{\underset{..}{Br}}\cdot \rightleftarrows H\text{—}Br + \_\_\_\_\_\_$$

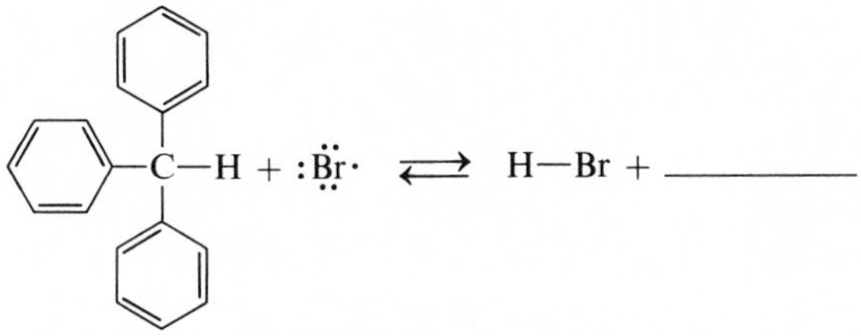

3.10 LEWIS ACID-BASE THEORY

A base can accept a proton because a base bears an unshared electron pair. The electron pair can form a coordinate covalent bond with the proton that has an empty orbital. In actuality, any compound or ion that has a stable, empty orbital can react with a base to form a coordinate covalent bond. An example is boron trifluoride, BF_3. Boron is in group III of the periodic table, and it therefore has three valence electrons. The formula

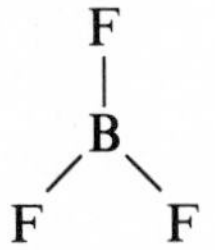

F
|
B
/ \
F F

is planar because the three bonding orbitals in boron are sp^2 hybrids. There is an empty, unused p orbital available on boron, and it can react with a base such as a fluoride ion to form BF_4^-. (Note that in BF_4^- the boron is sp^3 hybridized.)

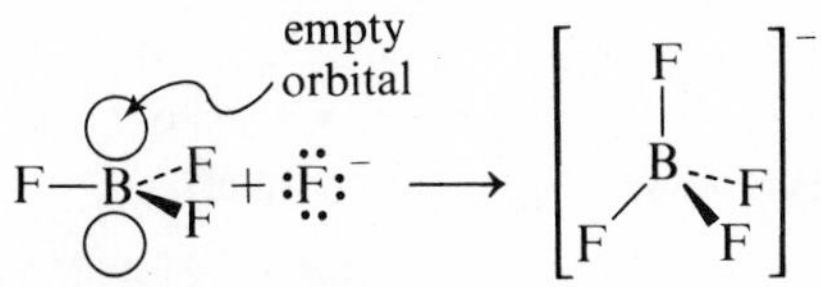

empty orbital

$$F\text{—}BF_2 + :\ddot{\underset{..}{F}}:^- \longrightarrow [BF_4]^-$$

The resulting ion has a formal charge of -1 on boron.

G. N. Lewis recognized that this class of reaction represents a kind of acid-base reaction, so he proposed the following new definition of acids and bases:

An acid is an electron pair acceptor.
A base is an electron pair donor.

These new definitions include those of Brønsted and Lowry; that is, a proton donor is a Lewis acid as well as a Brønsted–Lowry acid. A Brønsted–Lowry base, to be a proton acceptor, must also be an electron pair donor. Thus Brønsted–Lowry bases are also Lewis bases. The Lewis definition also includes a wide variety of compounds as acids that the Brønsted–Lowry definition does not include. Some examples are carbocations, BF_3, $AlCl_3$, $ZnCl_2$, and $FeCl_3$. All these species possess a vacant orbital that can accept an electron pair to form a coordinate covalent bond. For example, using Cl^- as a base:

$$\begin{array}{c} :\ddot{F}: \\ | \\ :\ddot{\underset{..}{F}}-B \\ | \\ :\underset{..}{F}: \end{array} + :\ddot{\underset{..}{Cl}}:^- \longrightarrow \begin{array}{c} :\ddot{F}: \\ | \\ :\ddot{\underset{..}{F}}-B^- -\ddot{\underset{..}{Cl}}: \\ | \\ :\underset{..}{F}: \end{array}$$

$$\begin{array}{c} :\ddot{Cl}: \\ | \\ :\ddot{\underset{..}{Cl}}-B \\ | \\ :\underset{..}{Cl}: \end{array} + :\ddot{\underset{..}{Cl}}:^- \longrightarrow \begin{array}{c} :\ddot{Cl}: \\ | \\ :\ddot{\underset{..}{Cl}}-B^- -\ddot{\underset{..}{Cl}}: \\ | \\ :\underset{..}{Cl}: \end{array}$$

We can summarize the Brønsted–Lowry and Lewis definitions as follows:

1. A Brønsted–Lowry acid is also a Lewis acid.
2. A Brønsted–Lowry base is also a Lewis base.
3. A Lewis base is also a Brønsted–Lowry base.
4. *A Lewis acid is not necessarily a Brønsted–Lowry acid.*

Problem 3.12
Explain why the carbocation R_3C^+ and $AlCl_3$ can be classified as Lewis acids. Write equations for the reaction of each with ammonia, $:NH_3$.

Problem 3.13
Give an example of (a) a Lewis acid that is also a Brønsted–Lowry acid and (b) a Lewis acid that is not a Brønsted–Lowry acid.

NEW TERMS

Acid Brønsted–Lowry definition: a substance that is capable of releasing a hydrogen ion (3.1). Lewis definition: an electron pair acceptor (3.10).

Acid dissociation constant (K_a) The constant

$$K_a = \frac{[H_3O^+]\,[\text{conjugate base}]}{[\text{acid}]}$$

for the reaction of an acid with water (3.3):

$$\text{Acid} + H_2O \rightleftarrows H_3O^+ + \text{conjugate base}$$

Acid strength The extent to which the equilibrium

$$HX + H_2O \rightleftarrows H_3O^+ + X^-$$

favors the products $H_3O^+ + X^-$. In other words, the extent to which the acid HX dissociates. A strong acid dissociates to a great extent; a weak acid dissociates to a slight extent (3.2).

Activation energy (E_a) The amount of energy that reacting molecules must acquire to achieve the transition state (3.4).

Base Brønsted–Lowry definition: a substance that is capable of reacting with a hydrogen ion (3.1). Lewis definition: an electron pair donor (3.10).

Base dissociation constant (K_b) The constant

$$K_b = \frac{[OH^-][\text{conjugate acid}]}{[\text{base}]}$$

for the reaction of a base with water (3.3):

$$\text{Base} + H_2O \leftrightarrows OH^- + \text{conjugate acid}$$

Carbanion An ion in which a carbon atom bears an unshared pair of electrons and has a formal charge of -1 (3.9).

Carbocation An ion in which a carbon atom has only six electrons in its valence shell and therefore has a formal charge of $+1$ (3.8).

Catalyst A substance that increases the rate of a chemical reaction without being consumed itself in the reaction (3.5).

Chain reaction A set of reactions in which the product of one reaction is a reactant in another reaction (3.7).

Conjugate acid The molecule or ion formed when a Brønsted–Lowry base accepts a hydrogen ion (3.1).

Conjugate base The molecule or ion that remains after a Brønsted–Lowry acid has lost a hydrogen ion (3.1).

Free energy (ΔG) The energy available to do work at a constant temperature and pressure. If the temperature is 25°C, the pressure is 1.0 atmosphere, and all concentrations are 1.0 mole per liter, the free energy is designated ΔG° (3.4).

Free radical An atom or molecule that has an unpaired electron (3.7).

Heterolytic cleavage A form of bond breakage in which both electrons of the bonding pair remain on one atom, the more electronegative one (3.8).

Homolytic cleavage A form of bond breakage in which one electron of the bonding pair remains on each atom involved in the bond (3.7).

Hydronium ion H_3O^+ (3.1).

Proton A hydrogen ion, H^+ (3.1).

Transition state The configuration of reacting molecules represented by the top of the energy barrier in a plot of ΔG° versus reaction coordinate (Fig. 3.1); the transition state is also called the activated complex (3.4).

Water constant (K_w) $K_w = [H_3O^+][OH^-] = 1 \times 10^{-14}$ at 25°C (3.3). $K_w = K_a \cdot K_b$ for a conjugate acid-base pair.

ADDITIONAL PROBLEMS

3.14 Identify the Brønsted–Lowry acids and bases in the following compounds. In each, give the conjugate base and acid, respectively. Which are neither Brønsted–Lowry acids nor Brønsted–Lowry bases? Which are both Brønsted–Lowry acids and Lewis acids? Which are both Brønsted–Lowry bases and Lewis bases?

[a] H_2O [b] $AlCl_3$ [c] HF
[d] HSO_4^- [e] NO_3^- (f) H_2S
(g) BF_3 (h) HI (i) HSO_3^-
(j) SH^- (k) $H_2PO_4^-$

[3.15] Write equations for the reactions of the base, water, with each of the Brønsted–Lowry acids in Problem 3.14. Identify the conjugate acid-base pairs in each equation.

[3.16] Write equations for the reaction of the acid, water, with each of the Brønsted–Lowry bases in Problem 3.14. Identify the conjugate acid-base pairs in each equation.

3.17 Calculate the K_b for each of the following bases.

[a] $HC_2O_4^-$ [b] F^- [c] OH^- (d) H_2O
(e) HSO_4^-

3.18 Calculate the equilibrium constant, K, for the homolytic dissociation of methane: $CH_4 \longrightarrow CH_3\cdot + \cdot H$, at 25°C if $\Delta G°$ for this reaction is 10.0 kcal/mole. (R = 1.987 cal/mole-degree.)

[3.19] The reaction of hydrogen with oxygen is very slow unless a flame or a catalyst is present. Explain in terms of Figure 3.1(b) how each of these agents accelerates the reaction.

3.20 Compare the free energy versus reaction coordinate diagrams of the following pairs of acids. Assume that each is dissolved in water, and that equilibrium is reached rapidly.

[a] $CH_3C(=O)-OH$ and HCN (b) $CH_3C(=O)-OH$ and C_6H_5-OH (phenol, drawn as a benzene ring bearing —OH)

3.21 Draw the Lewis dot structure for each of the following, and show formal charges where applicable.

[a] the methyl free radical
[b] the ethyl cation
[c] the phenyl carbanion (phenyl group = C_6H_5-, drawn as a benzene ring with one open bond)
[d] the product(s) of homolytic cleavage of the Br_2 molecule
[e] the products of heterolytic cleavage of the Br_2 molecule
(f) the hydroxyl free radical
(g) the methyl carbocation
(h) the ethyl carbanion
(i) the product(s) of homolytic cleavage of the Cl_2 molecule
(j) the products of heterolytic cleavage of the Cl_2 molecule

[3.22] Explain with examples how Lewis acids differ from Brønsted–Lowry acids. Give examples of Lewis acids that are *not* Brønsted–Lowry acids.

[3.23] Do Lewis bases differ from Brønsted–Lowry bases? Explain.

3.24 Predict the products of the following acid-base equilibria.

[a] $HF + H_2O \rightleftarrows$

[b] $AlBr_3 + Br^- \rightleftarrows$

[c] $FeCl_3 + Cl^- \rightleftarrows$

[d] $CH_3^+ + H_2O \rightleftarrows$

[e] $CH_3^- + H_2O \rightleftarrows$

(f) $HI + H_2O \rightleftarrows$

(g) $BCl_3 + Cl^- \rightleftarrows$

(h) $FeBr_3 + Br^- \rightleftarrows$

(i) $CH_3CH_2^+ + HOCH_3 \rightleftarrows$

(j) $CH_3CH_2^- + HOCH_3 \rightleftarrows$

[3.25] Where possible, predict which of the equilibria in Problem 3.24 occur to a large extent to the right.

3.26 Which of the reactions in Problem 3.24 can be defined as acid-base reactions *only* in terms of the Lewis acid-base theory?

[3.27] Which of the reactions in Problem 3.24 can be defined as acid-base reactions *only* in terms of the Brønsted–Lowry acid-base theory? Explain.

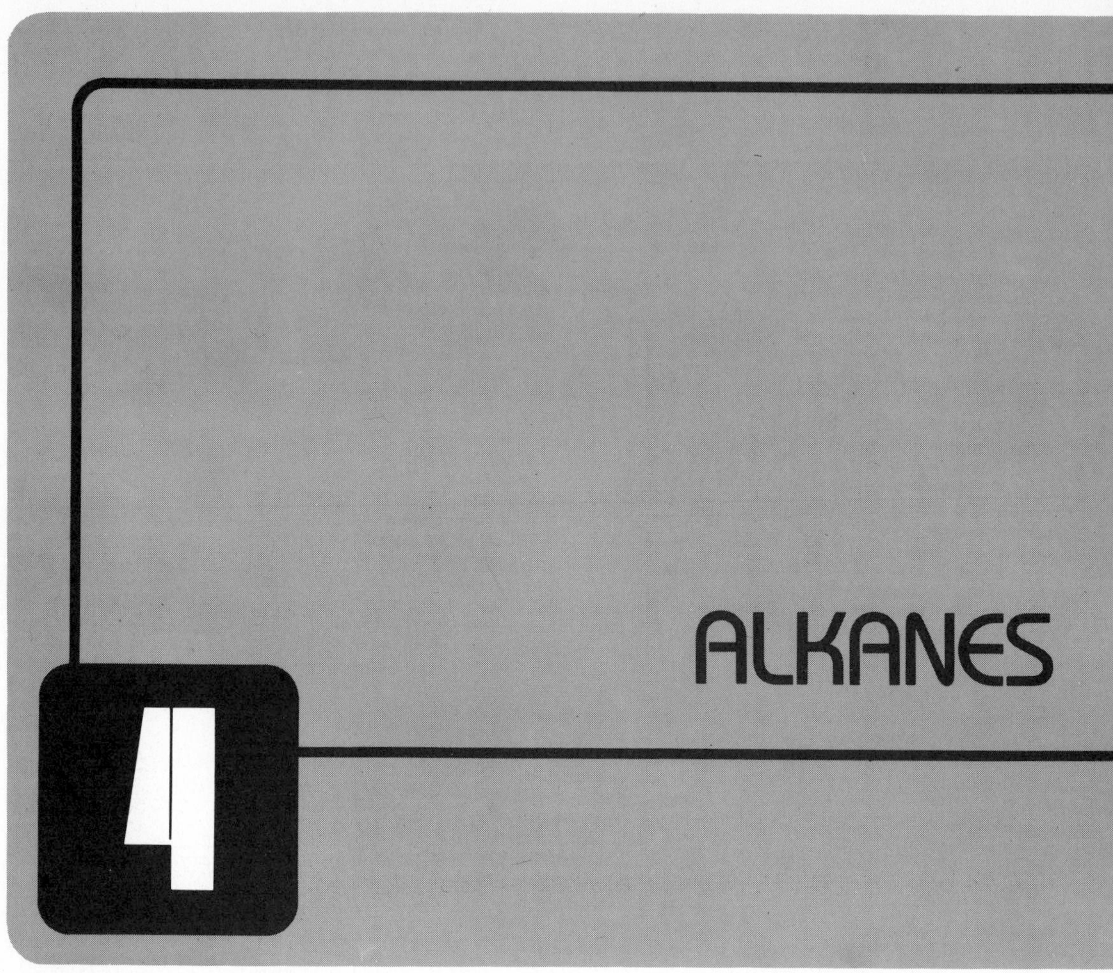

4.1 STRUCTURE

From the standpoint of structure, the alkanes, or saturated hydrocarbons, are the simplest organic compounds. They contain only the elements carbon and hydrogen, only single bonds, and no rings (they are acyclic). There is apparently no limit to the number of carbon atoms that can be linked together to form stable molecules: polyethylene molecules have carbon chains that contain thousands of carbon atoms.

The series of alkanes is an example of a **homologous series**. In a homologous series of compounds each member differs from the one preceding it by an additional $-CH_2-$ group. The general formula of the alkanes is

$$H\!\left(CH_2\right)_n H \quad \text{or} \quad C_nH_{2n+2}, \text{ where } n = 1, 2, 3, \ldots$$

Table 4.1 shows the first several members of the alkane series.

Table 4.1 The Simple Alkanes

Molecular Formula	Condensed Formula	Name
CH_4	CH_4	Methane
C_2H_6	CH_3CH_3	Ethane
C_3H_8	$CH_3CH_2CH_3$	Propane
C_4H_{10}	$CH_3(CH_2)_2CH_3$	Butane
C_5H_{12}	$CH_3(CH_2)_3CH_3$	Pentane
C_6H_{14}	$CH_3(CH_2)_4CH_3$	Hexane
C_7H_{16}	$CH_3(CH_2)_5CH_3$	Heptane
C_8H_{18}	$CH_3(CH_2)_6CH_3$	Octane
C_9H_{20}	$CH_3(CH_2)_7CH_3$	Nonane
$C_{10}H_{22}$	$CH_3(CH_2)_8CH_3$	Decane
$C_{11}H_{24}$	$CH_3(CH_2)_9CH_3$	Undecane
$C_{12}H_{26}$	$CH_3(CH_2)_{10}CH_3$	Dodecane
$C_{15}H_{32}$	$CH_3(CH_2)_{13}CH_3$	Pentadecane
$C_{20}H_{42}$	$CH_3(CH_2)_{18}CH_3$	Eicosane

We have seen in Chapter 1 that carbon is tetravalent. The combination of a 2*s* with three 2*p* orbitals gives four sp^3 hybrid orbitals which are oriented toward the corners of a regular tetrahedron. The overlap of an sp^3 hybrid orbital with the 1*s* orbital of a hydrogen atom (Fig. 1.5) gives a strong C—H bond. The structure of methane is

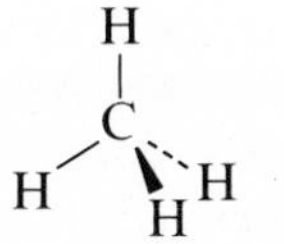

H
|
C
H H H

In ethane, bonding between carbon atoms occurs by the overlap of two sp^3 hybrid orbitals, one from each carbon (Fig. 1.6). We arrive at the structure of ethane, C_2H_6, by joining hydrogen atoms to each carbon atom according to the octet rule: There must be no more than four covalent bonds (eight electrons) around each carbon atom.

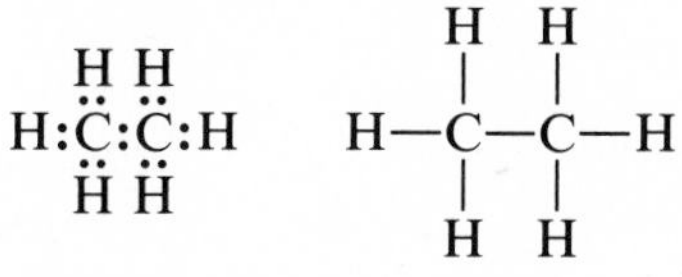

```
  H H          H   H
  ·· ··        |   |
H:C:C:H    H—C—C—H
  ·· ··        |   |
  H H          H   H
```

As in methane, the four bonds around carbon are tetrahedrally oriented. The following drawings represent ball-and-stick models (Fig. 2.1) of ethane:

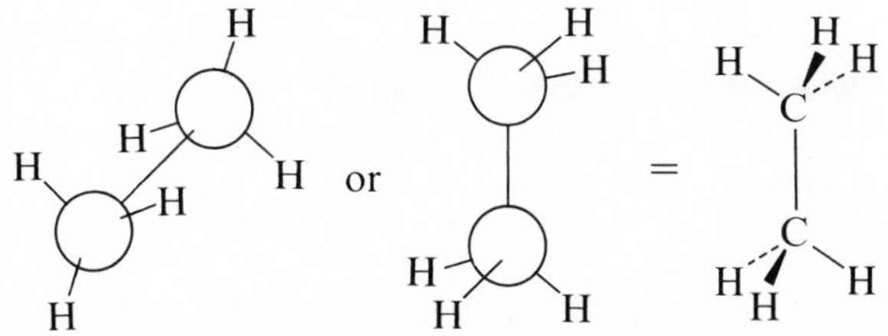

H H H H H or H H H H H H H = H H H C C H H H

Electron diffraction and spectroscopic measurements show that the bond angles in ethane are 109.5°, the C—H bond length is 1.10 Å* (the same as in methane), and the C—C bond length is 1.53 Å. These values are nearly the same for all the alkanes.

The next member of the alkane series is propane, C_3H_8. Propane boils at $-42°C$ and is therefore a gas at room temperature. It resembles methane and ethane in its chemical and physical properties. Butane, C_4H_{10}, is also a gas. It resembles methane, ethane, and propane in its properties, but differs in one respect: Butane exists as two isomers, called butane and isobutane (Sec. 2.3).

$CH_3CH_2CH_2CH_3$ (butane) $CH_3CH(CH_3)CH_3$ (isobutane)

All the higher members of the alkane series exhibit isomerism.

Problem 4.1
Draw the structural formulas of all the isomers of C_5H_{12} and C_6H_{14}.

4.2 CONFORMATION AND CONFORMERS

The structure of ethane can be drawn in two extreme conformations. A **conformation** is a particular spatial arrangement of the atoms in a molecule. Conformations, or **conformers**, *differ only in the rotation of their atoms about single bonds.* For this reason there are an infinite number of conformations possible for a given molecule. In one of these, the **eclipsed** conformation, the C—H bonds of one carbon atom are directly behind those of the other carbon atom if we look along the C—C bond axis. In the **staggered** conformation we can see all six C—H bonds if we look along the C—C bond axis (Figure 4.1).

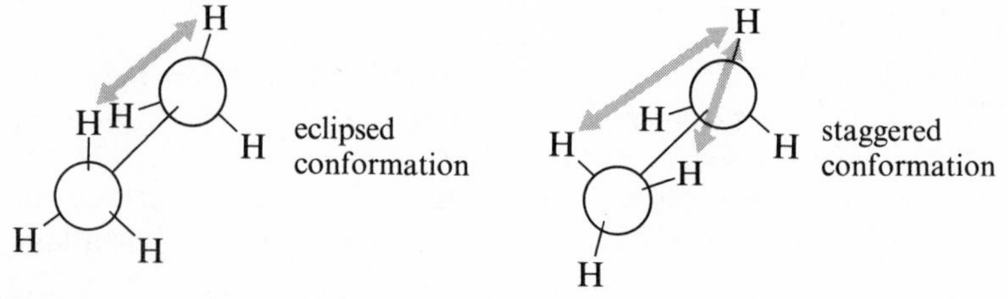

eclipsed conformation

staggered conformation

Newman projection formulas are useful in representing these conformations. The C—C bond is perpendicular to the plane of the page, and the circle represents a plane that separates the front and back carbons.

* The **Angstrom unit**, Å, equals 10^{-8} cm.

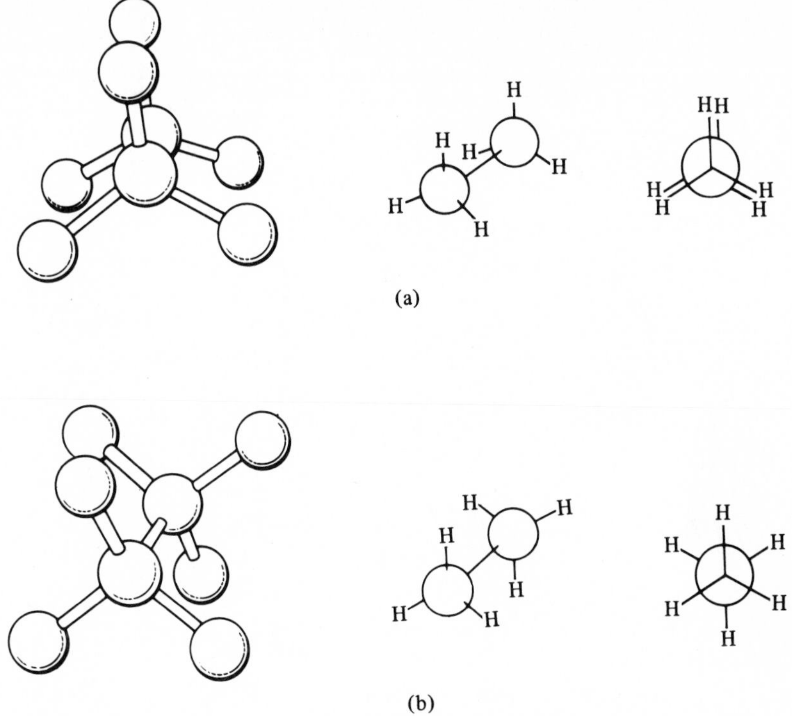

Figure 4.1. Ethane in the (a) eclipsed conformation and (b) in the staggered conformation.

eclipsed (H's on back carbon cannot be seen.)

staggered (H's on back carbon are visible.)

The carbon-carbon single bond is symmetrical about the bond axis. Rotation about the C—C bond should therefore be free. There is, however, a slight resistance to rotation. The free energy of ethane is lowest in the staggered conformation and increases as it approaches the eclipsed conformation (Fig. 4.2). The energy required to convert a molecule from one staggered conformation to another is very small compared with the average kinetic energy of molecules under ordinary conditions. Therefore, near room temperature, rotation about the single bond in ethane is essentially free.

Problem 4.2

Draw Newman projection formulas of all the staggered and eclipsed conformations of butane that result from rotation about the bond that joins the second and third carbons.

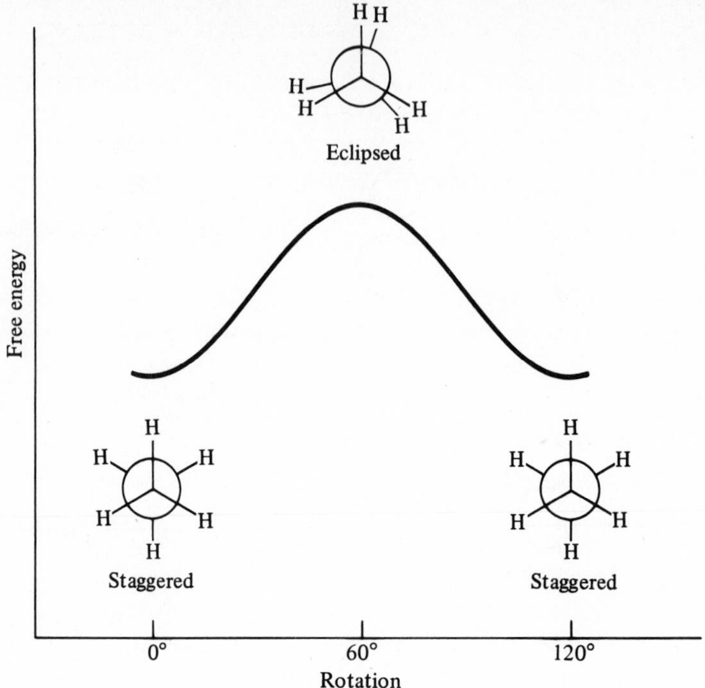

Figure 4.2. Free energy changes that accompany the rotation of the carbon–carbon single bond in ethane.

4.3 ALKYL GROUPS

Let us consider a molecule that bears a functional group; for example, $CH_3CH_2CH_2OH$. Because one hydrogen atom has been replaced by the OH group, the general formula of the hydrocarbon part is C_nH_{2n+1}. We name the hydrocarbon part of such a molecule by replacing the ending *-ane* of the alkane with *-yl*. In the above example, the $CH_3CH_2CH_2—$ group is called propyl (from propane). The general name of a hydrocarbon group derived from an alkane is *alkyl*.

The names of a few common alkyl groups are as follows:

$CH_3—$	methyl
$CH_3CH_2—$	ethyl
$CH_3CH_2CH_2—$	propyl
$CH_3CH_2CH_2CH_2—$	butyl
$CH_3(CH_2)_5CH_2—$	heptyl

Problem 4.3

Name the following alkyl groups.

(a) $CH_3CH_2CH_2CH_2CH_2—$ (b) $CH_3(CH_2)_4CH_2—$

(c) $CH_3(CH_2)_6CH_2—$

Problem 4.4
Although there are no isomers of propane, propyl derivatives such as the chlorides and the alcohols exhibit isomerism. Draw the structures of these isomers.

4.4 CLASSIFICATION OF CARBON

We can classify any given carbon atom in a molecule according to the number of other carbon atoms bonded to it. (Note that methane, CH_4, is in a class by itself.) We identify the following carbon atoms in a molecule:

Primary (1°) carbon atom has a bond to only one other carbon.

Secondary (2°) carbon atom has bonds to two other carbons.

Tertiary (3°) carbon atom has bonds to three other carbons.

Quaternary (4°) carbon atom has bonds to four other carbons.

As an example, we can classify the carbon atoms in the following molecule as shown.

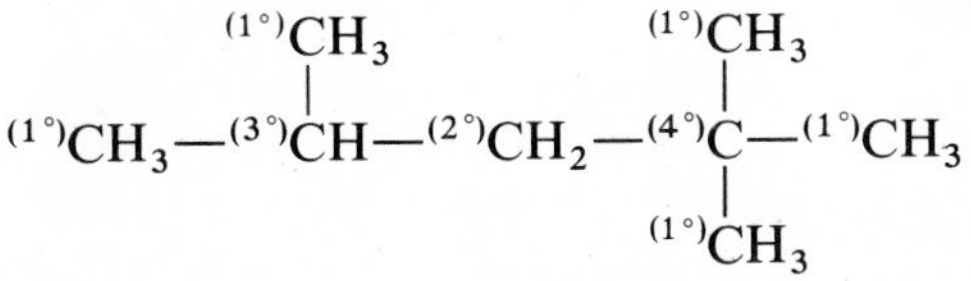

$$
\begin{array}{ccccccccc}
 & & ^{(1°)}CH_3 & & & & ^{(1°)}CH_3 & & \\
 & & | & & & & | & & \\
^{(1°)}CH_3 & — & ^{(3°)}CH & — & ^{(2°)}CH_2 & — & ^{(4°)}C & — & ^{(1°)}CH_3 \\
 & & & & & & | & & \\
 & & & & & & ^{(1°)}CH_3 & &
\end{array}
$$

Problem 4.5
Label the carbon atoms in each of the following molecules as 1°, 2°, 3°, or 4°.

(a) CH_3CH_3

(b)
$$
\begin{array}{ccccc}
 & & CH_3 & & \\
 & & | & & \\
CH_3 & — & C & — & CH_3 \\
 & & | & & \\
 & & CH_3 & &
\end{array}
$$

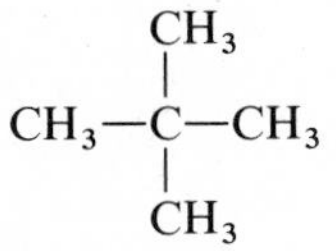

(c) [cyclopentane ring]

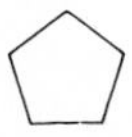

(d) [cyclohexane ring]–CH_3

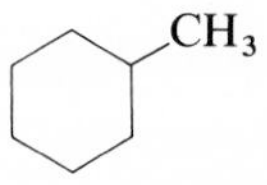

Two different alkyl groups are possible with a three-carbon chain—the propyl group and the isopropyl group (see Problem 4.4):

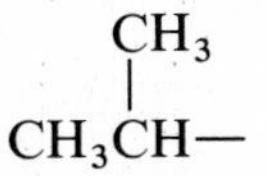

$$
\begin{array}{c}
CH_3 \\
| \\
CH_3CH—
\end{array}
$$

isopropyl group

$CH_3CH_2CH_2—$

propyl group

The two butanes—butane and isobutane—similarly give rise to four alkyl groups:

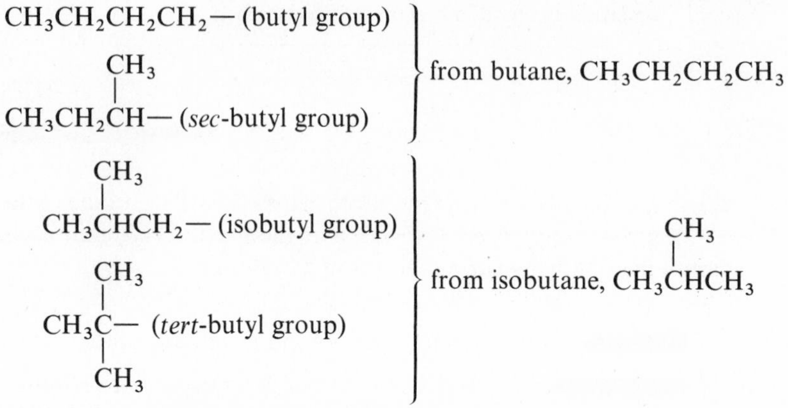

$$\left.\begin{array}{l} CH_3CH_2CH_2CH_2\text{—}\ \text{(butyl group)} \\ \\ \quad\quad CH_3 \\ \quad\quad | \\ CH_3CH_2CH\text{—}\ (\textit{sec}\text{-butyl group}) \end{array}\right\} \text{from butane, } CH_3CH_2CH_2CH_3$$

$$\left.\begin{array}{l} \quad CH_3 \\ \quad | \\ CH_3CHCH_2\text{—}\ \text{(isobutyl group)} \\ \quad CH_3 \\ \quad | \\ CH_3C\text{—}\ (\textit{tert}\text{-butyl group}) \\ \quad | \\ \quad CH_3 \end{array}\right\} \text{from isobutane, } \begin{array}{c} CH_3 \\ | \\ CH_3CHCH_3 \end{array}$$

The prefixes iso-, neo-, *sec*-, and *tert*- have precise meanings. The stem of the word with which they are used refers to the *total* number of carbon atoms in the group; propyl and isopropyl groups have three carbon atoms, and all the butyl groups have four carbon atoms.

Iso- means that the end farthest from the incomplete valence bond has the *iso structure.*

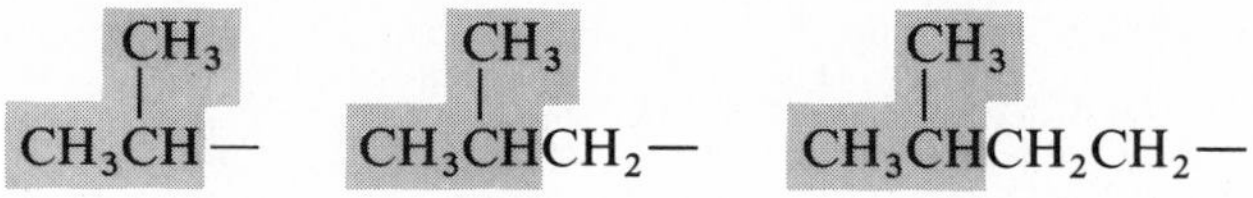

$$\begin{array}{c} CH_3 \\ | \\ CH_3CH\text{—} \end{array} \qquad \begin{array}{c} CH_3 \\ | \\ CH_3CHCH_2\text{—} \end{array} \qquad \begin{array}{c} CH_3 \\ | \\ CH_3CHCH_2CH_2\text{—} \end{array}$$

iso structures

Neo- means that the end farthest from the incomplete valence bond has the *neo structure.*

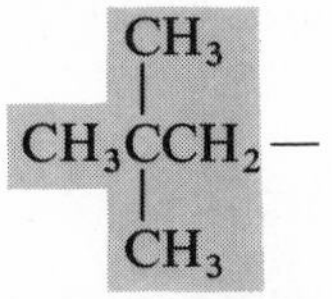

$$\begin{array}{c} CH_3 \\ | \\ CH_3CCH_2\text{—} \\ | \\ CH_3 \end{array} \qquad\qquad \begin{array}{c} CH_3 \\ | \\ CH_3CCH_2CH_2\text{—}Cl \\ | \\ CH_3 \end{array}$$

neo structure 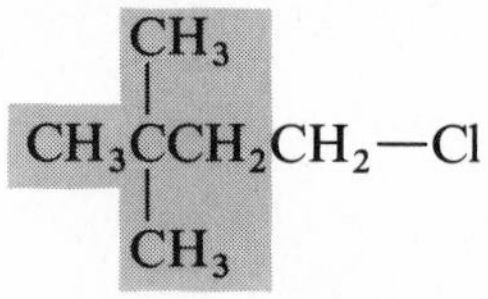   neohexyl chloride

Sec- is always italicized and separated from the rest of the name by a hyphen. It means that the incomplete valence bond is at a particular *secondary* carbon. Some *sec*- structures are

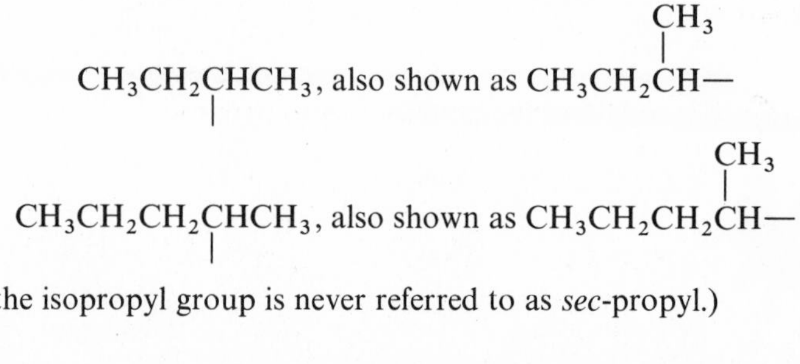

$$CH_3CH_2\underset{|}{C}HCH_3\text{, also shown as } \begin{array}{c} CH_3 \\ | \\ CH_3CH_2CH\text{—} \end{array}$$

$$CH_3CH_2CH_2\underset{|}{C}HCH_3\text{, also shown as } \begin{array}{c} CH_3 \\ | \\ CH_3CH_2CH_2CH\text{—} \end{array}$$

(Note that the isopropyl group is never referred to as *sec*-propyl.)

Tert- is also always italicized and separated from the rest of the name by a hyphen. It means that the incomplete valence bond is at a tertiary carbon atom. Some *tert-* structures are

$$\begin{array}{c} CH_3 \\ | \\ CH_3C- \\ | \\ CH_3 \end{array} \qquad \begin{array}{c} CH_3 \\ | \\ CH_3CH_2C- \\ | \\ CH_3 \end{array}$$

Problem 4.6

Name all the iso-, neo-, *sec-*, and *tert-* alkyl groups shown in Section 4.4.

4.5 NOMENCLATURE: COMMON NAMES

As organic chemistry developed into a science, chemists became painfully aware of the problem of giving a unique name to every organic compound. At first, the chemist who discovered a compound would give it whatever name he thought appropriate. It was not long before chemists realized that there would soon be too many names to remember, and more important, that there was little or no connection between the name of a compound and its structure. Many names dating from this early period are still in use. These are called *common* or *trivial* names. Examples are acetone $\left(CH_3\overset{O}{\overset{\|}{C}}CH_3\right)$, acetylene ($HC{\equiv}CH$), and the names of the first four alkanes (Table 4.1).

4.6 SYSTEMATIC NOMENCLATURE: IUPAC SYSTEM

An international, systematic nomenclature system now exists and is used by organic chemists throughout the world. It was proposed by the International Union of Pure and Applied Chemistry (*IUPAC*). This systematic nomenclature is not only universal in its scope, but it is also easy to learn and use. The following rules are used for naming alkanes:

1. The general name for the saturated hydrocarbons is *alkane*. All specific names have the ending *-ane* (meth*ane*, eth*ane*, prop*ane*, but*ane*, pent*ane*, hex*ane*, hept*ane*, oct*ane*, non*ane*, dec*ane*, etc.).
2. In naming a saturated hydrocarbon, select the longest *continuous* chain of carbon atoms in the molecule. The name of that "parent" chain is the basis for the name of the compound.

3. Assign numbers to the carbon atoms of the parent chain. Begin numbering at the end nearer to a substituent group. If there is a substituent group equally near to both ends, count from the end having the next nearer substituent.
4. Write the name of each substituent group in alphabetical order (ignore the prefixes di-, tri-, etc.) preceded by the number of its position on the parent chain. If the same group occurs more than once, use the prefixes di-, tri-, tetra-, etc., *and one number for each group.* Numbers are separated from each other by commas, and from the rest of the name by hyphens.

An example will help to clarify these rules.

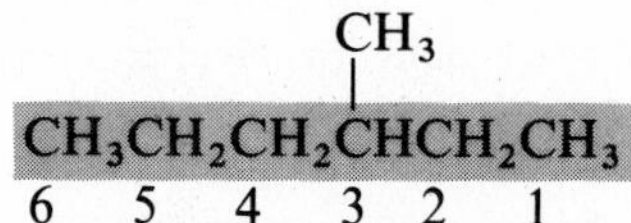

The longest continuous chain (shaded) has six carbon atoms, and the parent name is thus hexane (rules 1 and 2). We number the chain as shown because the substituent methyl group is on the third carbon from that end (it is on the fourth carbon if we number from the other end) (rule 3). Because the methyl substituent is attached to carbon number three, the name is 3-methylhexane (rule 4). Table 4.2 gives other examples.

Table 4.2 Systematic Names of Some Alkanes

CH_3 on C3, CH_3 on C5: $CH_3CH_2CHCH_2CHCH_2CH_3$ 3,5-dimethylheptane	This chain can be numbered from either end.
$CH_2CH_2CH_3$ on C4 (upper), CH_2CH_3 on C2 (lower): $CH_3CHCH_2CHCH_3$ 3,5-dimethyloctane	Be careful to use the longest *continuous* chain, even if it is drawn as it is here.
CH_3 above and below C2 and C5: $CH_3CCH_2CH_2CCH_2CH_3$ 2,2,5,5-tetramethylheptane	Use one number for *each* substituent, even if more than one substituent occurs on the same carbon atom of the parent chain.
CH_3CH_2 and CH_3 above: $CH_3CH_2CHCHCH_3$ 3-ethyl-2-methylpentane	Separate numbers and names with hyphens, and name in alphabetical order.

Problem 4.7

Give the systematic name of each of the following compounds.

(a) $CH_3CH_2CH_2CH_3$

(b) 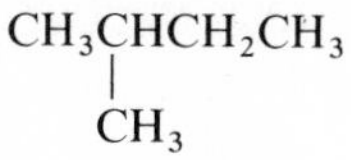
$$\begin{array}{l} CH_3\overset{}{C}HCH_2CH_3 \\ \quad\;\;\, | \\ \quad\; CH_3 \end{array}$$

(c)
$$\begin{array}{c} CH_3 \\ | \\ CH_3CCH_3 \\ | \\ CH_3 \end{array}$$

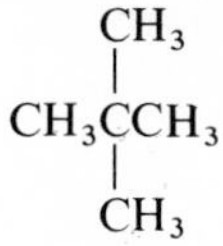

(d) 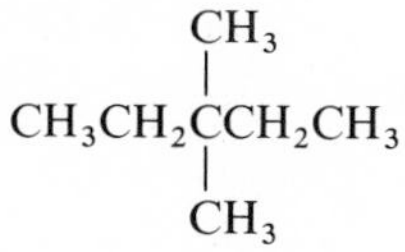
$$\begin{array}{c} CH_3 \\ | \\ CH_3CH_2CCH_2CH_3 \\ | \\ CH_3 \end{array}$$

Problem 4.8

Write the structural formula of each of the following compounds.

(a) 2,2,4,4-tetramethylpentane,

(b) 3,5-diethylheptane,

(c) 3-ethyl-2,3-dimethyl-4-propylnonane,

(d) 4,6-dipropylnonane

Generally systematic names do not include trivial names. Some trivial names occur so often, however, that the IUPAC has adopted them. The IUPAC recognizes the following special trivial names for alkyl groups:

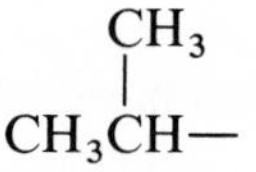

$$\begin{array}{c} CH_3 \\ | \\ CH_3CH- \end{array}$$
isopropyl-

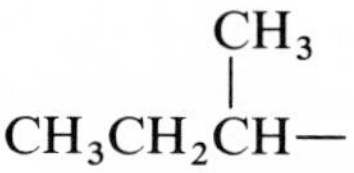

$$\begin{array}{c} CH_3 \\ | \\ CH_3CH_2CH- \end{array}$$
sec-butyl-

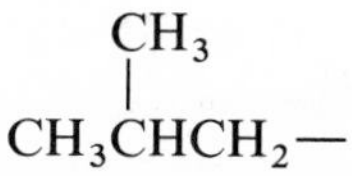

$$\begin{array}{c} CH_3 \\ | \\ CH_3CHCH_2- \end{array}$$
isobutyl-

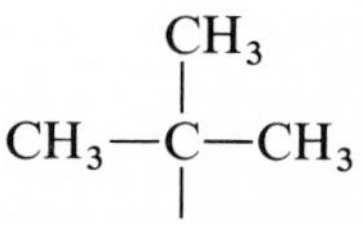

$$\begin{array}{c} CH_3 \\ | \\ CH_3-C-CH_3 \\ | \end{array}$$
tert-butyl-

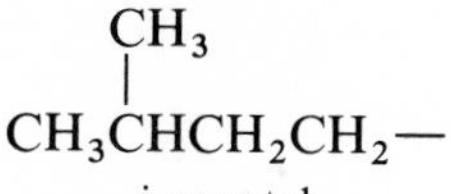

$$\begin{array}{c} CH_3 \\ | \\ CH_3CHCH_2CH_2- \end{array}$$
isopentyl-

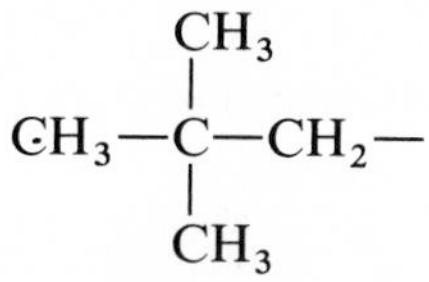

$$\begin{array}{c} CH_3 \\ | \\ CH_3-C-CH_2- \\ | \\ CH_3 \end{array}$$
neopentyl-

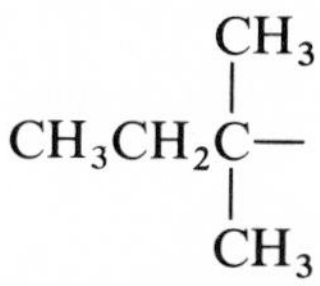

$$\begin{array}{c} CH_3 \\ | \\ CH_3CH_2C- \\ | \\ CH_3 \end{array}$$
tert-pentyl-

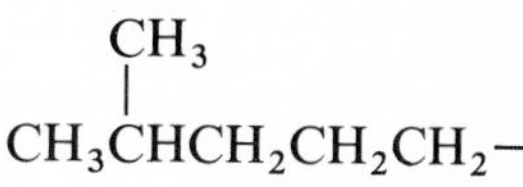

$$\begin{array}{c} CH_3 \\ | \\ CH_3CHCH_2CH_2CH_2- \end{array}$$
isohexyl-

The above are the only trivial names for alkyl groups permitted by the IUPAC. The use of these names for substituent groups must follow IUPAC rules. We may use them only when they are not part of the longest continuous chain. For example,

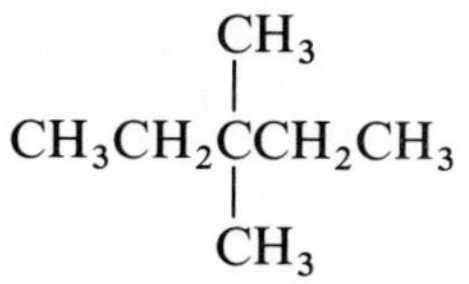

$$\begin{array}{c} CH_3 \\ | \\ CH_3CH_2CCH_2CH_3 \\ | \\ CH_3 \end{array}$$

is named 3,3-dimethylpentane, and not *tert*-pentylethane.

The prefixes iso-, neo-, *sec*-, and *tert*- may never be used as part of the parent name (that is, the name of the longest continuous chain). In the IUPAC system, these prefixes are restricted to use with the alkyl substituents given above.

Problem 4.9

Give the systematic name of each of the following.

(a) 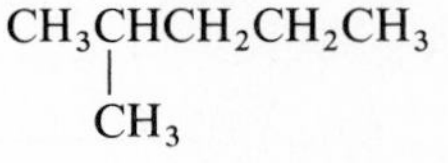
```
CH3CHCH2CH2CH3
   |
   CH3
```

(b) 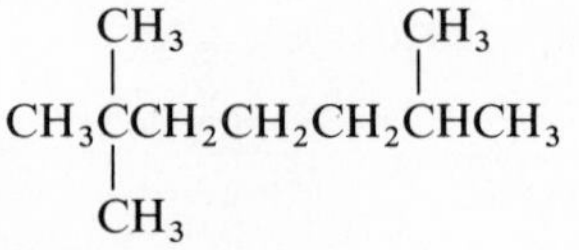
```
     CH3            CH3
     |              |
  CH3CCH2CH2CH2CHCH3
     |
     CH3
```

(c) 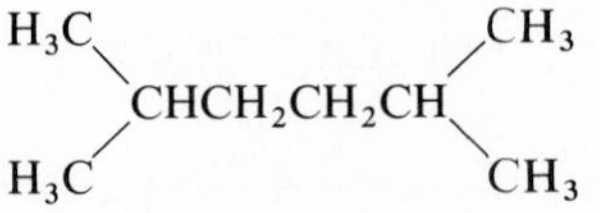
```
H3C               CH3
   \             /
    CHCH2CH2CH
   /             \
H3C               CH3
```

(d) 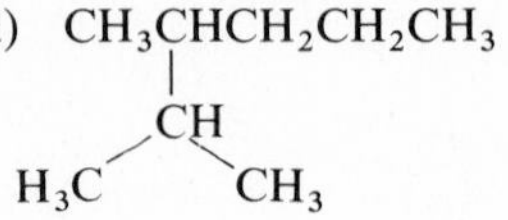
```
CH3CHCH2CH2CH3
   |
   CH
  /  \
H3C   CH3
```

Problem 4.10

Draw the structural formulas of the following compounds.

(a) 4-isopropylheptane

(b) 5-isopropyl-5-*tert*-butyldecane

(c) *sec*-butyl alcohol

(d) *tert*-pentyl alcohol

4.7 PHYSICAL PROPERTIES

Members of a homologous series such as the alkanes are usually related in their physical and chemical properties. Table 4.3 lists the first several members of the unbranched alkanes with their boiling points, melting points, and densities.

The alkanes show a strong regularity in boiling and melting points: The lower members, methane through butane, are gases under ordinary conditions. In chains of up to about 20 carbon atoms, the unbranched alkanes are liquids; only those with more than 20 carbon atoms are solids at room temperature.

Table 4.3 Physical Properties of Unbranched Alkanes

Alkane	Molecular Weight	Boiling Point, °C	Melting Point, °C	Density, g/ml (20°C)
Methane	16	−182.5	−164	0.466 (−164°)
Ethane	30	−88.6	−183	0.572 (−108°)
Propane	44	−42.2	−190	0.585 (−45°)
Butane	58	−0.5	−138	0.601 (0°)
Pentane	72	36.1	−130	0.626
Hexane	86	69.0	−95	0.660
Heptane	100	98.4	−90.6	0.684
Octane	114	126	−56.5	0.704
Nonane	128	151	−53.7	0.718
Decane	142	174	−30	0.730
Undecane	156	196	−26.5	0.741
Dodecane	170	215	−12	0.751
Pentadecane	212	271	10	0.769
Eicosane	282	343	36.8	0.789

Boiling Point

Boiling points of alkanes increase in a regular manner along the series. The boiling point is the temperature at which the vapor pressure of a substance is equal to atmospheric pressure. The vapor pressure of a substance is greater when intermolecular attractions are small. Therefore, the lower the intermolecular attraction, the greater will be the vapor pressure and the lower will be the boiling point. Because alkanes have the lowest boiling and melting points of all classes of organic compounds, we conclude that attractive forces between alkane molecules are very weak. The attractive forces between alkane molecules are weak due to the nonpolar character of alkane molecules. The C—C and C—H bonds are quite nonpolar, and there is no possibility of hydrogen bonding. Only weak intermolecular attractions exist.

Melting Point

Melting points of alkanes also increase as molecular size increases. In this case, however, the melting points oscillate as they increase. The melting points of the even-numbered alkanes lie on a smooth curve, and those of the odd-numbered alkanes lie on a smooth curve. The curve for the odd-numbered ones lies slightly lower than the one for the even-numbered ones. This difference between even- and odd-numbered chains is due to different packing arrangements in their crystal lattices. Recall that carbon chains are zigzag:

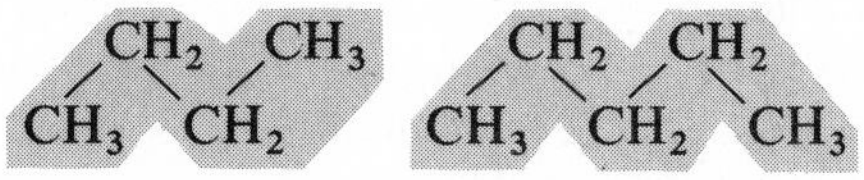

```
    CH2   CH3          CH2   CH2
   /   \  /           /   \  /   \
CH3    CH2         CH3    CH2    CH3
```

Solubility

The alkanes are insoluble in water. This is a result of their low bond polarities and their inability to form hydrogen bonds with water molecules. The liquid alkanes are completely soluble in each other, and alkanes are soluble in solvents of low polarity such as benzene, carbon tetrachloride, ether, and chloroform.

Density

Alkanes are the least dense of all the classes of organic compounds. Their densities increase with molecular size but reach a maximum near 0.8 g/ml. Thus, they are all less dense than water.

Problem 4.11

Plot boiling point versus number of carbon atoms for all the alkanes in Table 4.3. Make similar plots for melting point versus number of carbon atoms and for density versus number of carbon atoms.

4.8 COMMERCIAL PREPARATION OF ALKANES: AN ELECTIVE TOPIC

elective topic

The major source of alkanes is petroleum, a complex mixture of hydrocarbons. If petroleum is distilled, the hydrocarbons of lowest boiling point distill first (Fig. 4.3). As these substances are removed, higher boiling substances distill. A mixture collected in this way over a narrow temperature range is called a *fraction*. Each fraction is a mixture of hydrocarbons that boil over a relatively narrow temperature range (Table 4.4). Many of these fractions—natural gas, gasoline, kerosene, and diesel fuel—are used without further purification. Less useful, high-boiling fractions that contain chains of 15–18 carbon atoms can be converted into more useful fractions by a process known as cracking.

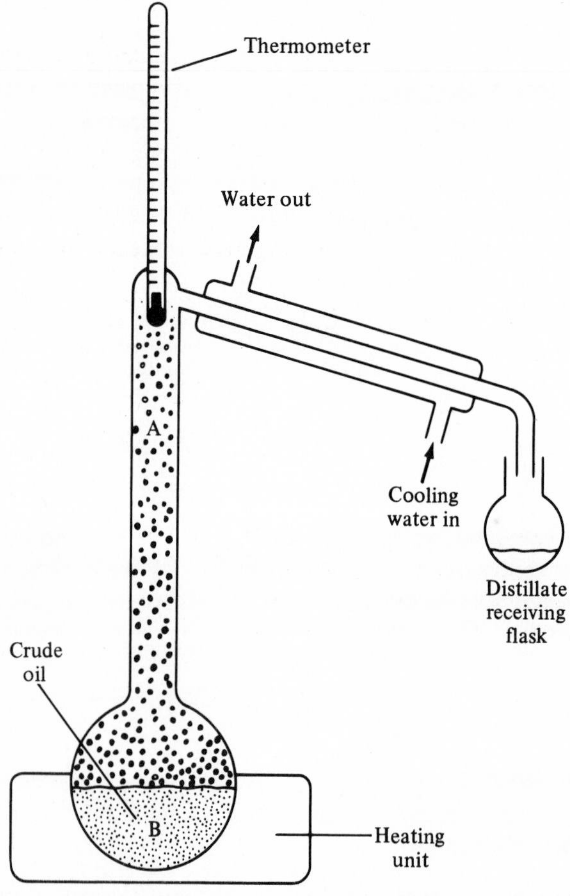

Figure 4.3. Schematic drawing of a simple laboratory distillation apparatus. The more volatile component, A (open circles), vaporizes to a greater extent than the less volatile component, B (solid circles). Near the top of the column, the vapor is nearly pure A. As it passes into the condenser, A vapor condenses to liquid A and is collected in the receiving flask.

Cracking

Much of our gasoline is made by a process called **cracking**. If a hydrocarbon is heated to 500°–1000°C in the absence of air, its molecules break apart homolytically. Under these conditions nearly every C—C and C—H bond can break to yield free radicals. These free radicals can react with other alkane molecules to produce new radicals, or they can recombine to form new, usually smaller molecules. This process is called **thermal cracking**. The cracking of propane serves to illustrate the complexity of the process:

$$CH_3CH_2CH_3 \begin{array}{l} \nearrow\ CH_3CH_2\cdot + \cdot CH_3 \\ \longrightarrow\ CH_3CH_2CH_2\cdot + \cdot H \\ \searrow\ CH_3\underset{\cdot}{C}HCH_3 + \cdot H \end{array}$$

followed by all the possible recombination and decomposition reactions

$$2CH_3CH_2\cdot \longrightarrow CH_3CH_2CH_2CH_3$$

$$2CH_3\cdot \longrightarrow CH_3CH_3$$

$$CH_3CH_2\cdot \longrightarrow CH_2{=}CH_2 + H\cdot$$

$$CH_3CH_2\cdot + CH_3CH_2CH_2\cdot \longrightarrow CH_3CH_2CH_2CH_2CH_3$$

$$CH_3CH_2\cdot + CH_3\underset{\cdot}{C}HCH_3 \longrightarrow CH_3CH_2\underset{\displaystyle CH_3}{\underset{|}{C}}HCH_3$$

$$2CH_3CH_2CH_2\cdot \longrightarrow CH_3CH_2CH_2CH_2CH_2CH_3$$

$$2CH_3\underset{\cdot}{C}HCH_3 \longrightarrow CH_3\underset{\displaystyle CH_3}{\underset{|}{C}}H{-}\underset{\displaystyle CH_3}{\underset{|}{C}}HCH_3$$

$$CH_3CH_2CH_2\cdot + CH_3\underset{\cdot}{C}HCH_3 \longrightarrow CH_3CH_2CH_2\underset{\displaystyle CH_3}{\underset{|}{C}}HCH_3$$

$$CH_3\cdot + CH_3CH_2CH_2\cdot \longrightarrow CH_3CH_2CH_2CH_3$$

$$CH_3\cdot + CH_3\underset{\cdot}{C}HCH_3 \longrightarrow CH_3\underset{\displaystyle CH_3}{\underset{|}{C}}HCH_3$$

$$CH_3CH_2CH_2\cdot \longrightarrow CH_3CH{=}CH_2 + H\cdot$$

$$2H\cdot \longrightarrow H_2$$

Actually, all the above reactions are not of equal importance. Cracking normally yields a much smaller number of products than the above possibilities suggest, although cracking even simple molecules gives complex mixtures of hydrocarbons. The process is particularly useful because the hydrocarbon mixtures are quite suitable as fuels, and separation into pure compounds is not necessary.

Table 4.4 Fractions from the Distillation of Crude Petroleum

Name of Fraction	Molecular Size	Approximate Boiling Range, °C
Natural gas	C_1–C_4	Below room temperature
Petroleum ether	C_5–C_6	20–60
Ligroin (light naphtha)	C_6–C_7	60–100
Gasoline	C_6–C_{12}	50–200
Kerosene	C_{12}–C_{18}	175–275
Gas oil (furnace oil, diesel oil)	above C_{18}	above 275
Lubricating oils		Nonvolatile liquids
Asphalt		Residue (nonvolatile solids)

Hydrocarbons can also be cracked at lower temperatures with the use of certain Lewis acid catalysts such as chromia-alumina (Cr_2O_3-Al_2O_3). This process is called **catalytic cracking** and probably involves carbocations. Catalytic cracking produces a greater proportion of branched-chain alkanes than thermal cracking.

The random nature of cracking makes it relatively useless in the laboratory, where chemists normally seek small amounts of pure compounds. On an industrial scale, however, every fraction is either removed as a useful product or else recycled through the cracking process to yield another mixture from which desirable products can be recovered.

4.9 LABORATORY PREPARATION OF ALKANES

Alkanes can be prepared in the laboratory in several ways. We shall consider three—the Wurtz reaction, the Corey-House Synthesis, and the reduction of alkyl halides.

Wurtz Reaction

The **Wurtz reaction** (named after A. Wurtz, its discoverer) is an example of a **coupling reaction**—a reaction in which two carbon chains are joined together to form a longer carbon chain. Coupling reactions always involve the formation of a new C—C bond. The Wurtz reaction occurs when an alkyl halide is heated in the presence of sodium metal. 1-Bromobutane reacts to form octane:

$$2CH_3CH_2CH_2CH_2Br + 2Na \xrightarrow{\text{boil}} CH_3CH_2CH_2CH_2CH_2CH_2CH_2CH_3 + 2NaBr$$

Chlorides or iodides can also be used, and in some cases zinc replaces sodium. The major limitation of the Wurtz reaction is that only even-numbered alkanes can be produced efficiently. The reaction of a mixture of two different alkyl halides with sodium yields three different alkanes. For example, a mixture of 1-bromopropane and 1-bromobutane yields a mixture of hexane, heptane, and octane:

$$3C_3H_7Br + 3C_4H_9Br + 6Na \xrightarrow{\text{boil}} C_6H_{14} + C_7H_{16} + C_8H_{18} + 6NaBr$$

It should not be too difficult to separate such a mixture by distillation because hexane, heptane, and octane have boiling points of 69°, 98°, and 126°C, respectively. The limitation, however, is that roughly two-thirds of the reactants would be was ed in producing two undesired products.

We can summarize the Wurtz reaction by the general equation:

$$2R{-}X + 2M \xrightarrow{\text{boil}} R{-}R + 2MX$$

(R is alkyl; X is Cl, Br, or I; and M is Na, K, or Zn.)

Corey–House Synthesis

Practically any alkane can be prepared by a variation of the Wurtz reaction known as the **Corey–House synthesis**. In this reaction, we couple two alkyl groups through the following sequence of reactions, all of which occur at or below room temperature:

$$RX + 2Li \longrightarrow RLi + LiX \quad (R = CH_3,\ 1^\circ \text{ or } 2^\circ \text{ alkyl};\ X = Cl, Br, \text{ or } I)$$

$$2RLi + CuI \longrightarrow R_2CuLi + LiI$$

$$R_2CuLi + R'X \longrightarrow R{-}R' + RCu + LiX \quad (R' = CH_3 \text{ or } 1^\circ \text{ alkyl})$$

An example is the synthesis of 3-methylheptane:

$$CH_3CH_2\overset{\overset{\displaystyle CH_3}{|}}{C}HBr + 2Li \longrightarrow CH_3CH_2\overset{\overset{\displaystyle CH_3}{|}}{C}HLi + LiBr$$

$$2CH_3CH_2\overset{\overset{\displaystyle CH_3}{|}}{C}HLi + CuI \longrightarrow (CH_3CH_2\overset{\overset{\displaystyle CH_3}{|}}{C}H)_2CuLi + LiI$$

$$(CH_3CH_2\overset{\overset{\displaystyle CH_3}{|}}{C}H)_2CuLi + ClCH_2CH_2CH_2CH_3$$

$$\longrightarrow CH_3CH_2\overset{\overset{\displaystyle CH_3}{|}}{C}HCH_2CH_2CH_2CH_3 + CH_3CH_2\overset{\overset{\displaystyle CH_3}{|}}{C}HCu + LiCl$$

This reaction takes advantage of the fact that RLi is not as reactive as RNa; that is, RLi is more covalent (less ionic) than RNa. Because of this lower reactivity, RLi does not react with unreacted RX in the first step. (If we had used RX with sodium, the RNa produced would have reacted immediately with unreacted RX to form the

Wurtz product, R—R.) Once formed, RLi is separated and allowed to react with CuI to produce the more reactive R_2CuLi, which does react with alkyl halides. In the third step we bring about coupling with R′X.

The major limitation of the Corey–House synthesis is that we cannot use tertiary alkyl halides in the first step, and we *must* use only primary alkyl halides or methyl halides in the third step. Other alkyl halides (2° or 3°) in the third step yield mainly alkenes and little coupling product.

Problem 4.12
We can use the Corey–House synthesis to prepare 3-methylpentane from ethyl chloride and *sec*-butyl chloride. Of the two ways that we could use these starting compounds, which is preferable? [Explain.]

Problem 4.13
What advantage does the Corey–House synthesis have over the Wurtz synthesis?

Reduction of Alkyl Halides

Reduction is defined as a process in which an atom gains electrons. Organic reactions normally involve only covalent bonds, and electron gain or loss (oxidation) is relative. Substitution of a less electronegative atom such as hydrogen for a more electronegative one such as chlorine results in an increase in the electron density of the carbon atom:

$$\overset{\delta+}{C}-\overset{\delta-}{Cl} \xrightarrow{\text{reduction}} \overset{\delta-}{C}-\overset{\delta+}{H}$$

low electron density (C of C—Cl) ... high electron density (C of C—H)

Because most nonmetallic elements are more electronegative than hydrogen, hydrogenation usually amounts to reduction. For simplicity, most organic chemists use the terms reduction and hydrogenation to mean the same thing.

We can produce alkanes by the reduction of alkyl halides. In this reaction, hydrogen replaces chlorine, and the alkane contains the same number of carbon atoms as the alkyl halide reactant. An example is the formation of butane from butyl chloride:

$$CH_3CH_2CH_2CH_2Cl + Zn + CH_3CO_2H \xrightarrow{\Delta} CH_3CH_2CH_2CH_3 + Cl^- + Zn^{2+} + CH_3CO_2^-$$

The reaction is brought about by boiling the alkyl halide with a mixture of zinc metal and acetic acid. We can summarize the reaction by the general equation

$$2R-X + Zn + CH_3CO_2H \xrightarrow{\Delta} R-H + Zn^{2+} + X^- + CH_3CO_2^-$$

where R is an alkyl group and X is Cl, Br, or I.

Alkyl halides can also be reduced by the important and versatile reducing agent lithium aluminum hydride, $LiAlH_4$. An example is the reduction of 1-bromodecane:

$$4CH_3CH_2CH_2CH_2CH_2CH_2CH_2CH_2CH_2CH_2Br + LiAlH_4 \longrightarrow 4C_{10}H_{22} + LiBr + AlBr_3$$

We can summarize this reaction by the general equation

$$\boxed{4R{-}X + LiAlH_4 \longrightarrow 4R{-}H + LiX + AlX_3}$$

Problem 4.14
How does reduction of alkyl halides differ from both the Wurtz and Corey–House methods of preparing alkanes?

Problem 4.15
Assume that you have any necessary organic starting materials of four carbons or fewer (other than the desired product) and any inorganic compounds, and write equations to show how you would prepare (a) hexane, (b) 3,4-dimethylhexane, and (c) propane.

Problem 4.16
For which of the compounds of Problem 4.15 could you use *both* of the coupling reactions described in this section? Explain. Write equations for those reactions that you did not give in Problem 4.15.

Problem 4.17
Write equations for all the reactions that could occur during the cracking of ethane.

4.10 REACTIONS OF ALKANES

The bulk of our alkanes are burned as fuels. We have seen fit to use only the energy stored in their chemical bonds and to waste the atoms themselves. By burning alkanes, we use only the energy and expel the carbon and hydrogen atoms into the atmosphere as CO_2 and H_2O.

Alkanes are quite unreactive. They do not react with strong acids or bases at ordinary temperatures. They are not oxidized or reduced by the ordinary oxidizing or reducing agents. Combustion, halogenation, and nitration—three rather drastic reactions—are the only typical reactions of alkanes.

Combustion

Oxidation is any process that reduces the electron density of an atom. Oxidation of carbon occurs whenever we substitute a more electronegative element for a less electronegative one. Because oxygen is quite electronegative, reaction of an organic compound with oxygen is oxidation.

All the alkanes burn in the presence of excess oxygen to give CO_2 and water as the final products. We call this burning process **combustion**

$$CH_4 + 2O_2 \xrightarrow{\text{heat}} CO_2 + 2H_2O \ (\Delta H^\circ = -192 \text{ kcal/mole})$$

$\Delta H°$, the **heat of reaction**, is the heat that must be added to convert one mole of reactant to products at 0°C, 1.0 atmosphere pressure, and a concentration of 1.0 mole per liter. As with $\Delta G°$, $\Delta H°$ is negative if heat is evolved; that is, $\Delta H°$ is considered to be a reactant.

If there is insufficient oxygen, alkanes burn incompletely to give carbon monoxide, CO, and elemental carbon. In either case, oxidation occurs because oxygen replaces hydrogen.

We do not fully understand the mechanism of combustion of alkanes. The reaction is very rapid and passes through various intermediates that are difficult to study under combustion conditions.

The heat of reaction for combustion can be measured with great precision. We know, for example, that $\Delta H°$ of combustion of butane is -687.4 kcal/mole, whereas the $\Delta H°$ of combustion of isobutane is -685.4 kcal/mole. Both react according to the same equation:

$$C_4H_{10} + 6\tfrac{1}{2}O_2 \longrightarrow 4CO_2 + 5H_2O$$

The difference in $\Delta H°$ between these two isomers (2.0 kcal/mole) shows that the branched alkane is more stable than the unbranched alkane; that is, isobutane has 2.0 kcal/mole less energy content than butane (Fig. 4.4). Later (Section 4.13), we shall use heats of combustion to compare the stabilities of molecules.

Halogenation

Methane reacts directly with chlorine or bromine in a process called **halogenation**. The reaction involves the substitution of a halogen atom for a hydrogen atom, and is also an oxidation reaction:

$$CH_4 + Cl_2 \longrightarrow CH_3Cl + HCl$$

The reaction is slow in the dark at room temperature, but very fast when we heat the mixture or expose it to light. For this reason we write the equation with the special conditions, light or heat (abbreviated Δ), over the arrow:

$$CH_4 + Cl_2 \xrightarrow{\text{light or } \Delta} CH_3Cl + HCl$$

The product, chloromethane, is capable of reacting further, and the reaction can be made to go to the completely chlorinated compound tetrachloromethane. We designate this series of reactions in which the product of one step reacts further in a second step, and so forth, as follows:

$$CH_4 \xrightarrow[\substack{\text{or light} \\ (-HCl)}]{Cl_2/\text{heat}} CH_3Cl \xrightarrow[\substack{\text{or light} \\ (-HCl)}]{Cl_2/\text{heat}} CH_2Cl_2 \xrightarrow[\substack{\text{or light} \\ (-HCl)}]{Cl_2/\text{heat}} CHCl_3 \xrightarrow[\substack{\text{or light} \\ (-HCl)}]{Cl_2/\text{heat}} CCl_4$$

The symbols above and below the arrows indicate the reagents used, the conditions, and the by-products that are removed (HCl, in this case).

Bromine reacts in the same way although it is less reactive than chlorine. Iodine does not react. Fluorine is extremely reactive and usually gives the completely fluorinated product CF_4.

Chlorine and bromine react with the higher alkanes. If a small amount of halogen is allowed to react with excess alkane, monohalogenation results. If an excess of

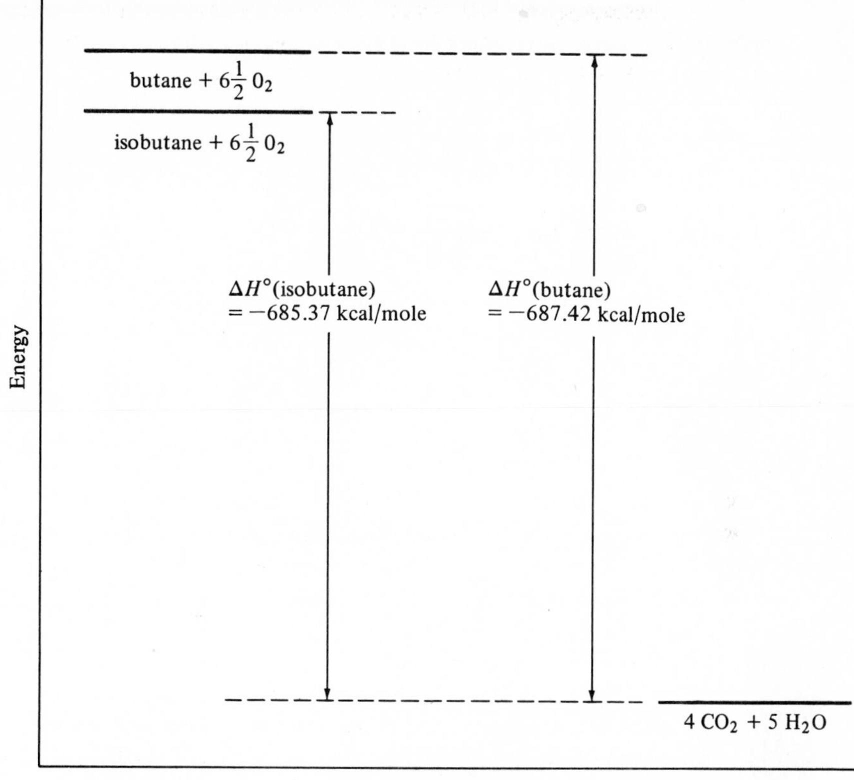

Figure 4.4. Reaction coordinate versus energy for the combustion of butane and isobutane.

halogen is used, especially at higher temperatures, polyhalogenation or even complete halogenation occurs. Chlorine is quite reactive and tends to give polychlorinated products. Bromine is less reactive than chlorine, and bromination is therefore easier to control.

In most cases, halogenation yields mixtures of isomers that must be separated into pure compounds. The simplest example is the chlorination of propane:

$$CH_3CH_2CH_3 + Cl_2 \xrightarrow[\text{heat}]{\text{light or}} CH_3CH_2CH_2Cl + CH_3\overset{\overset{\displaystyle Cl}{|}}{C}HCH_3$$
$$+ HCl + \text{polychlorinated products}$$

We summarize the halogenation of alkanes by the general equation

$$R{-}H + X_2 \xrightarrow{\text{light or heat}} R{-}X + HX$$

where X is Cl or Br.

The reaction of fluorine with alkanes produces so much heat that the energy released is more than enough to break carbon-carbon bonds. For example, the reaction

$$CH_3CH_2CH_3 + F_2 \longrightarrow CH_3\overset{\displaystyle F}{\overset{|}{C}}HCH_3 + HF$$

has a heat of reaction, $\Delta H°$, of -112 kcal/mole. The energy required to break a carbon–carbon bond is ~83 kcal/mole. Direct fluorination of higher alkanes, therefore, usually leads to decomposition of the alkane into smaller molecules that are completely fluorinated. This makes the products of direct fluorination difficult to predict, and the reaction is difficult to control. It should be noted that fluorine is an extremely dangerous chemical. For these reasons fluorine gas is not a routine laboratory stock item.

Many halogenated alkanes are produced in large quantities in the United States. Table 4.5 lists some of them, their uses, and their toxic effects on humans.

Nitration

The reaction of alkanes with nitric acid vapors at 400°–500°C produces nitroalkanes, R—NO_2:

$$R—H + HO—NO_2 \xrightarrow[\text{vapor}]{400°\text{–}500°} R—NO_2 + H_2O$$

Table 4.5 Some Commercially Important Halogenated Alkanes

Formula, Common Name	Uses	Effects on Humans
H H—C—Br H methyl bromide	Rodent poison, soil and grain fumigant	Toxic, causes skin injury
H H—C—Cl Cl carbon tetrachloride	Solvent, ingredient in paint and grease removers	Moderately toxic, narcotic effect
Cl H—C—Cl Cl chloroform	Solvent, used in the purification of penicillin	Toxic, causes burning sensation on skin, anesthetic (no longer used)
Cl Cl—C—Cl Cl carbon tetrachloride	Solvent for oils, fats, waxes, varnishes; dry cleaning agent	Toxic, causes skin burns and liver damage

Table 4.5 *(Continued)*

Formula, Common Name	Uses	Effects on Humans
CCl_2F_2 (Cl—C—Cl with F above and below) dichlorodifluoromethane (Freon-12)	Refrigerant, propellant in aerosol containers	Slightly toxic
CCl_3F (F—C—Cl with Cl above and below) trichlorofluoromethane (Freon-11)	Refrigerant	
CH_3CH_2Cl (H—C—C—Cl with H above and below each C) ethyl chloride	Manufacture of tetraethyl-lead for gasoline	Very slightly toxic, local anesthetic—causes numbness when sprayed on skin
CH_2ClCH_2Cl (H—C—C—H with H above, Cl below each C) ethylene dichloride	Additive to leaded gasoline	Toxic, causes liver enlargement and reduces blood pressure
CH_2BrCH_2Br (H—C—C—H with H above, Br below each C) ethylene dibromide	Scavenger for lead in aircraft fuels	Mild narcotic, causes skin blisters and irritation of mucous lining
$CHCl_2CHCl_2$ (H—C—C—H with Cl above and below each C) tetrachloroethane	Cleaning fluid, used in synthesis of other chemicals	Toxic, skin irritant
$CF_3CHClBr$ (F—C—C—H; F above and below first C; Cl above, Br below second C) halothane	Inhalation anesthetic	Nonflammable, nontoxic

The introduction of a nitro group, $—NO_2$, into a molecule is called **nitration**, and is also an oxidation process. Because the reaction requires such high temperatures, nitration causes much decomposition. For example, in the nitration of methane, only 20% of the methane ends up as nitromethane. The remainder is oxidized to CO_2 and water:

$$CH_4 + HO—NO_2 \xrightarrow{475^\circ} \begin{cases} \xrightarrow{20\%} CH_3NO_2 + H_2O \\ \xrightarrow{80\%} CO_2 + H_2O + \text{nitrogen oxides} \end{cases}$$

Much C—C and C—H bond breaking occurs during nitration of the higher alkanes. For example, nitration of butane produces nitromethane, nitroethane, 1-nitropropane, 2-nitropropane, 1-nitrobutane, and 2-nitrobutane. Oxidation to CO_2 and water also occurs.

We can generalize the simple nitration of alkanes by the equation

$$R—H + HO—NO_2 \xrightarrow{\Delta} R—NO_2 + H_2O$$

Because of the drastic conditions required and the extensive decomposition that normally occurs, nitration of alkanes is not a useful laboratory method. Nitration of alkanes is carried out on a large scale commercially, and the products are used widely as solvents and as raw materials in the production of drugs, explosives, and insecticides.

4.11 REACTIVITY OF ALKANES

The reactions of alkanes described above show that alkanes are quite unreactive because they react only under vigorous conditions. When enough energy is supplied to cause them to react, the reactions are undiscriminating; either random decomposition occurs, or else the reaction is difficult to control. Combustion and nitration, for example, usually result in complete C—C bond breakage, and halogenation usually leads to the introduction of several halogen atoms into the molecule.

We shall see in the following chapters that alkyl groups are usually unaffected during organic reactions because nearly all the functional groups are more reactive than alkyl groups.

Problem 4.18

Complete and balance the equation for (a) the complete combustion of pentane, (b) the monochlorination of ethane, (c) the hexachlorination of ethane, (d) the nitration of ethane.

Problem 4.19

(a) What conditions would you employ for the laboratory preparation of chloromethane that would minimize polychlorination? (b) What conditions would you employ in the laboratory preparation of tetrachloromethane?

Saturated hydrocarbons in which the carbon atoms form a ring are called *cyclo-alkanes* (Sec. 2.6). A ring of carbon atoms has two fewer hydrogen atoms than the corresponding alkane:

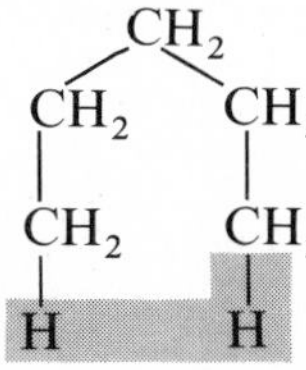

pentane (C_nH_{2n+2}) 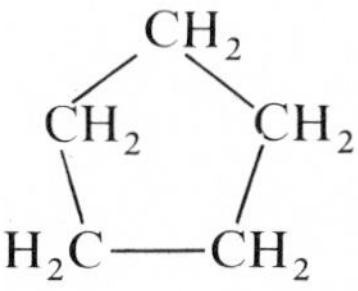 cyclopentane (C_nH_{2n}) 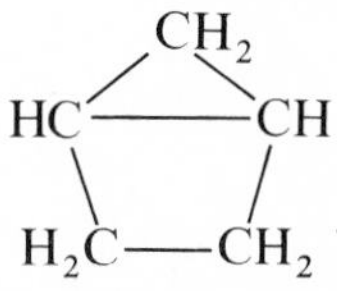 bicyclopentane (C_nH_{2n-2})

The general formula for a cycloalkane is therefore C_nH_{2n}. Compounds that have two rings have the general formula C_nH_{2n-2}.

The smallest possible cycloalkane contains three carbon atoms. The most common naturally occurring cycloalkanes have five or six carbon atoms (Figure 4.5).

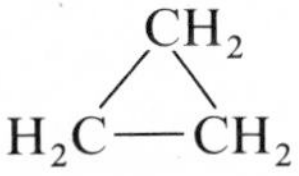

cyclopropane 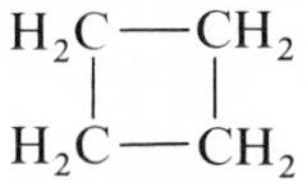 cyclobutane 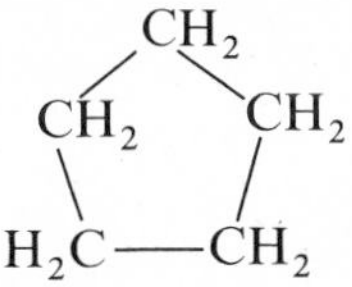 cyclopentane 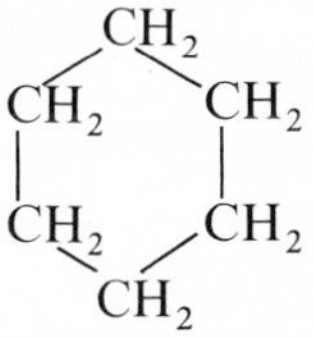 cyclohexane

The systematic names of cycloalkanes contain the prefix, cyclo-, attached to the alkane name as shown above. We name substituents as before:

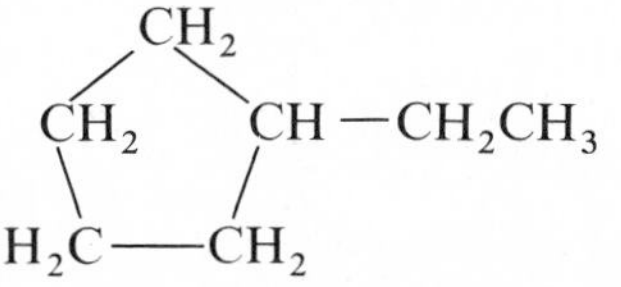

ethylcyclopentane

When the ring has two or more substituents, we designate the position of one of them as position one, and number the ring carbons in the direction that results in the lowest possible numbers in the name. For example, the compound

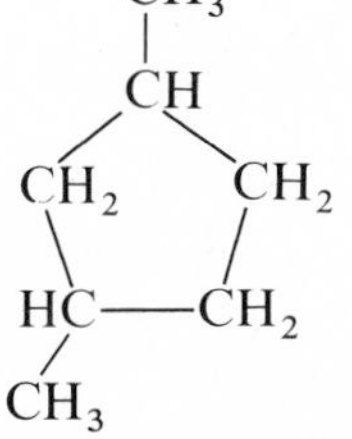

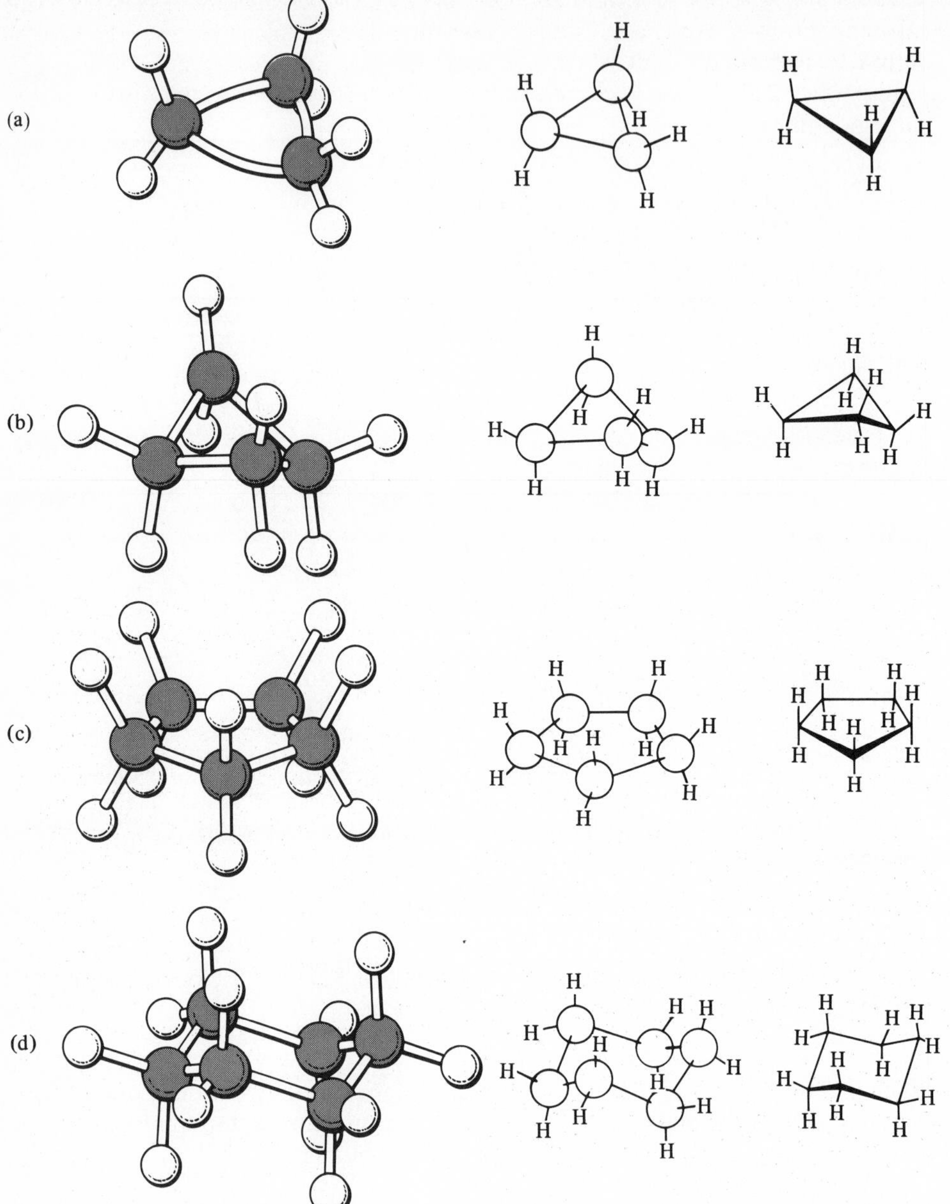

Figure 4.5. Cycloalkanes: (a) cyclopropane; (b) cyclobutane; (c) cyclopentane; (d) cyclohexane.

is 1,3-dimethylcyclopentane and *not* 1,4-dimethylcyclopentane. If the ring itself is a substituent on a larger chain, then we treat the ring as an alkyl group (called *cyclo-alkyl-*). For example,

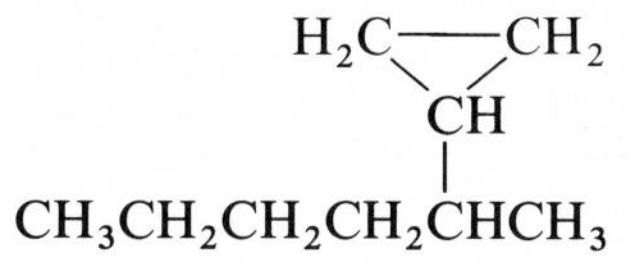

$$\begin{array}{c} H_2C\text{——}CH_2 \\ \diagdown\ \diagup \\ CH \\ | \\ CH_3CH_2CH_2CH_2CHCH_3 \end{array}$$

has the name 2-cyclopropylhexane.

Problem 4.20
Give the systematic name of each of the following compounds.

(a) 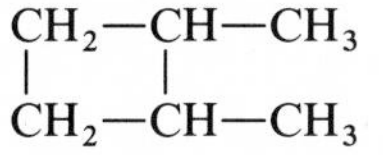
$$\begin{array}{l} CH_2\text{—}CH\text{—}CH_3 \\ |\qquad\ \ | \\ CH_2\text{—}CH\text{—}CH_3 \end{array}$$

(b) 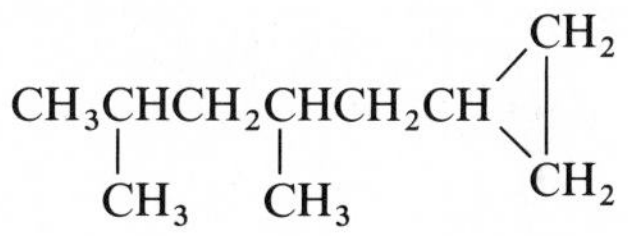
$$\begin{array}{l} \qquad\qquad\qquad\qquad\qquad\quad CH_2 \\ CH_3CHCH_2CHCH_2CH\ \ \ | \\ \qquad\qquad\qquad\qquad\qquad\quad CH_2 \\ \quad\ |\qquad\quad | \\ \quad CH_3\quad\ CH_3 \end{array}$$

Problem 4.21
Draw the structure of each of the following compounds.
(a) 1,1-dimethylcyclopentane, (b) 1,3-diethyl-2,4-dipropylcyclobutane,
(c) methylcyclononane

4.13 STRAIN ENERGY IN CYCLOALKANES: AN ELECTIVE TOPIC

elective topic

The carbon atoms in cyclopropane lie in a plane. The formula of cyclopropane shows that the bond angles are much smaller than the tetrahedral angle of 109.5°. Moreover, the hydrogens are in an eclipsed conformation. We should therefore expect cyclopropane to be highly strained.

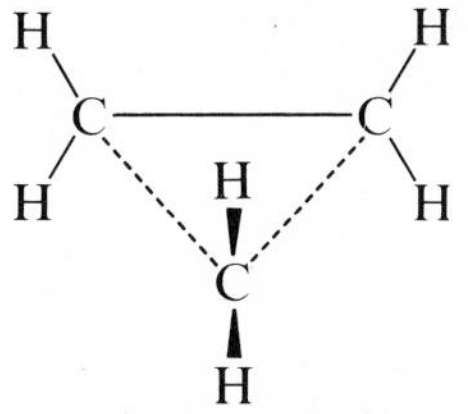

$$\begin{array}{ccc} H & & H \\ \diagdown & & \diagup \\ C & \text{———} & C \\ \diagup & H & \diagdown \\ H & | & H \\ & C & \\ & | & \\ & H & \end{array}$$

We can measure the amount of **strain energy** in molecules by comparing heats of combustion. When we burn a hydrocarbon in excess oxygen, heat is released. The amount of heat released is a measure of the energy that was stored in the molecules in the form of bond energy, bond angle distortion energy, and energy of repulsion due to crowding of bonds in eclipsed conformations. The greater the amount of heat released

($-\Delta H°$), the greater the amount of energy in the molecules. Note that the **heat of combustion** (ΔH_C) is defined as the amount of heat **released** ($-\Delta H°$). The heat of combustion is thus a positive quantity. We cannot compare heats of combustion, directly, however, because large molecules have more bonds and therefore more stored energy. We can compare molecules only if we use the quantity ΔH_C *per* CH_2 *unit.* We obtain the heat of combustion per CH_2 unit by dividing ΔH_C (in energy units per mole) by the number of CH_2 units in the molecule:

$$\frac{\Delta H_C}{n} = \text{heat of combustion per mole per } CH_2 \text{ unit in } (CH_2)_n.$$

If compound A has a greater ΔH_C per CH_2 unit than compound B, then molecules of compound A have a greater amount of strain energy (potential energy) than molecules of compound B (see Fig. 4.4).

Table 4.6 lists the heats of combustion for the first several cycloalkanes. We see in Table 4.6 that cyclopropane has the largest amount of strain energy (the largest ΔH_C per CH_2 unit). We expect cyclopropane to be highly strained because its bond angles are highly distorted from 109.5° and because the hydrogens are eclipsed. Cyclobutane has less bond angle distortion, and its heat of combustion per CH_2 unit is smaller. The cyclobutane molecule is "folded" or "puckered" (Figure 4.5(b)).

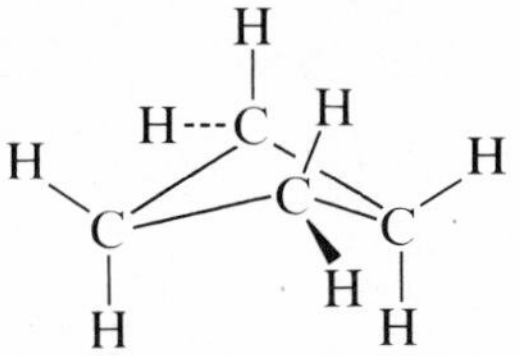

Although puckering strains the bond angles slightly, it also allows the hydrogens to be staggered. The result is a net relaxation of the molecule and a lower energy. The carbon atoms in cyclopentane are nearly planar. Although there is little bond angle strain, a

Table 4.6 Heats of Combustion of Cycloalkanes

Cycloalkane $(CH_2)_n$	ΔH_C	ΔH_C per CH_2 Unit
Cyclopropane, $(CH_2)_3$	499.83	166.6
Cyclobutane, $(CH_2)_4$	655.86	164.0
Cyclopentane, $(CH_2)_5$	793.52	158.7
Cyclohexane, $(CH_2)_6$	944.48	157.4
Cycloheptane, $(CH_2)_7$	1108.2	158.3
Cyclooctane, $(CH_2)_8$	1269.2	158.6
Cyclononane, $(CH_2)_9$	1429.5	158.8
Cyclodecane, $(CH_2)_{10}$	1586.0	158.6
Cyclopentadecane, $(CH_2)_{15}$	2362.5	157.5
Alkane, (C_nH_{2n+2})		157.4

slight puckering (Figure 4.5(c)) partially relieves the strain due to eclipsed hydrogens. Cyclohexane is the smallest cycloalkane that is completely strain-free (Figure 4.5(d)).

4.14 CONFORMATIONS OF CYCLOHEXANE

Cyclohexane exists in two extreme conformations—the *chair* and the *boat*. Both conformations are puckered, and the bond angles are exactly 109.5°. Thus neither exhibits any bond angle strain. Figure 4.6 shows the boat form. Although the boat form has no bond angle strain, it is strained because some of its hydrogens are eclipsed. The chair form (Fig. 4.7) is more stable than the boat form because it is free of eclipsing strain—its hydrogens are completely staggered. The chair and boat forms can interconvert into one another, although equilibrium favors the chair form.

The chair form of cyclohexane has two types of hydrogens. Six of the hydrogens are arranged around the ring's circumference. These are called **equatorial** hydrogens (Fig. 4.8). Bonds to the other six hydrogens, called the **axial** hydrogens, are perpendicular to the ring; three are above the plane of the ring and three are below. At room

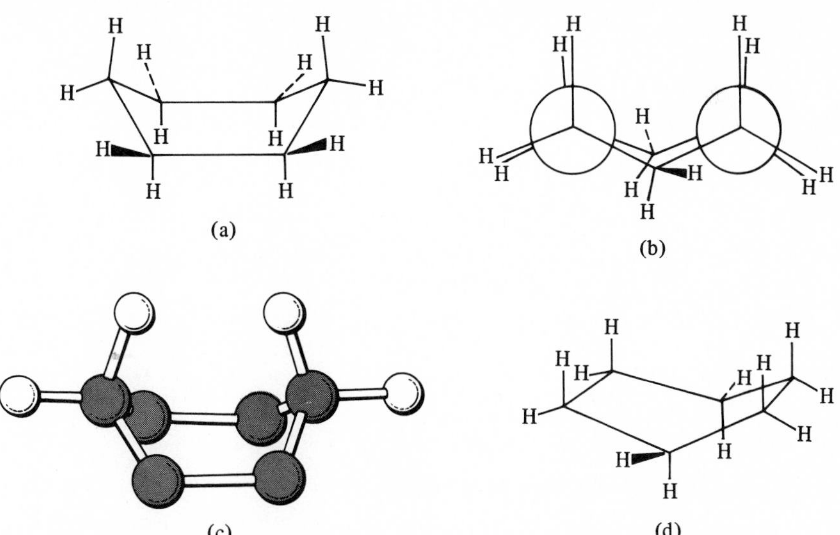

Figure 4.6. Representations of the boat conformation of cyclohexane: (a) line drawing; (b) Newman projection formula; (c) ball-and-stick model with only the 1,4-hydrogens shown; (d) line drawing of the Newman projection formula (b).

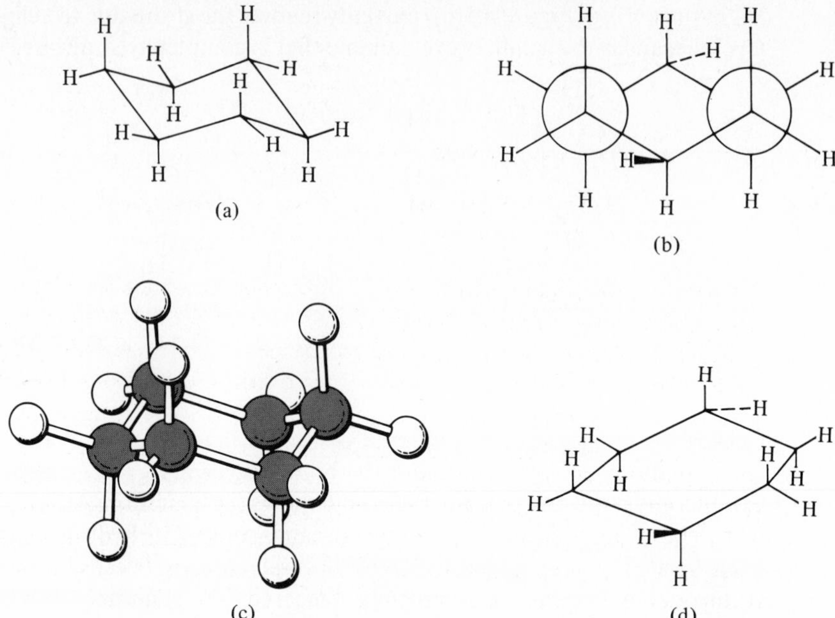

Figure 4.7. Representations of the chair conformation of cyclohexane: (a) line drawing; (b) Newman projection formula; (c) ball-and-stick model; (d) line drawing of the Newman projection formula (b).

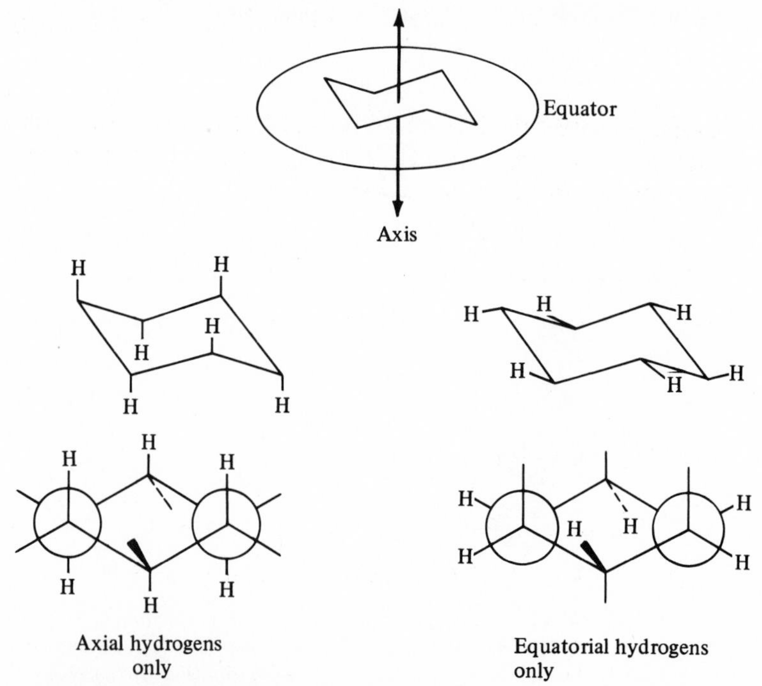

Figure 4.8. Representations that show axial and equatorial positions.

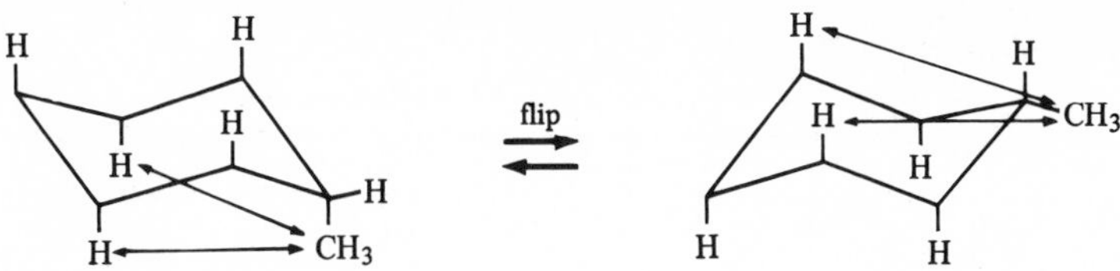

Figure 4.9. Methylcyclohexane in the axial and equatorial conformations (only axial hydrogens are shown).

temperature the ring flips back and forth rapidly between chair forms, and the hydrogens are indistinguishable. Repulsions between axial hydrogens are greater than repulsions between equatorial hydrogens. For this reason, a substituent group has more room when it is equatorial than when it is axial. A substituent will therefore spend a greater part of its time in an equatorial position than in an axial position (Fig. 4.9).

Problem 4.22
Draw Newman projection formulas for 1,2-dimethylcyclohexane when (a) both methyl groups are in axial positions, (b) both methyl groups are in equatorial positions, and (c) one methyl group is axial and one is equatorial. Construct a model of each of the above and verify your drawings.

Problem 4.23
Repeat Problem 4.22 using 1,3-dimethylcyclohexane and 1,4-dimethylcyclohexane.

4.15 CIS–TRANS ISOMERISM IN CYCLOALKANES

Cycloalkanes exhibit another structural feature—***cis–trans*** **isomerism** (Sec. 2.5). The ring prevents complete rotation about any C—C single bond; 1,2-dimethylcyclopropane can thus exist in two isomeric forms:

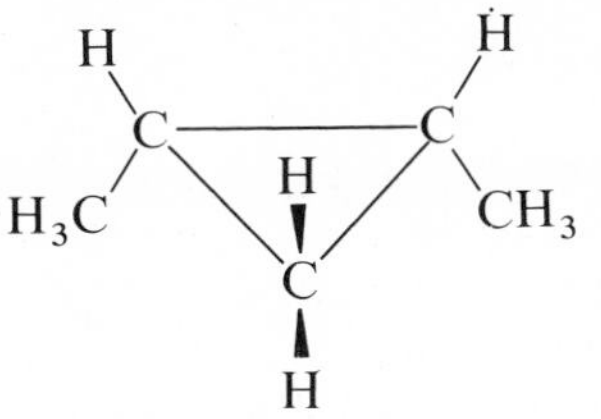

H H H H
C C
H
H_3C CH_3 shown as H_3C H CH_3
C
H
H

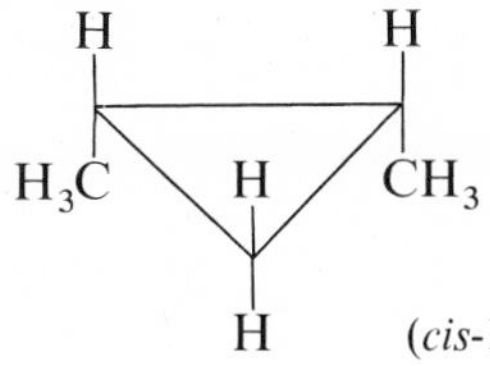

(*cis*-1,2-dimethylcyclopropane)

and

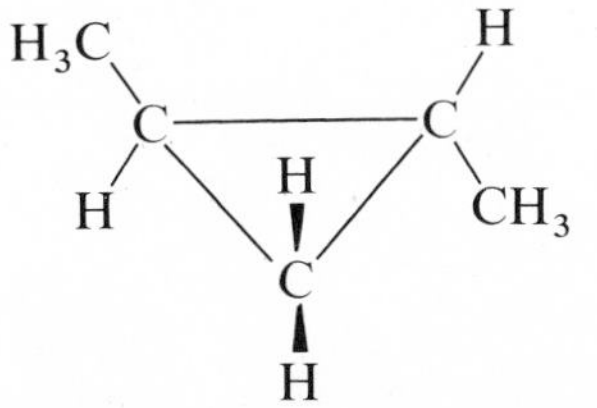

H_3C H CH_3 H
C C
H
H CH_3 shown as H H CH_3
C
H
H

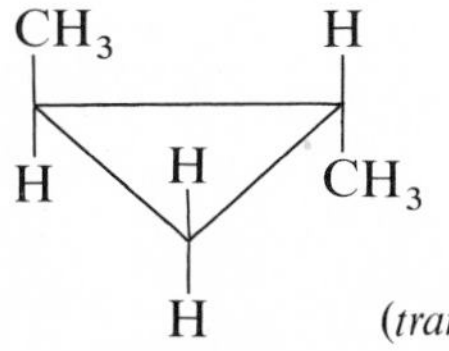

(*trans*-1,2-dimethylcyclopropane)

Higher cycloalkanes also exhibit *cis–trans* isomerism.

Problem 4.24

Draw the *cis* and *trans* isomers of (a) 1,2-dimethylcyclobutane, (b) 1,3-dimethylcyclobutane, (c) 1,3-dimethylcyclopentane. Construct models of the above molecules and verify your drawings.

cis–trans Isomerism in the cyclohexanes is best shown using Newman projection formulas. 1,2-dimethylcyclohexane has two isomers,

trans: CH_3 CH_3 flip $\rightleftarrows$ CH_3 CH_3

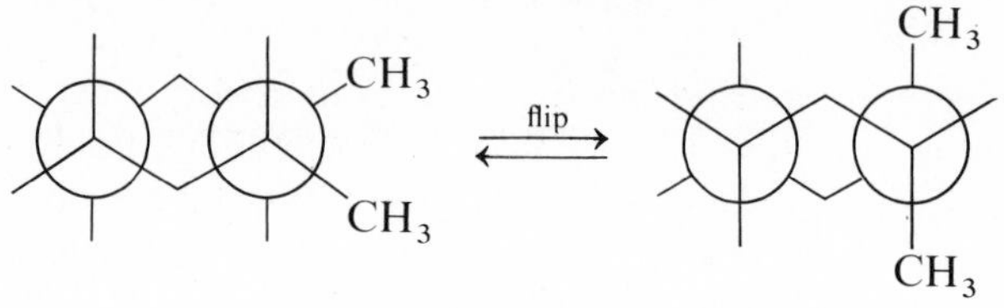

and

cis: CH_3 CH_3 flip $\rightleftarrows$ CH_3 CH_3

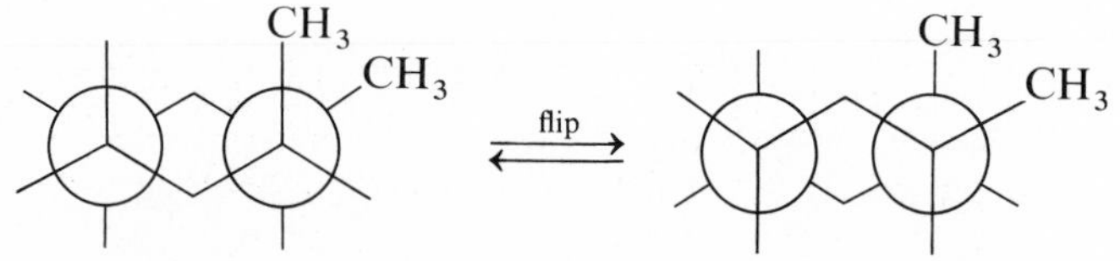

The two *trans* forms are conformers and interconvert readily. They therefore exist in equilibrium with each other. At equilibrium, the diequatorial conformer predominates because it experiences fewer axial-axial repulsions between methyl groups than does the diaxial conformer. The two *cis* forms are equivalent.

Problem 4.25

Which is the more stable conformation of (a) *trans*-1,2-dimethylcyclohexane? (b) *cis*-1,2-dimethylcyclohexane? (c) Which of the two isomers, *cis* or *trans*, is more stable? Explain.

4.16 PREPARATION OF CYCLOALKANES

Cyclopropane is the only cycloalkane that can be prepared by the Wurtz reaction. In this case zinc is used:

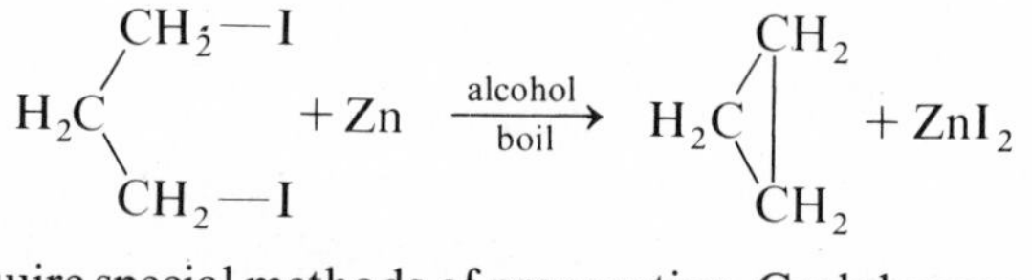

$$H_2C(CH_2{-}I)(CH_2{-}I) + Zn \xrightarrow[\text{boil}]{\text{alcohol}} \text{cyclo-}(CH_2)_3 + ZnI_2$$

Higher alkanes require special methods of preparation. Cyclohexane, for example, can be prepared by the reaction of benzene (Sec. 2.6) with hydrogen:

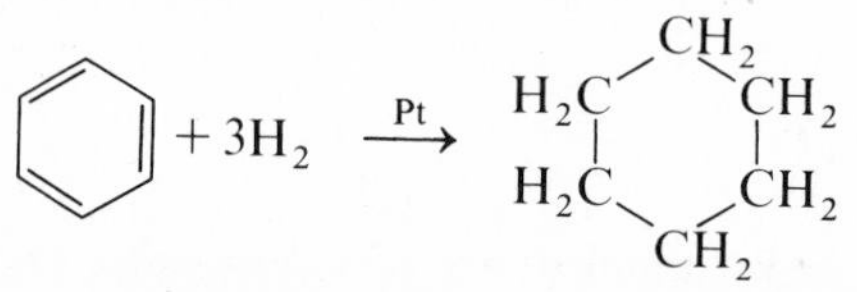

$$C_6H_6 + 3H_2 \xrightarrow{\text{Pt}} \text{cyclo-}(CH_2)_6$$

4.17 REACTIONS OF CYCLOALKANES

The reactions of the higher cycloalkanes are, in general, the same as those of the alkanes. Cyclopropane and cyclobutane are different because their molecules are strained. These two compounds, especially cyclopropane, tend to relieve bond strain by opening the ring. Examples of ring-opening reactions of cyclopropane are

$$\text{cyclopropane} \xrightarrow{HBr} CH_2(H)-CH_2-CH_2Br \quad \text{(1-bromopropane)}$$

$$\text{cyclopropane} \xrightarrow[25°C,\ dark]{Br_2} BrCH_2-CH_2-CH_2Br \quad \text{(1,3-dibromopropane)}$$

$$\text{cyclopropane} \xrightarrow[80°C,\ Ni]{H_2} CH_3-CH_2-CH_3 \quad \text{(propane)}$$

Cyclobutane is under less strain than cyclopropane and undergoes ring opening only with hydrogen and at higher temperatures:

$$\begin{array}{c} H_2C-CH_2 \\ | \quad\ \ | \\ H_2C-CH_2 \end{array} + H_2 \xrightarrow[200°C]{Ni} \begin{array}{c} H_2C-CH_3 \\ | \\ H_2C-CH_3 \end{array} \quad \text{(butane)}$$

The C—C bonds in cyclopropane (and to a lesser extent in cyclobutane) seem to be weaker than C—C bonds in other alkanes. We explain this weakness by the poor overlap of sp^3 orbitals in this molecule: The orbitals are not pointed in the best direction for maximum overlap (Fig. 4.10).

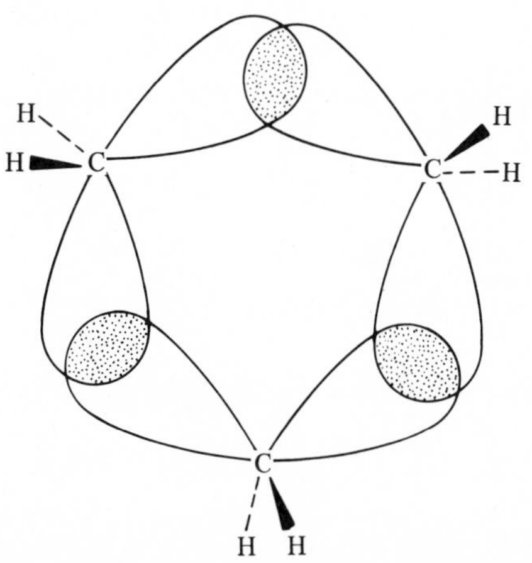

Figure 4.10. Orbital overlap in the cyclopropane molecule.

Problem 4.26

Suggest a method for the preparation of methylcyclopropane using the Wurtz reaction.

Problem 4.27

Predict the products of the reactions of cyclopropane with each of the following.

(a) HCl, (b) Cl_2 (dark, 25°C),
(c) D_2/Ni (80°C)(D = deuterium, an isotope of hydrogen)

4.18 OCCURRENCE OF ALKANES AND CYCLOALKANES

Petroleum is the major source of the world's alkanes. Geologists believe that petroleum was formed from tiny plants and animals that lived in the sea several hundred million years ago. At that time the sea covered much of the present land masses. These organisms thrived and died in enormous numbers, and then sank to the bottom where they mixed with mud and sand and formed layers of sediment. Later, more mud and sand covered these layers, which finally turned into rock. With the passing of time, the sea

Thujone, found in the essential oils of many plants

α–Pinene, major component of oil of turpentine

Cyclobuxine, obtained from Buxus sempervirens L., Buxaceae

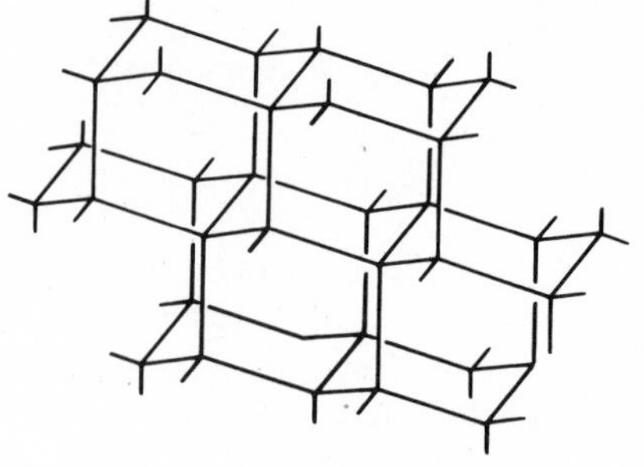

Diamond, a partial structure

Figure 4.11. Some naturally occurring compounds that contain ring structures.

receded and the earth's crust buckled under the influence of severe physical forces. The intense heat and pressure converted the organic substances into oil in the layers deep beneath the surface of the earth.

Crude petroleum is taken out of the earth by drilling and tapping large underground pools. Crude petroleum consists of thousands of compounds, most of which are alkanes. Some crude oils are dark and viscous; others are clear and fluid. Their appearance varies from one oil field to another.

Methane gas erupts from the earth in marshes and swamps and is called *marsh gas*. Methane is a product of the anaerobic (occurring without oxygen) fermentation of cellulose by microorganisms. This reaction occurs when plants become covered over with water in marshy areas. The gas is often visible as it rises to the water's surface. Sewage sludge also forms methane upon fermentation. It may be possible to use this process as a source of energy.

Alkanes and cycloalkanes also occur in many plants. Heptane is found in the turpentine of certain pines. Nonacosane ($C_{29}H_{60}$) has been found in cabbage leaves, and apples contain C_{27} and C_{29} alkanes. Beeswax contains heptacosane ($C_{27}H_{56}$).

Cycloalkane ring systems occur naturally in wide variety. We shall examine some of them in the chapters that follow. A few examples of naturally occurring compounds that contain rings are shown in Figure 4.11.

4.19 ANALYSIS AND IDENTIFICATION

Chemical analysis of the alkanes is difficult because of their inertness in the presence of most chemical reagents. They are not attacked by strong acids or bases or by strong oxidizing or reducing agents. For this reason we shall find that chemists can detect any functional group chemically without the interference of the alkyl parts of the molecule.

NEW TERMS

Ångstrom unit A unit of length equal to 10^{-8} cm (4.1).

Axial hydrogens The six hydrogen atoms of cyclohexane that align themselves perpendicular to the plane of the ring (Fig. 4.8) (4.14).

Catalytic cracking Cracking process carried out in the presence of a catalyst and at lower temperatures than those required in thermal cracking (4.8).

Classification of carbon atoms A *primary* (1°) carbon atom has bonds to only one other carbon atom; a *secondary* (2°) carbon atom has bonds to two other carbon atoms; a tertiary (3°) carbon atom has bonds to three other carbon atoms; a quaternary (4°) carbon atom has bonds to four other carbon atoms (4.4).

Combustion The process of burning with oxygen (4.10).

Conformation A particular spatial arrangement of the atoms in a molecule. Conformations differ from each other only in the rotation of their atoms about single bonds. Conformations are also called *conformers* (4.2).

Coupling reaction A chemical reaction in which two carbon chains are joined together to form a longer carbon chain. Examples are the **Wurtz reaction** and the **Corey-House synthesis** (4.9).

Eclipsed conformation The conformation in which the C—H bonds of one carbon atom are in the same plane as the C—H bonds of an adjacent carbon atom (4.2).

Equatorial hydrogens The six hydrogen atoms of cyclohexane that radiate out around the ring's circumference (Fig. 4.8) (4.14).

Halogenation The chemical combination of a compound with a halogen (fluorine, chlorine, bromine, or iodine) (4.10).

Heat of combustion (ΔH_C) The heat energy released when a compound is burned completely. (4.13)

Heat of reaction (ΔH°) The heat that must be added to convert one mole of reactant to products at 0°C, 1.0 atm. pressure, and a concentration of 1.0 mole per liter (4.10).

Homologous series A series of compounds in which each member differs from the one preceding it by an additional $—CH_2—$ group (4.1).

Newman projection formula A method of representing eclipsed and staggered conformations in two dimensions (Fig. 4.1) (4.2).

Nitration The introduction of a nitro group ($—NO_2$) into a compound, usually by reaction with nitric acid (4.10).

Oxidation A process in which an atom loses electrons or loses electron density (4.10).

Reduction A process in which an atom gains electrons or gains electron density (4.9).

Staggered conformation The conformation in which the C—H bonds of one carbon atom are staggered with the C—H bonds of an adjacent carbon atom (Fig. 4.1) (4.2).

Strain energy Chemical potential energy resulting from the distortion of bond angles or the eclipsing of bonds (4.13).

Thermal cracking A process for converting alkane mixtures into mixtures of different compositions by heating to 500°–1000°C in the absence of air (4.8).

ADDITIONAL PROBLEMS

4.28 Draw the line-dash-wedge formula (see Sec. 2.1) of hexane.

[4.29] Draw Newman projection formulas for all the staggered conformations of 2,3-dimethylbutane.

4.30 Give systematic names for each of the following.

[a] $CH_3\underset{\displaystyle CH_3}{\underset{|}{C}}H{—}\underset{\displaystyle CH_3}{\underset{|}{C}}HCH_3$,

[b] $CH_3\overset{\displaystyle CH_3}{\overset{|}{\underset{\displaystyle CH_3}{\underset{|}{C}}}}CH_2CH_3$,

[c] $CH_3CH_2CH_2\underset{\displaystyle CH_3}{\underset{|}{C}}HCH_3$,

(d) $CH_3CH_2\underset{\displaystyle CH_3}{\underset{|}{C}}H{—}\underset{\displaystyle CH_3}{\underset{|}{C}}HCH_3$,

(e) $CH_3CH_2\overset{\displaystyle CH_3}{\overset{|}{\underset{\displaystyle CH_3}{\underset{|}{C}}}}CH_2CH_2CH_3$,

(f) $CH_3CH_2\underset{\displaystyle CH_2CH_3}{\underset{|}{C}}HCH_2CH_3$

4.31 Draw the structural formula of each of the following compounds.

[a] 2,2,3,3-tetramethylpentane
[b] 3-ethyl-4-methylhexane
[c] heptane
[d] 4-neopentyloctane
(e) 2,2,4-trimethylpentane
(f) 3-ethyl-3-methylpentane
(g) octane
(h) 1,4-dineopentylcyclohexane

[4.32] Explain why alkanes are insoluble in water.

4.33 Show how you could prepare [a] 2,5-dimethylhexane by the Wurtz reaction and by the reduction of an alkyl halide; (b) octane by the Wurtz reaction and by the reduction of an alkyl halide.

4.34 Write a balanced equation for each reaction below.

[a] preparation of neopentyl chloride by direct chlorination
[b] mononitration of cyclohexane
[c] complete combustion of cyclobutane
(d) preparation of chlorocyclohexane by direct chlorination
(e) preparation of nitroethane by nitration
(f) complete combustion of nonane

[4.35] Explain why nitration of alkanes tends to give more decomposition than halogenation.

4.36 Give the systematic name of each of the following.

[a]

```
              CH2
             /   \
CH3CH2CH          CHCH2CH3
             \   /
              CH2
```

[b]

```
                 CH3
            CH2  |
           /   \ |
       H2C       C—CH3
         \      /
        H2C——C—CH3
             |
             CH3
```

(c)

```
     CH2CH2CH3
     |
     CH
    /   \
 CH2     CHCH2CH2CH3
    \   /
     CH2
```

(d)

```
          H2C——CH2
          /       \
   H3C—C           C—CH3
      /  \        / \
 CH3CH2   \      /   CH2CH3
           CH2
```

4.37 Draw the structural formula of each of the following.

[a] cyclopropylcyclopropane,
[b] 3-butyl-1-methylcyclohexane,
[c] *tert*-butylcyclohexane,
(d) cyclopentylcyclohexane,
(e) 1-isobutyl-1-ethylcyclobutane,
(f) *sec*-butylcyclohexane

[4.38] For those compounds in Problem 4.37 that exist in more than one conformation, tell which conformer is more stable.

[4.39] *trans*-1,4-Di-*tert*-butylcyclohexane exists in only one conformation. Draw its Newman projection formula and explain why the other conformation does not exist to any appreciable extent. Verify your explanation using models.

4.40 Predict the product of the Wurtz reaction of [a] 2,2-dimethyl-1,3-diiodopropane with Zn in alcohol (b) 1,3-dibromohexane with Zn in alcohol

4.41 Write equations to show how you could prepare each of the following from cyclopropane and any needed inorganic reagents.

[a] 1-chloropropane, [b] 1,3-dichloropropane,
[c] propane, (d) 1-iodopropane,
(e) 1,3-dibromopropane, (f) 1-deuteriopropane

4.42 Write equations for a reasonable laboratory method of preparing each of the following compounds starting with compounds that have four carbons or fewer and any solvents and inorganic reagents.

[a] pentane, [b] 2,4-dimethylpentane,
(c) 2,4-dimethylhexane, (d) isobutane

[4.43] Write equations for an alternative laboratory method of preparing each of the compounds of Problem 4.42 using the same restrictions of Problem 4.42. Compare each of these methods with the one you gave for Problem 4.42; that is, tell which is better and explain why.

[4.44] Explain why nitration of an alkane to form a nitroalkane is considered an oxidation reaction.

UNSATURATED HYDROCARBONS: ALKENES

5.1 CHARACTERISTICS OF ALKENES

Alkenes are hydrocarbons that contain a carbon–carbon double bond. Acyclic alkenes occur widely and have the general formula C_nH_{2n}. As a group the alkenes have chemical properties that distinguish them from the alkanes. These distinguishing chemical properties are due to the presence of the double bond.

1. The alkenes are *unsaturated*; that is, they contain fewer hydrogen atoms than the alkanes. The term unsaturated refers to the fact that unsaturated compounds can react with hydrogen to produce saturated molecules (alkanes) that cannot react further with hydrogen:

$$C_nH_{2n} + H_2 \longrightarrow C_nH_{2n+2}$$

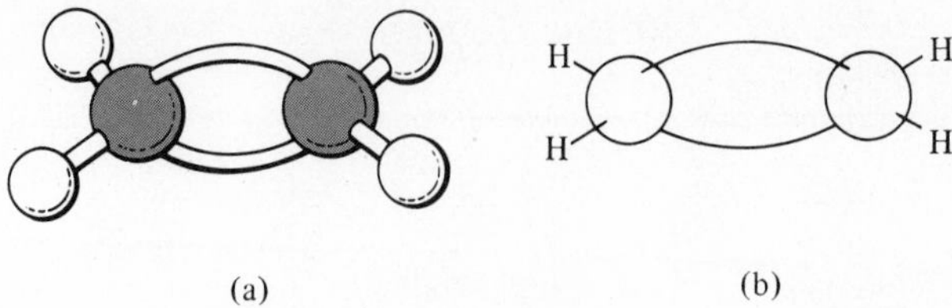

Figure 5.1. Ethene (a) ball-and-stick model and (b) circle and line drawing.

2. Unlike the alkanes, alkenes react readily, usually at mild temperatures, with various oxidizing and reducing agents, acids, free radicals, and a variety of other reagents.
3. Alkenes contain a carbon–carbon double bond. X-ray diffraction analysis shows that the atoms a, b, c, d, and the two carbons atoms of

$$\begin{matrix} a & & c \\ & C{=}C & \\ b & & d \end{matrix}$$

all lie in a plane (Figure 5.1).
4. The isolation of *cis–trans* isomers (Sec. 2.5) indicates that rotation does not occur about the double bond:

$$\begin{matrix} a & & c \\ & C{=}C & \\ b & & d \end{matrix} \quad \not\rightarrow \quad \begin{matrix} a & & d \\ & C{=}C & \\ b & & c \end{matrix}$$

We shall try to explain the gross features of alkenes in this chapter. Before proceeding to discuss them, however, we shall learn how to name them.

5.2 NOMENCLATURE: IUPAC SYSTEM

The simplest alkene is C_2H_4; its common name is ethylene. Other members have similar common names, but we will not be using them.

In the IUPAC system, the class name of compounds that have a carbon–carbon double bond is *alkene* (Table 5.1). Rules for naming alkenes include those for alkanes, with the following modifications:

1. Select as parent the longest continuous chain *that contains both carbons of the double bond.* For the parent name we use the name of the corresponding alkane and change the ending *-ane* to *-ene*. The simple members of the series are ethene, propene, butene, pentene, hexene, heptene, octene, nonene, and decene. For the compound

$$\begin{matrix} & CH{=}CH_2 \\ & | \\ CH_3CH_2CH_2CH_2 & CHCH_2CH_2CH_3 \end{matrix}$$

the parent chain is heptene (shown by shading).

Table 5.1 Some Simple Unbranched Alkenes

Name	Formula
Ethene	$CH_2{=}CH_2$
Propene	$CH_3CH{=}CH_2$
1-Butene	$CH_3CH_2CH{=}CH_2$
2-Butene	$CH_3CH{=}CHCH_3$
1-Pentene	$CH_2{=}CHCH_2CH_2CH_3$
3-Hexene	$CH_3CH_2CH{=}CHCH_2CH_3$

2. Number the parent chain beginning at the end nearer to the double bond, and designate the position of the double bond by using the number of the first carbon of the double bond. For example, the parent chain in the above compound has the name 1-heptene.
3. Name and indicate the location of each substituent group exactly as for the alkanes. Again, in the above compound, the substituent group is propyl, and it is at the 3- position. The complete systematic name is 3-propyl-1-heptene.
4. Number substituted cycloalkenes in the direction around the ring that gives the position 1,2- to the double bond, and the lowest numbers to the substituent groups. (Note that this rule often makes it unnecessary to give the position of the double bond because it is always 1,2-.) For example,

   (cyclohexene ring numbered 1–6, with $-CH_3$ at position 3)

   is 3-methylcyclohexene, and *not* 6-methylcyclohexene.
5. Two substituent groups that have special names are **vinyl-** and **allyl-**

   $$\begin{matrix} H & & H \\ & C{=}C & \\ H & & \end{matrix},$$

   or $CH_2{=}CH-$, is the vinyl group, and

   $$\begin{matrix} H & & H \\ & C{=}C & \\ H & & CH_2- \end{matrix},$$

   or $CH_2{=}CHCH_2-$, is the allyl group. For example,

   (cyclopentene)$-CH{=}CH_2$ is 4-vinylcyclopentene, and (cyclopentene)$-CH_2CH{=}CH_2$

   is 3-allylcyclopentene.

Other examples in Table 5.2 will clarify these rules.

Table 5.2 Systematic Names of Some Alkenes

$CH_3CH_2CH_2CHCH_2CH_2CH_3$ with $CH{=}CH_2$ on C-3 3-propyl-1-hexene	*Rule 1*: identify the longest continuous chain that contains the double bond; *Rule 2*: number from the end nearer the double bond; *Rule 3*: name and locate substituents.
$CH_3CH_2CCH_2CH_3$ with $=CH_2$ on the central C 2-ethyl-1-butene	*Rule 1*: parent chain must include *both* carbons of the double bond: *Rule 3*: name and indicate substituents.
Cyclopentene ring with CH_2CH_3 at C-1 (numbered 1–5) (*not* numbered with CH_2CH_3 at C-2) 1-ethylcyclopentene	*Rule 4*: number the ring so that the double bond is at position 1,2, and the substituent has the lowest number.
Cyclopropyl–$CHCH_2CH{=}CH_2$ (ring: CH_2–CH_2–CH) 3-cyclopropylpropene, or allylcyclopropane	*Rule 5*: allyl group (in this case either name is acceptable).

Problem 5.1
Give the structural formula of each of the following compounds.
(a) 2-pentene, (b) cyclopropene, (c) 3-methylcyclobutene,
(d) 1-cyclobutylcyclobutene

Problem 5.2
Give the systematic name of each of the following compounds.

(a) $CH_3CH_2CH{=}CHCH_2CH_3$

(b) $CH_3CH_2CH_2CHCH_3$ with $CH_2CH{=}CH_2$ on C-2

(c) Cyclohexene ring–$CH_2CH{=}CH_2$

(d) $CH_3CHCH{=}CHCH_3$ with CH_3 on C-2

(e) $CH_2{=}CH$–cyclopentene ring–$CH{=}CH_2$

Problem 5.3
Explain what is *wrong* with each of the following names.
(a) 3-pentene, (b) 3-vinylpentane, (c) 2-methylcyclobutene,
(d) 2-methyl-3-butene

The structural features of the carbon–carbon double bond are summarized as follows and in Figure 5.2:

1. All six atoms in ethene lie in a plane.
2. The carbon–carbon bond distance is 1.34 Å (recall that the C—C single bond length in ethane is 1.54 Å).
3. Bond angles are very close to 120° (recall that the bond angles in ethane are 109.5°).

These structural features of ethene are in total agreement with the structure arrived at using the concepts of orbital hybridization presented in Section 1.13. Using those concepts, we determine the structure of ethene as described in the following discussion.

Sigma Bonding

First we join the atoms in the molecule with single bonds to arrive at the skeletal structure:

```
H         H
 \       /
  Ċ — Ċ
 /       \
H         H
```

Three bonds are needed for each carbon atom; therefore carbon must utilize three sp^2 hybrid orbitals, and one electron is left in a p orbital on each carbon atom. The C—H and C—C single bonds are formed by orbital overlap, as shown in Figures 1.6(a) and 1.7. We call such bonds **sigma bonds**, or **σ bonds**

Pi Bonding

Each of the two remaining $2p$ orbitals contains one electron. We form a **pi (π) molecular orbital** by combining these two p orbitals as shown in Figure 5.3. The resulting bond is called a **π-bond** and contains a pair of electrons; it resembles two sausages, one on either side of the plane of the molecule. Note that one π bond includes both lobes.

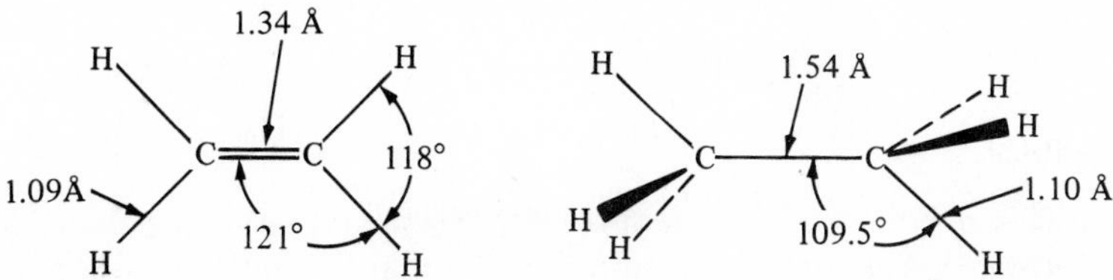

Figure 5.2. Structural features of ethene and ethane.

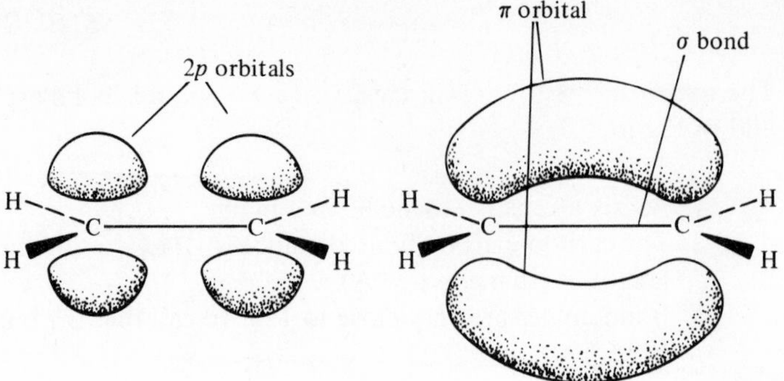

Figure 5.3. Formation of a π orbital by the combination of $2p$ orbitals of adjacent carbon atoms.

π Bonding requires that the molecule maintain the planar structure. Rotation about the C—C bond would break the π bond:

rotation

We can estimate the energy required for such rotation as follows. The energy required to dissociate a carbon–carbon double bond is 146 kcal/mole,

$$H_2C{=}CH_2 \longrightarrow 2\ H_2C{:};\ \Delta H^\circ = 146\ \text{kcal/mole}$$

The bond dissociation energy of the C—C bond in ethane is 84 kcal/mole,

$$CH_3{-}CH_3 \longrightarrow 2CH_3\cdot,\ \Delta H^\circ = 84\ \text{kcal/mole}$$

We conclude that the energy necessary to break only the π bond is approximately $146 - 84 = 62$ kcal/mole,

$$H_2C{=}CH_2 \longrightarrow H_2\dot{C}{-}\dot{C}H_2,\ \Delta H^\circ = 62\ \text{kcal/mole}$$

This π bond energy is approximately the energy needed to interconvert *cis–trans* isomers. The interconversion requires that the molecule achieve a transition state in which the π bond has dissociated into two mutually perpendicular p orbitals.

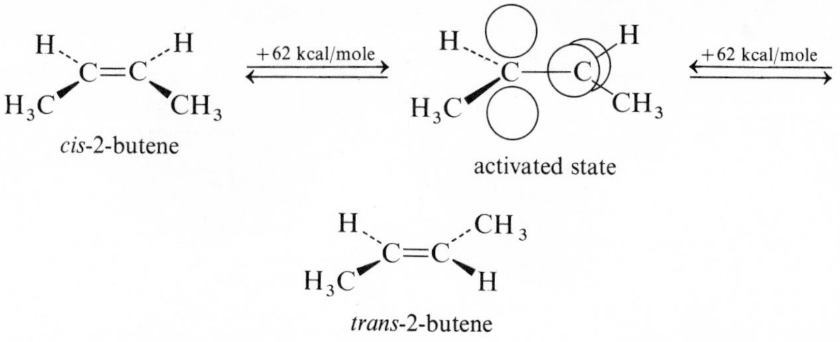

$$\underset{\textit{cis}\text{-2-butene}}{\overset{H\quad\quad H}{\underset{H_3C\quad\quad CH_3}{C{=}C}}} \underset{}{\overset{+62\text{ kcal/mole}}{\rightleftharpoons}} \underset{\text{activated state}}{\overset{H\quad\quad H}{\underset{H_3C\quad\quad CH_3}{C{-}C}}} \overset{+62\text{ kcal/mole}}{\rightleftharpoons}$$

$$\underset{\textit{trans}\text{-2-butene}}{\overset{H\quad\quad CH_3}{\underset{H_3C\quad\quad H}{C{=}C}}}$$

We can also account for the shorter bond distance (1.34 Å) of the C=C double bond on the basis of the π bond. Orbital overlap in the π bond adds to the overall degree of bonding. The extra electron pair draws the two carbon atoms closer together than they are in a single bond. We can also account for bond angles of near 120° by the geometry of sp^2 hybrid orbitals (see Fig. 1.8).

The above discussion shows that we can account neatly for the structure of the double bond in terms of orbital hybridization and the concept of σ and π bonding. As we shall see in Sections 5.4 and 5.5, this σ-π bonding model of the double bond also allows us to understand the physical and chemical properties of the alkenes.

Problem 5.4

The energy required to rupture a π bond is approximately 62 kcal/mole; the energy required to rupture the accompanying σ bond is approximately 84 kcal/mole. Explain this difference in bond energies between σ and π bonds. In other words, why is a σ bond stronger than a π bond?

Problem 5.5

Draw the structure of *cis*- and *trans*-1,2-dichloroethene. Designate the σ and π bonds.

Problem 5.6

Which of the following compounds exhibit *cis–trans* isomerism? Draw structures of the *cis–trans* isomers. Confirm your structures by constructing models.

(a) propene, (b) 2-butene,
(c) cyclopentene, (d) 1,2-dimethylcyclopentane

5.4 PHYSICAL PROPERTIES OF ALKENES

The presence of a carbon–carbon double bond in a molecule does not seem to alter its boiling and melting points or its density significantly (Table 5.3). In fact, the physical properties of the alkenes are comparable in many ways to those of their alkane counterparts. As is the case with the alkanes, melting and boiling points of alkenes increase with increasing molecular size. Alkenes are insoluble in water, and very soluble in nonpolar solvents such as alkanes, alkenes, benzene, ether, and chloroform. They are less dense than water.

Table 5.3 Physical Properties of Some Alkenes

Name	Structure	Melting Point, °C	Boiling Point, °C	Density, g/ml @ 20°C
Ethene	$CH_2{=}CH_2$	−169	−104	
Propene	$CH_2{=}CHCH_3$	−185	−47	
1-Butene	$CH_2{=}CHCH_2CH_3$	−130	−6	
cis-2-Butene	(H)(H)C=C(CH₃ below left as H_3C)(CH₃): $H, H / C{=}C / H_3C, CH_3$	−139	4	
trans-2-Butene	$H_3C, H / C{=}C / H, CH_3$	−106	1	
2-Methylpropene	$CH_2{=}C(CH_3)CH_3$	−141	−7	
Cyclobutene	$H_2C{-}CH$, $H_2C{-}CH$ (ring, CH=CH)		2	
Cyclobutane	$H_2C{-}CH_2$, $H_2C{-}CH_2$ (ring)	−80	13	
1-Pentene	$CH_2{=}CHCH_2CH_2CH_3$	−138	30	0.641
cis-2-Pentene	$H, H / C{=}C / H_3C, CH_2CH_3$	−151	37	0.655
trans-2-Pentene	$H, CH_2CH_3 / C{=}C / H_3C, H$		36	0.647
3-Methyl-1-butene	$CH_2{=}CHCH(CH_3)_2$	−135	25	0.648
2-Methyl-2-butene	$CH_3C(CH_3){=}CHCH_3$	−123	39	0.660
Cyclopentene	H_2C, CH, CH_2, CH, H_2C (ring, CH=CH)	−93	46	0.774
Cyclopentane	H_2C, CH_2, CH_2, CH_2, H_2C (ring)	−94	49	0.746

Table 5.3 *(Continued)*

Name	Structure	Melting Point, °C	Boiling Point, °C	Density, g/ml @ 20°C
1-Hexene	$CH_2{=}CHCH_2CH_2CH_2CH_3$	−99	64	0.673
2-Methyl-1-pentene	$CH_2{=}\underset{\underset{CH_3}{\vert}}{C}CH_2CH_2CH_3$	−136	62	0.682
4-Methyl-1-pentene	$CH_2{=}CHCH_2CH(CH_3)_2$	−154	54	0.665
1-Heptene	$CH_2{=}CH(CH_2)_4CH_3$	−119	93	0.697
1-Octene	$CH_2{=}CH(CH_2)_5CH_3$	−102	123	0.716
1-Nonene	$CH_2{=}CH(CH_2)_6CH_3$	−82	146	0.731
1-Decene	$CH_2{=}CH(CH_2)_7CH_3$	−66	171	0.741

The presence of the double bond confers a small polarity on alkene molecules because the π bond contains a high electron density. A propene molecule, for example, is therefore slightly polar. This polarity is represented by the use of vectors as in (a):

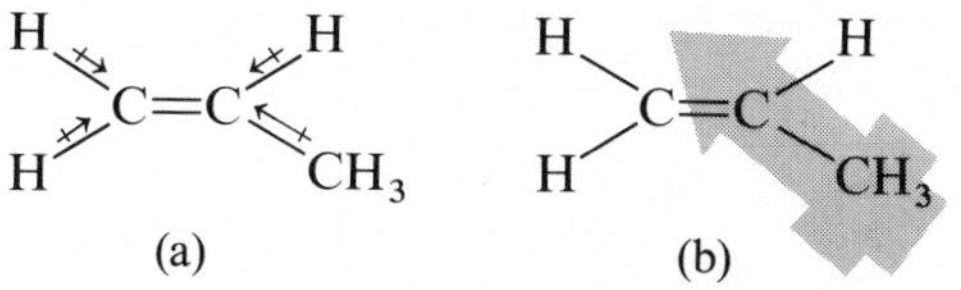

(a) (b)

The resultant of all the vectors is shown by the large shaded arrow in (b). (The vector, $\leftrightarrow$, represents a polar bond in which the arrowhead is the negative end and the cross at the other end is the positive end of the bond.)

Let us look at *cis*- and *trans*-2-butene. In *trans*-2-butene, the polarities of the various bonds all cancel, and the molecule considered as a unit is nonpolar:

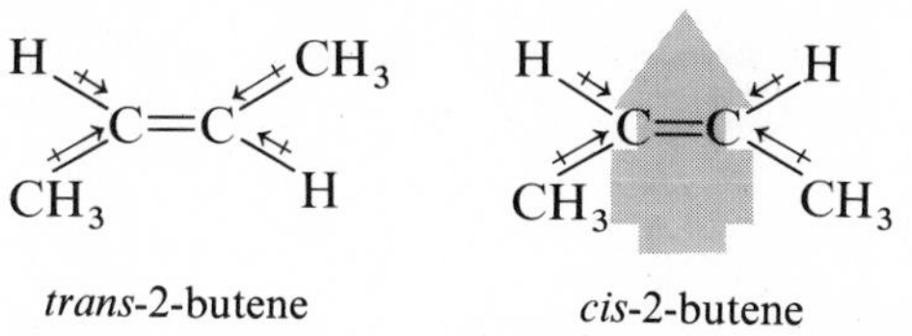

trans-2-butene *cis*-2-butene

The bond polarities in *cis*-2-butene do not cancel, and *cis*-2-butene is therefore a polar molecule. We can explain the small differences in physical properties of the two butenes on the basis of these differences in polarity: Because the *cis* isomer is polar, its molecules attract each other more strongly than do molecules of the *trans* isomer. Therefore, *cis*-2-butene boils at a higher temperature than *trans*-2-butene. The higher melting point of the *trans* isomer requires a different explanation: Melting requires the disruption of the crystal lattice of a solid. The greater symmetry of *trans*-2-butene allows its molecules to fit into a more regular and tighter crystal lattice than those of *cis*-2-butene.

The *trans* isomer, because it is more tightly packed than the *cis* isomer, melts at a higher temperature than the *cis* isomer.

Problem 5.7

Show both the individual bond polarities and the overall polarity of each of the following molecules.

(a) HCl, (b) chloromethane,
(c) 1,1-dichloroethene, (d) *cis*-1,2-dichloroethene,
(e) *trans*-1,2-dichloroethene, (f) 1,1,2,2-tetrachloroethene

Problem 5.8

Explain why alkanes and alkenes have lower densities than water.

5.5 PREPARATION OF ALKENES

As we saw in Section 4.8, cracking yields large quantities of alkenes, often as complex mixtures that are difficult to separate. Such separation is profitable only on a large industrial scale and is rarely used in the laboratory.

The most important laboratory methods of preparing alkenes involve elimination reactions. An **elimination reaction** is one in which a small molecule, AB, is removed from a compound:

$$-\underset{A}{\overset{|}{\underset{|}{C}}}-\underset{B}{\overset{|}{\underset{|}{C}}}- \longrightarrow \;\; >C=C< \; + AB$$

In this general equation we assume that the incomplete bonds are to hydrogen or other carbon atoms.

We shall look at two important elimination reactions—dehydration and dehydrohalogenation.

Dehydration

When we heat ethanol (CH_3CH_2OH, an alcohol) in the presence of a strong acid (H_2SO_4 or H_3PO_4), the alcohol molecule loses a water molecule, and ethene is formed:

$$H-\underset{H}{\overset{H}{\underset{|}{\overset{|}{C}}}}-\underset{OH}{\overset{H}{\underset{|}{\overset{|}{C}}}}-H \xrightarrow[\text{heat}]{H_2SO_4} \underset{H}{\overset{H}{>}}C=C\underset{H}{\overset{H}{<}} + H_2O$$

The reaction does not occur unless acid is present. The reaction occurs by the acid–base reaction of ethanol (a Brønsted–Lowry base) with the acid:

Step 1:

$$CH_3{-}CH_2{-}\ddot{\underset{\cdot\cdot}{O}}H + H{-}\ddot{\underset{\cdot\cdot}{O}}{-}\underset{\underset{:\underset{\cdot\cdot}{O}}{\|}}{\overset{\overset{:\ddot{O}}{\|}}{S}}{-}\ddot{\underset{\cdot\cdot}{O}}{-}H \rightleftarrows CH_3{-}CH_2{-}\underset{\underset{H}{|}}{\overset{+}{\ddot{O}}}{-}H + :\overset{-}{\ddot{\underset{\cdot\cdot}{O}}}{-}\underset{\underset{:\underset{\cdot\cdot}{O}:}{\|}}{\overset{\overset{:\ddot{O}:}{\|}}{S}}{-}OH$$

[Here and elsewhere in the text we shall use a curved arrow to indicate the movement of the electron pair of an atom during a chemical reaction (see also Sec. 3.7 and 3.8). This convention represents (a) the rupture of a bond to one atom and the formation of a bond to another, (b) dissociation to form ions, or (c) the reaction of ions to form a new covalent bond (Fig. 5.4).]

In a second step, the conjugate acid $CH_3CH_2OH_2{}^+$ loses a water molecule, and in a third step, the carbocation loses a proton on the adjacent carbon atom to form ethene and H_3O^+:

Step 2:

$$H{-}\underset{\underset{H}{|}}{\overset{\overset{H}{|}}{C}}{-}\underset{\underset{H}{|}}{\overset{\overset{H}{|}}{C}}{-}\underset{\underset{H}{|}}{\overset{+}{\ddot{O}}}{-}H \longrightarrow H{-}\underset{\underset{H}{|}}{\overset{\overset{H}{|}}{C}}{-}\overset{+}{C}\begin{smallmatrix}H\\ \\ H\end{smallmatrix} + H_2O$$

Step 3:

$$\begin{smallmatrix}H\\ \\ H\end{smallmatrix}\ddot{O}: + H{-}\underset{\underset{H}{|}}{\overset{\overset{H}{|}}{C}}{-}\overset{+}{C}\begin{smallmatrix}H\\ \\ H\end{smallmatrix} \longrightarrow H_3O^+ + \begin{smallmatrix}H\\ \\ H\end{smallmatrix}C{=}C\begin{smallmatrix}H\\ \\ H\end{smallmatrix}$$

(a)

$$R{-}\ddot{\underset{\cdot\cdot}{O}}H + H{-}X \longrightarrow R{-}\underset{\cdot\cdot}{\overset{\overset{H}{|\,+}}{O}}H + :X^-$$

A new O–H bond is formed, and the H–X bond is broken.

(b)

$$H{-}X \longrightarrow H^+ + :X^-$$

A covalent bond dissociates to form ions.

(c)

$$X:^- + H^+ \longrightarrow X{-}H$$

Combination of ions to form a covalent bond.

Figure 5.4. The use of curved arrows to show (a) the rupture of a bond to one atom and the formation of a new bond to another atom, (b) dissociation of a covalent bond to form ions, and (c) the combination of ions to form a new covalent bond.

Step 3 is also a Brønsted–Lowry acid–base reaction. The acid $CH_3\overset{+}{C}H_2$ donates a hydrogen ion to the base H_2O, after losing a water molecule itself.

If the hydroxyl group is attached to a primary carbon (Sec. 4.4), only one product is formed:

$$R{-}CH_2CH_2{-}OH \xrightarrow[\text{heat}]{H_2SO_4} R{-}CH{=}CH_2 + H_2O$$

Alcohols in which the hydroxyl group is attached to a secondary carbon may undergo dehydration to yield a mixture of alkenes because hydrogens on different carbon atoms can be eliminated:

$$CH_3CH_2\underset{\displaystyle OH}{\underset{|}{C}}HCH_3 + H_2SO_4 \xrightarrow{\Delta} \begin{cases} CH_3CH{=}CHCH_3 \ \text{(major product)} \\ \quad (\textit{cis} \text{ and } \textit{trans}) \\ CH_3CH_2CH{=}CH_2 \ \text{(minor product)} \end{cases}$$

When more than one alkene is possible, the reaction usually favors the most highly substituted one. That is, the order of preference is

$$R_2C{=}CR_2 > R_2C{=}CHR > RHC{=}CHR > R_2C{=}CH_2 > RHC{=}CH_2 > H_2C{=}CH_2$$

Alkyl groups attached to the $\rangle C{=}C\langle$ unit stabilize the double bond. The above order of alkene formation therefore means that the reaction yields the most stable alkene in the greatest quantity.

Dehydrohalogenation

Alkyl halides react with strong bases to yield alkenes. HX is eliminated in this reaction, but it appears as the salt and water:

$$CH_3{-}\underset{\displaystyle Br}{\underset{|}{C}}H{-}CH_3 + Na^+OH^- \xrightarrow[\Delta]{\text{alcohol}} CH_3{-}CH{=}CH_2 + H_2O + Na^+Br^-$$

The strong base, OH^-, undergoes an acid-base reaction with the hydrogen atom on the second carbon from the halogen:

$$CH_3{-}\underset{\displaystyle Br}{\underset{|}{C}}H{-}\underset{\displaystyle H}{\overset{\displaystyle H}{\overset{|}{\underset{|}{C}}}}{-}H + :\ddot{\underset{..}{O}}H^- \longrightarrow CH_3{-}CH{=}CH_2 + H_2O + Br^-$$

Mixtures of products can also occur. For example, 2-bromobutane yields a mixture of 1-butene and 2-butene:

$$\underset{\displaystyle\underset{Br}{|}}{CH_3CH_2CHCH_3} + K^+OH^- \xrightarrow[\Delta]{\text{ethanol}} \underset{\substack{(\textit{cis} \text{ and } \textit{trans-})\\ \text{(major product)}}}{CH_3CH{=}CHCH_3} + \underset{\text{(minor product)}}{CH_2{=}CHCH_2CH_3}$$

As with dehydration, the more substituted (most stable) alkene predominates.

We can summarize dehydration and dehydrohalogenation reactions in the following general equations:

Dehydration:

$$-\overset{|}{\underset{\underset{H}{|}}{C}}-\overset{|}{\underset{|}{C}}-OH + \text{acid} \xrightarrow{\Delta} \;\rangle C{=}C\langle\; + H_2O$$

$$(\text{acid} = H_2SO_4 \text{ or } H_3PO_4)$$

Dehydrohalogenation:

$$-\overset{|}{\underset{\underset{H}{|}}{C}}-\overset{|}{\underset{|}{C}}-X + OH^- \xrightarrow[\Delta]{\text{alcohol}} \;\rangle C{=}C\langle\; + H_2O + X^-$$

$$(X = Cl, Br, \text{ or } I)$$

Problem 5.9

Write equations including necessary reagents and conditions for the preparation of (a) propene from $CH_3CH_2CH_2OH$, (b) propene from $CH_3CH_2CH_2Cl$.

Problem 5.10

Predict the products of the following elimination reactions. When more than one alkene forms, predict which would be the major product. (Write balanced equations.)

(a) dehydration of $CH_3CH_2CH_2CH_2OH$

(b) dehydrobromination of 1-bromobutane

(c) dehydration of $\underset{\displaystyle\underset{OH}{|}}{CH_3C}(CH_3)_2$

(d) dehydrochlorination of 2-chloro-2-methylpropane

(e) dehydration of $CH_3CH_2CH_2\underset{\displaystyle\underset{OH}{|}}{C}HCH_3$

(f) dehydroiodination of 3-iodo-3-methylhexane

Just as the principal method for preparing alkenes is elimination,

$$\underset{\substack{| \quad |\\ A \quad B}}{-\overset{|}{C}-\overset{|}{C}-} \xrightarrow{\text{Elimination}} \;\rangle C{=}C\langle \; + \text{A—B}$$

the principal reaction of alkenes is **addition**, the reverse of elimination:

$$\rangle C{=}C\langle \; + \text{A—B} \xrightarrow{\text{addition}} -\overset{|}{\underset{A}{C}}-\overset{|}{\underset{B}{C}}-$$

In this section we shall examine the addition reactions shown in Table 5.4. We can explain the first four of them in terms of one general reaction mechanism. Because alkenes undergo addition reactions readily—often at room temperature—we shall see that some addition reactions serve as simple tests for unsaturation.

General Mechanism

The double bond in an alkene is rich in electrons. It should therefore behave as a Brønsted–Lowry base toward acidic reagents.

$$\rangle C{=}C\langle \; + \text{H—A} \rightleftarrows \rangle \overset{+}{C}-C\langle^{H} \; + \; :A^-$$

(base$_1$) (acid$_2$) (acid$_1$) (base$_2$)

The π bond breaks to form a new σ bond to the proton and leaves an empty $2p$ orbital on the adjacent carbon (a carbocation). We can write the above equation using curved arrows to represent the shift of electrons:

$$\rangle C{=}C\langle \; + \text{H—A} \rightleftarrows \rangle \overset{+}{C}-C\langle^{H} \; + :A^-$$

In the second step, the very reactive carbocation combines with the anion $:A^-$:

$$\rangle \overset{+}{C}-C\langle^{H} \; + \; :A^- \longrightarrow {}^{A}\rangle C-C\langle^{H}$$

Let us see how this general mechanism applies to actual examples.

Table 5.4 Addition Reactions of Alkenes

1. *Hydration*

$$CH_2{=}CH_2 + H_2O \xrightarrow{\text{acid}} \underset{\mid}{CH_2}(H){-}\underset{\mid}{CH_2}(OH)$$

2. *Hydrohalogenation*

$$CH_2{=}CH_2 + HX \longrightarrow CH_2(H){-}CH_2(X) \quad (X = Cl, Br, \text{ or } I)$$

3. *Halogenation*

$$CH_2{=}CH_2 + X_2 \longrightarrow CH_2(X){-}CH_2(X) \quad (X = Cl \text{ or } Br)$$

4. *Polymerization*

$$n\,CH_2{=}CH_2 \xrightarrow{\text{In}} {+}CH_2{-}CH_2{+}_n$$

(In = a free radical, an acid, or a base initiator)

5. *Hydrogenation*

$$CH_2{=}CH_2 + H_2 \xrightarrow{\text{catalyst}} CH_2(H){-}CH_2(H)$$

forms alkanes

(catalyst = Ni, Pt, or Pd)

6. *Glycol Formation*

$$CH_2{=}CH_2 + KMnO_4 + H_2O \xrightarrow[25^\circ C]{} CH_2(OH){-}CH_2(OH)$$

7. *Ozonolysis*

$$CH_2{=}CH_2 + O_3 \longrightarrow \text{(ozonide: } CH_2{-}O{-}CH_2 \text{ bridged by } O{-}O) \xrightarrow[H_2O]{Zn} 2\ H_2C{=}O$$

Hydration

Alkenes react with water in the presence of a strong mineral acid, usually sulfuric acid, to produce alcohols. Ethene, for example, yields ethanol:

$$CH_2{=}CH_2 + H_2O \xrightarrow[\Delta]{H_3O^+} CH_3CH_2OH$$

Most of the world's supply of ethanol is produced by this reaction. We can apply the above general mechanism to this reaction:

Step 1: $CH_2{=}CH_2 + H{-}\overset{+}{\ddot{O}}{-}H$ (with H on O) $\rightleftarrows$ $\overset{+}{C}H_2CH_3 + {:}\ddot{O}{-}H$ (with H on O)

Step 2: $H{-}\ddot{O}{:}$ (with H on O) $+ \overset{+}{C}H_2CH_3 \rightleftarrows H{-}\overset{+}{\ddot{O}}{-}CH_2CH_3$ (with H on O)

Step 3: $H{-}\ddot{O}{:}$ (with H on O) $+ H{-}\overset{+}{\ddot{O}}{-}CH_2CH_3$ (with H on O) $\rightleftarrows H_3O^+ + HOCH_2CH_3$

This reaction is the reverse of the dehydration reaction that we examined in Section 5.5.

When we hydrate other alkenes, an interesting pattern emerges. Propene, for example, reacts with water to give only 2-propanol. 1-Propanol ($CH_3CH_2CH_2OH$) is not formed.

$$CH_3CH{=}CH_2 + H_2O \xrightarrow{H_3O^+} CH_3\underset{\underset{OH}{|}}{C}HCH_3$$

(2-propanol)

2-Methylpropene reacts similarly to give only 2-methyl-2-propanol. 2-Methyl-1-propanol, $CH_3CH(CH_3)CH_2OH$, is not formed.

$$CH_3\overset{\overset{CH_3}{|}}{C}{=}CH_3 + H_2O \xrightarrow{H_3O^+} CH_3\overset{\overset{CH_3}{|}}{\underset{\underset{OH}{|}}{C}}CH_3$$

(2-methyl-2-propanol)

The general rule that tells us the direction of addition was expressed in 1870 by the Russian chemist V. Markovnikov, and is known as **Markovnikov's Rule**: *When an acid, HA, adds to an alkene, the hydrogen atom of HA adds to that carbon atom of the double bond that already bears the greater number of hydrogen atoms.*

greater number of hydrogen atoms

$$CH_3{-}\overset{\overset{H}{|}}{C}{=}CH_2 \qquad CH_3{-}\overset{\overset{CH_3}{|}}{C}{=}CH_2$$

A—H A—H

In most cases only the Markovnikov product is formed.

It is possible to explain Markovnikov's rule in terms of our general mechanism. Step 1 of the general mechanism is the reaction of the π bond with the proton of HA. The π bond could break in either of two ways. Propene could thus give either a secondary or a primary carbocation:

$$CH_3{-}CH{=}CH_2 + H{-}A \rightleftarrows CH_3{-}\overset{+}{C}H{-}CH_3 + :A^-$$

(a secondary carbocation)

or

$$CH_3{-}CH{=}CH_2 + A{-}H \rightleftarrows CH_3{-}CH_2{-}\overset{+}{C}H_2 + :A^-$$

(a primary carbocation)

Carbocation stability depends on the groups attached to the positive carbon. An alkyl group tends to furnish electrons more effectively than a hydrogen atom, and alkyl carbocation stability therefore increases in the order

$$R_3C^+ > R_2HC^+ > RH_2C^+ > H_3C^+$$

(tertiary carbocation) > (secondary carbocation) > (primary carbocation) > (methyl carbocation)

We can now explain Markovnikov's rule: *Addition of HA occurs to give the most stable carbocation.* In general, this means that the H^+ becomes attached to the less alkyl-substituted carbon because the positive charge is better accommodated (more stable) on the more alkyl-substituted carbon atom.

$$CH_3{-}CH{=}CH_2$$

+ charge more stable here (at CH); H^+ (at CH_2)

Problem 5.11
What is the major product of the acid-catalyzed hydration of 2-methyl-2-butene?

Hydrohalogenation

Alkenes react readily with hydrogen halides, HCl, HBr, and HI, to give alkyl halides in accordance with Markovnikov's rule.

$$CH_3CH{=}CH_2 + HCl \longrightarrow CH_3\underset{\displaystyle Cl}{\underset{|}{C}}HCH_3$$

This reaction usually occurs readily at room temperature. It follows the general mechanism given above.

$$\textbf{Step 1:}\quad CH_3CH{=}CH_2 + H{-}\ddot{\underset{..}{Cl}}: \longrightarrow CH_3{-}\overset{+}{C}H{-}CH_3 + :\ddot{\underset{..}{Cl}}:^-$$

$$\textbf{Step 2:}\quad CH_3{-}\overset{+}{C}H{-}CH_3 + :\ddot{\underset{..}{Cl}}:^- \longrightarrow CH_3\underset{\displaystyle Cl}{\underset{|}{C}}HCH_3$$

Here again, the addition reaction—hydrohalogenation—is the reverse of the elimination reaction—dehydrohalogenation.

Halogenation

Alkenes react with chlorine or bromine, usually in CCl_4 solution, to form the dihalide.

$$CH_3CH{=}CH_2 + Br_2 \longrightarrow CH_3\underset{\displaystyle Br}{\underset{|}{C}}H{-}\underset{\displaystyle Br}{\underset{|}{C}}H_2$$

This reaction occurs rapidly and usually completely at room temperature, and is useful as a simple test for unsaturation. Bromine is often used because it has a deep reddish-brown color which disappears when the bromine is consumed. The dibromides produced are colorless. The test involves adding a few drops of Br_2 dissolved in CCl_4 to the sample to be tested. If the bromine color disappears immediately, the sample is unsaturated.

Because halogen molecules are nonpolar they do not dissociate spontaneously in solution, and they cannot follow the first step of the general mechanism outlined above. Instead, halogenation occurs through the following mechanism:

$$\textbf{Step 1:}\quad \begin{matrix}\diagdown\ \diagup\\ C\\ \|\\ C\\ \diagup\ \diagdown\end{matrix} + :\ddot{\underset{..}{Br}}{-}\ddot{\underset{..}{Br}}: \rightleftarrows \left[\begin{matrix}\diagdown\ \diagup\\ C\\ \|\\ C\\ \diagup\ \diagdown\end{matrix}\ \overset{\delta+}{:\ddot{\underset{..}{Br}}}{-}\overset{\delta-}{\ddot{\underset{..}{Br}}:}\right] \rightleftarrows \begin{matrix}\diagdown\ |\\ C\\ |\\ C\\ \diagup\ |\end{matrix}\!\!>\ddot{\underset{..}{Br}}^+ + :\ddot{\underset{..}{Br}}:^-$$

$$\textbf{Step 2:}\quad :\ddot{\underset{..}{Br}}:^- + \begin{matrix}\diagdown\ |\\ C\\ |\\ C\\ \diagup\ |\end{matrix}\!\!>\ddot{\underset{..}{Br}}^+ \longrightarrow \begin{matrix} & | & \\ :\ddot{\underset{..}{Br}}{-} & C & {-}\\ & | & \\ {-} & C & {-}\ddot{\underset{..}{Br}}:\\ & | & \end{matrix}$$

Dissociation of the bromine molecule occurs when it collides with the electron-rich alkene. The π electrons in the alkene cause the Br—Br bond to polarize and eventually to dissociate into ions. In step 2, the bromine anion attacks one of the carbon atoms of the cyclic intermediate to produce the dibromide. Note that the cyclic intermediate is very reactive. Its carbon atoms bear considerable positive charge because the positively charged bromine atom polarizes the C—Br bonds.

Polymerization

Alkene molecules can react with each other in the presence of certain compounds (initiators) to form long chains called **polymers** (*poly-* = many, *mer-* = unit).

$$n \;\; \gt C{=}C\lt \;\xrightarrow{\text{initiator}}\; \left(-\overset{|}{\underset{|}{C}}-\overset{|}{\underset{|}{C}}- \right)_n$$

Initiator molecules end up as part of the final polymer molecules and are therefore not catalysts.

Alkenes may polymerize by three mechanisms that involve either free radicals, cations, or anions.

*Free radical polymerization mechanism:**

$$A\cdot + \gt C{=}C\lt \longrightarrow A-\overset{|}{\underset{|}{C}}-\overset{|}{\underset{|}{C}}\cdot \xrightarrow{\gt C=C\lt} A-\overset{|}{\underset{|}{C}}-\overset{|}{\underset{|}{C}}-\overset{|}{\underset{|}{C}}-\overset{|}{\underset{|}{C}}\cdot \longrightarrow \text{etc.}$$

Cationic polymerization mechanism:

$$B^+ + \gt C{=}C\lt \longrightarrow B-\overset{|}{\underset{|}{C}}-\overset{|}{\underset{|}{C}}^+ \xrightarrow{\gt C=C\lt} B-\overset{|}{\underset{|}{C}}-\overset{|}{\underset{|}{C}}-\overset{|}{\underset{|}{C}}-\overset{|}{\underset{|}{C}}^+ \longrightarrow \text{etc.}$$

Anionic polymerization mechanism:

$$D{:}^- + \gt C{=}C\lt \longrightarrow D-\overset{|}{\underset{|}{C}}-\overset{|}{\underset{|}{C}}{:}^- \xrightarrow{\gt C=C\lt} D-\overset{|}{\underset{|}{C}}-\overset{|}{\underset{|}{C}}-\overset{|}{\underset{|}{C}}-\overset{|}{\underset{|}{C}}{:}^- \longrightarrow \text{etc.}$$

Polyethylene is prepared by heating ethene (ethylene) under pressure with a small amount of oxygen:

$$n\,CH_2{=}CH_2 \xrightarrow[1000\text{–}1200\text{ atm}]{O_2,\ 100\text{–}400°C} \left(CH_2-CH_2 \right)_n \qquad (\Delta H° = -22 \text{ kcal/mole ethene})$$

The reaction is exothermic because the energy released in forming the new carbon–carbon σ bond exceeds the energy required to break the π bond of ethene.

Problem 5.12

Account for the $\Delta H°$ of the above reaction on the basis of the π bond broken and the new σ bond formed. How does your calculation compare with the observed value of -22 kcal/mole for $\Delta H°$?

The above reaction proceeds through a free radical mechanism. The oxygen molecule has two unpaired electrons and serves to initiate the reaction

$$\cdot\ddot{\underset{..}{O}}-\ddot{\underset{..}{O}}\cdot + CH_2{=}CH_2 \longrightarrow \cdot\ddot{\underset{..}{O}}-\ddot{\underset{..}{O}}-CH_2-CH_2\cdot \xrightarrow{CH_2=CH_2} \left(CH_2CH_2 \right)_n$$

* See Sec. 3.7 for curved arrow notation.

We show polymer structures with open valences, $\lbrace CH_2CH_2 \rbrace_n$, although the chain ends are actually occupied by initiator molecules. Because polymer chains are usually quite long (n = hundreds or sometimes thousands), the chemical identity of the ends does not affect the properties appreciably. Many polymers have commercial or industrial uses; some of the more important ones are described below.

Polymerization of chloroethene (common name, vinyl chloride) by a process similar to the last equation above gives *poly(vinyl chloride)*. The polymer has a high molecular weight (n = 24,000), and it is a hard, rigid plastic useful for making pipes and phonograph records. Cyanoethene (common name, acrylonitrile) polymerizes to give polyacrylonitrile, which is sold as *Orlon*.

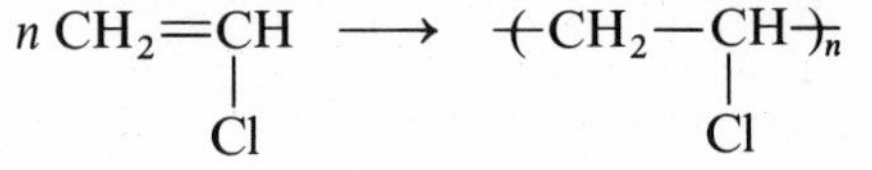

$$n\,CH_2{=}\underset{\displaystyle Cl}{\underset{|}{CH}} \longrightarrow \lbrace CH_2{-}\underset{\displaystyle Cl}{\underset{|}{CH}} \rbrace_n$$

vinyl chloride — poly(vinyl chloride)

$$n\,CH_2{=}\underset{\displaystyle CN}{\underset{|}{CH}} \longrightarrow \lbrace CH_2{-}\underset{\displaystyle CN}{\underset{|}{CH}} \rbrace_n$$

acrylonitrile — Orlon

Orlon is used in making fabrics. *Teflon*, $\lbrace CF_2{-}CF_2 \rbrace_n$, is chemically inert and very resistant to wear. These properties make it useful in the manufacture of bearings and cooking utensils.

Poly(methyl methacrylate)

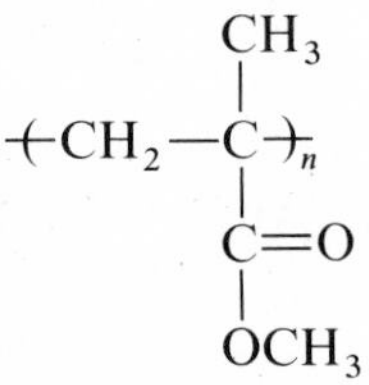

$$\lbrace CH_2{-}\overset{\displaystyle CH_3}{\overset{|}{\underset{\displaystyle \underset{\displaystyle OCH_3}{\underset{|}{C{=}O}}}{\underset{|}{C}}}} \rbrace_n$$

has good optical transparency and is sold under the names Plexiglas and Lucite.

Polyisobutylene is produced in a cation mechanism. 2-Methylpropene (common name, isobutylene) polymerizes when it is treated with a strong acid such as H_2SO_4.

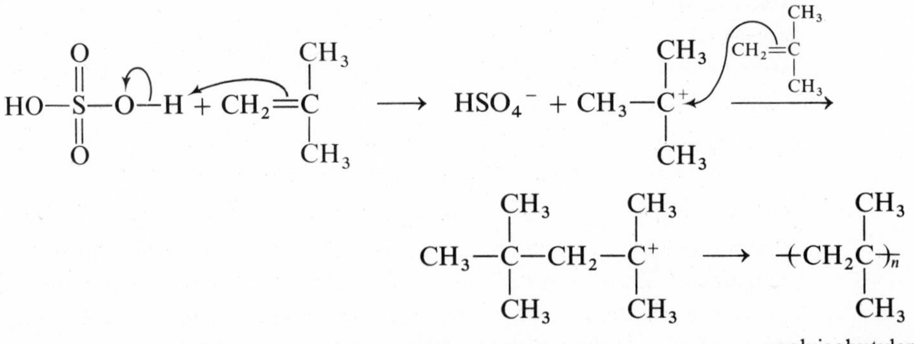

$$HO{-}\overset{\displaystyle O}{\overset{\|}{\underset{\displaystyle O}{\underset{\|}{S}}}}{-}O{-}H + CH_2{=}\overset{\displaystyle CH_3}{\overset{|}{\underset{\displaystyle CH_3}{\underset{|}{C}}}} \longrightarrow HSO_4^- + CH_3{-}\overset{\displaystyle CH_3}{\overset{|}{\underset{\displaystyle CH_3}{\underset{|}{C^+}}}} \xrightarrow{CH_2=C(CH_3)_2}$$

$$CH_3{-}\overset{\displaystyle CH_3}{\overset{|}{\underset{\displaystyle CH_3}{\underset{|}{C}}}}{-}CH_2{-}\overset{\displaystyle CH_3}{\overset{|}{\underset{\displaystyle CH_3}{\underset{|}{C^+}}}} \longrightarrow \lbrace CH_2\overset{\displaystyle CH_3}{\overset{|}{\underset{\displaystyle CH_3}{\underset{|}{C}}}} \rbrace_n$$

polyisobutylene

Although 2-methylpropene reacts in this mechanism, ethene does not. We can explain the difference on the basis of carbocation stability. The polymerization of 2-methylpropene involves tertiary carbocations as intermediates, whereas primary carbocations would have to be involved in the polymerization of ethene. Primary carbocations are so unstable that they are rarely formed under ordinary conditions. In other words, a primary carbocation is so strong an acid that the equilibrium

$$\underset{(\text{base}_1)}{CH_2{=}CH_2} + \underset{(\text{acid}_2)}{H_2SO_4} \rightleftarrows \underset{(\text{acid}_1)}{CH_3{-}CH_2^+} + \underset{(\text{base}_2)}{HSO_4^-}$$

lies completely to the left.

The anionic mechanism is not as widely used to prepare polymers as the other two, although one example is polyacrylonitrile prepared from cyanoethene. The strong base sodium amide (sodamide), $NaNH_2$, is used to initiate the polymerization.

$$H_2\ddot{N}{:}^- + CH_2{=}\underset{\displaystyle CN}{\underset{|}{CH}} \xrightarrow{NH_3} H_2N{-}CH_2{-}\underset{\displaystyle CN}{\underset{|}{\bar{\ddot{C}}H}} \xrightarrow{CH_2=CH(CN)}$$

$$H_2NCH_2\underset{\displaystyle CN}{\underset{|}{C}}HCH_2\underset{\displaystyle CN}{\underset{|}{\bar{\ddot{C}}H}} \longrightarrow \text{+}CH_2\underset{\displaystyle CN}{\underset{|}{C}}H\text{+}_n$$

polyacrylonitrite

Styrene ($C_6H_5{-}CH{=}CH_2$) undergoes polymerization by all three mechanisms. The product, polystyrene, is sold under various names including Styrofoam, a polymer which is allowed to solidify as a foam to produce buoyant and insulating materials.

Problem 5.13
Give the formulas of the alkenes used to make Teflon and poly(methyl methacrylate).

Hydrogenation

Alkenes react with hydrogen gas in the presence of finely divided metals (Ni, Pt, or Pd) to form alkanes.

$$CH_2{=}CH_2 + H_2 \xrightarrow[\text{pressure}]{\text{Ni, 25°C}} CH_3{-}CH_3$$

This reaction does not occur through the cation mechanism described above. The reaction depends on the high affinity of hydrogen gas for certain metals—Ni, Pt, Pd. The hydrogen is adsorbed on the metal surface along with alkene molecules. When a hydrogen molecule and an alkene molecule lie next to each other, their closeness and the weakening of their bonding permit a rearrangement of electrons to form new bonds

(Fig. 5.5). These hydrogenation reactions usually result in addition of the two hydrogens to the same side of the double bond. This is to be expected because the H_2 and alkene molecules lie side by side on the catalyst surface at the moment of reaction.

Hydrogenation is characteristic of alkenes. It is not a useful test for unsaturation because hydrogenation requires elaborate equipment to handle gases under pressure. It is useful, however, for determining how many double bonds are in a molecule. For example, a compound whose formula is C_7H_{12} (C_nH_{2n-2}) may have two double bonds, one double bond and a ring, or two rings:

$CH_2{=}CHCH_2CH_2CH_2CH{=}CH_2$, (ring: CH=CH, CH_2, $CHCH_3$, CH_2CH_2), or (bicyclic: CH_2—CH, CH_2—CH, with CH_2, CH_2, CH_2 bridge)

(methylcyclohexene)—CH_3 + 1 mole H_2 $\xrightarrow[25°C]{Ni}$ (cyclohexane)—CH_3 (C_7H_{14})

(bicyclo[3.2.0]heptane) + H_2 $\xrightarrow[25°C]{Ni}$ no reaction (under mild conditions)

Problem 5.14

A compound, C_5H_8, was hydrogenated at 25°C to give C_5H_{10}. Give all the structural formulas possible for the original compound.

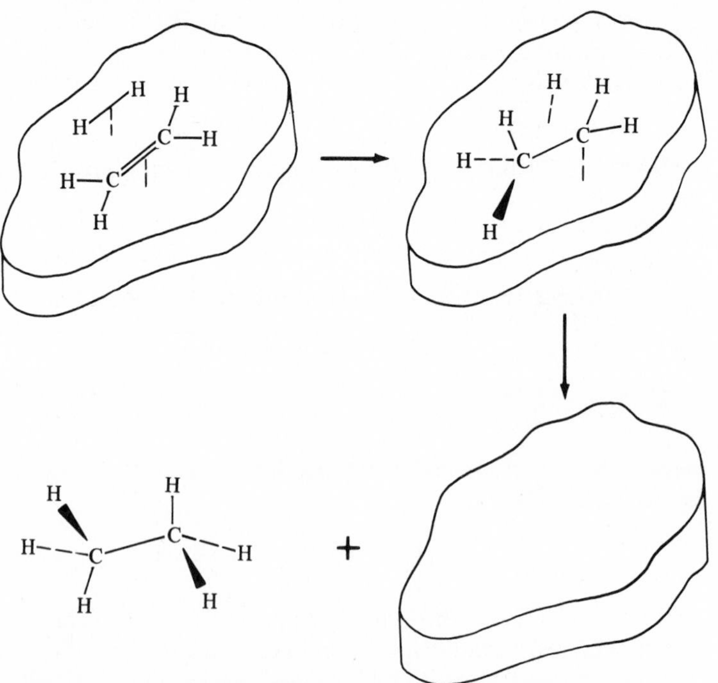

Figure 5.5. Schematic drawing of the hydrogenation of an alkene on a metal surface.

Problem 5.15

Draw the structural formula of the major isomer produced in the reaction

$$\text{(1,2-dimethylcyclopentene)} + H_2 \xrightarrow[\text{Pressure}]{\text{Ni, 25°C}} ?$$

Glycol Formation

If we treat an alkene with an aqueous solution of potassium permanganate at room temperature, we obtain a dihydroxyl compound called a glycol.

$$\rangle C{=}C\langle + MnO_4^- + 2H_2O \longrightarrow -\underset{OH}{\underset{|}{\overset{|}{C}}}-\underset{OH}{\underset{|}{\overset{|}{C}}}- + MnO_2\downarrow + 2OH^-$$

(a glycol)

Potassium permanganate is a strong oxidizing agent, and if we heat the mixture, the glycol will be oxidized further. The above reaction is a convenient test for unsaturation because a distinct color change occurs when MnO_4^- (deep purple solution) is reduced to MnO_2 (a black, insoluble solid). To test for an alkene we simply have to add a few drops of aqueous $KMnO_4$ solution to the sample. If the color changes, the sample is unsaturated.

Ozonolysis

The reaction of ozone, O_3, with alkenes is an oxidation reaction in which the carbon-carbon double bond is completely broken. The reaction is usually carried out by bubbling ozone (generated by passing O_2 through an electrical discharge) through a solution of the alkene.

We may write the reaction as follows:

$$\underset{B}{\overset{A}{\rangle}}C{=}C\underset{E}{\overset{D}{\langle}} + O_3 \longrightarrow \left[\underset{B}{\overset{A}{\rangle}}C\begin{smallmatrix}\diagup O \diagdown \\ \diagdown O{-}O \diagup\end{smallmatrix}C\underset{E}{\overset{D}{\langle}}\right]$$

(an ozonide)

$$\xrightarrow[H_2O]{Zn} \underset{B}{\overset{A}{\rangle}}C{=}O + O{=}C\underset{E}{\overset{D}{\langle}}$$

(A, B, D, E are alkyl or aryl groups or H)

Ozonides are often explosive and need not be isolated. Instead, the reaction mixture is treated with zinc metal and water to decompose the ozonide to the carbonyl compounds.

Ozonolysis is useful for locating the position of the double bond in a molecule. We do so by isolating and identifying the carbonyl compounds (aldehydes or ketones) that are produced. For example, let us examine the products of ozonolysis of 1-butene, 2-butene, and 2-methylpropene:

$$\underset{\text{(1-butene)}}{CH_2{=}CHCH_2CH_3} + O_3 \longrightarrow \left[\text{ozonide: } CH_2 \text{ and } CHCH_2CH_3 \text{ joined through } O \text{ and } O{-}O \right] \xrightarrow[H_2O]{Zn} \underset{\text{(formaldehyde)}}{HCH{=}O} + \underset{\text{(propanal)}}{HCCH_2CH_3(=O)}$$

$$\underset{\text{(2-butene)}}{CH_3CH{=}CHCH_3} \xrightarrow{\text{ozonolysis}} \underset{\text{(ethanal)}}{2CH_3CH{=}O}$$

$$\underset{\text{(2-methylpropene)}}{CH_3C(CH_3){=}CH_2} \xrightarrow{\text{ozonolysis}} \underset{\text{(acetone)}}{CH_3C(CH_3){=}O} + \underset{\text{(formaldehyde)}}{HCH{=}O}$$

Each alkene above gives a different set of ozonolysis products. Identification of these products allows us to determine the alkene that gave rise to them.

Problem 5.16

Ozonolysis of an alkene, C_6H_{12}, gave only acetone. Give the structure of the original alkene and write equations for the ozonolysis.

5.7 CHEMICAL TESTS

It is often necessary to determine which functional groups are present in an unknown compound. Many methods are available to the chemist for making such determinations; they include chemical tests and measurements of physical properties such as melting and boiling points, densities, and so forth, as well as more sophisticated physical determinations of mass spectra, ultraviolet, visible, and infrared spectra, and nuclear magnetic resonance spectra. In this section we shall be concerned only with simple chemical tests.

A simple chemical test is one in which we observe an immediate visual or other sensory response when we add a given test reagent to our sample. Many of the reactions that we have studied satisfy this requirement. We shall outline them below.

Bromine test. Bromination is a good simple test for unsaturation because the deep red-brown color of bromine is lost as the reaction proceeds:

$$\underset{\text{(red-brown)}}{C_nH_{2n} + Br_2} \xrightarrow[\text{room temperature}]{CCl_4} \underset{\text{(colorless)}}{C_nH_{2n}Br_2}$$

Baeyer Test

Cold, aqueous potassium permanganate is also a good simple test for unsaturation because the purple color of the permanganate ion (MnO_4^-) disappears as it is reduced to the insoluble, black manganese dioxide (MnO_2).

$$>C{=}C< + \underset{\text{(purple)}}{MnO_4^-} \xrightarrow[\text{room temperature}]{H_2O\text{, rapid at}} \underset{\text{(colorless)}}{-\underset{OH}{\overset{|}{\underset{|}{C}}}-\underset{OH}{\overset{|}{\underset{|}{C}}}-} + \underset{\text{(black solid)}}{MnO_2\downarrow}$$

Solubility in Cold, Concentrated H_2SO_4

A different but equally useful test for functional groups that are Lewis bases is cold, concentrated sulfuric acid. Sulfuric acid reacts with nearly all unsaturated compounds and compounds that contain oxygen atoms. The reaction is a Brønsted–Lowry acid–base reaction that yields ions which are soluble in cold, concentrated sulfuric acid:

$$>C{=}C< + H_2SO_4 \rightleftarrows \underbrace{-\underset{H}{\overset{|}{\underset{|}{C}}}-\overset{|}{\underset{+}{C}}- + HSO_4^-}_{\text{soluble in } H_2SO_4}$$

$$-C{\equiv}C- + H_2SO_4 \rightleftarrows \underbrace{-\underset{H}{\underset{|}{C}}{=}\overset{+}{C}- + HSO_4^-}_{\text{soluble in } H_2SO_4}$$

$$R-\ddot{\underset{\cdot\cdot}{O}}H + H_2SO_4 \rightleftarrows \underbrace{R-\underset{H}{\underset{|}{\overset{+}{\ddot{O}}}}H + HSO_4^-}_{\text{soluble in } H_2SO_4}$$

$$C_nH_{2n+2} + H_2SO_4 \longrightarrow \text{no reaction (does not dissolve)}$$

$$\text{Cycloalkanes} + H_2SO_4 \longrightarrow \text{no reaction (does not dissolve)}$$

Table 5.5 Simple Chemical Tests for Alkanes and Cycloalkanes*

Compound	Bromine Test, Br_2/CCl_4	Baeyer Test, $KMnO_4$, Dilute, Aqueous	H_2SO_4, Cold, Concentrated
Alkanes and cycloalkanes (except cyclopropane)	no reaction	no reaction	insoluble
Cyclopropane	red-brown → colorless in presence of $AlCl_3$	no reaction	soluble
Alkenes	red-brown → colorless	purple solution → black precipitate	soluble

* All tests are conducted at room temperature unless otherwise stated.

Saturated and aromatic hydrocarbons are not soluble in cold, concentrated sulfuric acid. The test therefore involves adding a small amount of sample to cold, concentrated H_2SO_4 and noting whether it dissolves. If so, then the sample contains a basic functional group. Table 5.5 gives the simple chemical tests that we have examined thus far in this text.

Others, however, are not useful as simple chemical tests. An example of a reaction that is not a useful simple chemical test is hydrogenation of an alkene:

$$C_nH_{2n} + H_2 \xrightarrow{Ni} C_nH_{2n+2}$$

This reaction requires elaborate equipment to handle gaseous hydrogen, and we observe reaction by the loss in pressure of H_2. There is no simple color change.

Problem 5.17

Write equations for all the reactions described in Table 5.5. Do not balance, but merely give the important reagents and products.

Problem 5.18

Describe simple chemical tests that could be used to distinguish between the compounds in each of the following pairs.

(a) 1-pentene and cyclopentane (b) 1-pentene and pentane

NEW TERMS

Addition reaction The combination of a double bond with a molecule A—B to produce.

```
  |   |
 —C—C—
  |   |
  A   B
```

Addition is the reverse of elimination (5.6).

Allyl group The group $CH_2{=}CH{-}CH_2{-}$ (5.2).

Baeyer test A simple chemical test in which aqueous $KMnO_4$ solution is added to a compound suspected of having a C=C double bond. A positive test consists of disappearance of the purple color of the $KMnO_4$ and formation of a black MnO_2 precipitate (5.7).

Dehydration The elimination of a molecule of water from an alcohol to produce an alkene (5.5):

$$-\underset{H}{\underset{|}{\overset{|}{C}}}-\underset{OH}{\underset{|}{\overset{|}{C}}}- \xrightarrow[\text{heat}]{H_2SO_4} \;\rangle C{=}C\langle \; + H_2O$$

Dehydrohalogenation The elimination of HCl, HBr, or HI from an alkyl halide to produce an alkene (5.5):

$$-\underset{H}{\underset{|}{\overset{|}{C}}}-\underset{X}{\underset{|}{\overset{|}{C}}}- + KOH \xrightarrow{\text{alcohol}} \;\rangle C{=}C\langle \; + KX + H_2O$$

Elimination reaction A reaction in which a double bond is formed in a molecule by the loss of an atom or group from each of two adjacent atoms (5.5):

$$-\underset{A}{\underset{|}{\overset{|}{C}}}-\underset{B}{\underset{|}{\overset{|}{C}}}- \longrightarrow \;\rangle C{=}C\langle \; + AB$$

Glycol A dihydroxyl compound (5.6).

Hydration The addition of a water molecule to an alkene to produce an alcohol (5.6):

$$\rangle C{=}C\langle \; + H_2O \xrightarrow{\text{acid}} -\underset{H}{\underset{|}{\overset{|}{C}}}-\underset{OH}{\underset{|}{\overset{|}{C}}}-$$

Hydrogenation The addition of hydrogen to an alkene to produce an alkane (5.6):

$$\rangle C{=}C\langle \; + H_2 \xrightarrow[\text{or Pd}]{\text{Ni, Pt}} -\underset{H}{\underset{|}{\overset{|}{C}}}-\underset{H}{\underset{|}{\overset{|}{C}}}-$$

Hydrohalogenation The addition of HCl, HBr, or HI to an alkene to produce an alkyl halide (5.6):

$$\rangle C{=}C\langle \; + HX \longrightarrow -\underset{H}{\underset{|}{\overset{|}{C}}}-\underset{X}{\underset{|}{\overset{|}{C}}}-$$

Markovnikov's rule When an acid HA adds to an alkene, the hydrogen atom of HA adds to that carbon atom of the double bond that already bears the greater number of hydrogen atoms (5.6).

Ozonolysis The reaction of $\rangle C{=}C\langle$ with ozone (O_3) to produce an ozonide which can be reduced and cleaved to two carbonyl groups. Ozonolysis cleaves the alkene at the double bond (5.6):

$$\rangle C{=}C\langle \; + O_3 \longrightarrow \rangle C\underset{O-O}{\overset{O}{\diagup\!\diagdown}}C\langle \longrightarrow \;\rangle C{=}O + O{=}C\langle$$

Pi (π) bond The bond formed by the overlap of two *p* orbitals on adjacent atoms. The electrons in a π bond lie in regions above and below the plane of the atoms involved in the π bond. This region in which π electrons exist is called a **π molecular orbital** (5.3).

Polymerization The combination of many molecules of a compound to form long-chain molecules (**polymers**) (5.6).

Sigma (σ) bond A bond in which the electron pair lies in the region between the two nuclei (Figs. 1.6 and 1.7) (5.3).

Vinyl group The group $CH_2{=}CH{-}$ (5.3).

ADDITIONAL PROBLEMS

5.19 Draw structural formulas and give IUPAC names for all the isomers of [a] C_4H_8 and (b) C_5H_{10}.

5.20 Draw the structural formula of each of the compounds below.
[a] 2,5-dimethyl-3-hexene, [b] *trans*-2-butene,
[c] *trans*-3,4-dimethylcyclobutene, (d) 3,4-dimethyl-3-hexene,
(e) *cis*-2-pentene, (f) *cis*-3,4-dimethylcyclopentene

[5.21] Explain the planar nature of the carbon–carbon double bond on the basis of the σ–π model.

[5.22] Explain the general chemical features of the carbon–carbon double bond on the basis of the σ–π model.

5.23 Draw the σ–π structure of each of the following compounds.
[a] *cis*-2-butene, [b] cyclobutene,
(c) *trans*-2-butene, (d) cyclopentene

5.24 Which of the following compounds exhibit *cis–trans* isomerism? Draw their structures.
[a] 1-pentene, [b] 2-pentene,
[c] 2,3-dimethyl-2-pentene, (d) 1-butene,
(e) 2-butene, (f) 2,3-dimethyl-2-butene

5.25 Which of the following compounds have a zero polarity? Demonstrate by drawing vector diagrams of each compound.

[a] CH_2Cl_2

[b] Cl(upper left), H(upper right), C=C, H(lower left), Cl(lower right)

[c] Cl(upper left), H(upper right), C=C, Cl(lower left), H(lower right)

[d] Cl(upper left), Cl(upper right), C=C, H(lower left), H(lower right)

(e) CF_2Cl_2

(f) F(upper left), H(upper right), C=C, H(lower left), Cl(lower right)

(g) F(upper left), H(upper right), C=C, F(lower left), H(lower right)

(h) Br(upper left), F(upper right), C=C, F(lower left), Br(lower right)

[5.26] Give the structures of all the isomeric alcohols whose molecular formula is $C_5H_{11}OH$.

5.27 Write equations to show how you could prepare [a] 2-methylpropene and (b) 2,3-dimethyl-2-butene by dehydration and by dehydrobromination.

[5.28] Write equations for the reactions that occur when 2-methyl-2-butene is treated with each of the following reagents.

(a) H_2/Ni, (b) Br_2,
(c) HCl, (d) cold, aqueous $KMnO_4$,
(e) ozone followed by Zn/H_2O

5.29 Repeat Problem 5.28 using 2-butene instead of 2-methyl-2-butene.

[5.30] Write equations for the polymerization of vinyl chloride initiated by oxygen.

[5.31] Compound A, C_4H_6, gives C_4H_{10} on hydrogenation. Compound B, C_5H_8, gives C_5H_{10} on hydrogenation. On the basis of this information, write all the possible structures of A and B.

5.32 A compound whose molecular formula is C_6H_{10} reacts with H_2/Ni to give C_6H_{12}. Ozonolysis of the original followed by Zn/H_2O reduction of the ozonide gives the dialdehyde

$$\overset{\displaystyle O}{\overset{\|}{HC}}CH_2CH_2CH_2CH_2\overset{\displaystyle O}{\overset{\|}{C}}H$$

What is the structure of the original compound?

[5.33] A compound whose molecular formula is C_8H_{14} reacts with H_2 in the presence of Ni to give C_8H_{16}. Ozonolysis of the original compound followed by Zn + H_2O reduction of the ozonide gives the diketone

$$CH_3\underset{\underset{\displaystyle O}{\|}}{C}CH_2CH_2CH_2CH_2\underset{\underset{\displaystyle O}{\|}}{C}CH_3$$

What is the structure of the original compound?

5.34 Each of the following reactions produces one isomer in preference to others. Give the structural formula of the major isomer produced in each of these reactions.

[a] (cyclohexene) + Br_2 $\xrightarrow{CCl_4}$ [b] (cyclohexene) + D_2 $\xrightarrow[\text{Pressure}]{\text{Ni, 25°C}}$

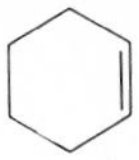

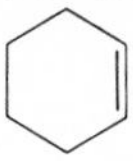

(*Note*: *D* is deuterium, an isotope of hydrogen.)

(c) 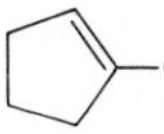 (cyclopentene)$-CH_3$ + HI $\longrightarrow$

6 UNSATURATED HYDROCARBONS: DIENES AND ALKYNES

6.1 DIENES AND POLYENES: NOMENCLATURE

Many compounds have more than one double bond. Compounds that have two double bonds are called **dienes**. Those that have three double bonds are called trienes. We name **polyenes** (compounds that have many double bonds) by replacing the ending of the alkane name, *-ne*, with *-diene*, *-triene*, *-tetraene*, and so forth to give the names alkadiene, alkatriene, and alkatetraene. We denote the position of each double bond by one number—the number of the lower-numbered carbon atom of the double bond. For example, $CH_2{=}C{=}CH{-}CH_3$ is 1,2-butadiene; $CH_2{=}CH{-}CH{=}CH_2$ is 1,3-butadiene; and $CH_3{-}CH{=}CH{-}CH_2{-}CH{=}CH_2$ is 1,4-hexadiene. We number from the end nearer to a double bond. We name cyclic polyenes in the same way except that one of the double bonds must be numbered 1, and the numbering must be in the direction that gives the number 2 to the other carbon of the same double bond *and* gives the lowest possible numbers to the other double bonds. For example.

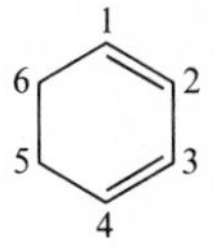

[Structure: 1,3-cyclohexadiene with ring atoms numbered 1–6; double bonds at C1=C2 and C3=C4]

must be numbered as shown and not

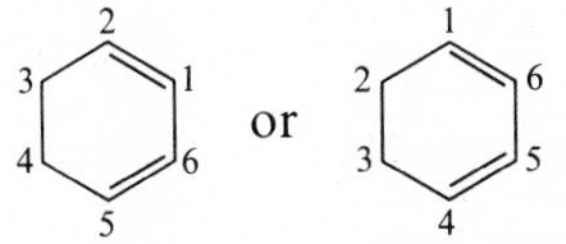

[Structures: the same ring numbered with double bonds at C1=C2 and C5=C6 (left), or C1=C6 and C4=C5 (right)]

Problem 6.1

Give the systematic name of each of the following:

(a) 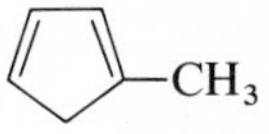 [Structure: cyclopentene ring with a CH_3 on a double-bond carbon]

(b) 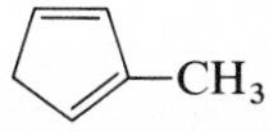 [Structure: cyclopentene ring with CH_3 on a ring carbon; structure differs from (a)]

(c) 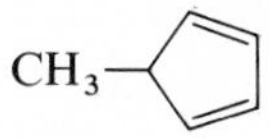 [Structure: CH_3 attached to a cyclopentadiene ring]

(d) 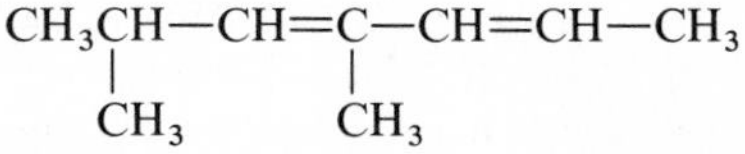 $CH_3CH(CH_3)-CH=C(CH_3)-CH=CH-CH_3$

6.2 CLASSIFICATION OF DIENES

We classify dienes as cumulated, conjugated, or isolated. **Cumulated double bonds** share one carbon atom. **Conjugated double bonds** are separated by one single bond. **Isolated double bonds** are separated by more than one single bond.

Cumulated double bonds: $>C=C=C<$

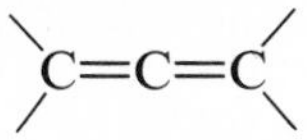

Conjugated double bonds: $>C=\overset{|}{C}-\overset{|}{C}=C<$

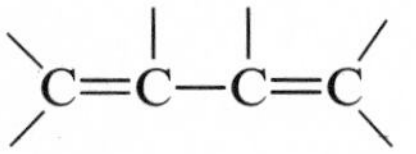

Isolated double bonds: $>C=\overset{|}{C}\text{+}\overset{|}{\underset{|}{C}}\text{+}_n\overset{|}{C}=C<$ $(n \geq 1)$

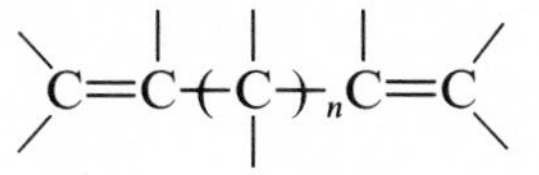

We classify dienes in this way because each class has its own set of properties.

6.3 STABILITY OF DIENES

Table 6.1 lists heats of hydrogenation of some alkenes and alkadienes. The heat of hydrogenation is the energy involved in the reaction

$$R_2C=CR_2 + H_2 \xrightarrow{Ni} R_2CH-CHR_2$$

Table 6.1 Heats of Hydrogenation (ΔH_h°) of Some Polyenes

Name	Formula	ΔH_h°, kcal/mole	ΔH_h° per Double Bond
1,2-Propadiene	$CH_2{=}C{=}CH_2$	−71.3	−35.7
1,2-Butadiene	$CH_2{=}C{=}CHCH_3$	−69.9	−35.0
1,3-Butadiene	$CH_2{=}CHCH{=}CH_2$	−57.1	−28.6
1,2-Pentadiene	$CH_2{=}C{=}CHCH_2CH_3$	−70.4	−35.2
1,3-Pentadiene	$CH_2{=}CHCH{=}CHCH_3$	−54.2	−27.1
1,4-Pentadiene	$CH_2{=}CHCH_2CH{=}CH_2$	−60.8	−30.4
1,5-Hexadiene	$CH_2{=}CH(CH_2)_2CH{=}CH_2$	−60.5	−30.3
Propene	$CH_2{=}CHCH_3$	−30.1	−30.1
1-Butene	$CH_2{=}CHCH_2CH_3$	−30.3	−30.3
1-Pentene	$CH_2{=}CHCH_2CH_2CH_3$	−30.1	−30.1

By comparing heats of hydrogenation we can estimate the relative stabilities of molecules. Heats of hydrogenation are negative, and therefore heat is released in the reaction. If two dienes have the same carbon skeleton, then the larger the heat of hydrogenation, the larger the energy content (potential energy) of the alkene. In Figure 6.1 we see that 1,2-butadiene has 12.8 kcal/mole more energy than 1,3-butadiene.

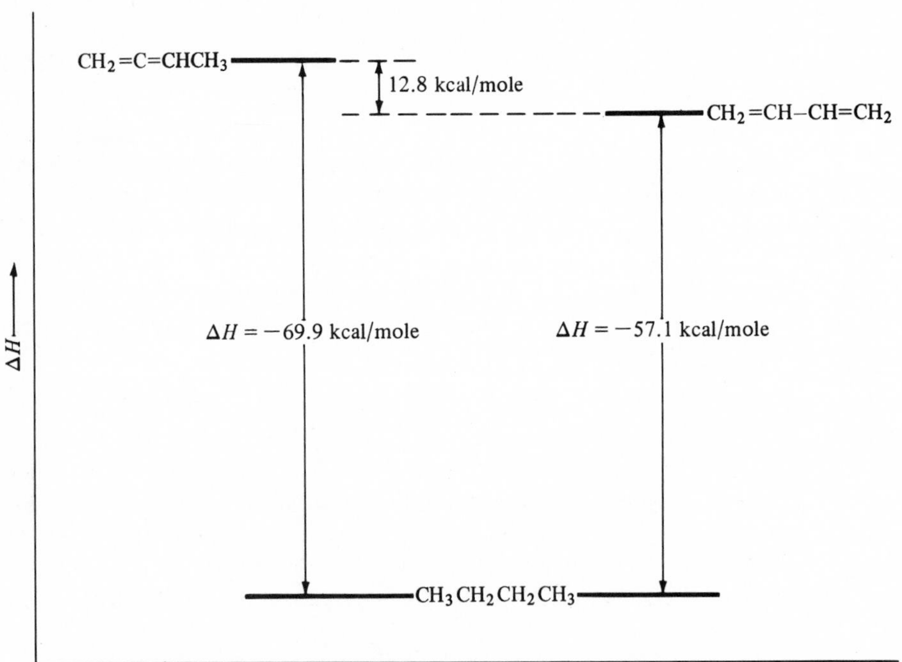

Figure 6.1. Heats of hydrogenation as a measure of the potential energy of a molecule.

Problem 6.2

Draw an energy diagram (such as in Fig. 6.1) to show the hydrogenation of 1,2-pentadiene, 1,3-pentadiene, and 1,4-pentadiene.

Table 6.1 also shows some alkenes. We see that alkenes have heats of hydrogenation of about 30 kcal/mole. From the values of heats of hydrogenation per double bond we can get an idea of the relative stabilities of cumulated, conjugated, and isolated double bonds. Compounds containing cumulated double bonds have heats of hydrogenation per double bond that are greater than those of alkenes. We conclude that molecules containing cumulated double bonds have higher potential energies per double bond than alkenes. Using the same logic, we see that molecules containing conjugated double bonds have lower potential energies per double bond than alkenes. Compounds whose molecules have isolated double bonds have potential energies per double bond that are roughly equal to those of alkenes.

6.4 ELECTRONIC STRUCTURES OF CONJUGATED DIENES

We can explain the unusual stabilities of conjugated dienes in terms of their electronic structures. The closeness of the two π orbitals allows a slight amount of overlap between

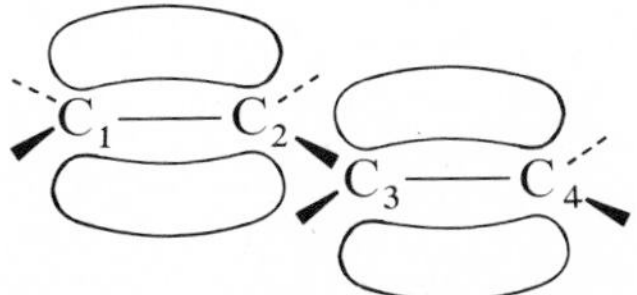

these two orbitals. This overlap is such that the electrons in these orbitals can, to a slight extent, move over the entire four-carbon system. Electrons behave like waves, and extending the wavelength of a wave lowers its energy. We express this idea quantitatively in Planck's equation for the energy of a light quantum:

$$E = hf = \frac{hc}{\lambda}$$

where E = energy of the quantum, f = frequency of the wave, λ = wavelength of the wave, c = velocity of light, and h = a universal constant called Planck's constant. If we think of the electron as a wave that is localized to one double bond, it has a wavelength equal to λ_1. If it is delocalized over both double bonds of a conjugated system, it has a longer wavelength, λ_2, and consequently a lower energy.

|←λ_1→|
C═C—C═C
|←——λ_2——→|

Thus a conjugated system of double bonds has a lower energy than a system of isolated double bonds where the two π orbitals are not close enough to each other to overlap.

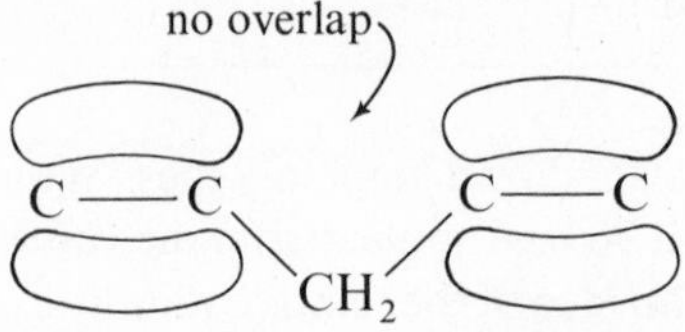

Cumulated double bonds cannot overlap, although they are even closer to each other than conjugated double bonds. The reason is that they lie at right angles to each other:

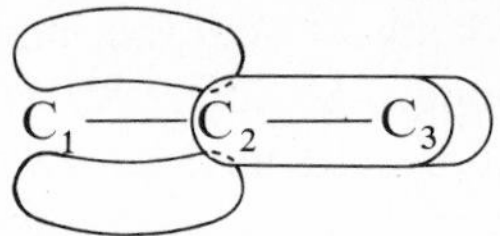

Viewing the system along the C—C—C bond axis, we see that orbital overlap is impossible:

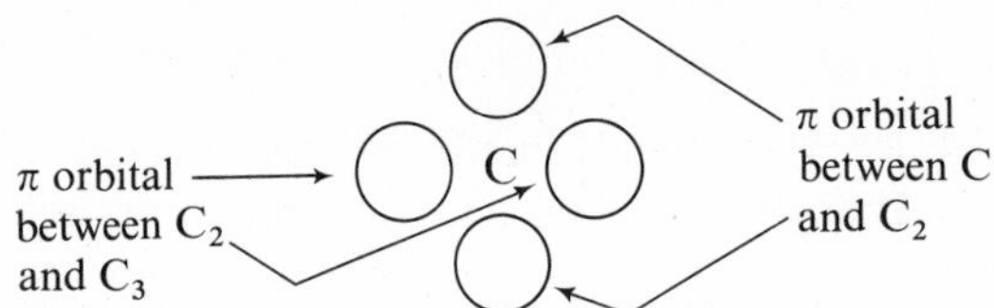

6.5 PREPARATION OF DIENES

Reactions used to prepare alkenes can also be used to prepare dienes. For example, the dehydration of 1,4-butanediol or the dehydrobromination of 1,4-dibromobutane gives 1,3-butadiene:

$$HOCH_2CH_2CH_2CH_2OH \xrightarrow[\Delta]{H_2SO_4} CH_2{=}CHCH{=}CH_2 + 2H_2O$$

$$BrCH_2CH_2CH_2CH_2Br + 2KOH \xrightarrow[\Delta]{\text{alcohol}} CH_2{=}CHCH{=}CH_2 + 2KBr + 2H_2O$$

Problem 6.3

Write equations to show how you could prepare 2,3-dimethyl-1,3-butadiene by (a) dehydration and (b) dehydrohalogenation.

6.6 REACTIONS OF DIENES

Isolated dienes behave as though the double bonds have no chemical effect on each other; each double bond undergoes its reactions independently of the other.

Conjugated dienes exhibit an interesting behavior in addition reactions: The reaction of one mole of hydrogen chloride with one mole of 1,3-butadiene gives two products:

$$CH_2{=}CH{-}CH{=}CH_2 + HCl$$

$$\xrightarrow{25°C} \underset{\substack{\text{(3-chloro-1-butene)} \\ \text{(major product)}}}{CH_3{-}\overset{\overset{\displaystyle Cl}{|}}{C}H{-}CH{=}CH_2} + \underset{\substack{\text{(1-chloro-2-butene)} \\ \text{(minor product)}}}{CH_3{-}CH{=}CH{-}CH_2Cl}$$

The second product, 1-chloro-2-butene, is surprising. It involves an apparent anti-Markovnikov addition and a migration of the double bond. We can interpret this set of products using a slight modification of the general mechanism for addition reactions that we used in Section 5.6:

Step 1:

$$CH_2{=}CH{-}CH{=}CH_2 + HCl \rightleftarrows \underset{\text{I}}{CH_3{-}\overset{+}{C}H{-}CH{=}CH_2} + Cl^-$$

We can picture the **allylic*** carbocation, I, by making use of the molecular orbital picture.

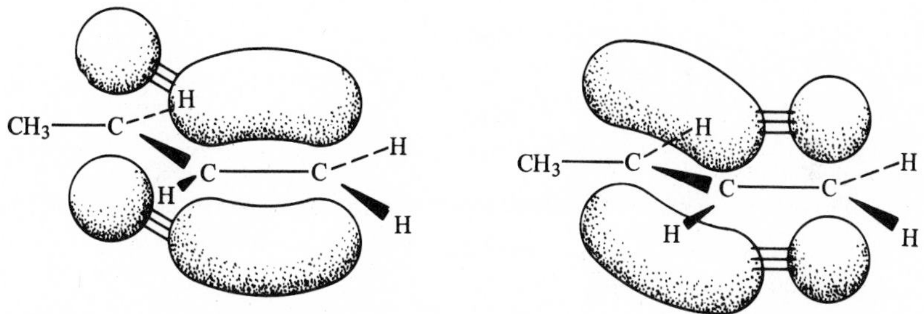

The π orbital and the empty p orbital overlap. The electron pair is therefore delocalized over three carbon atoms rather than over just two carbon atoms. The greater space available to the bonding π electrons reduces their energy (Sec. 6.4). We can represent the actual structure of I by the composite hybrid structure

$$\underset{4}{CH_3}{-}\underset{3}{\overset{\delta+}{CH}}{\cdots}\underset{2}{CH}{\cdots}\underset{1}{\overset{\delta+}{CH_2}}$$

in which the dashed lines represent partial bonds and $\delta+$ represents a partial positive charge (the sum of the $\delta+$'s = 1).

* **Allylic** refers to any group that has the C=C—C— structural unit; the term allylic means "allyl-like."

In the second step of our mechanism the chloride ion can react with either C_1 or C_3:

Step 2:

$$CH_3{-}\overset{\delta+}{CH}{\cdots}CH{\cdots}\overset{\delta+}{CH_2} + Cl^- \xrightarrow{a} CH_3{-}\underset{|}{\overset{Cl}{CH}}{-}CH{=}CH_2$$

$$\xrightarrow{b} CH_3{-}CH{=}CH{-}CH_2Cl$$

Reaction at C_3 gives 3-chloro-1-butene; reaction at C_1 gives 1-chloro-2-butene. Because this ion (I) is stabilized by electron delocalization, it is more stable than a tertiary carbocation. It is called an allylic carbocation because it is related to the allyl group, $CH_2{=}CH{-}CH_2{-}$.

We can view this reaction formally as addition to either the 1,2-carbons or the 1,4-carbons:

1,2- *Addition*

$$CH_2{=}CH{-}CH{=}CH_2 + H{-}Cl \longrightarrow \underset{H}{CH_2}{-}\underset{Cl}{CH}{-}CH{=}CH_2$$

1,4- *Addition*

$$CH_2{=}CH{-}CH{=}CH_2 + H{-}Cl \longrightarrow \underset{H}{CH_2}{-}CH{=}CH{-}\underset{Cl}{CH_2}$$

Therefore, we refer to addition to a conjugated diene as **1,2-addition** and **1,4-addition**.

Conjugated dienes undergo other addition reactions that give 1,2- and 1,4-products. Some examples are

$$CH_2{=}CH{-}CH{=}CH_2 + HBr \xrightarrow{40^\circ C} \underset{\text{(minor product)}}{CH_3\overset{Br}{C}HCH{=}CH_2} + \underset{\text{(major product)}}{CH_3CH{=}CHCH_2Br}$$

$$CH_2{=}CH{-}CH{=}CH_2 + Br_2 \xrightarrow{-15^\circ C} \underset{(54\%)}{\overset{Br}{CH_2}{-}\overset{Br}{C}HCH{=}CH_2} + \underset{(46\%)}{\overset{Br}{C}H_2CH{=}CH\overset{Br}{C}H_2}$$

In general, 1,4- addition predominates because the 1,4- product is more stable; that is, it is a more highly substituted alkene than the 1,2- product. Lower temperatures favor the 1,2- product.

Problem 6.4
Write equations to show 1,2- and 1,4-addition of (a) Cl_2 and (b) H_2O (in acid solution) to 1,3-cyclopentadiene.

6.7 DIELS–ALDER REACTION

Conjugated dienes undergo another kind of 1,4- addition reaction, named the **Diels–Alder reaction** after its discoverers. It is a "cycloaddition" reaction that occurs in one step without the development of ions. An example is the reaction of 1,3-butadiene with ethene. When these two compounds are heated together they react to form cyclohexene.

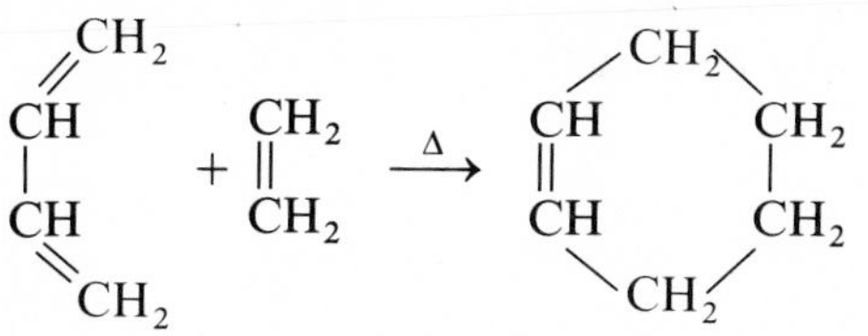

$$CH_2{=}CH{-}CH{=}CH_2 + CH_2{=}CH_2 \xrightarrow{\Delta} \text{cyclo-}(CH_2CH{=}CHCH_2CH_2CH_2) \quad (20\%)$$

This is actually a poor example of the Diels–Alder reaction because it gives a low yield even when the mixture is heated to 200°C. However, it serves to define the reaction. The two reactants are called **diene** and **dienophile**. In this example, 1,3-butadiene is the diene and ethene is the dienophile. We can diagram the electronic shifts using arrows and skeletal line formulas:

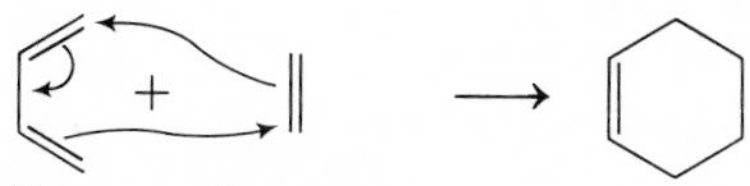

diene + dienophile →

Most examples of the Diels–Alder reaction give much better yields than the one above. Reaction is aided when the dienophile bears an electronegative functional group such as

$$\underset{}{-\overset{\overset{O}{\|}}{C}-OH}, \quad -\overset{\overset{O}{\|}}{C}-OR, \quad \text{or} \quad -\overset{\overset{O}{\|}}{C}-R$$

Some excellent examples of Diels–Alder reactions follow.

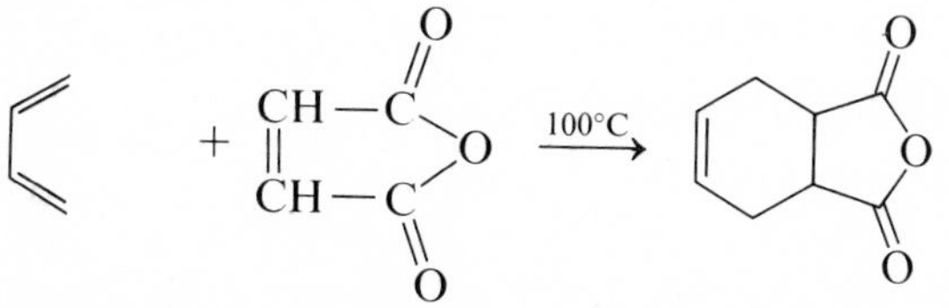

$$\text{1,3-butadiene} + \text{maleic anhydride} \xrightarrow{100°C} \text{product} \quad (100\%)$$

1,3-butadiene　　maleic anhydride

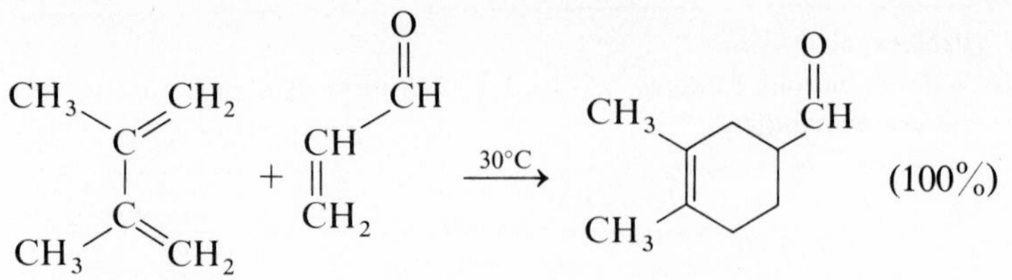

2,3-dimethyl-1,3-butadiene + propenal $\xrightarrow{30°C}$ (100%)

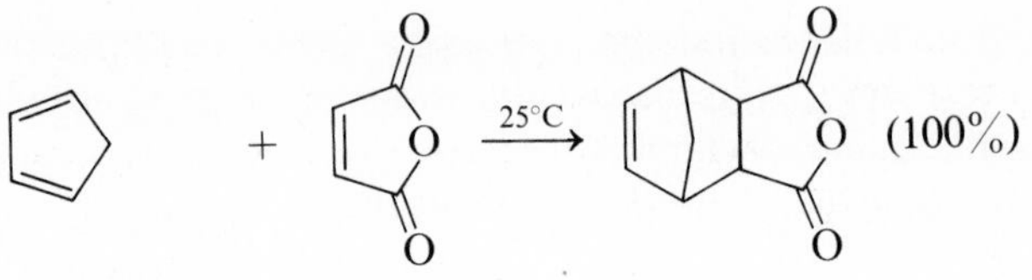

1,3-cyclopentadiene + maleic anhydride $\xrightarrow{25°C}$ (100%)

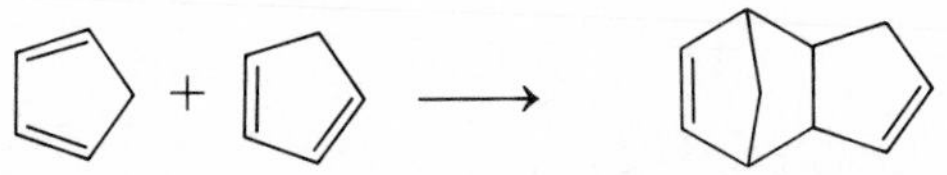

"dicyclopentadiene"

Problem 6.5
Label the diene and dienophile in each of the reactions shown above.

Problem 6.6
Using line formulas, give an equation for the Diels–Alder reaction of each of the following pairs of compounds. Label the diene and dienophile in each equation.

(a) 1,3-butadiene + propene
(b) 2,3-dimethyl-1,3-butadiene + maleic anhydride
(c) 1,3-cyclohexadiene + propenal

6.8 ALKYNES

Alkynes are hydrocarbons whose molecules have a carbon–carbon triple bond. The simplest member, $HC{\equiv}CH$, has the common name acetylene. The IUPAC rules are the same as for alkenes. The ending *-ane* of the alkanes is changed to *-yne*. Thus the simpler members of the alkyne series are ethyne, propyne, butyne, pentyne, hexyne, heptyne, octyne, nonyne, and decyne. We locate the position of the triple bond with one number just as we do for alkenes. Some examples of IUPAC systematic names of alkynes are listed below.

$CH_3{-}C{\equiv}CH$ propyne

$CH_3{-}C{\equiv}C{-}CH_3$ 2-butyne

$CH_3{-}C{\equiv}C{-}CH_2CH(CH_3)CH_3$ 5-methyl-2-hexyne

$HC{\equiv}C{-}C{\equiv}CH$ 1,3-butadiyne

Problem 6.7
Give the systematic name of each of the following compounds.

(a) $HC{\equiv}C{-}CH_2{-}CH_3$,

(b) $$\begin{array}{l} CH_3{-}CH_2{-}\underset{\displaystyle |}{C}H{-}CH_3 \\ \qquad\qquad\quad C{\equiv}CH \end{array}$$

6.9 STRUCTURE OF ALKYNES

We can visualize the structure of the triple bond most easily by using the σ-π molecular orbital approach. To construct the triple bond in acetylene, we first join all the atoms with single (σ) bonds. We need two sigma bonds for each carbon atom, and we construct them from *sp* hybrid orbitals. This leaves two half-filled *p* orbitals on each carbon. The p_y orbitals on each carbon combine to form one π orbital in the *xy* plane, and the p_z orbitals on each carbon combine to form a π orbital in the *xz* plane.

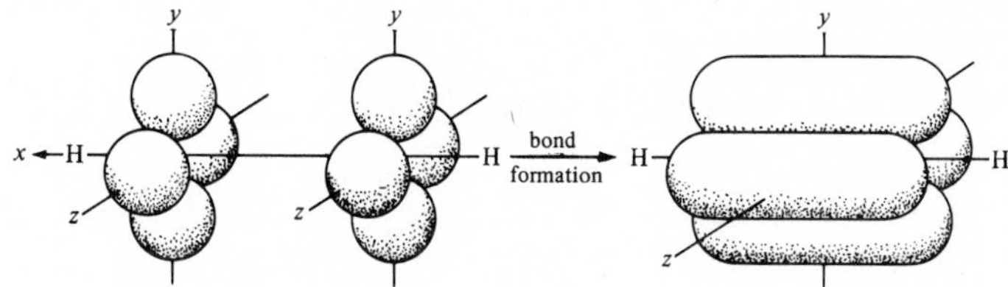

Thus the triple bond consists of two π bonds. The π orbital in the *xy* plane has lobes above and below the *xz* plane (overlap of two p_y orbitals), and the π orbital in the *xz* plane has lobes in front of and behind the *xy* plane (overlap of two p_z orbitals). The net effect is that the triple bond behaves as though it were cylindrical:

H—C—C—H

In the above formula of acetylene we used *sp* hybrid orbitals to construct the σ bonds. This suggests that the bond angles should be 180°. Electron diffraction analysis bears out this prediction: the acetylene molecule is linear.

6.10 PREPARATION OF ALKYNES

Alkynes are prepared by the same type of elimination reactions that are used to prepare alkenes. For example, either of the following dehydrohalogenation reactions yields acetylene.

$$CH_3{-}\underset{\displaystyle Cl}{\overset{\displaystyle Cl}{C}}H + 2\,NaNH_2 \xrightarrow[NH_3]{\text{liquid}} HC{\equiv}CH + 2\,NaCl + 2\,NH_3$$

$$Cl{-}CH_2{-}CH_2{-}Cl + 2\,NaNH_2 \xrightarrow[NH_3]{\text{liquid}} HC{\equiv}CH + 2\,NaCl + 2\,NH_3$$

Dehalogenation of 1,1,2,2-tetrahaloalkanes also yields alkynes:

$$R-\underset{Br}{\overset{Br}{\underset{|}{\overset{|}{C}}}}-\underset{Br}{\overset{Br}{\underset{|}{\overset{|}{C}}}}H + 2Zn \longrightarrow R-C\equiv CH + 2ZnBr_2$$

These methods are summarized in Table 6.2.

The most important member of the series is acetylene, which can be made by special methods not applicable to other alkynes. In one method, coke and lime are heated to 2500°C in a furnace to produce calcium carbide. Calcium carbide reacts rapidly with water at room temperature to yield acetylene:

$$\underset{\text{coke}}{3C} + \underset{\text{lime}}{CaO} \xrightarrow{2500°C} \underset{\text{calcium carbide}}{CaC_2} + CO$$

$$CaC_2 + 2H_2O \xrightarrow{25°C} Ca(OH)_2 + HC\equiv CH$$

Another industrial process involves heating methane (natural gas) to 1500°C for a fraction of a second:

$$2CH_4 \xrightarrow[<1\text{ second}]{1500°C} HC\equiv CH + 3H_2$$

Mixtures of acetylene and air burn at a very high temperature (2800°C). This property makes acetylene an important welding fuel.

Table 6.2 Methods of Preparing Alkynes

Dehydrohalogenation:

$$-\underset{H}{\overset{H}{\underset{|}{\overset{|}{C}}}}-\underset{X}{\overset{X}{\underset{|}{\overset{|}{C}}}}- + 2NaNH_2 \xrightarrow[NH_3]{\text{liquid}} -C\equiv C- + 2NaX + 2NH_3$$

$$-\underset{X}{\underset{|}{C}}H\underset{X}{\underset{|}{C}}H- + 2NaNH_2 \xrightarrow[NH_3]{\text{liquid}} -C\equiv C- + 2NaX + 2NH_3$$

Dehalogenation:

$$-\underset{X}{\overset{X}{\underset{|}{\overset{|}{C}}}}-\underset{X}{\overset{X}{\underset{|}{\overset{|}{C}}}}- + 2Zn \longrightarrow -C\equiv C- + 2ZnX_2$$

Alkynes undergo addition reactions in much the same way as alkenes. Moreover, additions obey Markovnikov's rule:

$$CH_3{-}C{\equiv}CH + 1\text{ mole HCl} \longrightarrow CH_3{-}\underset{\underset{Cl}{|}}{C}{=}CH_2 \xrightarrow[\text{HCl}]{\text{1 mole}} CH_3{-}\overset{\overset{Cl}{|}}{\underset{\underset{Cl}{|}}{C}}{-}CH_3$$

$$CH_3{-}C{\equiv}CH + 2\text{ moles }Br_2 \longrightarrow CH_3{-}\overset{\overset{Br}{|}}{\underset{\underset{Br}{|}}{C}}{-}\overset{\overset{Br}{|}}{\underset{\underset{Br}{|}}{CH}}$$

Hydration of alkynes, however, gives an unexpected result:

$$CH_3{-}C{\equiv}CH + H_2O \xrightarrow[HgSO_4]{H_2SO_4} CH_3{-}\overset{\overset{O}{\|}}{C}{-}CH_3$$

The reaction requires not only an acid, but also an Hg^{2+} salt. We can explain the formation of a ketone as follows. The addition of one molecule of water gives 2-propen--2-ol.

$$CH_3{-}C{\equiv}CH + H_2O \longrightarrow \underset{\text{an enol}}{CH_3{-}\overset{\overset{OH}{|}}{C}{=}CH_2}$$

2-Hydroxypropene, a vinylic alcohol, is called an **enol** because it contains a hydroxyl group bonded to an unsaturated carbon (a C=C double bond). Enol is a contraction of the endings *-ene* and *-ol*. Enols rearrange readily, especially in the presence of traces of acid, to the carbonyl compound called the **keto form**

$$\underset{\text{enol form}}{CH_3{-}\overset{\overset{O-H}{|}}{C}{=}\underset{\underset{H}{|}}{C}{-}H} \rightleftarrows \underset{\text{keto form}}{CH_3{-}\overset{\overset{O}{\|}}{C}{-}\overset{\overset{H}{|}}{\underset{\underset{H}{|}}{C}}{-}H}$$

Keto-enol forms are constitutional isomers (Sec. 2.4) that are in equilibrium; they cannot be separated from each other. We call such isomers **tautomers** and the process of interconversion **tautomerization**. We shall describe tautomers and tautomerization more fully in Section 12.12. At this point we only need realize that the keto form is the more stable, and that hydration of an alkyne yields predominantly the ketone. Table 6.3 summarizes the reactions of alkynes.

Table 6.3 Reactions of Alkynes

Hydrohalogenation (Section 6.11):

$$\text{R—C}\equiv\text{CH} + \text{HX (1 mole)} \longrightarrow \text{R—}\overset{\overset{\text{X}}{|}}{\text{C}}\text{=CH}_2 \quad \text{(Markovnikov addition)}$$

$$\text{R—}\overset{\overset{\text{X}}{|}}{\text{C}}\text{=CH}_2 + \text{HX (1 mole)} \longrightarrow \text{R—}\overset{\overset{\text{X}}{|}}{\underset{\underset{\text{X}}{|}}{\text{C}}}\text{—CH}_3$$

(X = Cl, Br, or I)

Halogenation (Section 6.11):

$$\text{R—C}\equiv\text{CH} + 2\text{X}_2 \longrightarrow \text{R—}\overset{\overset{\text{X}}{|}}{\underset{\underset{\text{X}}{|}}{\text{C}}}\text{—}\overset{\overset{\text{X}}{|}}{\underset{\underset{\text{X}}{|}}{\text{CH}}}$$

(X = Cl or Br)

Hydration (Section 6.11):

$$\text{R—C}\equiv\text{CH} + \text{H}_2\text{O} \xrightarrow[\text{HgSO}_4]{\text{H}_2\text{SO}_4} \left[\text{R—}\overset{\overset{\text{OH}}{|}}{\text{C}}\text{=CH}_2\right] \rightleftarrows \text{R—}\overset{\overset{\text{O}}{\|}}{\text{C}}\text{—CH}_3$$

enol form — keto form

Acidity (Section 6.12):

$$\text{R—C}\equiv\text{CH} + \text{NaNH}_2 \xrightarrow[\text{NH}_3]{\text{liquid}} \text{R—C}\equiv\text{C:}^- + \text{Na}^+ + \text{NH}_3$$

$$\text{R—C}\equiv\text{CH} + \text{NaOH} \longrightarrow \text{no reaction}$$

Silver and Copper Acetylides (Section 6.13):

$$\text{R—C}\equiv\text{CH} + \text{Ag(NH}_3)_2{}^+\text{OH}^- \xrightarrow{\text{H}_2\text{O}} \text{R—C}\equiv\text{CAg}\downarrow + \text{H}_2\text{O} + 2\text{NH}_3$$

$$\text{R—C}\equiv\text{CH} + \text{Cu(NH}_3)_2{}^+\text{OH}^- \xrightarrow{\text{H}_2\text{O}} \text{R—C}\equiv\text{CCu}\downarrow + \text{H}_2\text{O} + 2\text{NH}_3$$

Problem 6.8

Predict the products of the following reactions.

(a) 2-butyne + 2 moles HBr $\longrightarrow$

(b) 2-butyne + 2 moles Cl_2 $\longrightarrow$

(c) 1-butyne + water (in the presence of H_2SO_4 and $HgSO_4$) $\longrightarrow$

6.12 ACIDITY OF ALKYNES

1-Alkynes undergo a reaction that alkenes do not undergo: they are relatively strong acids (compared with other hydrocarbons) and react with bases to form salts called

acetylides. Sodium amide reacts with acetylene in liquid ammonia solution to give ammonia and sodium acetylide:

$$\underset{}{HC{\equiv}CH} + \underset{\text{sodium amide}}{NaNH_2} \xrightarrow[NH_3]{\text{liquid}} \underset{\text{sodium acetylide}}{HC{\equiv}CNa} + NH_3$$

Sodium hydroxide is not a strong enough base to convert acetylene to its conjugate base. We can summarize the reactions of acetylene with the two bases $:\ddot{N}H_2^-$ and $:\ddot{O}H^-$ using equilibrium equations. With the stronger base, $:\ddot{N}H_2^-$, the reaction occurs largely to the right: with the weaker base, $:\ddot{O}H^-$, the reaction occurs only slightly to the right. These equilibria are represented with arrows of unequal length:

$$HC{\equiv}CH + \underset{\text{stronger base}}{:\ddot{N}H_2^-} \rightleftharpoons \underset{\text{weaker base}}{HC{\equiv}C:^-} + :NH_3$$

$$HC{\equiv}CH + \underset{\text{weaker base}}{:\ddot{O}H^-} \rightleftharpoons \underset{\text{stronger base}}{HC{\equiv}C:^-} + H_2\ddot{O}:$$

The relative positions of equilibrium allow us to rank the conjugate bases in the following order of base strengths:

$$\text{Base strength: } :\ddot{N}H_2^- > HC{\equiv}C:^- > :\ddot{O}H^-$$

We can also rank the conjugate acids:

$$\text{Acid strength: } H_2O > HC{\equiv}CH > NH_3$$

Alkenes and alkanes do not undergo such acid-base reactions, and we conclude that they are weaker acids than ammonia. Table 6.4 gives approximate acid dissociation constants of these compounds.

Table 6.4 Some Acid Dissociation Constants

Acid	K_a	Conjugate Base
CH_3CH_3	$\sim 10^{-42}$	$CH_3{-}CH_2:^-$ (H above and below C)
$CH_2{=}CH_2$	$\sim 10^{-36}$	$CH_2{=}\ddot{C}^-{-}H$
$:NH_3$	$\sim 10^{-36}$	$:\ddot{N}H_2^-$
$HC{\equiv}CH$	$\sim 10^{-26}$	$HC{\equiv}C:^-$
H_2O	10^{-16}	$:\ddot{O}H^-$

We can explain the relative acidities of alkanes, alkenes, and alkynes on the basis of the hybridization state of the orbital that contains the electron pair.

$H-C\equiv C$: (*sp* orbital) $\qquad$ $H_2C=CH$: (*sp*² orbital) $\qquad$ H_3C-CH_2: (*sp*³ orbital)

sp orbital $\qquad$ *sp*^{2} orbital $\qquad$ *sp*^{3} orbital

The unshared electron pair of acetylene is in an *sp* hybrid orbital. This orbital has more *s* character than an *sp*^{2} or an *sp*^{3} orbital, and it is therefore closer to the carbon nucleus. (Recall that *s* orbitals are spherically symmetrical about the nucleus.) The potential energy of an electron pair is lowest in an *sp* orbital, higher in an *sp*^{2} orbital, and highest in an *sp*^{3} orbital. The base strengths of the anions depend on the availability of the electron pair. (Higher potential energy means that the electrons are farther from the nucleus and more available to react with an acid.) The base strengths of the anions are therefore $CH_3CH_2^- > CH_2{=}CH^- > HC{\equiv}C{:}^-$, and the acid strengths of the conjugate acids are in the reverse order, $HC{\equiv}CH > CH_2{=}CH_2 > CH_3CH_3$.

6.13 SILVER AND COPPER SALTS OF ALKYNES

Because of their acidity, 1-alkynes also undergo reactions with certain silver and copper salts.

$$CH_3{-}C{\equiv}CH + Ag(NH_3)_2{}^+OH^- \xrightarrow{H_2O} CH_3{-}C{\equiv}CAg\downarrow + H_2O + 2NH_3$$

$$CH_3{-}C{\equiv}CH + Cu(NH_3)_2{}^+OH^- \xrightarrow{H_2O} CH_3{-}C{\equiv}CCu\downarrow + H_2O + 2NH_3$$

These acetylides have more covalent character than sodium acetylides. They are insoluble in water and precipitate from solution as they are formed. These reactions occur rapidly at room temperature and therefore serve as simple visual tests to distinguish 1-alkynes from alkynes that have an interior triple bond, as well as to distinguish 1-alkynes from alkenes and polyenes. Table 6.3 summarizes the reactions of alkynes.

Problem 6.9
Write equations for the reactions of 2-butyne and 1-butyne with (a) $NaNH_2$ in liquid NH_3 solution, (b) $Ag(NH_3)_2OH$ in water solution, and (c) $Cu(NH_3)_2OH$ in water solution.

6.14 SIMPLE CHEMICAL TESTS

We can now expand Table 5.5 to include dienes and alkynes and the new tests described in the present chapter. Table 6.5 lists these new simple chemical tests.

Table 6.5 Simple Chemical Tests for Dienes and Alkynes*

Compound	Bromine Test Br_2/CCl_4	Baeyer Test, $KMnO_4/H_2O$	H_2SO_4, Cold, Concentrated	$Ag(NH_3)_2OH$
Dienes	red-brown → colorless	purple solution → black precipitate	soluble	no reaction
1-Alkynes	red-brown → colorless	purple solution → black precipitate	soluble	precipitate forms
Other Alkynes	red-brown → colorless	purple solution → black precipitate	soluble	no reaction

* All tests are conducted at room temperature unless otherwise stated.

Problem 6.10
Which simple chemical tests would make it possible to distinguish between the compounds in the following pairs?
(a) 1,3-butadiene and 1-butene, (b) 1-butyne and 2-butyne

6.15 OCCURRENCE OF UNSATURATED HYDROCARBONS: AN ELECTIVE TOPIC

elective topic

Unsaturated compounds occur widely in the plant and animal kingdoms as well as in the earth as fossil fuels. It would be impossible and uninteresting to tabulate all the unsaturated compounds found in nature. Instead we shall examine a few representative compounds, their structures, and their uses.

Fossil Fuels

Petroleum deposits contain varying amounts of ethene and other alkenes. Alkenes are also manufactured from petroleum by the cracking process. Ethene (common name, ethylene) is used extensively in the manufacture of ethyl alcohol by hydration in the presence of sulfuric acid.

$$CH_2{=}CH_2 + H_2O \xrightarrow{\text{acid}} CH_3CH_2OH$$

Ethylene is also used in the manufacture of polyethylene (Sec. 5.6). An interesting property of ethylene is its ability to cause fruit to ripen.

Another important unsaturated fossil fuel is graphite. Coal is a crude, impure form of graphite. Graphite consists of molecular layers that contain the infinite network structure shown in Figure 6.2. X-ray diffraction analysis shows that all the C—C bond distances in the planar sheets are equal to 1.42 Å. The distance between layers is 3.35 Å. The structure described in Figure 6.2(a) represents one of many

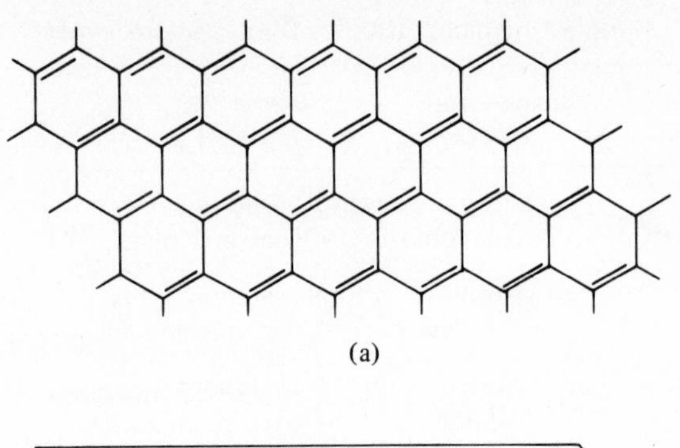

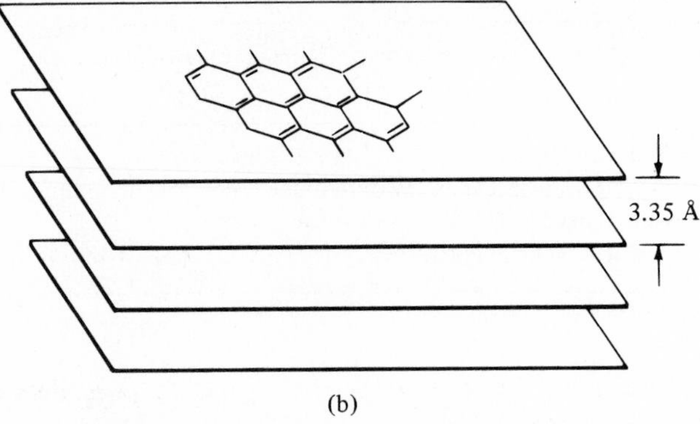

Figure 6.2. Graphite, shown as (a) molecular structure of a single layer and (b) layered structure.

possible resonance structures. Delocalization of electrons in the π system is so complete that graphite is an excellent conductor of electricity and heat. The large distance between sheets allows them to slide past each other, and graphite is therefore a good lubricant.

Terpenes

Terpenes are compounds that occur in plants and have the formula $(C_5H_8)_n$. They all contain the repeating unit

$$\begin{array}{c} \quad\;\; C \\ \quad\;\; | \\ C-C-C-C \end{array}$$

We may consider them to be formed from the apparently fundamental molecular unit *isoprene*,

$$\begin{array}{l} \qquad\quad\;\; CH_3 \\ \qquad\qquad | \\ CH_2{=}C{-}CH{=}CH_2 \end{array}$$

Many naturally occurring compounds besides terpenes appear to be made up of isoprene units.

Turpentine consists primarily of a mixture of two isomeric terpenes, α– and β–pinene.

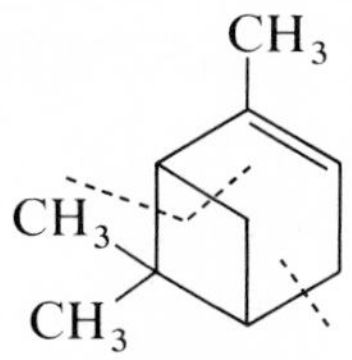

CH_3 CH_2 CH_3 CH_3 CH_3 CH_3

α–pinene 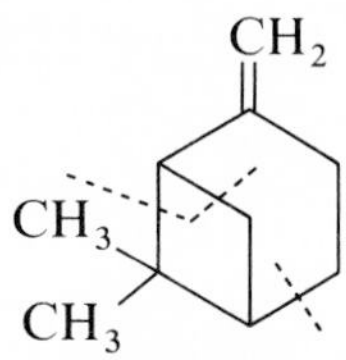 β–pinene

The mixture consists of about 60% α–pinene and about 40% β–pinene. Turpentine is used mainly as a paint solvent. The isoprene units (irrespective of double bonds) are indicated by the dashed lines in the formulas above.

Limonene, also a terpene, occurs in lemon oil and has the structure

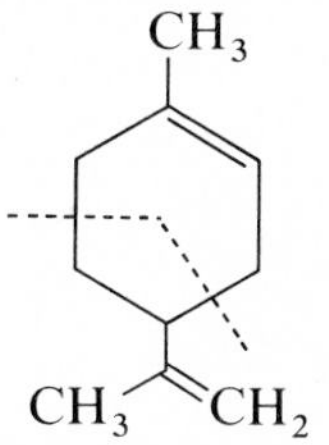

CH_3

CH_3 CH_2

The isoprene units are indicated with broken lines. If we draw limonene differently, its structural relation to α–pinene is obvious:

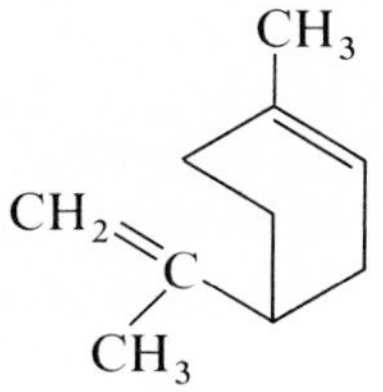

CH_3

$CH_2{=}C$

CH_3

Myrcene, a terpene, is found in oil of bay, verbena, and hop. Its structure is

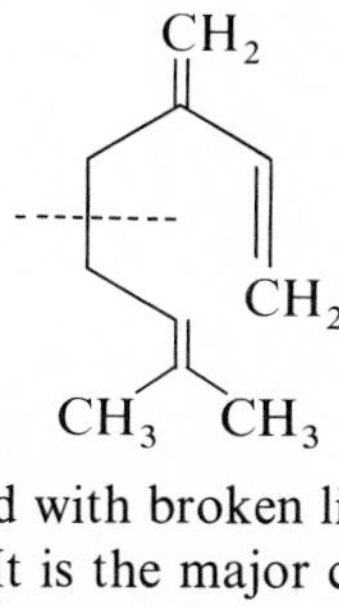

CH_2

CH_2

CH_3 CH_3

Again, isoprene units are indicated with broken lines.

Oleic acid is not a terpene. It is the major constituent of olive oil. It is an unsaturated carboxylic acid whose structure is

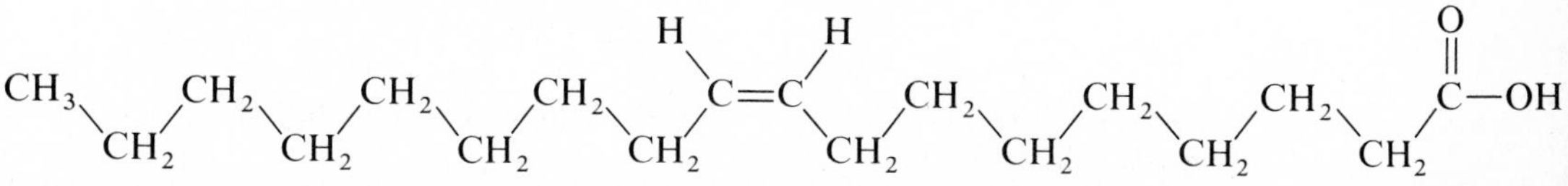

$$CH_3CH_2CH_2CH_2CH_2CH_2CH_2CH_2\text{–}CH{=}CH\text{–}CH_2CH_2CH_2CH_2CH_2CH_2CH_2\text{–}\overset{O}{\overset{\|}{C}}\text{–}OH$$

We may abbreviate this structure with a line formula

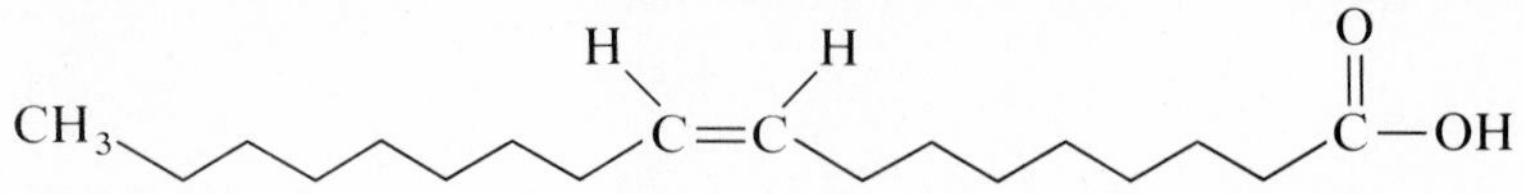

$$CH_3\text{/\/\/\/\/\/\/}\overset{H}{C}=\overset{H}{C}\text{/\/\/\/\/\/\/}\overset{O}{\overset{\|}{C}}-OH$$

where each line represents a single bond. Where the lines meet we understand that there is a carbon atom with enough hydrogens attached to use all four valence electrons of carbon. Oleic acid occurs in olive oil as the ester of glycerol.

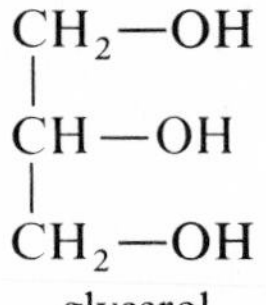

$$\begin{array}{l} CH_2-OH \\ | \\ CH-OH \\ | \\ CH_2-OH \end{array} \qquad \begin{array}{l} CH_2-O-\overset{O}{\overset{\|}{C}}-R \\ | \\ CH-O-\overset{O}{\overset{\|}{C}}-R \\ | \\ CH_2-O-\overset{O}{\overset{\|}{C}}-R \end{array}$$

glycerol

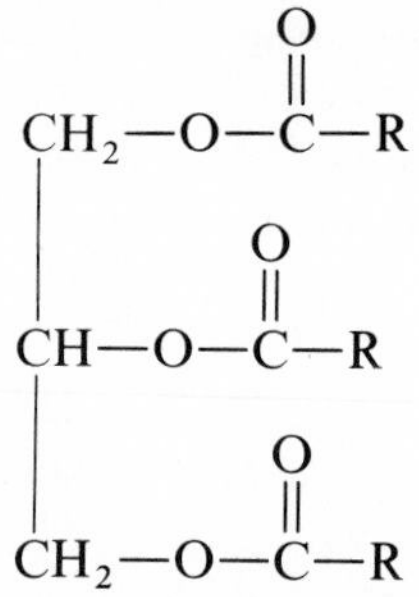

an oil

($HO-\overset{O}{\overset{\|}{C}}-R$ is oleic acid or another long-chain carboxylic acid.)

Carotenoids are an interesting group of compounds that occur in both plants and animals. Their color—usually yellow, orange, or red—is due to their long, conjugated polyene systems. Lycopene is found in ripe tomatoes and paprika. β–carotene is found in carrots. Their structures are similar and related:

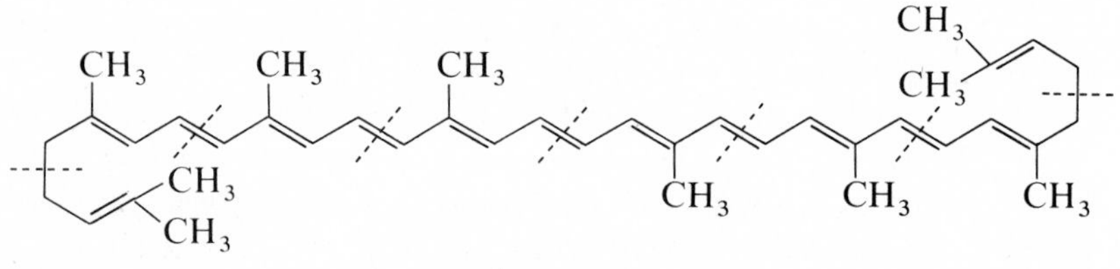

CH_3 CH_3 CH_3 CH_3 CH_3 CH_3 CH_3 CH_3 CH_3 CH_3

lycopene

CH_3 CH_3 CH_3 CH_3 CH_3 CH_3 CH_3 CH_3 CH_3 CH_3

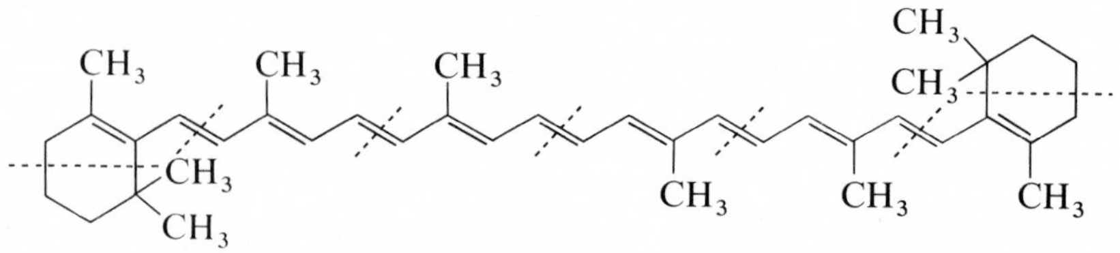

β–carotene

The isoprene units in these compounds are indicated with broken lines. In mammals, including humans, β–carotene is cleaved by an enzymatic oxidation reaction to give two molecules of retinal. Retinal in turn is reduced enzymatically to the alcohol called vitamin A_1.

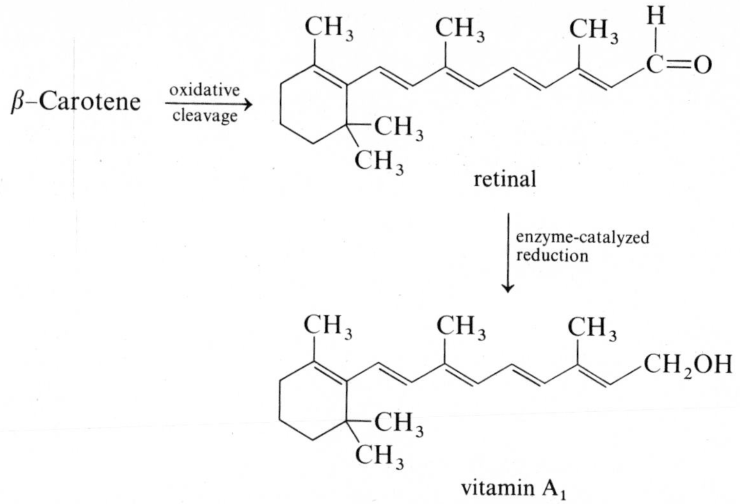

β-Carotene $\xrightarrow[\text{cleavage}]{\text{oxidative}}$ retinal $\xrightarrow{\text{enzyme-catalyzed reduction}}$ vitamin A_1

These compounds are believed to play a major role in vision. Retinal is converted biochemically (Fig. 6.3) to its isomer, neoretinal b (11-*cis*-retinal). The reaction is a conversion of the *trans*-11,12- double bond in retinal (all double bonds in retinal are *trans*) to the *cis* configuration. Neoretinal b then reacts with a protein called opsin, to give rhodopsin. Rhodopsin is stable only when the 11,12- double bond is *cis* because the *cis* form fits into the protein site exactly. Light induces the isomerization of the 11,12- double bond in rhodopsin to the *trans* configuration to give *trans*-rhodopsin, which is unstable. This process simultaneously gives rise to a nerve impulse. *trans*-Rhodopsin is not stable and hydrolyzes readily to give opsin and to regenerate retinal, which takes part in another cycle. The light-induced isomerization of rhodopsin to *trans*-rhodopsin triggers the nerve impulse from the optic nerve to the brain. The overall process is summarized in Figure 6.3.

Natural rubber is a polymer of isoprene. Two isomers of polyisoprene are possible, the all *cis* and the all *trans*. Both isomers occur in nature. The all *cis* isomer is the elastic substance known as rubber. It is called *hevea.* The all *trans* isomer is a hard, brittle solid used in making combs and other rigid objects. The all *trans* isomer is called *gutta-percha.*

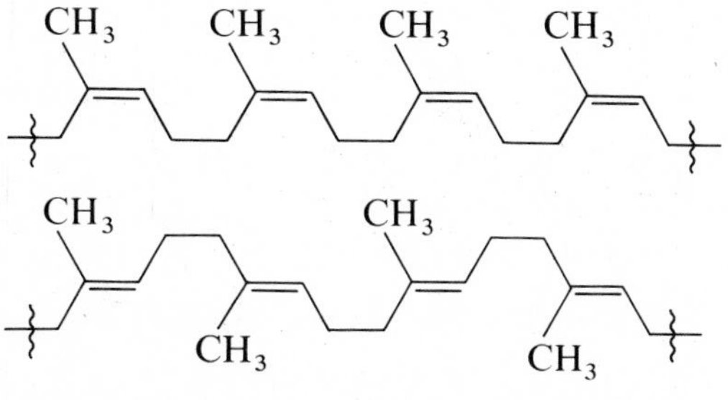

hevea (all *cis*)

gutta-percha (all *trans*)

Problem 6.11
Identify the isoprene units in hevea, gutta-percha, retinal, neoretinal b, and vitamin A_1.

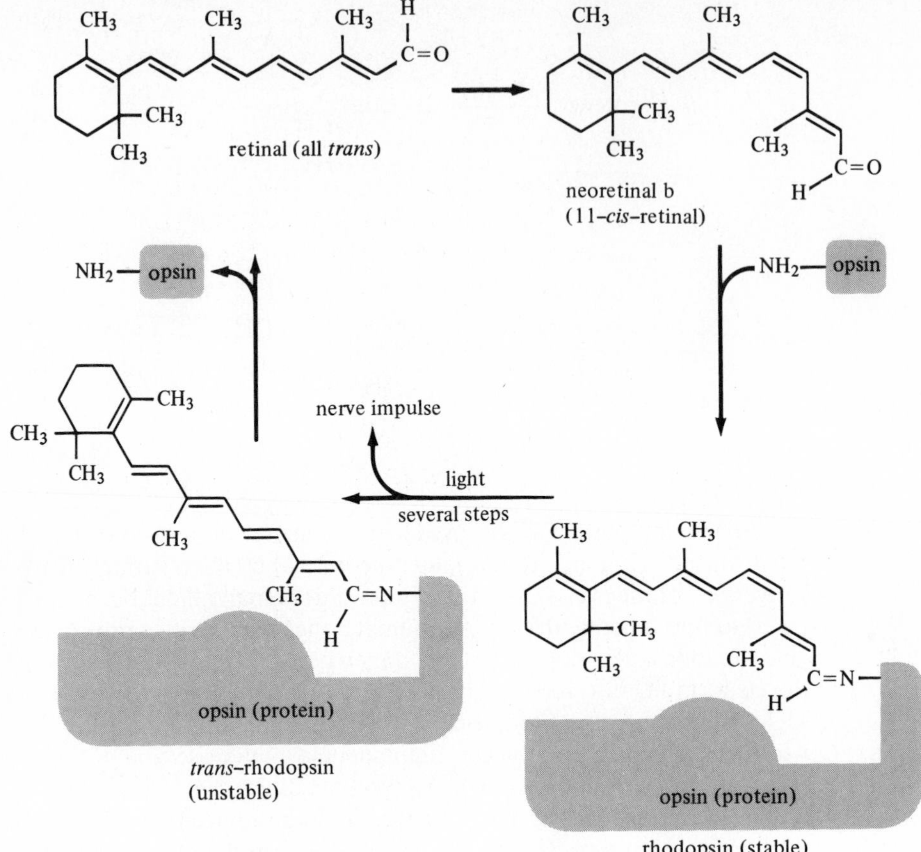

Figure 6.3. Summary of the chemistry of vision.

NEW TERMS

Acetylide The conjugate base of a 1-alkyne (6.12):

$$R{-}C{\equiv}C{:}^{-}$$

1,2-Addition Addition of a molecule A—B to one double bond of a conjugated diene (6.6):

$$\gt C{=}\overset{|}{C}{-}\overset{|}{C}{=}C\lt + A{-}B \longrightarrow -\overset{|}{\underset{A}{C}}{-}\overset{|}{\underset{B}{C}}{-}\overset{|}{C}{=}C\lt$$

1,4- Addition Addition of a molecule A—B to the first and fourth carbons of a conjugated diene (6.6):

$$\gt C{=}\overset{|}{C}{-}\overset{|}{C}{=}C\lt + A{-}B \longrightarrow -\overset{|}{\underset{A}{C}}{-}\overset{|}{C}{=}\overset{|}{C}{-}\overset{|}{\underset{B}{C}}-$$

Allylic Any group that has the C=C—C— structural unit and is therefore "allyl-like" (6.6).

Conjugated double bonds A system of double and single bonds in which the double and single bonds alternate (6.2):

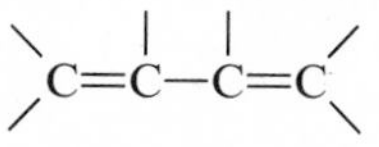

$$\gt C{=}\overset{|}{C}{-}\overset{|}{C}{=}C\lt$$

Cumulated double bonds A system of adjacent double bonds (6.2):

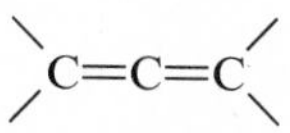

$$\gt C{=}C{=}C\lt$$

Diels–Alder reaction 1,4-Addition of a substituted alkene (called a **dienophile**) to a diene to form a cyclic compound (6.7):

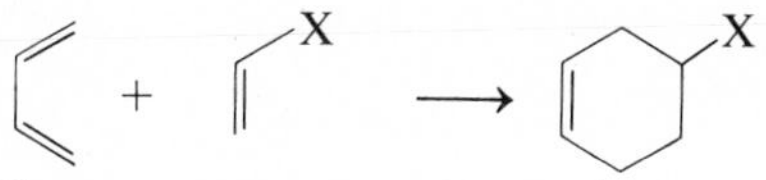

diene + dienophile (X) ⟶ cyclohexene bearing X

diene dienophile

Diene A hydrocarbon containing two carbon–carbon double bonds (6.1).

Enol A molecule in which a hydroxyl group is bonded to a C=C double bond:

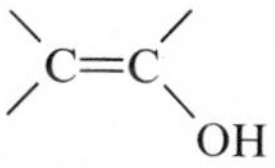

$$\gt C{=}C\lt^{OH}$$

Enols exist in equilibrium with the keto form (6.11).

Isolated double bonds A system in which double bonds are separated by two or more single bonds (6.2):

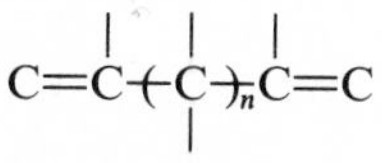

$$C{=}\overset{|}{C}{+}\underset{|}{\overset{|}{C}}{+}_{n}\overset{|}{C}{=}C \qquad (n \geq 1)$$

Keto form A molecule that has the structure

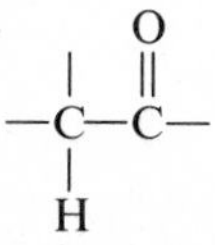

$$-\underset{H}{\overset{|}{\underset{|}{C}}}-\overset{O}{\overset{\|}{C}}-$$

Keto forms exist in equilibrium with the enol form (6.11):

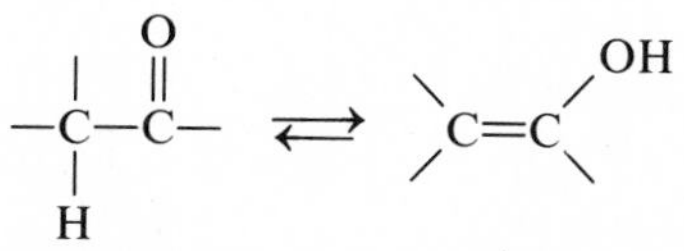

$$-\underset{H}{\overset{|}{\underset{|}{C}}}-\overset{O}{\overset{\|}{C}}- \rightleftarrows \gt C{=}C\lt^{OH}$$

keto enol

Polyene A hydrocarbon containing two or more carbon–carbon double bonds (6.1).

Tautomerization The process in which tautomers equilibrate with each other (6.11).

Tautomers Constitutional isomers that exist in equilibrium with each other; for example, keto-enol tautomers (6.11).

ADDITIONAL PROBLEMS

6.12 Give the systematic name of each of the following compounds.

[a] $CH_2{=}C{=}CHCH_3$ [b]

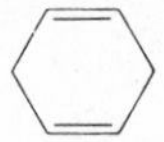

[c] $CH_3C{\equiv}CCH_2CH_2{-}$ (d) $CH_3CH{=}C{=}CHCH_3$

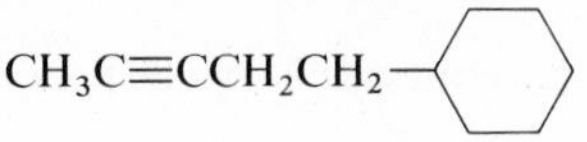

(e)  (f) 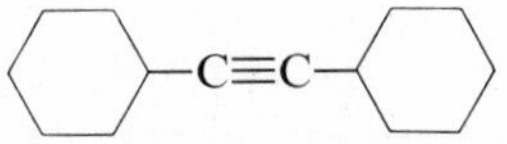 $-C{\equiv}C-$

[6.13] Write equations for the reactions that occur when 2,4-hexadiene is treated with one mole of the following reagents.

(a) HCl, (b) Br_2, (c) H_2O/H_3O^+

6.14 Repeat Problem 6.13 using 1,3-cyclohexadiene instead of 2,4-hexadiene.

6.15 What compounds would you react together to obtain the following Diels–Alder products?

[a] CN (b) 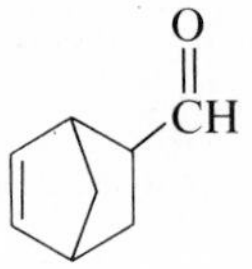 O ‖ CH

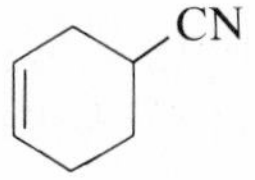

[6.16] Write equations to show how you would prepare acetylene starting with ethene and any inorganic reagents.

6.17 Write equations for the reactions that occur when 1-hexyne is treated with the following reagents. [Repeat using 3-hexyne.]

(a) 2 moles HCl, (b) 1 mole Br_2,
(c) 2 moles Br_2, (d) $H_2O/H_2SO_4/HgSO_4$,
(e) $NaNH_2/NH_3$

[6.18] Explain in terms of orbital hybridization why acetylene is a stronger acid than ethene.

6.19 Which simple chemical tests would make it possible to distinguish between the compounds in each of the following pairs?

[a] 1-pentene and 1-pentyne, [b] 1-butene and butane,
[c] cyclopentane and 2-pentene, (d) cyclopentene and 2-pentyne,
(e) 1-hexyne and 2-hexyne

6.20 Write equations to show the products of ozonolysis (followed by Zn/H_2O reduction) of [a] α-pinene, (b) β-pinene, [c] limonene, [d] myrcene, (e) lycopene, (f) β-carotene, [g] natural rubber

6.21 Draw the σ–π structure of [2-butyne] and 1-butyne.

6.22 Which of the following compounds have a zero dipole moment? Demonstrate by drawing vector diagrams of each compound.

[a] 1,3-butadiene, [b] 1-butyne, [c] 2-butyne,
[d] 1,2-butadiene, (e) propyne, (f) 3-hexyne,
(g) $Cl{-}C{\equiv}C{-}Cl$, (h) 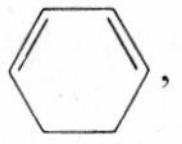 , (i)

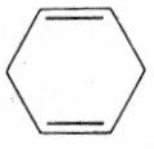

[6.23] Construct a table that includes all the compound classes and tests given in Tables 5.5 and 6.5.

[6.24] Compound A has the molecular formula C_6H_8. Hydrogenation of compound A yields a compound whose molecular formula is C_6H_{12}. Ozonolysis of compound A yields the compounds shown below. Give the structural formula of compound A.

$$\text{Compound A} + O_3 \longrightarrow \text{A ozonide}$$

$$\text{A ozonide} + \text{Zn} + H_2O \longrightarrow \overset{O}{\overset{\|}{HC}}-\overset{O}{\overset{\|}{CH}} + \overset{O}{\overset{\|}{HC}}CH_2CH_2\overset{O}{\overset{\|}{CH}}$$

6.25 Compound B has the molecular formula C_4H_6. Hydrogenation of compound B yields a compound whose molecular formula is C_4H_{10}. When compound B is treated with $Ag(NH_3)_2{}^+OH^-$ solution, a precipitate forms. Give the structural formula of compound B.

[6.26] Compound C has the molecular formula C_6H_{10}. Bromination of compound C yields $C_6H_{10}Br_4$. Treatment of compound C with $Cu(NH_3)_2{}^+OH^-$ solution gives no reaction. Ozonolysis of one mole of compound C yields two moles of $CH_3\overset{O}{\overset{\|}{C}}H$ and one mole of $\overset{O}{\overset{\|}{HC}}-\overset{O}{\overset{\|}{CH}}$. Give the structural formula of compound C.

7 AROMATIC HYDROCARBONS

7.1 UNUSUAL PROPERTIES OF BENZENE

We have seen that unsaturated compounds undergo addition reactions. In fact, addition is the most notable chemical property of unsaturated compounds.

There is an important class of hydrocarbons called **aromatic hydrocarbons** (Sec. 2.6) whose formulas suggest a high degree of unsaturation, although these compounds do not undergo addition reactions. (All other hydrocarbons are classed as **aliphatic**.) Benzene, C_6H_6, is the most common example of an aromatic compound. The formula C_6H_6 (C_nH_{2n-6}) suggests that benzene should undergo the addition reactions typical of unsaturated compounds. Instead, benzene reacts as follows:

$$C_6H_6 + Br_2 \xrightarrow[\text{dark}]{CCl_4} \text{no reaction}$$

$$C_6H_6 + Br_2 \xrightarrow{Fe} C_6H_5Br + HBr$$

$$C_6H_6 + KMnO_4 \xrightarrow{25°C} \text{no reaction}$$

$$C_6H_6 + H_2SO_4 \xrightarrow{\text{cold}} \text{no reaction, insoluble}$$

$$C_6H_6 + Ag(NH_3)_2{}^+OH^- \longrightarrow \text{no reaction}$$

When benzene reacts it undergoes substitution rather than addition.

In this chapter we shall examine the unique structure of benzene in order to explain its unusual chemical properties.

Problem 7.1

Complete the following equations.

(a) $C_6H_6 + Cl_2 \xrightarrow{\text{addition}}$ (b) $C_6H_6 + Cl_2 \xrightarrow{\text{substitution}}$

7.2 STRUCTURE OF BENZENE

X-ray diffraction techniques show that the benzene molecule is a planar hexagon in which all the carbons are equivalent and all the hydrogens are equivalent. All C—C bonds are 1.39 Å in length, and all C—H bonds are 1.10 Å in length. The C—C single bond length in alkanes is 1.53 Å, and C=C double bonds in alkenes are approximately 1.34 Å. The C—C bond in benzene is intermediate in length between a single and a double bond. The fact that the molecule is planar means that the angles are all 120°. Figure 7.1 compares the structures of benzene, ethane, and ethene.

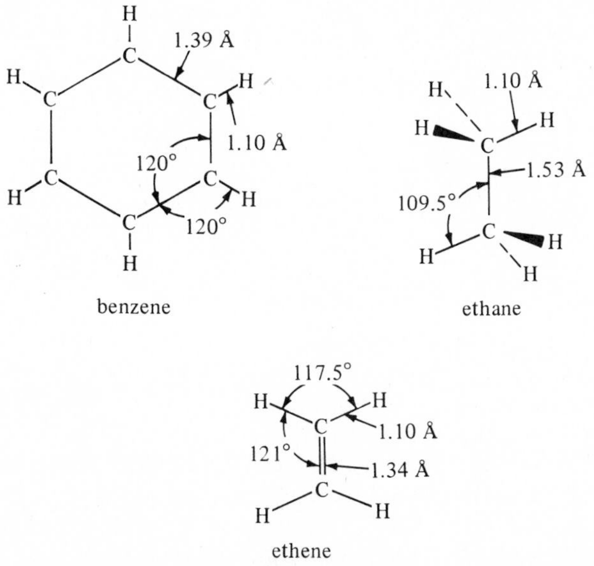

Figure 7.1. Comparison of the molecular parameters of benzene, ethane, and ethene.

To satisfy the octet rule we could draw the structure of benzene with alternating single and double bonds. This structure (we can call it 1,3,5-cyclohexatriene) requires unequal bond lengths, and it is therefore unsatisfactory. In addition, this structure predicts two isomers of 1,2-dibromobenzene:

1,3,5-cyclohexatriene   1,2-dibromo-1,3,5-cyclohexatriene   2,3-dibromo-1,3,5-cyclohexatriene

Only one isomer is known in which two bromine atoms are attached to adjacent carbon atoms in benzene. In addition, if benzene had the 1,3,5-cyclohexatriene structure it would undergo addition reactions, and it does not. Clearly the alternating single–double bond structure is unacceptable because it is not consistent with the known properties of benzene.

Problem 7.2

Assume that benzene has the 1,3,5-cyclohexatriene structure, and draw all the isomers of tribromobenzene, $C_6H_3Br_3$. Do the same using the symmetrical structure

Is it possible to decide which structure is preferable on this basis? (There are actually three known isomers of $C_6H_3Br_3$.)

7.3 HEAT OF HYDROGENATION OF BENZENE: ITS STABILITY

Although benzene does not add hydrogen as readily as ordinary alkenes or polyenes, it is possible to hydrogenate benzene under vigorous conditions to obtain cyclohexane.

$$C_6H_6 + 3H_2 \xrightarrow[\Delta]{Pt} C_6H_{12} \qquad \Delta H^\circ = -49.8 \text{ kcal/mole}$$

The heat of hydrogenation of benzene gives us a clue to its stability. If we compare the heat of hydrogenation of benzene with that of cyclohexene and 1,3-cyclohexadiene, we see a striking result (Table 7.1). The ΔH_h° of 1,3-cyclohexadiene is slightly less than twice that of cyclohexene. We have already seen this effect with conjugated polyenes (Sec. 6.3). The ΔH_h° value for benzene, however, is approximately half the expected value. The potential energy of benzene is therefore much lower than we expect for 1,3,5-cyclohexatriene (Fig. 7.2). The difference between the expected and the observed ΔH_h° is 36.0 kcal/mole. This difference between the energy of benzene and the calculated energy of the hypothetical 1,3,5-cyclohexatriene is called the **resonance energy** or **resonance stabilization energy** of benzene (Fig. 7.2). In addition to

Table 7.1 Heats of Hydrogenation, Δ_h°

Reaction	Observed ΔH_h°, kcal/mole	Expected $\Delta H_h^{\circ *}$
cyclohexene $+ H_2 \xrightarrow{Pt}$ cyclohexane	-28.6	-28.6
1,3-cyclohexadiene $+ 2H_2 \xrightarrow{Pt}$ cyclohexane	-55.4	-57.2
benzene $+ 3H_2 \xrightarrow{Pt}$ cyclohexane	-49.8	-85.8

* Expected ΔH_h° = (ΔH_h° of cyclohexene) × (number of double bonds). This quantity assumes that each double bond contributes to the total ΔH_h° as much as the one double bond in cyclohexene.

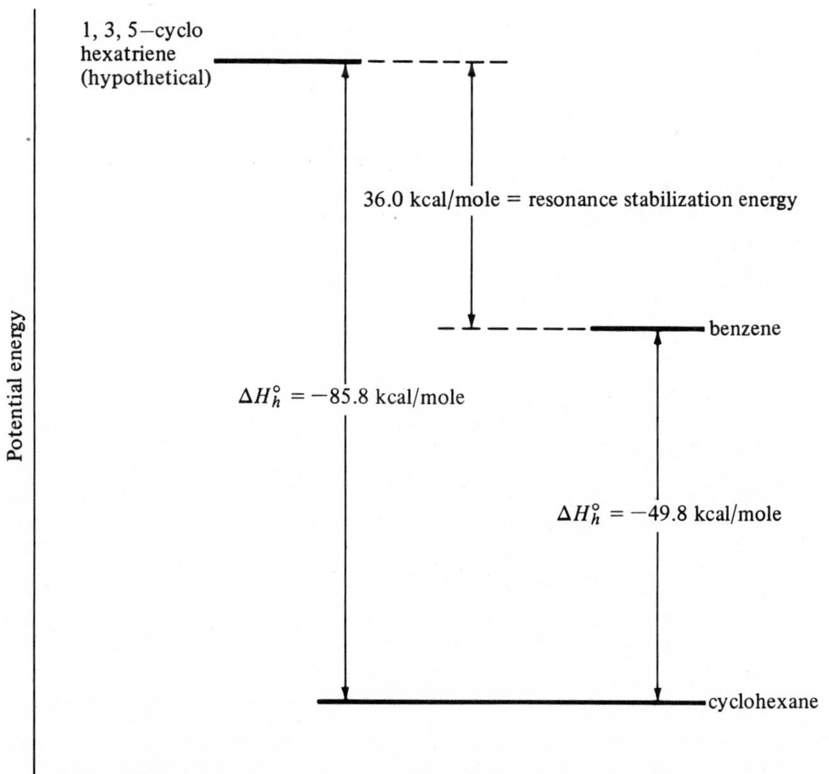

Figure 7.2. Comparison of observed ΔH_h° for benzene with the expected ΔH_h° calculated for the structure if it had three independent double bonds (the hypothetical 1,3,5-cyclohexatriene structure).

the other unusual properties of benzene, we must now account for this unusual stability.

Problem 7.3
Calculate the resonance stabilization energy of 1,3-cyclohexadiene.

7.4 EXPLANATION OF BENZENE STABILITY: RESONANCE

We can represent the structure of benzene as two equivalent valence bond structures.

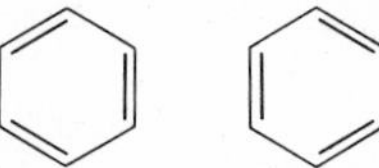

These structures differ only in the positions of the double bonds.

Whenever we can write more than one reasonable valence bond structure for a molecule, we know that the molecule has none of these structures. Instead, its structure is a *hybrid* of all of them. This approach to expressing the structure of a molecule as a hybrid or composite structure that we cannot draw is called **resonance**. The valence bond structures that we can draw are called **resonance structures**. It is easy to draw resonance structures if we follow a few rules.

Rules for Drawing Resonance Structures

(1) All the contributing structures must have their atoms in exactly the same relative positions. Contributing structures therefore differ only in the positions of their electrons. Such structures as butane and isobutane cannot contribute to a hybrid structure.

$$CH_3{-}CH_2{-}CH_2{-}CH_3 \qquad CH_3{-}\overset{\displaystyle CH_3}{\underset{\displaystyle H}{C}}{-}CH_3$$

(2) To arrive at different contributing structures we may move an electron pair (a) from an atom to a bond on the same atom, (b) from a bond to one of the atoms attached to the bond, or (c) from one bond to another on the same atom. These three possibilities are shown below using curved arrows to represent electron movement (Sec. 5.6). The double-headed arrow ↔ is used to refer to resonance structures.

We may not, however, move electrons any farther. For example, the two structures

$$\overset{-}{\ddot{C}}H_2-CH_2-\overset{+}{C}H_2 \quad \text{and} \quad \overset{+}{C}H_2-CH_2-\overset{-}{\ddot{C}}H_2$$

cannot contribute to a resonance hybrid because the electron pair is localized on a single atom and is not, in fact, capable of moving between C_1 and C_3. To move the electron pair from C_1 to C_3 requires removing them completely (ionizing them) from C_1.

(3) All the atoms involved in resonance must lie in the same plane. If they do not, then the *p* orbitals will not be able to overlap effectively (review Sec. 5.3 on orbital overlap).

(4) Contributing structures must have nearly the same energies. If one structure is much more stable than all the others, then that structure describes the molecule, and resonance is unimportant. For example, methane is amply described by structure I. The other structures shown are too energetic and do not contribute significantly to the actual structure.

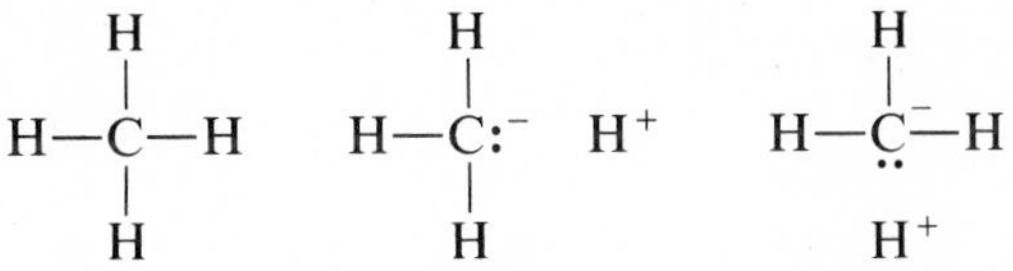

$$\begin{array}{ccc} \begin{array}{c} H \\ | \\ H-C-H \\ | \\ H \end{array} & \begin{array}{c} H \\ | \\ H-C\!:^{-} \quad H^{+} \\ | \\ H \end{array} & \begin{array}{c} H \\ | \\ H-\underset{\cdot\cdot}{\overset{-}{C}}-H \\ \\ H^{+} \end{array} \\ \text{I} & \text{II} & \text{III} \end{array}$$

(5) Contributing structures must have the same number of unpaired electrons. The structures

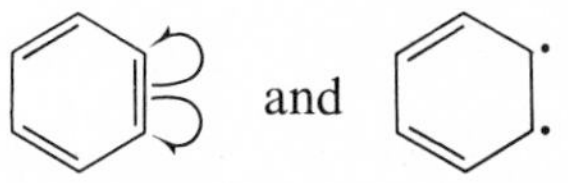

and

cannot contribute to a resonance hybrid because structures with unpaired electrons are much higher in energy than corresponding structures with electrons paired.

We shall make much use of the concept of resonance to describe molecules. At this point we must remind ourselves that contributing structures have no reality except on paper. Resonance is merely an attempt to describe, in terms of valence bond structures, a structure that we cannot describe adequately in any other way.

One final point must be made about the concept of resonance: *A hybrid structure has a lower energy than any of its contributing structures.* Therefore, any time we can draw resonance structures, the actual molecule will be more stable than we would expect on the basis of any of its contributing structures.

We can dismiss ionic structures such as (a) as unimportant because their energy is much higher than the two nonionic structures (b) and (c). Thus we can represent

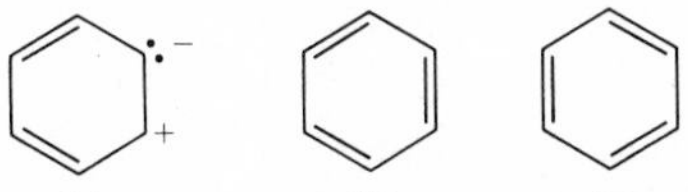

(a) (b) (c)

the structure of benzene as a hybrid of structures (b) and (c). We can express the benzene hybrid structure by a composite structure (d) that suggests the equality of all the C—C bonds.

(d)

Problem 7.4
Which of the following pairs of structures represent resonance structures? Explain each of your answers.

(a) $CH_3CH{=}CH_2$ and $CH_2{=}CHCH_3$
(b) $CH_3C{\equiv}CCH_3$ and $CH_2{=}CHCH{=}CH_2$
(c) $CH_3\overset{+}{C}H{-}CH{=}CH_2$ and $CH_3CH{=}CH{-}\overset{+}{C}H_2$
(d) H H and H H
H H H H
H H

Problem 7.5
One of the resonance structures of naphthalene is shown below. Draw two other structures.

The resonance stabilization energy of benzene represents the stabilization of benzene relative to a nonresonance structure. Actually, resonance stabilization energy is a hypothetical quantity because the structure 1,3,5-cyclohexatriene does not exist. However, resonance stabilization energy emphasizes the importance of using the concept of resonance to describe the structure of benzene. In other words, if we do not use resonance structures but instead use the structure

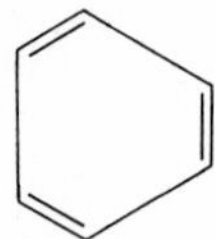

we shall be wrong by a factor of 36.0 kcal/mole. We are, of course, also wrong if we use this structure because it implies properties and reactivity that benzene does not possess.

7.5 MOLECULAR ORBITAL DESCRIPTION OF BENZENE

We can arrive at a more descriptive physical picture of the benzene resonance structure from the σ-π description. If we describe benzene using σ bonds we find that one p orbital is left on each carbon atom. These six p orbitals can combine to form six new **molecular orbitals**—orbitals that extend over several atoms. Some of these encompass the entire ring. The most stable of these molecular orbitals resembles two doughnuts, one above and one below the plane of the ring (Fig. 7.3). Electron delocalization reduces the electronic energy because delocalization means that the electrons have a longer electronic "wavelength," λ. A longer wavelength means a lower energy, according to Planck's equation (Sec. 6.4), $E = hc/\lambda$.

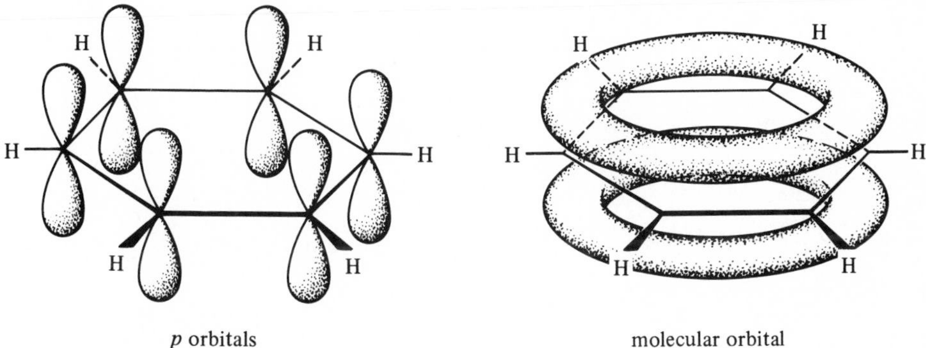

Figure 7.3. Atomic orbitals (p orbitals) combine to produce molecular orbitals.

7.6 AROMATICITY AND HÜCKEL'S RULE

We may now redefine aromatic compounds as compounds, like benzene, that have an unusually large resonance stabilization energy. The term aromatic originally referred to the aroma of many of the derivatives of benzene. It later came to refer to any compound that has the benzene conjugated ring system, and finally to any conjugated cyclic compound that is reluctant to undergo addition reactions. Table 7.2 lists some examples of common aromatic hydrocarbons.

An interesting compound is 1,3,5,7-cyclooctatetraene, which is not aromatic. That is, cyclooctatetraene adds Br_2 and other reagents readily, and it does not undergo substitution reactions; it is a typical polyene. Although we can write resonance structures for cyclooctatetraene, it has no appreciable resonance stabilization energy. The resonance approach, therefore, does not help us to understand why cyclooctatetraene is not aromatic while benzene and other compounds are aromatic. To decide whether a cyclic, conjugated compound is aromatic, the molecular orbital approach is more useful than the resonance approach.

In 1931, E. Hückel used molecular orbital theory to predict that any planar, monocyclic, completely conjugated hydrocarbon is aromatic if it has $(4n + 2)$ π

Table 7.2 Some Common Aromatic Hydrocarbons (One resonance structure is given for each compound.)

Name	Structure	Source, Uses, and Toxicity
Benzene		Coal, used as solvent and intermediate in synthesis
Naphthalene	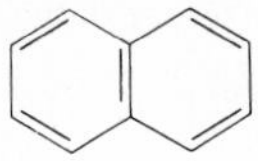	Coal tar, used in synthesis and as antiseptic (mothballs)
Anthracene	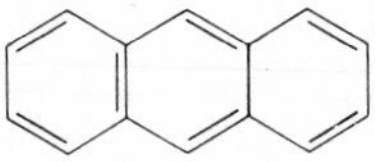	Coal tar, used in dye synthesis
Phenanthrene	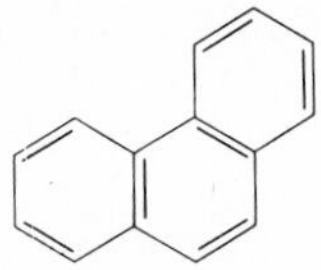	Coal tar, carcinogen, causes skin photosensitization
Pyrene	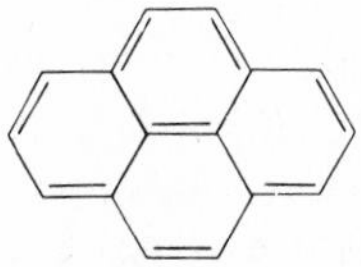	Coal tar, possible carcinogen
1,2-Benzopyrene	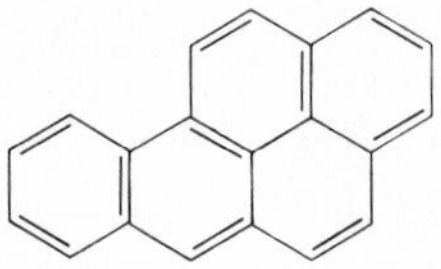	Coal tar, occurs in tobacco smoke, automobile exhausts, highly carcinogenic
10-Methyl-1,2-benzanthracene	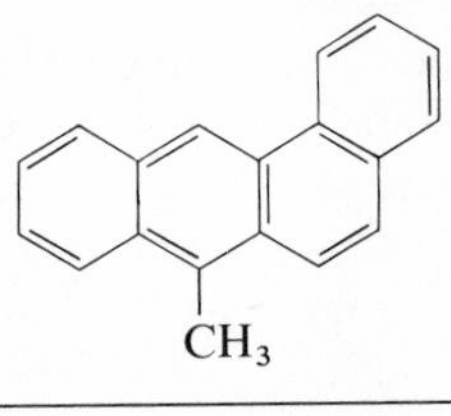 CH_3	Coal tar, extremely carcinogenic

electrons in its conjugated system (n is a positive integer ≥ 0). This $4n + 2$ rule has come to be known as **Hückel's rule**. Although we will not go into mathematical details, we can appreciate that Hückel based his prediction on calculations of the energies of electrons in the π system. If a molecule has $(4n + 2)\ \pi$ electrons in its cyclic, conjugated system, all those electrons will lie in molecular orbitals that are lower in energy than the original p orbitals (Fig. 7.3). Hückel's rule therefore allows us to predict whether a compound is aromatic. Benzene obeys Hückel's rule because it has three double bonds, and therefore six π electrons, and $6 = 4n + 2$ when $n = 1$. Note that n is not equal to the number of electrons; it is merely any integer.

Problem 7.6
Use Hückel's rule to predict which of the following compounds are aromatic:

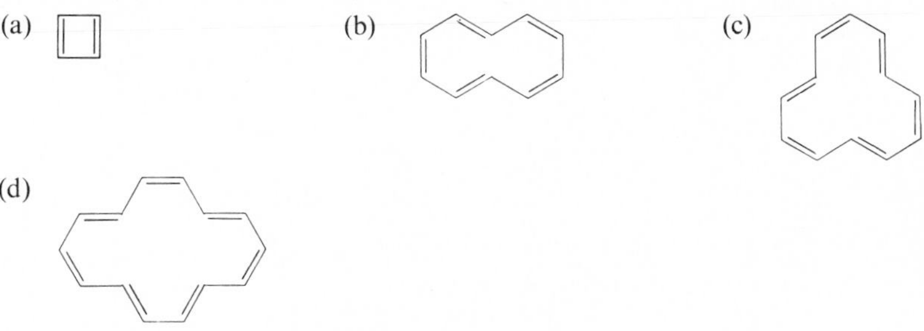

During the 1960s, F. Sondheimer set out to synthesize conjugated cyclic hydrocarbons in order to test Hückel's rule. The only such compounds known at that time were benzene and 1,3,5,7-cyclooctatetraene. Sondheimer synthesized most of the conjugated cyclic hydrocarbons (called **annulenes**) up to [24] annulene (the number in brackets is the number of carbons in the ring). Annulenes up to [30] annulene have been synthesized. [30] Annulene is not aromatic. Calculations show that an upper limit to Hückel's rule lies between [24] annulene and [26] annulene. Hückel's rule fails because of the inability of very large rings to maintain a planar conformation. Electron delocalization around the ring is possible only when the atoms of the ring all lie in a plane (rule 3, Sec. 7.3). Large rings are flexible and tend to bend out of the plane. Ring bending thus reduces the degree of electron delocalization in large rings.

Problem 7.7
Draw the structures of all the annulenes up to [22] annulene. Which of them obey Huckel's rule?

7.7 AROMATIC RINGS THAT CONTAIN ATOMS OTHER THAN CARBON

Many cyclic conjugated compounds that contain atoms other than carbon are also aromatic. Some of these are shown in Table 7.3. Purine and pyrimidine are essential building blocks of DNA and RNA (Sec. 17.15).

Table 7.3 Some Aromatic Compounds That Contain Atoms Other Than Carbon

pyridine	thiophene	pyrimidine	purine	pyrrole

7.8 AROMATIC IONS

Another class of organic species that exhibit aromatic character are certain cyclic, conjugated ions (Table 7.4). The compound

(7-bromo-1,3,5-cycloheptatriene)

exhibits properties characteristic of an ionic substance—high melting point, solubility in water, and insolubility in nonpolar solvents. As shown, 7-bromo-1,3,5-cycloheptatriene is not conjugated because the CHBr carbon is saturated. Dissociation of the bromide ion, however, produces a conjugated carbocation.

$\longleftrightarrow$ $\longleftrightarrow$ $\longleftrightarrow$ etc.

This cation satisfies Hückel's rule, and it is in fact aromatic.

Cyclopentadienyl anion and cyclopropenyl cation also obey Hückel's rule and are aromatic (Table 7.4).

Problem 7.8

Draw all the resonance structures of (a) the cycloheptatrienyl cation, (b) the cyclopentadienyl anion, and (c) the cyclopropenyl cation.

Problem 7.9

Why is the cyclopentadienyl cation not aromatic?

Table 7.4 Some Aromatic Ions

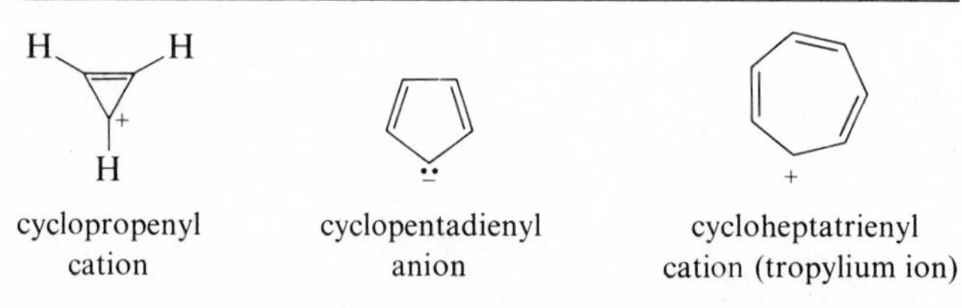

cyclopropenyl cation	cyclopentadienyl anion	cycloheptatrienyl cation (tropylium ion)

We can name many benzene derivatives simply by naming the substituent as prefix followed by the word *benzene*.* We name the following derivatives as shown:

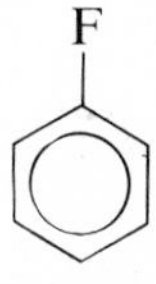

F Cl Br

fluorobenzene 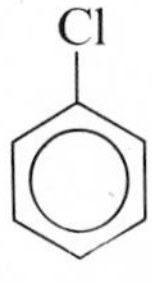 chlorobenzene 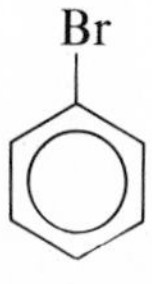 bromobenzene

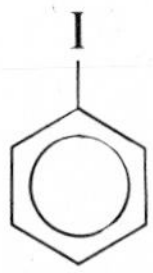

I NO_2 R

iodobenzene 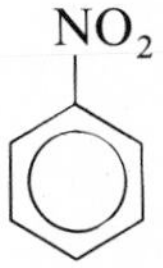 nitrobenzene 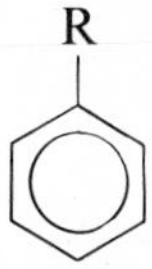 alkylbenzene

Some benzene derivatives have their own special names which may be used:

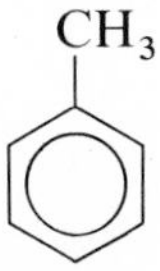

CH_3 CH_3 CH_3

toluene (or methylbenzene)

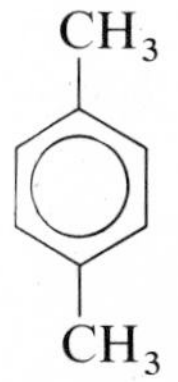

p-xylene (or *p*-dimethylbenzene)

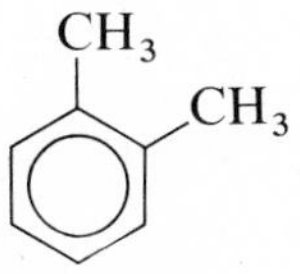

CH_3 CH_3 CH_3 CH_3

o-xylene (or *o*-dimethylbenzene)

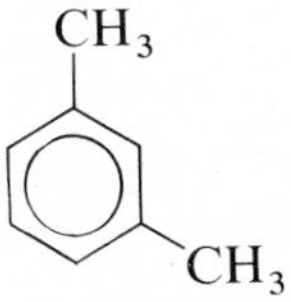

m-xylene (or *m*-dimethylbenzene)

For these compounds we may use either the special name or name them as derivatives of benzene.

If the ring bears only two groups, we may designate their positions with the terms *ortho*, *meta*, or *para*. Ortho groups are attached to the 1,2-positions, meta groups are

* We should recall that all the positions in benzene are equivalent. For example, there is only one chlorobenzene regardless of which carbon contains the chlorine atom.

attached to the 1,3-positions, and para groups are attached to the 1,4-positions. In systematic names, ortho, meta, and para are abbreviated *o*, *m*, and *p*.

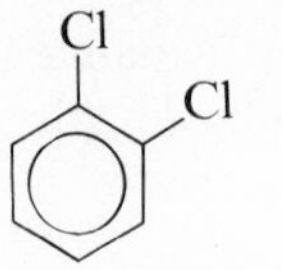

o-dichlorobenzene (1,2-dichlorobenzene)

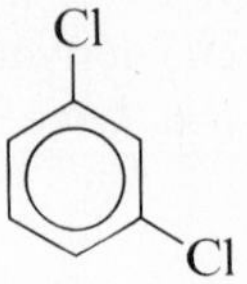

m-dichlorobenzene (1,3-dichlorobenzene)

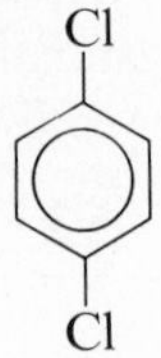

p-dichlorobenzene (1,4-dichlorobenzene)

When the ring contains more than two substituents, we must use numbers—one number for each substituent.

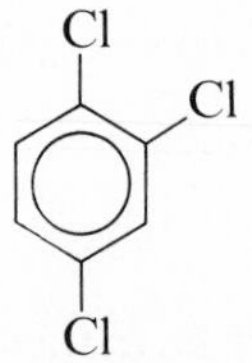

1,2,4-trichlorobenzene

However, when we can combine one substituent with the ring to make a new parent name, then we assume that that substituent is at position one, and it needs no number.

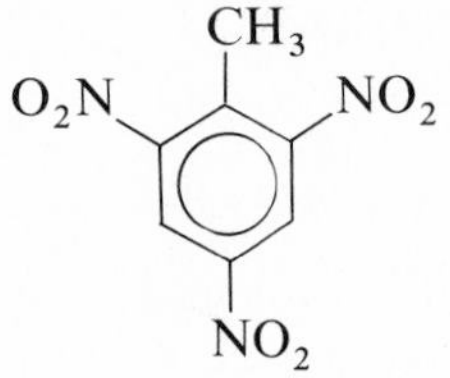

2,4,6-trinitrotoluene

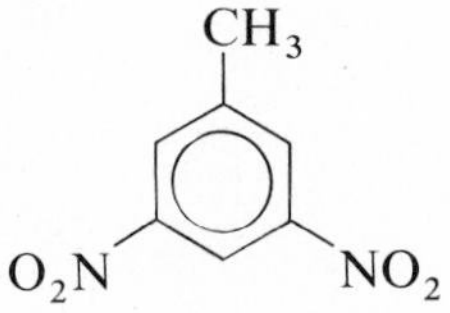

3,5-dinitrotoluene

When the benzene ring is itself a substituent, it is called a **phenyl** group.

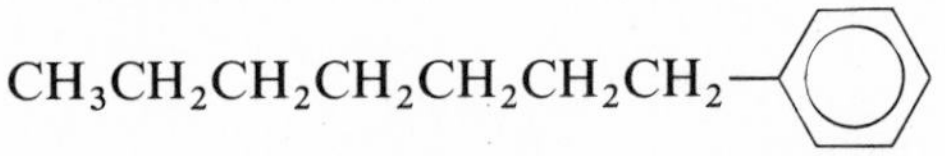

$CH_3CH_2CH_2CH_2CH_2CH_2CH_2-C_6H_5$

1-phenylheptane

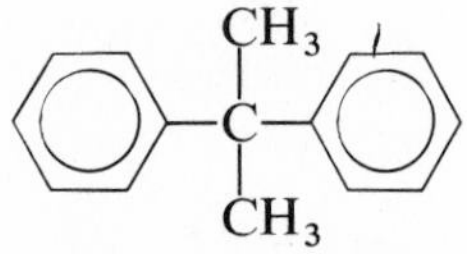

2,2-diphenylpropane

The grouping $C_6H_5-CH_2-$ has the special name *benzyl*.

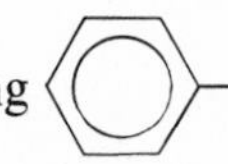

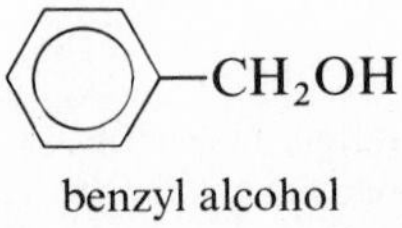

$C_6H_5-CH_2OH$

benzyl alcohol

$C_6H_5-CH_2Cl$

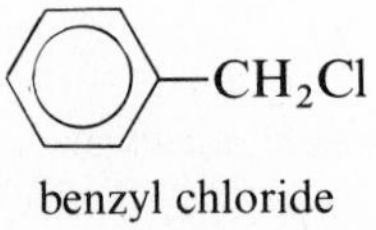

benzyl chloride

Problem 7.10

Give the structural formula of each of the following compounds.

(a) 2,4,6-trifluorotoluene
(b) *m*-xylene
(c) *p*-xylene
(d) 2,4-dinitrotoluene
(e) *p*-nitroethylbenzene
(f) *p*-chlorobenzyl chloride
(g) diphenylmethane

7.10 ADDITION VERSUS SUBSTITUTION IN BENZENE

We can now explain the reluctance of benzene to undergo addition reactions. Addition of a molecule of X—Y to benzene yields a carbocation:

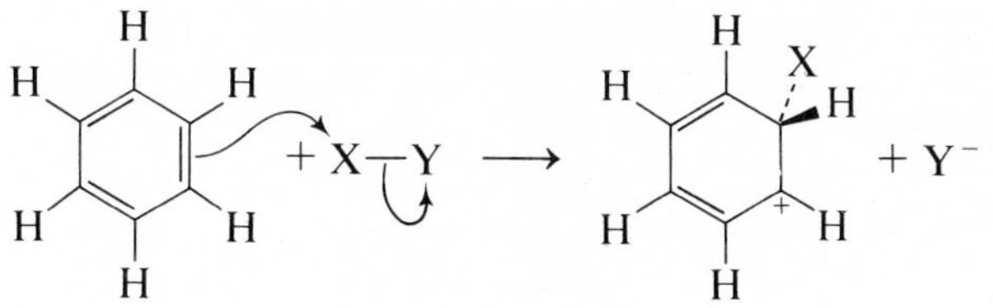

$$C_6H_6 + X\text{—}Y \longrightarrow C_6H_6X^+ + Y^-$$

Although this carbocation is a resonance hybrid

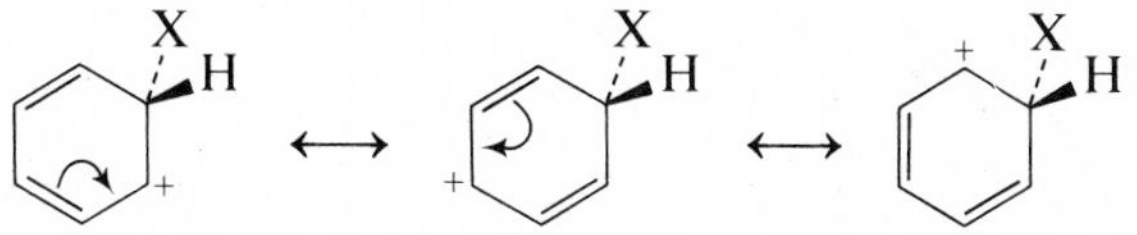

the π electrons are not delocalized over the entire ring but only over five carbon atoms. Addition of the Y^- ion would yield a 1,3-cyclohexadiene, which is not aromatic.

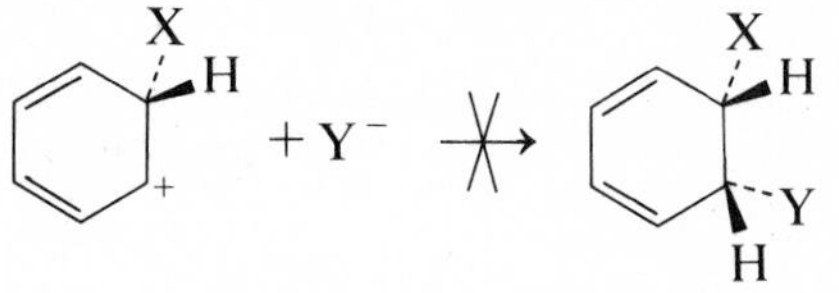

On the other hand, if the carbocation loses H^+, the ring becomes aromatic once again.

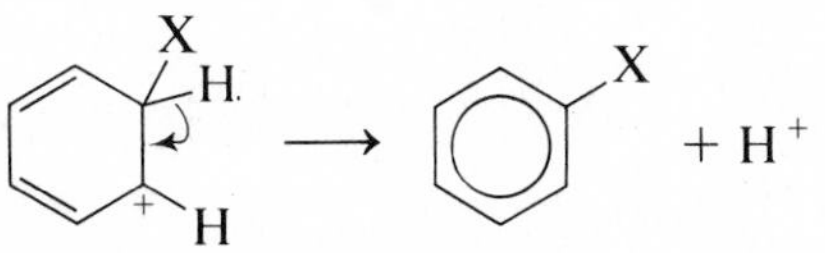

In other words, addition destroys aromaticity and requires an input of 36 kcal/mole—the resonance stabilization energy. Therefore another reaction—substitution—normally occurs that does not destroy aromaticity.

Problem 7.11
Explain why the resonance stabilization energy of 1,3-cyclohexadiene (Problem 7.3) is so much smaller than that of benzene.

7.11 SUBSTITUTION REACTIONS

Benzene undergoes a number of substitution reactions (Table 7.5). We shall be concerned with halogenation, nitration, sulfonation, alkylation, and acylation. Acylation is the introduction of an acyl group.

$$-\overset{\overset{\displaystyle O}{\|}}{C}-R$$

Table 7.5 Substitution Reactions of Benzene

Halogenation

$$C_6H_6 + X_2 \xrightarrow[\text{or } FeX_3]{Fe} C_6H_5X + HX \quad (X = Cl \text{ or } Br)$$

Nitration

$$C_6H_6 + HONO_2 \xrightarrow{H_2SO_4} C_6H_5NO_2 + H_2O$$

Sulfonation

$$C_6H_6 + SO_3 \xrightarrow{H_2SO_4} C_6H_5SO_3H$$

Desulfonation

$$C_6H_5SO_3H + 2\,H_2O \xrightarrow{H_3O^+} C_6H_6 + H_3O^+ + HSO_4^-$$

Friedel–Crafts Alkylation

$$C_6H_6 + R{-}X \xrightarrow{AlCl_3} C_6H_5R + HX \quad (X = Cl, Br, \text{ or } I)$$

(R may rearrange)

Friedel–Crafts Acylation

$$C_6H_6 + X-\overset{\overset{\displaystyle O}{\|}}{C}-R \xrightarrow{AlCl_3} C_6H_5-\overset{\overset{\displaystyle O}{\|}}{C}-R + HX \quad (X = Cl, Br, \text{ or } I)$$

Halogenation

Benzene reacts with chlorine or bromine in the presence of a small amount of iron salts.

$$C_6H_6 + X_2 \xrightarrow{FeX_3} C_6H_5X + HX \qquad (X = Cl \text{ or } Br)$$

The actual catalyst is $FeCl_3$ or $FeBr_3$. Recall that halogens react rapidly with iron to produce these salts:

$$2Fe + 3X_2 \longrightarrow 2FeX_3 \qquad (X = Cl, Br)$$

Light does not speed up this reaction, which therefore differs from the halogenation of alkanes. For this reason we can cause an alkylbenzene such as toluene to react either in the ring or in the alkyl side chain by simply controlling the conditions:

$$C_6H_5CH_3 + Cl_2 \xrightarrow[\text{dark}]{Fe} p\text{-}ClC_6H_4CH_3$$

$$C_6H_5CH_3 + Cl_2 \xrightarrow[\text{(no Fe)}]{\text{light}} C_6H_5CH_2Cl$$

We shall see later why ring chlorination occurs mainly in the para position.

This reaction follows the mechanism outlined in Section 7.10. Reaction of the halogen molecule with FeX_3, a Lewis acid, produces the positive halogen atom.

$$FeX_3 + :\ddot{X}-\ddot{X}: \longrightarrow FeX_4^- + :\ddot{X}:^+$$

The positive ion, X^+, then attacks the electron-rich benzene ring.

$$:\ddot{X}^+ + C_6H_6 \longrightarrow [C_6H_6X]^+ \xrightarrow{FeX_4^-} C_6H_5X + FeX_3 + HX$$

We call this kind of substitution reaction, in which the attacking reagent seeks an electron-rich group, an *electrophilic substitution* reaction.

Nitration

Benzene reacts with a mixture of nitric and sulfuric acids to produce nitrobenzene.

$$C_6H_6 + HONO_2 \xrightarrow{H_2SO_4} C_6H_5NO_2 + H_2O$$

Sulfuric acid catalyzes the reaction.

Here again, the positive ion, an electrophilic reagent, is produced by reaction of nitric acid with the catalyst:

$$H_2SO_4 + HONO_2 \rightleftarrows H_2\overset{+}{O}{-}NO_2 + HSO_4^-$$

$$H_2\overset{+}{O}{-}NO_2 \rightleftarrows H_2O + NO_2^+$$

$$H_2O + H_2SO_4 \longrightarrow H_3O^+ + HSO_4^-$$

$$C_6H_6 + \overset{+}{N}O_2 \longrightarrow C_6H_6NO_2^+ \xrightarrow{HSO_4^-} C_6H_5NO_2 + H_2SO_4$$

Sulfonation

Benzene reacts with a solution of sulfur trioxide in concentrated sulfuric acid (this solution is called fuming sulfuric acid) to produce benzenesulfonic acid.

$$C_6H_6 + SO_3 \xrightarrow{H_2SO_4} C_6H_5SO_3H$$

Unlike halogenation and nitration, **sulfonation** is reversible. Benzenesulfonic acid reacts with steam to re-form benzene and sulfuric acid.

$$C_6H_5SO_3H + 2\,H_2O \xrightarrow[H_3O^+]{\Delta} C_6H_6 + H_3O^+ + HSO_4^-$$

Water converts the SO_3 that is produced into the unreactive HSO_4^- and H_3O^+. This **desulfonation** reaction is useful for introducing a deuterium atom into the ring.

$$C_6H_6 + SO_3 \xrightarrow{H_2SO_4} C_6H_5SO_3H \xrightarrow[D_3O^+]{D_2O} C_6H_5D + D_3O^+ + HSO_4^-$$

In sulfonation, SO_3 is the electrophile and it reacts directly with the benzene ring.

H, $\ddot{O}$:, S^{2+}, SO_3^-, H, H

$\rightleftarrows$ SO_3^- H $\xrightarrow{H_2SO_4}$ SO_3H + HSO_4^-

Friedel–Crafts Alkylation

Benzene reacts with alkyl halides in the presence of Lewis acid catalysts to yield alkyl benzenes:

$$C_6H_6 + Cl\text{—}CH(CH_3)CH_3 \xrightarrow{AlCl_3} C_6H_5CH(CH_3)CH_3 + HCl$$

This reaction bears the names of its discoverers, C. Friedel and J. M. Crafts. It is a very useful reaction, but it has some limitations. The alkyl halide often rearranges during the reaction. For example, 1-bromopropane gives mainly isopropyl benzene.

$$C_6H_6 + BrCH_2CH_2CH_3 \xrightarrow{AlCl_3} C_6H_5CH(CH_3)CH_3 + HBr$$

For this reason we may suspect (correctly) that a carbocation is involved. $AlCl_3$ is a strong Lewis acid, and it reacts with the alkyl bromide to produce a carbocation which can rearrange.

$$Cl_3Al + :\ddot{Br}\text{—}CH_2CH_2CH_3 \longrightarrow [Cl_3Al\text{—}\ddot{Br}:]^- + \overset{+}{C}H_2CH_2CH_3$$

$$\rightleftarrows CH_3\overset{+}{C}HCH_3$$

The more stable carbocation, an electrophile, is now able to attack the benzene ring according to the general mechanism described in Section 7.10.

$$C_6H_6 + \overset{+}{C}H(CH_3)_2 \longrightarrow [C_6H_6CH(CH_3)CH_3]^+ \xrightarrow{AlCl_3Br^-} C_6H_5CH(CH_3)CH_3 + HBr$$

Friedel–Crafts Acylation

A similar reaction occurs when we mix an acyl halide and benzene in the presence of a Lewis acid catalyst.

$$C_6H_6 + Cl{-}\overset{\overset{\large O}{\|}}{C}{-}CH_2CH_3 \xrightarrow{AlCl_3} C_6H_5{-}\overset{\overset{\large O}{\|}}{C}{-}CH_2CH_3 + HCl$$

Rearrangements do not occur in this reaction. Reduction of the ketone yields propylbenzene, which cannot be prepared by the Friedel–Crafts alkylation reaction.

$$C_6H_5{-}\overset{\overset{\large O}{\|}}{C}{-}CH_2CH_3 \xrightarrow[\text{aqueous HCl}]{\text{Zn/Hg}} C_6H_5{-}CH_2CH_2CH_3 + Zn^{2+} + H_2O + Cl^-$$

Friedel–Crafts acylation occurs through the general mechanism (Sec. 7.10).

$$Cl{-}\underset{\underset{\large Cl}{|}}{\overset{\overset{\large Cl}{|}}{Al}} + :\ddot{Cl}{-}\overset{\overset{\large O}{\|}}{C}{-}R \rightleftharpoons AlCl_4^- + O{=}\overset{+}{C}{-}R$$

$$C_6H_6 + O{=}\overset{+}{C}{-}R \longrightarrow \left[C_6H_6(\text{H})(\text{C(=O)R})\right]^+$$

$$\xrightarrow{AlCl_4^-} C_6H_5{-}\overset{\overset{\large O}{\|}}{C}{-}R + HCl + AlCl_3$$

Problem 7.12
What is the geometry of the acylium ion, $R{-}\overset{+}{C}{=}\ddot{O}{:}$? Draw the important resonance structures.

Problem 7.13
Predict the product of the following Friedel–Crafts alkylation.

$$C_6H_6 + Cl{-}\overset{\overset{\large CH_3}{|}}{CH}{-}\underset{\underset{\large CH_3}{|}}{\overset{\overset{\large CH_3}{|}}{C}}{-}CH_3 \xrightarrow{AlCl_3}$$

If the benzene ring already bears a substituent, that substituent will determine the positions in the ring at which further substitution will occur. This effect is called **orientation**. The following examples illustrate orientation.

$$C_6H_5CH_3 + HONO_2 \xrightarrow{H_2SO_4} p\text{-}CH_3C_6H_4NO_2 + o\text{-}CH_3C_6H_4NO_2 \ (\text{major products}) \quad m\text{-}CH_3C_6H_4NO_2 \ (\text{minor product})$$

$$C_6H_5NO_2 + HONO_2 \xrightarrow{H_2SO_4} m\text{-}C_6H_4(NO_2)_2 \ (\text{major product}) + \left\{\begin{matrix}\text{small amounts of}\\ \text{ortho and para}\\ \text{dinitrobenzenes}\end{matrix}\right\}$$

Orientation depends *only* on the substituent already on the ring. It does not depend on the entering group. We can classify substituent groups into two categories: those that direct entering groups to ortho-para positions and those that direct entering groups to the meta position. Table 7.6 summarizes this effect.

We can understand orientation in terms of the resonance-stabilized carbocation intermediate

(three resonance structures of the cyclohexadienyl cation bearing X and H on the same carbon, with the + charge at the ortho, para, and ortho positions, connected by ⟷)

A substituent already on the ring affects the stability of the intermediate carbocation.

Table 7.6 Orientation in Aromatic Substitution Reactions

Ortho-Para-Directing Groups	Meta-Directing Groups
$—NH_2$	$—NO_2$
—OH, —OR	$—SO_3H$
—R (alkyl) C_nH_{2n+1}	$—\overset{O}{\overset{\|}{C}}—X$ (X = any atom or group)
—X (halogen)	—CN

An electron-releasing group will stabilize the carbocation; an electron-withdrawing group will destabilize the carbocation.

Let us look at the carbocations that are formed when attack occurs at the ortho, para, and meta positions. In the examples that follow, G is a substituent group and X—Y is the reagent.

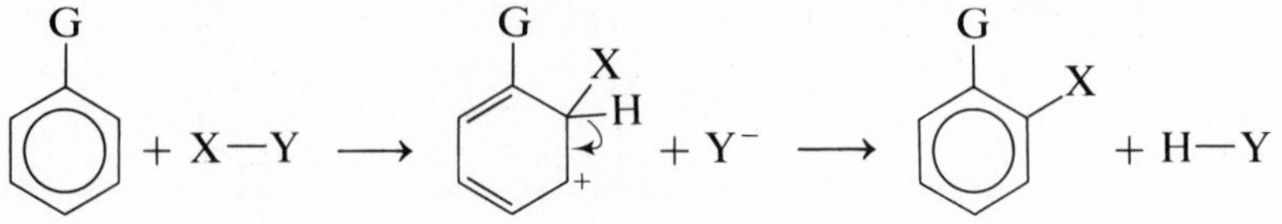

G, X, H, + X—Y ⟶ + Y^- ⟶ + H—Y

When G is electron-releasing (shown as G →), ortho, para, and meta attack produce carbocations that have the following resonance structures:

ortho attack ⟶ I ⟷ II ⟷ III (unusually stable)

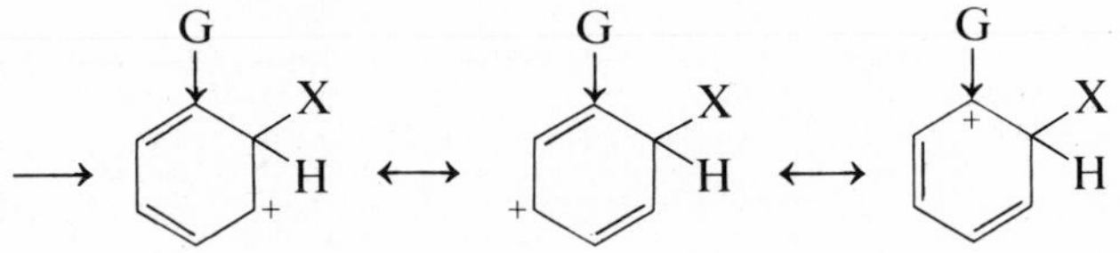

para attack ⟶ IV ⟷ V ⟷ VI

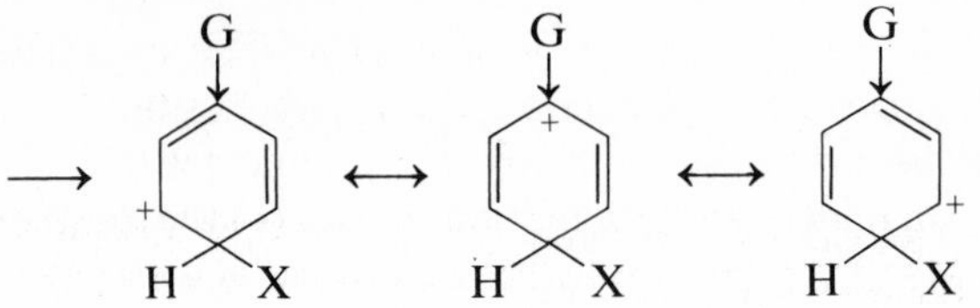

(unusually stable)

meta attack ⟶ VII ⟷ VIII ⟷ IX

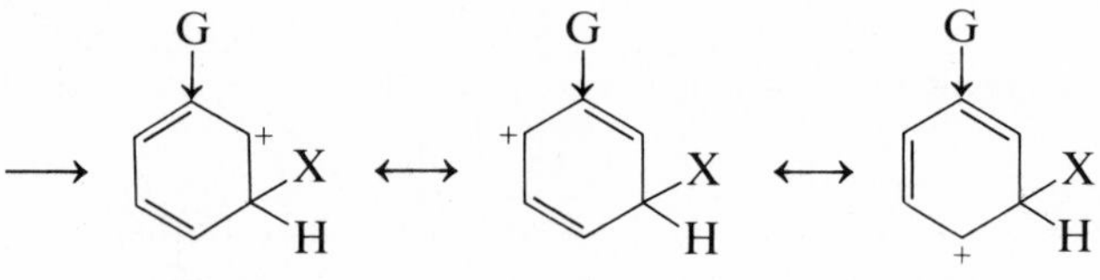

Ortho and para attack each produce a carbocation that is unusually stable because one of its contributing structures (III or V) has the positive charge adjacent to the electron-releasing substituent, G. Meta attack gives a carbocation that is not unusually stable. Thus an electron-releasing substituent directs attack at the ortho and para positions because in doing so, it allows the reaction to proceed through the path of lower energy.

If the substituent group is electron-attracting (shown as G ←), the situation is reversed:

ortho attack ⟶ I ⟷ II ⟷ III (unusually unstable)

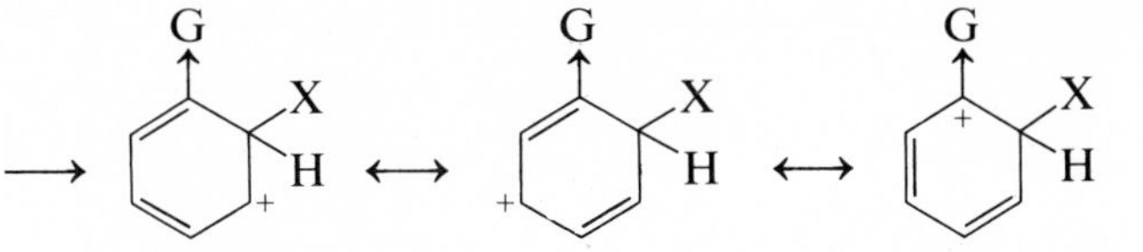

para attack ⟶ IV ⟷ V (unusually unstable) ⟷ VI

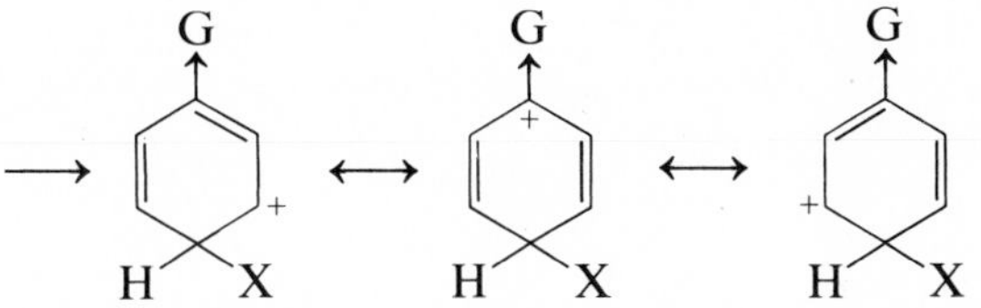

meta attack ⟶ VII ⟷ VIII ⟷ IX

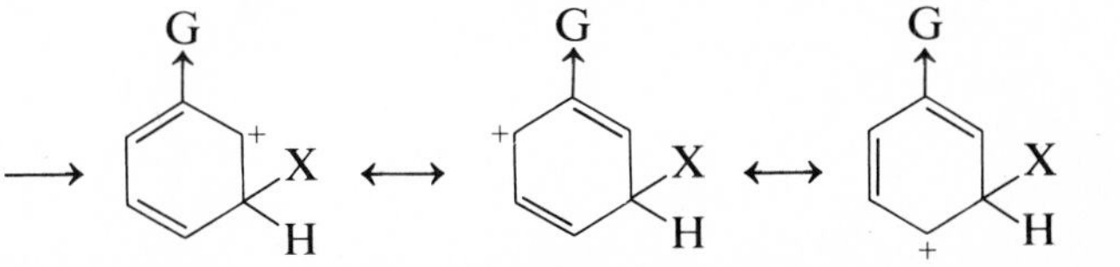

In this reaction, ortho or para attack yield an unstable carbocation; in each case one contributing structure is unusually unstable (III or V) because the positive charge is adjacent to the electron-attracting group. Thus meta attack gives the most stable carbocation (the one that is *least destabilized*).

How do we recognize a functional group as electron-releasing or electron-attracting? Again we can use resonance theory. The amino group is electron-releasing because we can write resonance structures in which the electron pair on nitrogen is released into the ring:

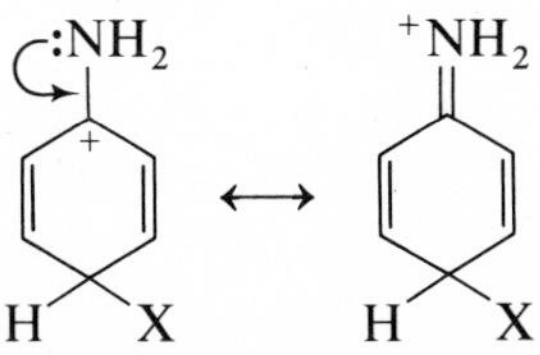

$:NH_2$ ⟷ $^+NH_2$

A substituent is electron-releasing if the atom joined to the ring bears an unshared electron pair. The $—\ddot{O}H$ and $—\ddot{X}:$ (halogen) groups are thus electron-releasing. Alkyl groups do not bear electron pairs; however, they do release electrons. The evidence for their electron-releasing ability is that alkyl groups stabilize carbocations (Sec. 5.6).

Meta-directing groups have resonance structures in which the atom attached to the ring bears a positive charge:

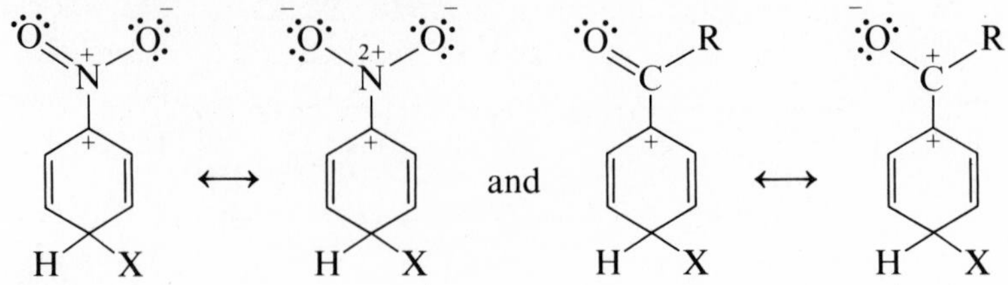

[resonance structures of nitro-substituted and acyl-substituted arenium ions (H, X on ring carbon), with $\longleftrightarrow$ and "and"]

Problem 7.14

Predict whether each group below is electron-attracting or electron-releasing.

(a) $-\ddot{N}(CH_3)_2$ (b) $-C(CH_3)_3$ (c) $-C(=O)-\ddot{N}H_2$

(d) $-C(=\ddot{N}H)-CH_3$ (e) $-C(=\ddot{S}:)-CH_3$

Problem 7.15

Predict the major products of each of the following reactions (assume that only monosubstitution occurs in each case).

(a) isopropylbenzene + $HONO_2$ $\xrightarrow{H_2SO_4}$

(b) phenol + Cl_2 $\xrightarrow{Fe}$ (c) $C_6H_5CN + Br_2$ $\xrightarrow{Fe}$

(d) $C_6H_5\overset{O}{\overset{\|}{C}}CH_3 + HONO_2$ $\xrightarrow{H_2SO_4}$ (e) $C_6H_5CCl_3 + SO_3$ $\xrightarrow{H_2SO_4}$

7.13 REACTIONS OF ALKYLBENZENES

The benzene ring affects many of the reactions of the alkyl group in alkylbenzenes. We shall survey a few of these reactions.

Ozonolysis

The ring remains intact during ozonolysis, and undergoes ozonolysis only under very vigorous conditions. For this reason we can ozonize phenylalkenes; for example, 1,2-diphenylethene gives two moles of benzaldehyde.

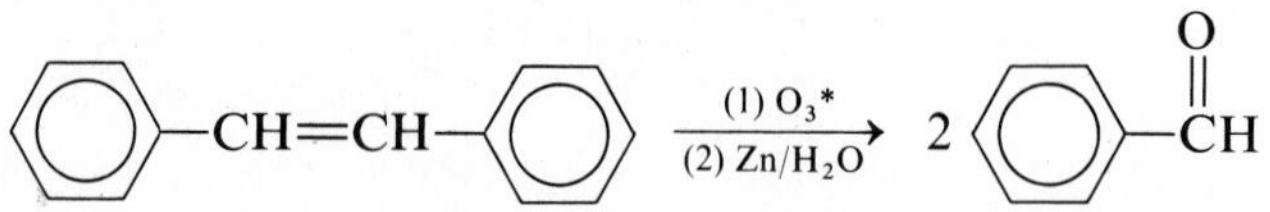

$$C_6H_5-CH{=}CH-C_6H_5 \xrightarrow[(2)\ Zn/H_2O]{(1)\ O_3^*} 2\ C_6H_5-\overset{O}{\overset{\|}{C}}H$$

* This notation means that the product of ozonolysis [Step (1)] is then treated with Zn + H_2O [Step (2)].

Oxidation

Hot, alkaline potassium permanganate solution oxidizes alkylbenzenes to benzoic acid regardless of the nature of the alkyl group. This reagent does not attack the ring.

$$C_6H_5{-}R \xrightarrow{KMnO_4,\ OH^-,\ heat} C_6H_5{-}\overset{O}{\overset{\|}{C}}{-}O^- + MnO_2\downarrow + \cdots$$

$$\xrightarrow{H_3O^+} C_6H_5{-}\overset{O}{\overset{\|}{C}}{-}OH$$

Oxidation degrades any alkyl substituent down to one carbon atom—the carboxyl group:

$$C_6H_5{-}CH(CH_3)_2 \xrightarrow[(2)\ H_3O^+]{(1)\ KMnO_4,\ OH^-,\ heat} C_6H_5{-}\overset{O}{\overset{\|}{C}}{-}OH$$

Other, nonalkyl substituents are generally not affected:

$$o\text{-}Cl{-}C_6H_4{-}CH_3 \xrightarrow[(2)\ H_3O^+]{(1)\ KMnO_4,\ OH^-,\ heat} o\text{-}Cl{-}C_6H_4{-}CO_2H$$

Because benzoic acids are well known, this reaction is especially useful in determining the structures of isomeric compounds.

Problem 7.16

Two isomeric alkyl benzenes have the formula C_8H_{10}. Isomer A, on permanganate oxidation, gives benzoic acid (melting point 122°C). Isomer B, on permanganate oxidation, gives *m*-phthalic acid,

$$m\text{-}C_6H_4(CO_2H)_2$$ (melting point 330°C).

Identify isomers A and B.

Halogenation

Toluene reacts with chlorine or bromine in the presence of light to give mono-, di-, or trihalogenation:

$$C_6H_5CH_3 \xrightarrow[\text{light}]{Cl_2} C_6H_5CH_2Cl + HCl \xrightarrow[\text{light}]{Cl_2} C_6H_5CHCl_2 + HCl \xrightarrow[\text{light}]{Cl_2} C_6H_5CCl_3 + HCl$$

In this reaction, toluene behaves like a typical alkane. Reaction does not occur in the ring as long as iron or other catalysts are absent.

With other alkyl groups, the reaction occurs preferentially on the carbon attached to the ring:

$$C_6H_5CH_2CH_3 + Cl_2 \xrightarrow{\text{light}} C_6H_5CH(Cl)CH_3 + C_6H_5CH_2CH_2Cl + HCl$$

(major product)

$$C_6H_5CH(CH_3)_2 + Br_2 \xrightarrow{\text{light}} C_6H_5C(CH_3)_2Br + HBr$$

(major product)

Elimination

Reactions of the side chain that produce a carbon–carbon double bond normally give as the major product an alkene that is conjugated with the ring:

$$C_6H_5CH_2CH(Cl)CH_2CH_3 + KOH \xrightarrow{\text{ethanol}} C_6H_5CH{=}CHCH_2CH_3$$

(major product)

As with other alkene-forming reactions, these reactions yield the more stable (conjugated) alkene.

Problem 7.17

Predict the products of the following reactions.

(a) propylbenzene + Br_2 $\xrightarrow{\text{light}}$

(b) propylbenzene + Br_2 $\xrightarrow{\text{Fe}}$

(c) 1-phenyl-2-bromo-2-methylpropane + KOH $\xrightarrow{\text{ethanol}}$

7.14 ANALYSIS OF AROMATIC HYDROCARBONS

Because aromatic hydrocarbons have such a high carbon content (low hydrogen content), they burn with a yellow sooty flame. Alkanes and alkenes also burn with a yellow flame but produce little or no soot. The presence of soot is thus a reliable test for most benzene derivatives.

The great stability of the aromatic ring makes specific chemical testing difficult. We can distinguish benzene from alkenes and alkynes, however, by the failure of benzene to dissolve in cold, concentrated sulfuric acid. In this test benzene behaves like an alkane.

Benzene differs from both alkanes and unsaturated hydrocarbons in its reaction with bromine. We can summarize the differences as follows:

$$\text{Alkane} + Br_2/CCl_4 \xrightarrow{\text{light}} \text{alkyl bromide} + HBr \text{ (only in the presence of light)}$$

$$\text{Alkene} + Br_2/CCl_4 \longrightarrow \text{dibromoalkane (no light or catalyst required)}$$

$$\text{Benzene} + Br_2/CCl_4 \xrightarrow{\text{Fe}} \text{bromobenzene} + HBr \text{ (catalyst required)}$$

Problem 7.18

Expand Tables 5.5 and 6.5 to include benzene.

Problem 7.19

Explain how you would distinguish between the compounds in the following pairs. (In some cases reactions more complicated than simple chemical tests may be necessary.)

(a)  and $CH_2{=}CH{-}C{\equiv}C{-}CH{=}CH_2$

(b) $CH_2{=}CH{-}C{\equiv}C{-}CH{=}CH_2$ and $HC{\equiv}C{-}CH{=}CH{-}CH{=}CH_2$

(c) 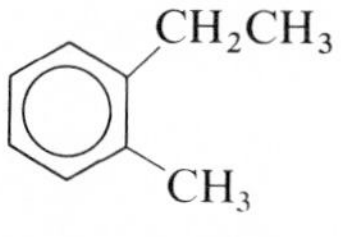 CH_2CH_3 CH_3 and $CH_2CH_2CH_3$

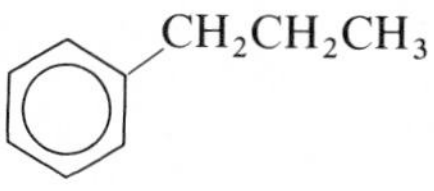

7.15 OCCURRENCE OF AROMATIC COMPOUNDS: AN ELECTIVE TOPIC

elective topic

The most abundant source of aromatic hydrocarbons on earth is coal. Coal is the remains of freshwater plants that lived 250 to 350 million years ago. Once these plants—trees, leaves, limbs—had been covered with water, bacterial action turned them into peat. When the oxygen supply was gone, decomposition stopped. Time and pressure converted these carbon-rich vegetable deposits into coal, an impure form of graphite (Sec. 6.15).

Such deposits are still forming, although large peat beds are not common. A typical one lies in the Dismal Swamp region of Virginia and North Carolina. No period of coal formation has been as prolific or extensive as the one that took place 250 to 350 million years ago.

If coal is heated to 1000–1300°C in the absence of oxygen, it decomposes to give a hard residue called *coke* and a mixture of more volatile substances that condense as *coal tar*. The process is called **destructive distillation** because in the process molecules decompose and re-form new ones. Distillation of coal tar yields a variety of aromatic hydrocarbons including (in addition to benzene, toluene, the xylenes, and the compounds shown in Table 7.2) the compounds shown in Table 7.7.

Many aromatic hydrocarbons are carcinogenic. As such, they are an important problem in our society. Two major sources of aromatic hydrocarbons in the air we

Table 7.7 Aromatic Compounds from Coal Tar

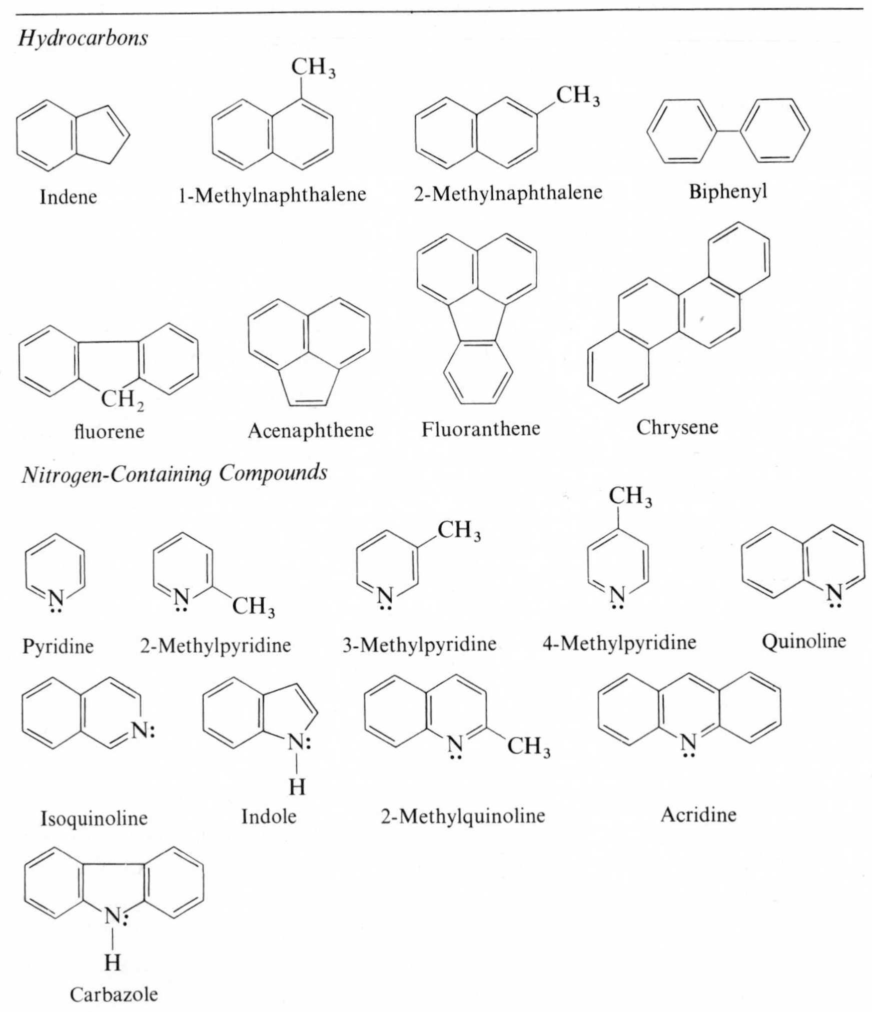

Oxygen-Containing Compounds

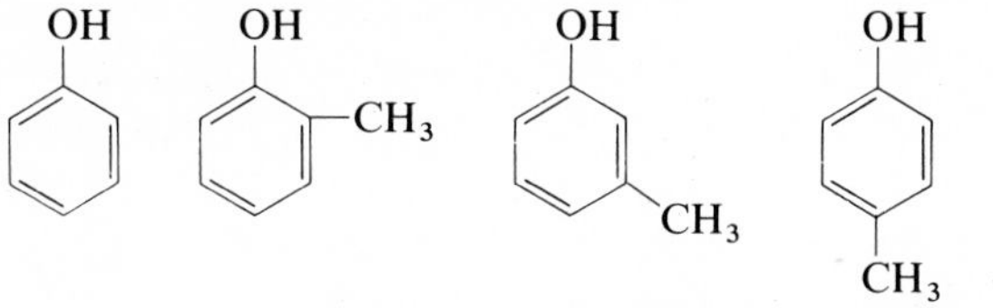

Phenol — *o*-Cresol — *m*-Cresol — *p*-Cresol

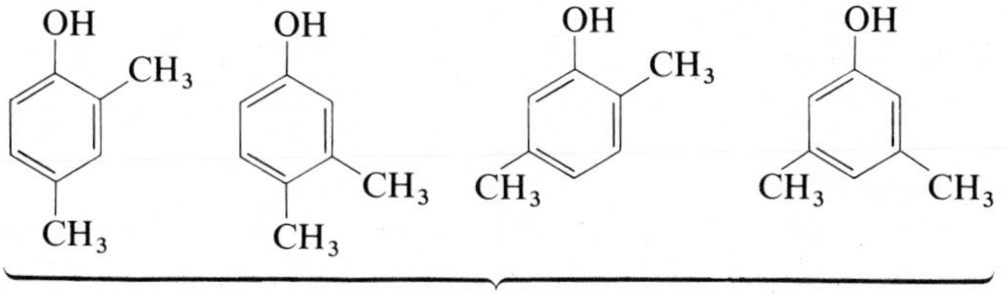

Dimethylphenols

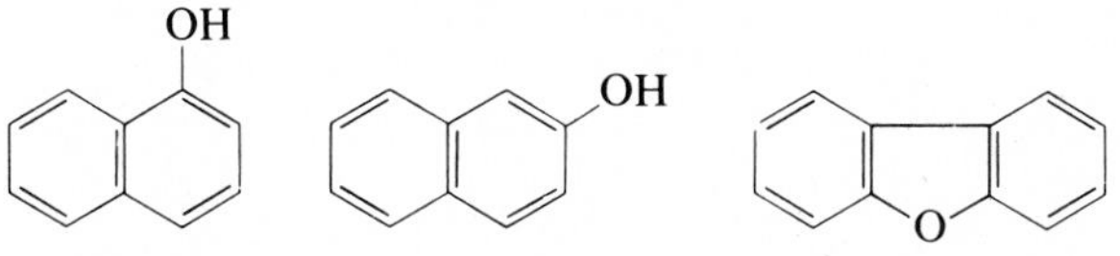

1-Naphthol — 2-Naphthol — Diphenylene oxide

breathe are automobile exhaust and cigarette smoke. In a burning cigarette, cellulose and other organic compounds burn incompletely. The burning tip of a cigarette reaches a temperature of 700–800°C. Because there is very little oxygen within this tip, most of the organic compounds do not burn but instead undergo destructive distillation to yield many of the aromatic compounds shown in Table 7.7. One of the most dangerous compounds found in cigarette smoke is 1,2-benzopyrene, an extremely potent carcinogen:

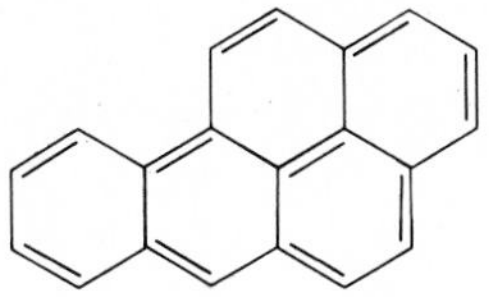

NEW TERMS

Annulene Any completely conjugated, monocyclic hydrocarbon (7.6).

Destructive distillation Decomposition of a substance by heating in the absence of oxygen (7.15).

Desulfonation Reaction of a benzenesulfonic acid with aqueous acid to remove the sulfonic acid group (7.11):

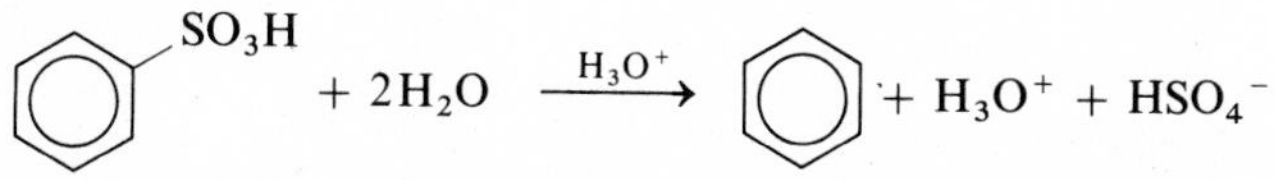

$$C_6H_5SO_3H + 2H_2O \xrightarrow{H_3O^+} C_6H_6 + H_3O^+ + HSO_4^-$$

Friedel–crafts acylation Reaction of a benzene ring with an acyl halide, in the presence of $AlCl_3$, to produce a phenyl ketone (7.11):

$$C_6H_6 + R-\overset{O}{\overset{\|}{C}}-Cl \xrightarrow{AlCl_3} C_6H_5-\overset{O}{\overset{\|}{C}}-R + HCl$$

Friedel–crafts alkylation Reaction of a benzene ring with an alkyl halide, in the presence of $AlCl_3$, to produce an alkyl benzene (7.11):

$$C_6H_6 + R-Cl \xrightarrow{AlCl_3} C_6H_5-R + HCl$$

Halogenation See also Section 4.10 (7.11).

$$C_6H_6 + X_2 \xrightarrow[\text{or } FeX_3]{Fe} C_6H_5-X + HX\ (X = Cl \text{ or } Br)$$

Heat of hydrogenation (ΔH_h°) ΔH° for the addition of H_2 to an unsaturated compound.

Hückel's rule Any planar, monocyclic, completely conjugated hydrocarbon is aromatic if it has $(4n + 2)$ π electrons in its conjugated system (n is any positive integer ≥ 0) (7.6).

Molecular orbital An orbital (σ or π) that extends over two or more atoms (7.5).

Nitration See also Section 4.10 (7.11).

$$C_6H_6 + HONO_2 \xrightarrow{H_2SO_4} C_6H_5-NO_2 + H_2O$$

Orientation If a benzene ring bears a substituent, that substituent will determine the positions in the ring where further substitution will occur. This effect is called orientation (7.12).

Oxidation Hot, alkaline $KMnO_4$ oxidation of the aliphatic substituents on a benzene ring to the corresponding carboxyl group (7.13):

$$C_6H_5-R \xrightarrow[\text{heat, } OH^-]{KMnO_4} C_6H_5-\overset{O}{\overset{\|}{C}}-O^-$$

Ozonolysis See section 5.6 (7.13).

Phenyl The group C_6H_5- (7.9).

Resonance Whenever we can write more than one reasonable valence bond structure for a molecule, then the true structure of that molecule is a hybrid of all of these valence bond structures. Resonance is the representation of the structure of a molecule as a hybrid or composite of several valence bond structures (7.11).

Resonance stabilization energy The difference between the observed heat of hydrogenation of a conjugated compound and the heat of hydrogenation calculated by assuming that there is no interaction between the double bonds. The calculated ΔH_h° is obtained by multiplying the ΔH_h° of one isolated double bond by the number of double bonds in the conjugated compound. Resonance stabilization energy is also called **resonance energy** (7.3).

Sulfonation Reaction of the benzene ring with SO_3 to produce a benzenesulfonic acid (7.11):

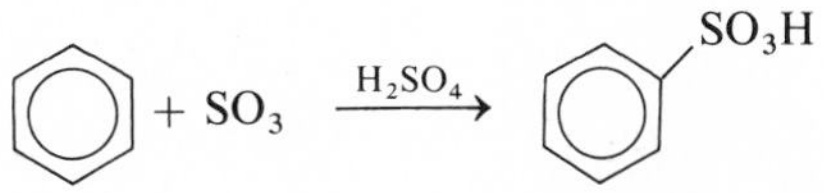

$$C_6H_6 + SO_3 \xrightarrow{H_2SO_4} C_6H_5SO_3H$$

ADDITIONAL PROBLEMS

7.20 Write equations for the bromination reactions of cyclohexane, cyclohexene, and benzene that show how these compounds differ chemically.

[7.21] Explain why the C—C bonds of benzene are all equal and intermediate in length between a single and a double bond.

[7.22] Use resonance stabilization energy to explain why benzene undergoes substitution reactions rather than addition reactions.

7.23 Calculate the number of π electrons in each of the hydrocarbons in Table 7.2. Do all of them obey Hückel's rule? Explain why those that do not may still be classified as aromatic. (Hint: What are the restrictions of Hückel's rule?)

7.24 Explain why cyclopentadienyl anion is aromatic whereas cyclopentadienyl cation is not.

7.25 Name the following compounds.

[a] CH_3 / Cl (benzene ring with CH_3 and Cl in the meta positions)

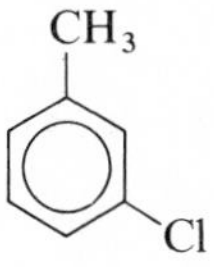

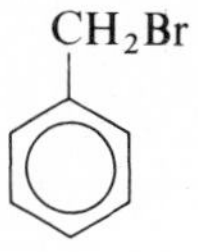

[b] CH_2Br (benzene ring with a CH_2Br substituent)

[c] O_2N / NO_2 / NO_2 (benzene ring with three NO_2 groups, 1,3,5-positions)

[d] CH_3 / CH_3 (benzene ring with two CH_3 groups in the meta positions)

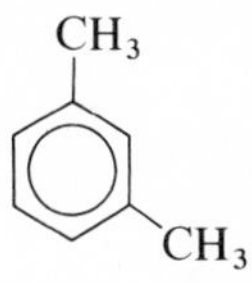

(e) 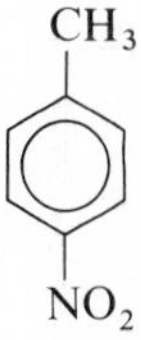 CH_3 / NO_2 (benzene ring with CH_3 and NO_2 in the para positions)

(f) 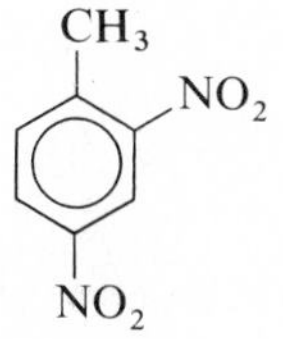 CH_3 / NO_2 / NO_2 (CH_3 at the top, one NO_2 ortho to it and one NO_2 para to it)

(g) 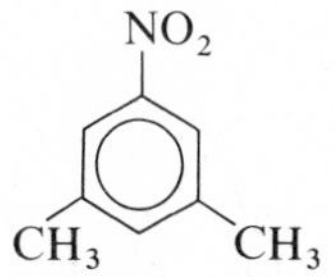 NO_2 / CH_3 / CH_3 (NO_2 at the top, CH_3 groups at the two lower ring positions, meta to the NO_2)

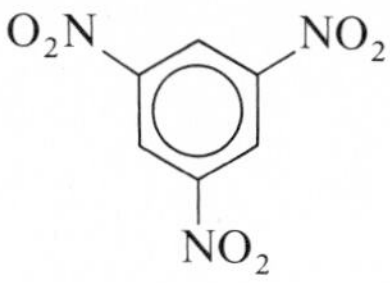

(h) 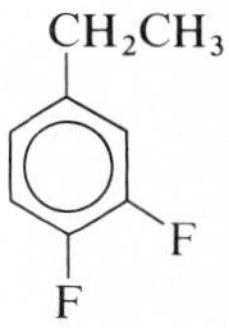 CH_2CH_3 / F / F (CH_2CH_3 at the top, one F para to it and one F on the adjacent lower-right position)

7.26 On the basis of the equilibrium between nitric acid and sulfuric acid given in Section 7.11 (nitration mechanism), which is the stronger acid, nitric or sulfuric?

7.27 Write equations to show how you would prepare each of the following compounds starting with either benzene or toluene and any aliphatic and inorganic compounds.

[a] bromobenzene, [b] benzenesulfonic acid,

[c] benzoic acid, (d) nitrobenzene,

(e) ethylbenzene, (f) 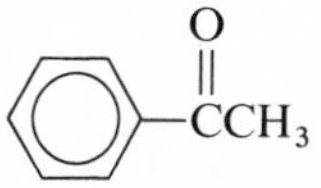 C_6H_5—$\overset{O}{\overset{\|}{C}}CH_3$

7.28 Write equations to show how you would prepare each of the following compounds starting with benzene and any nonaromatic and inorganic compounds.

[a] *p*-chloronitrobenzene,

[b] Cl—C₆H₄—C(=O)CH₃ (para), $Cl-C_6H_4-\overset{O}{\overset{\|}{C}}CH_3$,

(c) *m*-chloronitrobenzene,

(d) Cl—C₆H₄—C(=O)CH₃ (meta), $\overset{O}{\overset{\|}{C}}CH_3$

7.29 A compound whose formula is C_9H_{12} reacts with hot $KMnO_4$ to produce benzoic acid. What are the possible structures for the compound? Describe reactions that would make it possible to determine which of the possible structures is correct.

[7.30] Compound A, whose molecular formula is $C_{10}H_{12}$, undergoes the following reactions:

(a) $A + Br_2/CCl_4 \longrightarrow C_{10}H_{12}Br_2$

(b) $A + KMnO_4 \xrightarrow[(2)\ H_3O^+]{(1)\ OH^-,\ \Delta}$ benzene-1,3-dicarboxylic acid (CO_2H, CO_2H)

(c) $A + H_2 \xrightarrow{Ni} C_{10}H_{14} \xrightarrow[light]{Br_2} C_{10}H_{13}Br \xrightarrow[alcohol]{KOH} C_{10}H_{12}(B)$

Compound B is an isomer of A. Give the structural formula of compound A.

7.31 Cite simple chemical tests that would make it possible to distinguish between the compounds in the following pairs. (Describe what you would observe.)

[a] $C_6H_5-C{\equiv}C-C_6H_5$ and $C_6H_5-C_6H_4-C{\equiv}CH$

[b] $C_6H_5-CH{=}CHCH_3$ and $C_6H_5-CH_2CH{=}CH_2$

(c) C_6H_5–cyclopentyl and $C_6H_5-CH{=}CHCH_2CH_2CH_3$

(d) 1,2-dimethylbenzene (CH_3, CH_3) and 1,2-dimethylcyclohexane (CH_3, CH_3) (1,2-dimethylcyclohexane)

[7.32] We saw in Section 6.6 that conjugated dienes undergo 1,2- and 1,4-addition. For example, the addition of HCl to 1,3-butadiene gives both $CH_3CHClCH{=}CH_2$ and $CH_3CH{=}CHCH_2Cl$. Those reactions involve a carbocation intermediate. Use the concept of resonance to explain why conjugated dienes give both 1,2- and 1,4- addition products. (Hint: Examine the intermediate carbocation.)

MOLECULAR SHAPES: CHIRALITY AND OPTICAL ACTIVITY

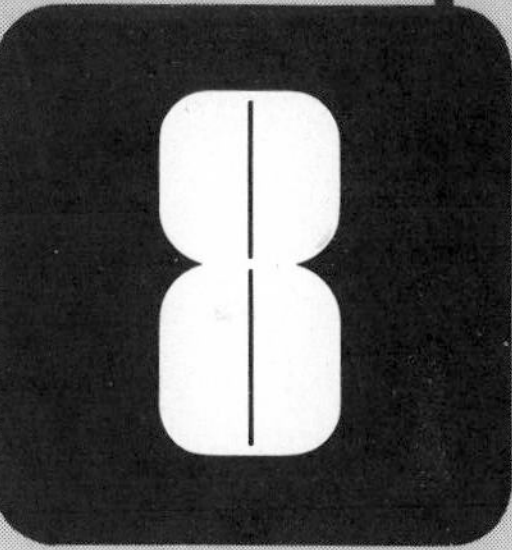

8.1 REVIEW OF ISOMERISM

Isomers are compounds that have the same molecular formula but different structures and therefore different properties.

In Chapter 2 we identified two kinds of isomers—constitutional isomers (Sec. 2.4) and stereoisomers (Sec. 2.5).

Molecules are **constitutional isomers** of one another if their atoms are bonded to each other in different sequences. Two examples of constitutional isomers are the butanes and the xylenes:

$$CH_3CH_2CH_2CH_3 \qquad \underset{\displaystyle CH_3}{CH_3\overset{|}{C}HCH_3}$$

butane — isobutane

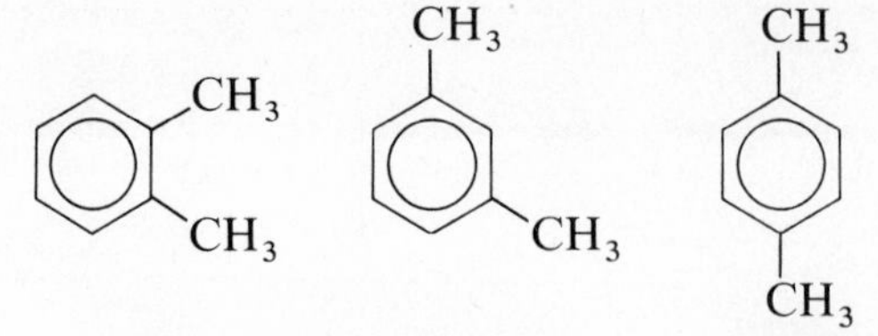

o-xylene m-xylene p-xylene

Two molecules are **stereoisomers** of one another if their atoms are bonded to each other in the same sequence (that is, if they are not constitutional isomers) and if they differ in the orientation of those atoms in space (ignoring differences due to rotation about single bonds). Examples of stereoisomers are the *cis–trans* isomers of 2-butene and the *cis–trans* isomers of 1,2-dimethylcyclopropane:

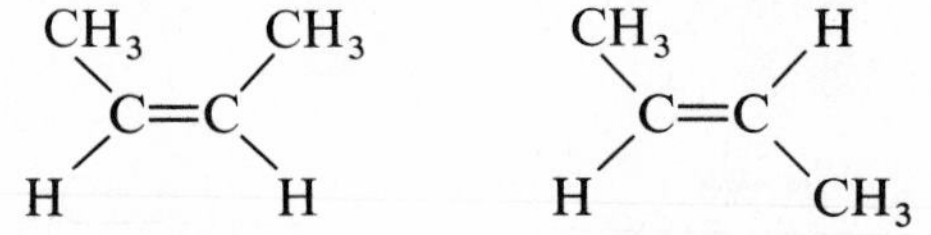

cis-2-butene *trans*-2-butene

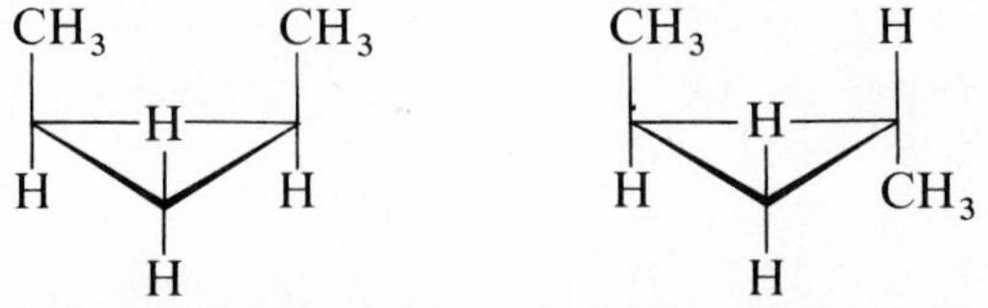

cis-1,2-dimethylcyclopropane *trans*-1,2-dimethylcyclopropane

Conformers (Sec. 4.2) are molecules that differ only in the rotation of their atoms about single bonds. Conformers are not usually considered to be isomers although they are sometimes referred to as conformational isomers. The eclipsed and staggered forms of ethane and the chair and boat conformations of cyclohexane are examples of conformers. Conformers normally interconvert readily at moderate temperatures.

ethane (eclipsed conformation)

ethane (staggered conformation)

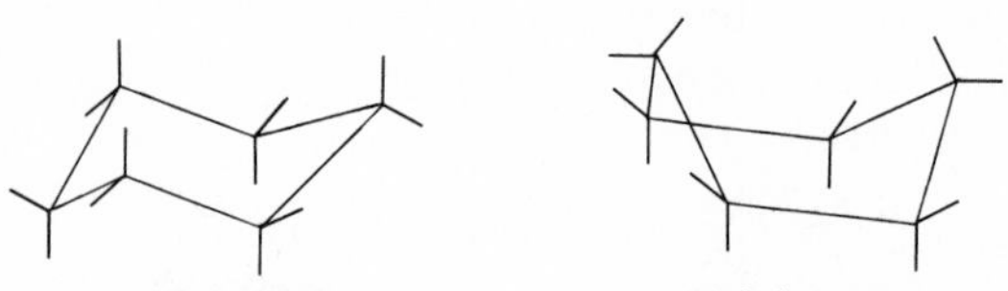

cyclohexane (chair form)

cyclohexane (boat form)

We shall see below that there are two kinds of stereoisomers, but before we can describe them we shall have to examine the phenomenon of plane-polarized light.

Problem 8.1

Label each pair of compounds as constitutional isomers, stereoisomers, conformers, nonisomers, or identical.

(a) [Newman projection, all H] and [Newman projection, all H]

(b) 1-methylcyclopentene (CH_3 on ring double bond) and methylenecyclopentane ($=CH_2$)

(c) $CH_3\overset{O}{\overset{\|}{C}}-OH$ and $H\overset{O}{\overset{\|}{C}}-OCH_3$

(d) $(CH_3)(Cl)C=C(F)(Br)$ and $(Cl)(CH_3)C=C(F)(Br)$

8.2 PLANE-POLARIZED LIGHT

Ordinary visible light consists of electromagnetic waves. Each electromagnetic wave lies in a plane. It is possible to "filter" ordinary light so that the light that emerges consists only of electromagnetic waves in a single plane. Light filtered in this way is called **plane-polarized** (Fig. 8.1).

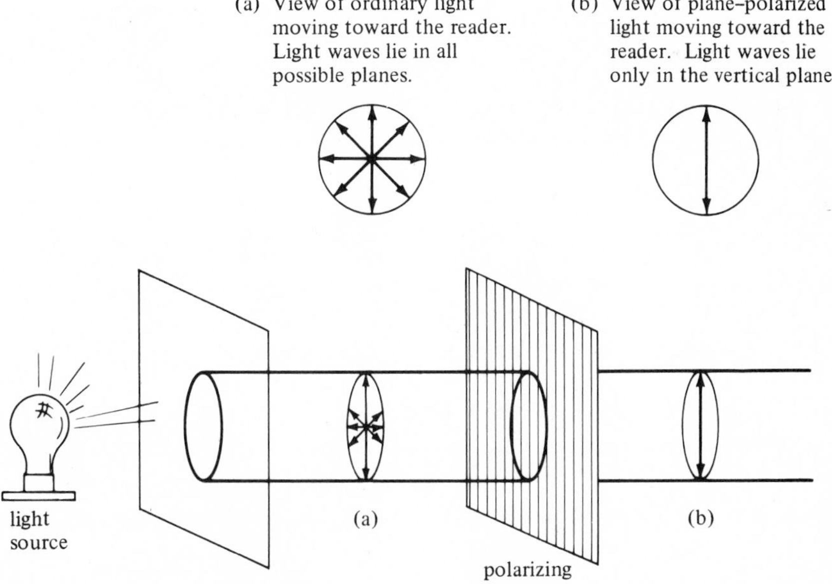
(a) View of ordinary light moving toward the reader. Light waves lie in all possible planes.
(b) View of plane-polarized light moving toward the reader. Light waves lie only in the vertical plane.
light source
(a)
(b)
polarizing filter

Figure 8.1. Ordinary visible light (a) consists of waves that vibrate in all possible planes. Plane-polarized light (b) consists of waves that vibrate in only one plane.

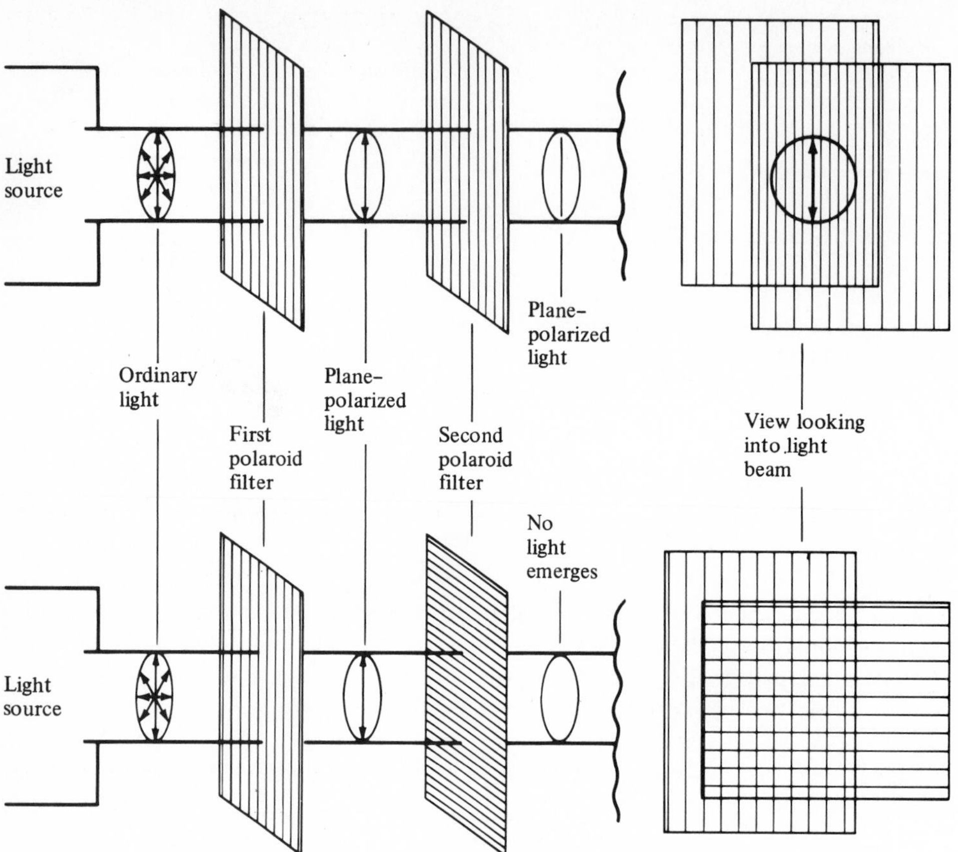

Figure 8.2. When light passes through two consecutive Polaroid filters whose polarizing axes are parallel, it emerges plane-polarized as if only one filter had been used. If the polarizing axes of the two filters are perpendicular to each other, no light emerges from the second one.

One kind of polarizing filter is the plastic sheet used to make Polaroid sun glasses. If we pass the light through two such Polaroid filters, we can get two different results. If the polarizing axes of the two filters are aligned parallel to each other, light passes through both of them as though there were only one filter. If we align them so that their polarizing axes are perpendicular to each other, then no light will pass through the second one. Figure 8.2 illustrates these situations.

8.3 OPTICAL ACTIVITY

Quartz crystals have an unusual effect on plane-polarized light: When plane-polarized light passes through the crystal, the light that emerges is still plane-polarized, but its plane of polarization has rotated (Fig. 8.3). Many naturally occurring liquids

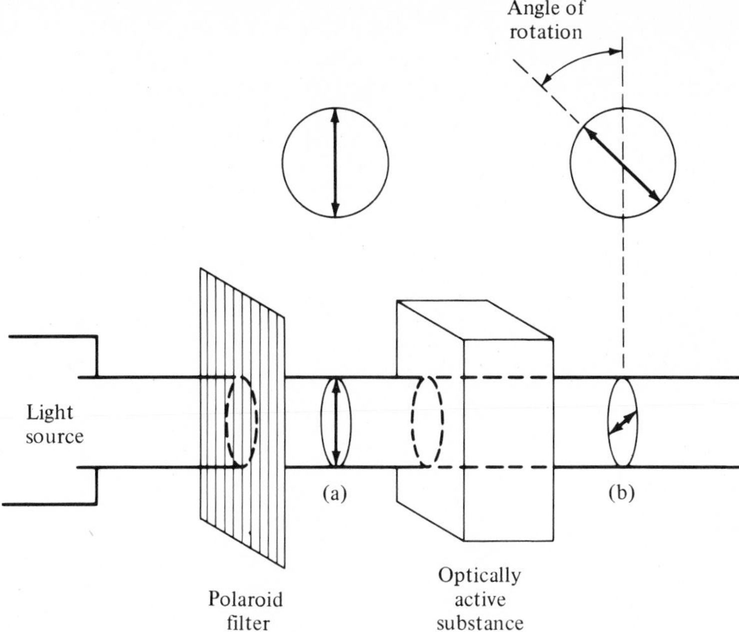

Figure 8.3. When plane-polarized light (a) passes through an optically active substance it emerges (b) with its plane of polarization rotated.

such as turpentine also have the power to rotate the plane of polarization of plane-polarized light. Substances that have this power are called *optically active* substances. The phenomenon is called **optical activity**.

In the mid-nineteenth century, Louis Pasteur separated two forms of sodium ammonium tartrate crystals by hand. The two forms are visibly different but they are related in the same way that the right hand is related to the left—one is the mirror image of the other (Fig. 8.4). Ammonium tartrate crystals were known to be optically active. Pasteur found that when he dissolved each kind of crystal in water, the water solutions were optically active! He concluded that, because the crystals no longer exist when they are dissolved, the *molecules themselves* must exist in right- and left-handed forms. Pasteur called such molecules, or objects, *dissymmetric*. Today, the more common term is **chiral**.

Chiral objects are capable of existing in two forms that are related as a right and left hand are related (Fig. 8.4). That is, chiral objects have a right-handed and a left-handed form. Examples of common chiral objects are hands, gloves, feet, shoes, ears, and screws. Examples of **achiral** (nonchiral) objects are noses, hammers, screw-drivers, socks, spoons, and forks.

Problem 8.2

Suggest how you could measure the angle of rotation of an optically active substance using a setup like the one shown in Figures 8.2 and 8.3. (Check your answer with Fig. 8.9.)

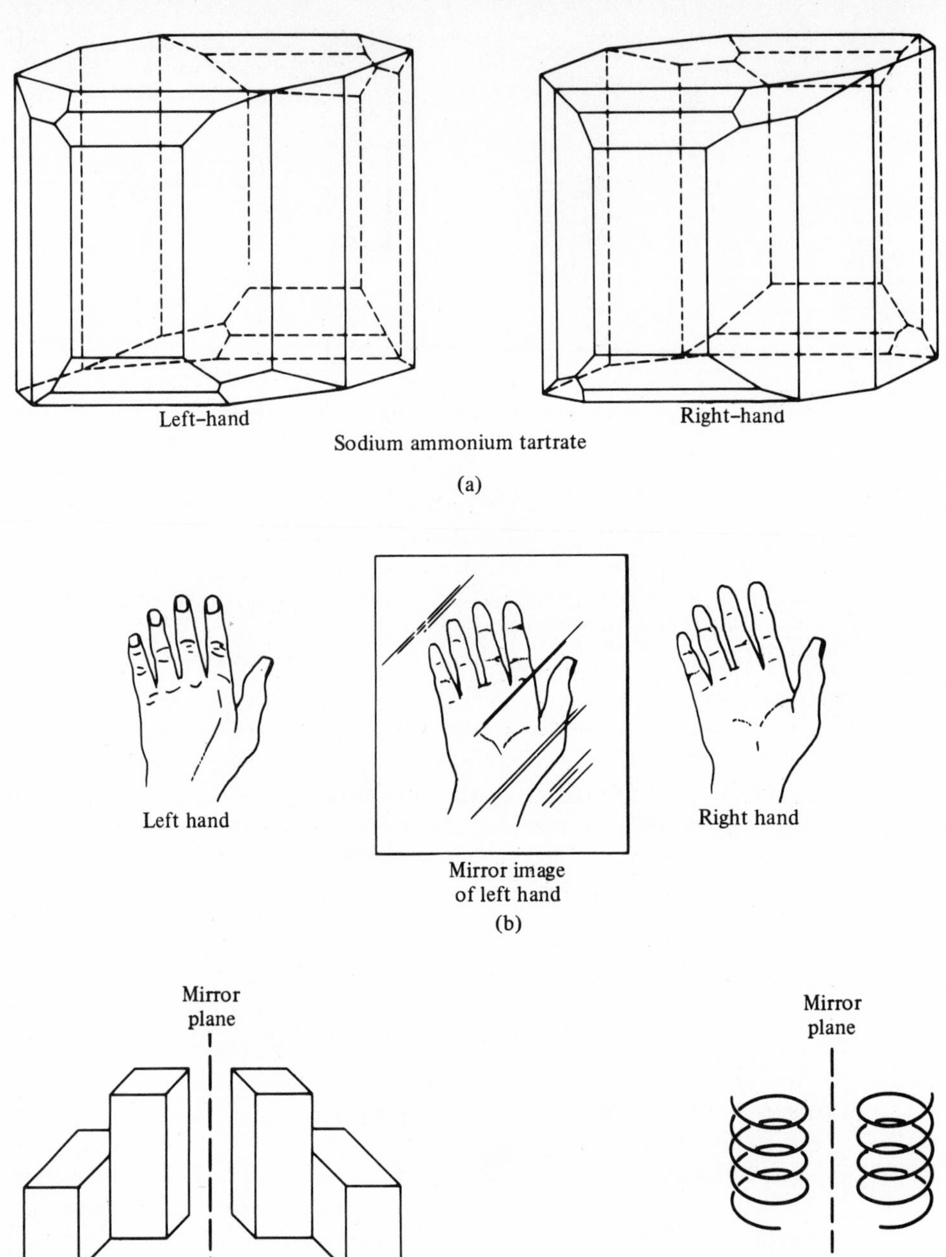

Figure 8.4. The two forms of sodium ammonium tartrate crystals (a) are related in the way that a right hand is related to a left hand (b). Two pairs of simple objects that are similarly related are shown in (c) and (d).

Problem 8.3
Classify the following objects as chiral or achiral.

(a) coffee cup,
(b) kitchen knife,
(c) plain wedding ring,
(d) plain pencil,
(e) classroom desk with writing arm,
(f) water pitcher,
(g) wine bottle,
(h) fountain pen

8.4 CONDITIONS FOR CHIRALITY

How do we decide whether an object is chiral or achiral? We can answer the question in two ways: A chiral object has a mirror image that is somehow different from the object itself. This difference is the difference between a right hand and a left hand. Right and left hands are not superposable. By **superposable** we mean that in our mind we place one object upon the other so that they occupy the same space and so that every part of one object coincides with the corresponding part of the other object. We can demonstrate that a right and a left hand are nonsuperposable by trying to fit a left-handed glove on a right hand. The left hand is, however, superposable with the mirror image of the *right* hand.

We can now define a chiral object as an object that is not superposable on its own mirror image. An achiral object is superposable on its mirror image. Figure 8.5 shows some common objects and their mirror images.

Another way of deciding whether an object is chiral or achiral is to determine whether it has a plane of symmetry. Chiral objects have no planes of symmetry. A **plane of symmetry** divides an object into two halves that are mirror images of each other. A coffee cup, for example, has a plane of symmetry that passes through its handle (Fig. 8.6). If an object has a plane of symmetry, it is not chiral. There are other elements of symmetry that a chiral object may not possess, although, the plane of symmetry criterion works for all molecules of interest to us in this text.

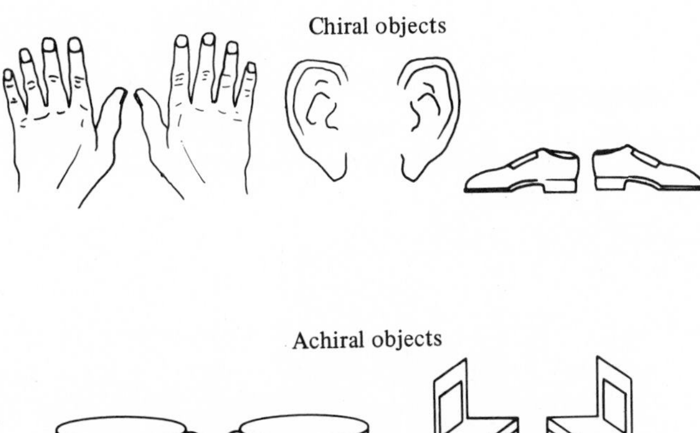

Figure 8.5. Some objects and their mirror images.

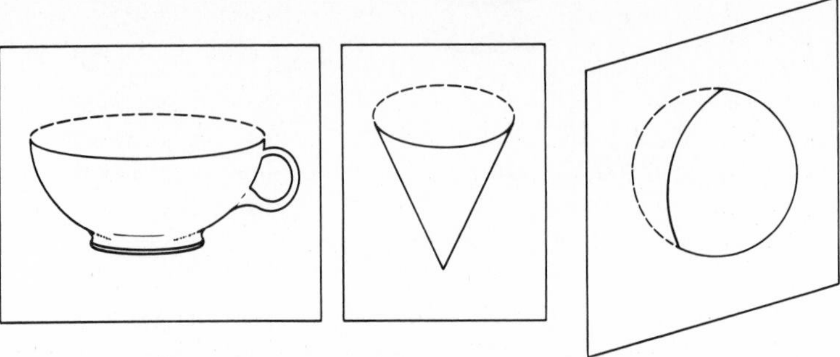

Figure 8.6. Archiral objects possess a plane of symmetry.

Problem 8.4

Demonstrate whether each object in Problem 8.3 has a plane of symmetry. Looking at each object in front of a mirror, determine which have nonsuperposable mirror images. Do your observations coincide with your answers to Problem 8.3?

Problem 8.5

Repeat Problem 8.4 for the following molecules, using models. Which are chiral? (Recall that in using the superposability criterion you may mentally rotate or otherwise move the mirror image in an attempt to make it superposable.)

(a) CH_4

(b)

(c) *cis*-1,2-dichloroethene

(d) *trans*-1,2-dichloroethene

(e) cyclohexane

8.5 ENANTIOMERS AND RACEMIC MIXTURES

The two forms of a chiral molecule are called **enantiomers**. Enantiomers have identical properties in all respects except in their interaction with plane-polarized light or in their interaction with another chiral object. For example, both enantiomers of sodium ammonium tartrate have the same melting point, density, solubility, color, and reactivity toward acids and bases. They differ, however, in the direction in which they rotate the plane of polarization of plane-polarized light. Both rotate the light to exactly the same extent (same angle), but one rotates the plane clockwise (+) while the other rotates the plane counterclockwise (−).

When enantiomers react with a pure chiral substance, the two enantiomers do not show the same reactivity toward the chiral substance. We shall examine this type of reaction later; however, we can explain such an interaction by the hand-glove analogy: a left-handed glove will fit (react with) a left hand better than it will a right hand.

A mixture of exactly equal amounts of two enantiomers is not optically active because the effect of each enantiomers on plane-polarized light is exactly offset by the other enantiomer. An equimolar mixture of a pair of enantiomers is called a **racemic mixture**. Racemic mixtures are optically inactive even though they contain chiral molecules.

8.6 CHIRALITY AND OPTICAL ACTIVITY: AN EXPLANATION

We may now ask why chiral molecules exhibit optical activity. Although this is a difficult question, we can answer it qualitatively as follows: Plane-polarized light interacts to some extent with *all* molecules. Such a light beam encounters an extremely large number of molecules as it passes through a sample. The direction and magnitude of the rotation that a single molecule produces will depend partly on its orientation when the light beam strikes it.

Let us examine an achiral molecule such as 2-bromopropane. In a sample of 2-bromopropane, all the molecules are identical. For every orientation of a given molecule relative to the light beam there is a high probability that the beam will strike another molecule in the opposite (mirror image) orientation, as shown in Figure 8.7. We must keep in mind that even the smallest sample will contain billions or trillions of molecules. The result of all these encounters between the plane-polarized light beam and countless identical achiral molecules in countless orientations is a zero net observed rotation of the plane of polarization.

When plane-polarized light passes through a sample of one enantiomer of 2-bromobutane (a chiral molecule), the situation is different. Molecules with the opposite (mirror-image) orientation are not present.

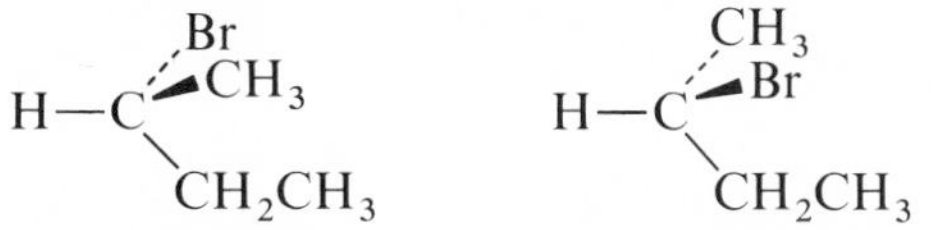

Br, CH_3, H—C, CH_2CH_3
(sample contains this enantiomer)

CH_3, Br, H—C, CH_2CH_3
(this enantiomer is not present)

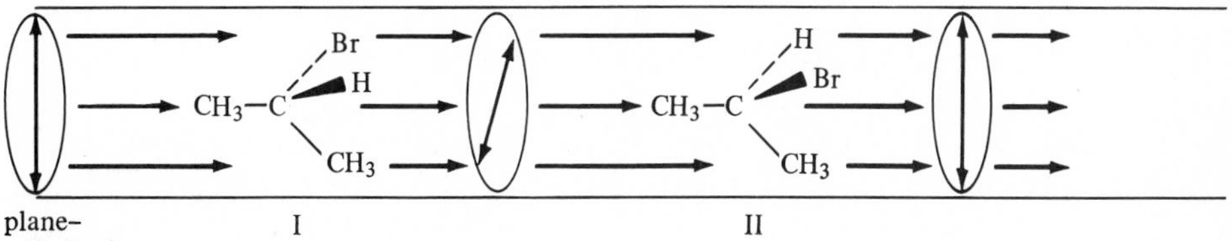

Br, H, CH_3—C, CH_3 — I

H, Br, CH_3—C, CH_3 — II

plane-polarized light

(II is obtained by rotating I.)

Figure 8.7. When a beam of plane-polarized light encounters an achiral molecule (2-bromopropane), it experiences a slight rotation that is exactly cancelled when it encounters an identical molecule (II) in its mirror-image orientation. The result is no observed rotation.

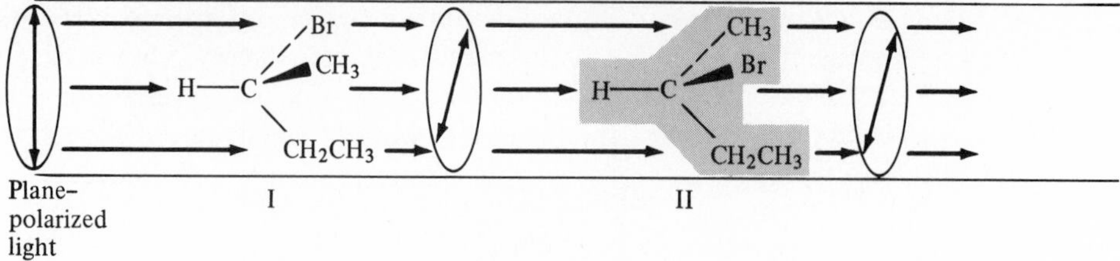

Figure 8.8. When a beam of plane-polarized light encounters a chiral molecule such as a pure enantiomer of 2-bromobutane, the light also experiences a slight rotation. In this case, however, there are no mirror-image molecules, II, present to cancel the rotation exactly. The net effect is an overall observed rotation.

Thus there is no exactly opposite orientation of a molecule that will exactly cancel the rotation of any other orientation. The result is an overall rotation of the plane of the polarized light beam. This situation is diagramed in Figure 8.8.

8.7 MEASUREMENT OF OPTICAL ACTIVITY

We measure optical rotation with a **polarimeter** (Fig. 8.9). A polarimeter uses the components shown in Figure 8.2.

Several factors affect the optical activity (rotation) of a given substance—its molecular structure, its concentration in the light beam, the length of the light beam through the sample, the temperature, and the wavelength of the light used. Ordinarily, the sodium D line (589.6 nm) and a temperature of 25°C are used. Concentration is given in grams of chiral solute per 100 ml of solvent, and the path length is 10 cm (1.0 decimeter).

Measurements of optical activity are given in terms of **specific rotation** ($[\alpha]_\lambda^t$):

$$[\alpha]_\lambda^t = \frac{100\alpha}{l \cdot c}$$

where α = observed rotation in degrees, t = temperature, λ = wavelength of light used, l = length of sample cell (length of the light path through the sample) in decimeters, and c = concentration of solute in grams per 100 milliliters of solvent. $[\alpha]_\lambda^t$, the specific rotation, is the rotation of a one decimeter solution that contains 100 g solute per 100 ml solvent; that is, specific rotation is the rotation of a pure sample one decimeter thick. The specific rotation is an exact physical property of a substance just as melting point, boiling point, and density are.

Any substance that rotates plane-polarized light in a clockwise direction is called **dextrorotatory**. Substances that rotate plane-polarized light in a counterclockwise direction are called **levorotatory**. We designate dextrorotatory substances by the prefix (+)- and levorotatory substances by the prefix (−)- placed before the name.

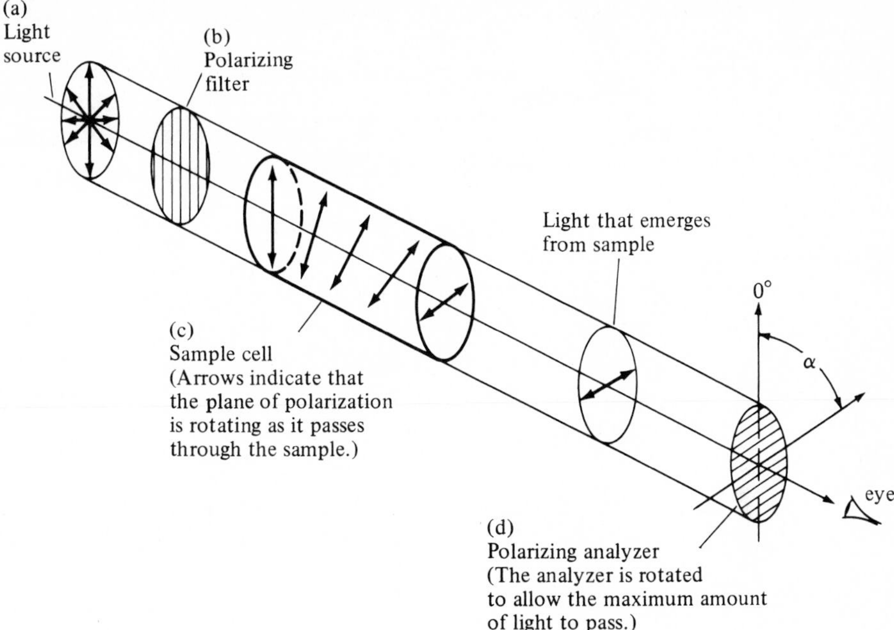

Figure 8.9. Schematic drawing of a polarimeter. Light from the light source, (a), is polarized by the polarizing filter, (b). The light that enters the sample cell, (c), is polarized (plane is vertical). The sample causes the plane of polarization to rotate in a clockwise direction (+) as seen by the observer. Light emerges from the sample with its plane rotated by $\alpha°$. The angle is measured by rotating the polarizing analyzer (d) until the light is brightest to the observer.

Enantiomers always have specific rotations that are equal in magnitude but opposite in sign.

Problem 8.6
Which of the following properties differ for each member of a pair of enantiomers?
(a) solubility in water,
(b) reaction with Cl_2,
(c) behavior in a polarimeter,
(d) taste (recall that the molecules that comprise taste buds are nearly all chiral)

Problem 8.7
Calculate the specific rotation, $[\alpha]_D^{25}$, for a substance if a solution of 2.5 g of the substance in 100 ml alcohol exhibits a rotation of $+3.7°$ at 25°C in a one decimeter cell (D line of sodium is used).

Problem 8.8
A substance has a specific rotation of $-12°$. What rotation would you observe for a solution of this substance in a sample tube 1.0 meter long if the concentration of the sample is 0.2 g per 100 ml of solvent?

Many types of structural features give rise to chirality in molecules. The most common occurs in a carbon atom (or other tetrahedral atom) that has four *different* groups attached to it. Such a carbon atom is called an **asymmetric** carbon atom or a **chiral** carbon atom. A molecule that has a chiral carbon atom has no plane of symmetry.*

Figure 8.10 shows some simple derivatives of methane. All but one have at least one plane of symmetry. Only the compound with four different atoms attached to carbon has no symmetry elements whatever, and that is why it is called an asymmetric carbon (asymmetric = no symmetry).

A compound that has a chiral carbon is not superposable on its mirror image (Figure 8.11).

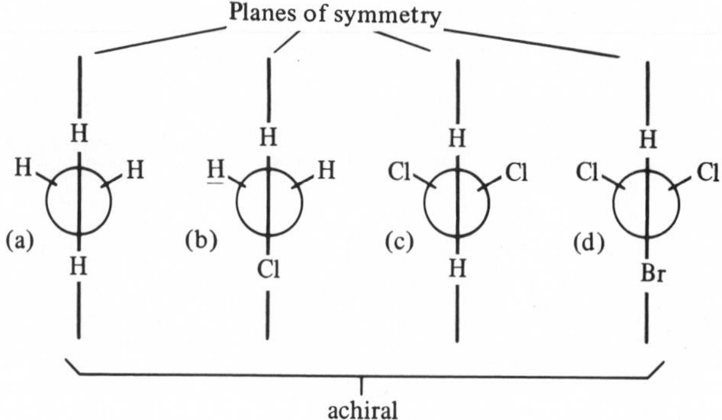

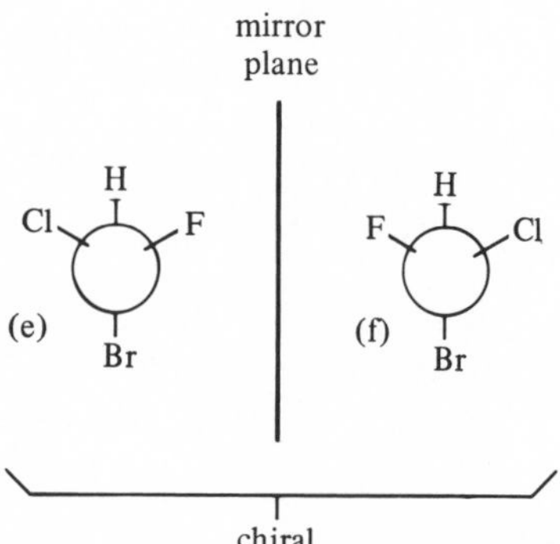

Figure 8.10. Compounds (a), (b), (c), and (d) all have at least one plane of symmetry vertical and perpendicular to the plane of the page. Compounds (e) and (f) have no plane of symmetry. They are related as an object and its nonsuperposable mirror image.

* We shall see that molecules that have two chiral atoms *may* have a plane of symmetry.

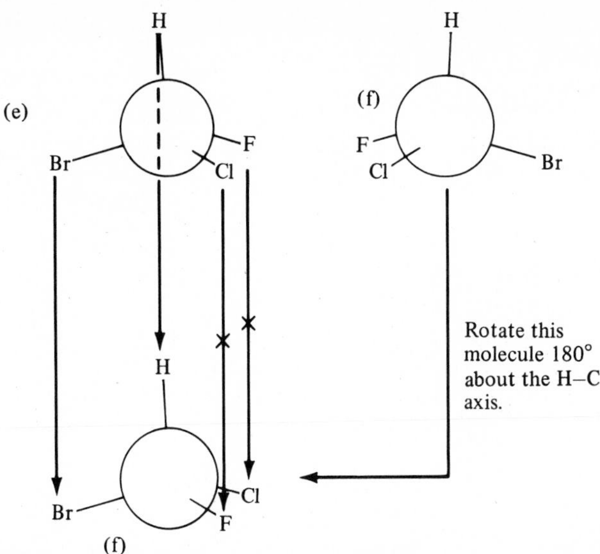

Figure 8.11. The pair of enantiomers, (e) and (f), of Figure 8.10 are not superposable; when we superpose the H and Br atoms, the Cl and F atoms do not coincide.

Problem 8.9
Which of the following compounds are chiral? Circle the chiral atom in each.

(a) $CH_3\underset{\displaystyle OH}{\underset{|}{C}}HCH_2CH_3$

(b) $CH_3\overset{O}{CH}{-}\underset{\displaystyle CH_3}{\underset{|}{C}}{-}CH_3$ (O bridging CH and C)

(c) cyclopentane ring bearing CH_3

(d) $C_6H_5{-}\underset{\displaystyle Cl}{\underset{|}{C}}H{-}C_6H_4{-}NO_2$

Problem 8.10
Draw structural formulas for the two enantiomers of each of the chiral molecules in Problem 8.9.

8.9 MOLECULES THAT CONTAIN TWO CHIRAL ATOMS: DIASTEREOMERS

There is one kind of stereoisomer that we have not yet discussed—stereoisomers that are not enantiomers.

If a molecule has two different chiral carbon atoms, then four stereoisomers are possible. Let us take the compound 2, 3-dibromopentane as an example:

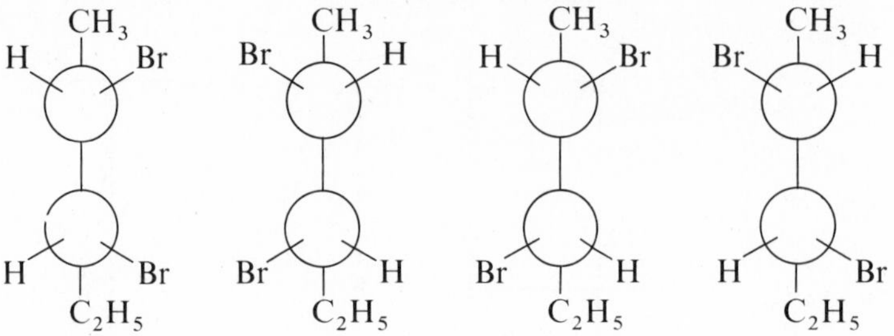

A B C D

a pair of enantiomers a pair of enantiomers

A and B are enantiomers because B is the nonsuperposable mirror image of A. C and D are also enantiomers. But what of the pairs AC, AD, BC, and BD? They are not enantiomeric pairs. Molecules that are stereoisomers of one another but not enantiomers of one another are called **diastereomers**.

Because diastereomers are not enantiomers, they have totally different sets of properties. Two diastereomers will have different melting points, boiling points, densities, and solubilities, and they will have different chemical reactivities toward most reagents.

8.10 MESO COMPOUNDS

A special kind of diastereomer exists when a molecule has an even number of chiral atoms and possesses a plane of symmetry. This occurs when one half of the molecule is the mirror image of the other. In this case the molecule is achiral because it has a plane of symmetry *even though it has chiral atoms.* An achiral molecule that has chiral atoms is called a **meso compound**. 2,3-Dibromobutane is an example of a meso compound:

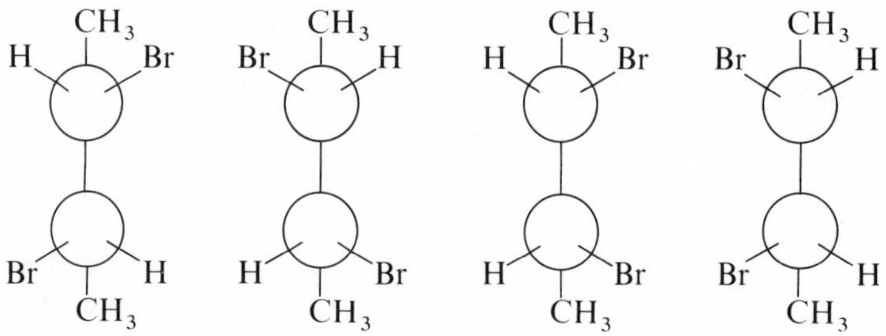

E F G H

enantiomers identical

Here again, E and F are enantiomers. G and H, however, are identical. We may superpose H on G if we rotate it 180° without lifting it out of the plane of the page.

Compound G is a meso compound because it has chiral atoms *and* a plane of symmetry.

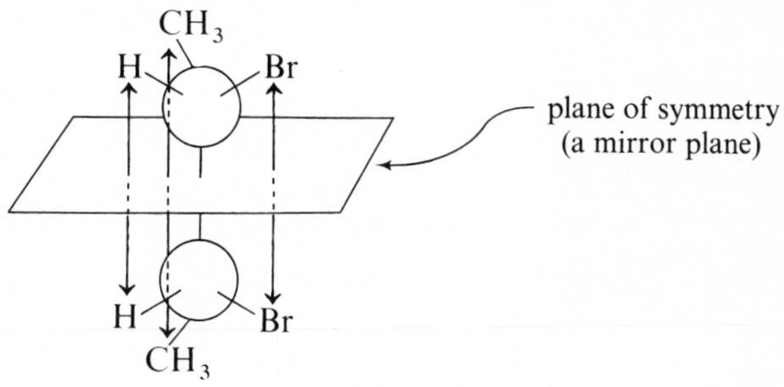

CH_3, H, Br, plane of symmetry (a mirror plane), H, Br, CH_3

G

Compound G is a diastereomer of A and B.

Problem 8.11

Draw structures of all the stereoisomers of each compound below. Label each enantiomeric pair, all diastereomers, and each meso compound.

(a) 2-bromo-3-chlorobutane

(b) 1,2-diphenyl-1,2-ethanediol

8.11 SUMMARY OF ISOMERISM

We may now classify any pair of isomers as one of three different kinds—constitutional isomers, enantiomers, or diastereomers. Figure 8.12 shows the relationships among these classes of isomers.

We must note that this definition of diastereomers includes stereoisomers that have no chirality. *Cis–trans* isomers are achiral diastereomers; they are not constitutional isomers because they do not differ in the sequential arrangement of their atoms. They are therefore stereoisomers. Because they are not enantiomers, we must classify them as diastereomers:

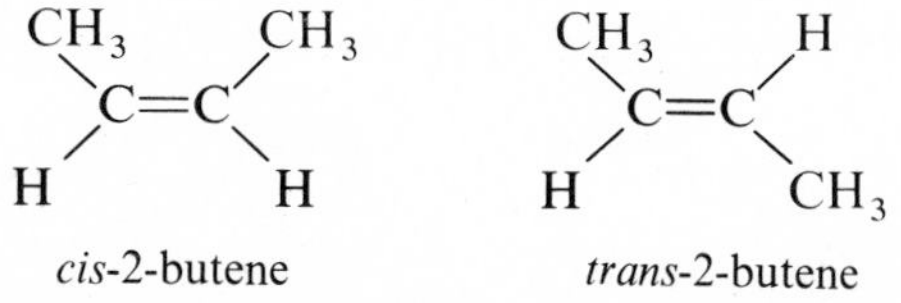

$CH_3(H)C{=}C(H)CH_3$ *cis*-2-butene

$CH_3(H)C{=}C(CH_3)H$ *trans*-2-butene

Diastereomers

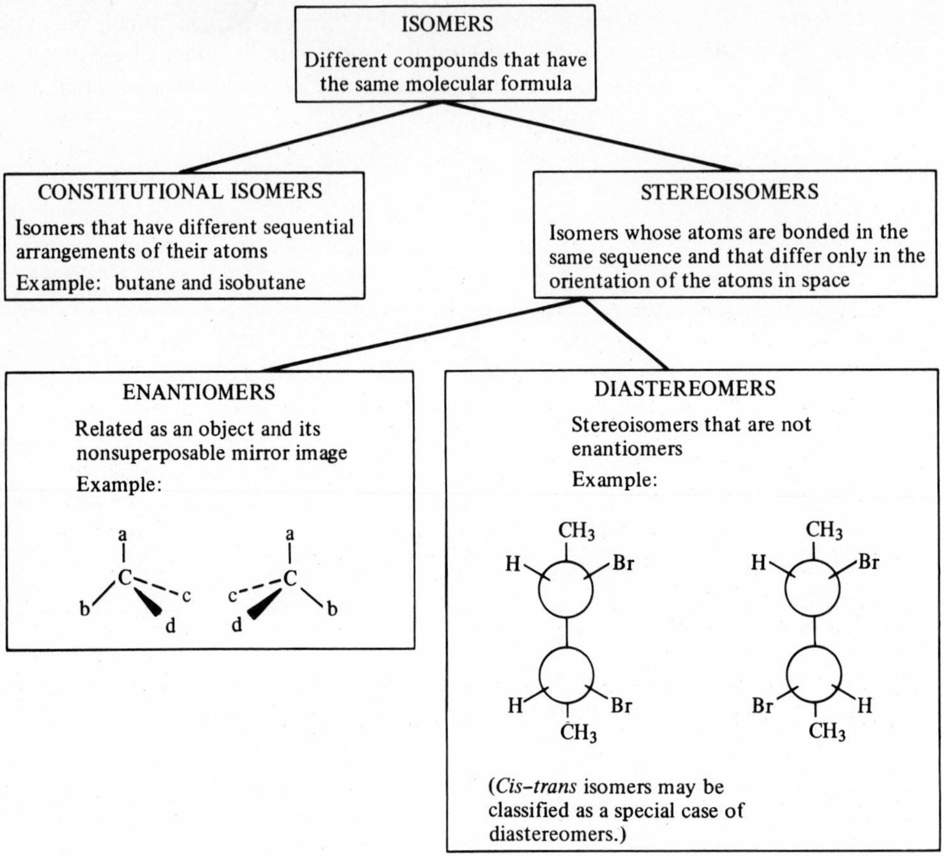

ISOMERS

Different compounds that have the same molecular formula

CONSTITUTIONAL ISOMERS

Isomers that have different sequential arrangements of their atoms

Example: butane and isobutane

STEREOISOMERS

Isomers whose atoms are bonded in the same sequence and that differ only in the orientation of the atoms in space

ENANTIOMERS

Related as an object and its nonsuperposable mirror image

Example:

DIASTEREOMERS

Stereoisomers that are not enantiomers

Example:

(*Cis–trans* isomers may be classified as a special case of diastereomers.)

Figure 8.12. Classification of isomers.

Problem 8.12

Classify each of the following pairs of structures as enantiomers, diastereomers, constitutional isomers, or identical. Which are meso compounds, and which are *cis–trans* isomers?

(a) and

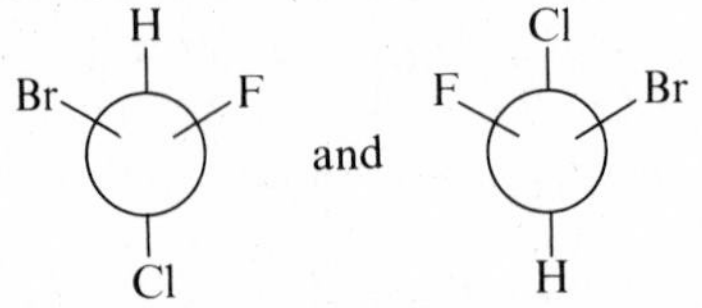

(b) and

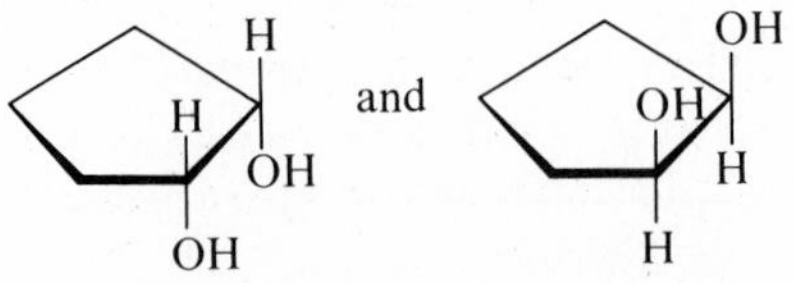

(c) 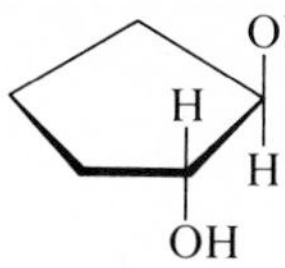 OH H H OH and H OH OH H

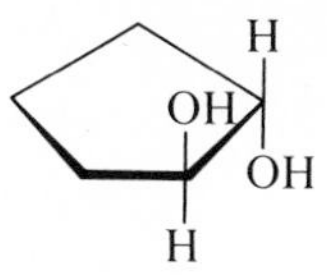

(d) 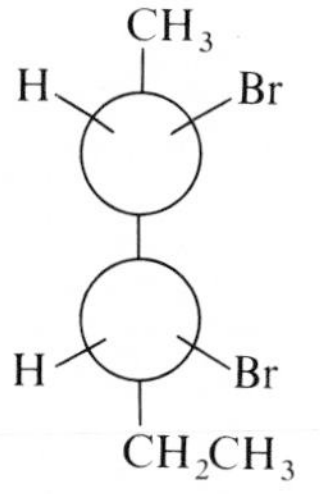 CH_3 H Br H Br CH_2CH_3 and CH_3 Br H Br H CH_2CH_3

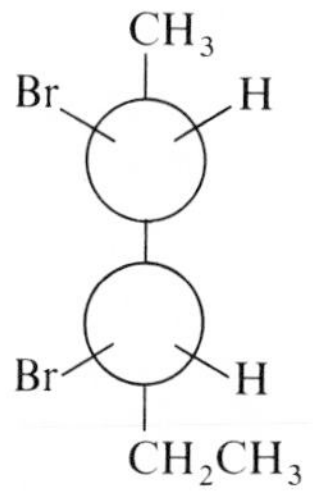

(e) 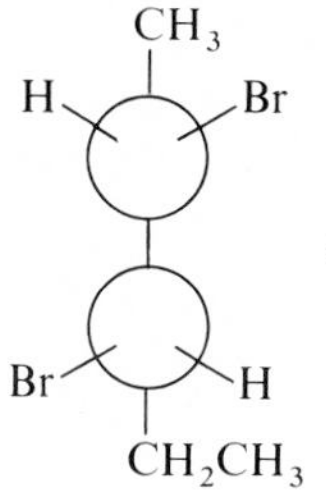 CH_3 H Br Br H CH_2CH_3 and CH_2CH_3 H Br Br H CH_3

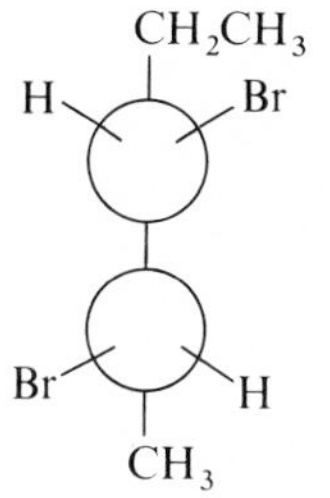

(f) 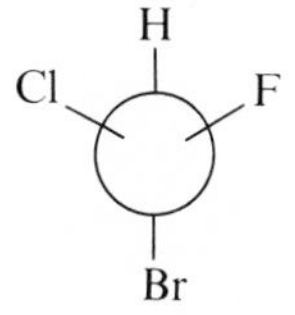 H Cl F Br and Cl Br H F

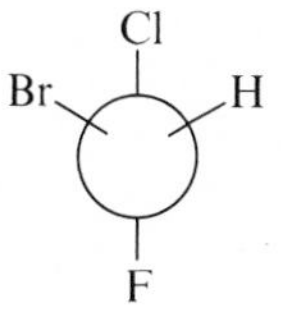

8.12 SEPARATION OF ENANTIOMERS: RESOLUTION

The process of separating a racemic mixture into its pure enantiomers is called **resolution**. We cannot separate enantiomers from one another by distillation or other physical methods because enantiomers always have identical physical properties. Pasteur separated sodium ammonium tartrate enantiomers by hand, but this method is rarely possible.

One method of resolution of enantiomers depends on the fact that diastereomers have different properties, and they can usually be separated by ordinary physical means—distillation or crystallization. To separate a pair of enantiomers, then, we must react them with a chiral reagent which converts them into diastereomers, separate the diastereomers, and then decompose the separate diastereomers into the original enantiomers.

We can separate a racemic carboxylic acid in this way by reacting it with an optically active base (a pure enantiomer). The salts formed are diastereomers:

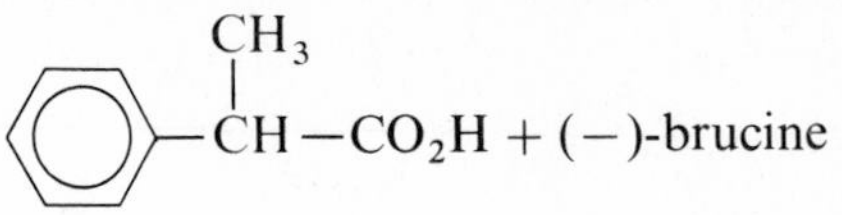

$$C_6H_5-\underset{|}{\overset{CH_3}{CH}}-CO_2H + (-)\text{-brucine}$$

(racemic mixture) (the pure levorotatory enantiomer occurs naturally; see text.)

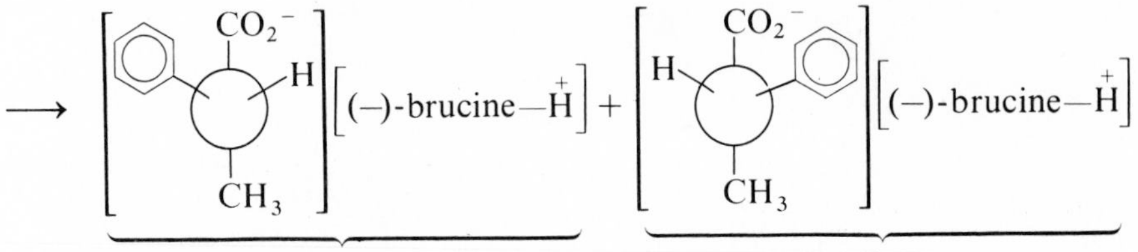

$$\longrightarrow \underbrace{[C_6H_5C(H)(CH_3)CO_2^-][(-)\text{-brucine}-\overset{+}{H}]}_{\text{salt A}} + \underbrace{[C_6H_5C(H)(CH_3)CO_2^-][(-)\text{-brucine}-\overset{+}{H}]}_{\text{salt B}}$$

Salt A and salt B are diastereomers and therefore have entirely different physical properties, including solubilities in water. By slowly cooling a saturated solution of the mixture of salt A and salt B, one of them—the less soluble one—will precipitate out of solution. This salt can then be filtered out. The original acid can be recovered by reacting the salt with a strong mineral acid:

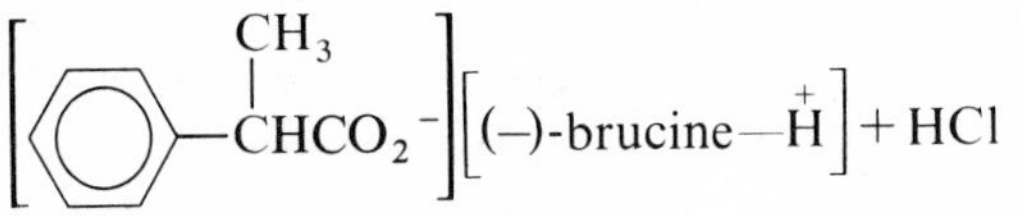

$$[C_6H_5-\overset{CH_3}{CH}CO_2^-][(-)\text{-brucine}-\overset{+}{H}] + HCl$$

pure diastereomer

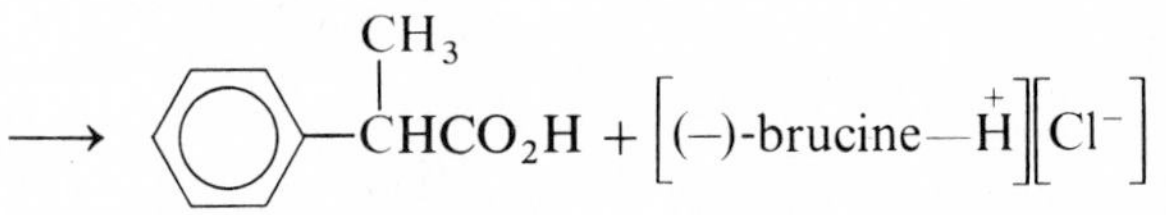

$$\longrightarrow C_6H_5-\overset{CH_3}{CH}CO_2H + [(-)\text{-brucine}-\overset{+}{H}][Cl^-]$$

pure enantiomer

Brucine has the structure

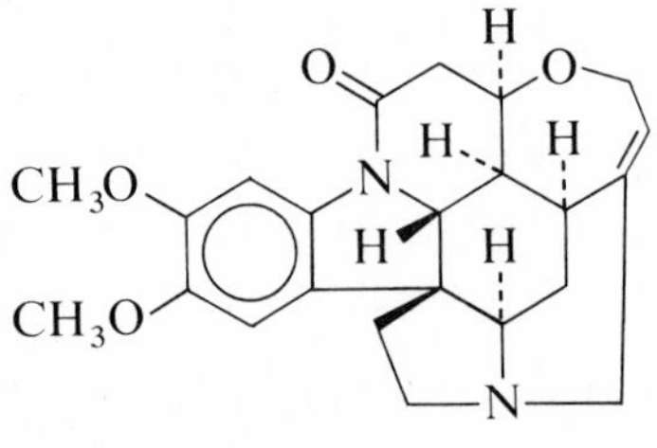

It occurs in Strychnos seeds (Strychnos nux-vomica) as the pure enantiomer ($[\alpha]_D^{25} = -127°$). We shall encounter many naturally occurring chiral plant and animal products throughout our study of organic chemistry.

Still another method of resolving enantiomers depends on the fact that enantiomers react at different rates with a pure chiral reagent. Pasteur, who discovered the method described above, also pioneered the method of selective reactivity. He discovered that the mold Penicillium glaucum preferentially decomposes (+)-tartaric acid when it is allowed to grow in the presence of racemic tartaric acid. In this way, Pasteur was able to recover pure, unreacted (−)-tartaric acid. This method is possible because the reaction caused by the mold involves chiral molecules in the mold.

Problem 8.13
Identify the chiral carbons in the structure of brucine.

Problem 8.14
Draw the structures of all the stereoisomers of tartaric acid,

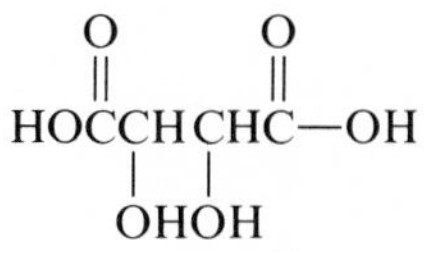

$$HO\overset{O}{\overset{\|}{C}}\underset{OH}{\underset{|}{C}H}\underset{OH}{\underset{|}{C}H}\overset{O}{\overset{\|}{C}}{-}OH$$

Label the enantiomeric pairs and meso form, if any.

8.13 SYNTHESIS OF CHIRAL MOLECULES

Reactions between achiral substances may produce chiral molecules. However, any such reaction that uses achiral reactants will lead to a racemic mixture. For example, the reaction of 1-butene with HCl yields 2-chlorobutane, which is chiral:

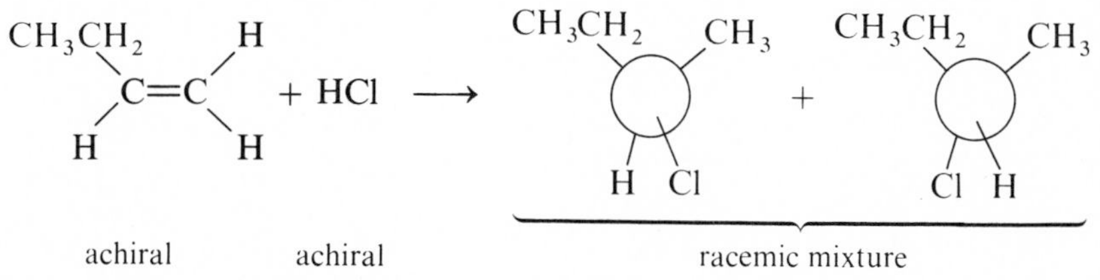

$$\underset{\text{achiral}}{CH_3CH_2(H)C{=}CH_2} + \underset{\text{achiral}}{HCl} \longrightarrow \underbrace{CH_3CH_2\text{–}CH(Cl)\text{–}CH_3 \;(\text{H, Cl}) + CH_3CH_2\text{–}CH(Cl)\text{–}CH_3 \;(\text{Cl, H})}_{\text{racemic mixture}}$$

We can understand why a racemic mixture results if we recall that the reaction proceeds through a carbocation intermediate (Sec. 5.6) and that carbocations are planar trigonal structures (the carbon is sp^2-hybridized):

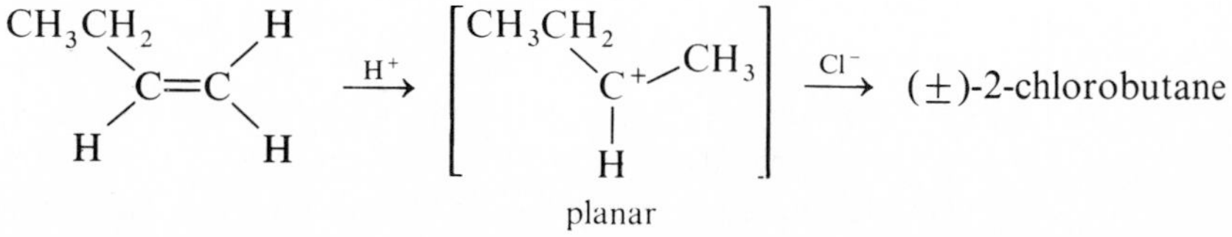

$$CH_3CH_2(H)C{=}CH_2 \xrightarrow{H^+} \underset{\text{planar}}{\left[CH_3CH_2\text{–}\overset{+}{C}H\text{–}CH_3 \right]} \xrightarrow{Cl^-} (\pm)\text{-2-chlorobutane}$$

The Cl^- can attack the carbocation from either side of the plane with equal ease, and therefore the two enantiomers are produced in equal amounts.

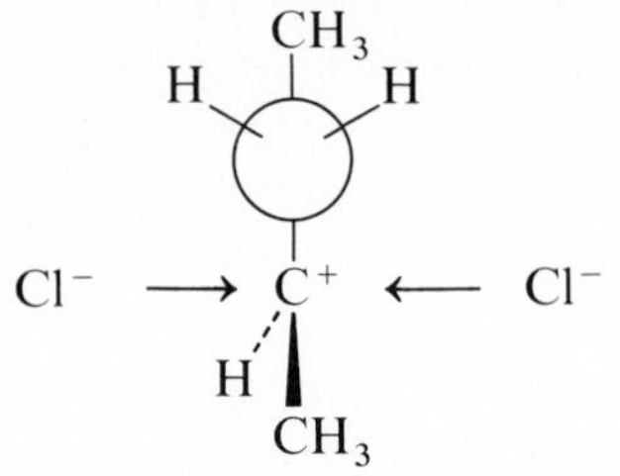

If one of the reagents is chiral, however, the reaction will not give equal amounts of products.

diastereomers

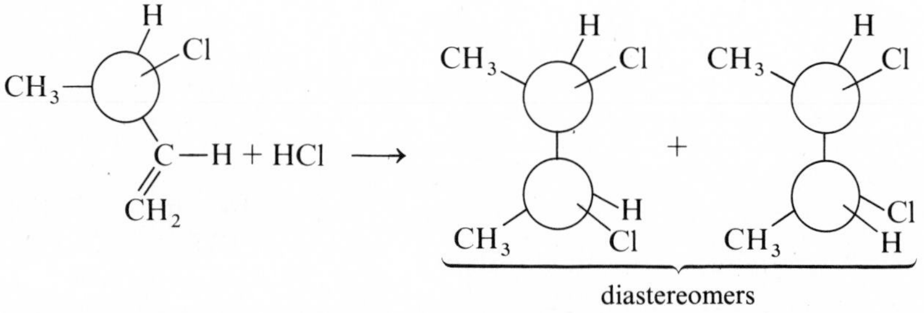

In this example the two products are diastereomers, and they are produced in unequal amounts. We explain the unequal amounts of diastereomeric products as follows: Although the intermediate carbocation is planar at the sp^2-hybridized carbon, the approaching Cl^- encounters different environments on the two sides of the plane. On one side, it finds a chlorine on the adjacent carbon; on the other, it finds a hydrogen. Even if the C_2—C_3 bond rotates, the approach will not be the same from the two sides. (This can be verified with models.) Hence, because approach from one side of the molecule is favored, one isomer will be formed at a faster rate, and that isomer will predominate.

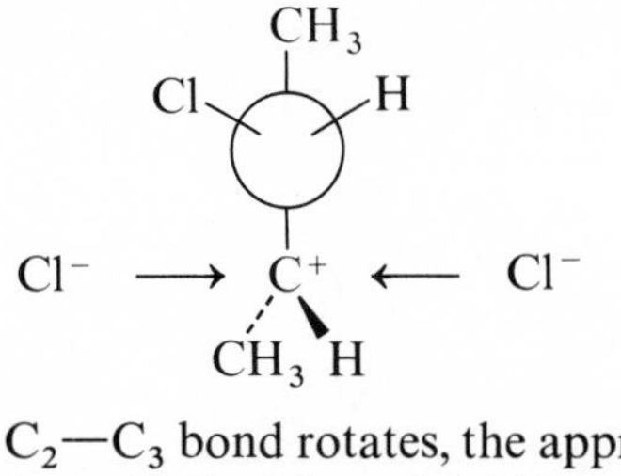

Most biological systems are chiral. For this reason, reactions in biological systems are usually highly specific in the stereoisomers that they produce or consume.

Problem 8.15

Draw the structural formulas of the products of each of the reactions below, and indicate whether each reaction would yield achiral products, a racemic mixture, a mixture of unequal amounts of products, or a single chiral product.

(a) $CH_3CH{=}CH_2 + HCl \longrightarrow$

(b) $CH_3CH{=}CHCH_3 + HCl \longrightarrow$

(c) [Newman projection: front carbon bearing CH_3 (up), CH_3CH_2 (left), CH_2OH (right); rear H] $+ CH_3\overset{O}{\overset{\|}{C}}{-}OH \xrightarrow{H^+}$ (ester)

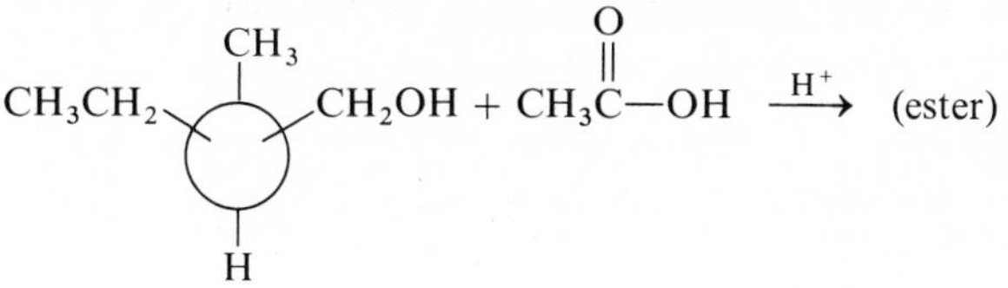

(d) [Newman projection: front carbon bearing CH_3 (up), CH_3CH_2 (left), CH_2OH (right); rear H] $+ (\pm){-}CH_3\overset{F}{\overset{|}{C}}H\overset{O}{\overset{\|}{C}}{-}OH \xrightarrow{H^+}$ (ester)

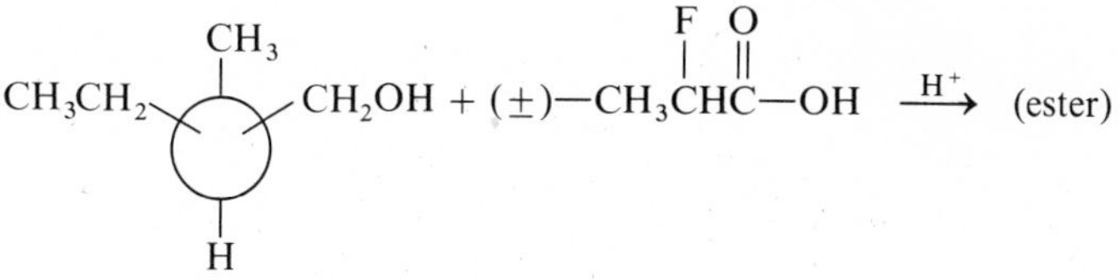

(e) $(\pm){-}CH_3CH_2\overset{CH_3}{\overset{|}{C}}HCH_2OH + (\pm){-}CH_3\overset{F}{\overset{|}{C}}H\overset{O}{\overset{\|}{C}}{-}OH \xrightarrow{H^+}$ (ester)

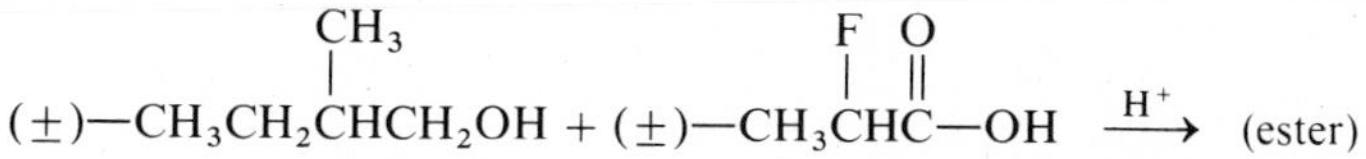

8.14 *R-S* NOMENCLATURE FOR REPRESENTING CONFIGURATIONS

How are we to describe the configuration of a particular chiral molecule without drawing its structure? The *Cahn–Ingold–Prelog* nomenclature system, named after its originators, does just this. In their system, we arrange the four groups attached to the chiral carbon in descending order of precedence (a, b, c, d) by applying the rules given below. We then view the chiral center with the lowest-ranking group (d) on the side opposite the viewer (group ranks are shown in parentheses):

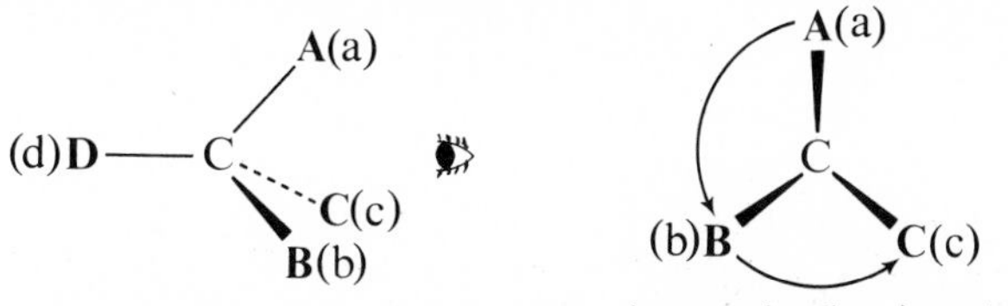

A(a)

(d)**D**—C

C(c)

B(b)

A(a)

C

(b)**B** C(c)

(as seen by the viewer)

counterclockwise = *S*

In the above example, **A** has the highest rank (a); **B** has the next highest (b); **C** has the next (c); and **D** has the lowest rank (d). We now go from high to low rank ($a \rightarrow b \rightarrow c$). To do so we move in a counterclockwise direction, and therefore we assign to this structure the ***S*** configuration (*S* = sinister = left). In the structure below,

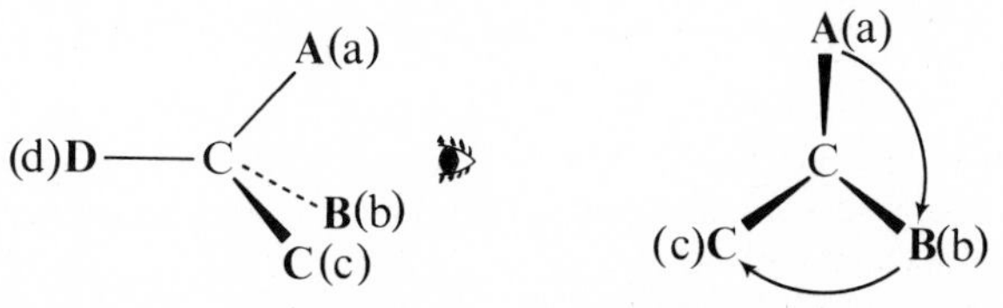

A(a)

(d)**D**—C

B(b)

C(c)

A(a)

C

(c)**C** **B**(b)

(as seen by the viewer)

clockwise = *R*

we go from highest to lowest rank by moving in a clockwise direction. We therefore assign to this structure the ***R*** configuration (R = rectus = right).

To establish group ranks we use the following rules.

1. Of the atoms attached directly to the chiral carbon atom, the one with the *highest atomic number has the highest rank*. For example, I > Br > Cl > F > O > N > C > H.
2. If two atoms attached to the chiral carbon atom are the same, we determine rank by going to the next atom away from the chiral carbon atom. For example,

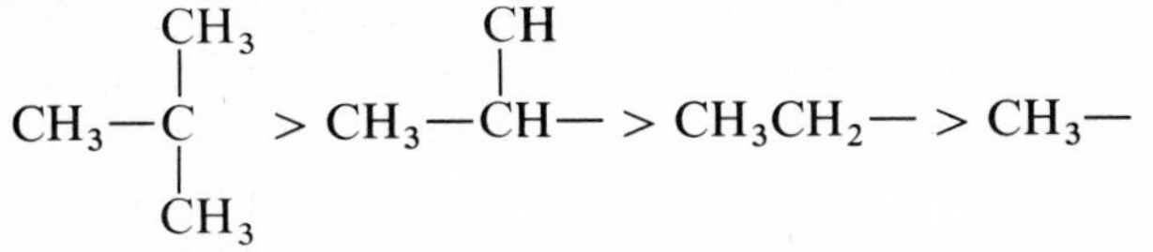

$$(CH_3)_3C{-} > CH_3{-}\underset{}{\overset{CH}{CH}}{-} > CH_3CH_2{-} > CH_3{-}$$

   Ethyl has a higher rank than methyl because the ethyl group has CHH attached to the first carbon, whereas the methyl carbon has only hydrogens (HHH), and C has precedence over H. Isopropyl is of higher rank than ethyl because it has two carbons attached to the first carbon and ethyl has only one. If there is no difference at the second atom in the chain, we go to the next and so forth to the first point of difference.
3. We treat a double bond as though each atom of the double bond were bonded to two atoms:

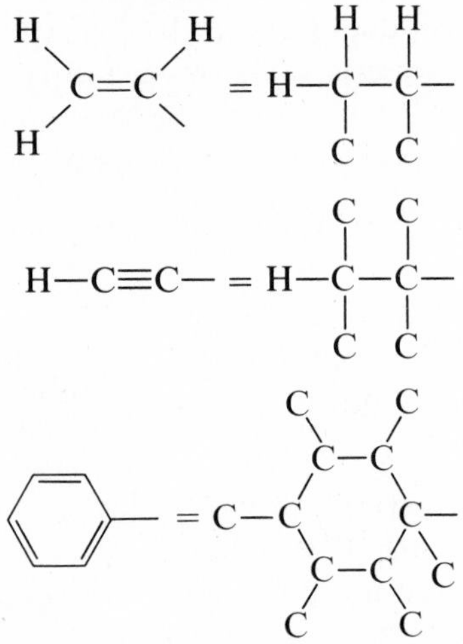

$$H_2C{=}CH{-} \;=\; H{-}\overset{H}{\underset{C}{C}}{-}\overset{H}{\underset{C}{C}}{-}$$

$$H{-}C{\equiv}C{-} \;=\; H{-}\overset{C}{\underset{C}{C}}{-}\overset{C}{\underset{C}{C}}{-}$$

$$\text{Phenyl}{-} \;=\; C{-}C_6 \text{ ring, each ring carbon bearing a duplicate C}$$

   Thus

$$\text{Phenyl}{-} > HC{\equiv}C{-} > CH_3{-}\overset{CH_3}{\underset{CH_3}{C}}{-}$$

$$> CH_2{=}CH{-} > CH_3{-}\overset{CH_3}{CH}{-} > CH_3CH_2{-}$$

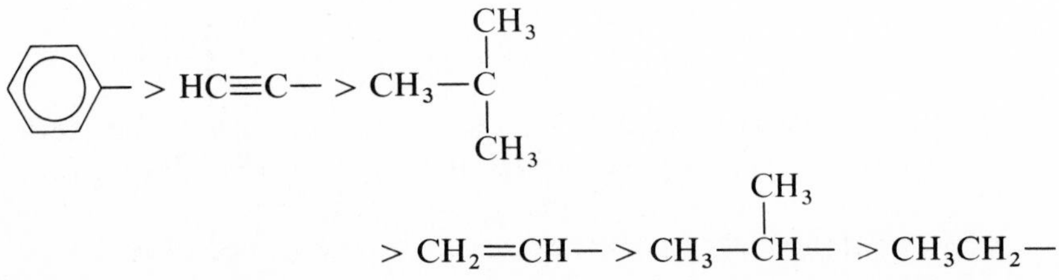

Additional rules exist for more complicated structures, but we shall not need them.

Let us take as an example the structure

OH | C---H, C=O, OH (phenyl attached to C)

The four atoms attached to the chiral carbon are H, O of the hydroxyl group, C of the phenyl group and C of the carboxyl group. We assign the lowest rank to H and the highest to O. We decide between the two carbons by considering the next atom. Phenyl has other carbons, carboxyl has oxygens. We therefore assign the priorities

$$-OH > -CO_2H > C_6H_5- > H.$$

The molecule is already shown with the lowest-ranked group, H, behind the page. To proceed in the sequence OH → CO_2H → C_6H_5

OH | C---H, C=O, OH (phenyl attached to C)

we move in a clockwise direction; therefore the molecule has the *R* configuration.

We use the letters *R* and *S* in front of the name, separated from it by a hyphen.

Problem 8.16

Tell whether each of the structural formulas below has the *R* or *S* configuration.

(a) CH_3CH_2—C (CH_3, OH, H)

(b) C_6H_5—C (CH_3, H, CN)

(c) HC≡C—C (H, Cl, CH=CH_2)

Before we leave this topic, we must emphasize that the *R–S* designation does not tell us how a molecule will rotate plane-polarized light. The *R–S* system is strictly a nomenclature system that allows us to describe with the symbol *R* or *S* the configuration of a structure as drawn.

As long as bonds to the chiral atom are not broken, any chemical reaction of a chiral substance will give a product that has the same configuration as the starting compound. An example is the reaction of an optically active acid with methyl alcohol to produce the ester:

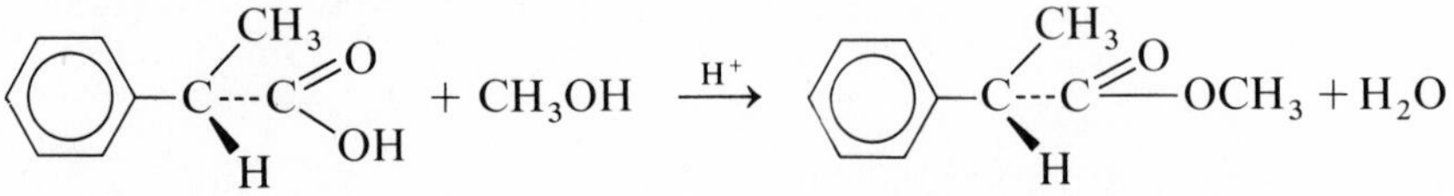

$$C_6H_5\text{—}C(CH_3)(H)\text{—}C(=O)OH + CH_3OH \xrightarrow{H^+} C_6H_5\text{—}C(CH_3)(H)\text{—}C(=O)\text{—}OCH_3 + H_2O$$

(*S* configuration) (*S* configuration)

These two compounds have the same *relative configuration.*

Until recently, however, it was impossible to know which enantiomer of a pair had a particular *R–S* configuration. In 1951, a method was devised for determining the **absolute configuration** of a substance using an X-ray diffraction technique. By this method, the following enantiomers with the indicated optical rotations were shown to have the configurations pictured here.

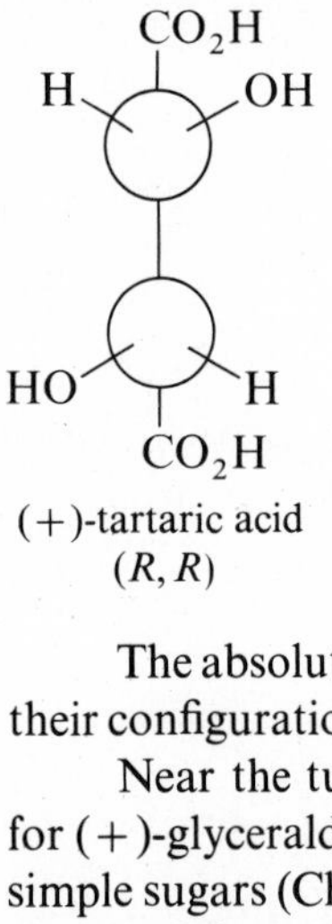

(+)-tartaric acid (*R*, *R*) — Newman projection: front carbon CO_2H, H, OH; rear carbon HO, H, CO_2H

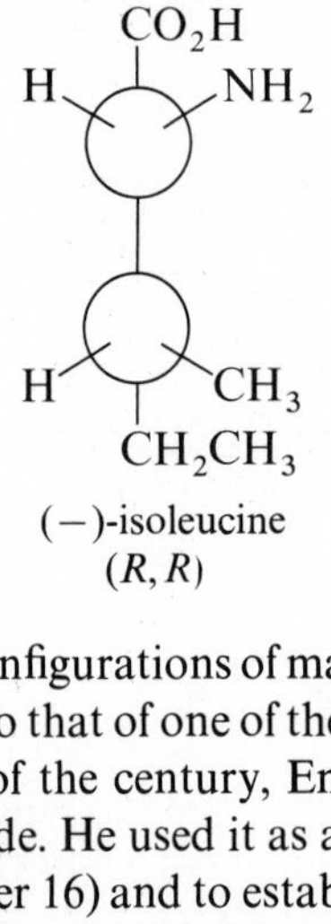

(−)-isoleucine (*R*, *R*) — Newman projection: front carbon CO_2H, H, NH_2; rear carbon H, CH_3, CH_2CH_3

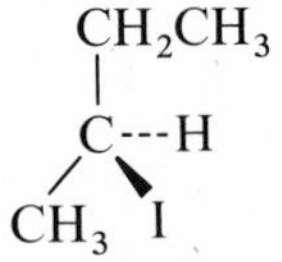

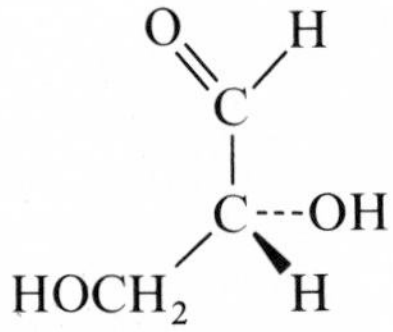

(+)-2-iodobutane (*S*) — CH_2CH_3, C---H, CH_3, I

(+)-glyceraldehyde (*R*) — O=C—H, C---OH, $HOCH_2$, H

The absolute configurations of many molecules were then determined by relating their configurations to that of one of the compounds of known absolute configuration.

Near the turn of the century, Emil Fischer suggested the above configuration for (+)-glyceraldehyde. He used it as a starting compound to synthesize many of the simple sugars (Chapter 16) and to establish their relative configurations. His arbitrary choice of the correct structure of (+)-glyceraldehyde was fortunate. Fischer gave it the designation D and called it D-(+)-glyceraldehyde. D stands for dextrorotatory. The enantiomer is designated L-(−)-glyceraldehyde. L stands for levorotatory. We shall examine the D–L designations in Chapter 16.

All proteins are made up of various combinations of twenty amino acids (Chapter 17). Amino acids have the general structure

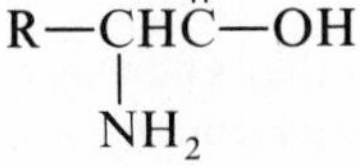

$$R\text{—}\underset{NH_2}{\underset{|}{C}}H\text{—}\overset{O}{\overset{\|}{C}}\text{—}OH$$

where *R* may contain various combinations of alkyl, aryl, and other functional groups. It is interesting that all naturally occurring amino acids obtained from proteins (except glycine, which is achiral) have the same absolute configuration at the chiral carbon. Note also that amino acids are related* to (−)-glyceraldehyde and not to (+)-glyceraldehyde, as are the naturally occurring sugars.

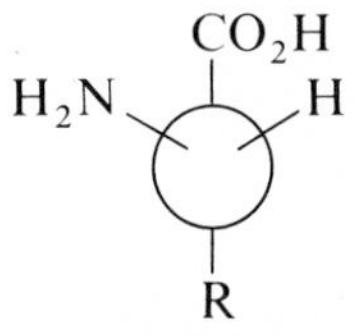

CO_2H
H_2N H
R

generalized amino acid (*S*)

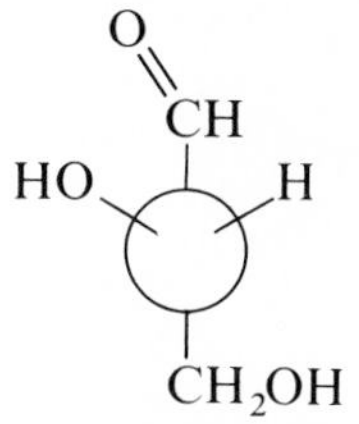

O
CH
HO H
CH_2OH

(−)-glyceraldehyde (*S*)

Table 8.1 lists a few amino acids with their specific rotations.

Problem 8.17
Judging by the structures shown in this section, is there any relationship between the direction of rotation of plane-polarized light by a sample and its *R*—*S* configuration?

Table 8.1 Some Natural Amino Acids

Name	Structure	$[\alpha]_D^{25}$
Alanine	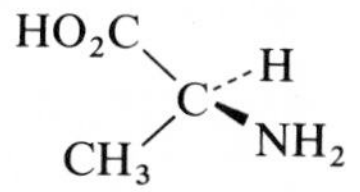 HO_2C, H, NH_2, CH_3 on C	+8.5°
Leucine	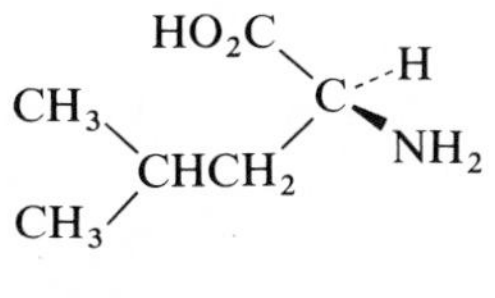 HO_2C, H, NH_2 on C; $(CH_3)_2CHCH_2$	−10.8°
Proline	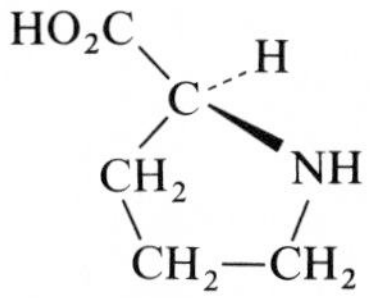 HO_2C, H on C; ring C–CH_2–CH_2–CH_2–NH	−85.0°
Cysteine	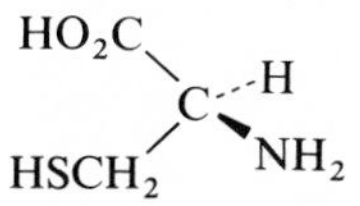 HO_2C, H, NH_2, $HSCH_2$ on C	+6.5

* We see how they are related if we draw the carbon skeleton vertically and if we substitute the amino group for the hydroxyl group on the chiral carbon atom.

Problem 8.18
Give the R—S designation of all the chiral atoms whose structures are shown in Sections 8.9 and 8.10.

Problem 8.19
Give the R—S designation of the chiral atoms in the structures of Problem 8.11.

8.16 CHIRAL MOLECULES THAT CONTAIN NO CHIRAL ATOMS

Many molecules are chiral even though they have no chiral atoms. These molecules are chiral because their atoms are linked into a chiral arrangement. Some examples are shown.

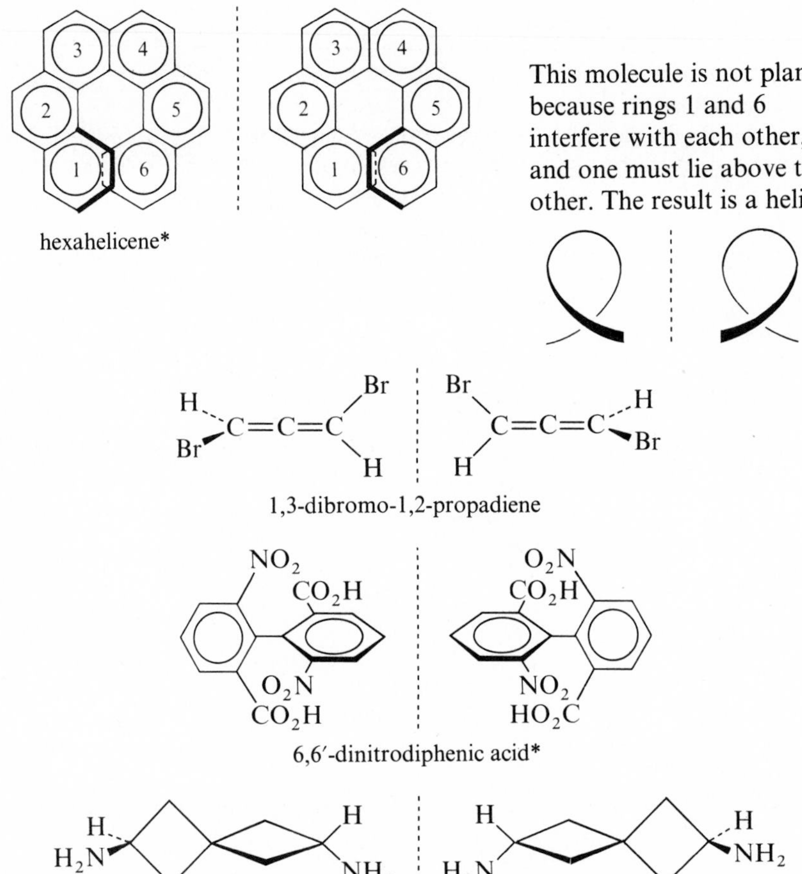

* The student is not expected to know these names.

Problem 8.20
Construct models of the above structures and verify that the model and its mirror image are not superposable.

8.17 MOLECULES THAT POSSESS CHIRAL ATOMS OTHER THAN CARBON

Amines that have three different groups attached to nitrogen are chiral. All known examples of such amines are optically inactive, however, because the two enantiomeric forms interconvert rapidly at room temperature. The energy barrier in this interconversion is only about 5–10 kcal/mole; this interconversion is about as rapid as the interconversion of conformers.

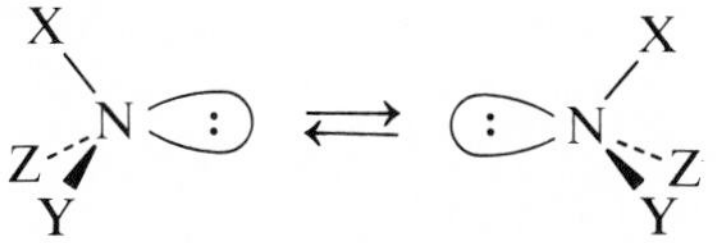

Analogous phosphorus compounds are separable, however, because the energy barrier between the two forms is about 30 kcal/mole.

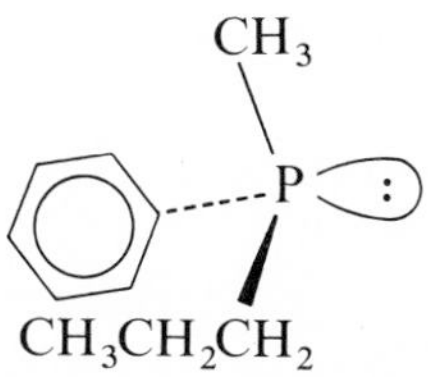

$[\alpha]_D^{20} = +16.8°$

Chiral ammonium salts, in contrast to amines, do not interconvert, and they exhibit optical activity.

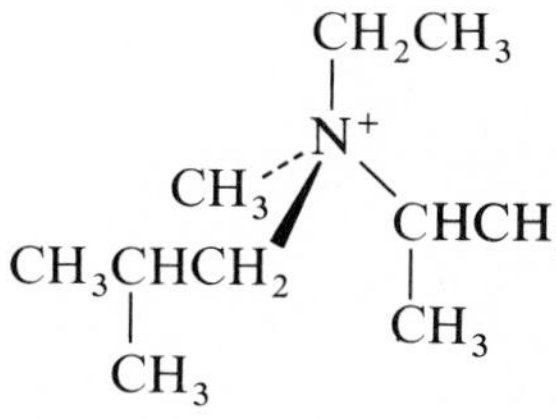

Problem 8.21
Certain metal ions (Ni^{2+}, Pt^{2+}, Pd^{2+}, Cu^{2+}, Au^{3+}) form square planar complexes. An example is

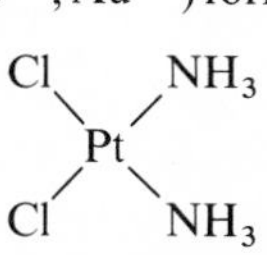

Could such structures exhibit optical isomerism under any circumstances? Explain.

8.18 *E-Z* NOMENCLATURE FOR *Cis-Trans* ISOMERS

Some alkenes cannot be designated *cis* or *trans* because no two groups are alike. An example is the alkene

$$\begin{array}{ccc} CH_3 & & CH_2CH_3 \\ & C{=}C & \\ H & & CH(CH_3)_2 \end{array}$$

We name such stereoisomers as well as all other *cis–trans* isomers by an application of the Cahn–Ingold–Prelog *R–S* system (Sec. 8.14). We assign group ranks to the two groups bonded to each carbon atom of the double bond. In the above example, the assignments are as shown below.

$$\begin{array}{ccc} (a)\ CH_3 & & CH_2CH_3\ (b) \\ \text{-----------} & C{=}C & \text{-----------} \\ (b)\ H & & CH(CH_3)_2\ (a) \end{array}$$

(a) groups are on opposite sides.

If the two groups of higher rank are on the same side of a line that goes through the double bond (dashed line above), the isomer is labeled *Z* (from the German *zusammen*, together). If the higher-ranked groups are on opposite sides, the isomer is labeled *E* (from the German *entgegen*, opposite). The above example is designated *E* because the methyl and isopropyl groups are on opposite sides. Some examples will illustrate the *E–Z* method of nomenclature.

$$\begin{array}{ccc} (a)\ Cl & & CH_2CH_2CO_2H\ (a) \\ & C{=}C & \\ (b)\ H & & H\ (b) \end{array}$$

(*Z*)-5-chloro-4-pentenoic acid

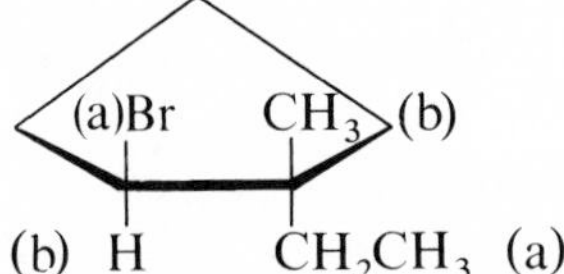

E-2-bromo-1-ethyl-1-methylcyclopentane

Problem 8.22

(a) Give the systematic name, including the *E—Z* designation, of the compound

$$\begin{array}{ccc} CH_3 & & H \\ & C{-}C & \\ H & & CH_3 \end{array}$$

(b) Give the structural formula of *Z*-3-methyl-2-pentene.

NEW TERMS

Absolute configuration The actual orientation in space of groups bonded to the chiral atom of a particular stereoisomer (8.15).

Achiral A term that means not chiral; that is, an object that possesses a plane of symmetry (8.3).

Asymmetric atom A tetrahedral atom that is bonded to four different groups or atoms; also called a **chiral** atom (8.8).

Chiral A term that describes objects that can exist in two forms related as a right and left hand. Chiral objects possess no plane of symmetry (8.3).

Dextrorotatory A term that describes an optically active substance that rotates the plane of polarization of plane-polarized light in a clockwise direction (8.7).

Diastereomers Stereoisomers that are not enantiomers of one another (see Fig. 8.12) (8.9).

Enantiomers Stereoisomers that are related as an object and its nonsuperposable mirror image. The right- and left-handed forms of a chiral molecule are enantiomers (8.5).

E–Z nomenclature A nomenclature system used to describe alkene isomers that cannot be called *cis* or *trans* because no two groups attached to the C=C are alike. The *E–Z* system also applies to ordinary cis-trans isomers.

Levorotatory A term that describes an optically active substance that rotates the plane of polarization of plane-polarized light in a counterclockwise direction (8.7).

Meso compound A molecule that possesses an even number of chiral atoms and also possesses a plane of symmetry. Meso compounds are achiral even though they have chiral atoms (8.10).

Optical activity Rotation of the plane of polarization of plane-polarized light as it passes through an **optically active** substance (8.3).

Plane of symmetry A plane that divides an object into two halves that are mirror images of each other (8.4).

Plane-polarized light Light whose electromagnetic waves vibrate in a single plane.

Polarimeter An instrument used to measure the angle of rotation of plane-polarized light by an optically active substance (8.7).

Racemic mixture A mixture of exactly equal amounts of two enantiomers. A racemic mixture is optically inactive (8.5).

Resolution The process of separating a racemic mixture into its pure enantiomers (8.12).

R–S nomenclature A nomenclature system used to describe the spatial configuration of a particular chiral molecule (8.14).

Specific rotation $[\alpha]_\lambda^t = 100\alpha/l \cdot c$ where $[\alpha]_\lambda^t$ is the specific rotation, t is the temperature, λ is the wavelength of polarized light used, α is the measured rotation, l is the path length of light through the sample and c is the concentration of sample in grams per 100 ml solvent (8.7).

Superposable Two objects are superposable if we can imagine placing one object upon the other so that they occupy the same space and so that every part of one object coincides with the corresponding part of the other object (8.4).

ADDITIONAL PROBLEMS

8.23 Which of the following compounds are capable of existing as a pair of enantiomers?

[a] CCl_3F, [b] CCl_2F_2,
[c] CCl_2FBr, (d) $CHClFBr$,

(e) 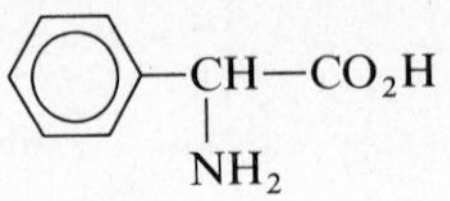 C_6H_5—CH—CO_2H with NH_2 on CH

(f) 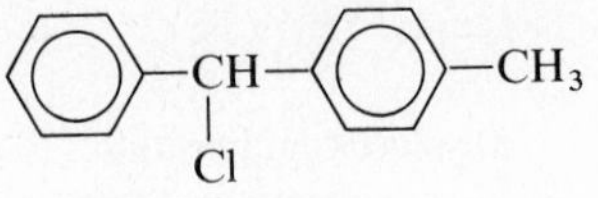 C_6H_5—CH(Cl)—C_6H_4—CH_3

8.24 Classify each of the following pairs of structures as enantiomers, diastereomers, constitutional isomers, conformers, the same molecule, or different, nonisomeric compounds.

[a] H, C---OH, CH_3, CH_2CH_3 and H, C---CH_2CH_3, HO, CH_3

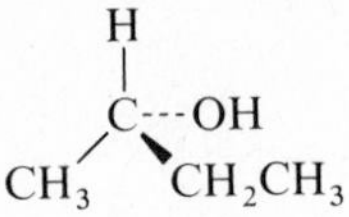

(b) H, C---OH, CH_3, CH_2CH_3 and H, C---OH, CH_3CH_2, CH_3

[c] H, C---OH, CH_3, CH_2CH_3 and CH_3, C---OH, CH_3CH_2, H

(d) H, C---H, CH_3, Cl and H, C---CH_3, Cl, H

[e] Newman projections: front carbon Cl, H, OH; rear carbon H, OH, CH_3 and front carbon Cl, HO, H; rear carbon HO, H, CH_3

(f) Newman projections: front carbon CH_3, H, Cl; rear carbon H, Cl, CH_3 and front carbon CH_3, Cl, H; rear carbon Cl, H, CH_3

[g] Newman projections: front carbon CH_3, H, Cl; rear carbon H, Cl, CH_3 and front carbon Cl, H, CH_3; rear carbon H, Cl, CH_3

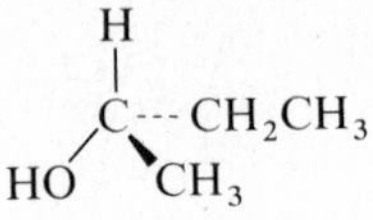

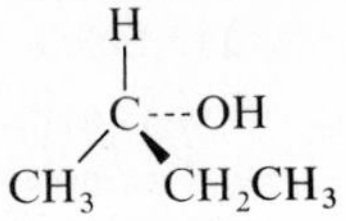

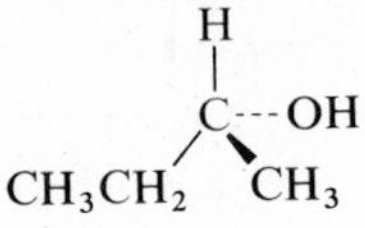

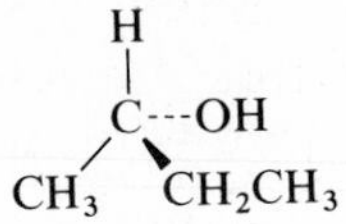

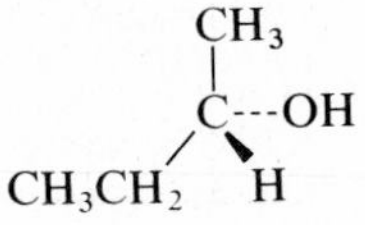

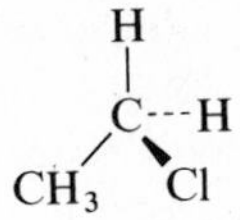

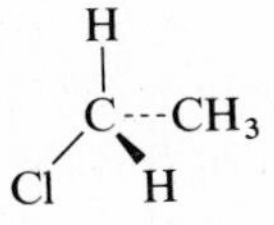

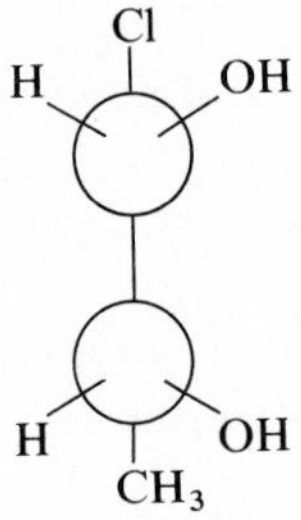

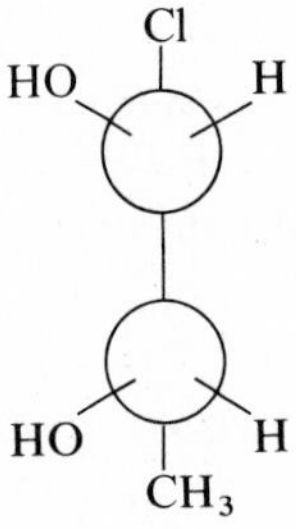

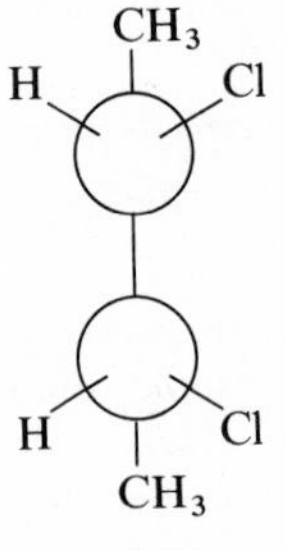

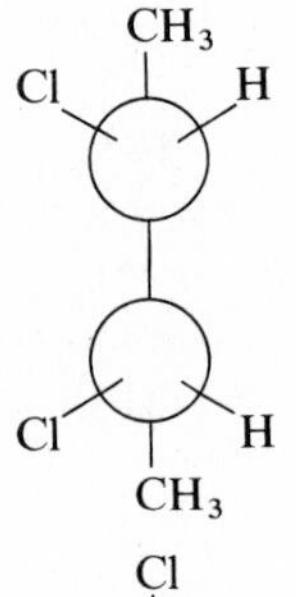

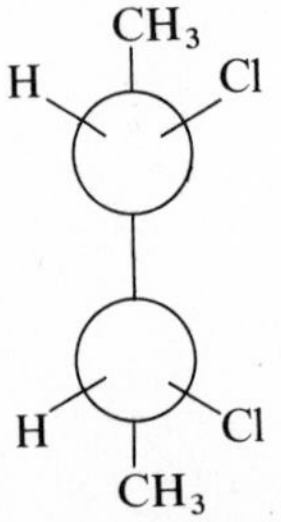

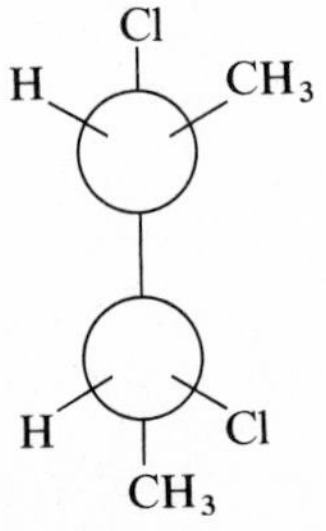

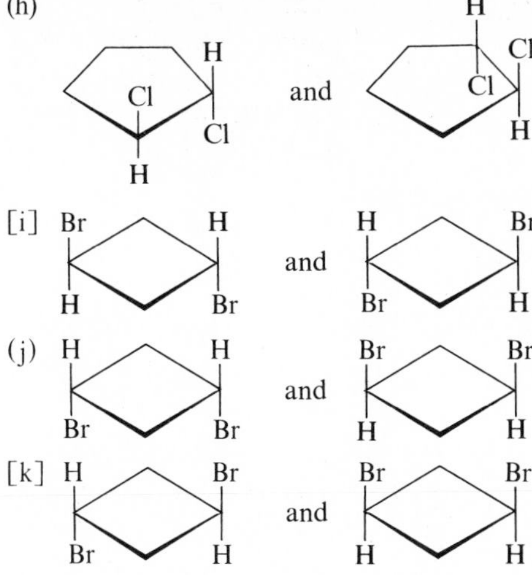

[8.25] Explain the difference between the terms chirality and optical activity.

[8.26] Of the structures in Problem 8.24, identify all those that are (a) chiral, (b) achiral, (c) meso compounds, (d) *cis–trans* isomers.

8.27 Two isomeric alcohols have the formula C_4H_9OH. They have identical melting and boiling points and densities. Give the structural formulas of the two alcohols.

[8.28] If the alcohols of Problem 8.27 had different properties, what could their structural formulas be?

8.29 Calculate the rotation using the sodium D line at 25°C of a solution of 5.0 g alanine (Table 8.1) dissolved in one liter of water, and using a 10 cm cell.

8.30 Give the *R–S* designation of the chiral atoms in:

[a] H; F; CH_3; Cl

(b) CH_3; CH_3CH_2; $CH{=}CH_2$; F

[c] CH_3; Cl; H

(d) CH_3; H; OH; HO; H; CH_3

[8.31] Identify the chiral atoms in the structure below. Label each chiral atom *R* or *S*.

OH H O OH H H

8.32 Give the *E–Z* designation of the following compounds:

[a] (structure: cyclohexane with H, CH_3, CH_2CH_3, CH_3) [b] (structure: CH_3, CH_3, C=C, H, cyclohexane) (c) Cl, H, C=C, Br, CH_3

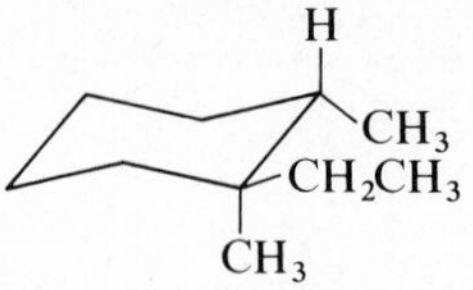

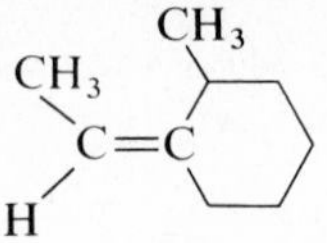

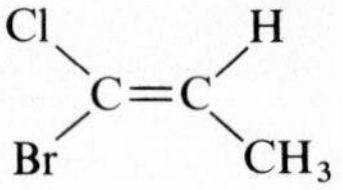

[8.33] Give the *E–Z* designations of the isomers in Problem 8.24 to which *E–Z* designations apply.

[8.34] Draw the structural formulas of all the stereoisomers of 2,3-dihydroxybutane and 2,3-dihydroxypentane. Indicate which are enantiomers, diastereomers, and meso compounds, if any.

8.35 Draw the structural formulas of the products of the following reactions.

(a) $CH_3CH{=}CHCH_3 + H_2O \xrightarrow{H^+}$ (alcohol)

[b] $CH_3CH{=}CHCH_3 + Br_2 \longrightarrow (C_4H_8Br_2)$

(c) 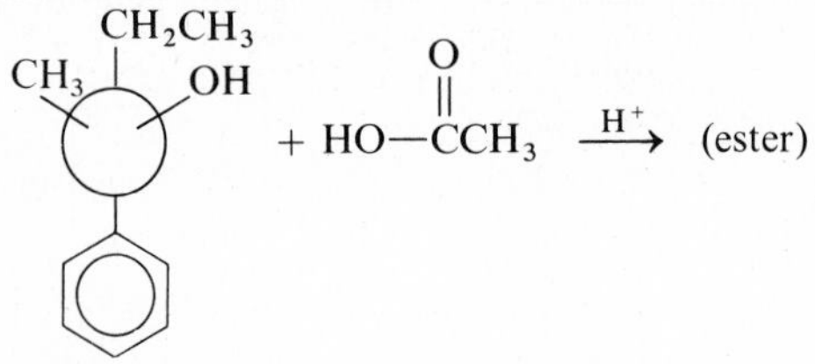 (Newman projection: CH_2CH_3, CH_3, OH; phenyl) $+ HO{-}\overset{O}{\overset{\|}{C}}CH_3 \xrightarrow{H^+}$ (ester)

[d] (Newman projection: CH_2CH_3, CH_3, OH; phenyl) $+ (\pm){-}HO{-}\overset{O}{\overset{\|}{C}}{-}\overset{CH_3}{\overset{|}{C}H}{-}C_6H_5 \xrightarrow{H^+}$ (ester)

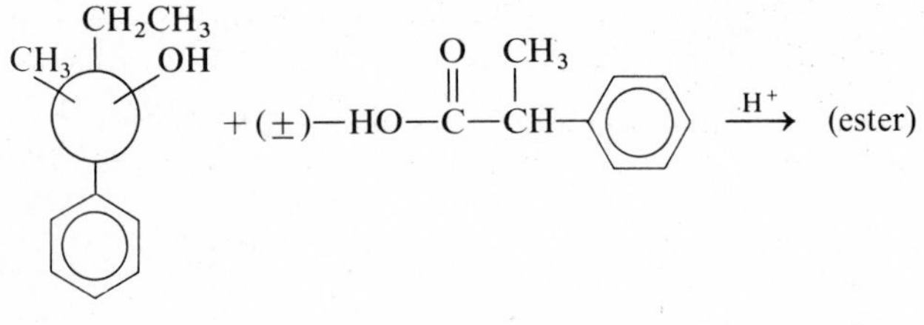

8.36 Which of the following compounds are chiral? Which are optically active?

[a] H, H, C=C=C, Br, Br [b] CH_3, H, C=C=C, H, CH_3 [c] (biphenyl with Br, NO_2, Br, NO_2)

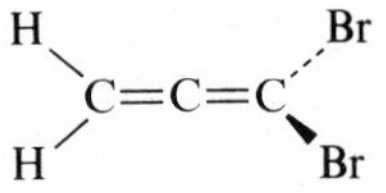

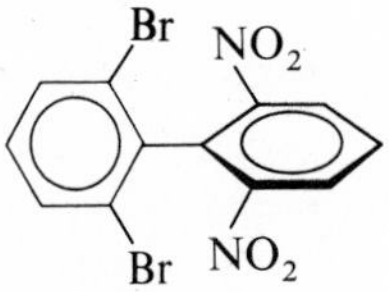

(d) 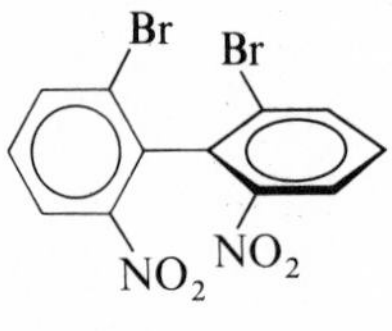 (biphenyl with Br, Br, NO_2, NO_2) (e) 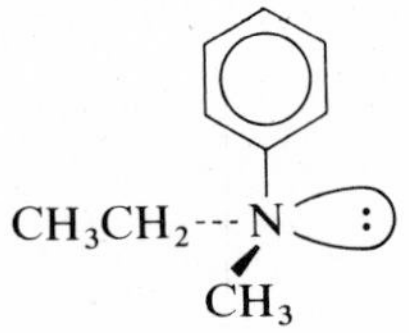 (phenyl, CH_3CH_2---N with lone pair, CH_3) (f) 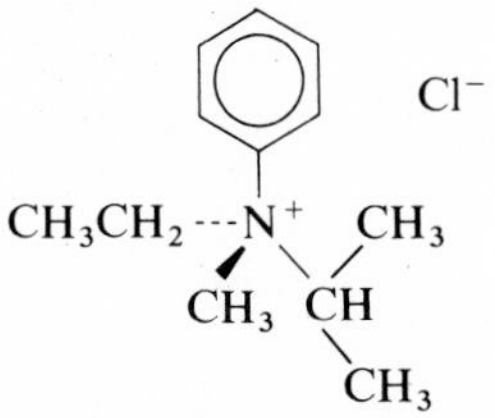 (phenyl, CH_3CH_2---N^+, CH_3, CH_3, CH, CH_3) Cl^-

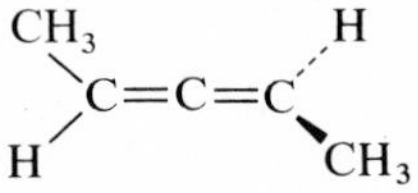

ORGANIC HALOGEN COMPOUNDS

We have mentioned organic halogen compounds throughout the first five chapters (a halogen is any element from group VII of the periodic table). Organic halogen compounds are important in synthesis and as useful end products—notably plastics, insecticides, and the Freons. These compounds are also important because of their effects on our physiology and the environment.

Organic halogen compounds also provide an interesting study of two important reactions—substitution and elimination.

9.1 STRUCTURE AND PROPERTIES

Carbon–halogen bonds are polar, especially C—F and C—Cl bonds:

$$\overset{\delta+}{-\!\!\underset{|}{\overset{|}{C}}}\!-\overset{\delta-}{\ddot{\underset{\cdot\cdot}{X}}}\!: \qquad (X = F, Cl, Br, \text{ or } I)$$

Table 9.1 Properties of Some Organic Halogen Compounds

Formula	Name	Boiling Point, °C	Density, g/ml @ 20°C
CH_3F	Methyl fluoride	−78.6	0.8774 @ −78.6°C
CH_3Cl	Methyl chloride	−24.2	0.991 @ −25°C
CH_3Br	Methyl bromide	3.6	1.732 @ 0°C
CH_3I	Methyl iodide	42.5	2.279
CH_2Cl_2	Methylene chloride	40.1	1.336
$CHCl_3$	Chloroform	61.3	1.498
CCl_4	Carbon tetrachloride	76.8	1.595
CCl_2F_2	Dichlorodifluoromethane (Freon-12)	−28	1.486 @ −30°C
C_6Cl_6	Hexachlorobenzene	326	2.044
$C_6H_6Cl_6$	α-1,2,3,4,5,6-Hexachlorocyclohexane*	288 (decomposes)	1.87
$C_6H_6Cl_6$	β-1,2,3,4,5,6-Hexachlorocyclohexane†	sublimes	1.89

* α = 1,2,4- chlorines are equatorial; 3,5,6- chlorines are axial.

† β = 1,3,5- chlorines are equatorial; 2,4,6- chlorines are axial.

In spite of the polarity of carbon–halogen bonds, organic halogen compounds are not soluble in water. Generally, hydrogen bonding is far more important than polarity in determining water solubility, and organic halogen compounds form only very weak hydrogen bonds with water.

The large atomic weights of Cl, Br, and I impart high densities to their compounds. Table 9.1 lists some organic halogen compounds along with some of their properties.

The boiling points of organic halogen compounds are somewhat higher than the boiling points of the corresponding hydrocarbons. We can explain these higher boiling points by the polarity of organic halogen compounds.

9.2 NOMENCLATURE

Simple halogenated alkanes may be called **alkyl halides**. Thus CH_3Cl is methyl chloride, $CH_3CHClCH_3$ is isopropyl chloride, and $BrC(CH_3)_3$ is *tert*-butyl bromide. The systematic IUPAC system simply treats the halogen atom as a substituent. Thus $CH_3CHClCH_2CH_3$ is 2-chlorobutane. Table 9.2 shows some examples of systematic IUPAC names.

Problem 9.1

Give the systematic name of each of the following compounds.

(a) $CH_3CH_2CH_2F$

(b) [cyclohexane with H, Cl on one carbon and H, Cl on the adjacent carbon]

(c) $CH_2{=}C(Cl)CH_2CH_2Cl$

(d) $H(BrCH_2)C{=}C(H)CH_2F$ (H above, $BrCH_2$ below on the left carbon; H above, CH_2F below on the right carbon)

(e) Bromocyclopentane (cyclopentane ring with Br)

Problem 9.2

Draw the structural formula of each of the following compounds and give their systematic names.

(a) allyl iodide, (b) vinyl iodide, (c) *m*-dichlorobenzene,
(d) *m*-chlorotoluene, (e) hexachlorobenzene

Table 9.2 Some Systematic IUPAC Names

Compound	Systematic IUPAC Name
CH_3CH_2Br	Bromoethane
$CH_3CH(Cl)CH_3$	2-Chloropropane
$CH_3C(CH_3)_2Cl$	2-Chloro-2-methylpropane
C_6H_5—Cl (benzene ring—Cl)	Chlorobenzene
$CH_2{=}CHF$	Fluoroethene
$CH_2{=}CHCH_2Br$	3-Bromopropene
CH_2Cl_2	Dichloromethane
$CHCl_3$	Trichloromethane
CCl_4	Tetrachloromethane

9.3 PREPARATION OF ORGANIC HALOGEN COMPOUNDS

We have already encountered several methods of preparing organic halogen compounds. These are listed in Table 9.3.

We have discussed reactions 1–3 in previous chapters, and we shall discuss only reactions 4 and 5 in the following paragraphs.

Alkyl Halides From Alcohols

Alcohols react with hydrogen halides (HX, where X = Cl, Br, or I) to produce the corresponding alkyl halides.

$$R{-}OH + \text{concentrated } HX \longrightarrow R{-}X + H_2O$$

Example: $(CH_3CH_2)_3COH + HCl \longrightarrow (CH_3CH_2)_3CCl + H_2O$

Two other reagents are commonly used to prepare alkyl halides from alcohols—phosphorus tribromide (PBr_3) and thionyl chloride ($SOCl_2$):

$$3ROH + PBr_3 \longrightarrow 3RBr + P(OH)_3$$

$$ROH + SOCl_2 \longrightarrow RCl + SO_2 + HCl$$

Table 9.3 Methods of Preparing Organic Halogen Compounds (R = alkyl, Ar = aryl)

1. *Halogenation (Substitution)* (X = Cl or Br)

   (a) Alkanes: $R{-}H + X_2 \xrightarrow[\text{heat}]{\text{light or}} R{-}X + HX$ (Sec. 4.10)

   (b) Aromatic hydrocarbons: $Ar{-}H + X_2 \xrightarrow{\text{Fe}} Ar{-}X + HX$ (Sec. 7.10)

2. *Halogenation (Addition)* (X = Cl or Br)

   (a) Alkenes: $\rangle C{=}C\langle + X_2 \longrightarrow -\underset{X}{\underset{|}{\overset{|}{C}}}-\underset{X}{\underset{|}{\overset{|}{C}}}-$ (Sec. 5.7)

   (b) Alkynes: $-C{\equiv}C- + 2X_2 \longrightarrow -\underset{X}{\underset{|}{\overset{X}{\overset{|}{C}}}}-\underset{X}{\underset{|}{\overset{X}{\overset{|}{C}}}}-$ (Sec. 6.10)

3. *Hydrohalogenation* (X = Cl, Br, or I)

   (a) Alkenes: $\rangle C{=}C\langle + HX \xrightarrow[\text{addition}]{\text{Markovnikov}} -\underset{H}{\underset{|}{\overset{|}{C}}}-\underset{X}{\underset{|}{\overset{|}{C}}}-$ (Sec. 5.7)

   (b) Alkynes: $-C{\equiv}C- + HX \xrightarrow[\text{addition}]{\text{Markovnikov}} -\underset{H}{\underset{|}{\overset{H}{\overset{|}{C}}}}-\underset{X}{\underset{|}{\overset{X}{\overset{|}{C}}}}-$ (Sec. 6.10)

4. *Reaction of Alcohols with HX or* PX_3 (X = Cl, Br, or I)

   $R{-}OH + HX \longrightarrow R{-}X + H_2O$ (Sec. 9.3)

   $R{-}OH + PX_3 \longrightarrow R{-}X + P(OH)_3$

5. *Halogen Exchange* (X and X′ = F, Cl, Br, or I)

   $R{-}X + X'^- \longrightarrow R{-}X' + X^-$ (Sec. 9.3)

Halogen Exchange

Organic fluorides and iodides are not as easy to prepare by direct methods as are chlorides and bromides. Both of these classes of halides can be prepared by **halogen exchange**.

Mercury or antimony fluorides are normally used to prepare an alkyl fluoride from the corresponding chloride or bromide:

$$2\,RCl + HgF_2 \longrightarrow 2\,RF + HgCl_2$$

Iodides are prepared by reaction of an alkyl chloride or bromide with sodium iodide:

$$R\text{—}Cl + NaI \xrightarrow{\text{acetone}} RI + NaCl$$

Halogen exchange is an equilibrium reaction. In this latter case, the equilibrium shifts in favor of the alkyl iodide because NaCl is less soluble than NaI in acetone solution.

Problem 9.3

Write equations for the preparation of the following compounds starting from an alcohol or an organic bromide and any inorganic chemicals.

(a) 1-fluoropropane,
(b) 1-iodopropane,
(c) 2-chloropropane,
(d) benzyl bromide from a hydrocarbon,
(e) bromobenzene from benzene

9.4 REACTIONS OF ALKYL HALIDES: SUBSTITUTION

Alkyl halides undergo two important types of reaction—substitution and elimination. We shall consider elimination reactions in Section 9.5.

Alkyl halides are very reactive toward a class of molecules or ions called nucleophiles. A **nucleophile** is a molecule or ion that has an unshared electron pair. Because of its available electron pair, a nucleophile is a base, and it seeks a positive center (hence the name nucleophile or "nucleus-loving"). Typical nucleophiles are F^-, Cl^-, Br^-, I^-, RO^-, ArO^-, OH^-, H_2O, NH_3, CN^-, NH_2^-, and RS^-.

When a nucleophile reacts with an alkyl halide, the nucleophile displaces the halide ion. We therefore call such reactions **nucleophilic substitution reactions** abbreviated S_N. If $Nu:^-$ is a nucleophile, the S_N reaction is

$$Nu:^- + R\text{—}\ddot{\underset{..}{X}}: \longrightarrow Nu\text{—}R + :\ddot{\underset{..}{X}}:^-$$

Nucleophilic substitution reactions are usually carried out in water, alcohol, or acetone. Charged nucleophiles are usually in the form of the sodium salt. Some typical S_N reactions follow.

Nucleophilic Reagent		Product
$RX + NaI$	$\xrightarrow[\text{exchange}]{\text{halogen}}$	$RI + NaX$
$RX + NaOH$	$\longrightarrow$	$ROH + NaX$
$RX + NaOR'$	$\longrightarrow$	$ROR' + NaX$
$RX + NaNH_2$	$\longrightarrow$	$RNH_2 + NaX$
$RX + NaSH$	$\longrightarrow$	$RSH + NaX$
$RX + Na{:}C{\equiv}CR'$	$\longrightarrow$	$RC{\equiv}CR' + NaX$
$RX + {:}NH_3$	$\longrightarrow$	$R{-}NH_3^+ + X^-$

Problem 9.4

Write equations for the preparation of each of the following from an alkyl bromide.

(a) $CH_3CH_2CH_2SH$

(b) $C_6H_5{-}CH_2NH_2$

(c) $C_6H_5{-}CH_2C{\equiv}CCH_3$

(d) $CH_3CH_2CH_2CH_2OH$

(e) $C_6H_5{-}CH_2I$

9.5 REACTIONS OF ALKYL HALIDES: ELIMINATION

Alkyl halides react with strong bases to lose the elements H and X. The most common solvents are alcohols:

$$-\underset{\underset{H}{|}}{\overset{|}{C}}-\underset{|}{\overset{|}{C}}-X + NaOH \xrightarrow{CH_3CH_2OH} \;\rangle C{=}C\langle\; + H{-}O{-}H + NaX$$

(X = Cl, Br, or I)

As we noted in Section 5.5, the major product is the most highly substituted alkene. The hydroxide ion, a strong base, is also a nucleophile. Thus we see that elimination and nucleophilic substitution are competing reactions because nucleophiles are also bases. It is not difficult to decide which reaction will occur, however, because the structure of the halide and the basicity of the base (nucleophile) affect the outcome in predictable ways. The reactivities of alkyl halides in the two reactions using a strong base are:

Substitution: primary > secondary (tertiary does not react)

Elimination: tertiary > secondary > primary

Some of the reasons for the above reactivity sequence are presented in Sections 9.6 and 9.7. For the moment, we simply state that with a strongly basic nucleophile, if the alkyl halide is primary, the predominant reaction is substitution. If the alkyl halide is tertiary, the predominant reaction depends on the conditions: With a strong base and a solvent of low polarity, tertiary alkyl halides undergo mainly elimination; with a weakly basic nucleophile and a polar solvent, tertiary alkyl halides undergo mainly substitution. Some specific examples will illustrate this generalization.

Conditions: Strongly basic nucleophile ($CH_3CH_2O^-$) and nonpolar solvent (CH_3CH_2OH)

$$CH_3CH_2ONa + \underset{\text{(primary)}}{CH_3CH_2Br} \begin{cases} \xrightarrow[(90\%)]{\text{substitution}} CH_3CH_2OCH_2CH_3 + NaBr \\ \xrightarrow[(10\%)]{\text{elimination}} CH_2{=}CH_2 + CH_3CH_2OH + NaBr \end{cases}$$

$$CH_3CH_2ONa + \underset{\text{(secondary)}}{CH_3\underset{Br}{CH}CH_3} \begin{cases} \xrightarrow[(21\%)]{\text{substitution}} CH_3\underset{OCH_2CH_3}{CH}CH_3 + NaBr \\ \xrightarrow[(79\%)]{\text{elimination}} CH_3CH{=}CH_2 + CH_3CH_2OH + NaBr \end{cases}$$

$$CH_3CH_2ONa + \underset{\text{(tertiary)}}{CH_3\overset{CH_3}{\underset{Br}{C}}CH_3} \begin{cases} \xrightarrow[(9\%)]{\text{substitution}} CH_3\overset{CH_3}{\underset{OCH_2CH_3}{C}}CH_3 + NaBr \\ \xrightarrow[(91\%)]{\text{elimination}} CH_3\overset{CH_3}{C}{=}CH_2 + CH_3CH_2OH + NaBr \end{cases}$$

Conditions: Weakly basic nucleophiles (H_2O and CH_3CH_2OH) and polar solvent (water or water-containing solvent mixture)

$$CH_3\overset{CH_3}{\underset{Br}{C}}CH_3 \xrightarrow[20\%\ H_2O]{80\%\ CH_3CH_2OH} \begin{cases} \xrightarrow[(83\%)]{\text{substitution}} CH_3\overset{CH_3}{\underset{OH}{C}}CH_3 + CH_3\overset{CH_3}{\underset{OCH_2CH_3}{C}}CH_3 + H_3O^+ + Br^- \\ \xrightarrow[(17\%)]{\text{elimination}} CH_3\overset{CH_3}{CH}{=}CH_2 + H_3O^+ + Br^- \end{cases}$$

Problem 9.5

Predict the products of the following reactions and designate the major product in each case.

(a) 2-bromo-2-methylpropane + NaOH $\xrightarrow{CH_3CH_2OH}$

(b) 1-bromobutane + $NaNH_2$ $\xrightarrow[NH_3]{\text{liquid}}$

(c) benzyl chloride + $NaOCH_2CH_3$ $\xrightarrow{CH_3CH_2OH}$

(d) allyl iodide + NaSH $\xrightarrow{CH_3CH_2OH}$

9.6 MECHANISMS OF SUBSTITUTION AND ELIMINATION REACTIONS: AN ELECTIVE TOPIC

S_N2 and E2 mechanisms

elective topic

Alkyl halides present two positive centers to a basic (nucleophilic) reagent—the carbon bearing the halogen atom and the hydrogen on the adjacent carbon:

$$\text{H}-\overset{|}{\underset{|}{\text{C}}}-\overset{\text{X}}{\overset{|}{\text{C}}}-\quad \text{(a), (b)}\quad :\text{Nu}^-$$

Reaction (a) involves substitution, and reaction (b) involves elimination:

$$\text{(a)}\quad \text{Nu}:^- + \text{C}-\ddot{\text{X}}: \xrightarrow{\text{substitution}} \text{Nu}-\text{C} \quad \text{H} + :\ddot{\text{X}}:^-$$

$$\text{(b)}\quad \text{Nu}:^- + \text{H}-\overset{|}{\underset{|}{\text{C}}}-\overset{|}{\underset{|}{\text{C}}}-\ddot{\text{X}}: \xrightarrow{\text{elimination}} \text{NuH} + \text{C}{=}\text{C} + :\ddot{\text{X}}:^-$$

Because both of these reactions depend on the collision of two molecules—alkyl halide and base (nucleophile)—these mechanisms are called **bimolecular**. The bimolecular nucleophilic substitution reaction is abbreviated **S_N2**, and the bimolecular elimination reaction is abbreviated **E2**.

Effect of Alkyl Halide Structure

We can explain the dependence of these reactions on alkyl halide structure quite easily. The S_{N2} reaction requires that the nucleophile attack *from the back side* of the carbon bearing the halogen:

$$\mathrm{Nu{:}^-} \; \mathrm{H_2C(R){-}X} \longrightarrow \left[\overset{\delta-}{\mathrm{Nu}}\cdots\mathrm{C(H_2)(R)}\cdots\overset{\delta-}{\mathrm{X}}\right] \longrightarrow \mathrm{Nu{-}CH_2R} + \mathrm{X^-}$$

(transition state)

In the transition state, both the entering nucleophile and the departing halide are partially bonded to the carbon atom. The carbon atom at this point is highly crowded because it is bonded to five atoms. With secondary alkyl halides, the transition state is even more crowded, and with tertiary alkyl halides crowding is so severe that the transition state cannot form. Crowding in the transition state thus accounts for the reactivity order of alkyl halides as $1° > 2°$ in S_N2 reactions (Figure 9.1).

In the E2 reaction, on the other hand, the nucleophile attacks a hydrogen atom on an adjacent carbon (Fig. 9.1d). Crowding of the halogen-bearing carbon is not a

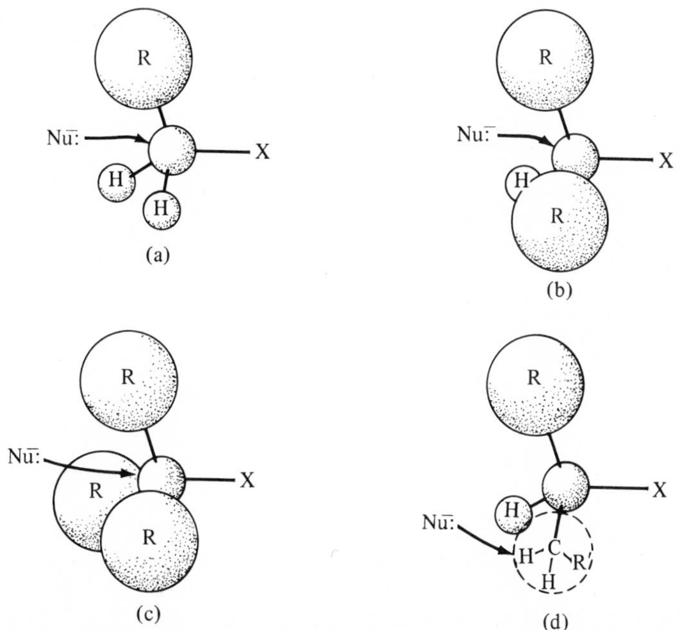

Figure 9.1. Effect of an alkyl halide's structure on its reaction with a nucleophile. In (a) the alkyl halide is primary. In (b) the alkyl halide is secondary, and the two bulky R groups hinder the approach of the nucleophile. In (c) the alkyl halide is tertiary, and the three R groups prevent the approach of the nucleophile. The nucleophile is shown attacking the hydrogen on the adjacent carbon atom (d).

factor. Moreover, since there are more alkyl groups attached to the halogen-bearing carbon, there are more hydrogens available for reaction with the nucleophile. Thus the greater the number of alkyl groups, the greater the probability that the nucleophile will encounter a hydrogen rather than the halogen-bearing carbon atom. For this reason the order of reactivity of alkyl halides in E2 reactions is 3° > 2° > 1°.

$CH_3\underline{CH_2}CH_2{-}X$ has two reactive hydrogens

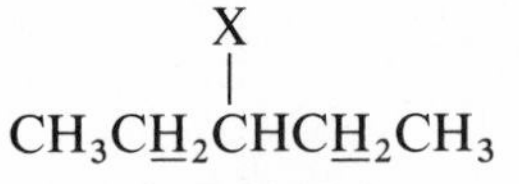

$CH_3\underline{CH_2}CH(X)\underline{CH_2}CH_3$ has four reactive hydrogens

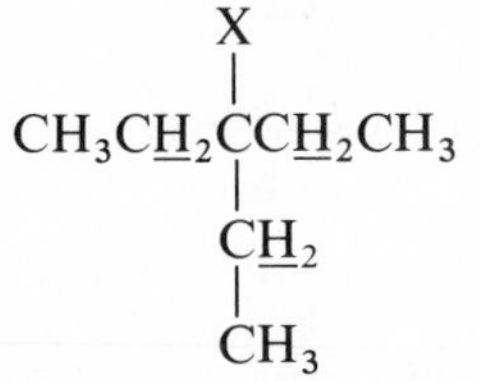

$CH_3\underline{CH_2}C(X)(\underline{CH_2}CH_3)\underline{CH_2}CH_3$ has six reactive hydrogens

Effect of Nucleophile: S_N1 and E1 mechanisms

S_N2 and E2 reactions occur when an alkyl halide reacts with a strongly basic nucleophile. When the nucleophile is less basic, and especially when water is used as solvent, another mechanism can operate—one that involves carbocations:

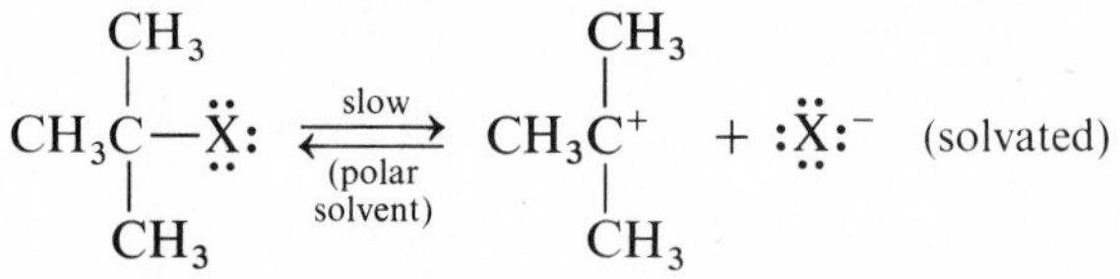

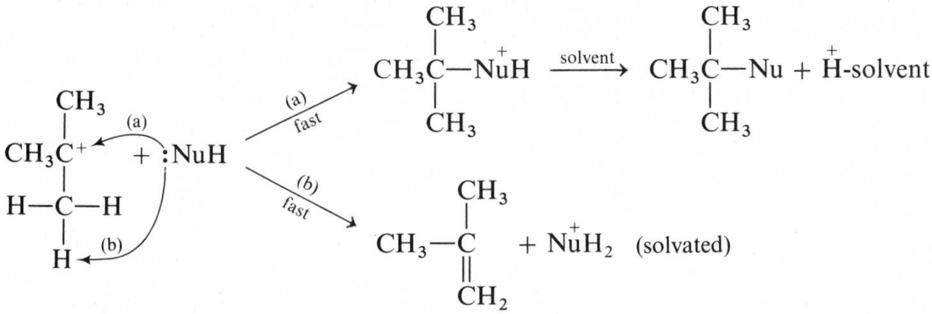

$$CH_3C(CH_3)_2{-}\ddot{X}\!: \xrightleftharpoons[\text{(polar solvent)}]{\text{slow}} CH_3C^+(CH_3)_2 + :\ddot{X}:^- \text{ (solvated)}$$

$$CH_3C^+(CH_3){-}CH_3 + :NuH \xrightarrow[\text{fast}]{(a)} CH_3C(CH_3)_2{-}\overset{+}{Nu}H \xrightarrow{\text{solvent}} CH_3C(CH_3)_2{-}Nu + \overset{+}{H}\text{-solvent}$$

$$CH_3C^+(CH_3){-}CH_3 + :NuH \xrightarrow[\text{fast}]{(b)} CH_3{-}C(CH_3){=}CH_2 + \overset{+}{Nu}H_2 \text{ (solvated)}$$

Because the stability of carbocations is in the order tertiary > secondary > primary (Sec. 5.6), the above mechanism predominates only with tertiary alkyl halides.

The slow step determines the rate of the overall reaction and is therefore called the **rate-determining** step. Because the reaction rate depends on the dissociation of the alkyl halide and the nucleophile does not play a role in this crucial step, we call the reaction mechanism **unimolecular**—only one molecule is involved in the rate-determining (slow) step. We therefore abbreviate the above substitution and elimination mechanisms $\mathbf{S_N1}$ and **E1**, respectively.

These mechanisms are important only for tertiary alkyl halides reacting in a very polar solvent such as water. In most cases the S_N1 reaction predominates over the E1 reaction.

Problem 9.6
What conditions would you employ to encourage a unimolecular reaction over a bimolecular reaction?

Problem 9.7

Which mechanism (S_N1 or S_N2) would you use if you wanted to prepare $CH_3\underset{\displaystyle OH}{\overset{\displaystyle CH_3}{\overset{|}{\underset{|}{C}}}}CH_2CH_3$ from the corresponding bromide? Explain your answer in terms of competition between substitution and elimination.

Problem 9.8
Predict the major product of each of the following reactions:

(a) 1-chloropentane + $NaOCH_2CH_3 \xrightarrow{CH_3CH_2OH}$

(b) *tert*-butyl bromide + $NaNH_2 \xrightarrow[NH_3]{\text{liquid}}$

(c) *tert*-butyl bromide + $H_2O \longrightarrow$

9.7 STEREOCHEMISTRY OF S_N AND E REACTIONS: AN ELECTIVE TOPIC

elective topic

Our discussions of substitution and elimination reactions suggest that each of these reactions occurs with its own stereochemistry. By the term **stereochemistry** we mean the three-dimensional description of the molecules as they react to form products. For example, the S_N2 mechanism requires that the nucleophile, $Nu:^-$, approach the carbon from the side opposite the departing halogen atom:

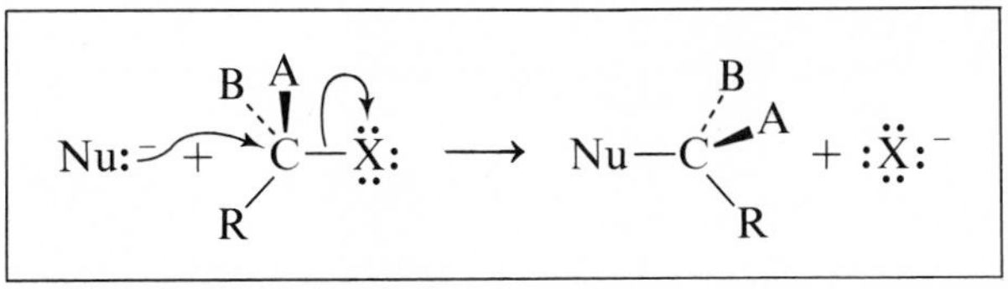

$$Nu:^- + \underset{R}{\overset{B\ \ A}{C}}-\ddot{\underset{..}{X}}: \longrightarrow Nu-\underset{R}{\overset{B}{C}}\!-A + :\ddot{\underset{..}{X}}:^-$$

An S_N2 reaction in which the halogen-bearing carbon has four different groups bonded to it will always produce the inverted product. Many examples of this stereochemical feature of S_N2 reactions have been observed. An example is the reaction of 2-bromooctane with hydroxide ion.

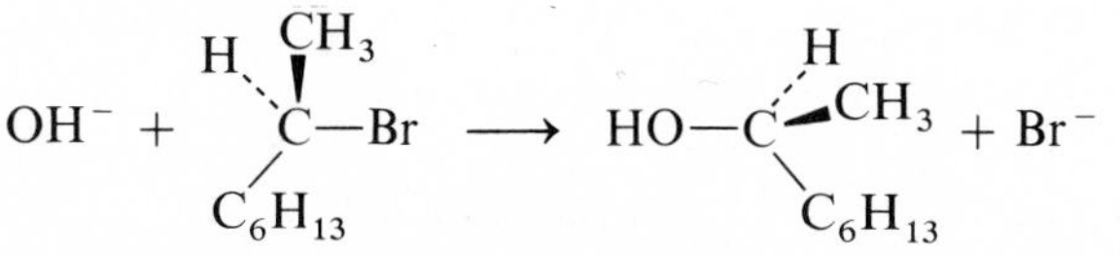

$$OH^- + \underset{C_6H_{13}}{\overset{H\ \ CH_3}{C}}-Br \longrightarrow HO-\underset{C_6H_{13}}{\overset{H}{C}}\!-CH_3 + Br^-$$

If we carry out the reaction with the pure enantiomer shown, the product is also a pure enantiomer but has the inverted configuration.

Problem 9.9
Give the *R–S* designation of the reactant and product in the above equation.

Problem 9.10
Draw the structure of the product of the S_N2 reaction of (*R*)-2-bromobutane with $NaOCH_2CH_3$.

E2 reactions also follow a definite stereochemical path. Because the C—H and C—X bonds break at the same time that the C=C bond forms, the H—C—C—X bonds must lie in a plane as shown with the H—C bond parallel to the C—X bond:

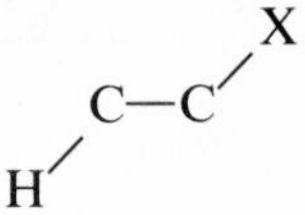

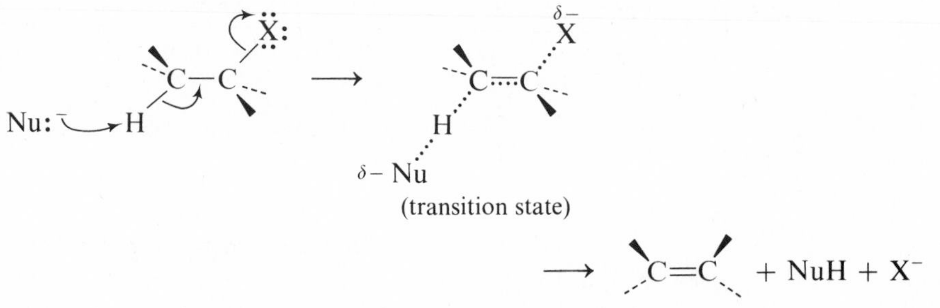

Nu:⁻ → H—C—C—X: → H···C⋯C···X (δ−) with δ− Nu (transition state) → C=C + NuH + X^-

In this way, the electron pair of the C—H bond displaces the X^- ion to form the double bond.

An interesting example illustrates this stereochemistry: We would expect the dehydrohalogenation of *trans*-2-bromo-1-methylcyclopentane to yield the more highly substituted alkene, I (Sec. 5.5):

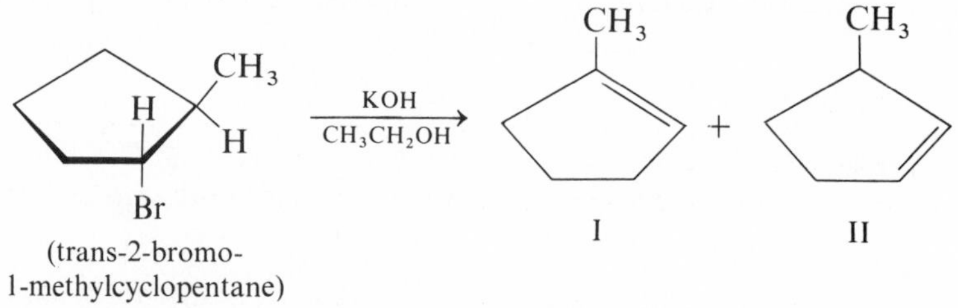

(trans-2-bromo-1-methylcyclopentane) $\xrightarrow[CH_3CH_2OH]{KOH}$ I + II

Instead, the reaction yields the less substituted alkene, II, as the major product. In the reactant (a *trans* isomer) only one hydrogen atom is on the side opposite the bromine:

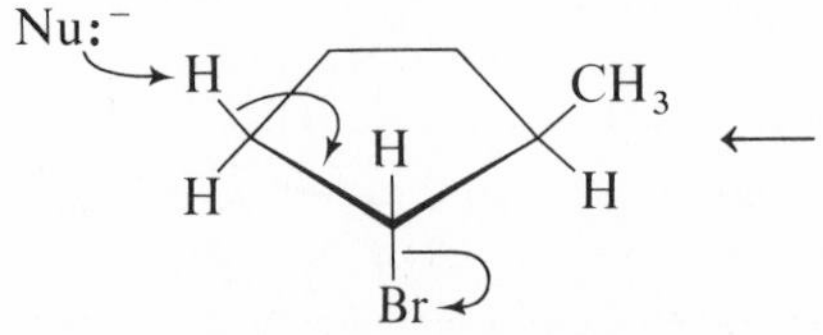

Nu:⁻ → H ... Br ← This hydrogen is on the same side of the ring as the Br, and cannot react.

In the *cis* isomer, the hydrogen on C_1 is opposite the bromine and the reaction gives the expected product, I.

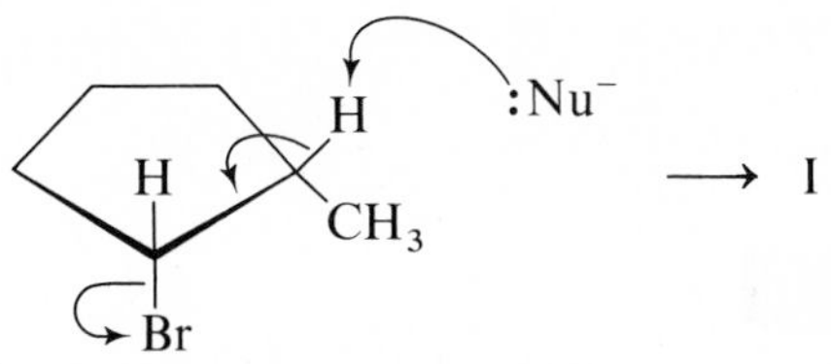

H :Nu⁻
H
CH₃ ⟶ I
Br

(*cis*-2-bromo-1-methylcyclopentane)

S_N1 and E1 reactions proceed through carbocations. Carbocations are planar, and the nucleophile can approach either side of the carbocation with equal ease:

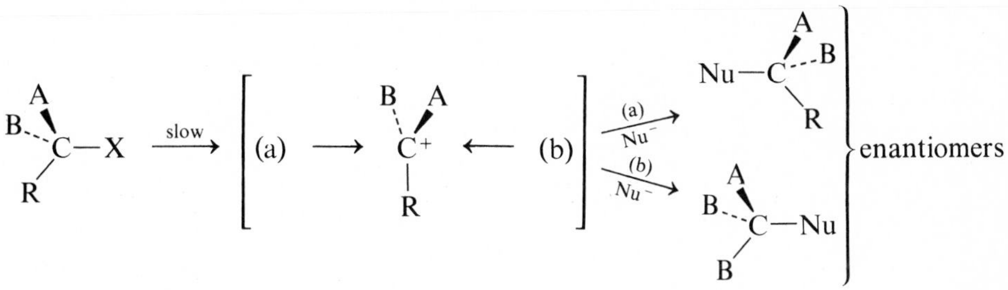

$$B,A,R\text{-}C\text{-}X \xrightarrow{\text{slow}} \left[(a) \longrightarrow C^{+} \longleftarrow (b)\right] \begin{cases} \xrightarrow[Nu^-]{(a)} Nu\text{-}C(A)(B)(R) \\ \xrightarrow[Nu^-]{(b)} (A)(B)(B)C\text{-}Nu \end{cases} \text{enantiomers}$$

The reaction is unimolecular because the rate-determining step (slow step) involves only the alkyl halide. The nucleophile reacts in a later, fast step. In the E1 reaction, any available hydrogen on the adjacent carbon can be removed because rotation about the R—C bond can occur in the carbocation:

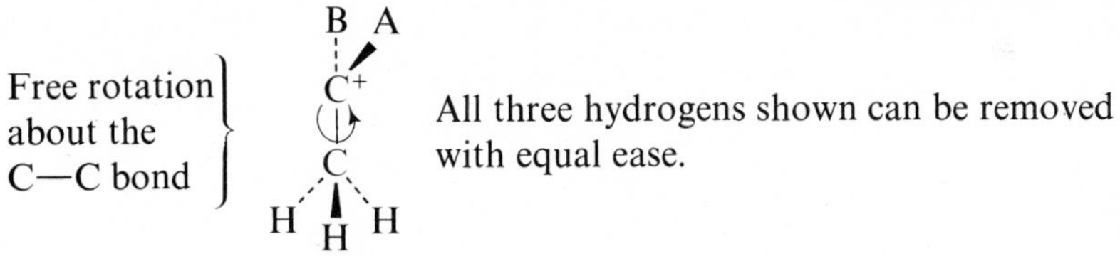

Free rotation about the C—C bond } B A C⁺ C H H H — All three hydrogens shown can be removed with equal ease.

Problem 9.11

Draw the structure and show the stereochemistry of the product of each of the following reactions.

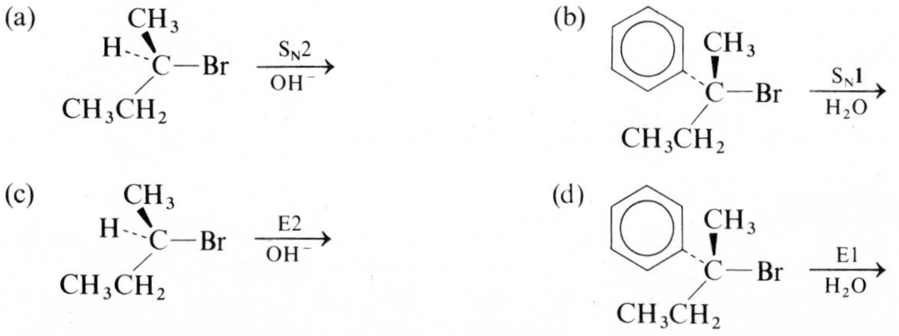

(a) $H,CH_3,CH_3CH_2\text{-}C\text{-}Br \xrightarrow[OH^-]{S_N2}$

(b) $C_6H_5,CH_3,CH_3CH_2\text{-}C\text{-}Br \xrightarrow[H_2O]{S_N1}$

(c) $H,CH_3,CH_3CH_2\text{-}C\text{-}Br \xrightarrow[OH^-]{E2}$

(d) $C_6H_5,CH_3,CH_3CH_2\text{-}C\text{-}Br \xrightarrow[H_2O]{E1}$

9.8 HALIDES OF UNUSUAL REACTIVITY

So far we have discussed only saturated primary, secondary, and tertiary alkyl halides. Some halides have unusually high or unusually low reactivities because of the presence of a carbon–carbon double bond.

Allyl Halides

Allyl halides are unusually reactive. If we add an alcoholic solution of silver nitrate to an allyl halide, we obtain an immediate precipitation of silver halide:

$$CH_2{=}CHCH_2Cl + AgNO_3 \xrightarrow[H_2O\text{-alcohol}]{\text{(immediate)}} CH_2{=}CHCH_2{-}OH + AgCl\downarrow + \cdots$$

Saturated alkyl halides react much more slowly because they produce much less stable carbocations. Allylic carbocations are stabilized by resonance (Sec. 7.3 and Problem 7.32):

$$CH_2{=}CH{-}\overset{+}{C}H_2 \longleftrightarrow \overset{+}{C}H_2{-}CH{=}CH_2$$

Saturated carbocations are not stabilized in this way.

We can now include allylic carbocations in our order of carbocation stability:

Carbocation stability: allylic $\geq$ tertiary $>$ secondary $>$ primary $>$ methyl

Benzyl Halides

Benzyl halides are as reactive as allylic halides toward $AgNO_3$. The benzyl carbocation is stabilized by resonance in much the same way as allylic carbocations:

$$C_6H_5{-}\overset{+}{C}H_2 \longleftrightarrow \overset{+}{C_6H_5}{=}CH_2 \longleftrightarrow {}^{+}C_6H_5{=}CH_2 \longleftrightarrow C_6H_5^{+}{=}CH_2$$

We can therefore place the benzylic carbocation along with the allylic carbocation in our stability order:

$$\text{Carbocation stability: } \left.\begin{matrix}\text{allylic}\\ \text{or}\\ \text{benzylic}\end{matrix}\right\} \geq \text{tertiary} > \text{secondary} > \text{primary} > \text{methyl}$$

Vinyl Halides

Vinyl halides are much less reactive than saturated alkyl halides toward $AgNO_3$. In fact, they do not react appreciably over long periods of time even when heated.

The reasons for the low reactivity of vinyl halides are that resonance does not stabilize the vinyl carbocation, $CH_2{=}\overset{+}{C}H$, and resonance *does* stabilize the vinyl halide itself.

$$\underbrace{H_2C{=}CHCl}_{\text{resonance stabilized}} \not\rightleftarrows \underbrace{H_2C{=}\overset{+}{C}H}_{\text{not resonance-stabilized}} + Cl^-$$

The resonance structures of the vinyl halide molecule are:

$$\underset{\text{(I)}}{CH_2{=}CH{-}\ddot{\underset{\cdot\cdot}{Cl}}:} \longleftrightarrow \underset{\text{(II)}}{\bar{\ddot{C}}H_2{-}CH{=}\underset{\cdot\cdot}{Cl}:^+}$$

Resonance structure (II) accounts for a C—Cl bond that has partial double bond character and is therefore stronger than the C—Cl bonds in ordinary saturated alkyl halides.

Aryl Halides

Compounds in which the halogen is attached directly to a benzene ring are as unreactive as vinyl halides.

$$\underset{\text{aryl chloride}}{C_6H_5{-}\ddot{\underset{\cdot\cdot}{Cl}}:} \qquad \underset{\text{vinyl chloride}}{H_2C{=}CH{-}\ddot{\underset{\cdot\cdot}{Cl}}:}$$

The reasons are the same as for the vinyl halides: the phenyl carbocation is not stabilized by resonance, and the C—X bond has partial double bond character due to resonance structures such as

$$C_6H_5{-}\ddot{\underset{\cdot\cdot}{Cl}}: \longleftrightarrow \bar{C}_6H_5{=}\ddot{Cl}:^+ \longleftrightarrow$$

Problem 9.12
Draw all the remaining important resonance structures of chlorobenzene.

In summary, allyl and benzyl halides are very reactive in both S_N1 and S_N2 reactions, and vinyl and aryl halides are unreactive. The order of carbocation stability is

> Carbocation stability:
>
> $$\left\{\begin{matrix}\text{allylic}\\ \text{benzylic}\end{matrix}\right\} \geq \text{tertiary} > \text{secondary} > \text{primary} > CH_3^+ > \left\{\begin{matrix}\text{vinyl}\\ \text{aryl}\end{matrix}\right\}$$

We can categorize the S_N1 and S_N2 reactions of the various organic halides as follows:

Halides that undergo only S_N2 reactions:

Methyl halides, 1° and 2° alkyl halides except allylic and benzylic halides: CH_3X, $R{-}CH_2{-}X$, $R{-}\underset{\displaystyle R'}{\underset{|}{C}}H{-}X$

Halides that may undergo either S_N1 or S_N2 reactions:

1° and 2° allylic and benzylic halides: $Ar{-}CH_2{-}X$, $Ar\underset{\displaystyle R}{\underset{|}{C}}H{-}X$

$CH_2{=}CHCH_2{-}X$, $CH_2{=}CH\underset{\displaystyle R}{\underset{|}{C}}H{-}X$

Table 9.4 Summary of S_N and E Reactions

Structure of Alkyl Halide	S_N2	E2	S_N1	E1
1° (except allylic and benzylic halides)	S_N2 favored over E2		No reaction	No reaction
2° (except allylic and benzylic halides)	E2 favored over S_N2		No reaction	No reaction
1° and 2° allylic and benzylic halides	Favored	No reaction	Favored	No reaction
3° alkyl, allylic, and benzylic halides	No reaction	Favored	S_N1 favored over E1	
Vinyl and aryl halides	No reaction	No reaction	No reaction	No reaction
* * * * * * *				
Nucleophile Strength	Large effect	Large effect	No effect	No effect
Increasing Solvent Polarity	Little effect	Little effect	Large effect	Large effect

Halides that undergo only S_N1 reactions:

3° alkyl, allylic, or benzylic halides

$$R-\overset{R}{\underset{R}{C}}-X,\quad Ar-\overset{R}{\underset{R}{C}}-X,\quad CH_2=CH-\overset{R}{\underset{R}{C}}-X$$

Table 9.4 summarizes the facts on S_N1, S_N2, E1, and E2 reactions.

9.9 GRIGNARD REAGENTS

One of the most important synthetic reactions of organic halides is the reaction with magnesium metal. Nearly all organic halides react with metallic magnesium in dry ether as solvent to give a very reactive compound called the **Grignard reagent**, named after its discoverer, the French chemist Victor Grignard.

$$R-X + Mg \xrightarrow[(X\,=\,Cl,\,Br,\,or\,I)]{ether} R-MgX$$

alkyl magnesium halide
(a Grignard reagent)

Grignard reagents behave as though the alkyl or aryl group were a carbanion:

$$\overset{\delta-}{R}-\overset{\delta+}{Mg}X$$

The negative character of the carbon atom attached to Mg in the Grignard reagent makes this reagent a strong base and a strong nucleophile. Examples of the reactions of Grignard reagents as bases and nucleophiles are as follows.

Reaction as Base

$$C_6H_5-MgBr + H-O-H \longrightarrow C_6H_5-H + MgBrOH$$

(base) (acid) (acid) (base)

$$C_6H_5-MgBr + HO-CH_2CH_3 \longrightarrow C_6H_5-H + MgBr\,OCH_2CH_3$$

(base) (acid) (acid) (base)

$$CH_3MgBr + HC\equiv CR \longrightarrow CH_4 + RC\equiv C{:}MgBr$$

(base) (acid) (acid) (base)

The general reaction is

$$RMgX + HA \longrightarrow R-H + MgXA$$

where HA is an acid.

Reaction as Nucleophile

$$\overset{\delta-}{C}H_3\overset{\delta+}{C}H_2MgBr + :\ddot{O}=\overset{\delta+}{C}=\overset{\delta-}{\ddot{O}}: \longrightarrow CH_3CH_2-\underset{\underset{O}{\|}}{C}-\overset{-}{\ddot{O}}:\overset{+}{M}gBr$$

$$\xrightarrow{H_3O^+} CH_3CH_2-C(=O)OH + Mg^{2+} + Br^-$$

The general reaction is

$$RMgX + CO_2 \longrightarrow R\overset{O}{\overset{\|}{C}}OMgX$$

$$RCO_2MgX + HA \longrightarrow RCO_2H + MgXA$$

In the following chapters, we shall encounter many other reactions of Grignard reagents as both bases and nucleophiles.

Problem 9.13
Explain why dry ether is necessary in the preparation and use of Grignard reagents.

Problem 9.14
Given that the reactions above occur essentially completely as shown, compare the basicity of the Grignard reagent with the bases OH^-, RO^-, and $RC{\equiv}C{:}^-$.

Problem 9.15
Write equations to show how you would prepare (a) methylmagnesium iodide and (b) phenylmagnesium bromide starting only with hydrocarbons, any inorganic compounds, and any solvents.

9.10 HALOGENATED COMPOUNDS—MANUFACTURE AND USES: AN ELECTIVE TOPIC

elective topic

Most halogenated hydrocarbons are prepared commercially by direct halogenation. On an industrial scale, complex mixtures of products are readily separated, and most by-products are either sold or used to make other useful products. Examples are the chlorination of methane, benzene, and biphenyl:

$$CH_4 + Cl_2 \longrightarrow CH_3Cl + CH_2Cl_2 + CHCl_3 + CCl_4 + HCl$$

$$C_6H_6 + Cl_2 \xrightarrow{Fe} C_6H_5Cl + HCl + \text{polychlorinated benzenes}$$

$$\text{biphenyl} + \text{excess } Cl_2 \xrightarrow{Fe} \text{decachlorobiphenyl} + \left\{\begin{array}{l}\text{partially}\\\text{chlorinated}\\\text{biphenyls}\end{array}\right.$$

(biphenyl)

Chlorobenzene is used to prepare DDT (*d*ichloro*d*iphenyl*t*richloroethane):

$$CCl_3CHO + C_6H_5Cl \xrightarrow{H_2SO_4} (ClC_6H_4)_2CH{-}CCl_3 + H_2O$$

DDT

This reaction also produces other isomers in which one or both chlorines are at ortho or meta positions. Commercial DDT contains all these isomers because purification to the all-para isomer is expensive and unnecessary.

Another commercially important group of compounds is the chlorinated cyclohexanes. These are misnamed *benzene hexachlorides* because they are prepared by the chlorination of benzene.

$$C_6H_6 + 3\,Cl_2 \longrightarrow C_6H_6Cl_6$$

1,2,3,4,5,6-hexachlorocyclohexane

Several isomers of this compound are possible because the chlorine atoms may be in either axial or equatorial positions. Lindane, an important insecticide, has axial chlorines at the 1,2,3- carbons and equatorial chlorines at the 4,5,6- carbons:

lindane

Aldrin, another insecticide, is made by the Diels–Alder reaction of hexachlorocyclopentadiene with norbornadiene:

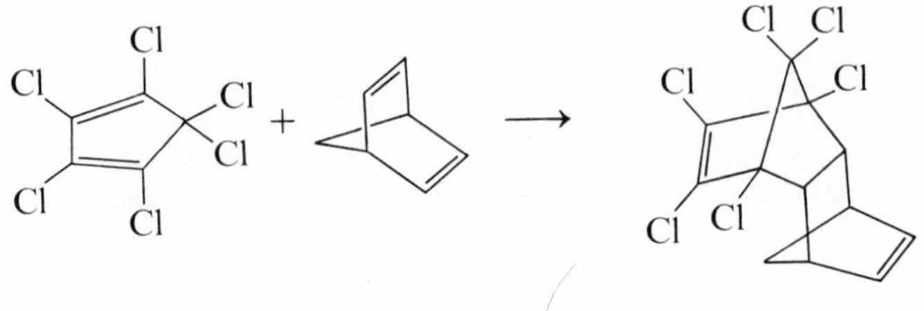

(norbornadiene) aldrin

Chlordane is prepared by the Diels–Alder reaction,

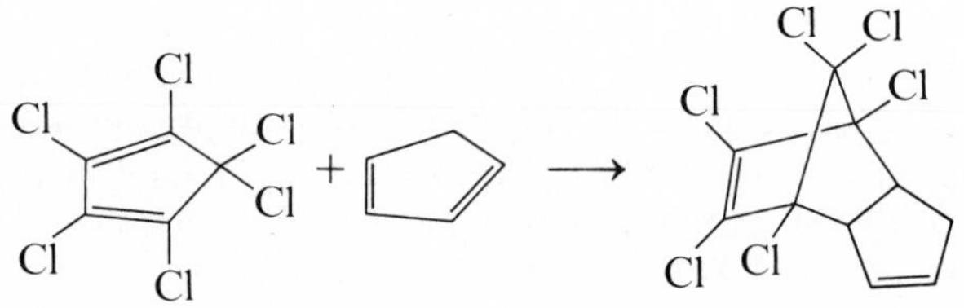

followed by addition of chlorine to the unsubstituted double bond,

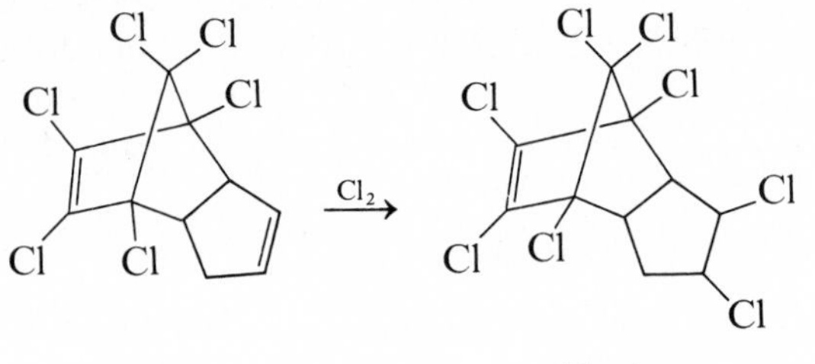

chlordane

Table 4.5 lists some other commercially important organic halogen compounds.

Nearly all chlorinated hydrocarbons are poisonous to many animals including humans. [Halothane (Table 4.5) is a notable exception.] Pesticides like DDT, aldrin, lindane, and chlordane are not only toxic when first used, but continue to persist in the environment because they resist the usual atmospheric or microbial decomposition pathways that eventually destroy or transform most other organic substances. In addition to their great stability, they are insoluble in water and soluble in nonpolar solvents including the fatty tissues of animals. They tend to concentrate in such fatty tissues and thus are not only carried along in individual animals but are transferred from one animal to another in the food chain. Such insecticides are commonly found in the liver, brain, gonads, and in mammals' milk fat.

In addition to the above disadvantages associated with the use of chlorinated hydrocarbons, insects have—through natural selection—become increasingly resistant to pesticides. As a result, it has become necessary to use larger and larger concentrations of insecticides to accomplish the same effect.

Teflon is an example of another useful but less dangerous compound. The utility of polyfluorinated polymers lies in the great stability of the C—F bond. Teflon is prepared by polymerizing tetrafluoroethene:

$$nCF_2{=}CF_2 \longrightarrow \text{+}CF_2CF_2\text{+}_n$$

This polymer is inert to chemical agents and undergoes wear at a very slow rate. Its chemical inertness also imparts an oily feel to the plastic. Teflon is used in the manufacture of bearings, in cooking utensils, and in artificial body parts in medicine.

9.11 ANALYSIS AND IDENTIFICATION OF ORGANIC HALOGEN COMPOUNDS

Organic halides can be made to react with alcoholic solutions of silver nitrate. Vinyl and aryl halides react so slowly at room temperature that they are practically inert. Allylic and benzylic halides react almost immediately to give a silver halide precipitate. Other halides react at varying rates according to the order: allyl = benzyl > tertiary > secondary > primary > methyl > vinyl = aryl.

Halides that react slowly with $AgNO_3$ can be analyzed by the *sodium fusion technique*: We heat a small piece of sodium in a test tube over a gas burner (caution!) When the sodium has melted and begun to vaporize, we add a small amount of the sample. This vigorous reaction converts organic halides into Na salts. When the reaction mixture cools we add water, acidify with nitric acid, and add a few drops of aqueous $AgNO_3$. Halide ions will precipitate immediately as silver halide:

$$RX + Na \xrightarrow{\Delta} NaX + H_2O + \text{other decomposition products}$$

$$NaX + AgNO_3 \xrightarrow{H_3O^+} AgX\downarrow + NaNO_3$$

Nitrogen and sulfur compounds also give a positive test in sodium fusion.

$$\text{Nitrogen-containing organic compounds} \xrightarrow{Na} NaCN + \cdots$$

$$NaCN + AgNO_3 \xrightarrow{H_3O^+} AgCN\downarrow + NaNO_3$$

$$\text{Sulfur-containing organic compounds} \xrightarrow{Na} Na_2S$$

$$Na_2S + AgNO_3 \xrightarrow{H_3O^+} Ag_2S\downarrow + NaNO_3$$

We can expel both the cyanide and sulfide ions from the reaction mixture by acidifying and heating before we add the $AgNO_3$ solution. This treatment converts ionic cyanides to gaseous HCN, and ionic sulfides to gaseous H_2S:

$$NaCN + H^+ \xrightarrow{\Delta} HCN\uparrow + Na^+$$

$$2NaS + H^+ \xrightarrow{\Delta} H_2S\uparrow + 2Ha^+$$

Problem 9.16

Show how you could distinguish between the compounds in each of the following pairs.

(a) $CH_3CH{=}CHCH_2Cl$ and $CH_2{=}CHCH_2CH_2Cl$

(b) $CH_3CH{=}CHCH_2Cl$ and $CH_3CH_2CH{=}CHCl$

(c) $CH_3CH_2CH_2CH_2CH_3$ and $CH_3CH_2CH_2CH_2Cl$

(d) CH_3CH_2Cl and CH_3CH_2SH

NEW TERMS

Alkyl halide Any simple halogenated alkane (9.2).

Bimolecular reaction A reaction that occurs by the collision of two molecules or ions (9.6).

E1 reaction An elimination reaction in which the rate-determining step is unimolecular (9.6).

E2 reaction An elimination reaction in which the rate-determining step is bimolecular (9.6).

Grignard reaction Reaction of a Grignard reagent with an acid HA, or with a carbonyl or other polar group (9.9).

Grignard reagent A compound RMgX in which R is alkyl or aryl and X is Cl, Br, or I. Grignard reagents are prepared by the reaction of an alkyl or aryl halide with magnesium in dry ether (9.9).

$$R{-}X + Mg \xrightarrow{\text{ether}} RMgX$$

Halogen exchange Reaction of an alkyl halide RX with the salt of a different halide X′ to produce RX′. This reaction is used primarily to prepare alkyl fluorides and alkyl iodides (9.3).

Nucleophile A molecule or ion that has an unshared electron pair (a Lewis base). (9.4).

Nucleophilic substitution reaction (S_N) Reaction of a nucleophile with an organic compound RX that leads to the substitution of the nucleophile for X (9.4):

$$:Nu^- + RX \longrightarrow R{-}Nu + :X^-$$

Rate-determining step The slowest step in a reaction sequence (9.6).

S_N1 reaction A nucleophilic substitution reaction in which the rate-determining step is unimolecular (9.6).

S_N2 reaction A nucleophilic substitution reaction in which the rate-determining step is bimolecular (9.6).

Stereochemistry The three-dimensional description of a molecule or molecules as they react to form products (9.7).

Unimolecular reaction A reaction involving only one molecule or ion (9.6).

ADDITIONAL PROBLEMS

[9.17] 1-Chlorobutane is quite insoluble in water, whereas butyl alcohol is fairly soluble. Explain.

9.18 Draw the structure of the following compounds.

[a] *Z*-1,2-difluoroethene,
[b] *trans*-1-chloro-2-butene,
[c] *m*-dichlorobenzene,
[d] 2,3-difluorocyclopentene,
(e) *E*-1,2-dichloropropene,
(f) *trans*-1,4-dichloro-2-butene,
(g) 2,4-dibromotoluene,
(h) 1,3-diiodocyclopentene

9.19 Give the systematic name of each of the following:

[a] I—C_6H_4—I (para-diiodobenzene structure)

[b] CCl_3CH_3

[c] $CHCl_2CCl_2CHCl_2$

[d] Cl-substituted cyclopentene (Cl on the double-bond carbon)

(e) Cl—C_6H_4—Br

(f) CH_2ClCH_2Cl

(g) $CCl_3CH_2CCl_3$

(h) Br-substituted cyclopentene

9.20 Write equations for the preparation of each of the following by direct halogenation.

[a] $CH_3CCl_2CH_3$

[b] CH_2ClCH_3

[c] CH_2ClCH_2Cl

[d] C_6H_5Br (bromobenzene)

[e] CCl_4

(f) CH_3CHBr_2

(g) $CH_3CHBrCH_3$

(h) $CH_3CHClCHClCH_3$

(i) CH_3—C_6H_4—Br (para)

(j) CH_3Br

9.21 Of the methods given in this chapter, which would be best for preparing each of the following:

[a] CH_3CHFCH_3

[b] C_6H_5—CH_2I

[c] $CH_3CHClCH_3$

(d) C_6H_5—CH_2F

(e) C_6H_5—$CH(CH_3)$—I

(f) C_6H_5—CH_2CH_2Cl

9.22 Give the products of each of the following nucleophilic substitution reactions.

[a] $CH_2{=}CHCH_2Br + NaCN \longrightarrow$

[b] $CH_2{=}CHCH_2Br + NH_3 \longrightarrow$

[c] C_6H_5—$CH_2Br + NaSCH_2CH_3$

(d) C_6H_5—$CH_2Cl + NaCN \longrightarrow$

(e) C_6H_5—$CH_2I + NaOH \longrightarrow$

(f) $CH_2{=}CHCH_2Br + NaOCH_2CH_3 \longrightarrow$

9.23 Predict the major product of each of the following reactions.

[a] $CH_3CH_2CH_2CH_2Br + NaOH \xrightarrow{H_2O}$

[b] $(CH_3)_3CBr + NaOH \xrightarrow{CH_3CH_2OH}$

(c) C_6H_5—$CH_2CH_2CH_2I + NaOH \xrightarrow{H_2O}$

(d) C_6H_5—$CH_2C(CH_3)(CH_2CH_3)$—$I + NaOH \xrightarrow{CH_3CH_2OH}$

*9.24 Predict the mechanism—S_N1 or S_N2—for the reaction of each of the following compounds with an alcoholic solution of NaOH.

[a] $CH_3CH{=}CHCH_2Br$

[b] $CH_2{=}C(CH_3)CH_2Br$

[c] $CH_2{=}CHC(CH_2CH_3)(CH_3)Br$

(d) $(CH_3)_2C{=}CHCH_2Br$

(e) $CH_3CH{=}C(CH_3)CH_2Br$

(f) C_6H_5—$C(CH_2CH_3)(CH_3)Br$

*[9.25] For each of the compounds of Problem 9.24, predict the mechanism—S_N1 or S_N2—if the reactions are carried out by heating in a solution of alcohol containing water.

* 9.26 For each of the reactions in Problem 9.24, predict whether the corresponding elimination reaction occurs, and if so, whether it would be E1 or E2.

*[9.27] For each of the reactions in Problem 9.25, predict whether the corresponding elimination reaction occurs, and if so, whether it would be E1 or E2.

*[9.28] Predict the stereochemistry of the products of each of the reactions in Problems 9.24 and 9.25.

9.29 Which of the following molecules are chiral?

[a] CH_4

[b] CH_2Cl_2

[c] C_6H_5—$CHCl_2$ (Cl above CHCl)

[d] C_6H_5—$CHBrCl$ (Cl above CHBr)

(e) CH_3CH_3

(f) $CH_3CHClCH_2CH_3$

(g) $(C_6H_5)_2CBr_2$

(h) $(C_6H_5)_2CBrCl$

[9.30] For each of the chiral molecules in Problem 9.29, draw the structures of the two enantiomers.

[9.31] Explain why chlorobenzene is unreactive in nucleophilic substitution reactions.

[9.32] Grignard reagents must be prepared in dry ether solution. Explain why water interferes with the preparation of Grignard reagents. What other types of compounds interfere with the preparation of Grignard reagents?

*[9.33] Why do chlorinated hydrocarbons persist so long in the environment?

9.34 Give simple chemical tests that would make it possible to distinguish between the compounds in each of the following pairs.

[a] chlorobenzene and toluene

[b] chlorobenzene and benzyl chloride

[c] vinyl chloride and benzyl chloride

(d) vinyl chloride and butane

(e) bromobenzene and C_6H_5–CHCl–C_6H_5

(f) vinyl bromide and *p*-chlorobenzyl bromide

*[9.35] In the first step of the synthesis of chlordane (Sec. 9.10), the Diels–Alder reaction can give two enantiomeric products. Draw their structures.

---

* From an Elective Topic.

10 ALCOHOLS AND PHENOLS

10.1 STRUCTURE OF ALCOHOLS AND PHENOLS

Both alcohols and phenols have a hydroxyl group attached to a hydrocarbon group. They differ in the nature of the hydrocarbon. In **alcohols**, the OH group is bonded to an aliphatic carbon atom; in **phenols**, the OH group is bonded to an aromatic ring. As we shall see, this small structural difference produces a great difference between alcohols and phenols. Table 10.1 shows some common alcohols and phenols.

$$R_3C{-}OH$$

Alcohol (R = H or alkyl)

$$R_5C_6{-}OH$$

Phenol (R = H or any other group)

Table 10.1 Some Common Alcohols and Phenols

Alcohols*		Phenols*	
CH_3OH Methanol (Methyl alcohol)	$CH_2{=}CHCH_2OH$ 2-Propenol-1 (Allyl alcohol)	OH Phenol (Carbolic acid)	OH, CH_3 *o*-Cresol
CH_3CH_2OH Ethanol (Ethyl alcohol)	$C_6H_5-CH_2OH$ Benzyl alcohol	OH 1-Naphthol	OH 2-Naphthol
$CH_2(OH)-CH_2(OH)$ 1,2-Ethanediol (Ethylene glycol)	$CH_2(OH)-CH(OH)-CH_2(OH)$ 1,2,3-Propanetriol (Glycerol)	OH, OH Resorcinol	OH, O_2N, NO_2, NO_2 2,4,6-Trinitrophenol (Picric acid)

* Common names are given in parentheses.

10.2 CLASSIFICATION OF ALCOHOLS

We classify alcohols according to the carbon to which the OH is bonded. In *primary* (1°) *alcohols*, the OH is bonded to a primary carbon (Sec. 4.4). In *secondary* (2°) *alcohols*, the OH is bonded to a secondary carbon. In *tertiary* (3°) *alcohols*, the OH is bonded to a tertiary carbon. There are no quaternary alcohols. The simplest alcohol, CH_3OH, is in a class by itself.

We shall see in Section 10.11 that primary, secondary, and tertiary alcohols exhibit different chemical reactivities.

Problem 10.1

Classify each of the following alcohols as 1°, 2°, or 3°.

(a) CH_3CH_2OH, (b) $CH_3CH(CH_3)CH_2OH$, (c) $C_6H_5-CH_2OH$,

(d) $CH_3C(CH_3)_2-OH$, (e) $C_6H_5-C(CH_3)(CH_3)-CH(OH)-C(CH_3)(CH_3)-C_6H_5$

Problem 10.2
Classify the alcohols in Table 10.1 as 1°, 2°, or 3°.

10.3 COMMON ALCOHOLS

Common names usually consist of the name of the alkyl group followed by the word alcohol, thus *alkyl alcohol.* Some examples are given in Table 10.1. The names used most often, in addition to those shown in Table 10.1, are shown below.

$CH_3CH_2CH_2OH$
propyl alcohol

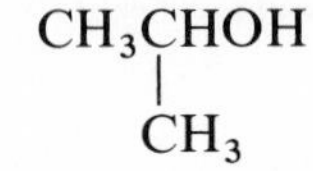

$CH_3CH(CH_3)OH$ (CH₃CHOH with CH₃ branch)
isopropyl alcohol

$CH_3CH_2CH_2CH_2OH$
butyl alcohol

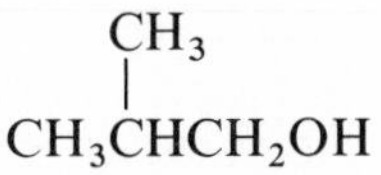

$CH_3CH(CH_3)CH_2OH$
isobutyl alcohol

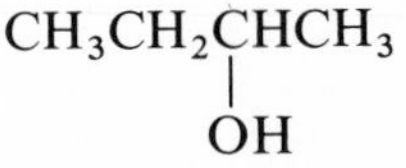

$CH_3CH_2CH(OH)CH_3$
sec-butyl alcohol

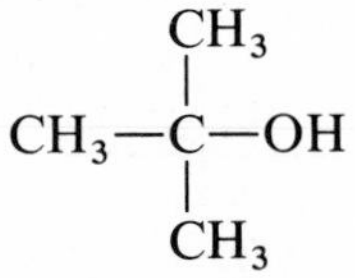

$CH_3-C(CH_3)_2-OH$
tert-butyl alcohol

10.4 NOMENCLATURE OF ALCOHOLS

We form the systematic names of alcohols by using the parent name, *alkanol*, preceded by a number that gives the position of the OH group on the chain. We form the alkanol name by replacing the final *-e* of the alkane with *-ol*.

Examples:

CH_3OH
methanol

CH_3CH_2OH
ethanol

$CH_3CH_2CH_2OH$
1-propanol

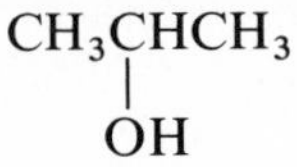

$CH_3CH(OH)CH_3$
2-propanol

$CH_3CH_2CH(OH)CH_2CH_2CH_3$
3-hexanol

If the alcohol contains substituent groups, we name them as described in Section 4.6. We number the chain beginning at the end nearer to the OH group. If a compound contains both an alkene and a hydroxyl group, the hydroxyl group determines the numbering.

Examples:

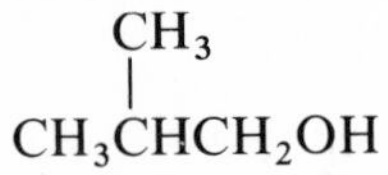

$CH_3CH(CH_3)CH_2OH$
2-methyl-1-propanol

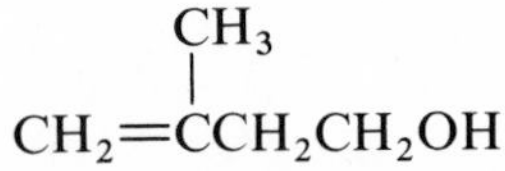

$CH_2{=}C(CH_3)CH_2CH_2OH$
3-methyl-3-buten-1-ol
(*not* 2-methyl-1-buten-4-ol)

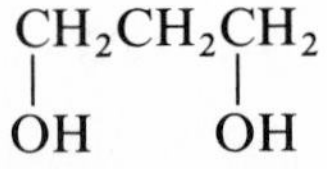

$CH_2(OH)CH_2CH_2(OH)$
1,3-propanediol

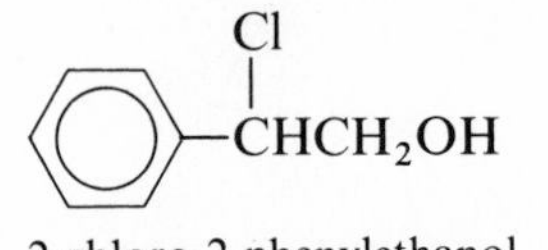

$C_6H_5-CH(Cl)CH_2OH$
2-chloro-2-phenylethanol

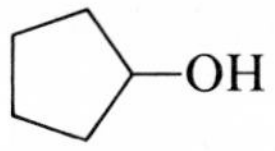

cyclopentanol 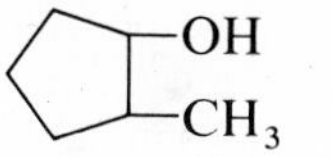 2-methylcyclopentanol

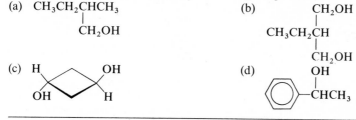

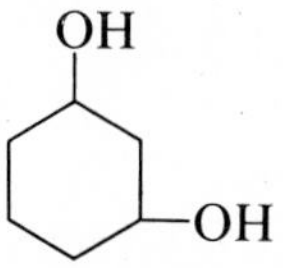

1,3-cyclohexanediol

Problem 10.3

Give the structural formula of each of the following compounds.

(a) 4,4-dimethyl-1-pentanol, (b) 2,3-dimethyl-2-butanol,
(c) 1,3,5-cyclohexanetriol, (d) *trans*-3-methylcyclobutanol

Problem 10.4

Give the systematic name of each of the following compounds.

(a) $CH_3CH_2CHCH_3$ with CH_2OH on C-3 (b) CH_3CH_2CH bearing two CH_2OH groups

(c) cyclobutane with H/OH on C-1 and OH/H on C-3 (d) C_6H_5–CH(OH)CH_3

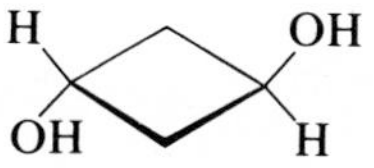

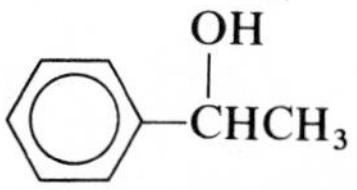

10.5 NOMENCLATURE OF PHENOLS

The simplest phenol, C_6H_5OH, has the name *phenol.* The following phenols have special names:

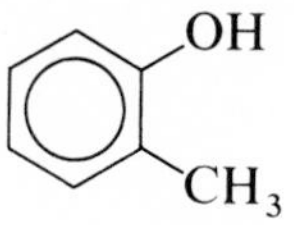

o-cresol 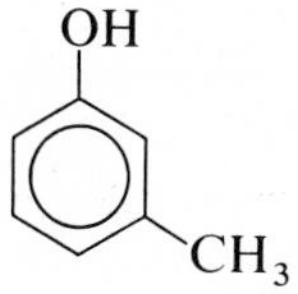 *m*-cresol  *p*-cresol

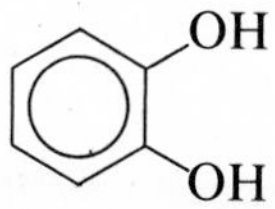

catechol 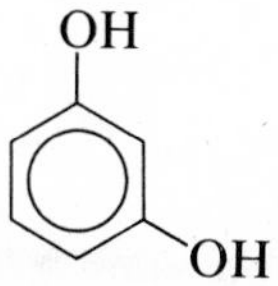 resorcinol  hydroquinone

We must name other phenols as derivatives of phenol and not as derivatives of any of the specially named compounds shown above. We always assign position 1 to the ring carbon that bears the OH group. Some examples will illustrate the systematic nomenclature system for phenols.

4-chloro-2-methylphenol

3-ethylphenol or *m*-ethylphenol

6-chloro-3-methyl-2-nitrophenol

2-chlorophenol or *o*-chlorophenol

2,4,6-tribromophenol

2,6-dinitrophenol

Problem 10.5
Give the structural formula of each of the following compounds.

(a) 2,4-dimethylphenol, (b) resorcinol,
(c) *m*-cresol, (d) *p*-chlorophenol

Problem 10.6
Give the name of each of the following compounds.

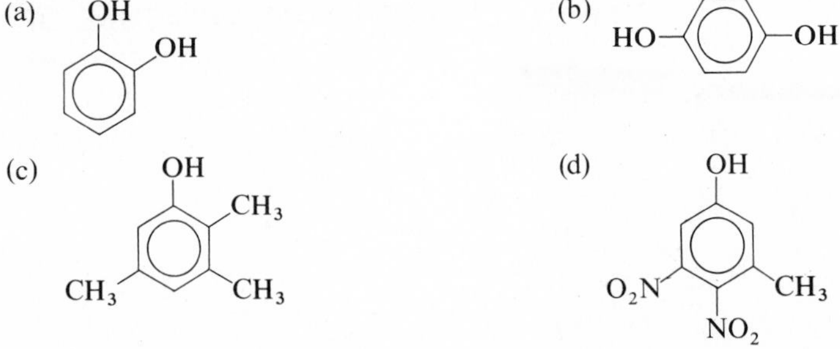

10.6 PHYSICAL PROPERTIES OF ALCOHOLS AND PHENOLS

Boiling Point

Like water, alcohols and phenols have higher boiling points than hydrocarbons of comparable molecular weights (Table 10.2). The high boiling points can be attributed to hydrogen bonding (Sec. 1.8). Due to hydrogen bonding, the separation of molecules from one another during boiling requires more energy than the separation of hydrocarbon molecules.

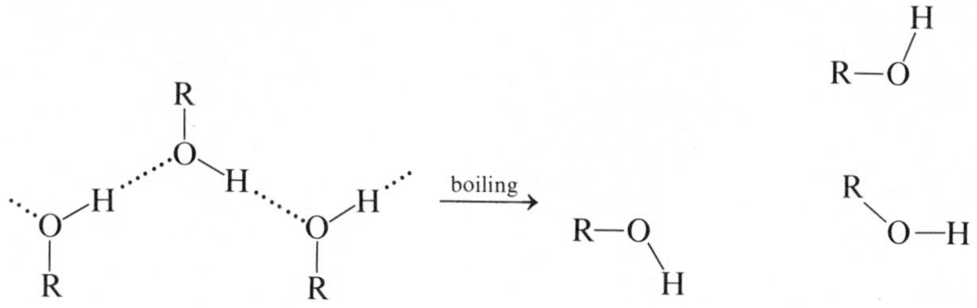

$$\cdots O(R)\!-\!H\cdots O(R)\!-\!H\cdots O(R)\!-\!H\cdots \xrightarrow{\text{boiling}} R\!-\!O\!-\!H \quad R\!-\!O\!-\!H \quad R\!-\!O\!-\!H$$

liquid alcohol — alcohol vapor

Solubility

The lower alcohols are soluble in water. Phenols are only slightly soluble, but they are more soluble than the corresponding hydrocarbons.

We can explain water solubility on the basis of hydrogen bonding between water molecules and alcohol (or phenol) molecules.

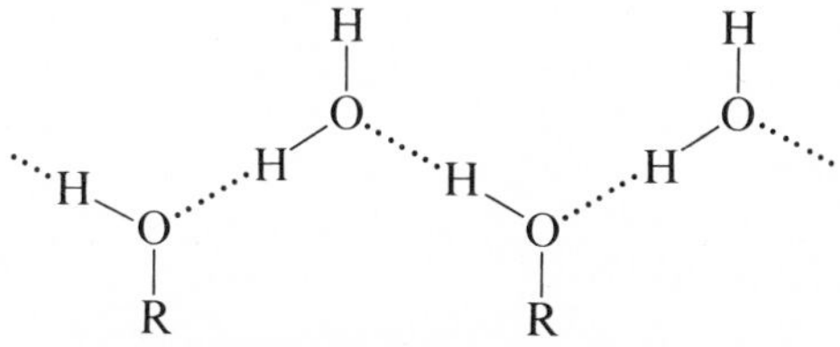

$$\cdots H\!-\!O(R)\cdots H\!-\!O(H)\cdots H\!-\!O(R)\cdots H\!-\!O(H)\cdots$$

The water solubility of alcohols decreases as the number of carbon atoms in the alcohol increases. Alcohols that have more than five carbon atoms are only slightly soluble in water. As alcohol molecules become more "hydrocarbon-like," they become less soluble in water and more soluble in hydrocarbons. This is an example of the rule *like dissolves like.*

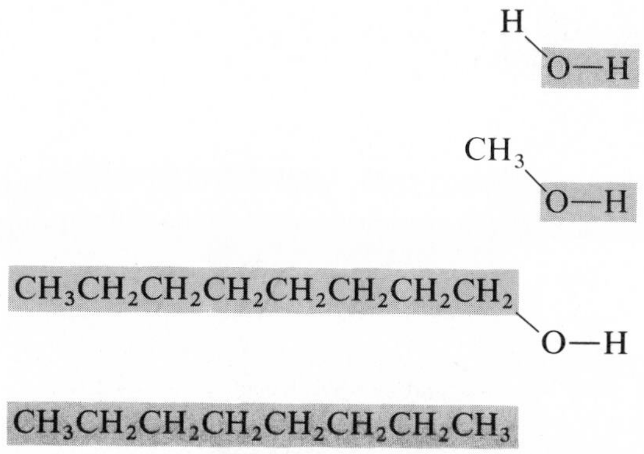

H—O—H

CH_3—O—H Resembles water more than it does methane.

$CH_3CH_2CH_2CH_2CH_2CH_2CH_2CH_2$—$O$—$H$ Resembles octane more than it does water.

$CH_3CH_2CH_2CH_2CH_2CH_2CH_2CH_3$

Table 10.2 Properties of Some Alcohols and Phenols

Name	Formula	Boiling Point, °C	Density, g/ml at 20°C	Solubility, g/100 g H_2O at 25°C
ALCOHOLS				
Methanol	CH_3OH	64.5	0.792	∞
Ethanol	CH_3CH_2OH	78.5	0.789	∞
1-Propanol	$CH_3CH_2CH_2OH$	97.8	0.804	∞
2-Propanol	$CH_3CH(OH)CH_3$	82.5	0.789	∞
1-Butanol	$CH_3CH_2CH_2CH_2OH$	117.7	0.810	7.9
2-Butanol	$CH_3CH(OH)CH_2CH_3$	99.5	0.808	12.5
1-Pentanol	$CH_3(CH_2)_3CH_2OH$	137.9	0.817	2.3
1-Hexanol	$CH_3(CH_2)_4CH_2OH$	156.5	0.819	0.6
1-Octanol	$CH_3(CH_2)_6CH_2OH$	195	0.827	0.05
1-Decanol	$CH_3(CH_2)_8CH_2OH$	228	0.829	insoluble
1,2-Ethanediol	$CH_2(OH){-}CH_2(OH)$	197.5	1.115	∞
1,2,3-Propanetriol	$CH_2(OH){-}CH(OH){-}CH_2(OH)$	290	1.260	∞

Water solubility is also greater when a compound has more than one hydroxyl group per molecule. Examples are the dihydrobenzenes—catechol, resorcinol, and hydroquinone (Table 10.2). This is another example of the rule that like dissolves like; the more OH groups in the molecule, the more "water-like" the molecule, and the greater the opportunity for hydrogen bonding.

Density

In general, alcohols are less dense than water. Only those that have two or more hydroxyl groups per molecule have densities greater than one. Phenols are generally denser than water. They also have higher densities if they have more OH groups.

Problem 10.7
On one graph, plot boiling point versus molecular weight for the unbranched alkanes (Table 4.3) and the unbranched alcohols (Table 10.2). What trend do you see? Explain.

Table 10.2 (*continued*)

Name	Formula	Boiling Point, °C	Density, g/ml at 20°C	Solubility, g/100 g H_2O at 25°C
PHENOLS				
Phenol	C_6H_5—OH	182	1.072	6.7
o-Cresol	*o*-$CH_3C_6H_4$—OH	192	1.047	3.1
m-Cresol	*m*-$CH_3C_6H_4$—OH	203	1.034	2.5
p-Cresol	CH_3—C_6H_4—OH	203	1.035	2.4
Catechol	*o*-HOC_6H_4—OH	204	1.371	45.1
Resorcinol	*m*-HOC_6H_4—OH	277	1.285	229
Hydroquinone	HO—C_6H_4—OH	286	1.358	5.9

10.7 ALCOHOLS AND PHENOLS AS ACIDS AND BASES

The hydroxyl group is potentially an acid and a base, and it is therefore *amphoteric*. It acts as an acid if it donates a hydrogen ion to a base; it acts as a base if it accepts a hydrogen ion from an acid (Sec. 3.1):

$$\underset{\text{acid}}{R{-}\ddot{\underset{..}{O}}{-}H} + :B \rightleftarrows \underset{\substack{\text{conjugate}\\ \text{base}}}{R{-}\ddot{\underset{..}{O}}:^-} + H{-}B^+$$

$$\underset{\text{base}}{R{-}\ddot{\underset{..}{O}}{-}H} + HA \rightleftarrows \underset{\substack{\text{conjugate}\\ \text{acid}}}{R{-}\underset{\substack{|\\ H}}{\ddot{O}^+}{-}H} + A^-$$

The anion $R—O^-$ is called an **alkoxide ion** (methoxide, ethoxide, and so forth), and the cation ROH_2^+ is called an **oxonium ion**.

Alcohols are such weak acids that we normally regard them as neutral. This is because the alkoxide ion, RO^-, is a very strong base—stronger, in fact, than the hydroxide ion. The following equilibrium lies almost entirely to the left:

$$\underset{\text{weaker base}}{R—OH + OH^-} \rightleftharpoons \underset{\text{stronger base}}{R—O^-} + H_2O$$

To cause an alcohol to react as an acid we must find a base that is stronger than the OH^- ion. The NH_2^- ion is such a base. The equilibrium

$$R—OH + \underset{\text{sodium amide}}{NaNH_2} \rightleftharpoons R—ONa + NH_3$$

lies almost entirely to the right, which shows that NH_2^- is a stronger base than $R—O^-$. In this reaction we must use a solvent other than water (liquid NH_3, for example) because the amide ion, NH_2^-, would react with the water to produce the weaker base, the OH^- ion:

$$\underset{\text{stronger base}}{NH_2^-} + H_2O \rightleftharpoons NH_3 + \underset{\text{weaker base}}{OH^-}$$

On the basis of the above equilibria we can rank the base strengths of these three ions:

$$NH_2^- > R—O^- > OH^-$$

Problem 10.8
Phenol reacts completely with NaOH to form sodium phenoxide:

$$C_6H_5—OH + NaOH \longrightarrow C_6H_5—ONa + H_2O$$

Where would you place the phenoxide ion in the ranking of relative basicities given above?

Problem 10.9
Write the net ionic equation for the reaction of Problem 10.8. (The net ionic equation excludes all "spectator" ions—ions that occur on both sides of the equation.)

Table 10.3 lists the acid dissociation constants of the four acids that we have discussed above.

Problem 10.10
Write the equations for the acid dissociations described by the K_as of Table 10.3.

Problem 10.11
On the basis of K_as rank the compounds of Table 10.3 according to their relative acidities.

Table 10.3 Acid Dissociation Constants

Acid	Conjugate Base	K_a
H_2O	OH^-	10^{-16}
CH_3CH_2OH	$CH_3CH_2O^-$	10^{-18}
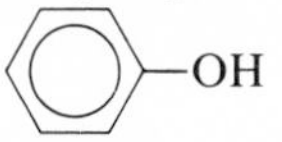 C_6H_5—OH	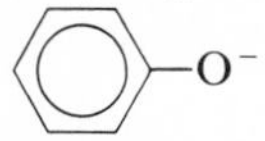 C_6H_5—O^-	10^{-10}
NH_3	NH_2^-	10^{-35}

Phenols are generally much stronger acids than alcohols. We can explain this difference by the weaker basicity of **phenoxide ions** compared with alkoxide ions. The phenoxide ion is a weak base because the negative charge is delocalized over the entire molecule. Resonance structures illustrate this:

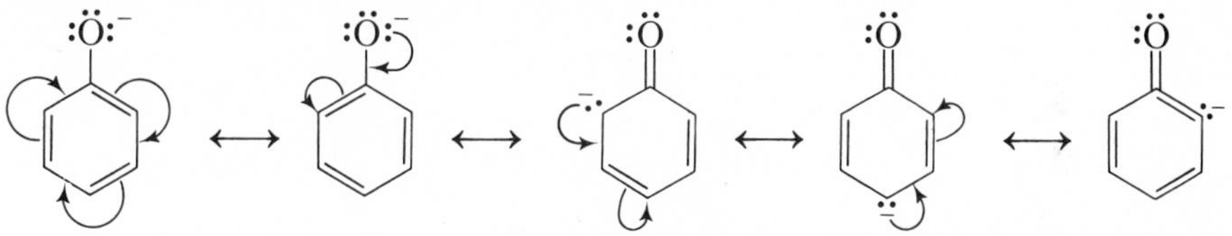

In alkoxide ions, the entire negative charge is localized on the oxygen atom:

$$CH_3{-}CH_2{-}\ddot{\underset{\cdot\cdot}{O}}{:}^-$$

Problem 10.12
Make a free energy versus reaction coordinate diagram (Fig. 3.1) for the dissociation of ethanol (as an acid) in water (as a base), and for the similar acidic dissociation of phenol in water. Explain how resonance affects the energies of reactants and products in both reactions.

Problem 10.13
p-Cresol and 1-pentanol have nearly the same solubilities in water (Table 10.2). Predict their relative solubilities in dilute aqueous NaOH solution.

Alcohols and phenols behave as bases only in the presence of strong acids such as H_2SO_4, H_3PO_4, HCl, HBr, and HI. The reaction of ethanol with concentrated H_2SO_4 is given by the equation

$$\underset{\text{base}}{CH_3CH_2{-}\ddot{\underset{\cdot\cdot}{O}}{-}H} + H_2SO_4 \longrightarrow \underset{\text{conjugate acid}}{CH_3CH_2{-}\overset{\cdot\cdot}{O}^{+}(H){-}H} + HSO_4^-$$

Because concentrated H_2SO_4 is so highly polar, it dissolves the ions that are produced. For this reason, solubility in cold, concentrated sulfuric acid is a good visual test for alcohols, phenols, and in fact for most compounds that contain oxygen atoms.

Problem 10.14
Write an equation for the reaction of phenol with sulfuric acid.

10.8 INDUSTRIAL PREPARATION OF ALCOHOLS: AN ELECTIVE TOPIC

Methanol

elective topic

The most important method for the manufacture of methanol involves the reaction of carbon monoxide with hydrogen in the presence of a special catalyst.

$$CO + 2H_2 \xrightarrow[300\text{–}375^\circ C,\ 4000\text{–}5200\ \text{psi}]{ZnO,\ Cr_2O_3} CH_3OH$$

The CO—H_2 mixture is produced by passing steam over coke:

$$C + H_2O \longrightarrow CH + H_2$$

In another process, methane is catalytically oxidized with atmospheric oxygen.

$$2CH_4 + O_2 \xrightarrow[\text{catalyst}]{450\ \text{psi},\ 450\text{–}470^\circ C} 2CH_3OH$$

Problem 10.15
Suggest two reasons why a laboratory chemist would not use either of the above methods to prepare a small quantity of methanol.

Ethanol

Ethanol was first made in ancient times by the fermentation of fruit juices under the action of yeast. The reaction is now known to involve eleven enzymes acting in combination to catalyze the conversion of sugars to ethanol and CO_2. The old name for this enzyme mixture is zymase.

$$C_6H_{12}O_6 \xrightarrow{\text{zymase}} 2CH_3CH_2OH + 2CO_2$$

In this process, ethanol builds up to a concentration of about 12% and is then removed by distillation.

Problem 10.16
Many home recipes for making wine call for simply allowing raw, unpasteurized grape juice to stand in a barrel without adding yeast. Account for the formation of wine under these conditions. (Hint: If the juice is pasteurized first, it will not form wine unless yeast is added.)

Only a small amount of the ethanol used in the United States is produced by fermentation. The most important industrial method for its production is the hydration of ethylene (Sec. 5.6). This process involves the reaction of ethylene with 98%

sulfuric acid to form the intermediate ethyl hydrogen sulfate, which then reacts with water in a second step to produce the alcohol.

$$CH_2{=}CH_2 + H_2SO_4 \xrightarrow[150\text{–}200\text{ psi}]{50\text{–}80°C} CH_3CH_2{-}O{-}\underset{\underset{O}{\|}}{\overset{\overset{O}{\|}}{S}}{-}OH$$

$$CH_3CH_2{-}O{-}\underset{\underset{O}{\|}}{\overset{\overset{O}{\|}}{S}}{-}OH + H_2O \longrightarrow H_2SO_4 + CH_3CH_2OH$$

The overall reaction amounts to hydration of the ethylene:

$$CH_2{=}CH_2 + H_2O \xrightarrow{H_2SO_4} CH_3CH_2OH$$

Distillation of ethanol from its water solutions always produces a mixture that contains, at most, 95% ethanol. Further distillation cannot yield a purer alcohol because the 95% mixture boils at a slightly lower temperature than pure ethanol. Pure ethanol, called **absolute alcohol**, can be prepared by several methods. One of them involves treating the 95% mixture with lime, CaO, which reacts with the water to form $Ca(OH)_2$.

Problem 10.17
Write and balance the equation for the reaction of CaO with water.

The "proof' of an alcoholic beverage is simply twice the percentage of alcohol. For example, a whiskey that is marked 86 proof contains 43% ethanol.

Ethanol used to make beverages is highly taxed by the government. Much of the ethanol used for other purposes is **denatured**—rendered unfit to drink. There is no tax on denatured alcohol. Alcohol is denatured by adding to it a small amount of a toxic substance. Many different denaturing agents are used including methanol, camphor, gasoline, benzene, and nicotine.

Problem 10.18
What is the proof of a beverage that is 52% ethanol?

Isopropyl Alcohol

Isopropyl alcohol (2-propanol) is also manufactured by hydration. The overall equation is

$$CH_3CH{=}CH_2 + H_2O \xrightarrow{H_2SO_4} CH_3\underset{\underset{OH}{|}}{C}HCH_3$$

Isopropyl alcohol is used as a solvent and as rubbing alcohol.

Hydration of Alkenes

We have discussed hydration of alkenes in Sections 5.6 and 10.8.

Hydroboration

Acid-catalyzed hydration of alkenes obeys Markovnikov's rule (Sec. 5.6). We can also add the elements H—OH to an alkene in the reverse direction.

$$CH_3-\underset{|}{\overset{CH_3}{\overset{|}{C}}}=CH_2 \xrightarrow[\text{addition of H—OH}]{\text{Markovnikov}} CH_3-\underset{OH}{\underset{|}{\overset{CH_3}{\overset{|}{C}}}}-\underset{H}{CH_2}$$

$$CH_3-\overset{CH_3}{\overset{|}{C}}=CH_2 \xrightarrow[\text{addition of H—OH}]{\text{reverse Markovnikov}} CH_3-\underset{H}{\underset{|}{\overset{CH_3}{\overset{|}{C}}}}-\underset{OH}{\underset{|}{CH_2}}$$

We bring about reverse Markovnikov addition of H—OH to an alkene by reacting the alkene with diborane, $(BH_3)_2$. Diborane is a very reactive gas that explodes on contact with air. It is generated when needed by mixing boron trifluoride (BF_3) with sodium borohydride ($NaBH_4$) in the presence of the alkene.

$$BF_3 + 3\,NaBH_4 \longrightarrow 2(BH_3)_2 + 3\,NaF$$

Alkenes react with diborane in three steps to yield the trialkylborane as the final product:

$$2R-\overset{H}{\overset{|}{C}}=CH_2 + (BH_3)_2 \longrightarrow R-\underset{H}{\underset{|}{\overset{H}{\overset{|}{C}}}}-CH_2-BH_2$$

(an alkylborane)

$$R-\overset{H}{\overset{|}{C}}=CH_2 + RCH_2CH_2-BH_2 \longrightarrow RCH_2CH_2-\overset{H}{\overset{|}{B}}-CH_2CH_2R$$

(a dialkylborane)

$$R-\overset{H}{\overset{|}{C}}=CH_2 + (RCH_2CH_2)_2BH \longrightarrow RCH_2CH_2-\underset{CH_2CH_2R}{\underset{|}{B}}-CH_2CH_2R$$

(a trialkylborane)

Oxidation of the trialkylborane with hydrogen peroxide in basic solution yields the alcohol.

$$(RCH_2CH_2)_3B + H_2O_2 \xrightarrow{OH^-} 3RCH_2CH_2OH + B(OH)_3$$

In the addition reactions above we note that the $H{-}B\langle$ undergoes apparent anti-Markovnikov addition to the double bond. The reason for this apparent reversal is that boron is less electronegative than hydrogen, so the bond is polarized as shown:

$$\overset{\delta^-}{H}{-}\overset{\delta^+}{B}\langle$$

Hydrogen therefore behaves like an anion ($H{:}^-$) rather than a cation (H^+).

Hydrolysis of Alkyl Halides: S_N and E Reactions

We discussed S_N1 and S_N2 reactions of alkyl halides in Section 9.6. We summarize the use of those reactions for the preparation of alcohols as follows.

1. Primary alkyl halides react through the S_N2 mechanism to yield primary alcohols as the major product.

$$RCH_2CH_2X + NaOH \longrightarrow RCH_2CH_2OH + NaX$$

2. Secondary alkyl halides react through the S_N2 mechanism to yield secondary alcohols; however, E2 elimination to form the alkene is the major reaction.

$$RCH(X)CH_2R' + NaOH \xrightarrow{S_N2} RCH(OH)CH_2R' + NaX$$

$$RCH(X)CH_2R' + NaOH \xrightarrow{E2} RCH{=}CHR' + NaX$$

(major product)

3. Tertiary alkyl halides yield alcohols through the S_N1 reaction. The S_N1 reaction usually predominates over the E1 elimination, and the alcohol is the major product.
4. No S_N2 reaction occurs with tertiary alkyl halides. If we use conditions that favor the second-order reaction, only E2 elimination occurs.
5. S_N1 and E1 reactions do not occur with ordinary primary or secondary alkyl halides. S_N1 and E1 reactions are favored by using a polar solvent, usually water, and a weak nucleophile (base). S_N2 and E2 reactions are favored by using a strong nucleophile (base); the solvent is relatively unimportant.

Problem 10.19
Give the major organic product of each reaction below.

(a) $CH_3CH{=}CH_2 \xrightarrow[H_2SO_4]{H_2O}$

(b) $CH_3CH{=}CH_2 \xrightarrow{(BH_3)_2} \xrightarrow[OH^-]{H_2O_2}$

(c) $CH_3CH(Cl)CH_3 \xrightarrow[ethanol]{KOH}$

(d) $CH_3CH_2CH_2Cl \xrightarrow[ethanol]{KOH}$

(e) $(CH_3)_3CCl \xrightarrow[ethanol]{KOH}$

(f) $(CH_3)_3CCl \xrightarrow[H_2O]{dilute\ KOH}$

10.10 LABORATORY PREPARATION OF PHENOLS

Reaction of an aryl halide with OH^- is not a suitable laboratory method for phenol synthesis unless the benzene ring contains nitro groups or other strongly electron-withdrawing groups ortho or para to the halide. Note in the following examples that lower reaction temperatures are required as more nitro groups are present in the ring.

$$\text{2-chloronitrobenzene (Cl, } NO_2\text{)} + OH^- \xrightarrow{130°C} \text{2-nitrophenol (OH, } NO_2\text{)}$$

$$\text{1-chloro-2,4-dinitrobenzene (Cl, } NO_2, NO_2\text{)} + OH^- \xrightarrow{100°C} \text{2,4-dinitrophenol (OH, } NO_2, NO_2\text{)}$$

$$\text{1-chloro-2,4,6-trinitrobenzene (Cl, } O_2N, NO_2, NO_2\text{)} + OH^- \xrightarrow{35°C} \text{2,4,6-trinitrophenol (OH, } O_2N, NO_2, NO_2\text{)}$$

The electron-withdrawing nitro groups apparently make the benzene ring more positive and therefore more susceptible to attack by a nucleophile.

Reaction With Active Metals

Alcohols, like water, react with sodium, potassium, and other alkali metals to produce hydrogen gas and an alkoxide:

$$2R{-}OH + 2Na \longrightarrow 2R{-}ONa + H_2\uparrow$$

The evolution of hydrogen gas is a test for the hydroxyl or other acidic group.

Alkyl Halides From Alcohols

We discussed the reaction of alcohols with hydrogen halides in Section 9.3. Both S_N1 and S_N2 mechanisms occur (Sec. 9.6).

S_N1 Mechanism (with 3° alcohols)

$$R{-}\ddot{\underset{\cdot\cdot}{O}}H + HX \rightleftharpoons R{-}\overset{\cdot\cdot\,+}{\underset{\underset{H}{|}}{O}}H + X^-$$

$$R{-}\overset{+}{O}H_2 \underset{}{\overset{\text{slow}}{\rightleftharpoons}} R^+ + H_2O$$

$$R^+ + X^- \xrightarrow{\text{fast}} R{-}X$$

S_N2 Mechanism (with 1° and 2° alcohols)

$$R{-}\ddot{\underset{\cdot\cdot}{O}}H + HX \rightleftharpoons R{-}\overset{+}{O}H_2 + X^-$$

$$X^- + R{-}\overset{+}{O}H_2 \rightleftharpoons [\overset{\delta^-}{X}\cdots R\cdots\overset{\delta^-}{O}H_2] \longrightarrow X{-}R + H_2O$$

(activated state)

The S_N reaction of alcohols with HCl is the basis of the **Lucas test** used to determine whether an alcohol is primary, secondary, or tertiary. In the Lucas test, we mix the alcohol with a solution of $ZnCl_2$ (a powerful Lewis acid) in concentrated hydrochloric acid. Most alcohols dissolve in this strong acid solution because they form oxonium ions. (Tertiary alcohols require only H_3O^+; primary and secondary alcohols require the stronger Lewis acid $ZnCl_2$.)

$$3^\circ \quad R{-}OH + H_3O^+ \rightleftharpoons R{-}\underset{\underset{H}{|}}{\overset{+}{O}}{-}H + H_2O$$

(an oxonium ion)

$$\rightleftharpoons R^+ + H_2O$$

a carbocation

$$1^\circ \text{ or } 2^\circ \quad R{-}OH + ZnCl_2 \rightleftharpoons R{-}\underset{\underset{H}{|}}{\overset{+}{O}}{-}ZnCl_2$$

(an oxonium ion)

Reaction of the carbocation or oxonium ion with the Cl^- ion yields an alkyl halide that is insoluble in this mixture and that precipitates out as a visible, separate layer. The Lucas test is based on the different reaction rates of primary, secondary, and tertiary alcohols:

3°	$R{-}OH \xrightarrow[ZnCl_2]{HCl} R{-}Cl$	(Alkyl chloride layer forms immediately.)
2°	$R{-}OH \xrightarrow[ZnCl_2]{HCl} R{-}Cl$	(Solution becomes cloudy after a minute or so, and the alkyl chloride layer is clearly visible within 5 minutes.)
1°	$R{-}OH \xrightarrow[ZnCl_2]{HCl} R{-}Cl$	(No visible sign of reaction even after 30 minutes or more.)

Allyl and benzyl alcohols react very rapidly in the Lucas test because allylic and benzylic carbocations are very stable (Sec. 9.8). Phenols, on the other hand, do not react because phenyl carbocations are extremely unstable and do not form.

Problem 10.20
Predict how each of the following compounds would react in the Lucas test.

(a) $CH_3C(CH_3)_2CH_2OH$ (drawn as CH_3CCH_2OH with CH_3 above and below the central C)

(b) $C_6H_5{-}CH_2OH$ (benzene ring—CH_2OH)

(c) $CH_2{=}CHCH_2OH$

(d) $C_6H_5{-}OH$ (benzene ring—OH)

We may also convert alcohols into alkyl halides by using thionyl chloride ($SOCl_2$) or phorphorus trihalide (PX_3, where X = Cl, Br, or I). In each of these reactions the intermediate in an ester.

$$R{-}OH + Cl{-}\overset{\overset{O}{\|}}{S}{-}Cl \longrightarrow HCl + R{-}O{-}\overset{\overset{O}{\|}}{S}{-}Cl \quad \text{(an alkyl chlorosulfite)}$$

$$\longrightarrow R{-}Cl + SO_2$$

$$R{-}OH + X{-}\underset{\underset{X}{|}}{P}{-}X \longrightarrow HCl + R{-}O{-}\underset{\underset{X}{|}}{P}{-}X \quad \text{(an alkyl chlorophosphite)}$$

$$\longrightarrow R{-}X + HO{-}\underset{\underset{OH}{|}}{P}{-}OH$$

phosphorous acid

The reaction with $SOCl_2$ goes essentially to completion due to the formation of gaseous products—HCl and SO_2—which are easily removed. When we use PX_3, we can remove the alkyl halide by distillation because phosphorous acid boils at a very high temperature.

Carboxylic Esters

Alcohols react with a variety of acids and acid derivatives to form **esters**. Alcohols react directly with carboxylic acids in the presence of a mineral acid catalyst to produce carboxylic esters. The reaction is an equilibrium and may be forced to the right by removing the water as it forms:

$$\underset{}{R{-}OH} + \underset{\text{carboxylic acid}}{HO{-}\overset{\overset{\displaystyle O}{\|}}{C}{-}R'} \overset{H^+}{\rightleftharpoons} \underset{\text{carboxylic ester}}{R{-}O{-}\overset{\overset{\displaystyle O}{\|}}{C}{-}R'} + H_2O$$

Inorganic Esters

When a primary alcohol is mixed with pure sulfuric acid at 0°C, the product is an alkyl hydrogen sulfate:

$$R{-}OH + \underset{\text{sulfuric acid}}{HO{-}\overset{\overset{\displaystyle O}{\|}}{\underset{\underset{\displaystyle O}{\|}}{S}}{-}OH} \overset{0°C}{\rightleftharpoons} \underset{\text{alkyl hydrogen sulfate}}{R{-}O{-}\overset{\overset{\displaystyle O}{\|}}{\underset{\underset{\displaystyle O}{\|}}{S}}{-}OH} + H_2O$$

Alkyl hydrogen sulfates are very reactive. If the mixture of alcohol and H_2SO_4 is heated, the reaction yields the ether or the alkene depending on the temperature:

$$2CH_3CH_2OH \xrightarrow{H_2SO_4/140°C} CH_3CH_2{-}O{-}CH_2CH_3 + H_2O$$

$$CH_3CH_2OH \xrightarrow{H_2SO_4/180°C} CH_2{=}CH_2 + H_2O$$

Nitric acid reacts with alcohols to yield alkyl nitrates:

$$R{-}OH + \underset{\text{nitric acid}}{HO{-}N\begin{matrix}{/\!\!/}O\\ \diagdown O\end{matrix}} \longrightarrow \underset{\text{alkyl nitrate (a nitric ester)}}{R{-}O{-}N\begin{matrix}{/\!\!/}O\\ \diagdown O\end{matrix}} + H_2O$$

Nitric esters are explosive. A well-known example is the trinitric ester of glycerol.

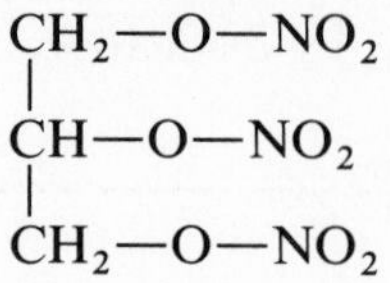

$$\begin{array}{l} CH_2-O-NO_2 \\ | \\ CH-O-NO_2 \\ | \\ CH_2-O-NO_2 \end{array} \quad \begin{array}{l}\text{glyceryl trinitrate} \\ \text{(nitroglycerin)}\end{array}$$

We prepare the trialkylphosphates by heating an alcohol with the acid halide $POCl_3$:

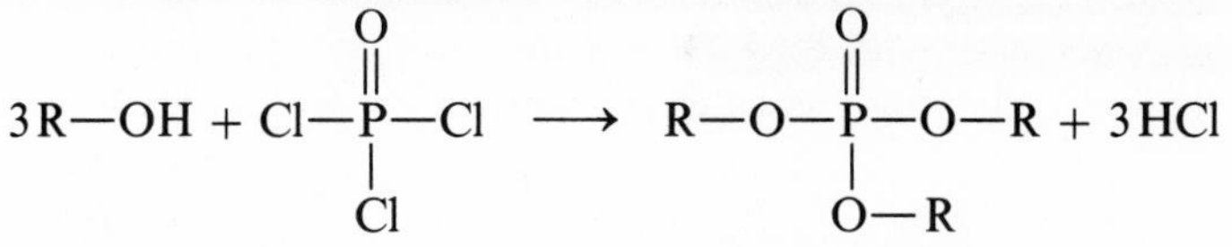

$$3R-OH + Cl-\overset{\overset{\displaystyle O}{\|}}{\underset{\underset{\displaystyle Cl}{|}}{P}}-Cl \longrightarrow R-O-\overset{\overset{\displaystyle O}{\|}}{\underset{\underset{\displaystyle O-R}{|}}{P}}-O-R + 3HCl$$

trialkyl phosphate

Alkyl dihydrogen phosphates and dialkyl hydrogen phosphates may also be prepared.

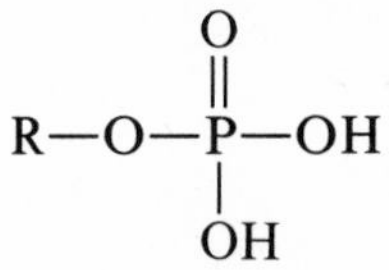

$$R-O-\overset{\overset{\displaystyle O}{\|}}{\underset{\underset{\displaystyle OH}{|}}{P}}-OH \qquad R-O-\overset{\overset{\displaystyle O}{\|}}{\underset{\underset{\displaystyle OH}{|}}{P}}-O-R$$

alkyl dihydrogen phosphate 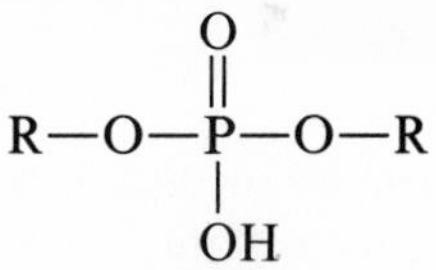 — dialkyl hydrogen phosphate

Phosphate esters are important in biochemical processes. A well-known example is deoxyribonucleic acid (DNA), which is a polymeric phosphate ester of the sugar, deoxyribose (a diol):

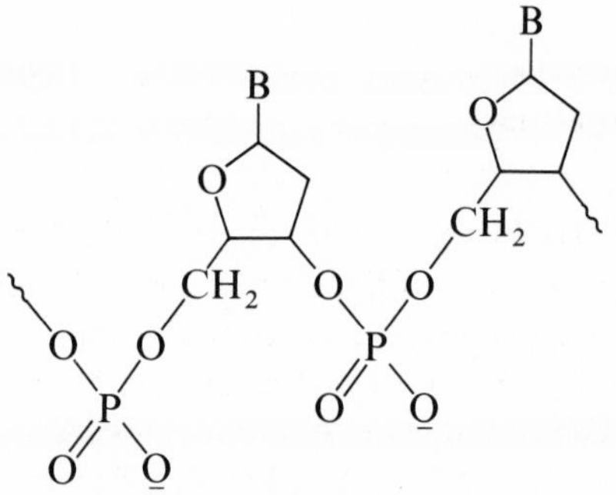

B
O
B
O
CH_2
CH_2
O O
O O
P
P
O O
O O

(B = a cyclic organic nitrogen group)

Problem 10.21
Draw the structural formula of the diol from which the DNA structure above is formed.

Oxidation

Primary alcohols are readily oxidized to aldehydes, and secondary alcohols are oxidized to ketones:

$$R{-}CH_2{-}OH \xrightarrow{\text{oxidize}} R{-}\overset{\overset{\displaystyle O}{\|}}{C}{-}H \quad \text{(an aldehyde)}$$

$$R{-}\underset{}{\overset{\overset{\displaystyle OH}{|}}{C}H}{-}R' \xrightarrow{\text{oxidize}} R{-}\overset{\overset{\displaystyle O}{\|}}{C}{-}R' \quad \text{(a ketone)}$$

Tertiary alcohols are not easily oxidized because oxidation requires the breakage of a carbon–carbon bond:

$$R{-}\underset{\underset{\displaystyle R''}{|}}{\overset{\overset{\displaystyle R'}{|}}{C}}{-}OH \xrightarrow{\text{oxidize}} \text{no reaction}$$

A convenient oxidizing agent for alcohols is chromic acid, H_2CrO_4. Chromic acid is orange-red; its reduction product is the green Cr^{3+} ion:

$$3R{-}CH_2{-}OH + \underset{\text{orange-red}}{2H_2CrO_4} + 6H_3O^+ \longrightarrow 3R{-}\overset{\overset{\displaystyle O}{\|}}{C}{-}H + \underset{\text{green}}{2Cr^{3+}} + 14H_2O$$

This color change allows us to use this reaction as a test for alcohols (1° or 2° only).

Aldehydes are oxidized far more easily than alcohols. For this reason the oxidation does not usually stop at the aldehyde, but instead goes on to give the carboxylic acid. In some cases the aldehyde may be removed as it forms. For example, many aldehydes have lower boiling points than the alcohol or the carboxylic acid; such an aldehyde can be distilled away from the oxidizing mixture as soon as it forms.

$$\underset{\text{bp } 78^\circ\text{C}}{CH_3CH_2OH} \xrightarrow{H_2CrO_4} \underset{\text{bp } 21^\circ\text{C}}{CH_3\overset{\overset{\displaystyle O}{\|}}{C}H} \xrightarrow{H_2CrO_4} \underset{\text{bp } 118^\circ\text{C}}{CH_3\overset{\overset{\displaystyle O}{\|}}{C}{-}OH}$$

Problem 10.22
Give the product of the reaction of 1-propanol with each of the following reagents.
(a) concentrated HBr, (b) H_2CrO_4, (c) H_2SO_4/0°C,
(d) Na, (e) $NaOH/H_2O$, (f) CH_3CO_2H/H^+

Problem 10.23
Repeat Problem 10.22 using 2-propanol instead of 1-propanol.

10.12 REACTIONS OF PHENOLS

Reaction With Alkali Metals

We have already seen that phenols react as acids with NaOH. Phenols also react with alkali metals to produce hydrogen gas.

$$2\,ArOH + 2\,Na \longrightarrow 2\,ArO^- + H_2\uparrow$$
(Ar = any aromatic ring)

Reaction With Bromine

The hydroxyl group renders the benzene ring of phenols very reactive. Bromine, for example, reacts with phenol at room temperature to give 2,4,6-tribromophenol even without an iron catalyst:

$$C_6H_5OH + 3\,Br_2 \xrightarrow[25°C]{H_2O} 2,4,6\text{-}Br_3C_6H_2OH + 3\,HBr$$

To obtain monobromination, we must carry out the reaction at a temperature near 0°C.

Once again, the color change serves as a visual test; bromine is red, and the products are only slightly colored.

Problem 10.24
Write equations for two reactions that convert phenol into sodium phenoxide.

Problem 10.25
Give a simple chemical test that is positive for alcohols and negative for phenols, and one that is negative for alcohols and positive for phenols.

10.13 OCCURRENCE AND USES OF ALCOHOLS AND PHENOLS: AN ELECTIVE TOPIC

elective topic

In addition to methanol, ethanol, and isopropyl alcohol (discussed in Sec. 10.8), ethylene glycol and glycerol are also widely used.

Ethylene glycol (1,2-ethanediol) is used as an antifreeze agent because of its complete solubility in water and its high boiling point (198°C).

$$\begin{array}{c} CH_2OH \\ | \\ CH_2OH \end{array} \qquad\qquad \begin{array}{c} CH_2OH \\ | \\ CHOH \\ | \\ CH_2OH \end{array}$$

ethylene glycol   glycerol

Glycerol (1,2,3-propanetriol) is also called glycerin. It is a by-product in the manufacture of soap. Soap is manufactured by boiling fats with sodium or potassium hydroxide. Fats are esters of glycerin and long-chain carboxylic acids.

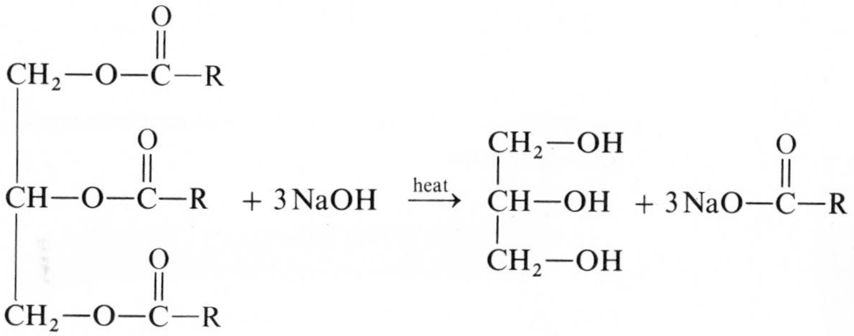

$$\begin{array}{l} CH_2{-}O{-}\overset{\overset{\large O}{\|}}{C}{-}R \\ | \\ CH{-}O{-}\overset{\overset{\large O}{\|}}{C}{-}R \\ | \\ CH_2{-}O{-}\overset{\overset{\large O}{\|}}{C}{-}R \end{array} + 3\,NaOH \xrightarrow{\text{heat}} \begin{array}{l} CH_2{-}OH \\ | \\ CH{-}OH \\ | \\ CH_2{-}OH \end{array} + 3\,NaO{-}\overset{\overset{\large O}{\|}}{C}{-}R$$

a fat   glycerol   a soap

Glycerol is a sweet-tasting, colorless, syrupy liquid. It is used as a solvent, as a moistening agent, as a softening agent in soaps and lotions, and as an additive to alter the properties of plastics.

Phenol is used as an antiseptic. It is commonly known as carbolic acid. Other phenols are widely used; hydroquinone, for example, is used as a photographic developer.

10.14 SULFUR ANALOGS OF ALCOHOLS AND PHENOLS: THIOLS AND THIOPHENOLS

Sulfur occurs immediately beneath oxygen in the periodic table, and the two elements have many properties in common.

Thiols or **mercaptans**, correspond to the alcohols with a sulfur atom in place of an oxygen. The formulas below show IUPAC names; common names are in parentheses.

CH_3CH_2SH	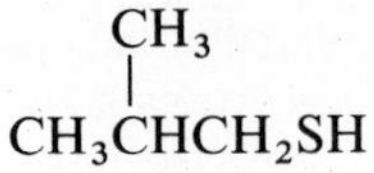 $CH_3CH(CH_3)CH_2SH$	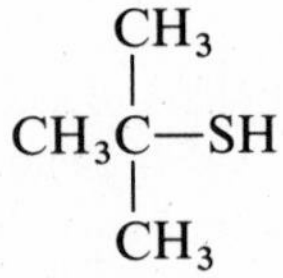 $CH_3C(CH_3)_2$—SH
ethanethiol (ethyl mercaptan)	2-methyl-1-propanethiol (isobutyl mercaptan)	2-methyl-2-propanethiol (*tert*-butyl mercaptan)

The SH group is often called the **sulfhydryl** group. Sulfur analogs of phenols are called **thiophenols**.

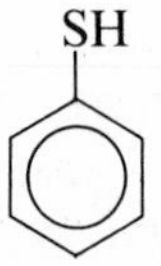

thiophenol   *p*-bromothiophenol

We can prepare thiols from alkyl halides by nucleophilic substitution using sodium hydrosulfide, NaSH.

$$R—Cl + SH^- \longrightarrow R—SH + Cl^-$$

In general, thiols are more acidic than alcohols just as H_2S is more acidic than H_2O.

Problem 10.26
Write the equation for the equilibrium that results when sodium ethoxide and ethanethiol are mixed together. Does the equilibrium favor reactants or products?

Thiols undergo an oxidation reaction that is not observed with alcohols. Oxidation of a thiol with hydrogen peroxide yields a **disulfide**:

$$2R—SH + H_2O_2 \longrightarrow \underset{\text{a disulfide}}{R—S—S—R} + 2H_2O$$

The O—H bond is stronger than the S—H bond; therefore oxidation of alcohols occurs at the weaker C—H bond rather than at the stronger O—H bond.

$$\mathrm{RCH_2{-}\ddot{\underset{..}{S}}{-}H \xrightarrow[(-H)]{\text{oxidize}} RCH_2{-}\ddot{\underset{..}{S}}\cdot \longrightarrow \underset{\text{a disulfide}}{R{-}CH_2{-}S{-}S{-}CH_2{-}R}}$$

$$\mathrm{\underset{\displaystyle H}{\underset{|}{RCH}}{-}\ddot{\underset{..}{O}}{-}H \xrightarrow[(-H)]{\text{oxidize}} R{-}\underset{\cdot}{C}H{-}\ddot{\underset{..}{O}}{-}H \xrightarrow[(-H)]{} \underset{\text{an aldehyde}}{R{-}\overset{\overset{\displaystyle O}{\|}}{C}H}}$$

The thiol ⇄ disulfide oxidation-reduction reaction is important in biochemical processes, especially in the formation of linkages that hold protein chains together in particular conformations. An example is the protein insulin, which has three interchain disulfide linkages. Interaction of random pairs of —SH groups would give "wrong" conformations of the chain. In the living system, the oxidation is catalyzed by the enzyme glutathion-insulin transhydrogenase, which not only catalyzes the oxidation but also brings together the "correct" thiol groups.

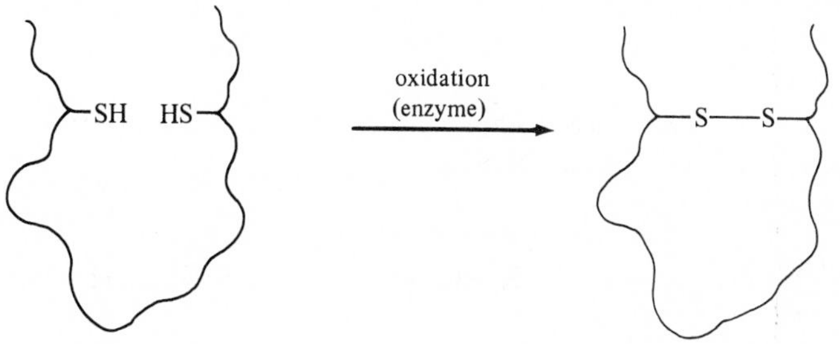

SH HS —oxidation (enzyme)→ S—S

reduced insulin — oxidized insulin

10.15 ANALYSIS AND IDENTIFICATION OF ALCOHOLS AND PHENOLS

Ferric Chloride Test

Throughout this chapter we have encountered many reactions that are useful tests for alcohols and phenols. In addition to the ones already discussed, an important test for phenols is the ferric chloride test. Phenols and enols—groups that have the $-\overset{|}{C}{=}\overset{|}{C}{-}OH$ group—react with aqueous $FeCl_3$ solutions to give pink, violet, or other colored solutions. The color depends on the phenol used, and arises from a complex formed between the phenol and the $FeCl_3$. Although the exact structure

Table 10.4 Simple Chemical Tests for Alcohols and Phenols

Test	Alcohols	Phenols
Sodium metal	Evolution of H_2	Evolution of H_2
Aqueous NaOH	Higher alcohols are insoluble	Soluble
Cold, concentrated H_2SO_4	Soluble	Soluble
Lucas test—concentrated HCl + $ZnCl_2$	3° alcohols → insoluble layer immediately 2° alcohols → insoluble layer within 5 min 1° alcohols are unreactive	No reaction
H_2CrO_4	1° and 2° alcohols → green color 3° alcohols are unreactive	Ring oxidation → green color
Br_2/H_2O	No reaction	Rapid reaction → loss of Br_2 color
$FeCl_3$ test	No reaction	Color change

of these complexes is not known, they undoubtedly contain a coordinate covalent bond between the phenolic OH and the iron:

$$\mathrm{H{-}\overset{+}{\ddot{O}}{-}\overset{-}{Fe}Cl_3}\quad(\text{O bonded to a benzene ring})$$

Alcohols do not give a color with ferric chloride solution.

Table 10.4 lists the chemical tests that can be used to identify alcohols and phenols.

NEW TERMS

Absolute alcohol 100% ethanol (10.8).

Alcohol A compound in which a hydroxyl group is bonded to an aliphatic hydrocarbon group (10.1).

Alkoxide ion The conjugate base (RO^-) of an alcohol (10.7).

Classification of alcohols Alcohols are classified according to the classification of the carbon atom (Sec. 4.4) to which the OH group is bonded. In primary (1°) alcohols, OH is bonded to a 1° carbon atom; in secondary (2°) alcohols, OH is bonded to a 2° carbon atom; in tertiary (3°) alcohols, OH is bonded to a 3° carbon atom (10.2).

Denatured alcohol Ethanol that has been rendered unfit to drink by the addition of a small amount of toxic substance (10.8).

Disulfides R—S—S—R in which R is alkyl or aryl (10.14).

Ester Product of the condensation reaction of an alcohol with an acid (10.11).

Hydroboration Reaction of an alkene with diborane $(BH_3)_2$ to produce the trialkylborane. Trialkylboranes can be oxidized with $H_2O_2 + OH^-$ to produce the corresponding alcohol. The overall reaction is anti-Markovnikov (10.9).

Lucas test Test used to distinguish among 1°, 2°, and 3° alcohols by observing the rate of reaction of an unknown alcohol with a solution of $ZnCl_2$ in concentrated HCl. The rates of formation of a second layer (the alkyl chloride) are in the order 3° > 2° > 1° (10.11).

Oxidation See Section 4.10 (10.11).

Oxonium ion The conjugate acid (ROH_2^+) of an alcohol (10.7).

Phenol A compound in which a hydroxyl group is bonded to an aromatic ring (10.1).

Phenoxide ion The conjugate base (ArO^-) of a phenol (10.7).

Sulfhydryl The —SH group (10.14).

Thiol R—SH in which R is alkyl. Thiols are also called **mercaptans** (10.14).

Thiophenols Ar-SH in which Ar is aryl (10.14).

ADDITIONAL PROBLEMS

10.27 Write structural formulas for each of the following compounds.

[a] 2,4-dichloro-1-hexanol
(b) *cis*-2-methylcyclopentanol
[c] 1-phenyl-1-propanol
(d) *cis*-1,3-cyclobutanediol
[e] *trans*-1,3-cyclobutanediol
(f) sodium *tert*-butoxide
[g] potassium *p*-nitrophenoxide
(h) 3-cyclohexenol
[i] 3-methyl-3-pentene-1-ol

[10.28] Classify the alcohols in Problem 10.27 as primary, secondary, or tertiary.

10.29 Name each of the following compounds.

[a] HO— —OH

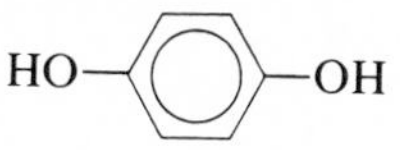

[b]
OH
|
$CH_3CH_2CHCH_2CH_3$

[c] Cl— —CH_2OH

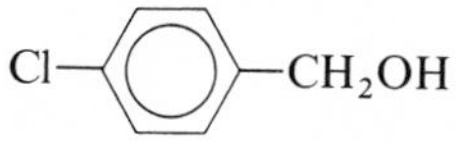

(d) Br, —OH

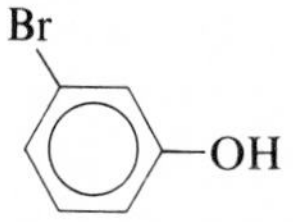

(e) Br, CH_3CH_2— —OH

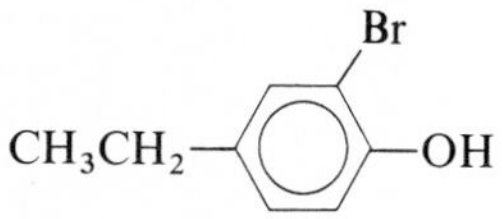

(f) OH

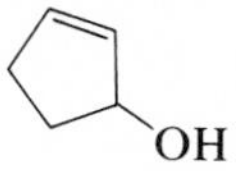

[10.30] Explain why diols are more soluble in water than alcohols of comparable molecular weight.

10.31 For each of the following pairs of compounds select the one with the higher boiling point and the one with the greater solubility in water. Explain your choices.

[a] $CH_3CH_2CH_2OH$ and $CH_3CH_2CH_2F$

[b] CH_3CHCH_3 and CH_3CHCH_3
with OH | and CH_3 | on the central carbons respectively

(c) glycerol and hexane

(d) glycerol and 1,2-propanediol

10.32 Write the net ionic equation for the reaction that might occur when we mix each of the following pairs of compounds. Show with arrows of unequal length whether the equilibrium occurs to a great extent to form reactants or products.

[a] C_6H_5—ONa + H_2O

(b) C_6H_5—SH + NaOH (aqueous)

[c] $CH_3CH_2CH_2SNa + CH_3CH_2CH_2OH$

(d) CH_3OH + NaOH (aqueous)

[e] $CH_3CH_2ONa + NH_3$

(f) C_6H_5—OH + CH_3CH_2ONa

[10.33] Explain why picric acid (2,4,6-trinitrophenol) is a stronger acid than phenol.

10.34 Write equations showing all the steps needed to convert 1-butene into each of the following compounds.

[a] $CH_3CH_2CH(OH)CH_3$

(b) $CH_3CH_2CH_2CH_2OH$

[c] $CH_3CH{=}CHCH_3$

(d) $CH_3CH_2CHClCH_3$

[e] $CH_3CH_2C(=O)CH_3$

(f) $CH_3CH_2CH_2CH{=}O$

10.35 Write equations showing all the steps needed to convert 2-phenylethanol into each of the following compounds.

[a] C_6H_5—CH_2CO_2H

[b] C_6H_5—$CH{=}CH_2$

[c] C_6H_5—$CH(OH)CH_3$

[d] C_6H_5—$C(=O)$—CH_3

(e) C_6H_5—CH_2CH_2Br

(f) C_6H_5—$CHClCH_3$

(g) C_6H_5—CH_2CH_2—O—$C(=O)CH_3$

10.36 Write equations showing all the steps needed to carry out each of the following transformations.

[a] C_6H_5—$CH_2OH \xrightarrow{?} C_6H_5$—$CH_2ONa$

(b) $Br-C_6H_4-OH \xrightarrow{?} Br-C_6H_4-ONa$

[c] $CH_3CH_2CH_2{-}SH \xrightarrow{?} CH_3CH_2CH_2{-}S{-}S{-}CH_2CH_2CH_3$

(d) $CH_3CH_2CH_2OH \xrightarrow{?} CH_3CH_2CH_2SH$

10.37 Give as many simple chemical tests as you can that would allow you to distinguish between the compounds in each of the following pairs. Write equations for each test and explain what you would observe.

[a] propyl alcohol and isopropyl alcohol
[b] propyl alcohol and 1-chloropropane
(c) phenol and chlorobenzene
[d] phenol and 2-hexanol
(e) allyl alcohol and *tert*-butyl alcohol
[f] *o*-cresol and *o*-xylene
(g) phenol and cyclohexanol

[10.38] Compound A is a liquid that is only slightly soluble in water. It reacts with sodium to evolve H_2, but does not dissolve in aqueous NaOH solution. Oxidation with chromic acid gives a carboxylic acid with five carbons. Write structures for all the possible compounds that A might be.

10.39 Compound B (C_7H_8O) is slightly soluble in water but dissolves in aqueous NaOH solution. It gives a positive $FeCl_3$ test and decolorizes bromine-water solution. Suggest a possible structure for compound B.

[10.40] Compound C ($C_5H_{12}O$) gives an immediate reaction in the Lucas test. What are the possible structures of C?

11 ETHERS AND EPOXIDES

Ethers are structurally related to water and the alcohols. The best known ether is the anesthetic called diethyl ether. Other ethers have a wide range of uses as solvents, refrigerants, artificial flavors, and drugs. We shall see that ethers as a class are relatively unreactive chemically.

11.1 STRUCTURE OF ETHERS

The ether functional group consists of an oxygen atom bonded to two hydrocarbon groups. The two groups may be saturated, unsaturated, or aromatic, or the oxygen may be bonded to two ends of a chain as in the cylic ethers. Some typical ethers are

$$CH_3CH_2{-}O{-}CH_2CH_3 \qquad CH_3{-}O{-}C_6H_5$$

diethyl ether — anisole

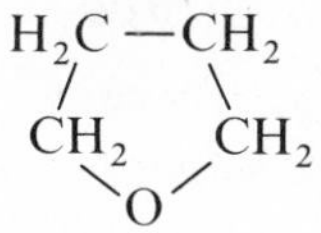

$$\begin{array}{c} H_2C-CH_2 \\ / \quad \backslash \\ CH_2 \quad CH_2 \\ \backslash \; O \; / \end{array}$$

tetrahydrofuran

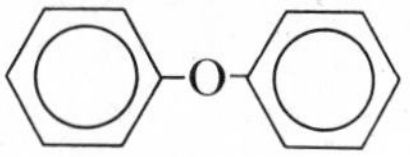

$C_6H_5-O-C_6H_5$

diphenyl ether

Ethers are isomeric with the alcohols. That is, diethyl ether is an isomer of butanol, $CH_3CH_2CH_2CH_2OH$. Both have the molecular formula $C_4H_{10}O$.

Problem 11.1
Draw the structural formulas of all the isomers of $C_4H_{10}O$. Which are ethers?

11.2 NOMENCLATURE OF ETHERS

Most ethers are named by following the names of the two alkyl groups with the word ether. Some examples are shown above and in Table 11.1. A few ethers, such as anisole, have special names. We shall be concerned with only a few of these special names.

In the systematic IUPAC name of an ether, R—O—R′, one of the two R groups—usually the larger—is the parent hydrocarbon. The other group including the oxygen (R—O—) is named as a substituent. We name the R—O— group *alkoxy*. Examples are shown below and in Table 11.1.

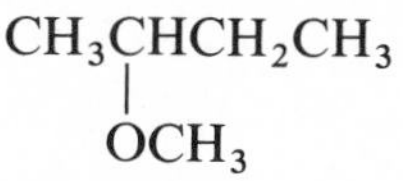

$CH_3CH(OCH_3)CH_2CH_3$

2-methoxybutane

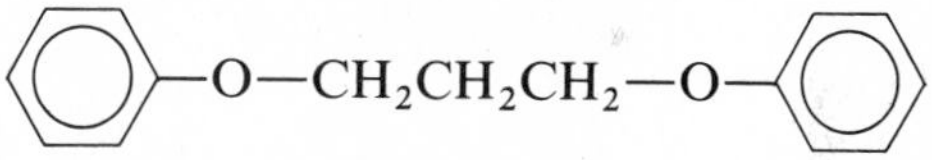

$C_6H_5-O-CH_2CH_2CH_2-O-C_6H_5$

1,3-diphenoxypropane

$CH_2(OCH_3)-CH(OCH_3)-CH_2(OCH_3)$

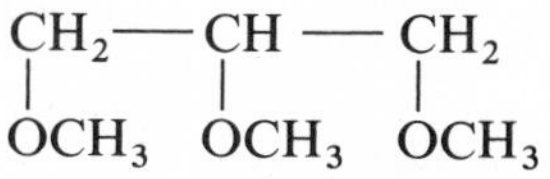

1,2,3-trimethoxypropane

Problem 11.2
Draw the structural formula of each of the following compounds.
(a) propoxybenzene, (b) 1,2-dimethoxyethane,
(c) *trans*-1,3-diethoxycyclobutane, (d) diisobutyl ether

Problem 11.3
Give two names for each of the compounds below.

(a) 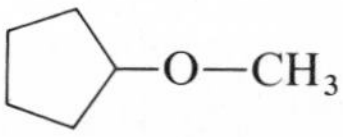 cyclopentyl$-O-CH_3$

(b) 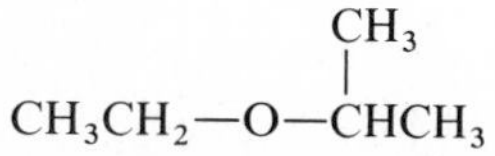 $CH_3CH_2-O-CH(CH_3)CH_3$

Table 11.1 Physical Properties of Some Ethers

Name[a]	Formula	Boiling Point, °C	Density, g/ml (20°)	Solubility, c/100 ml H_2O
Dimethyl ether (Methoxymethane)	$CH_3—O—CH_3$	−23.7	2.09[b]	3700[c]
Ethyl methyl ether (Methoxyethane)	$CH_3—O—CH_2CH_3$	7.9	0.726	soluble
Diethyl ether (Ethoxyethane)	$CH_3CH_2—O—CH_2CH_3$	34.6	0.7135	7.5
Ethyl propyl ether (Ethoxypropane)	$CH_3CH_2—O—CH_2CH_2CH_3$	64	0.747	soluble
Dipropyl ether (Propoxypropane)	$CH_3CH_2CH_2—O—CH_2CH_2CH_3$	91	0.736	0.25
Furan	(ring structure, O)	32	0.944	insoluble
Tetrahydrofuran	(ring structure, O)	66		very soluble
Dioxane	(ring structure, O O)	101	1.035	∞
Anisole (Methoxybenzene)	$CH_3—O—C_6H_5$	155	0.999	insoluble
Diphenyl ether (Phenoxybenzene)	$C_6H_5—O—C_6H_5$	259	1.073	very slightly soluble

[a] Systematic IUPAC names are in parentheses.
[b] This ether is gaseous, and the density is given in g/liter.
[c] Cubic centimeters of gaseous ether per 100 ml H_2O.

11.3 PHYSICAL PROPERTIES: HYDROGEN BONDING

Boiling Point

As we have seen in previous chapters, hydrogen bonding accounts for the physical properties of many compounds. Ethers lack the necessary hydrogen atom, and pure ethers therefore do not exhibit hydrogen bonding. As we might expect, ethers have boiling points lower than those of the isomeric alcohols. Table 11.1 lists physical properties of several ethers. Ethers have somewhat higher boiling points than hydrocarbons (Table 4.3) of comparable molecular weight because ether molecules are polar and therefore attract each other.

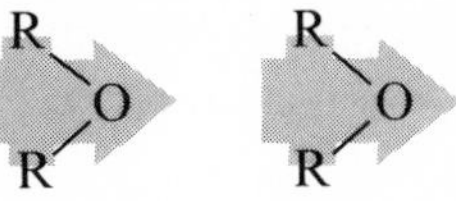

Problem 11.4

Verify that ethers have lower boiling points than alcohols by comparing the ethers in Table 11.1 with the alcohols in Table 10.2.

Solubility

Water solubility is a different matter. Ethers can form hydrogen bonds with water molecules:

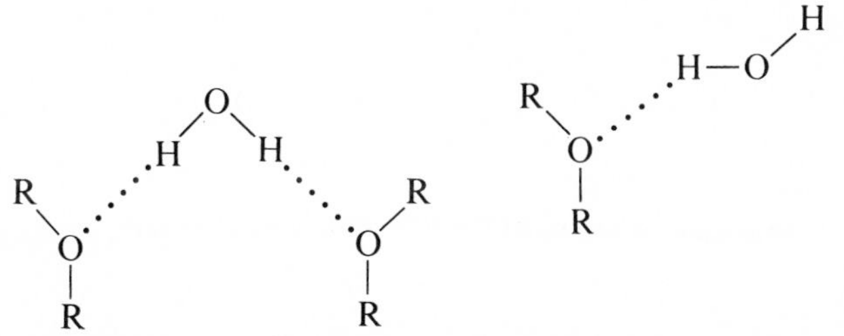

We therefore expect ethers to resemble the isomeric alcohols in their water solubilities, and they do.

We may summarize as follows: Ethers resemble the hydrocarbons of comparable molecular weight in their boiling points, and resemble the isomeric alcohols in their water solubilities. Table 11.2 illustrates these relationships.

Density

Ethers are generally less dense than the alcohols. Because they cannot form hydrogen bonds, their molecules do not form tight clusters, as do the molecules of alcohols and especially water. Ethers resemble the hydrocarbons in their densities.

Table 11.2 Comparison of an Ether, a Hydrocarbon, and an Alcohol

Compound	Formula	Molecular Weight	Boiling Point, °C	Solubility, g/100 ml H_2O
Pentane	$CH_3CH_2CH_2CH_2CH_3$	72	36.2	0.036
Diethyl ether	$CH_3CH_2O\ CH_2CH_3$	74	34.6	7.5
1-Butanol	$CH_3CH_2CH_2CH_2OH$	74	117.7	7.9

Problem 11.5
Would you expect pentane or diethyl ether to be more soluble in liquid ammonia (NH_3)? Explain.

11.4 PREPARATION OF ETHERS

Dehydration of Alcohols

Diethyl ether is manufactured by the dehydration of 95% ethanol with concentrated sulfuric acid.

$$CH_3CH_2{-}OH + HO{-}CH_2CH_3 \xrightarrow[125\text{–}140^\circ C]{H_2SO_4} CH_3CH_2{-}O{-}CH_2CH_3 + H_2O$$

If the reaction is carried out at higher temperatures, ethylene is the major product. At the higher temperature, dehydration occurs within a single molecule rather than between two molecules:

$$\underset{\displaystyle H\qquad OH}{CH_2{-}CH_2} \xrightarrow[180^\circ C]{H_2SO_4} CH_2{=}CH_2 + H_2O$$

This latter reaction is of course undesirable, because ethanol is manufactured from ethylene in the first place. This method is difficult to control in the laboratory, and some alkene is usually produced in the process.

With secondary or tertiary alcohols, alkenes tend to form more readily. This is because secondary or tertiary carbocations are produced, and these easily lose H^+ to form the alkene (Sec. 5.5):

$$CH_3{-}\overset{\displaystyle CH_3}{\underset{\displaystyle CH_3}{C}}{-}OH \underset{}{\overset{H_2SO_4}{\rightleftharpoons}} CH_3{-}\overset{\displaystyle CH_3}{\underset{\displaystyle CH_3}{C}}{-}OH_2^+ \overset{-H_2O}{\rightleftharpoons} CH_3{-}\overset{\displaystyle CH_3}{C^+}{-}CH_3 \;(\text{H–C–H, with } H{-}\ddot{O}{-}H \text{ removing } H^+)$$

$$\longrightarrow CH_3{-}\overset{\displaystyle CH_3}{C}{=}CH_2 + H_3O^+$$

For this reason, acid-catalyzed dehydrations of secondary and tertiary alcohols are usually carried out at temperatures as low as possible.

Like the Wurtz reaction (Sec. 4.9), the dehydration of alcohols is useful only for the synthesis of symmetrical ethers—ethers that have two identical groups attached to oxygen.

$$2R{-}OH \xrightarrow{H_2SO_4} R{-}O{-}R + H_2O$$
$$(R{-}OH = 1^\circ \text{ alcohol})$$

Attempts to make unsymmetrical ethers lead to mixtures:

$$R{-}OH + HO{-}R' \xrightarrow[-H_2O]{H_2SO_4} R{-}O{-}R + R{-}O{-}R' + R'{-}O{-}R'$$

Problem 11.6
What are two disadvantages of preparing ethers by the dehydration of alcohols?

Williamson Synthesis

A method that avoids the above disadvantages is the **Williamson synthesis**. In this reaction we react the sodium salt of an alcohol with an alkyl halide:

$$R{-}ONa + Cl{-}R' \longrightarrow R{-}O{-}R' + NaCl$$

The sodium alkoxide is prepared by reacting the alcohol with sodium metal (Sec. 10.11).

An example of this reaction is the formation of methyl *tert*-butyl ether:

$$CH_3{-}\underset{CH_3}{\overset{CH_3}{\underset{|}{\overset{|}{C}}}}{-}OH \xrightarrow[-\frac{1}{2}H_2]{Na} CH_3{-}\underset{CH_3}{\overset{CH_3}{\underset{|}{\overset{|}{C}}}}{-}ONa \xrightarrow{CH_3Br} CH_3{-}\underset{CH_3}{\overset{CH_3}{\underset{|}{\overset{|}{C}}}}{-}O{-}CH_3 + NaBr$$

We could have attempted the above synthesis using methanol and *tert*-butyl bromide:

$$CH_3OH \xrightarrow[-\frac{1}{2}H_2]{Na} CH_3ONa \xrightarrow{Br{-}C(CH_3)_3} CH_3{-}O{-}\underset{CH_3}{\overset{CH_3}{\underset{|}{\overset{|}{C}}}}{-}CH_3 + NaBr$$

This latter alternative, however, involving an S_N2 reaction, is not as good as the first because *tert*-butyl halides react readily with strong bases to give 2-methylbutene as the major product (an E2 elimination):

$$CH_3{-}\underset{Br}{\overset{CH_3}{\underset{|}{\overset{|}{C}}}}{-}CH_3 \xrightarrow[(E2)]{\text{strong base}} CH_3{-}\overset{CH_3}{\overset{|}{C}}{=}CH_2$$

Recall that S_N2 reactions do not occur with 3° alkyl halides. Even 2° alkyl halides undergo E2 reactions in preference to S_N2 reactions (Sec. 9.6). We must therefore be careful to select the best combination of alcohol and alkyl halide as reactants in the Williamson synthesis.

We can use the Williamson synthesis to prepare ethers in which one of the hydrocarbon groups is aryl. In this case, the aryl group must come from a phenol because aryl halides do not undergo nucleophilic substitutions:

$$R{-}ONa + Cl{-}C_6H_5 \not\longrightarrow R{-}O{-}C_6H_5 \quad \text{(no reaction)}$$

The preparation of anisole is an example:

$$C_6H_5{-}OH + Br{-}CH_3 + NaOH \xrightarrow{H_2O} C_6H_5{-}O{-}CH_3 + NaBr + H_2O$$

or, in general,

$$Ar{-}OH + R{-}X + NaOH \xrightarrow{H_2O} Ar{-}O{-}R + NaX + H_2O$$

Because phenols are acidic, we may simply mix the phenol and alkyl halide and add aqueous NaOH. Under these conditions the phenoxide ion is produced and the actual reaction is

$$C_6H_5{-}\ddot{\underset{..}{O}}:^- + CH_3Br \longrightarrow C_6H_5{-}\ddot{\underset{..}{O}}{-}CH_3 + Br^-$$

Problem 11.7

Write equations for the preparation of each of the following compounds.

(a) $CH_3{-}O{-}CH_3$

(b) $C_6H_5{-}O{-}CH_2{-}C_6H_5$

(c) $CH_3CH_2CH_2{-}O{-}C_6H_4{-}Cl$ (para)

Which of the above compounds can be prepared by both dehydration and the Williamson synthesis, and which can be prepared by only one of these methods (which one)? Explain.

11.5 REACTIONS OF ETHERS

Acids

The major reaction of ethers is with acids. Except for epoxides (Section 11.8), ethers do not react with bases.

Ethers react with hydrogen halides (HCl, HBr, or HI) to form alkyl halides:

$$R{-}O{-}R' + 2HX \longrightarrow R{-}X + R'{-}X + H_2O$$
$$(X = Cl, Br, or I)$$

We can explain this reaction with the following mechanism:

$$\underset{\text{base}}{R{-}\ddot{\underset{..}{O}}{-}R'} + H{-}\ddot{\underset{..}{Br}}: \rightleftarrows \underset{\text{conjugate acid}}{R{-}\overset{H}{\overset{|}{\underset{..}{O^+}}}{-}R'} + :\ddot{\underset{..}{Br}}:^-$$

$$:\ddot{\underset{..}{Br}}:^- + R{-}\overset{H}{\overset{|}{\underset{..}{O^+}}}{-}R' \xrightarrow{S_N2} R{-}Br + R'{-}\ddot{\underset{..}{O}}H$$

Of course, the alcohol R′OH reacts further through the same path:

$$R'-\ddot{O}H + H-Br \rightleftarrows R'-\overset{H}{\overset{|}{O^+}}-H + Br^-$$

$$Br^- + R'-OH_2^+ \longrightarrow R'-Br + H_2O$$

We call this reaction **ether cleavage**. If we examine the above mechanism closely, we see that it is the exact reverse of the reaction of an alkyl halide with water to form an alcohol, or the reaction of an alkyl halide with an alcohol to form an ether (Sec. 9.6).

Acid cleavage of aryl ethers stops at the phenol stage because phenols do not undergo nucleophilic substitution reactions.

$$C_6H_5-O-CH_2CH_3 \xrightarrow{HI} C_6H_5-OH + CH_3CH_2I$$

Problem 11.8
Predict the products of the acid cleavage of the following ethers using HI as the acid.
(a) ethyl isopropyl ether,
(b) anisole,
(c) diphenyl ether

Problem 11.9
Explain why ethers do not react with neutral solutions of sodium halides to produce alkyl halides according to the equation

$$R-O-R' + Cl^- \not\longrightarrow R-Cl + R'-Cl$$

Peroxides

Ethers react with atmospheric oxygen to form **peroxides** and **hydroperoxides**.

$$R-O-\overset{|}{\underset{|}{C}}-H + O_2 \longrightarrow R-O-\overset{|}{\underset{|}{C}}-O-O-H$$

a hydroperoxide

$$\longrightarrow R-O-\overset{|}{\underset{|}{C}}-O-O-\overset{|}{\underset{|}{C}}-O-R$$

a peroxide

These products are extremely explosive, and their formation is a hazard. Ether that has stood for long periods of time tends to form dangerous quantities of peroxides. Purification of such an ether sample by distillation removes the ether and leaves the less volatile peroxides in the residue. Near the end of the distillation process, the temperature rises and a major explosion can occur. This explosive property of ethers coupled with their high flammability makes them fairly dangerous chemicals.

We can easily test for the presence of peroxides and hydroperoxides in ethers by shaking a small sample of the ether with aqueous KI solution. Peroxides oxidize iodide ion to I_2. The characteristic iodine color indicates that peroxides are present.

$$R-O-O-H + 4I^- + 2H_2O \longrightarrow R-H + 2I_2 + 4OH^-$$

We remove peroxides and hydroperoxides from ethers by stirring the ether with ferrous sulfate:

$$R{-}O{-}O{-}H + 2Fe^{2+} + 2H_2O \longrightarrow R{-}H + 2Fe^{3+} + 4OH^-$$

11.6 NOMENCLATURE OF EPOXIDES

The general class of three-membered cyclic ethers is commonly called **epoxides** although the IUPAC name is **oxiranes**. In the IUPAC system, we assign the number 1 to the non-carbon atom in a ring. Common names of epoxides are based on the alkene from which they are made. The names are shown.

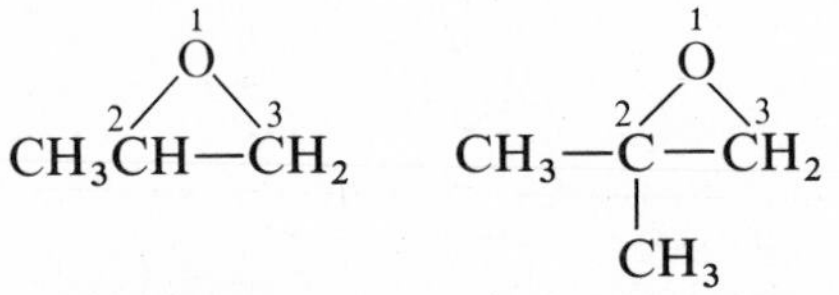

$$CH_3\overset{2}{C}H{-}\overset{3}{C}H_2 \ (\text{ring }\overset{1}{O}) \qquad CH_3{-}\underset{\displaystyle CH_3}{\overset{2}{C}}{-}\overset{3}{C}H_2 \ (\text{ring }\overset{1}{O})$$

2-methyloxirane (propylene oxide) 2,2-dimethyloxirane (isobutylene oxide)

The simplest epoxide is $CH_2{-}CH_2$ (bridged by O). It is commonly called ethylene oxide because it is manufactured by the oxidation of ethylene:

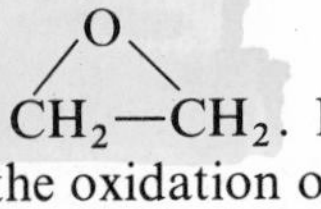

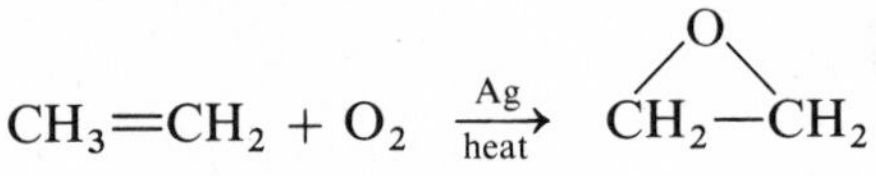

$$CH_3{=}CH_2 + O_2 \xrightarrow[\text{heat}]{\text{Ag}} CH_2{-}CH_2 \ (\text{ring O})$$

Problem 11.10
Give the structural formula of each of the following compounds.

(a) styrene oxide (styrene is $C_6H_5{-}CH{=}CH_2$)

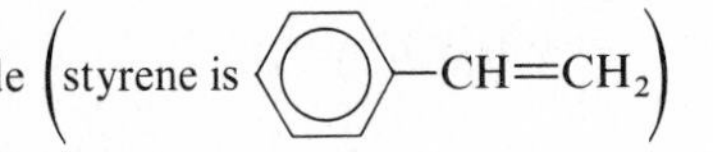

(b) 2,3-dimethyloxirane

11.7 PREPARATION OF EPOXIDES

Ethylene oxide is manufactured by two methods. We have already mentioned the first—air oxidation of ethylene. In the second method, ethylene and hypochlorous acid, HOCl (formed in chlorine-water solutions), react to give a **chlorohydrin**

$$CH_2{=}CH_2 + HO{-}Cl \longrightarrow \underset{\displaystyle OH}{CH_2}{-}\underset{\displaystyle Cl}{CH_2}$$

ethylene chlorohydrin

The chlorohydrin then reacts with NaOH in a reaction that resembles the Williamson synthesis:

$$\underset{\displaystyle OH\quad Cl}{CH_2{-}CH_2} \xrightleftharpoons{OH^-} \underset{\displaystyle O^-}{CH_2{-}\overset{\displaystyle Cl}{CH_2}} \xrightarrow[S_N2]{\text{internal}} \underset{\displaystyle O}{CH_2{-}CH_2} + Cl^-$$

Other epoxides may also be produced by the above methods. We can generalize this method by the following equations:

$$-\overset{|}{C}{=}\overset{|}{C}- + HOCl \longrightarrow -\underset{OH}{\overset{|}{C}}{-}\underset{Cl}{\overset{|}{C}}-$$

$$-\underset{OH}{\overset{|}{C}}{-}\underset{Cl}{\overset{|}{C}}- + NaOH \longrightarrow -\overset{|}{C}{-}\overset{|}{C}-\ (\text{epoxide, bridged by O}) + NaCl + H_2O$$

Problem 11.11

Write equations for the synthesis of the following epoxides.

(a) $CH_3CH{-}CHCH_3$ (epoxide, O bridging the two CH carbons)

(b) $C_6H_5{-}CH{-}CH{-}C_6H_5$ (epoxide, O bridging the two CH carbons)

11.8 REACTIONS OF EPOXIDES

Acid-Catalyzed Ring Opening

Epoxides react with acids or bases to open the ring.

With acids:

$$-\overset{|}{C}{-}\overset{|}{C}-\ (\text{epoxide}) + HX \longrightarrow -\underset{OH}{\overset{|}{C}}{-}\underset{X}{\overset{|}{C}}-$$

$$-\overset{|}{C}{-}\overset{|}{C}-\ (\text{epoxide}) + H_2O \xrightarrow{H^+} -\underset{OH}{\overset{|}{C}}{-}\underset{OH}{\overset{|}{C}}-$$

$$-\overset{|}{C}{-}\overset{|}{C}-\ (\text{epoxide}) + R{-}OH \xrightarrow{H^+} -\underset{OH}{\overset{|}{C}}{-}\underset{OR}{\overset{|}{C}}-$$

With bases:

$$-\underset{\diagdown O \diagup}{C-C}- + H_2O \xrightarrow{OH^-} -\underset{OH}{\overset{|}{C}}-\underset{OH}{\overset{|}{C}}-$$

$$-\underset{\diagdown O \diagup}{C-C}- + NH_3 \longrightarrow -\underset{OH}{\overset{|}{C}}-\underset{NH_2}{\overset{|}{C}}-$$

With acids, the mechanism is analogous to the acid cleavage of ethers:

$$\underset{\diagdown O \diagup}{CH_2-CH_2} + HBr \rightleftarrows \underset{\underset{H}{|}}{\underset{\diagdown O^+ \diagup}{CH_2-CH_2}} + Br^- \longrightarrow \underset{OH}{\underset{|}{CH_2}}-\overset{Br}{\overset{|}{CH_2}}$$

Nucleophiles other than halide ions may also react with the protonated epoxide to open the ring:

$$\underset{\underset{H}{|}}{\underset{\diagdown O^+ \diagup}{CH_2-CH_2}} + H-\ddot{O}-H \longrightarrow \underset{OH}{\underset{|}{CH_2}}-\overset{OH_2^+}{\overset{|}{CH_2}} \xrightarrow{-H^+} \underset{OH}{\underset{|}{CH_2}}-\underset{OH}{\underset{|}{CH_2}}$$

ethylene glycol

$$\underset{\underset{H}{|}}{\underset{\diagdown O^+ \diagup}{CH_2-CH_2}} + H-\ddot{O}-CH_3 \longrightarrow \underset{OH}{\underset{|}{CH_2}}-\overset{H\overset{+}{O}CH_3}{\overset{|}{CH_2}} \xrightarrow{-H^+} \underset{OH}{\underset{|}{CH_2}}-\underset{OCH_3}{\underset{|}{CH_2}}$$

(2-methoxyethanol)
(methyl cellosolve)

Methyl cellosolve is a good solvent because it is both an ether and an alcohol. It is used as an additive to jet fuels and as a solvent for plastics.

Base-Catalyzed Ring Opening

The highly strained three-membered ring makes epoxides more reactive than ordinary ethers. For example, unlike ordinary ethers, epoxides react with bases such as OH^- and NH_3 (see above in this section). We explain these reactions by the strain energy that is released when the ring opens. [We should recall that the cyclopropane ring also undergoes ring-opening reactions that are not typical of other saturated hydrocarbons

(Sec. 4.13 and 4.17).] Ethanolamine is a water-soluble base. It is used to absorb acidic gases such as CO_2 and H_2S in industrial processes.

Problem 11.12
Write equations for the reaction of each epoxide of Problem 11.11 with (a) H_2O, (b) HCl, (c) NH_3, (d) CH_3CH_2OH, (e) NaOH (aqueous).

Problem 11.13
The ring-opening reaction with NaOH is very fast with ethylene oxide, slow with oxetane (four-membered ring with O), and does not occur with tetrahydrofuran (five-membered ring with O). Explain the relative reactivities of these three cyclic ethers toward base-catalyzed ring opening.

11.9 DIETHYL ETHER AS AN ANESTHETIC

A surgeon named C. W. Long of Jefferson, Georgia, first used diethyl ether in an operation he performed in 1842. He did not publish his technique, and it went unnoticed until the dentist J. C. Warren rediscovered it in 1846 in the Massachusetts General Hospital. Warren gave this procedure the name anesthesia, which means insensibility, at the suggestion of Oliver Wendell Holmes.

Inhalation of diethyl ether vapor produces unconsciousness by depressing the activity of the central nervous system. It probably acts on cell membranes by altering their permeability and interfering with normal impulse transmission.

The use of diethyl ether as an anesthetic has declined in recent years because it is flammable, irritating to breathe, and acts slowly, and its effects disappear slowly. Neothyl, $CH_3OCH_2CH_2CH_3$, is an improved inhalation anesthetic.

NEW TERMS

Chlorohydrin An intermediate in the production of ethylene oxide (11.7):

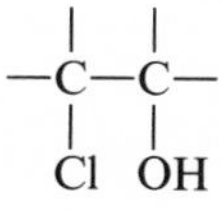

$$\begin{array}{c} |\quad | \\ -C-C- \\ |\quad | \\ Cl\ \ OH \end{array}$$

Epoxide A three-membered cyclic ether:

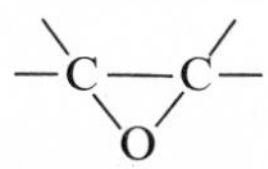

$$\begin{array}{c} \diagdown\qquad\diagup \\ -C—C- \\ \diagdown\ \diagup \\ O \end{array}$$

The systematic name of this ring system is **oxirane** (11.6).

Ether R—O—R′ in which R and R′ may be alkyl or aryl (11.1).

Ether cleavage Reaction of an ether with HCl, HBr, or HI to produce alkyl halides (11.5):

$$R—O—R' + 2HX \longrightarrow RX + R'X + H_2O$$

Hydroperoxide Any compound that has the —O—O—H group (11.5). Ether hydroperoxides have the structure:

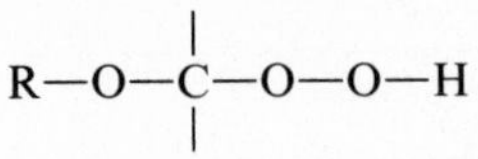

$$R{-}O{-}\overset{|}{\underset{|}{C}}{-}O{-}O{-}H$$

Peroxide Any compound that has the —O—O— group (11.5). Ether peroxides have the structure:

$$R{-}O{-}\overset{|}{\underset{|}{C}}{-}O{-}O{-}\overset{|}{\underset{|}{C}}{-}O{-}R$$

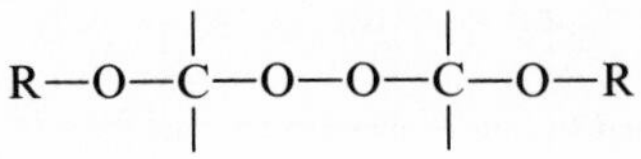

Williamson synthesis Ethers are prepared by reacting a sodium alkoxide with an alkyl halide (11.4):

$$R'{-}X + NaOR \longrightarrow R'{-}O{-}R + NaX$$

(X = Cl, Br, or I; R may be alkyl or aryl; R′ is 1° alkyl)

ADDITIONAL PROBLEMS

[11.14] Identify the ether groups in each of the following structural formulas.

(a) 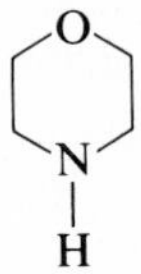 (morpholine)

(b) 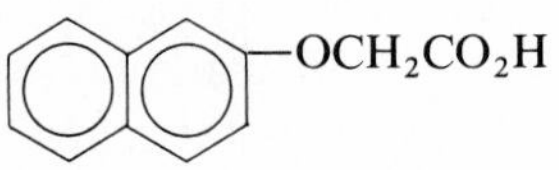 $-OCH_2CO_2H$

(c) brucine (see Sec. 8.12)

(d) 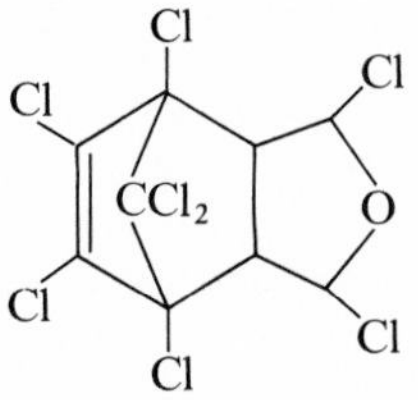 (isobenzan, an insecticide)

(e) 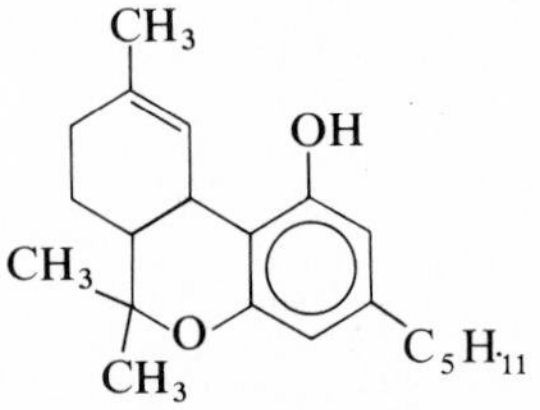 (tetrahydrocannabinol, active constituent of marijuana)

11.15 Draw the structural formulas of each of the following compounds.

[a] 1,2,3,4-tetramethoxybutane, [b] dibenzyl ether,
[c] phenoxybenzene, (d) 3-ethoxycyclopentene,
(e) dimethoxymethane

11.16 Give an acceptable name for each of the following compounds.

[a] C_6H_5—OCH_2CH_2O—C_6H_5

[b] CH_3—O—CH_2CH_2Cl

[c] CH_3—O—C_6H_4—O—CH_3

(d) $CH_3CH(CH_3)$—O—CH_2CH_3

(e) $CH_3CH(CH_3)CH_2$—O—$CH_2CH(CH_3)CH_3$

[11.17] For which of the compounds in Problem 11.16 can you assign *both* a common and an IUPAC name? Give those names that you did not give in Problem 11.16.

[11.18] Construct a table like Table 11.2 that compares the compounds propane, dimethyl ether, and ethanol. Does your table support the conclusions that we drew about the compounds in Table 11.2?

11.19 Write equations to show how you could prepare each of the following ethers.
[a] dipropyl ether, [b] diisopropyl ether,
[c] methyl isopropyl ether, (d) 1-phenoxypropane,
(e) dibenzyl ether

[11.20] Which of the compounds in Problem 11.19 could *not* be prepared conveniently by dehydration of an alcohol? Explain.

11.21 Complete each of the following equations.

[a] $CH_3OCH_3 \xrightarrow[\text{(excess)}]{HI}$

[b] Cl—C_6H_4—O—$CH_2CH_3 \xrightarrow[\text{(excess)}]{HBr}$

(c) C_6H_5—O—CH_2—$C_6H_5 \xrightarrow[\text{(excess)}]{HI}$

11.22 Write equations to show how you could prepare each of the following compounds starting with an alkene and any inorganic reagents.

[a] C_6H_5—CH—CH_2 (epoxide, O bridging CH and CH_2)

(b) cyclohexene oxide (cyclohexane ring fused with O epoxide)

[11.23] Bottle A and bottle B each contain a pure liquid. One of the liquids is tetrahydrofuran; the other is 2,3-dimethyloxirane. Explain how you could determine which compound is in bottle A and which is in bottle B.

11.24 An unknown compound C ($C_8H_{10}O$) is insoluble in water. Compound C reacts with excess HI to give an acidic compound, D (C_6H_6O), and ethyl iodide. Give the structural formulas of compounds C and D.

[11.25] Compound E ($C_8H_{10}O$) is insoluble in water. Compound E reacts with excess HI to give two neutral compounds, F and G, that contain no oxygen. Give structural formulas of compounds E, F, and G consistent with these results.

11.26 Compound H ($C_6H_{10}O$) is insoluble in water. Compound H reacts with aqueous NaOH to give a water-soluble compound, I ($C_6H_{12}O_2$). Give possible structural formulas for compounds H and I.

[11.27] Ethylene oxide reacts with NaOH to give not only ethylene glycol but also other, longer chain polyether alcohols which are commercially important solvents:

$$OH^- + \overset{\;\;O}{CH_2\!-\!CH_2} \xrightarrow{H_2O} \underset{\text{ethylene glycol}}{HOCH_2CH_2OH} + \underset{\text{diethylene glycol}}{HOCH_2CH_2OCH_2CH_2OH} +$$

$$\underset{\text{triethylene glycol}}{HOCH_2CH_2OCH_2CH_2OCH_2CH_2OH} + \cdots$$

Give a mechanism to explain how these polyether alcohols are produced.

ALDEHYDES AND KETONES

Aldehydes and ketones are among the most interesting compounds available to the organic chemist. They undergo a wide variety of chemical reactions. In addition to their use in synthesis, aldehydes and ketones have common practical uses; for example, formaldehyde is a disinfectant and preservative, and vanillin is a natural flavor. We find aldehydes and ketones in many vegetable products such as almonds and cinnamon, and in certain animal products such as sex hormones and musk. Many biochemically important compounds such as simple sugars possess aldehyde or ketone groups.

Aldehydes and ketones are only two classes of compounds that possess the carbonyl functional group. In this chapter we shall discuss the physical and chemical behavior of aldehydes and ketones in terms of the structure of the carbonyl group.

12.1 FUNCTIONAL GROUPS THAT CONTAIN THE CARBONYL GROUP

Oxidation of a primary or secondary alcohol results in the loss of two hydrogen atoms and the formation of a carbonyl group:

$$-\overset{\overset{\large OH}{|}}{\underset{\underset{\large H}{|}}{C}}- \xrightarrow[(-2H)]{\text{oxidation}} -\overset{\overset{\large O}{\|}}{C}-$$

carbonyl group

If the carbonyl group is bonded to at least one hydrogen, the compound is an **aldehyde**:

$$R-\overset{\overset{\large OH}{|}}{\underset{\underset{\large H}{|}}{C}}-H \xrightarrow{-2H} R-\overset{\overset{\large O}{\|}}{C}-H$$

an aldehyde

Examples:

$$H-\overset{\overset{\large O}{\|}}{C}-H \qquad CH_3\overset{\overset{\large O}{\|}}{C}-H$$

(formaldehyde) (acetaldehyde)

$$C_6H_5-\overset{\overset{\large O}{\|}}{C}-H$$

(benzaldehyde)

In a **ketone** the carbonyl group is bonded to two hydrocarbon groups:

$$R-\overset{\overset{\large OH}{|}}{\underset{\underset{\large H}{|}}{C}}-R' \xrightarrow{-2H} R-\overset{\overset{\large O}{\|}}{C}-R'$$

a ketone

Examples:

$$CH_3\overset{\overset{\large O}{\|}}{C}CH_3, \qquad CH_3\overset{\overset{\large O}{\|}}{C}-C_6H_5$$

(acetone) (acetophenone)

$$C_6H_5-\overset{\overset{\large O}{\|}}{C}-C_6H_5$$

(benzophenone)

Tables 12.1 and 12.2 list several aldehydes and ketones with their boiling points, sources, and uses.

Table 12.1 Some Common Aldehydes

Name	Boiling Point, °C	Formula	Source and Uses
Formaldehyde	−20	$H-\overset{\overset{O}{\|}}{C}-H$	Oxidation of methanol. Disinfectant, preservative, manufacture of resins.
Acetaldehyde	21	$CH_3-\overset{\overset{O}{\|}}{C}-H$	Oxidation of ethanol. Synthetic intermediate, plastics, synthetic rubber.
Propionaldehyde	49	$CH_3CH_2-\overset{\overset{O}{\|}}{C}-H$	Oxidation of propanol.
Butyraldehyde	75	$CH_3CH_2CH_2-\overset{\overset{O}{\|}}{C}-H$	Oxidation of butanol. Manufacture of plastics, solvents, plasticizers.
Benzaldehyde	179	$C_6H_5-\overset{\overset{O}{\|}}{C}-H$	Oil of almond. Manufacture of dyes, perfumes, and flavors.
Cinnamaldehyde	246	$C_6H_5-CH=CH-\overset{\overset{O}{\|}}{C}-H$	Oil of cinnamon. Manufacture of flavors and perfumes.
Vanillin	285	$HO-C_6H_3(OCH_3)-\overset{\overset{O}{\|}}{C}-H$	Vanilla, potato peels. Used as flavor in foods and beverages.
Citronellal	208	$(CH_3)_2C=CH-CH_2CH_2-CH(CH_3)-CH_2-\overset{\overset{O}{\|}}{C}-H$	Citronella oil, lemon. Used as perfume and insect repellent.

Table 12.2 Some Common Ketones

Name	Boiling Point, °C	Formula	Source and Uses
Acetone	56.5	$CH_3-\overset{\overset{\displaystyle O}{\|}}{C}-CH_3$	Fermentation and synthetic. Used as solvent for fats, oils, plastics, and as synthetic intermediate.
Methyl ethyl ketone	79.6	$CH_3-\overset{\overset{\displaystyle O}{\|}}{C}-CH_2CH_3$	Fermentation and synthetic. Used as solvent and in the manufacture of smokeless powder.
Methyl phenyl ketone (Acetophenone)	202	$CH_3-\overset{\overset{\displaystyle O}{\|}}{C}-$ (benzene ring)	Synthetic. Used as solvent, as synthetic intermediate, and in perfumes.
Diphenyl ketone	305	(benzene ring)$-\overset{\overset{\displaystyle O}{\|}}{C}-$(benzene ring)	Synthetic. Used in manufacture of perfumes, soaps, insecticides, hypnotics, and antihistamines.
Cyclohexanone	156	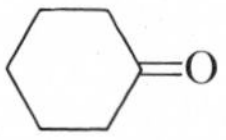 (cyclohexane ring)$=O$	Oxidation of cyclohexanol. Used as solvent for plastics and in synthesis.
Camphor	sublimes, melting point = 180°	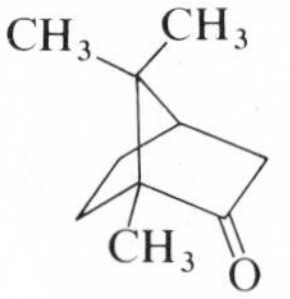 CH_3 CH_3 CH_3 O	Camphor tree. Used as plasticizer for cellulose esters and in manufacture of celluloid, explosives, moth repellent, and embalming fluids.
Testosterone	melting point = 155°	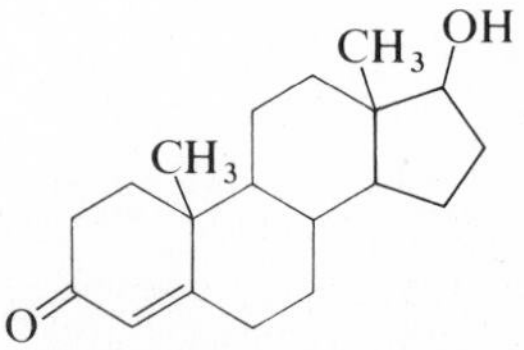 CH_3 OH CH_3 O	Male sex hormone.

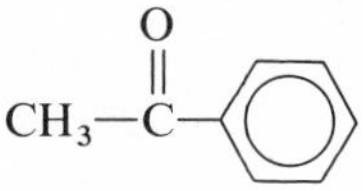

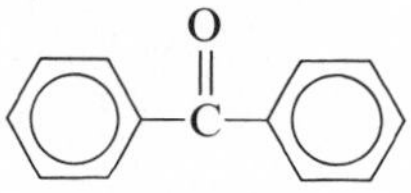

The carbonyl group occurs in other functional groups as well; the most important are carboxylic acids and their derivatives:

$$\mathrm{R{-}\overset{\overset{\large O}{\|}}{C}{-}OH}$$

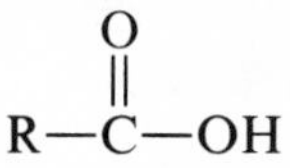

a carboxylic acid

We shall examine the carboxylic acids and their derivatives in Chapter 13.

Problem 12.1

Classify each of the following compounds as an aldehyde, a ketone, or neither.

(a) 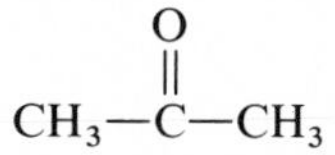 $\mathrm{CH_3{-}\overset{\overset{O}{\|}}{C}{-}CH_3}$

(b) $\mathrm{CH_3{-}\overset{\overset{O}{\|}}{C}{-}OCH_3}$

(c) $\mathrm{C_6H_5{-}\overset{\overset{O}{\|}}{C}{-}\underset{\underset{H}{|}}{\overset{\overset{CH_3}{|}}{C}}{-}CH_3}$ (benzene ring attached to the carbonyl carbon)

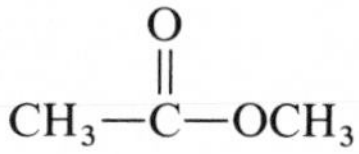

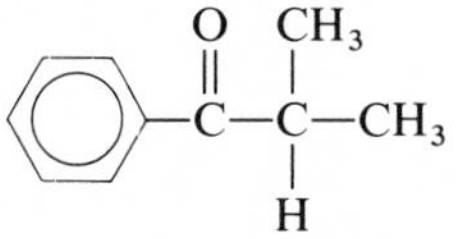

(d) 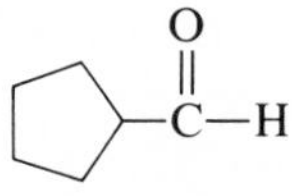 $\mathrm{C_5H_9{-}\overset{\overset{O}{\|}}{C}{-}H}$ (cyclopentane ring attached to the carbonyl carbon)

12.2 CARBONYL GROUP

The **carbonyl group** consists of a carbon–oxygen double bond.

$$\overset{\overset{:\ddot{O}}{\|}}{\mathrm{C}}$$

carbonyl group

This group is very reactive because of its polarity. Two effects operate to make the carbonyl group polar:

1. *Polarity of the* C—O *bond.* The electronegativity difference between carbon and oxygen makes the C—O sigma bond quite polar (Sec. 1.7).
2. *Resonance.* We can write the following resonance structures for the carbonyl group:

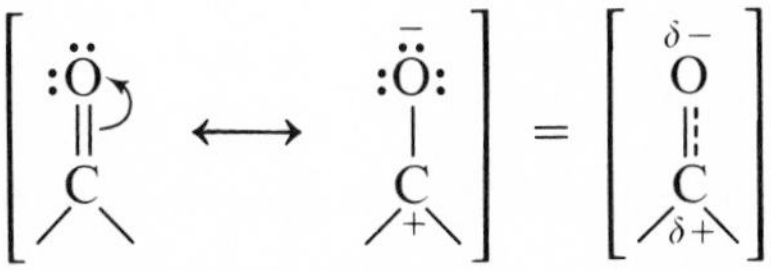

$$\left[\;\overset{\overset{:\ddot{O}}{\|}}{\mathrm{C}}\;\longleftrightarrow\;\overset{\overset{:\ddot{O}:^{-}}{|}}{\mathrm{C}^{+}}\;\right] = \left[\;\overset{\overset{\overset{\delta-}{O}}{\|}}{\mathrm{C}^{\delta+}}\;\right]$$

I   II   hybrid

Structure II contains formal charges. We know from physical measurements on a number of aldehydes and ketones that these two effects, bond polarity and resonance, combine to give the carbonyl group about 50% ionic character. That is, structures

I and II above contribute about equally to the resonance hybrid, and the carbonyl group in aldehydes and ketones is indeed quite polar.

Aldehydes and ketones can react with both acids and bases because of the polarity of the carbonyl group.

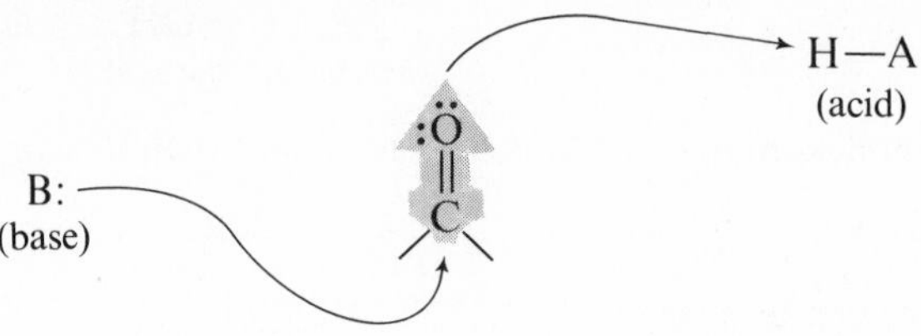

Problem 12.2

(a) Write the Lewis dot resonance structures of the conjugate acid of formaldehyde.

(b) Write the Lewis dot structure of the product formed when a hydroxide ion adds to a molecule of formaldehyde.

12.3 PHYSICAL PROPERTIES OF ALDEHYDES AND KETONES

Because the carbonyl group is so polar, aldehydes and ketones are able to form hydrogen bonds to water molecules and to other compounds capable of furnishing the necessary hydrogen atoms:

$$(R)(R')\overset{\delta+}{C}=\overset{\delta-}{O}\cdots\overset{\delta+}{H}-\overset{\delta-}{O}-H$$

Aldehydes and ketones, however, do not possess the necessary hydrogen atoms to form hydrogen bonds to each other. We can therefore make the following general statements about aldehydes and ketones:

1. Aldehydes and ketones are soluble in solvents that can furnish hydrogen atoms for hydrogen bonding. They thus resemble the alcohols in their water solubility.
2. Aldehydes and ketones have lower boiling points than alcohols or other hydrogen-containing compounds because hydrogen bonding among aldehyde molecules or ketone molecules is impossible. Aldehydes and ketones have higher boiling points than hydrocarbons of comparable molecular weights because the carbonyl group is polar. Polarity produces an attractive force between molecules that must be overcome during boiling.

$$(R)(R')\overset{\delta+}{C}=\overset{\delta-}{O} \qquad (R)(R')\overset{\delta+}{C}=\overset{\delta-}{O}$$

Table 12.3 Physical Properties of Aldehydes and Ketones

Compound	Formula	Boiling Point, °C	Solubility, g/100 ml H_2O
Formaldehyde	$CH_2{=}O$	−20	soluble
Methanol	CH_3OH	64.4	∞
Acetaldehyde	CH_3CHO	21	∞
Ethanol	CH_3CH_2OH	78	∞
Propionaldehyde	CH_3CH_2CHO	49	20
Acetone	$CH_3C(=O)CH_3$	56.5	∞
1-Propanol	$CH_3CH_2CH_2OH$	97	∞
Butyraldehyde	$CH_3CH_2CH_2CHO$	75.7	3.7
Methyl ethyl ketone	$CH_3C(=O)CH_2CH_3$	79.6	19
1-Butanol	$CH_3CH_2CH_2CH_2OH$	118	7.9

Table 12.3 lists some comparisons of boiling points and solubilities of some aldehydes, ketones, and alcohols of comparable molecular weight.

Problem 12.3

Which compound in each pair is more soluble in water and which has the higher boiling point?

(a) $C_6H_5-CH(OH)-CH_3$ and $C_6H_5-C(=O)-CH_3$

(b) $CH_3C(=O)CH_3$ and $CH_3CH(OH)CH_3$

(c) cyclohexanone (ring with =O) and cyclohexanol (ring with OH)

(d) $HOCH_2-C_6H_4-CH_2OH$ and $HC(=O)-C_6H_4-C(=O)H$

(e) $CH_3CH_2-O-CH_2CH_3$ and $CH_3C(=O)CH_2CH_3$

12.4 NOMENCLATURE: COMMON NAMES

Aldehydes and ketones are most often referred to by their common names. The common names of aldehydes are based on the related carboxylic acids. For example, formaldehyde can be oxidized to formic acid, and acetaldehyde can be oxidized to acetic acid:

$$\underset{\text{formaldehyde}}{H-\overset{\overset{\displaystyle O}{\|}}{C}-H} \xrightarrow{\text{oxidation}} \underset{\text{formic acid}}{H-\overset{\overset{\displaystyle O}{\|}}{C}-OH}$$

$$\underset{\text{acetaldehyde}}{CH_3-\overset{\overset{\displaystyle O}{\|}}{C}-H} \xrightarrow{\text{oxidation}} \underset{\text{acetic acid}}{CH_3-\overset{\overset{\displaystyle O}{\|}}{C}-OH}$$

Table 12.1 gives the common names of some simple aldehydes.

Ketones are often named by giving the names of the two hydrocarbon groups followed by the word ketone. Methyl ethyl ketone (Table 12.2) is an example. Some ketones have special names—acetone, acetophenone, and benzophenone are examples.

12.5 NOMENCLATURE: IUPAC

In the IUPAC systematic names of aldehydes, the ending *-e* of the parent alkane is replaced with the ending *-al*. The carbonyl carbon of aldehydes is always on the end of the chain, and it is always assigned the number 1. We need not include this number in the name of an aldehyde.

Some examples will illustrate how we name aldehydes.

Aldehyde	Common Name	Systematic Name
$HCHO$	Formaldehyde	Methanal
CH_3CHO	Acetaldehyde	Ethanal
CH_3CH_2CHO	Propionaldehyde	Propanal
$CH_3CH_2CH_2CHO$	Butyraldehyde	Butanal
$CH_3\underset{\underset{\displaystyle CH_3}{\vert}}{\overset{\overset{\displaystyle CH_3}{\vert}}{C}}CH_2\overset{\overset{\displaystyle CH_3}{\vert}}{C}HCHO$		2,4,4-Trimethylpentanal

We form systematic IUPAC names of ketones by replacing the *-e* of the parent alkane with the ending *-one*. We must designate the position of the carbonyl group of ketones except in the simplest cases where there is no ambiguity. Some examples are given below.

Ketone	Common Name	Systematic Name
$CH_3C(=O)CH_3$	Acetone	Propanone
$CH_3CH_2C(=O)CH_3$	Methyl ethyl ketone	Butanone
$CH_3CH_2C(=O)CH_2CH_3$	Diethyl ketone	3-Pentanone
$CH_3C(=O)CH_2CH_2CH_3$	Methyl propyl ketone	2-Pentanone
$CH_3C(=O)CH_2CH(C_6H_5)CH_3$		4-Phenyl-2-pentanone

Problem 12.4

Give the structural formula of each of the following compounds.

(a) octanal,
(b) 2-chlorobutanal,
(c) diisopropyl ketone,
(d) *sec*-butyl isobutyl ketone,
(e) 3-phenyl-4-octanone

Problem 12.5

Give the systematic IUPAC name and, if possible, an additional acceptable name for each ot the following compounds.

(a) cyclobutane=O

(b) $CH_3CH(CH_3)CHO$

(c) $C_6H_5-CH_2-C(=O)-CH_2-C_6H_5$

12.6 RELATIONSHIP TO OTHER FUNCTIONAL GROUPS: OXIDATION STATE

In Sections 4.9 and 4.10, we examined reduction and oxidation as they apply to covalent compounds. Oxidation (electron loss) occurs when an atom of higher electronegativity replaces one of lower electronegativity. The reverse process is reduction (electron gain).

$$\underset{\substack{\text{higher electron density}\\ \text{(reduced)}}}{H{-}\underset{\underset{H}{|}}{\overset{\overset{H}{|}}{C}}{-}H} \underset{\text{reduction}}{\overset{\text{oxidation}}{\rightleftarrows}} \underset{\substack{\text{lower electron density}\\ \text{(oxidized)}}}{H{-}\underset{\underset{H}{|}}{\overset{\overset{H}{|}}{C}}{-}OH}$$

In the example above, carbon is oxidized when oxygen replaces hydrogen because that substitution lowers the electron density at the carbon atom.

The most reduced state of carbon is in methane, and the most oxidized state of carbon is in carbon dioxide. We can say that the **oxidation state** of carbon depends on the number of bonds involving oxygen or other highly electronegative atoms such as the halogens or nitrogen. Table 12.4 lists some examples of compounds with the oxidation state of carbon shown.

We can use the concept of oxidation state to explain how we might convert one compound into another. For example, to convert an alcohol into an ether we do not need an oxidizing or reducing agent because alcohols and ethers have the same

Table 12.4 Oxidation States of Carbon in Its Compounds

Number of Bonds to O or X (Halogens)	Examples of Compounds			
0	CH_4 (alkanes)			
1	$CH_3{-}OH$, alcohols	$CH_3{-}O{-}CH_3$, ethers	$CH_3{-}Cl$ alkyl halides	
2	$CH_2{=}O$, aldehydes	$HO{-}CH_2{-}OCH_3$, hemiacetals	$CH_3O{-}CH_2{-}OCH_3$, acetals	CH_2Cl_2 alkylene dihalides
3	$H{-}\overset{\overset{O}{\|}}{C}{-}OH$, acids	$H{-}\overset{\overset{O}{\|}}{C}{-}OCH_3$, esters	$H{-}\underset{\underset{OCH_3}{|}}{\overset{\overset{OCH_3}{|}}{C}}{-}OCH_3$, ortho esters	$H{-}\underset{\underset{Cl}{|}}{\overset{\overset{Cl}{|}}{C}}{-}Cl$ trihalides
4	$O{=}C{=}O$, carbon dioxide	$HO{-}\overset{\overset{O}{\|}}{C}{-}OH$, carbonic acid	$R{-}O{-}\overset{\overset{O}{\|}}{C}{-}O{-}R$, carbonates	CCl_4 carbon tetrahalides

oxidation state. On the other hand, we need an oxidizing agent to convert an aldehyde into a carboxylic acid because this conversion requires raising the oxidation state of carbon.

We shall find that the concept of oxidation state is useful not only in ordinary organic reactions, but also in reactions that occur in living systems.

Problem 12.6

Classify the following compounds according to the oxidation state of the circled carbon atom.

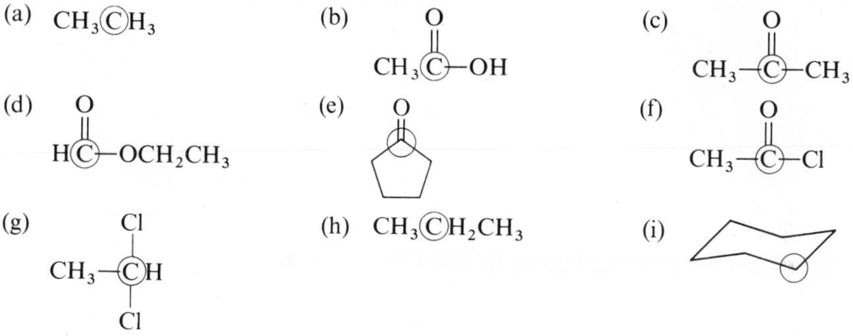

(a) $CH_3\text{Ⓒ}H_3$

(b) $CH_3\overset{O}{\overset{\|}{\text{Ⓒ}}}-OH$

(c) $CH_3-\overset{O}{\overset{\|}{\text{Ⓒ}}}-CH_3$

(d) $H\overset{O}{\overset{\|}{\text{Ⓒ}}}-OCH_2CH_3$

(e) cyclopentanone (circled carbonyl carbon)

(f) $CH_3-\overset{O}{\overset{\|}{\text{Ⓒ}}}-Cl$

(g) $CH_3-\text{Ⓒ}H(Cl)_2$ (Cl above and below the circled carbon)

(h) $CH_3\text{Ⓒ}H_2CH_3$

(i) cyclohexane (chair form, circled carbon)

Problem 12.7

For each of the following conversions indicate whether you would need an oxidizing agent, a reducing agent, or neither.

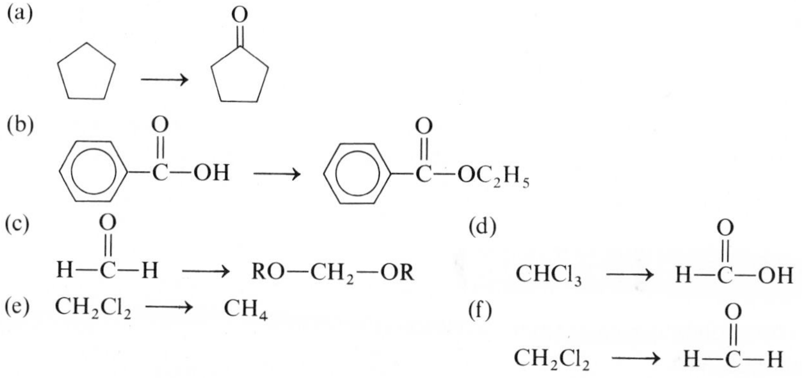

(a) cyclopentane $\longrightarrow$ cyclopentanone

(b) $C_6H_5-\overset{O}{\overset{\|}{C}}-OH \longrightarrow C_6H_5-\overset{O}{\overset{\|}{C}}-OC_2H_5$

(c) $H-\overset{O}{\overset{\|}{C}}-H \longrightarrow RO-CH_2-OR$

(d) $CHCl_3 \longrightarrow H-\overset{O}{\overset{\|}{C}}-OH$

(e) $CH_2Cl_2 \longrightarrow CH_4$

(f) $CH_2Cl_2 \longrightarrow H-\overset{O}{\overset{\|}{C}}-H$

12.7 LABORATORY PREPARATION OF ALDEHYDES AND KETONES

Oxidation of Alcohols

Aldehydes and ketones represent a higher oxidation state of carbon than alcohols. Therefore, one method of preparing aldehydes and ketones is by the oxidation of alcohols. The most common oxidizing agent for oxidizing alcohols to aldehydes and ketones is chromic acid, H_2CrO_4.

Oxidation of methanol yields formaldehyde:

$$CH_3OH + H_2CrO_4 + 4H^+ \longrightarrow CH_2O + Cr^{3+} + 4H_2O$$

Chromic acid is orange-red and the Cr^{3+} ion is green; we can easily follow the reaction by noting the color change.

Oxidation of primary alcohols yields aldehydes:

$$3R{-}CH_2OH + H_2CrO_4 \longrightarrow 3RCHO + 4H_2O + Cr^{3+}$$

Aldehydes are themselves more susceptible to oxidation than alcohols. They are easily oxidized to carboxylic acids. For this reason we must remove the aldehyde as it forms in order to prevent its oxidation. It is usually easy to remove the aldehyde by distillation because aldehydes have lower boiling points than the corresponding alcohols (Sec. 12.3). We need only carry out the reaction at a temperature above the boiling point of the aldehyde, and the aldehyde will distill out as it forms.

Oxidation of secondary alcohols yields ketones:

$$3R{-}\underset{}{\overset{OH}{\overset{|}{C}H}}{-}R' + H_2CrO_4 \longrightarrow 3R{-}\overset{O}{\overset{\|}{C}}{-}R' + 4H_2O + Cr^{3+}$$

Ketones lack the easily oxidized C—H bond of aldehydes. Ketones therefore resist further oxidation under normal reaction conditions.

Tertiary alcohols are not readily oxidized:
Oxidation of tertiary alcohols requires drastic conditions and involves breakage of carbon–carbon bonds. The resulting mixtures are complex and the reaction is not useful in synthesis.

Hydrolysis of Alkyl Dihalides

Alkyl dihalides with both halogens on the same carbon can be hydrolyzed to carbonyl compounds:

$$R{-}\underset{\underset{Cl}{|}}{\overset{Cl}{\overset{|}{C}H}} + 2OH^- \longrightarrow R{-}\overset{O}{\overset{\|}{C}H} + H_2O + 2Cl^-$$

This is a nucleophilic substitution reaction,

$$R{-}\underset{\underset{Cl}{|}}{\overset{Cl}{\overset{|}{C}}}{-}H + OH^- \longrightarrow R{-}\underset{\underset{OH}{|}}{\overset{Cl}{\overset{|}{C}}}{-}H + Cl^-$$

followed by elimination of H^+ and Cl^-:

$$R{-}\overset{\overset{Cl}{|}}{\underset{\underset{O-H}{|}}{C}}{-}H + OH^- \longrightarrow R{-}\underset{\underset{O}{\|}}{C}{-}H + H_2O + Cl^-$$

If the halogens are not on the end carbon, the reaction produces a ketone.

$$R{-}\overset{\overset{Cl}{|}}{\underset{\underset{Cl}{|}}{C}}{-}R' + 2OH^- \longrightarrow R{-}\overset{\overset{O}{\|}}{C}{-}R' + 2Cl^- + H_2O$$

This reaction is not an oxidation–reduction reaction because both the dihalide carbon and the carbonyl carbon are in the same oxidation state.

Friedel–Crafts Acylation

We can prepare aryl ketones by Friedel–Crafts acylation (Sec. 7.10).

$$C_6H_6 + Cl{-}\overset{\overset{O}{\|}}{C}{-}R \xrightarrow{AlCl_3} C_6H_5{-}\overset{\overset{O}{\|}}{C}{-}R + HCl$$

Problem 12.8

Write equations to show how each of the following compounds could be prepared, using methods outlined in this section.

(a) $C_6H_5{-}\overset{\overset{O}{\|}}{C}{-}CH_3$ (two methods),

(b) $CH_3\overset{\overset{O}{\|}}{C}CH_2CH_3$,

(c) $C_6H_5{-}\overset{\overset{O}{\|}}{C}{-}H$ (two methods),

(d) cyclopentanone (cyclopentane ring $=O$)

12.8 OXIDATION OF ALDEHYDES

We discussed the polar nature of the carbonyl group in Section 12.3. We can explain the reactions of aldehydes and ketones on the basis of this polarity.

In this and the following two sections we shall examine three general types of reactions: oxidation of aldehydes, additions to the carbonyl group, and reactions of the alpha hydrogens (the hydrogens on the carbon adjacent to the carbonyl group).

The C—H bond of aldehydes is very easily oxidized. Most aldehydes undergo spontaneous oxidation to the acid if exposed to the air.

$$2R{-}\overset{\overset{\displaystyle O}{\|}}{C}{-}H + O_2 \longrightarrow 2R{-}\overset{\overset{\displaystyle O}{\|}}{C}{-}OH$$

Ketones have no corresponding C—H bond and do not undergo oxidation readily. Under extreme oxidizing conditions C—C bond breakage occurs, and the reaction is not synthetically useful.

This difference in ease of oxidation makes it possible to distinguish between aldehydes and ketones. Three common laboratory reagents specifically oxidize aldehydes—Tollens's reagent, Fehling's reagent, and Benedict's reagent. Each of these oxidizing agents gives a visual change during the reaction, so we can use them as specific chemical tests for aldehydes.

Tollens's Reagent

Tollens's reagent is a water solution of silver nitrate and ammonia. This solution contains the complex ion $Ag(NH_3)_2{}^+$, which is a very mild oxidizing agent.

$$R{-}\overset{\overset{\displaystyle O}{\|}}{C}{-}H + 2Ag(NH_3)_2{}^+ + 2OH^-$$

$$\longrightarrow \underset{\text{carboxylic acid salt}}{R{-}\overset{\overset{\displaystyle O}{\|}}{C}{-}O^-NH_4{}^+} + \underset{\text{silver mirror}}{2Ag\downarrow} + 3NH_3 + H_2O$$

If we use a clean test tube the silver metal will coat the surface to give a "silver mirror." In fact, this reaction (employing formaldehyde solution) is used to manufacture mirrors.

Fehling's Reagent

When using **Fehling's reagent**, we oxidize the aldehyde with a basic water solution of cupric ion, Cu^{2+}. Cupric hydroxide is insoluble in water and cannot be used as such. In Fehling's solution, the Cu^{2+} ions are kept in solution in the form of a complex ion—cupric tartrate. The solution is prepared by adding sodium tartrate to the cupric hydroxide-water mixture.

$$R{-}\overset{\overset{\displaystyle O}{\|}}{C}H + 2Cu^{2+} + 5OH^- \xrightarrow[\text{tartrate}]{\text{sodium}} R{-}\overset{\overset{\displaystyle O}{\|}}{C}{-}O^- + Cu_2O\downarrow + 3H_2O$$

The original test solution is blue because of the complex Cu^{2+} ions; Cu_2O is a reddish-brown precipitate. The test consists of adding a small amount of Fehling's solution to the unknown and observing the disappearance of the blue color and the appearance of the reddish-brown precipitate.

Fehling's solution is unstable and a fresh mixture must be prepared each time it is used.

Benedict's Reagent

The use of **Benedict's reagent** involves the same oxidizing agent as Fehling's solution—Cu^{2+} ion. The difference is that the cupric ions are stabilized as the citrate complex ion instead of the tartrate complex ion. The solution is prepared by adding sodium citrate to the $Cu(OH)_2$–H_2O mixture. Benedict's test is visually the same as Fehling's test, but Benedict's solution has the advantage of being more stable and can be stored for long periods of time.

These reagents are used in clinical tests for glucose in urine. Glucose contains an aldehyde group, and the presence of glucose in the urine indicates one of several conditions including diabetes. Benedict's solution can detect as little as 0.25% glucose in urine.

12.9 ADDITION REACTIONS

Theory

The carbonyl group undergoes addition reactions much more readily than the more symmetrical carbon–carbon double bond of alkenes. The carbonyl group is more reactive due to its polarity. Nucleophilic reagents (ions or molecules that contain an unshared electron pair) attack the carbonyl carbon atom (the positive end of the carbonyl group) to form an ion with a negative charge on oxygen:

$$\mathrm{Nu{:}^-} + \mathrm{R}-\overset{\overset{\delta-}{:\ddot{O}}}{\underset{\delta+}{\overset{\|}{C}}}-\mathrm{R'} \longrightarrow \mathrm{R}-\overset{:\ddot{O}:^-}{\underset{\mathrm{Nu}}{\overset{|}{\underset{|}{C}}}}-\mathrm{R'}$$

Neutral nucleophiles react to give a product that has both a positive and a negative charge:

$$:\mathrm{NuH} + \mathrm{R}-\overset{:\ddot{O}^{\delta-}}{\underset{\delta+}{\overset{\|}{C}}}-\mathrm{R'} \longrightarrow \mathrm{R}-\overset{:\ddot{O}:^-}{\underset{^+\mathrm{NuH}}{\overset{|}{\underset{|}{C}}}}-\mathrm{R'}$$

Both of these intermediate products are very basic alkoxide ions, and they react readily with any proton-donating compound present. If the reaction is carried out in water, alcohol, or another proton-donating solvent, the second step (an acid-base reaction) is rapid:

$$\mathrm{R}-\overset{:\ddot{O}:^-}{\underset{\mathrm{Nu}}{\overset{|}{\underset{|}{C}}}}-\mathrm{R'} + \mathrm{H}-\mathrm{OH} \xrightarrow{\text{fast}} \mathrm{R}-\overset{:\ddot{O}H}{\underset{\mathrm{Nu}}{\overset{|}{\underset{|}{C}}}}-\mathrm{R'} + \mathrm{OH^-}$$

or

$$\text{R—}\overset{\text{:}\ddot{\text{O}}\text{:}^-}{\underset{^+\text{NuH}}{\text{C}}}\text{—R}' + \text{H—OH} \xrightarrow{\text{fast}} \text{R—}\overset{\text{:}\ddot{\text{O}}\text{H}}{\underset{^+\text{Nu—H}}{\text{C}}}\text{—R}' + \text{:}\ddot{\text{O}}\text{H}^- \longrightarrow \text{R—}\overset{\text{:}\ddot{\text{O}}\text{H}}{\underset{\text{Nu:}}{\text{C}}}\text{—R}' + H_2O$$

Neutral nucleophiles can be made more reactive by using a strong base catalyst. The base converts the less reactive neutral nucleophile, :NuH, to the more reactive anion, :Nu:$^-$.

$$\underset{\text{weak nucleophile}}{\text{:NuH + base}} \rightleftharpoons \underset{\text{strong nucleophile}}{\text{:Nu:}^-} + \text{H}^+\text{—base}$$

An example of such enhancement is the conversion of HCN (a weak nucleophile) to a cyanide ion (a strong nucleophile).

$$HCN + OH^- \rightleftharpoons CN^- + H_2O$$

Some addition reactions are catalyzed by acid. In this case we enhance the reactivity of the carbonyl group by converting it into a positive ion—the conjugate acid:

$$\text{R—}\overset{\text{:}\ddot{\text{O}}^{\delta-}}{\underset{\delta+}{\overset{\|}{\text{C}}}}\text{—R}' + \text{H—A} \rightleftharpoons \left[\text{R—}\overset{\text{:}\overset{+}{\text{O}}\text{H}}{\overset{\|}{\text{C}}}\text{—R}' \longleftrightarrow \text{R—}\overset{\text{:}\ddot{\text{O}}\text{H}}{\underset{+}{\text{C}}}\text{—R}'\right] + \text{A}^-$$

This protonated carbonyl group is much more reactive toward nucleophiles than the original carbonyl group. We normally use acid catalysis with weak nucleophiles.

In general, aldehydes are more reactive than ketones because aldehydes offer less hindrance to the attacking nucleophile: Ketones have two hydrocarbon groups attached to the carbonyl carbon, whereas aldehydes have only one.

Problem 12.9

The above discussion examines the effect of a base catalyst on the nucleophile and of an acid catalyst on the carbonyl group. What effects, if any, could a base catalyst have on the carbonyl group? What effects, if any, could an acid catalyst have on the nucleophile?

Addition of Alcohols

When we mix an alcohol with an aldehyde or ketone, an equilibrium is established with a compound called a **hemiacetal** or a **hemiketal**.

$$\text{R—}\overset{\text{O}}{\overset{\|}{\text{C}}}\text{—H} + \text{R}'\text{—OH} \rightleftharpoons \underset{\text{a hemiacetal}}{\text{R—}\overset{\text{OH}}{\overset{|}{\text{CH}}}\text{—OR}'}$$

$$R{-}\overset{\overset{\Large O}{\|}}{C}{-}R' + R''{-}OH \rightleftarrows R{-}\underset{\underset{\Large R'}{|}}{\overset{\overset{\Large OH}{|}}{C}}{-}OR''$$

a hemiketal

Hemiacetals are produced from aldehydes, and hemiketals are produced from ketones. Both of these reactions are reversible and it is not usually possible to isolate the hemiacetal or hemiketal from the reaction mixture. The reaction proceeds by the simple addition mechanism given above.

$$R{-}\overset{\overset{:O^{\delta-}}{\|}}{\underset{\delta+}{C}}{-}H + H\ddot{O}{-}R' \rightleftarrows R{-}\underset{\underset{HOR'^{+}}{|}}{\overset{\overset{:\ddot{O}:^{-}}{|}}{C}}{-}H$$

$$\xrightleftharpoons{H_2O} R{-}\underset{\underset{HOR'^{+}}{|}}{\overset{\overset{:\ddot{O}H}{|}}{C}}{-}H + OH^{-} \rightleftarrows R{-}\underset{\underset{:\ddot{O}R'}{|}}{\overset{\overset{:\ddot{O}H}{|}}{C}}{-}H + H_2O$$

A specific example is the reaction of propionaldehyde with ethanol:

$$CH_3CH_2\overset{\overset{\Large O}{\|}}{C}H + HOCH_2CH_3 \rightleftarrows CH_3CH_2\underset{\underset{OCH_2CH_3}{|}}{\overset{\overset{OH}{|}}{C}}H$$

If a small amount of dry HCl or other strong acid is present, further reaction occurs to give an **acetal** (a **ketal** if we use a ketone):

$$R{-}\underset{\underset{OR'}{|}}{\overset{\overset{OH}{|}}{C}}H + HOR' \xrightleftharpoons{HCl} R{-}\underset{\underset{OR'}{|}}{\overset{\overset{OR'}{|}}{C}}H + H_2O$$

an acetal

$$R{-}\underset{\underset{OR''}{|}}{\overset{\overset{OH}{|}}{C}}{-}R' + HOR'' \xrightleftharpoons{HCl} R{-}\underset{\underset{OR''}{|}}{\overset{\overset{OR''}{|}}{C}}{-}R' + H_2O$$

a ketal

Acetals and ketals can be isolated if we first neutralize the acid catalyst. In the absence of acid, both the forward and reverse reactions are very slow; for all practical purposes, we may say that the reaction stops.

The formation of acetal proceeds by the following mechanism:

$$\text{R—CH(}\ddot{\text{O}}\text{H)(}\ddot{\text{O}}\text{R}') + \text{HCl} \underset{}{\overset{-\text{Cl}^-}{\rightleftharpoons}} \text{R—CH(}\overset{+}{\text{O}}\text{H}_2\text{)(}\ddot{\text{O}}\text{R}') \rightleftharpoons \text{R—}\overset{+}{\text{C}}\text{—H(}\ddot{\text{O}}\text{—R}') + H_2O$$

(leaving group)

$$\left[\text{R—}\overset{+}{\text{C}}\text{H(}\ddot{\text{O}}\text{R}') \longleftrightarrow \text{R—CH}{=}\overset{+}{\text{O}}\text{R}'\right] + \text{H}\ddot{\text{O}}\text{—R}'$$

resonance hybrid

$$\rightleftharpoons \text{R—CH(}\overset{+}{\text{O}}\text{H—R}'\text{)(}\ddot{\text{O}}\text{R}') \overset{H_2O}{\rightleftharpoons} \text{R—CH(}\ddot{\text{O}}\text{R}'\text{)(}\ddot{\text{O}}\text{R}') + H_3O^+$$

The reaction requires acid because otherwise the very basic OH^- would have to be the leaving group. In the presence of acid the leaving group is water, a much weaker base.

Ketals and acetals are ethers in which both alkoxy groups are attached to the same carbon. Acetals and ketals are much more reactive than ordinary ethers.

Most hemiacetals are too unstable to isolate. Cyclic hemiacetals (formed from a molecule that has both an aldehyde and an alcohol group) are generally much more stable, especially if the ring contains five or six atoms.

$$HO\text{—}CH_2CH_2CH_2\text{—}CHO \rightleftharpoons \text{(ring: } CH_2\text{—}CH_2\text{—}CH_2\text{—}CH\text{—}OH \text{, closed by O)}$$

$$HO\text{—}CH_2CH_2CH_2CH_2\text{—}CHO \rightleftharpoons \text{(ring: } CH_2\text{—}CH_2\text{—}CH_2\text{—}CH_2\text{—}O\text{—}CH\text{—}OH\text{)}$$

Simple sugars such as glucose and fructose contain a carbonyl group. Glucose, for example, is a pentahydroxy aldehyde. Most simple sugars exist in equilibrium with the hemiacetal (Sec. 16.7, 16.8). The hemiacetal is usually the major component:

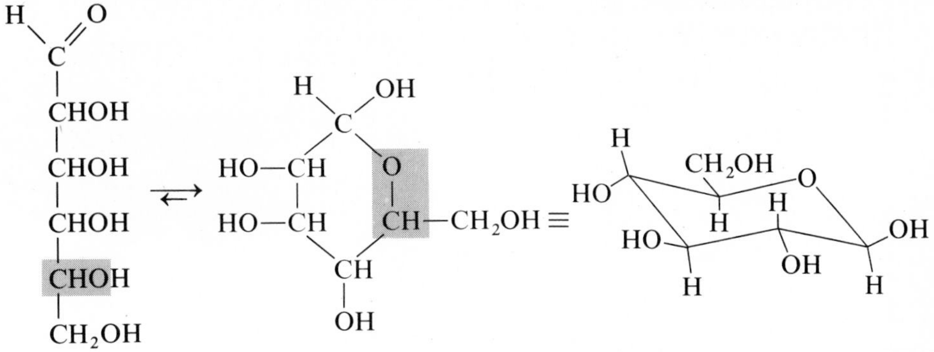

$$\begin{matrix} H\diagdown\!\!\overset{\displaystyle O}{\diagup\!\!\diagup} \\ C \\ | \\ CHOH \\ | \\ CHOH \\ | \\ CHOH \\ | \\ CHOH \\ | \\ CH_2OH \end{matrix} \rightleftarrows \text{(cyclic hemiacetal: ring of } C(H)(OH)\text{, O, CH—}CH_2OH\text{, CH(OH), HO—CH, HO—CH)} \equiv \text{(chair form with } CH_2OH\text{, OH, HO, H substituents)}$$

a simple sugar — cyclic hemiacetal form

For this reason, hemiacetals have an important role in biochemistry.

Problem 12.10

Write balanced equations for the reactions of ethanol with each of the following compounds. Show what happens in the absence of HCl and in the presence of HCl.

(a) 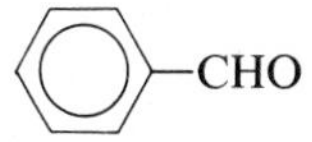 C_6H_5—CHO (benzaldehyde)

(b) 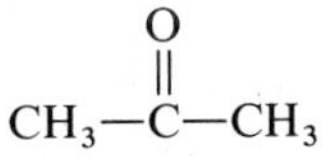 $CH_3-\overset{\overset{\displaystyle O}{\|}}{C}-CH_3$

Grignard Reaction

Aldehydes and ketones react with Grignard reagents (Sec. 9.9) to give alcohols.

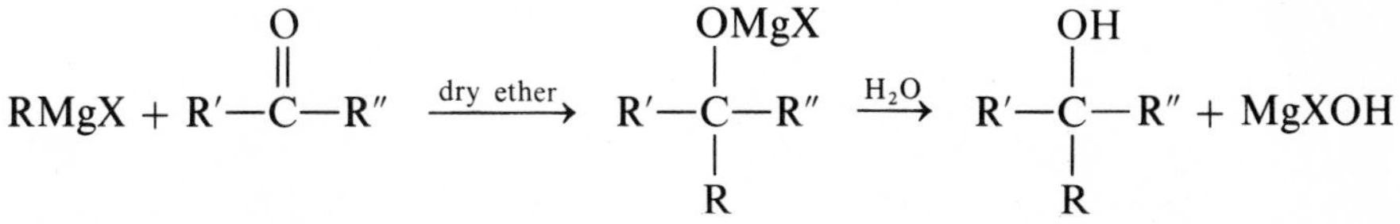

$$RMgX + R'-\overset{\overset{\displaystyle O}{\|}}{C}-R'' \xrightarrow{\text{dry ether}} R'-\overset{\overset{\displaystyle OMgX}{|}}{\underset{\underset{\displaystyle R}{|}}{C}}-R'' \xrightarrow{H_2O} R'-\overset{\overset{\displaystyle OH}{|}}{\underset{\underset{\displaystyle R}{|}}{C}}-R'' + MgXOH$$

In the first step, the hydrocarbon group of the Grignard reagent adds to the carbonyl carbon. The alkoxide salt thus produced is converted to the alcohol in a second step by reaction with water or another acid.

The Grignard reagent is a strong nucleophile because the C—Mg bond is highly polar.

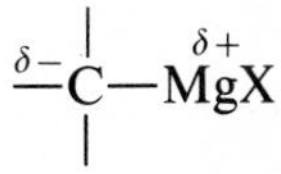

$$-\overset{|}{\underset{|}{\overset{\delta-}{C}}}-\overset{\delta+}{MgX}$$

The reaction occurs by attack of the anionic R group at the carbonyl carbon atom.

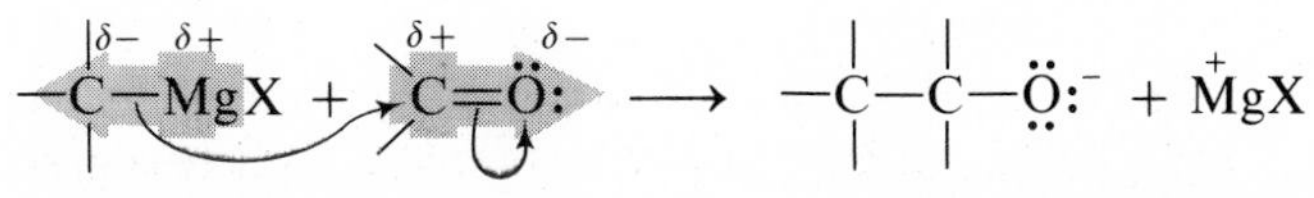

$$-\overset{|}{\underset{|}{\overset{\delta-}{C}}}-\overset{\delta+}{MgX} + \overset{\delta+}{C}=\overset{\delta-}{\ddot{O}}: \longrightarrow -\overset{|}{\underset{|}{C}}-\overset{|}{\underset{|}{C}}-\ddot{\underset{..}{O}}:^- + \overset{+}{MgX}$$

We can use the Grignard reaction to produce primary, secondary, or tertiary alcohols. If we use formaldehyde as the carbonyl compound, we obtain a primary alcohol.

$$RMgBr + H_2C{=}O \xrightarrow{\text{dry ether}} R{-}CH_2{-}OMgBr \xrightarrow{H_2O} RCH_2OH + MgBrOH$$

All other aldehydes yield secondary alcohols.

$$RMgBr + R'{-}\overset{\overset{\displaystyle O}{\|}}{C}H \xrightarrow[\text{ether}]{\text{dry}} R'{-}\overset{\overset{\displaystyle OMgBr}{|}}{C}H{-}R \xrightarrow{H_2O} R'{-}\overset{\overset{\displaystyle OH}{|}}{C}H{-}R + MgBrOH$$

Ketones yield tertiary alcohols.

$$RMgBr + R'{-}\overset{\overset{\displaystyle O}{\|}}{C}{-}R'' \xrightarrow[\text{ether}]{\text{dry}} R'{-}\underset{\underset{\displaystyle R}{|}}{\overset{\overset{\displaystyle OMgBr}{|}}{C}}{-}R'' \xrightarrow{H_2O} R'{-}\underset{\underset{\displaystyle R}{|}}{\overset{\overset{\displaystyle OH}{|}}{C}}{-}R'' + MgBrOH$$

The following examples illustrate the versatility of this reaction.

$$C_6H_5{-}MgBr + CH_2O \xrightarrow[\text{ether}]{\text{dry}} C_6H_5{-}CH_2OMgBr \xrightarrow{H_2O} C_6H_5{-}CH_2OH + MgBrOH$$

$$\text{cyclopentanone} + CH_3CH_2MgBr \xrightarrow[\text{ether}]{\text{dry}} \text{1-(BrMgO)-1-}(CH_2CH_3)\text{cyclopentane} \xrightarrow{H_2O} \text{1-(HO)-1-}(CH_2CH_3)\text{cyclopentane} + MgBrOH$$

$$C_6H_5{-}\overset{\overset{\displaystyle O}{\|}}{C}H + CH_3MgBr \xrightarrow[\text{ether}]{\text{dry}} C_6H_5{-}\overset{\overset{\displaystyle OMgBr}{|}}{C}HCH_3 \xrightarrow{H_2O} C_6H_5{-}\overset{\overset{\displaystyle OH}{|}}{C}HCH_3 + MgBrOH$$

Problem 12.11
Complete each of the following equations.

(a)

$$CH_3MgBr + CH_3\overset{\overset{\displaystyle O}{\|}}{C}CH_3 \xrightarrow[\text{ether}]{\text{dry}} ? \xrightarrow{H_2O} ?$$

(b)

$$\text{(cyclohexanone)} + C_6H_5\text{—}MgBr \xrightarrow[\text{ether}]{\text{dry}} ? \xrightarrow{H_2O} ?$$

(c) $CH_3MgBr + CH_2O \xrightarrow[\text{ether}]{\text{dry}} ? \xrightarrow{H_2O} ?$

(d)

$$CH_3\overset{\overset{\displaystyle CH_3}{|}}{C}HCHO + CH_3MgBr \xrightarrow[\text{ether}]{\text{dry}} ? \xrightarrow{H_2O} ?$$

Addition of HCN

Aldehydes and ketones* react with hydrogen cyanide to form **cyanohydrins** (compounds that have a cyano group and a hydroxyl group attached to the same carbon).

$$R\text{—}\overset{\overset{\displaystyle O}{\|}}{C}\text{—}R' + HCN \longrightarrow R\text{—}\underset{\underset{\displaystyle CN}{|}}{\overset{\overset{\displaystyle OH}{|}}{C}}\text{—}R'$$

a cyanohydrin

In this reaction the cyanide ion (strong nucleophile) attacks the carbonyl carbon:

$$R\text{—}\overset{\overset{\displaystyle :\ddot{O}}{\|}}{C}\text{—}R' + {}^{-}:C{\equiv}N: \rightleftarrows R\text{—}\underset{\underset{\displaystyle CN}{|}}{\overset{\overset{\displaystyle :\ddot{O}:^-}{|}}{C}}\text{—}R' \underset{-H^+}{\overset{+H^+}{\rightleftarrows}} R\text{—}\underset{\underset{\displaystyle CN}{|}}{\overset{\overset{\displaystyle OH}{|}}{C}}\text{—}R'$$

Because hydrogen cyanide is so toxic and volatile, the reaction is usually carried out by mixing the carbonyl compound with a water solution of NaCN or KCN (*also very toxic*) and then carefully adding mineral acid. The acid reacts with the cyanide salt to produce HCN.

* Highly hindered ketones fail to undergo this reaction.

Cyanohydrins are useful synthetic intermediates. The cyano group may be hydrolyzed to a carboxyl group, or it may be reduced and hydrolyzed to an aldehyde group:

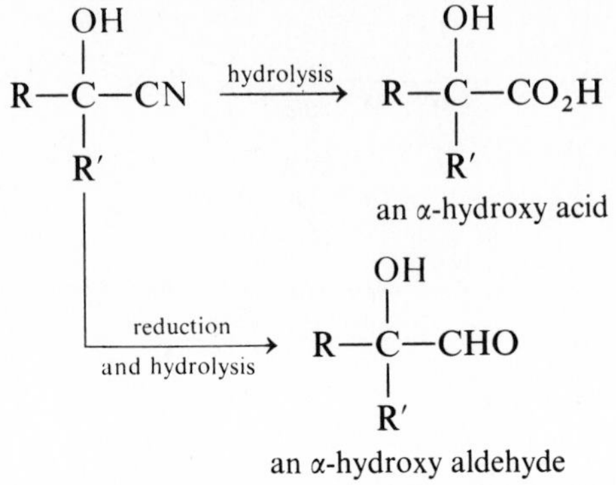

$$\underset{\underset{R'}{|}}{\overset{\overset{OH}{|}}{R-C-CN}} \xrightarrow{\text{hydrolysis}} \underset{\underset{R'}{|}}{\overset{\overset{OH}{|}}{R-C-CO_2H}}$$

an α-hydroxy acid

$$\xrightarrow[\text{and hydrolysis}]{\text{reduction}} \underset{\underset{R'}{|}}{\overset{\overset{OH}{|}}{R-C-CHO}}$$

an α-hydroxy aldehyde

Problem 12.12

Write equations for the reaction of HCN with (a) acetaldehyde, (b) formaldehyde, (c) acetone.

Addition of Amines and Their Derivatives

Ammonia and various other related nitrogen compounds are nucleophiles because they have an unshared electron pair on the nitrogen atom.

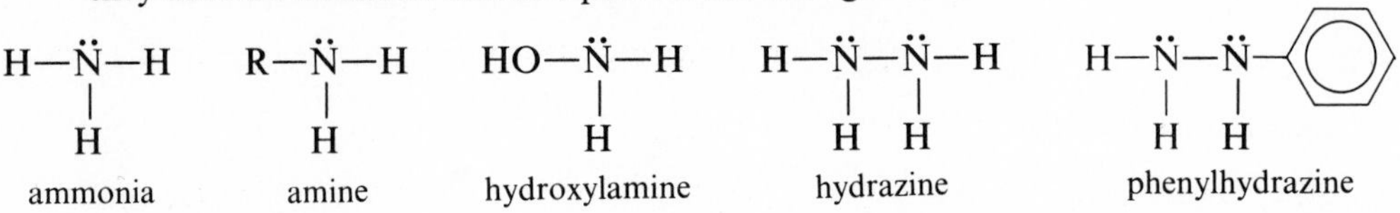

$$\underset{\text{ammonia}}{H-\ddot{N}(H)-H} \qquad \underset{\text{amine}}{R-\ddot{N}(H)-H} \qquad \underset{\text{hydroxylamine}}{HO-\ddot{N}(H)-H} \qquad \underset{\text{hydrazine}}{H-\ddot{N}(H)-\ddot{N}(H)-H} \qquad \underset{\text{phenylhydrazine}}{H-\ddot{N}(H)-\ddot{N}(H)-C_6H_5}$$

These compounds react as nucleophiles with aldehydes and ketones as above, except that in these cases the addition product reacts further to lose water and form an unsaturated compound. Ammonia is the simplest example. (In all the following R and R′ may be alkyl, aryl, or H.)

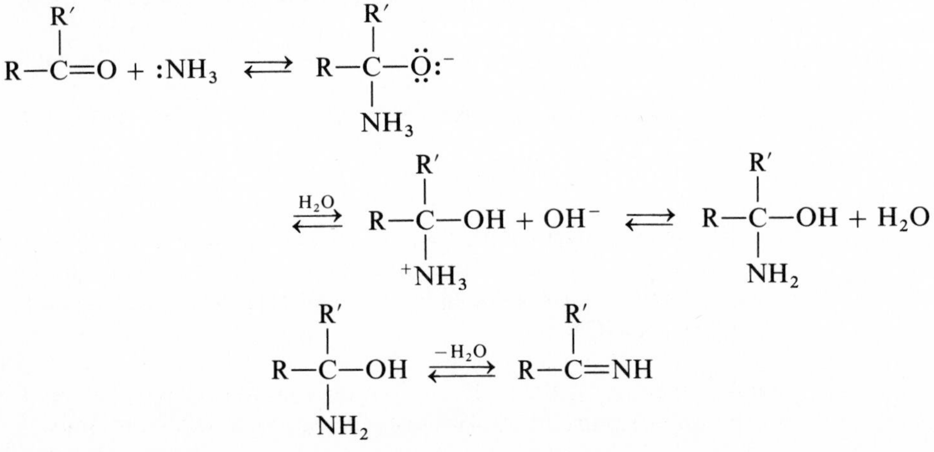

$$\overset{\overset{R'}{|}}{R-C=O} + :NH_3 \rightleftharpoons \underset{\underset{NH_3}{|}}{\overset{\overset{R'}{|}}{R-C-\ddot{\underset{..}{O}}:^-}}$$

$$\overset{H_2O}{\rightleftharpoons} \underset{\underset{^+NH_3}{|}}{\overset{\overset{R'}{|}}{R-C-OH}} + OH^- \rightleftharpoons \underset{\underset{NH_2}{|}}{\overset{\overset{R'}{|}}{R-C-OH}} + H_2O$$

$$\underset{\underset{NH_2}{|}}{\overset{\overset{R'}{|}}{R-C-OH}} \overset{-H_2O}{\rightleftharpoons} \overset{\overset{R'}{|}}{R-C=NH}$$

The overall general equation is

$$\begin{array}{c} R' \\ | \\ R{-}C{=}O \end{array} + NH_3 \longrightarrow \begin{array}{c} R' \\ | \\ R{-}C{=}NH \end{array} + H_2O$$

The other compounds shown above react in the same way.

$$\begin{array}{c} R' \\ | \\ R{-}C{=}O \end{array} + :\begin{array}{c} H \\ | \\ N{-}R' \\ | \\ H \end{array} \longrightarrow \begin{array}{c} R' \\ | \\ R{-}C{=}N{-}R' \end{array} + H_2O$$

an amine — a **schiff base**

$$\begin{array}{c} R' \\ | \\ R{-}C{=}O \end{array} + \begin{array}{c} R' \\ | \\ H{-}\ddot{N}{-}OH \end{array} \longrightarrow \begin{array}{c} R' \\ | \\ R{-}C{=}NOH \end{array} + H_2O$$

hydroxylamine — an **oxime**

$$\begin{array}{c} R' \\ | \\ R{-}C{=}O \end{array} + H_2\ddot{N}{-}\ddot{N}H_2 \longrightarrow \begin{array}{c} R' \\ | \\ R{-}C{=}N{-}NH_2 \end{array} + H_2O$$

hydrazine — a **hydrazone**

$$\begin{array}{c} R' \\ | \\ R{-}C{=}O \end{array} + H_2N{-}NH{-}C_6H_5 \longrightarrow \begin{array}{c} R' \\ | \\ R{-}C{=}N{-}NH{-}C_6H_5 \end{array} + H_2O$$

phenylhydrazine — a **phenylhydrazone**

The net effect of all these reactions is to convert C=O double bonds into C=N double bonds:

$$\begin{array}{c} O \\ \| \\ R{-}C{-}R' \end{array} + H_2NR'' \longrightarrow \begin{array}{c} R' \\ | \\ R{-}C{=}N{-}R'' \end{array} + H_2O$$

$$(R'' = H, \text{Alkyl}, OH, NH_2, NHC_6H_5)$$

Some of these derivatives, especially the oximes and phenylhydrazones, are useful in identifying aldehydes and ketones because they are solids whose melting points can be compared with those of derivatives of known structure.

Problem 12.13

Complete the following equations.

(a) $CH_3CHO + H_2NOH \longrightarrow$

(b) $C_6H_5{-}\overset{\overset{\displaystyle O}{\|}}{C}{-}CH_3 + H_2N{-}NH{-}C_6H_5 \longrightarrow$

Addition of Hydrogen

Aldehydes and ketones react with hydrogen gas in the presence of finely divided metals such as nickel, palladium, or platinum. Aldehydes (except formaldehyde) yield primary alcohols; ketones yield secondary alcohols.

$$\underset{\text{aldehyde}}{R{-}\overset{\overset{\displaystyle O}{\|}}{C}{-}H} + H_2 \xrightarrow{Ni} \underset{1^\circ\text{ alcohol}}{R{-}CH_2OH}$$

$$\underset{\text{ketone}}{R{-}\overset{\overset{\displaystyle O}{\|}}{C}{-}R'} + H_2 \xrightarrow{Ni} \underset{2^\circ\text{ alcohol}}{R{-}\overset{\overset{\displaystyle OH}{|}}{C}H{-}R'}$$

Hydrogenation occurs on the surface of the metal catalyst (Fig. 5.5, Sec. 5.6).

Metal Hydride Reductions

Certain **metal hydrides** also reduce the carbonyl group to the alcohol. The most commonly used metal hydrides are lithium aluminum hydride ($LiAlH_4$) and sodium borohydride ($NaBH_4$). These reductions occur by nucleophilic attack of the hydride (AlH_4^- or BH_4^-) at the carbonyl carbon.

$$R{-}\overset{\overset{\displaystyle O}{\|}}{C}{-}R' + H{-}AlH_3^- \longrightarrow \left[R{-}\overset{\overset{\displaystyle OAlH_3}{|}}{\underset{\underset{\displaystyle H}{|}}{C}}{-}R'\right]^-$$

At the end of the reduction the alcohol is present as the salts LiOCHRR′ and $Al(OCHRR')_3$, which must be hydrolyzed to obtain the alcohol:

$$(RR'CH{-}O)_3Al + 3H_2O \xrightarrow{H^+} 3R{-}\overset{\overset{\displaystyle OH}{|}}{\underset{\underset{\displaystyle H}{|}}{C}}{-}R' + Al(OH)_3$$

We may write the overall reaction as

$$R{-}\overset{\overset{\displaystyle O}{\|}}{C}{-}R' \xrightarrow{LiAlH_4} R{-}\overset{\overset{\displaystyle OH}{|}}{\underset{\underset{\displaystyle H}{|}}{C}}{-}R'$$

These metal hydrides do not attack the less polar C=C double bond, so we can use them to reduce a carbonyl group even when there is a C=C double bond present in the molecule. For example, we can reduce 2-butenal to 2-buten-1-ol:

$$CH_3CH{=}CH{-}\overset{\overset{\displaystyle O}{\|}}{C}H + LiAlH_4 \longrightarrow CH_3CH{=}CHCH_2OH$$

We can reduce an aldehyde or ketone all the way to the hydrocarbon by the use of zinc amalgam and hydrochloric acid. Zinc amalgam is a solution of zinc metal in mercury. The reaction is called a **Clemmensen reduction** and is often used to reduce ketones prepared by Friedel-Crafts acylation (Sec. 7.10).

$$C_6H_5{-}\overset{\overset{\displaystyle O}{\|}}{C}{-}CH_3 \xrightarrow[HCl]{Zn{-}Hg} C_6H_5{-}CH_2CH_3$$

The general reaction is

$$R{-}\overset{\overset{\displaystyle O}{\|}}{C}{-}R' \xrightarrow[HCl]{Zn{-}Hg} R{-}CH_2{-}R'$$

Problem 12.14

Write equations to show how you could convert benzophenone (diphenyl ketone) to (a) diphenylmethanol (two methods), (b) diphenylmethane.

12.10 ACIDITY OF ALDEHYDES AND KETONES: THE α HYDROGENS

Thus far we have examined the reactions of the carbonyl group itself. Now we shall see that the carbonyl group is so polar that it affects the carbon atom immediately adjacent to it. This adjacent carbon is often called the **alpha carbon**. Hydrogens attached to the alpha carbon are called **alpha hydrogens**.

$$\underset{\delta}{CH_3}\underset{\gamma}{CH_2}\underset{\beta}{CH_2}\underset{\alpha}{CH_2}{-}\overset{\overset{\displaystyle O}{\|}}{C}{-}H$$

Hydrocarbons are extremely weak acids. For example, the K_as of ethane, ethene, and acetylene are 10^{-42}, 10^{-36}, and 10^{-25}, respectively. Phenols and alcohols have higher acidities: C_6H_5—OH has a K_a of 10^{-10}, and CH_3CH_2OH has a K_a of 10^{-18}.

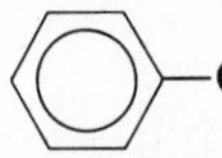

The α hydrogens of aldehydes and ketones are somewhat acidic. Most aldehydes and ketones have K_as of about 10^{-19} to 10^{-20}. Aldehydes and ketones are acidic because their conjugate bases are stabilized by resonance.

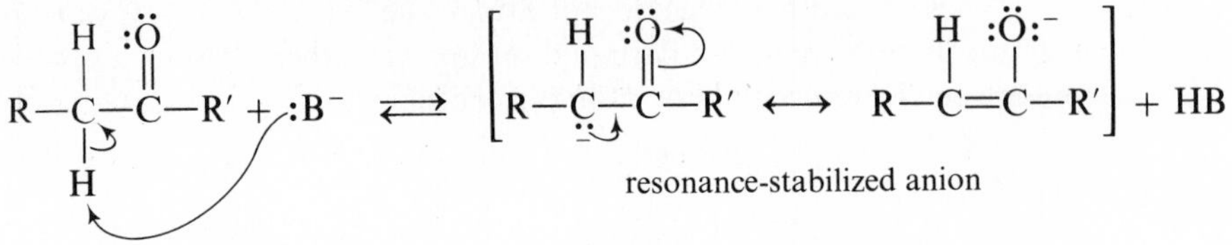

$$\mathrm{R{-}CH_2{-}C({=}\ddot{O}){-}R' + {:}B^- \rightleftarrows \left[R{-}\overset{H}{\underset{\cdot\cdot}{C}}{-}C({=}\ddot{O}){-}R' \longleftrightarrow R{-}\overset{H}{C}{=}C(\ddot{O}{:}^-){-}R' \right] + HB}$$

resonance-stabilized anion

This anion is called the **enolate ion**. We designate its resonance hybrid as

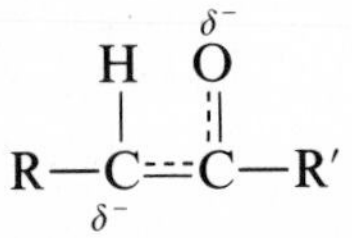

$$\mathrm{R{-}\underset{\delta^-}{\overset{H}{C}}{\cdots}C(\overset{\delta^-}{O}){-}R'}$$

Problem 12.15
Of the hydrocarbons, phenols, and alcohols, which most resemble aldehydes and ketones in acidity?

12.11 KETO-ENOL TAUTOMERISM

When an enolate ion accepts a proton, it may do so either at the alpha carbon or at the oxygen because both of these atoms bear part of the negative charge. Thus two different products can result:

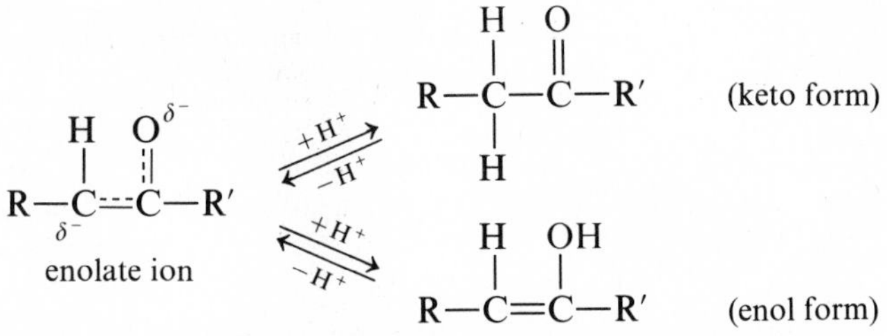

$$\underset{\text{enolate ion}}{\mathrm{R{-}\underset{\delta^-}{\overset{H}{C}}{\cdots}C(\overset{\delta^-}{O}){-}R'}} \quad \underset{-H^+}{\overset{+H^+}{\rightleftarrows}} \quad \mathrm{R{-}CH_2{-}C({=}O){-}R'} \quad \text{(keto form)}$$

$$\underset{-H^+}{\overset{+H^+}{\rightleftarrows}} \quad \mathrm{R{-}\overset{H}{C}{=}C(OH){-}R'} \quad \text{(enol form)}$$

If the proton becomes attached to the alpha carbon, the original carbonyl compound results. This molecule is called the *keto form*. If the proton becomes attached to the oxygen, a new molecule results. This new molecule is called the *enol form* (Sec. 6.11). Keto-enol forms are structural isomers in equilibrium; they cannot be separated from each other. The special term chemists use to describe structural isomers that are in equilibrium with each other and that interconvert readily is *tautomers*. Their interconversion is called *tautomerization* (Sec. 6.11).

Most aldehydes and ketones exist mainly in the keto form. In acetone, for example, the enol content is less than 1%. In acetaldehyde, the enol content is too small to measure.

Problem 12.16

The keto and enol forms of a compound cannot be considered to be resonance structures. Explain (see Sec. 7.3).

Problem 12.17

Draw the keto and enol forms of (a) cyclohexanone, (b) acetaldehyde, (c) acetophenone.

12.12 ALDOL CONDENSATION

Aldehydes and ketones that have alpha hydrogens undergo an interesting reaction called the **aldol condensation**.

$$2CH_3-\overset{\overset{\displaystyle O}{\|}}{C}-H \xrightarrow[5°C]{\text{dilute NaOH}} \underset{\text{an aldol}}{CH_3-\overset{\overset{\displaystyle OH}{|}}{C}H-CH_2-CHO}$$

The general equation for the aldol condensation is

$$2R-\underset{\underset{\displaystyle H}{|}}{\overset{\overset{\displaystyle R'}{|}}{C}}-\overset{\overset{\displaystyle O}{\|}}{C}-H \xrightarrow[\text{cold}]{\text{dilute }OH^-} R-\underset{\underset{\displaystyle H}{|}}{\overset{\overset{\displaystyle R'}{|}}{C}}-\underset{\underset{\displaystyle H}{|}}{\overset{\overset{\displaystyle OH}{|}}{C}}-\underset{\underset{\displaystyle R}{|}}{\overset{\overset{\displaystyle R'}{|}}{C}}-\overset{\overset{\displaystyle O}{\|}}{C}-H$$

The product is called an aldol because it has both an *ald*ehyde group and an alcoh*ol* group.

The aldol condensation occurs because of the acidity of the alpha hydrogen atom. The mechanism involves the following steps.

$$H\ddot{O}:^- + H-CH_2-\overset{\overset{\displaystyle O}{\|}}{C}-H \rightleftarrows H_2O + \left[^-:CH_2-\overset{\overset{\displaystyle :\ddot{O}}{\|}}{C}-H \longleftrightarrow CH_2=\overset{\overset{\displaystyle :\ddot{O}:^-}{|}}{C}-H \right]$$

enolate ion

$$CH_3-\overset{\overset{\displaystyle :\ddot{O}}{\|}}{C}-H + {}^-:CH_2-\overset{\overset{\displaystyle O}{\|}}{C}-H \rightleftarrows CH_3-\underset{\underset{\displaystyle H}{|}}{\overset{\overset{\displaystyle :\ddot{O}:^-}{|}}{C}}-CH_2-\overset{\overset{\displaystyle O}{\|}}{C}-H$$

$$H-O-H + CH_3-\overset{\overset{\displaystyle :\ddot{O}:^-}{|}}{C}H-CH_2-\overset{\overset{\displaystyle O}{\|}}{C}-H \rightleftarrows HO^- + CH_3-\overset{\overset{\displaystyle OH}{|}}{C}H-CH_2-\overset{\overset{\displaystyle O}{\|}}{C}-H$$

If the mixture is heated, dehydration occurs to give 2-butenal. The dehydration occurs because of the acidity of the remaining alpha hydrogens of aldol:

$$CH_3-\underset{}{\overset{OH}{\overset{|}{C}H}}-\underset{\underset{H}{|}}{C}H-\overset{O}{\overset{\|}{C}}-H + :\ddot{O}H^- \rightleftarrows CH_3-CH{=}CH-\overset{O}{\overset{\|}{C}}-H + OH^- + H_2O$$

2-butenal

The aldol condensation is a general reaction for aldehydes and ketones that have an alpha hydrogen. Propanal, for example, gives 3-hydroxy-2-methylpentanal.

$$2CH_3CH_2CHO \xrightarrow[\text{cold}]{\text{dilute NaOH}} CH_3CH_2\overset{OH}{\overset{|}{C}H}-\underset{\underset{CH_3}{|}}{C}HCHO$$

3-hydroxy-2-methylpentanal

Problem 12.18
Draw the structure of the product of the aldol condensation of each of the following compounds.

(a) $CH_3CH_2CH_2CHO$

(b) $CH_3\overset{O}{\overset{\|}{C}}CH_3$

(c) $C_6H_5-CH_2CHO$

12.13 CROSSED ALDOL CONDENSATION

Aldehydes or ketones that do not have an alpha hydrogen cannot undergo the aldol condensation. They may react, however, with enolate ions produced from an aldehyde or ketone that does have an alpha hydrogen. For example, benzaldehyde can react with the enolate ion of acetaldehyde:

$$C_6H_5-\overset{O}{\overset{\|}{C}}-H + {}^-:CH_2-\overset{O}{\overset{\|}{C}}-H \xrightarrow[\text{cold}]{\text{dilute OH}^-} C_6H_5-\overset{OH}{\overset{|}{C}H}-CH_2CHO \xrightarrow{\Delta} C_6H_5-CH{=}CH-CHO$$

This variation of the reaction in which two *different* carbonyl compounds react is called the **crossed aldol condensation**. Crossed aldol condensations are usually less useful than ordinary aldol condensations because they give a mixture of products. The above reaction, for example, gives 3-hydroxybutanal in addition to the product shown. 3-Hydroxybutanal is produced from the reaction of acetaldehyde with its own enolate ion.

Many biochemically important crossed aldol condensations are known. An example is the conversion of oxaloacetic acid into citric acid in the citric acid cycle.

$$\underset{\text{acetyl coenzyme A}}{\mathrm{CH_3\overset{\overset{\displaystyle O}{\|}}{C}{-}S{-}CoA^*}} + \underset{\text{oxaloacetic acid}}{\mathrm{O{=}\underset{\underset{\displaystyle CH_2CO_2H}{|}}{C}{-}CO_2H}} \xrightarrow{\text{citrate synthase}} \underset{\text{citric acid}}{\mathrm{HO{-}\overset{\overset{\displaystyle CH_2\overset{\overset{\displaystyle O}{\|}}{C}{-}OH}{|}}{\underset{\underset{\displaystyle CH_2CO_2H}{|}}{C}}{-}CO_2H}} + \mathrm{HSCoA^*}$$

Problem 12.19

What is the product of the crossed aldol condensation of benzaldehyde with propanal? What other products will form? [Show with equations.]

Problem 12.20

Show with structural formulas why benzaldehyde cannot undergo an aldol condensation with itself.

12.14 HALOFORM REACTION

Aldehydes and ketones that have alpha hydrogens react with halogens in the presence of a base. In this reaction, a halogen atom substitutes for the alpha hydrogen.

$$\mathrm{R{-}CH_2{-}\overset{\overset{\displaystyle O}{\|}}{C}{-}R'} + X_2 + OH^- \longrightarrow \mathrm{R{-}\underset{\underset{\displaystyle X}{|}}{CH}{-}\overset{\overset{\displaystyle O}{\|}}{C}{-}R'} + H_2O + X^-$$

(X = Cl, Br, or I)

If we use an excess of halogen, all the alpha hydrogens react:

$$\mathrm{R{-}\underset{\underset{\displaystyle X}{|}}{CH}{-}\overset{\overset{\displaystyle O}{\|}}{C}{-}R'} + X_2 + OH^- \longrightarrow \mathrm{R{-}\overset{\overset{\displaystyle X}{|}}{\underset{\underset{\displaystyle X}{|}}{C}}{-}\overset{\overset{\displaystyle O}{\|}}{C}{-}R'} + H_2O + X^-$$

* See Section 14.12 (Fig. 14.5) for the structural formula of acetyl coenzyme A.

This reaction is mechanistically similar to the aldol reaction in that the reactive intermediate is the enolate ion:

$$R-\underset{\underset{H}{|}}{CH}-\overset{\overset{O}{\|}}{C}-R' + :\ddot{O}H^- \rightleftharpoons R-\underset{..}{C}H-\overset{\overset{O}{\|}}{C}-R' + H_2O$$

$$R-\underset{..}{C}H-\overset{\overset{O}{\|}}{C}-R'' + :\ddot{X}-\ddot{X}: \longrightarrow R-\underset{\underset{X}{|}}{CH}-\overset{\overset{O}{\|}}{C}-R' + :\ddot{X}:^-$$

Because halogens are electron-withdrawing, the second alpha hydrogen is more acidic than the first. In other words, the enolate is stabilized by the presence of a halogen on the alpha carbon atom:

$$R-\underset{..}{\overset{\overset{Cl}{|}}{C}}-\overset{\overset{O}{\|}}{C}-R'$$

For this reason, all available alpha hydrogens usually react.

When the alpha carbon is a methyl group, a further reaction occurs to give haloform, CHX_3.

Haloform Reaction:

$$CH_3-\overset{\overset{O}{\|}}{C}-R + 3X_2 + 4OH^- \longrightarrow CHX_3 + R-\overset{\overset{O}{\|}}{C}-O^- + 3X^- + 3H_2O$$

$$(X = Cl, Br, \text{ or } I)$$

The reaction proceeds in steps to yield the trihalocarbonyl compound,

$$CH_3\overset{\overset{O}{\|}}{C}-R \xrightarrow[OH^-]{X_2} \underset{\underset{X}{|}}{CH_2}-\overset{\overset{O}{\|}}{C}-R \xrightarrow[OH^-]{X_2} \underset{\underset{X}{|}}{\overset{\overset{X}{|}}{C}H}-\overset{\overset{O}{\|}}{C}-R \xrightarrow[OH^-]{X_2} CX_3-\overset{\overset{O}{\|}}{C}-R$$

The presence of three electron-withdrawing halogens on the alpha carbon atom weakens the bond to the carbonyl group,

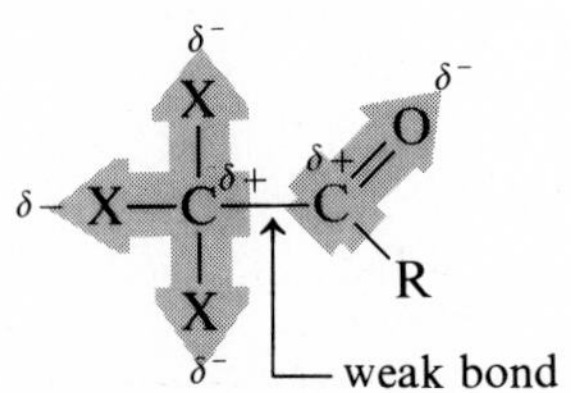

The C—C bond breaks when an OH^- ion attacks the carbonyl carbon:

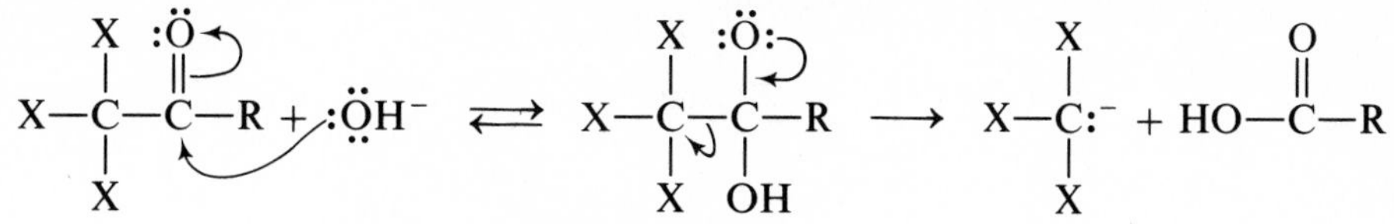

$$X-\underset{X}{\overset{X}{C}}-\overset{:\ddot{O}}{\overset{\|}{C}}-R + :\ddot{O}H^- \rightleftarrows X-\underset{X}{\overset{X}{C}}-\underset{OH}{\overset{:\ddot{O}:^-}{C}}-R \longrightarrow X-\underset{X}{\overset{X}{C}}:^- + HO-\overset{O}{\overset{\|}{C}}-R$$

$$^-:CX_3 + R-\overset{O}{\overset{\|}{C}}-OH \rightleftarrows \underset{\text{haloform}}{HCX_3} + R-\overset{O}{\overset{\|}{C}}-O^-$$

Iodoform Test

This reaction is useful as a test for the grouping $CH_3-\overset{O}{\overset{\|}{C}}-R$, in which R is H, alkyl, or aryl. Iodine is normally used because iodoform, CHI_3, is a brilliant yellow crystalline product that is easily detected. Alkaline solutions of halogens are good oxidizing agents, so compounds that have the grouping $CH_3-\overset{OH}{\overset{|}{C}}H-R$ also give a positive iodoform test. In this case, the first step of the test oxidizes the alcohol to a $CH_3-\overset{O}{\overset{\|}{C}}-R$ group:

$$CH_3-\overset{OH}{\overset{|}{C}}H-R + X_2 + 2OH^- \longrightarrow CH_3-\overset{O}{\overset{\|}{C}}-R + 2H_2O + 2X^-$$

Problem 12.21

Which of the following pairs of compounds can be distinguished by use of the iodoform test?

(a) $CH_3-\overset{O}{\overset{\|}{C}}-CH_3$ and $CH_3-\overset{O}{\overset{\|}{C}}-H$

(b) $C_6H_5-\overset{O}{\overset{\|}{C}}-CH_3$ and $C_6H_5-\overset{O}{\overset{\|}{C}}-CH_2CH_3$

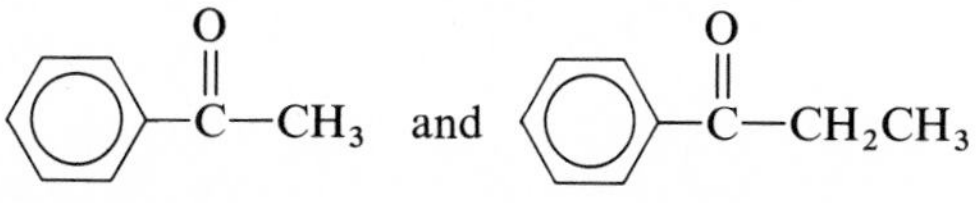

(c) $C_6H_5-\overset{O}{\overset{\|}{C}}-C_6H_5$ and $H-\overset{O}{\overset{\|}{C}}-H$

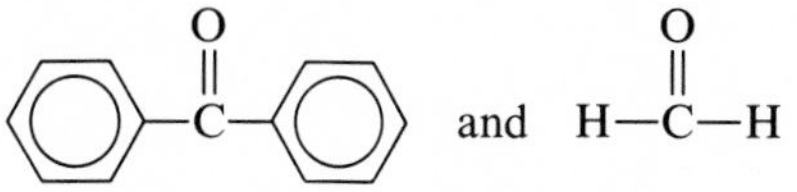

(d) CH_3CH_2OH and $CH_3\overset{O}{\overset{\|}{C}}CH_2CH_3$

(e) 2-pentanone and 3-pentanone

12.15 ANALYSIS AND IDENTIFICATION OF ALDEHYDES AND KETONES

Throughout this chapter we have examined various reactions that are suitable as simple chemical tests for aldehydes and ketones. Table 12.5 is a summary of those reactions and others that we can apply to aldehydes and ketones.

Table 12.5 Simple Chemical Tests for Aldehydes and Ketones

Test	Aldehydes	Ketones
Cold, concentrated H_2SO_4	Soluble	Soluble
Tollens's test, $Ag(NH_3)_2^+$	Ag mirror	No reaction
Fehling's test, Cu^{2+} tartrate	Red-brown Cu_2O precipitate	No reaction
Benedict's test, Cu^{2+} citrate	Red-brown Cu_2O precipitate	No reaction
Iodoform test, I_2 + NaOH (aqueous)	CHI_3 precipitates with acetaldehyde only	CHI_3 precipitates with methyl ketones only
Hydroxylamine, NH_2OH	Oxime precipitates	Oxime precipitates
Phenylhydrazine, $NH_2NH—C_6H_5$	Phenylhydrazone precipitates	Phenylhydrazone precipitates

Cold, concentrated sulfuric acid dissolves most compounds that contain oxygen atoms or unsaturated nonaromatic functional groups. Tollens's, Fehling's, and Benedict's tests are specific for aldehydes. The iodoform test is specific for the $CH_3—\overset{\overset{\displaystyle O}{\|}}{C}—R$ or $CH_3\overset{\overset{\displaystyle OH}{|}}{C}H—R$ groups where R = H, alkyl, or aryl. Hydroxylamine and phenylhydrazine react only with aldehydes and ketones to give precipitates. We may therefore use these latter reagents to distinguish aldehydes and ketones from other compounds.

NEW TERMS

Acetal

$$R—\underset{\underset{\displaystyle H}{|}}{\overset{\overset{\displaystyle OR'}{|}}{C}}—OR',$$

prepared by allowing an aldehyde to react with an alcohol R′OH in the presence of HCl (12.9).

Aldehyde

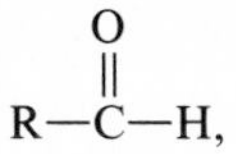

$$\overset{\displaystyle O}{\overset{\|}{\text{R—C—H}}},$$

where R is alkyl, aryl, or H (12.1).

Aldol condensation Condensation of two molecules of an aldehyde or ketone. The carbonyl carbon of one molecule becomes bonded to the alpha carbon of the other molecule (12.12):

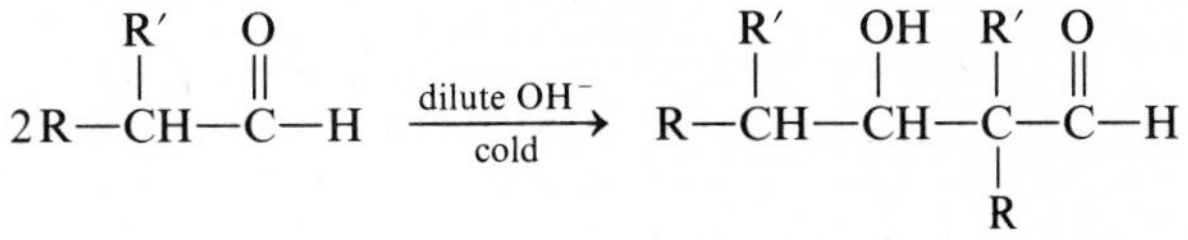

$$2\text{R—}\overset{\overset{\displaystyle R'}{|}}{\text{CH}}\text{—}\overset{\overset{\displaystyle O}{\|}}{\text{C}}\text{—H} \xrightarrow[\text{cold}]{\text{dilute OH}^-} \text{R—}\overset{\overset{\displaystyle R'}{|}}{\text{CH}}\text{—}\overset{\overset{\displaystyle OH}{|}}{\text{CH}}\text{—}\underset{\underset{\displaystyle R}{|}}{\overset{\overset{\displaystyle R'}{|}}{\text{C}}}\text{—}\overset{\overset{\displaystyle O}{\|}}{\text{C}}\text{—H}$$

Alpha carbon atom The carbon atom adjacent to a carbonyl atom (12.10).

Alpha hydrogen atom A hydrogen atom bonded to an alpha carbon atom (12.10).

Benedict's reagent A water solution of $Cu(OH)_2$ containing citrate ions to solubilize the Cu^{2+} ions. It serves the same purpose as Fehling's reagent but has a longer shelf life (12.8).

Carbonyl group (12.2)

$$\overset{\displaystyle O}{\overset{\|}{\text{—C—}}}$$

Clemmensen reduction Reduction of an aldehyde or ketone to the corresponding hydrocarbon by use of zinc amalgam and concentrated HCl (12.9):

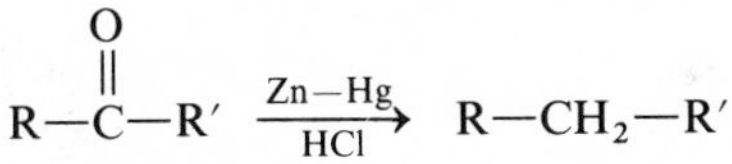

$$\text{R—}\overset{\overset{\displaystyle O}{\|}}{\text{C}}\text{—R}' \xrightarrow[\text{HCl}]{\text{Zn—Hg}} \text{R—CH}_2\text{—R}'$$

(R and R′ may be alkyl, aryl, or H.)

Crossed aldol condensation Aldol condensation between two different aldehydes or ketones (12.13).

Cyanohydrin

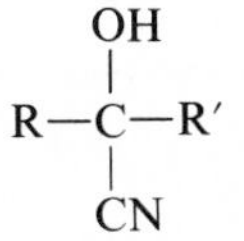

$$\text{R—}\underset{\underset{\displaystyle CN}{|}}{\overset{\overset{\displaystyle OH}{|}}{\text{C}}}\text{—R}'$$

(R and R′ may be alkyl, aryl, or H), prepared by reaction of an aldehyde or ketone with HCN (12.9).

Enolate ion A resonance-stabilized anion having the structural formula (12.10):

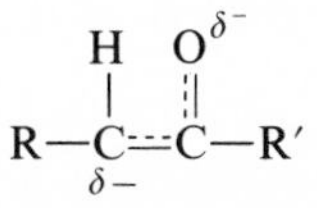

$$\text{R—}\underset{\delta-}{\overset{\overset{\displaystyle H}{|}}{\text{C}}}\text{---}\overset{\overset{\displaystyle O^{\delta-}}{\|}}{\text{C}}\text{—R}'$$

Fehling's reagent A water solution of $Cu(OH)_2$ containing tartrate ions to solubilize the Cu^{2+} ions. Fehling's reagent is a mild oxidizing agent and serves as a test for aldehydes (12.8).

Haloform reaction Reaction of a compound that has the grouping $CH_3\overset{O}{\overset{\|}{C}}—$, with $X_2 + NaOH$ ($X = Cl, Br, or I$), to yield a haloform (CHX_3) and the carboxylic acid with one less carbon atom (12.14):

$$CH_3\overset{O}{\overset{\|}{C}}—R \xrightarrow{X_2,\ OH^-} CHX_3 + R—\overset{O}{\overset{\|}{C}}—O^-$$

Hemiacetal

$$R—\overset{OH}{\overset{|}{C}}HOR',$$

prepared by allowing an aldehyde to react with an alcohol R′OH (12.9).

Hemiketal

$$R—\underset{R''}{\underset{|}{\overset{OH}{\overset{|}{C}}}}—OR',$$

prepared by allowing a ketone to react with an alcohol R′OH (12.9).

Hydrazone

$$R—\overset{R'}{\overset{|}{C}}=N—NH_2$$

(R and R′ may be alkyl, aryl, or H), prepared by reacting an aldehyde or ketone with hydrazine (NH_2NH_2) (12.9).

Iodoform test The haloform reaction is used as a test for the $CH_3\overset{O}{\overset{\|}{C}}—$ group when the halogen is I_2. In this test, yellow crystals of iodoform (CHI_3) precipitate out (12.14).

Ketal

$$R—\underset{R''}{\underset{|}{\overset{OR'}{\overset{|}{C}}}}—OR'$$

prepared by allowing a ketone $R—\overset{O}{\overset{\|}{C}}—R''$ to react with an alcohol R′OH in the presence of HCl (12.9).

Ketone

$$R—\overset{O}{\overset{\|}{C}}—R'$$

where R and R′ are alkyl or aryl (12.1).

Metal hydrides The most common ones are lithium aluminum hydride ($LiAlH_4$) and sodium borohydride ($NaBH_4$); used to reduce carbonyl compounds to alcohols (12.9).

Oxidation state The oxidation state of a carbon atom depends on the number of its bonds to oxygen or other highly electronegative atoms such as halogens or nitrogens (12.6).

Oxime

$$\begin{array}{c} R' \\ | \\ R{-}C{=}NOH \end{array}$$

(R and R′ may be alkyl, aryl, or H), prepared by reacting an aldehyde or ketone with hydroxylamine (NH_2OH) (12.9).

Phenylhydrazone

$$\begin{array}{c} R' \\ | \\ R{-}C{=}N{-}NH{-}C_6H_5 \end{array}$$

(R and R′ may be alkyl, aryl, or H), prepared by reacting an aldehyde or ketone with phenylhydrazine (12.9).

$$(C_6H_5{-}NHNH_2)$$

Schiff base

$$\begin{array}{c} R' \\ | \\ R{-}C{=}NR'' \end{array}$$

(R, R′ and R″ may be alkyl, aryl, or H), prepared by reaction of an aldehyde or ketone with 1° amine or NH_3 (12.9).

Tollens's reagent A water solution of $AgNO_3$ and ammonia; it is a mild oxidizing agent that oxidizes aldehydes to carboxylic acids and produces a "silver mirror" that serves to identify the unknown sample as an aldehyde (12.8).

ADDITIONAL PROBLEMS

[12.22] Write the Lewis dot structures of all the resonance forms of formaldehyde. Show formal charges.

12.23 Give the structural formula of each of the following compounds.

[a] dimethyl ketone
[b] butanal
[c] 4-phenyl-2-pentanone
[d] 2,2,4,4-tetramethyl-1,3-cyclobutanedione
(e) cyclohexanone
(f) dicyclopropyl ketone
(g) benzaldehyde
(h) 3,5-dinitroacetophenone

12.24 Give an acceptable name for each of the following structures.

[a] $CH_3CH_2CH_2CHO$ [b] Cl–⟨benzene ring⟩–CHO

[c] ⟨cyclohexane ring⟩–C(=O)–CH_3 [d] ⟨cyclohexanone ring with O, CH_3, CH_3⟩

(e) ⟨cyclopentenone ring⟩=O (f) CH_3CHCHO with CH_3 on the central C

(g) ⟨benzene ring⟩–CH_2–C(=O)–CH_2–⟨benzene ring⟩

(h) CH_2–CHO with OH on the CH_2

12.25 Predict which compound in each pair is the more soluble in water and which has the higher boiling point.

[a] ⟨cyclopentanone⟩ and ⟨cyclopentanol: ring C with OH and H⟩

[b] CH_3CH_2–C(=O)–CH_2CH_3 and $CH_3CH_2CHCH_2CH_3$ with OH on the central C

(c) CH_3CH_2–O–CH_2CH_3 and $CH_3CH_2CH_2CH_2OH$

(d) $CH_3CH_2CH_2CH_3$ and CH_3CCH_3 with =O on the central C

12.26 Which compound in each of the following pairs has carbon in the higher oxidation state? If both compounds are in the same oxidation state, explain.

[a] CH_3–C(=O)–CH_3 and CH_3CH_2–C(=O)–CH_3

[b] CH_3–C(=O)–CH_3 and CH_3–C(=O)–O–CH_3

[c] ⟨benzene ring with OH⟩–C(=O)–CH_3 and ⟨benzene ring with CH_3⟩–C(=O)–OH

[d] CH_3–⟨benzene ring⟩–C(=O)–H and H–C(=O)–⟨benzene ring⟩–C(=O)H

(e) $CH_3-\overset{\overset{O}{\|}}{C}-C_6H_5$ and $CH_3-O-C_6H_5$

(f) $CH_3-\underset{Cl}{\overset{Cl}{|}}{C}-CH_2CH_3$ and $CH_3-\overset{Cl}{CH}-\overset{Cl}{CH}-CH_3$

(g) $CH_3-\overset{\overset{O}{\|}}{C}-CH_2CH_3$ and $CH_3-\underset{Cl}{\overset{Cl}{C}}-CH_2CH_3$

12.27 Supply the reagent or reagents needed to carry out each of the following conversions.

[a] $CH_3CH_2OH \xrightarrow{?} CH_3CHO$

[b] $CH_3-\underset{Cl}{\overset{Cl}{C}}-C_6H_5 \xrightarrow{?} CH_3-\overset{\overset{O}{\|}}{C}-C_6H_5$

[c] $C_6H_6 \xrightarrow{?} C_6H_5-\overset{\overset{O}{\|}}{C}-CH_2CH_3$

[d] $HC{\equiv}C-CH_3 \xrightarrow{?} CH_3-\overset{\overset{O}{\|}}{C}-CH_3$

(e) cyclohexanol (chair, H axial, OH equatorial) $\xrightarrow{?}$ cyclohexanone (chair)

(f) $CH_3CH_2CHO \xrightarrow{?} CH_3CH_2CO_2H$

(g) $C_6H_5-CH{=}CH_2 \xrightarrow{?} C_6H_5-\overset{OH}{CH}-CH_3$

12.28 Supply the major products of each of the following reactions or equilibria.

[a] $C_6H_5-CH_2CHO + CH_3CH_2OH \rightleftarrows$

(b) $C_6H_5-CH_2CHO + CH_3CH_2OH \overset{H^+}{\rightleftarrows}$

[c] $C_6H_5-CH_2CHO + C_6H_5-MgBr \xrightarrow[\text{ether}]{\text{(1) dry}} \xrightarrow{\text{(2) } H_2O}$

(d) $C_6H_5-\overset{\overset{\large O}{\|}}{C}-CH_3 + C_6H_5-MgBr \xrightarrow[\text{ether}]{\text{(1) dry}} \xrightarrow{\text{(2) } H_2O}$

[e] $H-\overset{\overset{\large O}{\|}}{C}-H + C_6H_5-MgBr \xrightarrow[\text{ether}]{\text{(1) dry}} \xrightarrow{\text{(2) } H_2O}$

(f) $C_6H_5-CH_2CHO + NaCN \xrightarrow{H_3O^+}$

[g] $C_6H_5-\overset{\overset{\large O}{\|}}{C}-CH_3 + NaCN \xrightarrow{H_3O^+}$

(h) Product of (f) $\xrightarrow[\text{(hydrolysis)}]{H_2O/H^+}$

[i] Product of (g) $\xrightarrow[\text{(hydrolysis)}]{H_2O/H^+}$

(j) $C_6H_5-CH_2CHO + NH_3 \longrightarrow$

[k] $C_6H_5-CH_2CHO + CH_3CH_2NH_2 \longrightarrow$

(l) $C_6H_5-CH_2CHO + C_6H_5-NHNH_2 \longrightarrow$

[m] $C_6H_5-\overset{\overset{\large O}{\|}}{C}-CH_3 + NH_2NH_2 \longrightarrow$

(n) $C_6H_5-\overset{\overset{\large O}{\|}}{C}-CH_3 + NH_2OH \longrightarrow$

[o] $C_6H_5-CH_2CHO + H_2 \xrightarrow{Ni}$

(p) $C_6H_5-\overset{\overset{\large O}{\|}}{C}-CH_3 + H_2 \xrightarrow{Ni}$

[q] $C_6H_5-CH_2CHO + LiAlH_4 \longrightarrow$

(r) $C_6H_5-\overset{\overset{\large O}{\|}}{C}-CH_3 \xrightarrow[HCl]{Zn-Hg}$

[s] $C_6H_5-CH_2CHO \xrightarrow[\text{cold}]{\text{dilute } OH^-}$

(t) $C_6H_5-\overset{\overset{\large O}{\|}}{C}-CH_3 \xrightarrow[\text{cold}]{\text{dilute } OH^-}$

[u] $C_6H_5-CH_2CHO + I_2 \xrightarrow{OH^-}$

(v) $C_6H_5-\overset{O}{\overset{\|}{C}}-CH_3 + I_2 \xrightarrow{OH^-}$

12.29 Draw the structural formulas of the keto-enol tautomers if they exist for each of the following compounds. If they do not exist, explain why not.
[a] 2-phenylpropanal
[b] 2,2-dimethylpropanal
(c) cyclopentanone

12.30 Write equations for a reasonable laboratory preparation of each of the following compounds. Use the compound given as your only source of organic starting material. Use any solvents and inorganic reagents needed.
[a] 2-phenyl-2-propanol starting from benzene and acetone
(b) acetophenone starting from benzene and acetylene
[c] 1-methylcyclopentene starting from cyclopentanol and methanol

[12.31] Formaldehyde, on standing, forms a trimer whose structure is

(cyclic trimer: ring of alternating O and CH_2 — O, CH_2, O, CH_2, O, CH_2)

Write a reaction mechanism that accounts for the conversion of formaldehyde into this trimer.

12.32 Give a simple chemical test that would make it possible to distinguish between the compounds in each of the following pairs. Tell what you would *do* and what you would *observe*.

[a] cyclopentane and cyclopentanone ($C_5H_8{=}O$)

(b) $CH_3CH_2\overset{O}{\overset{\|}{C}}CH_2CH_3$ and $CH_3\overset{O}{\overset{\|}{C}}CH_2CH_2CH_3$

[c] $CH_3\overset{O}{\overset{\|}{C}}CH_2CH_2CH_3$ and cyclopentanone

(d) $CH_3-\overset{O}{\overset{\|}{C}}-C_6H_5$ and $CH_3-C_6H_4-CHO$

[e] $CH_3-\overset{OH}{\overset{|}{CH}}-C_6H_5$ and $CH_3-\overset{O}{\overset{\|}{C}}-C_6H_5$

(f) $CH_3CH_2CH_2CHO$ and $CH_3CH_2CH_2CH_2OH$

[12.33] Compound A (C_8H_8O) gives a negative Tollens's test, is soluble in cold, concentrated H_2SO_4, and gives CHI_3 when treated with I_2 and aqueous NaOH. Give the structural formula of A.

12.34 Compound B ($C_{13}H_{12}O$) reacts with sodium metal to liberate H_2, is insoluble in aqueous NaOH solution, reacts with H_2CrO_4 to form a green solution, and gives no reaction with Tollens's solution or with I_2 and aqueous NaOH. Give the structural formula of B.

12.35 Compound C (C_8H_8O) does not undergo an aldol condensation. It gives no reaction with I_2 and aqueous NaOH but does give a positive Fehling's test. Vigorous oxidation of C gives terephthalic acid ($HO_2C-C_6H_4-CO_2H$). Give the structural formula of C.

[12.36] When acetone is mixed with D_2O containing some NaOH and then distilled, the product contains $CH_3-\overset{\overset{O}{\|}}{C}-CH_2D$ and other, more highly deuterated products. Write a mechanism to show how deuterium becomes incorporated into the acetone molecule.

CARBOXYLIC ACIDS AND THEIR DERIVATIVES

The **carboxyl group**, $-\overset{\overset{\displaystyle O}{\|}}{C}-OH$ (also written $-CO_2H$), occurs widely in nature. It is often found in biological systems, and it has many synthetic uses.

Functional derivatives of **carboxylic acids** have various atoms or groups in place of the —OH group.

$$R-\overset{\overset{\displaystyle O}{\|}}{C}-X$$

All carboxylic acid derivatives contain the **acyl group**.

$$R-\overset{\overset{\displaystyle O}{\|}}{C}-$$

acyl group

Table 13.1 Carboxylic Acid Derivatives

Name	Structural Formula
Acyl halide (or acid halide)	$R{-}\overset{\overset{\displaystyle O}{\|}}{C}{-}X$ (X = F, Cl, Br, or I)
Carboxylic anhydride	$R{-}\overset{\overset{\displaystyle O}{\|}}{C}{-}O{-}\overset{\overset{\displaystyle O}{\|}}{C}{-}R$
Ester	$R{-}\overset{\overset{\displaystyle O}{\|}}{C}{-}O{-}R'$
Amide	$R{-}\overset{\overset{\displaystyle O}{\|}}{C}{-}NH_2$, $R{-}\overset{\overset{\displaystyle O}{\|}}{C}{-}NHR'$, $R{-}\overset{\overset{\displaystyle O}{\|}}{C}{-}NR_2'$

For this reason, carboxylic acid derivatives are often called *acyl compounds.* Table 13.1 lists the various carboxylic acid derivatives that we shall discuss in this chapter.

13.1 NOMENCLATURE OF CARBOXYLIC ACIDS

We form the systematic IUPAC name of a carboxylic acid by replacing the final *-e* of the name of the parent alkane with *-oic acid.* Because the carboxyl group is necessarily at the end of a chain, it always has the number 1. We do not have to give its number, however. Table 13.2 gives the systematic IUPAC names of a few simple carboxylic acids. For aromatic acids we normally use common names or names based on the parent compound, benzoic acid. Some examples follow.

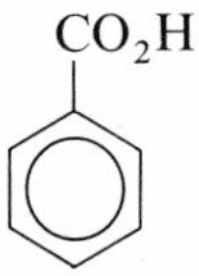

CO_2H CO_2H CO_2H
CO_2H
CO_2H

benzoic acid 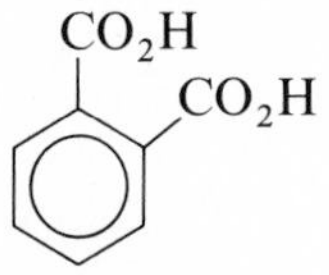 phthalic acid 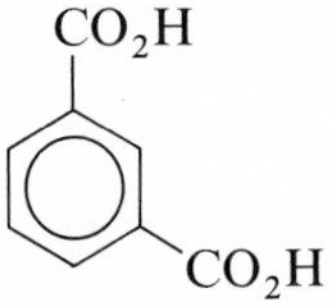 isophthalic acid

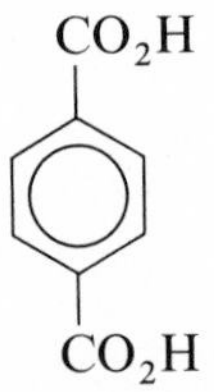

CO_2H CO_2H CO_2H
CH_3
CH_3
CO_2H

terephthalic acid 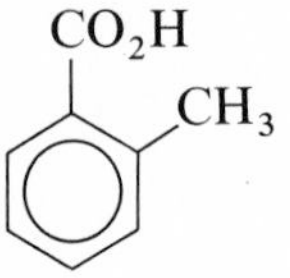 *o*-toluic acid 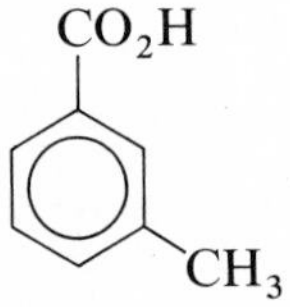 *m*-toluic acid

p-toluic acid 1-naphthoic acid 3,5-dinitrobenzoic acid

5-chloro-1-naphthoic acid

Many carboxylic acids have common names that derive from Latin or Greek words that refer to their source. Methanoic acid has the common name formic acid (from the Latin *formica*, or ant). Ethanoic acid is called acetic acid (from the Latin *acetum*, or vinegar). Butanoic acid is called butyric acid because it is responsible for the odor of rancid butter (the Latin *butyrum* means butter).

The common names of the simple carboxylic acids are important because they form the basis for the names of related compounds. Many of them are accepted by the IUPAC. It is useful to know the common names of at least the first four simple carboxylic acids.

The simple carboxylic acids are often called *fatty acids* because many of them are obtained by the hydrolysis of fats.

Table 13.2 Simple Carboxylic Acids

Formula	Systematic Name	Common Name	Melting Point, °C	Boiling Point, °C	Solubility, g/100 g H_2O, 25°C
HCO_2H	Methanoic acid	Formic acid	8	101	∞
CH_3CO_2H	Ethanoic acid	Acetic acid	16.6	118	∞
$CH_3CH_2CO_2H$	Propanoic acid	Propionic acid	−22	141	∞
$CH_3CH_2CH_2CO_2H$	Butanoic acid	Butyric acid	−8	164	∞
$CH_3CH_2CH_2CH_2CO_2H$	Pentanoic acid	Valeric acid	−34	187	5
$CH_3(CH_2)_4CO_2H$	Hexanoic acid	Caproic acid	−2	205	1
$CH_3(CH_2)_{10}CO_2H$	Dodecanoic acid	Lauric acid	44		0.006
$CH_3(CH_2)_{14}CO_2H$	Hexadecanoic acid	Palmitic acid	64	decomposes	0.0007
$CH_3(CH_2)_{16}CO_2H$	Octadecanoic acid	Stearic acid	69.4	383	0.0003

We name the **salts** of carboxylic acids by replacing the ending *-ic acid* with *-ate*. This applies to common names as well as to systematic names. Some examples will illustrate.

	Systematic Name	Common Name
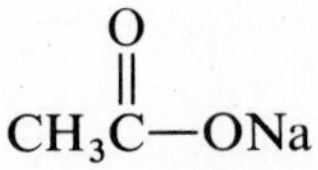 $CH_3C(=O)-ONa$	Sodium ethanoate	Sodium acetate
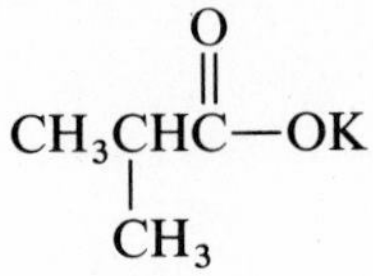 $CH_3CH(CH_3)C(=O)-OK$	Potassium 2-methyl propanoate	Potassium isobutyrate

Problem 13.1
Write the structural formula of each of the following compounds.
(a) heptanoic acid
(b) 2,3-dimethyloctanoic acid
(c) isobutyric acid
(d) sodium valerate

Problem 13.2
Give an acceptable name for each of the following compounds.

(a) 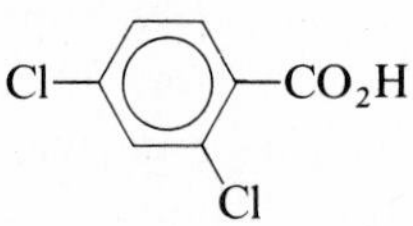 Cl—C_6H_3(Cl)—CO_2H (benzene ring with Cl para to CO_2H and a second Cl ortho to CO_2H)

(b) Cl—C_6H_3(Cl)—$C(=O)-ONa$ (benzene ring with Cl para and a second Cl ortho to the carboxylate group)

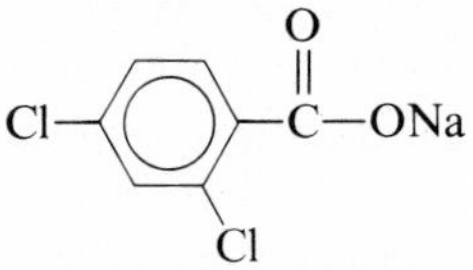

(c) $CH_3C(CH_3)_2CH_2CO_2H$

$$
\begin{array}{c}
CH_3 \\
| \\
CH_3CCH_2CO_2H \\
| \\
CH_3
\end{array}
$$

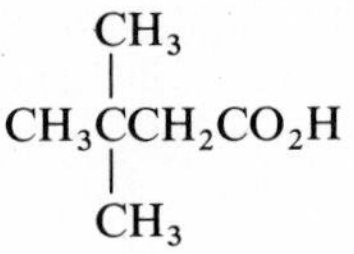

(d) 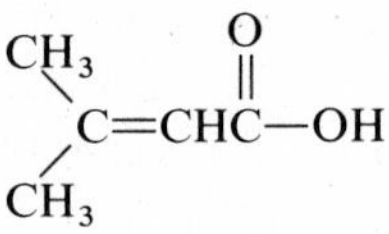 $(CH_3)_2C{=}CHC(=O)-OH$

13.2 PHYSICAL PROPERTIES: HYDROGEN BONDING

We see in Table 13.2 that the carboxylic acids have higher boiling points and water solubilities than either aldehydes, ketones, or alcohols having the same number of carbon atoms. We explain this increase in both boiling point and water solubility on the basis of very effective hydrogen bonding in the carboxylic acids. Determinations of the molecular weights of the lower carboxylic acids give values nearly double those we expect for RCO_2H. This suggests that carboxylic acids form dimers that are held together by hydrogen bonding:

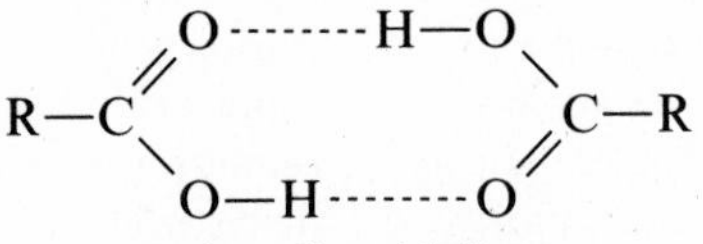

$$
\begin{array}{ccc}
 & O\cdots\cdots H-O & \\
R-C\!\!\!\!\! & & \!\!\!\!\!C-R \\
 & O-H\cdots\cdots O &
\end{array}
$$

carboxylic acid dimer

We can explain the unusually high water solubilities of carboxylic acids by hydrogen bonding. Carboxylic acid molecules, like water, can both furnish and accept hydrogens for hydrogen bonding.

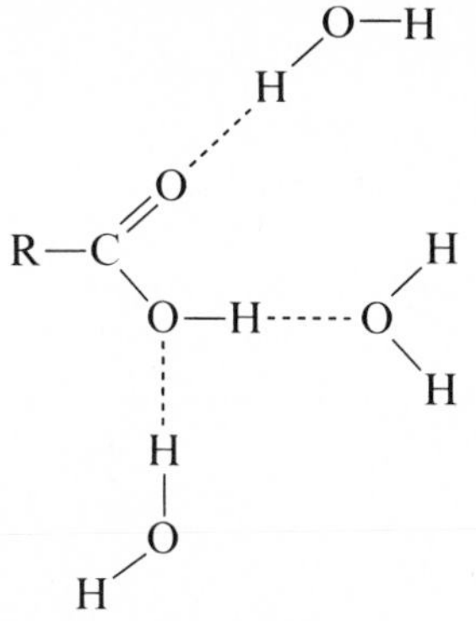

Problem 13.3
Explain why propanoic acid has *both* a higher boiling point and a higher water solubility than propanal (see propionaldehyde in Table 12.3).

Problem 13.4
Explain why butanoic acid has *both* a higher boiling point and a higher water solubility than 1-butanol (see Table 12.3).

13.3 ACIDITY OF CARBOXYLIC ACIDS

Carboxylic acids dissociate in water:

$$R-\overset{\overset{\displaystyle O}{\|}}{C}-OH + H_2O \rightleftharpoons R-\overset{\overset{\displaystyle O}{\|}}{C}-O^- + H_3O^+$$

The K_as of carboxylic acids are about 10^{-5}. They are therefore weak acids but stronger than phenols (K_as $\sim 10^{-10}$). Of course, the exact K_a value of a given carboxylic acid depends on the nature of R.

Table 13.3 lists some common carboxylic acids along with their acid dissociation constants, K_a.

Carbonic acid has a K_a that is intermediate between those of carboxylic acids and phenols. We can use this fact to distinguish chemically between carboxylic acids and phenols. We carry out the test by adding some of the unknown acid to a water solution of $NaHCO_3$. The reactions are as follows:

$$\underset{\text{(stronger acid)}}{RCO_2H} + HCO_3^- \rightleftharpoons RCO_2^- + \underset{\text{(weaker acid)}}{H_2CO_3}$$

$$(H_2CO_3 \rightleftharpoons H_2O + CO_2\uparrow)$$

$$\underset{\text{(weaker acid)}}{Ar-OH} + HCO_3^- \rightleftharpoons Ar-O^- + \underset{\text{(stronger acid)}}{H_2CO_3}$$

Table 13.3 K_as of Some Common Carboxylic Acids

Carboxylic Acid	Name	K_a (25°C)
HCO_2H	Formic acid	1.77×10^{-4}
CH_3CO_2H	Acetic acid	1.76×10^{-5}
$CH_3CH_2CH_2CO_2H$	Butyric acid	1.54×10^{-5}
$CH_3(CH_2)_3CO_2H$	Valeric acid	1.52×10^{-5}
CH_2ClCO_2H	Chloroacetic acid	1.40×10^{-3}
$CHCl_2CO_2H$	Dichloroacetic acid	3.32×10^{-2}
CCl_3CO_2H	Trichloroacetic acid	2.00×10^{-1}
$CH_3CHClCO_2H$	2-Chloropropanoic acid	1.47×10^{-2}
$CH_2ClCH_2CO_2H$	3-Chloropropanoic acid	1.47×10^{-3}
C_6H_5—CO_2H	Benzoic acid	6.46×10^{-5}
Cl—C_6H_4—CO_2H	*p*-Chlorobenzoic acid	1.04×10^{-4}
C_6H_4—CO_2H (Cl meta)	*m*-Chlorobenzoic acid	1.60×10^{-4}
CH_3—C_6H_4—CO_2H	*p*-Toluic acid	4.43×10^{-5}
NO_2—C_6H_4—CO_2H	*p*-Nitrobenzoic acid	3.93×10^{-4}
C_6H_5—OH	Phenol	1.1×10^{-10}
H_2CO_3	Carbonic acid	4.3×10^{-7}
CH_3CH_2OH	Ethanol	$\sim 10^{-18}$

Because carboxylic acids are stronger acids than carbonic acid, the equilibrium is displaced to the right; that is, equilibrium produces the weaker acid, H_2CO_3. H_2CO_3 is slightly soluble in water and forms CO_2 gas, which is clearly visible as bubbles. Phenols are generally weaker acids than carbonic acid, so the equilibrium favors the phenol (weaker acid). Because H_2CO_3 does not form in appreciable amounts, we see no CO_2 bubbles when we dissolve phenols in bicarbonate solutions.

13.4 EFFECT OF STRUCTURE ON ACID–BASE STRENGTH

We explain the acidity difference between phenols and carboxylic acids in terms of Brønsted–Lowry acid–base theory as follows. The stronger acid yields the weaker conjugate base.

$$\underset{\text{(stronger acid)}}{R{-}CO_2H} + H_2O \rightleftarrows H_3O^+ + \underset{\text{(weaker conjugate base)}}{R{-}CO_2^-}$$

$$\underset{\text{(weaker acid)}}{Ar{-}OH} + H_2O \rightleftarrows H_3O^+ + \underset{\text{(stronger conjugate base)}}{Ar{-}O^-}$$

The carboxylic acid is a stronger acid than phenol because the conjugate base of the carboxylic acid is a weaker base than the conjugate base of the phenol. That is, $R{-}CO_2^-$ is a weaker base than $Ar{-}O^-$. The base strength of a base depends on its ability to accommodate the electron pair left on it when the proton leaves. The carboxylate group, $R{-}\overset{\overset{\large O}{\|}}{C}{-}O^-$, is better able to stabilize those electrons than the phenoxide ion, $Ar{-}O^-$.

Problem 13.5
Choose the stronger acid in each pair below. (Refer to Table 13.3).
(a) valeric acid and benzoic acid
(b) butyric acid and *p*-nitrobenzoic acid
(c) *m*-chlorobenzoic acid and *p*-toluic acid

Problem 13.6
Choose the weaker conjugate base of each pair of acids in Problem 13.5.

Resonance

We can best explain stabilization of the carboxylate ion in terms of resonance.

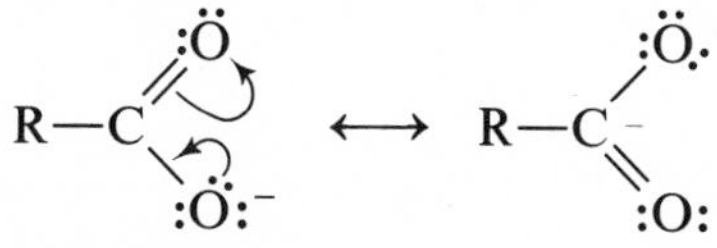

$$R{-}C\begin{matrix}\nearrow\!\!\!\!/\!\!/ :\ddot{O} \\ \searrow :\ddot{O}:^-\end{matrix} \longleftrightarrow R{-}C\begin{matrix}/ :\ddot{O}:^- \\ \backslash\!\!\backslash :O:\end{matrix}$$

We can represent the hybrid structure as

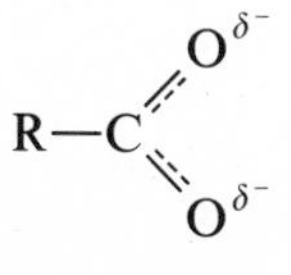

$$R{-}C\begin{matrix} \nearrow O^{\delta-} \\ \searrow O^{\delta-}\end{matrix}$$

Resonance stabilization of the carboxylate anion is especially important because the two contributing structures are equivalent. Resonance stabilization of the undissociated acid is less important because the two structures are not equivalent and therefore do not contribute equally to the resonance hybrid.

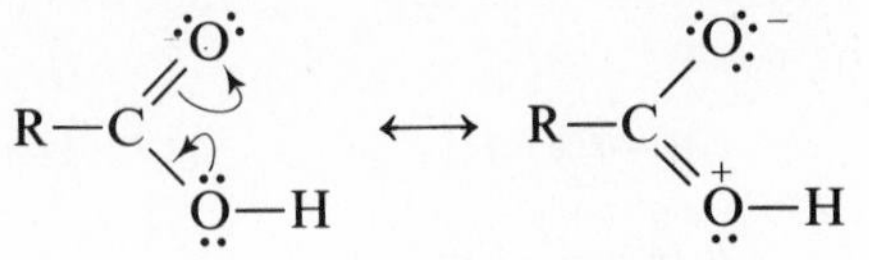

$$R-C(=\ddot{O}:)-\ddot{O}-H \longleftrightarrow R-C(-\ddot{O}:^{-})=\overset{+}{O}-H$$

Physical measurements of bond lengths in formic acid and formate ion (in sodium formate) bear this out.

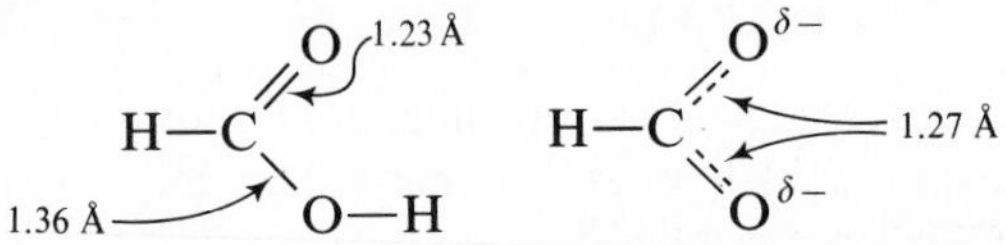

$$H-C(=O)-O-H \quad (C{=}O\ 1.23\ \text{Å};\ C{-}OH\ 1.36\ \text{Å}) \qquad H-C(O^{\delta-})(O^{\delta-}) \quad (1.27\ \text{Å})$$

In formic acid, the C=O bond is considerably shorter and therefore has more double bond character than the C—OH bond. In the anion, the two C—O bonds are identical in length and intermediate between the C=O and C—OH bond lengths in formic acid. The equivalence of the two C—O bonds in the anion indicates that the two resonance forms contribute equally.

Inductive Effect

We can also explain the variations in acid strength from one carboxylic acid to another. For example, acetic, chloroacetic, dichloroacetic, and trichloroacetic acids increase in acidity with each additional chlorine atom. Chlorine is very electronegative and its bond to carbon is a polar covalent bond (Sec. 1.7). The chlorine thus causes a partial positive charge to develop on the carbon to which it is attached. This effect is called the **inductive effect**. *The inductive effect is the polarization of a single bond (a sigma bond) due to a difference in electronegativity between the two atoms joined by the bond.* The positive charge on the alpha carbon induced by the chlorine is transmitted partly to the carbonyl carbon. This positive charge on the carbonyl carbon helps to stabilize the carboxylate anion by helping to neutralize its negative charge. We can summarize as follows: The greater the number of chlorine atoms, the greater the positive charge at the carbonyl carbon, the greater the stability of the anion, and the stronger the acid.

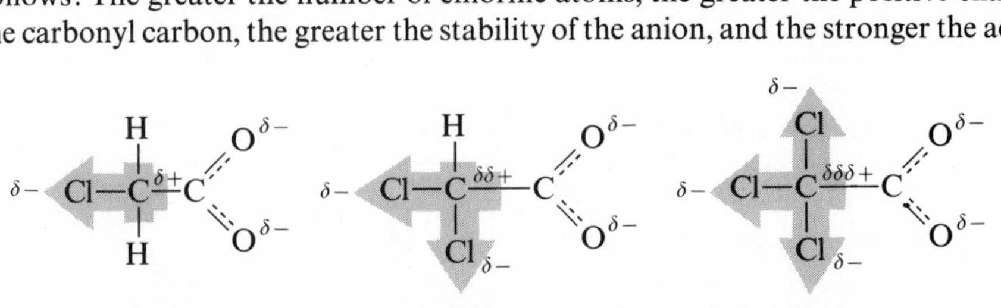

$$\overset{\delta-}{Cl}-CH_2-\overset{\delta+}{C}(O^{\delta-})(O^{\delta-}) \qquad \overset{\delta-}{Cl}-CH(\overset{\delta-}{Cl})-\overset{\delta\delta+}{C}(O^{\delta-})(O^{\delta-}) \qquad \overset{\delta-}{Cl}-C(\overset{\delta-}{Cl})_2-\overset{\delta\delta\delta+}{C}(O^{\delta-})(O^{\delta-})$$

K_a: 1.40×10^{-3} $\qquad$ 3.32×10^{-2} $\qquad$ 2.00×10^{-1}

The chlorine atoms stabilize the anion much more than they stabilize the undissociated acid because the acid is electrically uncharged. In other words, the effect of the chlorine is to stabilize the anion relative to the acid:

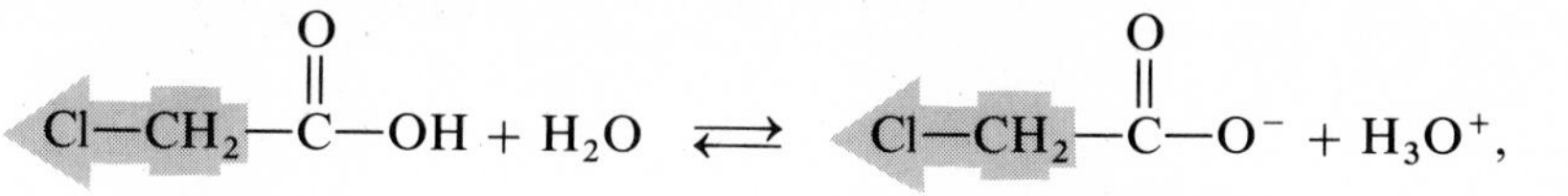

$$Cl{-}CH_2{-}\overset{\overset{\displaystyle O}{\|}}{C}{-}OH + H_2O \rightleftarrows Cl{-}CH_2{-}\overset{\overset{\displaystyle O}{\|}}{C}{-}O^- + H_3O^+,$$

(little or no stabilization by chlorine)

(stabilized by the electron-withdrawing effect of chlorine)

No such stabilization occurs in acetic acid because acetic acid does not contain an electron-withdrawing group near the carbonyl group. Chloroacetic acid therefore has a larger K_a than does acetic acid.

We see the same effect in the benzoic acids.

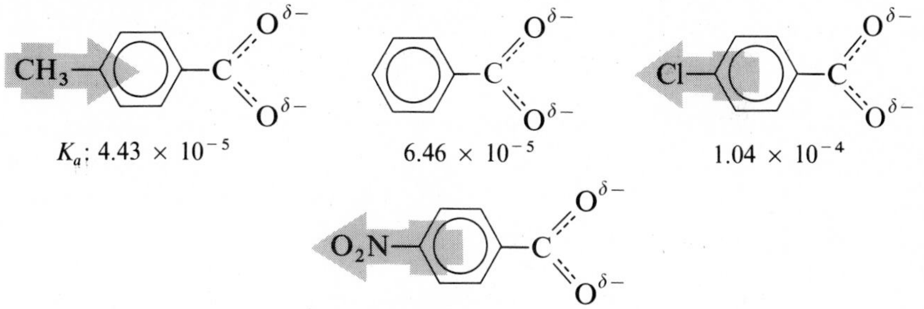

$CH_3{-}C_6H_4{-}C(O^{\delta-})_2$ K_a: 4.43×10^{-5}

$C_6H_5{-}C(O^{\delta-})_2$ 6.46×10^{-5}

$Cl{-}C_6H_4{-}C(O^{\delta-})_2$ 1.04×10^{-4}

$O_2N{-}C_6H_4{-}C(O^{\delta-})_2$ 3.93×10^{-4}

Both the chlorine atom and the nitro group withdraw electrons and thereby stabilize the carboxylate anion. *p*-Chlorobenzoic and *p*-nitrobenzoic acids are thus stronger acids than benzoic acid. Methyl is an electron-releasing group, and it therefore destabilizes the anion by adding negative charge to it. *p*-Toluic acid is thus a weaker acid than benzoic acid.

The inductive effect decreases with distance. For example, 3-chloropropanoic acid is a weaker acid than 2-chloropropanoic acid.

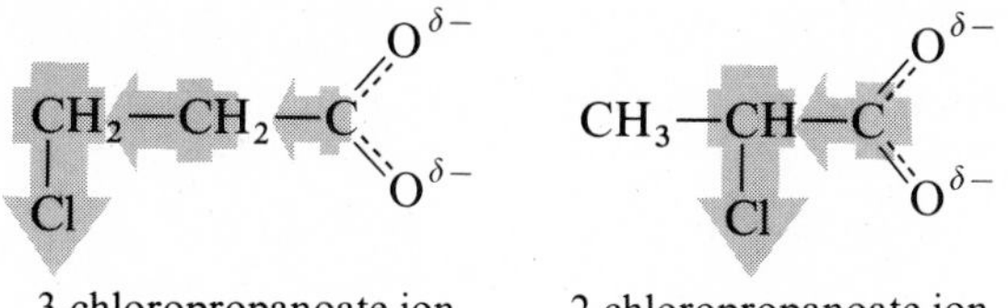

$\underset{\underset{\displaystyle Cl}{|}}{CH_2}{-}CH_2{-}C(O^{\delta-})_2$

3-chloropropanoate ion

K_a: 1.47×10^{-3}

$CH_3{-}\underset{\underset{\displaystyle Cl}{|}}{CH}{-}C(O^{\delta-})_2$

2-chloropropanoate ion

1.47×10^{-2}

Problem 13.7

Predict the stronger acid of each pair below (reason by analogy to compounds given in Table 13.3 and your knowledge of electronegativity).

(a) difluoroacetic acid and fluoroacetic acid

(b) $CH_3CH_2{-}C_6H_4{-}CO_2H$ and $CH_3CH_2{-}C_6H_4{-}OH$

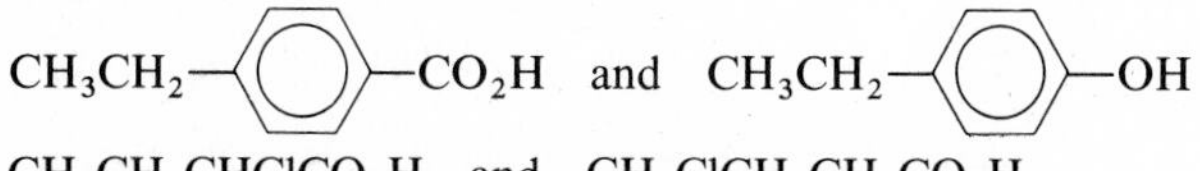

(c) $CH_3CH_2CHClCO_2H$ and $CH_2ClCH_2CH_2CO_2H$

13.5 PREPARATION OF CARBOXYLIC ACIDS: LABORATORY METHODS

Most of the methods of preparing carboxylic acids are familiar to us. They are summarized as follows.

Oxidation of Primary Alcohols (Sec. 10.11)

$$R{-}CH_2OH \xrightarrow{H_2CrO_4} R{-}\overset{\overset{\large O}{\|}}{C}{-}H \xrightarrow[\text{heat}]{H_2CrO_4} R{-}\overset{\overset{\large O}{\|}}{C}{-}OH$$

$$R{-}CH_2OH \xrightarrow[\text{heat}]{\text{(or directly) } H_2CrO_4} R{-}\overset{\overset{\large O}{\|}}{C}{-}OH$$

Oxidation of Aldehydes (Sec. 12.8, 10.11)

$$R{-}\overset{\overset{\large O}{\|}}{C}{-}H \xrightarrow{Ag(NH_3)_2^+} R{-}\overset{\overset{\large O}{\|}}{C}{-}O^- \xrightarrow{H_3O^+} R{-}\overset{\overset{\large O}{\|}}{C}{-}OH$$

(We may also use Cu^{+2} in Fehling's or Benedict's solutions)

Oxidation of Alkylbenzenes (Sec. 7.12)

$$C_6H_5{-}R \xrightarrow[OH^-,\ \text{heat}]{KMnO_4} C_6H_5{-}\overset{\overset{\large O}{\|}}{C}{-}O^- \xrightarrow{H_3O^+} C_6H_5{-}\overset{\overset{\large O}{\|}}{C}{-}OH$$

(R = any alkyl group)

Oxidation of Methyl Ketones: The Haloform Reaction (Sec. 12.14)

$$R{-}\overset{\overset{\large O}{\|}}{C}{-}CH_3 \xrightarrow{X_2,\ OH^-} R{-}\overset{\overset{\large O}{\|}}{C}{-}O^- + CHX_3$$

$$R{-}\overset{\overset{\large O}{\|}}{C}{-}O^- \xrightarrow{H_3O^+} R{-}\overset{\overset{\large O}{\|}}{C}{-}OH$$

(X = Cl, Br, or I)

Hydrolysis of Nitriles

In Section 12.9 we saw that cyanohydrins (prepared by the reaction of HCN with aldehydes or ketones) can be hydrolyzed to α-hydroxy carboxylic acids:

$$\underset{\text{cyanohydrin}}{R_2C(CN)(OH)} \xrightarrow[H_3O^+]{\text{(hydrolysis)}} \underset{\alpha\text{-hydroxy acid}}{R_2C(CO_2H)(OH)}$$

Ordinary nitriles (R—C≡N) can be prepared by reacting an alkyl halide with CN^-. This is an example of a nucleophilic substitution (S_N2) reaction (Sec. 9.6):

$$R{-}CH_2{-}X + CN^- \longrightarrow R{-}CH_2CN + X^-$$

$$R{-}CH_2CN \xrightarrow[\text{heat}]{H_3O^+} R{-}CH_2{-}\overset{O}{\overset{\|}{C}}{-}OH + NH_4^+$$

Because CN^- is a fairly strong base, this S_N2 reaction works well only with primary alkyl halides. Secondary and tertiary alkyl halides undergo elimination to produce the alkene as the major product (Section 9.5).

Carbonation of Grignard Reagents

Grignard reagents react with carbon dioxide to give magnesium carboxylate salts that yield carboxylic acids on acidification.

$$\overset{\delta-}{R}\,\overset{\delta+}{MgX} + O{=}C{=}O \xrightarrow[\text{ether}]{\text{dry}} R{-}C(=O){-}\overset{-}{O}\,\overset{+}{Mg}X \xrightarrow{H_3O^+} R{-}\overset{O}{\overset{\|}{C}}{-}OH + \overset{+}{Mg}X$$

This reaction works well with all R groups—aryl, 1°, 2°, 3° alkyl, and so forth.

Problem 13.8

Write equations to describe all the steps needed to carry out each of the following conversions.

(a) $CH_3CH_2CH_2OH \longrightarrow CH_3CH_2CO_2H$

(b) $C_6H_5{-}CH_2Cl \longrightarrow C_6H_5{-}CH_2CO_2H$ (two ways)

(c) $C_6H_5{-}CH_2Cl \longrightarrow C_6H_5{-}CO_2H$

(d) $C_6H_5CH_2\overset{O}{\overset{\|}{C}}-H \longrightarrow C_6H_5CH_2CO_2H$

(e) $C_6H_5CH_2\overset{O}{\overset{\|}{C}}CH_3 \longrightarrow C_6H_5CH_2CO_2H$

(f) $C_6H_5CH_2\overset{O}{\overset{\|}{C}}CH_3 \longrightarrow C_6H_5CO_2H$

(g) $C_6H_5CH_2\overset{O}{\overset{\|}{C}}-H \longrightarrow C_6H_5CH_2\underset{OH}{\underset{|}{C}}HCO_2H$

(h) $CH_3\overset{CH_3}{\overset{|}{\underset{CH_3}{\underset{|}{C}}}}Cl \longrightarrow CH_3\overset{CH_3}{\overset{|}{\underset{CH_3}{\underset{|}{C}}}}-CO_2H$

13.6 ACYL HALIDES

Figure 13.1 shows the functional derivatives (acyl derivatives) of the carboxylic acids.

We shall briefly review the structure, nomenclature, and preparation of these functional derivatives in the next four sections.

Problem 13.9
Which of the reactions of Figure 13.1 are oxidation-reduction? Explain.

Acyl halides have the general formula

$$R-\overset{O}{\overset{\|}{C}}-X$$

(X = F, Cl, Br, or I)

We form their names by substituting the ending *-yl halide* for the ending *-ic acid* in both the common and systematic names. Some examples will illustrate. Systematic names are in parentheses.

$$H-\overset{O}{\overset{\|}{C}}-Cl$$
formyl chloride
(methanoyl chloride)

$$CH_3\overset{O}{\overset{\|}{C}}-Br$$
acetyl bromide
(ethanoyl bromide)

$$C_6H_5-\overset{O}{\overset{\|}{C}}-Cl$$
benzoyl chloride

$$CH_3CH_2CH_2CH_2\overset{O}{\overset{\|}{C}}-Cl$$
valeryl chloride
(pentanoyl chloride)

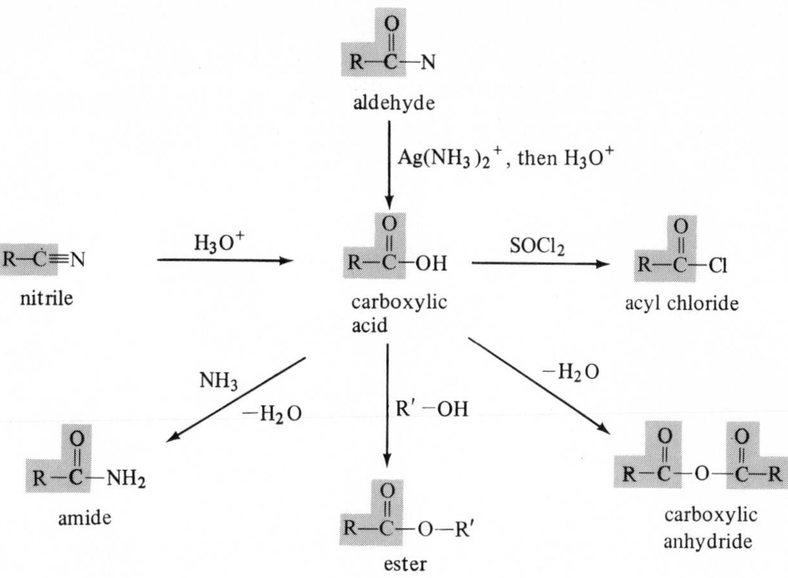

Figure 13.1. Functional derivatives of carboxylic acids and their interrelationships.

We prepare acyl chlorides by the reaction of a carboxylic acid with thionyl chloride ($SOCl_2$) or with phosphorus halides (PBr_3, PCl_3).

$$3\,R{-}\overset{\overset{\displaystyle O}{\|}}{C}{-}OH + PBr_3 \longrightarrow 3\,R{-}\overset{\overset{\displaystyle O}{\|}}{C}{-}Br + P(OH)_3$$

$$R{-}\overset{\overset{\displaystyle O}{\|}}{C}{-}OH + SOCl_2 \longrightarrow R{-}\overset{\overset{\displaystyle O}{\|}}{C}{-}Cl + SO_2 + HCl$$

Acyl halides are very reactive. They are, in fact, the most reactive acyl derivatives.

Problem 13.10

Give an acceptable name for

(a) $CH_3CH(CH_3)CH_2\overset{\overset{\displaystyle O}{\|}}{C}{-}Cl$

(b) $O_2N{-}C_6H_4{-}\overset{\overset{\displaystyle O}{\|}}{C}{-}Br$ (para-substituted benzene ring)

Problem 13.11

Write equations to show how you could prepare (a) propionyl bromide, (b) acetyl chloride, (c) benzoyl chloride.

13.7 CARBOXYLIC ANHYDRIDES

Carboxylic anhydrides have the formula

$$R-\overset{O}{\overset{\|}{C}}-O-\overset{O}{\overset{\|}{C}}-R$$

We call them anhydrides because they react with water to form the carboxylic acid:

$$R-\overset{O}{\overset{\|}{C}}-O-\overset{O}{\overset{\|}{C}}-R + H_2O \longrightarrow R-\overset{O}{\overset{\|}{C}}-OH + HO-\overset{O}{\overset{\|}{C}}-R$$

We name them by simply substituting the word *anhydride* for *acid*. (Systematic names are in parentheses.)

$$CH_3-\overset{O}{\overset{\|}{C}}-O-\overset{O}{\overset{\|}{C}}-CH_3$$

acetic anhydride
(ethanoic anhydride)

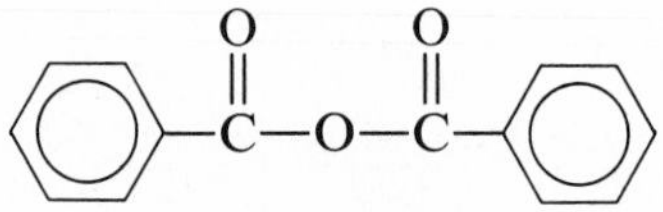

$$C_6H_5-\overset{O}{\overset{\|}{C}}-O-\overset{O}{\overset{\|}{C}}-C_6H_5$$

benzoic anhydride

We can prepare carboxylic anhydrides by the reaction of an acyl halide with a carboxylic salt.

$$R-\overset{O}{\overset{\|}{C}}-Cl + NaO-\overset{O}{\overset{\|}{C}}-R \longrightarrow R-\overset{O}{\overset{\|}{C}}-O-\overset{O}{\overset{\|}{C}}-R + NaCl$$

acyl chloride; sodium carboxylate

Problem 13.12
Give an acceptable name for

(a) $Cl-C_6H_4-\overset{O}{\overset{\|}{C}}-O-\overset{O}{\overset{\|}{C}}-C_6H_4-Cl$,

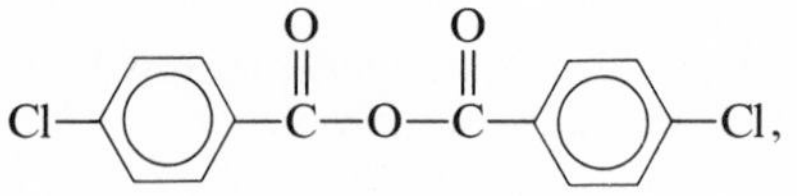

(b) $CH_3-\underset{CH_3}{\overset{CH_3}{C}}-CH_2\overset{O}{\overset{\|}{C}}-O-\overset{O}{\overset{\|}{C}}-CH_2-\underset{CH_3}{\overset{CH_3}{C}}-CH_3$

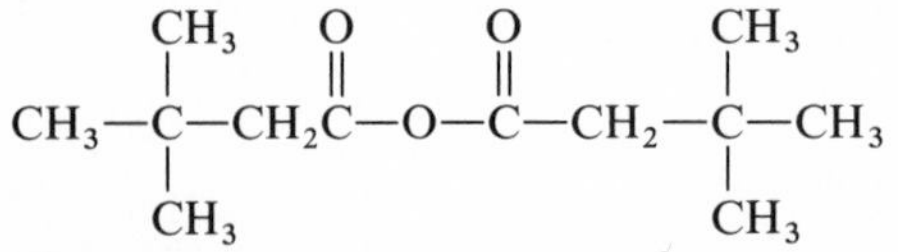

Problem 13.13
Write equations to show how you could synthesize each of the anhydrides of Problem 13.12 starting with the corresponding carboxylic acid. (Show the synthesis of all the necessary acyl intermediates.)

Esters have the general formula

$$\underbrace{R-\overset{\overset{\displaystyle O}{\|}}{C}}_{\text{acyl group}}-O-\underbrace{R'}_{\text{alcoholic group}}$$

We name esters by giving the name of the alcoholic R′ group followed by the name of the acyl group. Some examples will illustrate. (Systematic names are in parentheses.)

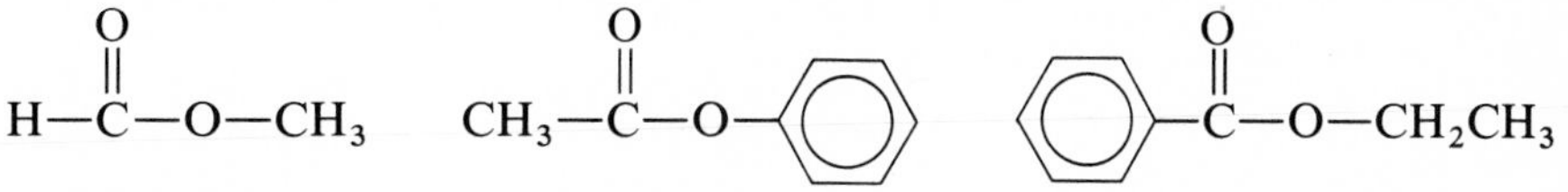

$$H-\overset{\overset{\displaystyle O}{\|}}{C}-O-CH_3 \qquad CH_3-\overset{\overset{\displaystyle O}{\|}}{C}-O-C_6H_5 \qquad C_6H_5-\overset{\overset{\displaystyle O}{\|}}{C}-O-CH_2CH_3$$

methyl formate (methyl methanoate) — phenyl acetate (phenyl ethanoate) — ethyl benzoate

Esters are prepared directly from the carboxylic acid, from the acyl halide, or from the carboxylic anhydride.

From the Carboxylic Acid

$$R-\overset{\overset{\displaystyle O}{\|}}{C}-OH + HO-R' \underset{}{\overset{H^+}{\rightleftarrows}} R-\overset{\overset{\displaystyle O}{\|}}{C}-O-R' + H_2O$$

We can shift the equilibrium to the right by using an excess of either alcohol or carboxylic acid, or by removing the water as it is produced.

From the Acyl Halide

$$R-\overset{\overset{\displaystyle O}{\|}}{C}-Cl + HO-R' \xrightarrow{OH^-} R-\overset{\overset{\displaystyle O}{\|}}{C}-OR' + H_2O + Cl^-$$

This reaction is not an equilibrium and is very useful for preparing esters in the laboratory.

From the Carboxylic Anhydride

$$R-\overset{\overset{\displaystyle O}{\|}}{C}-O-\overset{\overset{\displaystyle O}{\|}}{C}-R + HO-R' \longrightarrow R-\overset{\overset{\displaystyle O}{\|}}{C}-OR' + R-\overset{\overset{\displaystyle O}{\|}}{C}-OH$$

Here again, the reaction is not reversible. Anhydrides are often used to prepare esters in the laboratory.

Generally, esters are manufactured by direct reaction of an acid and an alcohol (first reaction above). Dimethyl terephthalate, used in the manufacture of Dacron, is manufactured by this process.

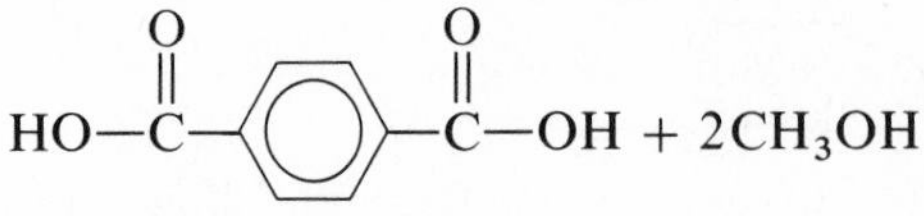

$$HO-\overset{\overset{\displaystyle O}{\|}}{C}-C_6H_4-\overset{\overset{\displaystyle O}{\|}}{C}-OH + 2CH_3OH$$

terephthalic acid

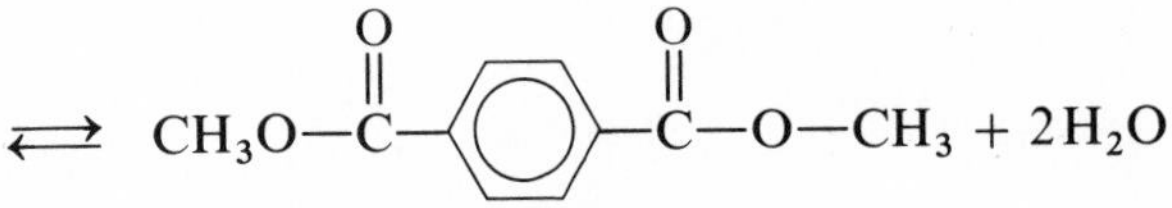

$$\rightleftarrows CH_3O-\overset{\overset{\displaystyle O}{\|}}{C}-C_6H_4-\overset{\overset{\displaystyle O}{\|}}{C}-O-CH_3 + 2H_2O$$

dimethyl terephthalate

Because methanol is quite inexpensive, the process uses 20–30 moles of methanol per mole of terephthalic acid to drive the equilibrium to the right.

Problem 13.14

Give an acceptable name for

(a) 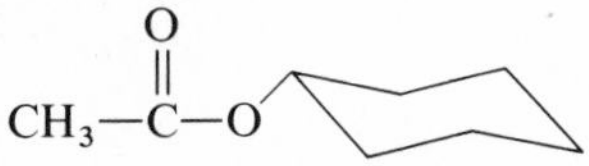 $CH_3-\overset{\overset{\displaystyle O}{\|}}{C}-O-C_6H_{11}$ (cyclohexyl)

(b) 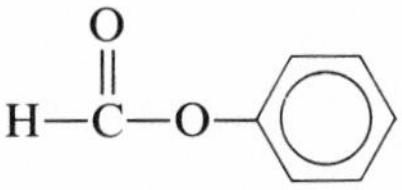 $H-\overset{\overset{\displaystyle O}{\|}}{C}-O-C_6H_5$

Problem 13.15

Write equations to show how you could prepare each of the esters in Problem 13.14.

13.9 AMIDES

Amides have the general formula

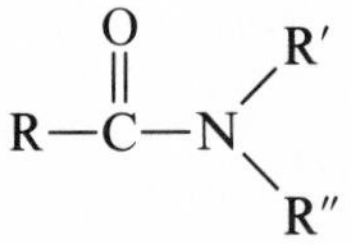

$$R-\overset{\overset{\displaystyle O}{\|}}{C}-N\begin{matrix}R' \\ R''\end{matrix}$$

(R′ and R″ may be H, alkyl, or aryl)

We name amides by replacing the ending *-ic acid* or *-oic acid* of the carboxylic acid with *-amide*. Some examples follow. (Systematic names are in parentheses.)

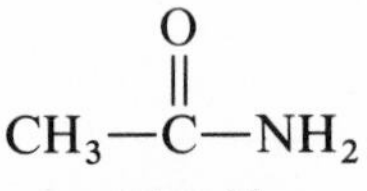

$$CH_3-\overset{\overset{\displaystyle O}{\|}}{C}-NH_2$$

acetamide
(ethanamide)

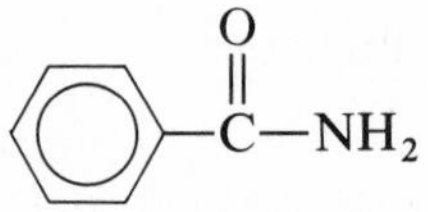

$$C_6H_5-\overset{\overset{\displaystyle O}{\|}}{C}-NH_2$$

benzamide

$$CH_3CH_2\overset{\overset{\displaystyle O}{\|}}{C}-NHCH_3$$

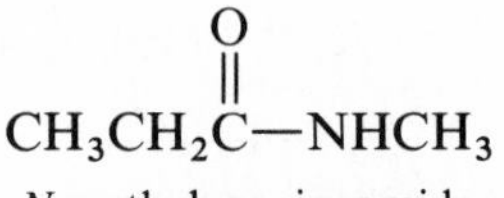

N-methylpropionamide
(*N*-methylpropanamide)

The prefix *N*- tells us that the alkyl group named is bonded to the nitrogen atom.

We can prepare amides from carboxylic acids, acyl chlorides, acid anhydrides, or esters.

From Carboxylic Acids

$$R{-}\overset{\overset{\displaystyle O}{\|}}{C}{-}OH + NH_3 \longrightarrow \underset{\text{ammonium salt}}{R{-}\overset{\overset{\displaystyle O}{\|}}{C}{-}\overset{-}{O}\overset{+}{N}H_4} \xrightarrow{\text{heat}} R{-}\overset{\overset{\displaystyle O}{\|}}{C}{-}NH_2 + H_2O$$

From Acyl Halides

$$R{-}\overset{\overset{\displaystyle O}{\|}}{C}{-}Cl + 2NH_3 \longrightarrow R{-}\overset{\overset{\displaystyle O}{\|}}{C}{-}NH_2 + \overset{+}{N}H_4\overset{-}{Cl}$$

From Carboxylic Anhydrides

$$R{-}\overset{\overset{\displaystyle O}{\|}}{C}{-}O{-}\overset{\overset{\displaystyle O}{\|}}{C}{-}R + NH_3 \longrightarrow R{-}\overset{\overset{\displaystyle O}{\|}}{C}{-}NH_2 + R{-}\overset{\overset{\displaystyle O}{\|}}{C}{-}OH$$

From Esters

$$R{-}\overset{\overset{\displaystyle O}{\|}}{C}{-}O{-}R' + NH_3 \longrightarrow R{-}\overset{\overset{\displaystyle O}{\|}}{C}{-}NH_2 + R'{-}OH$$

Problem 13.16
Give an acceptable name for

(a) $CH_3CH_2\overset{\overset{\displaystyle O}{\|}}{C}{-}NH_2$

(b) $\overset{\overset{\displaystyle O}{\|}}{C}{-}NH_2$ attached to a benzene ring bearing Cl in the meta position

Problem 13.17
Write equations to show how you could prepare each of the amides of Problem 13.16 starting from the corresponding carboxylic acid and any necessary reagents. Show the synthesis of all acyl intermediates.

---

elective topic

The general order of reactivity of acyl derivatives is:

$$\underset{\text{acyl chloride}}{R{-}\overset{\overset{O}{\|}}{C}{-}Cl} > \underset{\text{carboxylic anhydride}}{R{-}\overset{\overset{O}{\|}}{C}{-}O{-}\overset{\overset{O}{\|}}{C}{-}R} > \underset{\text{ester}}{R{-}\overset{\overset{O}{\|}}{C}{-}O{-}R'} > \underset{\text{amide}}{R{-}\overset{\overset{O}{\|}}{C}{-}NH_2}$$

We can use acyl chlorides to prepare anhydrides, esters, and amides by nucleophilic substitution reactions. Similarly, we can use anhydrides to prepare esters and amides but not acyl chlorides by S_N reactions, and we can use esters to prepare amides but not acyl chlorides or anhydrides. Finally, we cannot prepare any of the other acyl derivatives directly (except the carboxylic acid) by nucleophilic substitution reactions on amides.

We can generalize the reactions of acyl compounds by the following mechanism (Nu—H = nucleophile, L = leaving group):

$$R{-}\overset{\overset{:\ddot{O}}{\|}}{C}{-}L + :Nu{-}H \rightleftharpoons R{-}\underset{\underset{^{+}Nu{-}H}{|}}{\overset{\overset{:\ddot{O}:^{-}}{|}}{C}}{-}L$$

$$\rightleftharpoons R{-}\underset{\underset{Nu}{|}}{\overset{\overset{:\ddot{O}:^{-}}{|}}{C}}{-}\overset{+}{L}H \longrightarrow R{-}\overset{\overset{O}{\|}}{C}{-}Nu + LH$$

The above *substitution* reaction of acyl compounds differs from the *addition* reaction of aldehydes and ketones:

$$\underset{\text{aldehyde or ketone}}{R{-}\overset{\overset{:\ddot{O}}{\|}}{C}{-}R'} + :Nu{-}H \rightleftharpoons R{-}\underset{\underset{+NuH}{|}}{\overset{\overset{:\ddot{O}:^{-}}{|}}{C}}{-}R' \rightleftharpoons R{-}\underset{\underset{Nu}{|}}{\overset{\overset{OH}{|}}{C}}{-}R'$$

(R′ = H, alkyl, or aryl)

Aldehydes and ketones do not undergo substitution because to do so they would have to lose the very basic $:H^-$ ion or $:R^-$ ion.

$$R{-}\underset{\underset{Nu}{|}}{\overset{\overset{:\ddot{O}:^{-}}{|}}{C}}{-}R' \not\rightarrow R{-}\overset{\overset{:\ddot{O}}{\|}}{C}{-}Nu + :R'^{-}$$

$$R{-}\underset{\underset{Nu}{|}}{\overset{\overset{:\ddot{O}:^{-}}{|}}{C}}{-}H \not\rightarrow R{-}\overset{\overset{:\ddot{O}}{\|}}{C}{-}Nu + :H^{-}$$

These reactions do not occur.

In other words, acyl compounds are able to undergo substitution because they have a suitable leaving group which they can lose; aldehydes and ketones do not have a suitable leaving group, and therefore they undergo addition.

Some examples of nucleophilic substitution reactions of acyl derivatives will help clarify the general mechanism above.

Example 1: Esterification of a carboxylic acid.

Step 1:

$$R{-}C({=}\ddot{O}{:}){-}\ddot{O}H + HA \;\xrightarrow{\text{acid catalyst}}\; \rightleftharpoons \left[R{-}C({=}\overset{+}{\ddot{O}}H){-}\ddot{O}H \longleftrightarrow R{-}\overset{+}{C}({-}\ddot{O}H){-}\ddot{O}H \longleftrightarrow R{-}C({-}\ddot{O}H){=}\overset{+}{O}H \right] + A^-$$

a resonance hybrid

Step 2:

$$R{-}\overset{+}{C}({-}\ddot{O}H){-}OH + H\ddot{O}{-}R' \rightleftharpoons R{-}C({-}\ddot{O}H)({-}H\overset{+}{\ddot{O}}R'){-}\ddot{O}H$$

Step 3:

$$R{-}C({-}\ddot{O}H)({-}H\overset{+}{\ddot{O}}R'){-}OH \rightleftharpoons R{-}C({-}\ddot{O}H)({-}{:}\ddot{O}R'){-}\overset{+}{O}H_2$$

Step 4:

$$R{-}C({-}\ddot{O}H)({-}{:}\ddot{O}R'){-}\overset{+}{O}H_2 \rightleftharpoons \left[R{-}C({=}\overset{+}{\ddot{O}}H){-}\ddot{O}R' \longleftrightarrow R{-}C^+({-}\ddot{O}H){-}\ddot{O}R' \right] + H_2O$$

a resonance hybrid

Step 5:

$$R{-}C({=}\overset{+}{\ddot{O}}H){-}\ddot{O}R' + A^- \rightleftharpoons R{-}C({=}\ddot{O}{:}){-}OR' + HA$$

Isotopic labeling experiments have proved that the alcoholic oxygen atom substitutes for the OH group of the carboxylic acid:

$$C_6H_5{-}C({=}O){-}OH + CH_3{-}^{18}OH \xrightarrow{H^+} C_6H_5{-}C({=}O){-}^{18}OCH_3 + H_2O$$

That is, all the ^{18}O ends up in the ester and none in the water.

Example 2: Reaction of an acyl halide with water.

$$R-\overset{\overset{\ddot{O}:}{\|}}{C}-\ddot{Cl}: + :\ddot{O}(H)-H \rightleftharpoons R-\overset{:\ddot{O}:^-}{\underset{:\overset{+}{O}(H)-H}{C}}-\ddot{Cl}: \rightleftharpoons R-\overset{\overset{:\ddot{O}}{\|}}{C}-\overset{+}{\ddot{O}}(H)-H + :\ddot{Cl}:^-$$

$$\xrightarrow{-H^+} R-\overset{\overset{O}{\|}}{C}-OH$$

Example 3: Reaction of an acid anhydride with ammonia.

$$(RCO)_2O + :NH_3 \rightleftharpoons R-\overset{:\ddot{O}:^-}{C}(-\overset{+}{N}H_3)-\ddot{O}-C(=O)-R \rightleftharpoons R-\overset{:\ddot{O}:^-}{C}(-NH_2)-\overset{+}{:O}H-C(=O)-R \longrightarrow R-\overset{\overset{O}{\|}}{C}-NH_2 + R-\overset{\overset{O}{\|}}{C}-OH$$

In each of the above examples the leaving group is a very weak base: Example 1, $L = H_2O$; Example 2, $L = Cl^-$; Example 3, $L = R-CO_2H$. We can therefore explain the order of reactivity, acyl chloride > carboxylic anhydride > ester > amide, by the basicity of the leaving group. Acyl chlorides are the most reactive acyl derivatives because the chloride ion is an extremely weak base (an excellent leaving group). Amides are the least reactive because NH_3 is a poor leaving group (it is a relatively strong base). The order of leaving group ability is

$$Cl^- \qquad HO-\overset{\overset{O}{\|}}{C}-R \qquad HO-R' \qquad NH_3$$

← increasing effectiveness as leaving groups / decreasing base strength

Problem 13.18
Write a mechanism for the reaction,

$$CH_3\overset{\overset{O}{\|}}{C}-OCH_2CH_3 + NH_3 \longrightarrow CH_3\overset{\overset{O}{\|}}{C}-NH_2 + CH_3CH_2OH$$

13.11 NOMENCLATURE OF DICARBOXYLIC ACIDS

We name **dicarboxylic acids** systematically by adding the ending *-dioic acid* to the name of the parent hydrocarbon. Table 13.4 lists the first few simple dicarboxylic acids.

Table 13.4 Simple Dicarboxylic Acids

Formula	Systematic Name	Common Name
$HO-\overset{\overset{O}{\Vert}}{C}-\overset{\overset{O}{\Vert}}{C}-OH$	Ethanedioic acid	Oxalic acid
$HO-\overset{\overset{O}{\Vert}}{C}-CH_2-\overset{\overset{O}{\Vert}}{C}-OH$	Propanedioic acid	Malonic acid
$HO-\overset{\overset{O}{\Vert}}{C}-CH_2CH_2-\overset{\overset{O}{\Vert}}{C}-OH$	Butanedioic acid	Succinic acid
$HO-\overset{\overset{O}{\Vert}}{C}-CH_2CH_2CH_2-\overset{\overset{O}{\Vert}}{C}-OH$	Pentanedioic acid	Glutaric acid
$HO-\overset{\overset{O}{\Vert}}{C}-CH_2CH_2CH_2CH_2-\overset{\overset{O}{\Vert}}{C}-OH$	Hexanedioic acid	Adipic acid

13.12 PROPERTIES AND USES OF DICARBOXYLIC ACIDS

The simple dicarboxylic acids are colorless crystalline substances.

Although oxalic acid is poisonous, it occurs in plants and animals in the form of its salts. Potassium oxalate occurs in rhubarb. Calcium oxalate occurs in plant cells and in some stony deposits (calculi) found in the body.

Adipic acid is used in the manufacture of nylon (Chapter 18).

A number of dicarboxylic acids are important metabolic intermediates. Because they are relatively strong acids, they dissociate at neutral pH and therefore exist as ions. An advantage of having such charged compounds (at neutral pH) as intermediates is that the cell membranes retain them within the cells. Charged compounds do not readily cross membranes unless a specific mechanism exists for their transport. Indeed most metabolites are charged at physiological pH.

13.13 ANALYSIS AND IDENTIFICATION OF CARBOXYLIC ACIDS

The most characteristic property of carboxylic acids is their acidity. Even those that are insoluble in water are soluble in dilute aqueous NaOH solution. We can distinguish between carboxylic acids and phenols by the difference in their reaction with $NaHCO_3$ solutions: Carboxylic acids dissolve to yield bubbles of CO_2 whereas phenols do not (Sec. 13.3).

Problem 13.19

Describe simple tests that would make it possible to distinguish between the compounds in each of the following pairs.

(a) 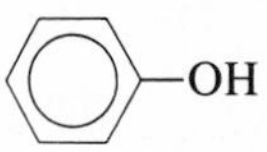 C_6H_5—OH and $CH_3CH_2CH_2CH_2OH$

(b) $CH_3CH_2CH_2CH_2OH$ and $CH_3CH_2CH_2CO_2H$

(c) $CH_3CH_2CH_2CO_2H$ and C_6H_5—OH

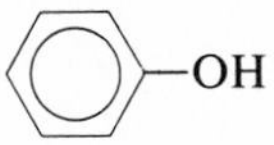

NEW TERMS

Acyl group

$$\mathrm{R{-}\overset{\overset{\displaystyle O}{\|}}{C}{-}}$$

functional group (R may be alkyl, aryl, or H) (Introduction).

Acyl halide

$$\mathrm{R{-}\overset{\overset{\displaystyle O}{\|}}{C}{-}X}$$

(X = Cl, Br, or I) (13.6).

Amide

$$\mathrm{R{-}\overset{\overset{\displaystyle O}{\|}}{C}{-}NR'_2}$$

(R′ may be alkyl, aryl, H, or any combination of them) (13.9).

Carboxyl group

$$\mathrm{{-}\overset{\overset{\displaystyle O}{\|}}{C}{-}OH}$$

functional group (Introduction).

Carboxylic acid

$$R-\overset{\overset{\displaystyle O}{\|}}{C}-OH$$

(R may be alkyl, aryl, or H) (Introduction).

Carboxylic anhydride

$$R-\overset{\overset{\displaystyle O}{\|}}{C}-O-\overset{\overset{\displaystyle O}{\|}}{C}-R \qquad (13.7).$$

Carboxylic salt

$$R-\overset{\overset{\displaystyle O}{\|}}{C}-OM$$

(M may be any metal ion) (13.1).

Dicarboxylic acid A compound whose molecules have two carboxyl groups (13.11).

Ester

$$R-\overset{\overset{\displaystyle O}{\|}}{C}-O-R'$$

(R′ may be alkyl or aryl) (13.8).

Esterification Conversion of a carboxylic acid into an ester (13.10).

Inductive effect The electronic effect produced by the polarization of a single bond (a sigma bond) due to differences in electronegativity between the two atoms joined by the bond (13.4).

ADDITIONAL PROBLEMS

13.20 Write the structural formula of each of the following compounds.

[a] propanoic acid
(b) adipic acid
[c] oxalic acid
(d) valeric acid
[e] 2-methyl-3-hexenoic acid
(f) phthalic anhydride
[g] diethyl phthalate
(h) caproamide
[i] dichloroacetic acid
(j) acetamide
[k] propenoic acid
(l) cyclohexyl pentanoate
[m] dimethyl adipate
(n) *N,N*-diethylacetamide
[o] methyl *o*-toluate
(p) *p*-toluic anhydride
[q] 3,5-dinitrobenzoyl bromide
(r) phenyl 1-naphthoate

13.21 Give an acceptable name for each of the following compounds.

[a] $C_6H_5-\overset{\overset{\displaystyle O}{\|}}{C}-Cl$

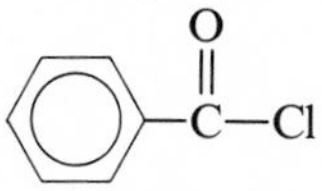

(b) $CH_3\underset{\underset{\displaystyle Cl}{|}}{C}HCH_2\underset{\underset{\displaystyle Cl}{|}}{C}HCO_2H$

[c] (cyclohexyl) $-O-\underset{\underset{\displaystyle O}{\|}}{C}-CH_3$

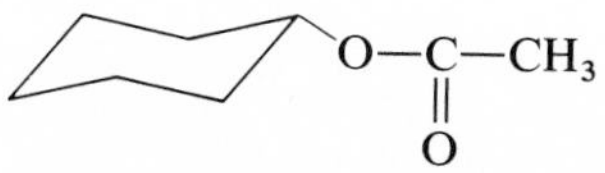

(d) 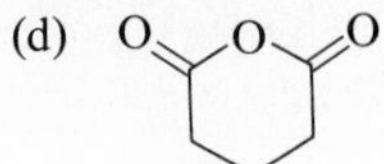

[e] $C_6H_5-\overset{\displaystyle O}{\overset{\|}{C}}-OK$

(f) $HO-\overset{\displaystyle O}{\overset{\|}{C}}-CH_2CH_2-\overset{\displaystyle O}{\overset{\|}{C}}-OH$

[g] $C_2H_5O-\overset{\displaystyle O}{\overset{\|}{C}}-CH_2-\overset{\displaystyle O}{\overset{\|}{C}}-OC_2H_5$

(h) $C_6H_{11}-O-\overset{\displaystyle O}{\overset{\|}{C}}-C_6H_4-\overset{\displaystyle O}{\overset{\|}{C}}-O-C_6H_{11}$

[i] $CH_3\overset{\displaystyle CH_3}{\underset{\displaystyle CH_3}{C}}-\overset{\displaystyle O}{\overset{\|}{C}}-Cl$

(j) $CH_3CH_2CO_2Na$

13.22 Which compound in each of the following pairs has the higher boiling point? Explain your choices.
[a] 1-butanol and propanoic acid
(b) propanoic acid and butanone
[c] butanoic acid and succinic acid

[13.23] Refer to Problem 13.22 and indicate which compound in each pair has the greater water solubility. Explain your choices.

13.24 Choose the stronger acid in each of the following pairs. Explain your choices.
[a] benzoic acid and *p*-nitrobenzoic acid
(b) benzoic acid and phenol
[c] benzoic acid and *p*-toluic acid
(d) 2-nitrobutanoic acid and 3-nitrobutanoic acid

[13.25] Explain why the C—O bonds in sodium formate are of equal length, whereas the C—O bonds in formic acid are of different lengths.

13.26 Write the equation for the reaction, if any, of butanoic acid with each of the following:
[a] aqueous NaOH,
(b) NH_3, then heat,
[c] $SOCl_2$,
(d) PBr_3,
[e] CH_3OH/HCl,
(f) H_3O^+.

13.27 Write equations for a laboratory synthesis of each of the following compounds. You may use any organic and/or inorganic reagents. Do not show any synthesis more than once.
[a] propanoic anhydride
[b] benzoyl chloride from benzoic acid
[c] benzamide from benzoic acid
[d] phthalic acid from *o*-xylene
(e) propanoic acid from butanone
(f) butanoic acid from 1-chloropropane
(g) butanoic acid from 1-butanol

13.28 Write equations to show how you could prepare each of the following acyl derivatives from the corresponding carboxylic acid and any necessary reagents.
[a] benzoic anhydride
[b] benzyl benzoate
(c) *o*-toluoyl chloride
(d) *o*-toluamide

*[13.29] Predict which of the following reactions will occur as written. Explain those that do not.

(a) $CH_3CO_2H + HCl \longrightarrow CH_3\overset{O}{\overset{\|}{C}}-Cl + H_2O$

(b) $CH_3\overset{O}{\overset{\|}{C}}-OC_2H_5 + NH_3 \longrightarrow CH_3\overset{O}{\overset{\|}{C}}-NH_2 + C_2H_5OH$

(c) $CH_3\overset{O}{\overset{\|}{C}}-O-\overset{O}{\overset{\|}{C}}CH_3 + NH_3 \longrightarrow CH_3\overset{O}{\overset{\|}{C}}-NH_2 + CH_3\overset{O}{\overset{\|}{C}}-ONH_4$

(d) $CH_3\overset{O}{\overset{\|}{C}}-Cl + H_2O \longrightarrow CH_3\overset{O}{\overset{\|}{C}}-OH + H_3O^+ + Cl^-$

*[13.30] Write a reasonable mechanism for the reaction

$$C_6H_5-\overset{O}{\overset{\|}{C}}-NH_2 + H_2O \longrightarrow C_6H_5-\overset{O}{\overset{\|}{C}}-O^- + NH_4{}^+$$

[13.31] Compound A, C_3H_7Cl, reacts with NaCN to give B, C_3H_6, as the major product. Compound B decolorizes Br_2/CCl_4 solution. Reaction of A with Mg in dry ether and subsequent treatment of that ether solution with CO_2 and then H_2O gives a compound, C, whose formula is $C_4H_8O_2$. When we add compound C to aqueous $NaHCO_2$ solution, bubbles are evolved. Give the structural formulas of A, B, and C, and write equations for all the reactions mentioned.

13.32 Compound E is neutral and has the formula $C_5H_6O_3$. If we boil E in water, we obtain compound F, $C_5H_8O_4$, which is soluble in aqueous NaOH solution and which liberates CO_2 when we dissolve it in aqueous $NaHCO_3$ solution. Heating dry F produces E. Give structural formulas of E and F, and write equations for all the reactions mentioned.

[13.33] Compound G, $C_9H_8O_2$, decolorizes Br_2/CCl_4 solution. It also reacts with H_2/Ni to give H, $C_9H_{10}O_2$. Both G and H dissolve in aqueous $NaHCO_3$ with the evolution of CO_2. Vigorous oxidation of either G or H with hot $KMnO_4$ solution gives benzoic acid. Give probable structural formulas for G and H, and write equations for all the reactions mentioned.

---

* These problems refer to an Elective Topic (Section 13.10).

14

FATS, OILS, WAXES, AND THEIR DERIVATIVES

Fats, oils, and waxes are naturally occurring acyl esters of glycerol. In mammals, fats serve as a source of chemical energy. When fats undergo oxidation to carbon dioxide and water they release more than twice as much energy per gram as carbohydrates or proteins. Waxes serve as protective coatings on the skin, fur, and feathers of animals and on the leaves of plants.

Fats and oils have always been an important part of our diet. We have also found many other uses for these compounds in soaps, drying oils, paints, and finishes of several kinds.

In this chapter we shall review the physical and chemical properties as well as the occurrence and uses of fats, oils, waxes, and their derivatives.

14.1 TRIACYLGLYCEROLS: FATS AND OILS

Most **fats** (solids) and **oils** (liquids) are esters of unbranched carboxylic acids that have even numbers of carbon atoms and the trihydroxy alcohol, glycerol. These esters are thus called **triacylglycerols** (an older name is glycerides).

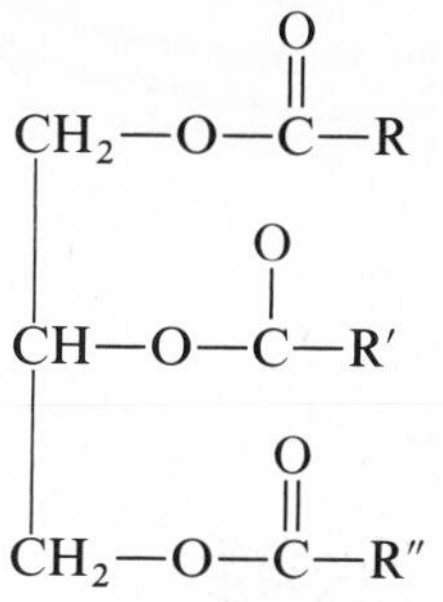

$$
\begin{array}{l}
\overset{O}{\|} \\
CH_2-O-C-R \\
| \qquad\quad\ \ \overset{O}{|} \\
CH-O-C-R' \\
| \qquad\quad\ \ \overset{O}{\|} \\
CH_2-O-C-R''
\end{array}
$$

a triacylglycerol
(or a triglyceride)

14.2 NOMENCLATURE

We name individual triacylglycerols, like other esters, by first giving the name *glyceryl* (the alcohol part) followed by the names of the acyl groups, called *alkanoate* (see Table 14.1). Some examples will illustrate.

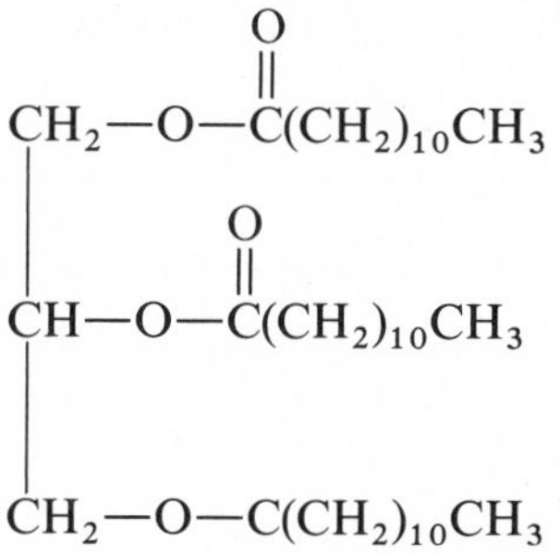

$$
\begin{array}{l}
\overset{O}{\|} \\
CH_2-O-C(CH_2)_{10}CH_3 \\
| \qquad\quad\ \ \overset{O}{\|} \\
CH-O-C(CH_2)_{10}CH_3 \\
| \\
CH_2-O-C(CH_2)_{10}CH_3
\end{array}
$$

glyceryl trilaurate
(from lauric acid)

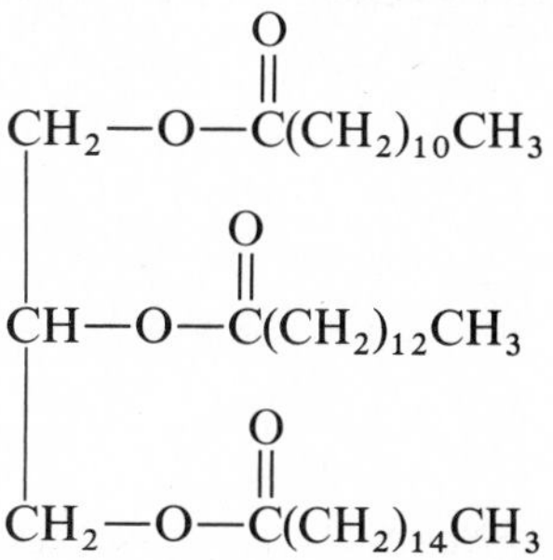

$$
\begin{array}{l}
\overset{O}{\|} \\
CH_2-O-C(CH_2)_{10}CH_3 \\
| \qquad\quad\ \ \overset{O}{\|} \\
CH-O-C(CH_2)_{12}CH_3 \\
| \qquad\quad\ \ \overset{O}{\|} \\
CH_2-O-C(CH_2)_{14}CH_3
\end{array}
$$

glyceryl lauromyristopalmitate
(from lauric, myristic,
and palmitic acids)

Problem 14.1
Write structural formulas for each of the following compounds.
(a) glyceryl trioleate
(b) glyceryl palmitolaurooleate

14.3 HYDROLYSIS OF FATS AND OILS: SAPONIFICATION

Hydrolysis of fats and oils yields glycerol and carboxylic acids. If we boil a fat or oil with aqueous sodium hydroxide until hydrolysis is complete, we obtain a mixture of glycerol and sodium salts of the carboxylic acids.

Table 14.1 Carboxylic Acids from Fats and Oils

		Melting Point, °C
Saturated Carboxylic Acids		
Lauric acid (dodecanoic acid)[a]	$CH_3(CH_2)_{10}CO_2H$	44
Myristic acid (tetradecanoic acid)	$CH_3(CH_2)_{12}CO_2H$	54
Palmitic acid (hexadecanoic acid)	$CH_3(CH_2)_{14}CO_2H$	63
Stearic acid (octadecanoic acid)	$CH_3(CH_2)_{16}CO_2H$	70
Unsaturated Carboxylic Acids[b]		
Oleic acid (9-octadecenoic acid)	$CH_3(CH_2)_7(H)C{=}C(H)(CH_2)_7CO_2H$	4
Linoleic acid (9,12-octadecadienoic acid)	$CH_3(CH_2)_4(H)C{=}C(H)CH_2(H)C{=}C(H)(CH_2)_7CO_2H$	−5
Linolenic acid (9,12,15-octadecatrienoic acid)	$CH_3CH_2(H)C{=}C(H)CH_2(H)C{=}C(H)CH_2(H)C{=}C(H)(CH_2)_7CO_2H$	−11

[a] Systematic IUPAC names are given in parentheses.
[b] Naturally occurring unsaturated carboxylic acids are generally *cis*.

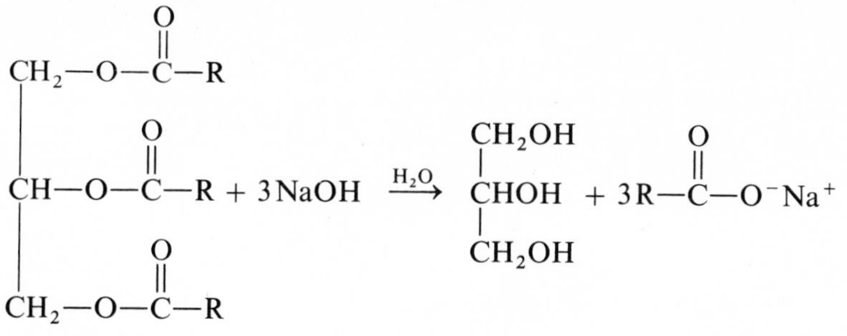

$$\begin{array}{l} \overset{\displaystyle O}{\overset{\|}{CH_2-O-C-R}} \\ | \\ \overset{\displaystyle O}{\overset{\|}{CH-O-C-R}} + 3NaOH \xrightarrow{H_2O} \begin{array}{c} CH_2OH \\ | \\ CHOH \\ | \\ CH_2OH \end{array} + 3R-\overset{\displaystyle O}{\overset{\|}{C}}-O^-Na^+ \\ | \\ \overset{\displaystyle O}{\overset{\|}{CH_2-O-C-R}} \end{array}$$

a fat or oil — glycerol — a soap

We precipitate these salts as soaps by adding sodium chloride to the mixture. After separating the soap which includes some glycerol, the remaining glycerol can be removed from the aqueous mixture by distillation.

The alkaline hydrolysis of fats and oils is called **saponification** (from the Latin *sapo*, which means soap).

Problem 14.2
Write an equation for the saponification of glyceryl trimyristate (see Table 14.1).

14.4 FATTY ACIDS

The carboxylic acids derived from fats and oils are often called **fatty acids**. The fatty acids found in fats and oils are usually unbranched and have even numbers of carbon atoms. Some of the acids are saturated and some have one or more alkene linkages. Table 14.1 lists the most common saturated and unsaturated carboxylic acids found in fats and oils. The most important unsaturated fatty acids have sixteen or eighteen carbons, and most of them occur in the *cis* configuration.

Problem 14.3
Name the following compounds.

(a) 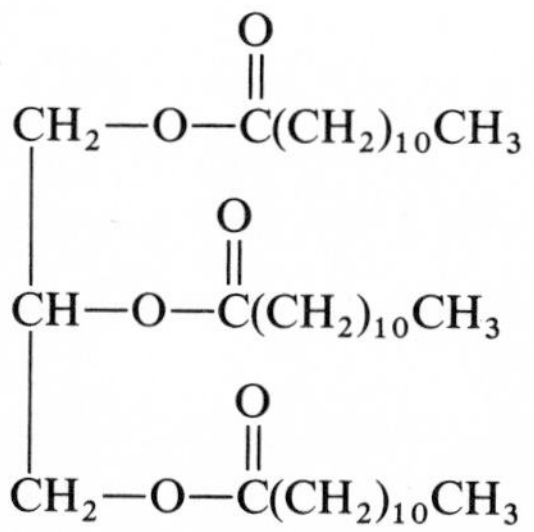
$$\begin{array}{l} CH_2-O-\overset{O}{\overset{\|}{C}}(CH_2)_{10}CH_3 \\ | \\ CH-O-\overset{O}{\overset{\|}{C}}(CH_2)_{10}CH_3 \\ | \\ CH_2-O-\overset{O}{\overset{\|}{C}}(CH_2)_{10}CH_3 \end{array}$$

(b) 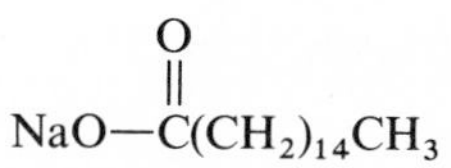 $NaO-\overset{O}{\overset{\|}{C}}(CH_2)_{14}CH_3$

Fats and oils have densities of slightly less than 1.0 g/ml. Solid fats are composed mainly of saturated fatty acids whereas oils contain greater percentages of unsaturated fatty acids. In fact, the melting point of a fat depends on the amount of unsaturation in the fatty acid parts of the triacylglycerol.

Vegetable oils tend to contain greater amounts of unsaturated fatty acids than animal fats. Table 14.2 lists the compositions of a few common fats and oils.

Solid shortenings such as Crisco and Spry are manufactured by partial hydrogenation of vegetable oils. By adding hydrogen atoms to some of the double bonds, a product of higher melting point results. Completely saturated fats are not convenient for this purpose because they are waxy and brittle.

Table 14.2 Compositions of Some Fats and Oils

	Fatty Acid Constant (%)					
	Saturated Fatty Acids				Unsaturated Fatty Acids	
Fat or Oil	Lauric (C_{12})	Myristic (C_{14})	Palmitic (C_{16})	Stearic (C_{18})	Oleic (C_{18})	Linoleic (C_{18})
Beef tallow		3–6	24–32	20–25	37–43	2–12
Olive oil			9	2	83	4
Corn oil		0.1–2	8–12	2–5	19–49	34–62
Peanut oil	Trace		9	3	56	26
Cottonseed oil		1–2	18–25	1–2	17–38	45–55

Beef tallow is obtained by melting down the fatty tissue of cattle. It is used chiefly in the manufacture of soap, candles, margarine, and lubricants.

Olive oil is manufactured by pressing olives. Spain is the major producer of olives although olives grow in all the Mediterranean countries. In addition to its use as a food, olive oil is used to produce soaps, textile lubricants, and cosmetics. It is sometimes used as a laxative and as an emollient (skin softener).

A by-product of the milling of corn is *corn oil.* It is used very widely in foods, in pharmacy as a vehicle for intramuscular injections, and in the manufacture of nonyellowing enamel paint.

Peanut oil is obtained by pressing shelled, skinned seeds. It is used in unaltered form as a food, and in the manufacture of shortening and margarine by hydrogenation. It is also used to make mayonnaise and confections as well as paints and soaps. Peanut oil is sometimes used as a vehicle for intramuscular injections and in ointments and liniments.

Cottonseed oil is obtained by pressing the seeds of the plant. In addition to its use in salad and cooking oils, it is used in the manufacture of shortening, margarine, lubricants, and soaps. It is used medically as an emollient laxative.

14.6 WAXES

Waxes are esters of unbranched carboxylic acids and unbranched 1-alkanols; both groups contain chains of sixteen or more carbon atoms. Waxes also contain varying amounts of hydrocarbons and alcohols. Waxes occur naturally in plants and animals and have a typical "waxy" feel.

Beeswax is obtained from the honeycombs of bees. It is a mixture of esters of C_{24} to C_{36} carboxylic acids and alcohols up to C_{36}. An example is

$$C_{30}H_{61}\text{—O—}\overset{\overset{\displaystyle O}{\|}}{C}\text{—}C_{25}H_{51}$$

Beeswax melts at 62–65°C. It is used in the production of wax paper, candles, artificial fruits and flowers, and shoe polish.

Carnuba wax exudes from the pores of the leaves of the Brazilian wax palm tree. It is hard and has a high polishing ability because of its high melting point, 82–85°C. It is a mixture of esters of C_{24} and C_{28} carboxylic acids and C_{32} and C_{34} 1-alkanols. Carnuba wax is used as a high-polish wax for automobiles, floors, and shoes, and for the manufacture of carbon papers and cosmetics such as depilatories (hair removers) and deodorant sticks. It is also used in pharmacy to coat tablets.

Spermaceti is extracted from the head of the sperm whale. It is chiefly cetyl palmitate,

$$C_{16}H_{33}\text{—O—}\overset{\overset{\displaystyle O}{\|}}{C}\text{—}C_{15}H_{31}$$

Spermaceti is white and melts at 42–50°C. It is used in ointments and in the manufacture of candles, soaps, cosmetics, and laundry wax. It is also an emollient.

Problem 14.4
How do waxes differ structurally from fats and oils?

14.7 SOAPS

When a **soap** is mixed with water it disperses in an interesting way. The soap molecule consists of a long hydrocarbon "tail" and an ionic "head." The ionic end

$$CH_3CH_2CH_2CH_2CH_2CH_2CH_2CH_2CH_2CH_2CH_2CH_2CH_2CH_2CH_2CH_2\overset{\overset{\displaystyle O}{\|}}{C}\text{—}O^-Na^+$$

is soluble in water but the hydrocarbon tail is not. As a result, soap forms **micelles** in which all the hydrocarbon tails are on the inside, away from the water, and all the carboxylate ions are on the surface (Fig. 14.1).

It is difficult to remove dirt with water alone because dirt particles are generally coated with fats or oils and are therefore not soluble in water. When soap is used, the

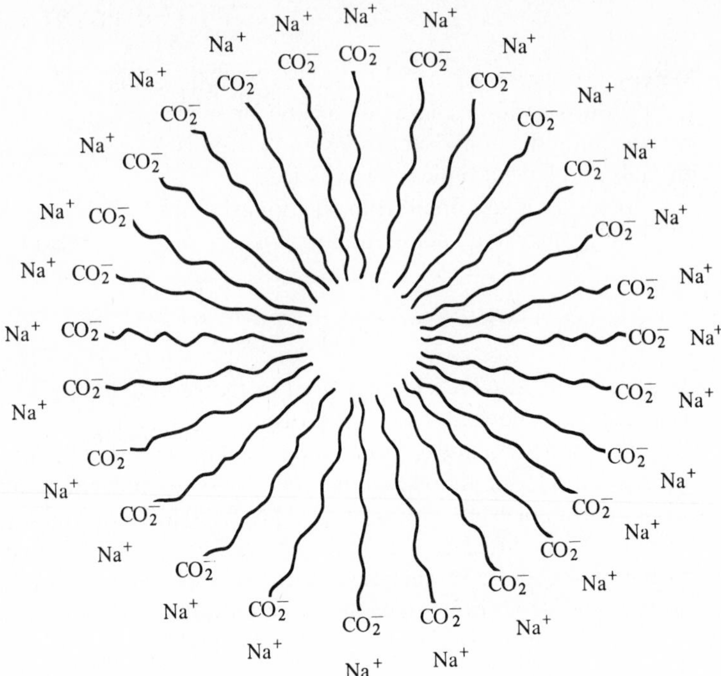

Figure 14.1. Cross-section of a soap micelle.

oily layer of the dirt particles dissolves in the hydrocarbon chains of the soap micelles and is therefore "solubilized" and carried away by the water (Fig. 14.2).

The exteriors of soap micelles are all negatively charged, so they repel each other and do not coalesce or precipitate.

Problem 14.5
Do you think soap would be helpful in removing an ink spot? Explain why or why not.

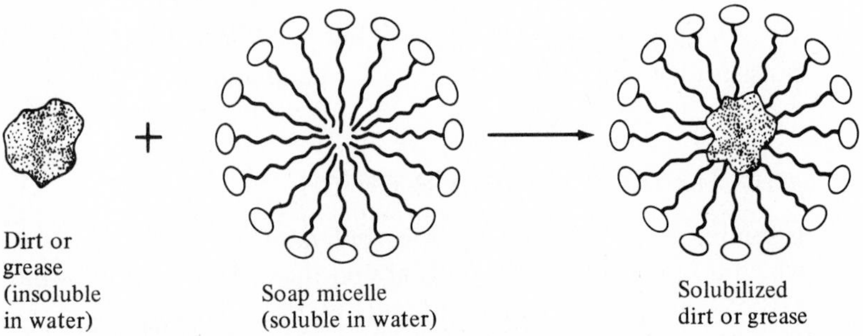

Figure 14.2. Action of soap on dirt and grease.

14.8 SYNTHETIC DETERGENTS

Soaps made from naturally occurring fats and oils contain unbranched acyl groups which are metabolized by certain bacteria. This biodegradation normally occurs in sewage treatment or after the soap solution has found its way into the ground.

Soaps have the disadvantage of leaving a scum when they are used with hard water. Hard water contains Ca^{2+}, Mg^{2+}, and Fe^{2+} ions which form insoluble salts with soaps.

$$\underset{\text{soluble}}{2\,R{-}\overset{\overset{\displaystyle O}{\|}}{C}{-}ONa} + Ca^{2+} \longrightarrow \underset{\text{insoluble}}{\left(R{-}\overset{\overset{\displaystyle O}{\|}}{C}{-}O\right)_2 Ca} + 2\,Na^{+}$$

To avoid this scum, chemists developed synthetic **detergents** in the 1930s. The first detergents were branched-chain alkylbenzenesulfonates,

$$R{-}C_6H_4{-}\overset{\overset{\displaystyle O}{\|}}{\underset{\underset{\displaystyle O}{\|}}{S}}{-}O^{-}K^{+}$$

(R = branched alkyl group)

The calcium, magnesium, and iron salts of these compounds are "soluble" in water.

These early detergents were not biodegradable, and they often appeared in rivers and lakes as foams. In the mid-1960s, new linear alkansulfonate detergents became available that were biodegradable because they were linear chains and resembled soap molecules:

$$CH_3(CH_2)_nSO_3K$$

Problem 14.6

Write the equation for the reaction of Ca^{2+} ions with soap prepared from glyceryl trimyristate.

14.9 REACTIONS OF SOAPS

Because soaps are the salts of strong bases and weak acids, aqueous soap solutions are alkaline and show a basic reaction with litmus. The carboxylate ion reacts with water to establish the equilibrium

$$R{-}\overset{\overset{\displaystyle O}{\|}}{C}{-}O^{-} + H_2O \rightleftharpoons R{-}\overset{\overset{\displaystyle O}{\|}}{C}{-}OH + OH^{-}$$

Soaps react with strong mineral acids to form the free fatty acid:

$$R{-}\overset{\overset{\displaystyle O}{\|}}{C}{-}ONa + HCl \longrightarrow R{-}\overset{\overset{\displaystyle O}{\|}}{C}{-}OH + Na^{+} + Cl^{-}$$

Long-chain fatty acids are insoluble in water, and they precipitate out of solution.

Problem 14.7
Write net ionic equations for the reactions of sodium oleate with (a) water, (b) concentrated HCl.

14.10 REACTIONS OF FATS AND OILS

Fats and oils are esters and undergo the reactions of ordinary esters. Fats and oils that contain unsaturated acyl groups also undergo the reactions of alkenes. Figure 14.3 outlines the reactions of triacylglycerols.

We can use bromination or permanganate oxidation to determine the amount of unsaturation in a fat or oil. We do this by measuring the amount of bromine or $KMnO_4$ consumed by a given weight of fat or oil.

Problem 14.8
What conclusion can you draw from the fact that one mole of a certain triacylglycerol reacts with two moles of Br_2 in CCl_4?

14.11 PHOSPHOLIPIDS

Compounds that have a glycerol molecule esterified by two fatty acids and one phosphoric acid molecule are called **phosphatidic acids**.

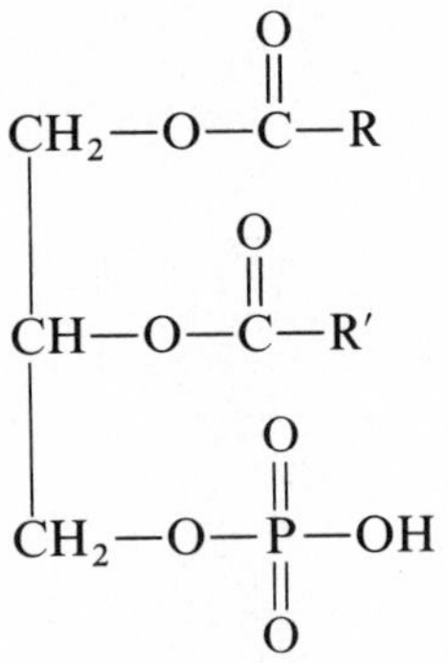

$$\begin{array}{l} CH_2{-}O{-}\overset{\overset{\displaystyle O}{\|}}{C}{-}R \\ | \\ CH{-}O{-}\overset{\overset{\displaystyle O}{\|}}{C}{-}R' \\ | \\ CH_2{-}O{-}\underset{\underset{\displaystyle O}{\|}}{\overset{\overset{\displaystyle O}{\|}}{P}}{-}OH \end{array}$$

a phosphatidic acid

Most phosphatidic acids contain one saturated acyl group and one unsaturated acyl group.

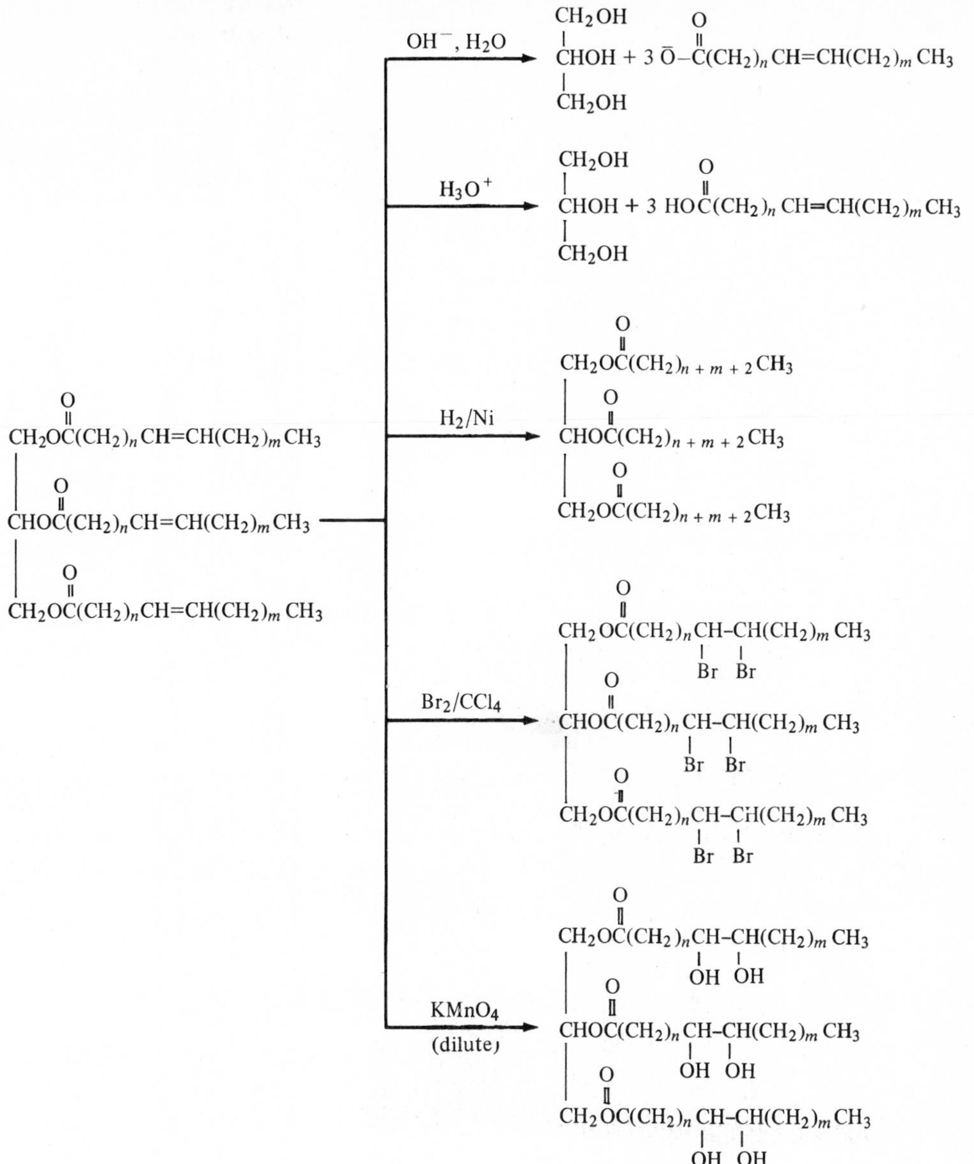

Figure 14.3. Reactions of triacylglycerols.

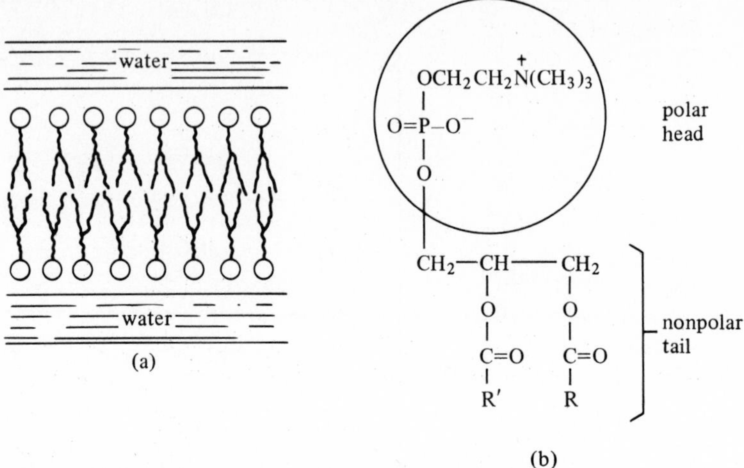

Figure 14.4. Schematic drawing of (a) a phospholipid bilayer and (b) the polar head and nonpolar tail of a phospholipid.

Free phosphatidic acids rarely occur. They usually occur with the phosphoric acid group esterified to another alcohol group (a **phosphatidyl ester**). An example is *phosphatidyl choline*, also known as *lecithin.*

$$
\begin{array}{l}
CH_2{-}O{-}\overset{\overset{\large O}{\|}}{C}(CH_2)_{14}CH_3 \\
| \\
CH{-}O{-}\overset{\overset{\large O}{\|}}{C}(CH_2)_{14}CH_3 \\
| \\
CH_2{-}O{-}\underset{\underset{\large O}{\|}}{\overset{\overset{\large O}{\|}}{P}}{-}OCH_2CH_2\overset{+}{N}(CH_3)_3
\end{array}
$$

phosphatidyl choline (lecithin)

Glyceryl esters that contain phosphoric acid groups have the general name **phospholipids**.*

Phospholipids form micelles when they are mixed with water. They also form **bilayers**, especially at the interface between two aqueous layers. Figure 14.4 shows such a bilayer. The hydrocarbon tails lie together with the polar heads on opposite sides of the bilayer. Membranes found in living systems appear to consist of such bilayer structures.

* **Lipids** are naturally occurring compounds that are soluble in hydrocarbons and insoluble in water. The term *lipid* includes fats, oils, waxes, natural hydrocarbons, and so forth.

Problem 14.9
Distinguish among the terms phospholipid, phosphatidic acid, and phosphatidyl ester.

14.12 METABOLISM OF FATS AND OILS

Fats provide more energy per gram than any other food. In addition to using fats as a source of energy, our bodies synthesize fats from other foods—carbohydrates and proteins. This combination of buildup and breakdown of foods in the body is called **metabolism**. We shall now look briefly at these two metabolic processes—**anabolism** (biosynthesis) and **catabolism** (biodegradation).

Most organic compounds burn in the presence of oxygen to yield carbon dioxide, water, and energy. Fats are no exception. The body cannot utilize this process of direct combustion, however, because the body must maintain a nearly constant temperature. Similarly, the buildup of molecules in the body cannot require high temperatures; that is, all biochemical reactions must occur at 37°C (98.6°F). Some of the energy is released during metabolism as heat (since no energy transformation is 100% efficient) and provides the means of maintaining normal body temperature in warm-blooded animals.

How do exothermic and endothermic reactions occur without evolving or absorbing large amounts of heat energy? The answer is that biochemical reactions are always coupled so that the energy released by one reaction is absorbed by another one that occurs at the same time. In other words, energy released by one reaction is absorbed in another reaction and transformed into chemical potential energy in the form of chemical bonds in the course of the second reaction. In an overall sense, food acts as a medium for transferring energy: Food (chemical potential energy in bonds) enters our bodies, where some of its energy is used to maintain our bodily functions, and the remainder is stored in other chemical bonds (for example, in fats). When we need extra energy we call on these stored molecules to release their energy.

Before we discuss the mechanism of fat buildup and breakdown, we should recall that fatty acids are unbranched and always contain an even number of carbon atoms. This suggests that the synthesis of fatty acids in the body must occur in steps of two carbon units. The early suggestion that fatty acids were built up in units of acetic acid was later tested by experiments that used acetic acid labeled with carbon-14 (written as *C). When a test animal was fed acetic acid with the carbon-14 label at the carboxyl group, the fatty acids produced in the animal contained the labels at alternate carbons beginning at the carboxyl group:

$$\underset{\text{fed to the animal}}{CH_3{*}CO_2H} \longrightarrow \underset{\text{produced in the animal}}{CH_3{*}CH_2(CH_2{*}CH_2)_nCH_2{*}CO_2H}$$

When the acetic acid was labeled at the methyl group, the fatty acids produced in the animal contained the labels at alternate carbons beginning with the alpha carbon (see Sec. 12.7 and 12.10 for the definition of the alpha carbon):

$$\underset{\text{fed to the animal}}{{*}CH_3CO_2H} \longrightarrow \underset{\text{produced in the animal}}{{*}CH_3CH_2({*}CH_2CH_2)_n{*}CH_2CO_2H}$$

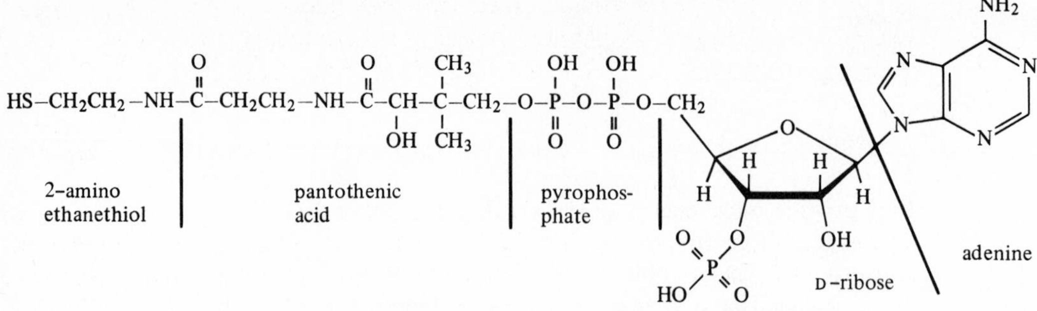

Figure 14.5. Structure of coenzyme A.

Both anabolism and catabolism of fatty acids involve a compound called coenzyme A (A = acetylation). Coenzyme A acts both as an acetyl group donor in anabolism and as an acetyl group acceptor in catabolism. Coenzyme A has the chemical structure shown in Figure 14.5. It undergoes its principal reactions at the SH group, so we can abbreviate coenzyme A with the symbol *CoA-SH*.

Catabolism

The breakdown of fatty acids in the body occurs through a process called β-oxidation (oxidation of the $-CH_2-$ group beta to the carbonyl). Under the influence of various **enzymes**,* fatty acid esters of CoA-SH undergo a series of reactions in which

* Enzymes are complex proteins that catalyze specific chemical reactions in living systems.

$$\underset{\text{stearic acid}}{CH_3(CH_2)_{14}CH_2CH_2\overset{O}{\overset{\|}{C}}-OH} + CoA-SH \underset{}{\overset{\text{(enzyme)}}{\rightleftharpoons}} CH_3(CH_2)_{14}CH_2CH_2\overset{O}{\overset{\|}{C}}-S-CoA + H_2O$$

$$\Big\updownarrow \; -2\text{ H (enzyme)}$$

$$CH_3(CH_2)_{14}\overset{OH}{\overset{|}{C}}HCH_2-\overset{O}{\overset{\|}{C}}-S-CoA \overset{H_2O\text{ (enzyme)}}{\rightleftharpoons} CH_3(CH_2)_{14}CH{=}CH\overset{O}{\overset{\|}{C}}-S-CoA$$

$$\Big\updownarrow \; -2\text{ H (enzyme)}$$

$$\underset{\beta\text{-keto thioester}}{CH_3(CH_2)_{14}\overset{O}{\overset{\|}{C}}-CH_2-\overset{O}{\overset{\|}{C}}-S-CoA} \overset{\text{CoA-SH/enzyme}}{\rightleftharpoons} \underset{\text{palmitic acid (CoA thioester)}}{CH_3(CH_2)_{14}\overset{O}{\overset{\|}{C}}-S-CoA} + \underset{\text{acetyl coenzyme A}}{CH_3\overset{O}{\overset{\|}{C}}-S-CoA}$$

Figure 14.6. Catabolism of stearic acid showing β-oxidation reaction.

the product is a fatty acid having two carbons fewer than the original. These reactions are outlined in Figure 14.6. The acetyl coenzyme A produced in this reaction sequence is used in many body processes including carbohydrate metabolism.

Anabolism

The buildup of fatty acids occurs through the reaction of acetyl coenzyme A with carbon dioxide to produce the half thioester of malonic acid, malonyl coenzyme A (see Fig. 14.7). This derivative then reacts with a second molecule of acetyl coenzyme A to produce a compound which loses CO_2 to yield a β-keto acid (acetoacetyl coenzyme A). Acetoacetyl coenzyme A (Fig. 14.7) is then reduced to the thioester of butanoic acid through the reversible sequence of reactions that are the reverse of those shown in Figure 14.6 for catabolism. Although anabolism and catabolism are reversible, the same pathways are not followed. The enzymes for each process are different, and are located in different parts of the cell.

The butanoyl coenzyme A reenters the cycle at point *a* (Fig. 14.7) to build up a chain of six carbon atoms. The process repeats to build up larger chains containing even numbers of carbon atoms.

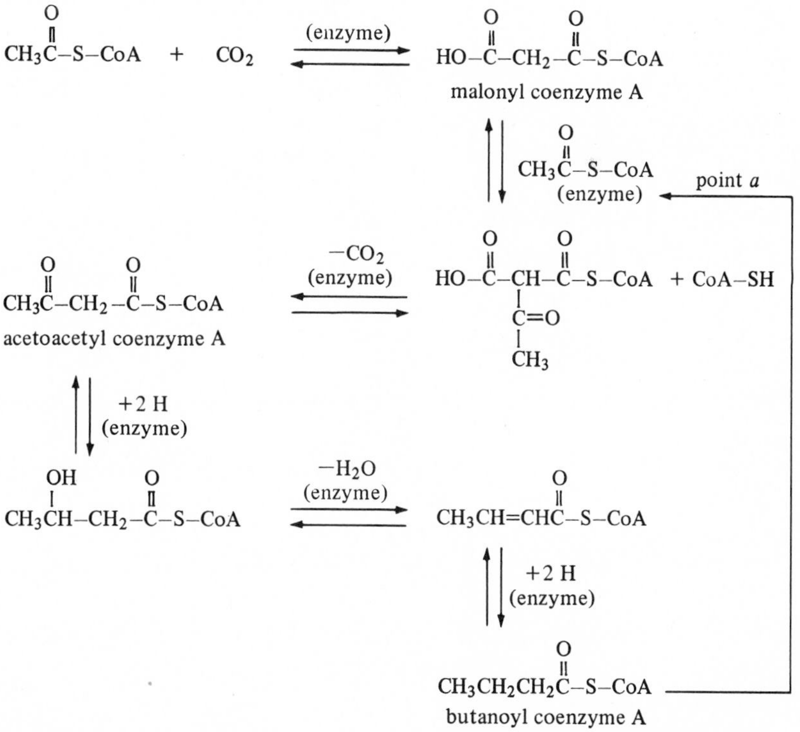

$$CH_3\overset{O}{\overset{\|}{C}}{-}S{-}CoA + CO_2 \underset{}{\overset{\text{(enzyme)}}{\rightleftharpoons}} HO{-}\overset{O}{\overset{\|}{C}}{-}CH_2{-}\overset{O}{\overset{\|}{C}}{-}S{-}CoA$$

malonyl coenzyme A

$$\Big\updownarrow \; CH_3\overset{O}{\overset{\|}{C}}{-}S{-}CoA \text{ (enzyme)} \leftarrow \text{point } a$$

$$CH_3\overset{O}{\overset{\|}{C}}{-}CH_2{-}\overset{O}{\overset{\|}{C}}{-}S{-}CoA \underset{}{\overset{-CO_2 \text{ (enzyme)}}{\rightleftharpoons}} HO{-}\overset{O}{\overset{\|}{C}}{-}\underset{\underset{CH_3}{|}}{\underset{C=O}{|}}{\overset{}{CH}}{-}\overset{O}{\overset{\|}{C}}{-}S{-}CoA + CoA{-}SH$$

acetoacetyl coenzyme A

$$\Big\updownarrow \; +2\,H \text{ (enzyme)}$$

$$CH_3\overset{OH}{\overset{|}{C}}H{-}CH_2{-}\overset{O}{\overset{\|}{C}}{-}S{-}CoA \underset{}{\overset{-H_2O \text{ (enzyme)}}{\rightleftharpoons}} CH_3CH{=}CH\overset{O}{\overset{\|}{C}}{-}S{-}CoA$$

$$\Big\updownarrow \; +2\,H \text{ (enzyme)}$$

$$CH_3CH_2CH_2\overset{O}{\overset{\|}{C}}{-}S{-}CoA$$

butanoyl coenzyme A

Figure 14.7. Anabolism reactions showing the buildup of a butanoic acid from an acetic acid unit.

Problem 14.10
Write equations to show the anabolism of butanoic acid into hexanoic acid.

Problem 14.11
Write equations to show the catabolism of hexanoic acid into butanoic acid.

14.13 ANALYSIS AND IDENTIFICATION OF FATS, OILS, WAXES, AND SOAPS: AN ELECTIVE TOPIC

elective topic

Saponification Number

We can saponify a fat or oil quantitatively. To do so we merely add a measured amount of potassium hydroxide to one gram of the fat or oil, and determine how much of the KOH is consumed. The **saponification number** is the number of milligrams of KOH required to saponify one gram of the fat or oil. (Remember that one mole of fat requires three moles of KOH.)

If a fat or oil has long chains, there will be fewer of them present in one gram than if the fat or oil has short chains. A low saponification number therefore indicates that the fat or oil has long chains.

Iodine Number

We can use the addition of iodine to an alkene double bond to determine the degree of unsaturation in a fat or oil. The reaction is carried out in alcohol solution. The **iodine number** is the number of grams of iodine consumed by 100 grams of the fat or oil. A high iodine number indicates a high degree of unsaturation in the oil or fat.

Table 14.3 lists some triacylglycerols with their saponification numbers and iodine numbers.

A sample calculation using saponification number follows.

Problem: What is the molecular weight of a fat that has a saponification number of 190?

Table 14.3

Fat or Oil	Saponification Number	Iodine Number
Beef tallow	190–195	35–40
Olive oil	187–196	79–90
Corn oil	187–196	109–133
Peanut oil	188–195	84–102
Cottonseed oil	190–195	103–111

Solution:

$$\text{Saponification number} = \text{S.N.} = \frac{\text{mg of KOH}}{\text{g of fat}}$$

Recall that 3 moles of KOH is required to saponify 1 mole of fat. Therefore,

$$3 \text{ moles KOH} \times 56.1 \text{ g KOH/mole} \times 1000 \text{ mg/g} = 168{,}300 \frac{\text{mg KOH}}{\text{mole fat}}$$

$$168{,}300\left(\frac{\text{mg KOH}}{\text{mole fat}}\right) \div \text{S.N.}\left(\frac{\text{mg KOH}}{\text{g fat}}\right) = \frac{168{,}300}{\text{S.N.}}\left(\frac{\text{g fat}}{\text{mole fat}}\right) = \frac{168{,}300}{190}$$

$$= 885.8 \text{ g/mole fat}$$

The molecular weight of the fat is therefore 886.

A sample calculation using iodine number follows.

Problem: How many double bonds does the above fat have per molecule if its iodine number (I.N.) is 57?

Solution:

$$\text{I.N.} = \frac{\text{g } I_2}{100 \text{ g fat}}$$

Recall that 1 mole of I_2 is required per double bond per mole of fat. Then,

$$\frac{253.8 \text{ g } I_2\text{/mole} \times n}{\text{molecular weight fat}} \times 100 = \text{I.N.}$$

n = the number of double bonds per molecule of fat. Then,

$$\frac{253.8 \text{ g } I_2\text{/mole} \times n}{885.8} \times 100 = 57$$

Rearranging,

$$n = \frac{57 \times 885.8}{253.8 \times 100} = 1.99 \cong 2$$

That is, the molecule of fat has two double bonds.

Problem 14.12

Arrange the fats and oils of Table 14.3 in order of (a) molecular weight and (b) degree of unsaturation.

NEW TERMS

Anabolism Biosynthesis (buildup) of molecules in the body (14.12).

Bilayer Organization of phospholipid molecules as shown in Figure 14.4 (14.11).

Catabolism Biodegradation (breakdown) of food molecules in the body (14.12).

Detergent Sulfonic salt $R—SO_3^-K^+$ in which R is a hydrocarbon group.

Enzyme Any complex protein that catalyzes a specific chemical reaction in living systems (14.12).

Fat Solid ester of unbranched carboxylic acids and glycerol (14.1).

Fatty acid Carboxylic acid derived from fats and oils; generally, aliphatic carboxylic acids are called fatty acids (14.4).

Iodine number The number of grams of iodine consumed by 100 grams of the fat or oil. The iodine number gives a measure of the amount of unsaturation in a fat or oil (14.13).

Lipid A naturally occurring compound soluble in hydrocarbons and insoluble in water. The term lipid includes fats, oils, waxes, and natural hydrocarbons (14.11).

Metabolism Buildup and breakdown of foods in the body (14.12).

Micelle A globular combination of soap molecules in which the hydrocarbon parts are on the inside and the carboxylate ions are on the surface. Micelles are formed when soap is mixed with water (14.7).

Oil A liquid ester of unbranched carboxylic acids and glycerol (14.1).

Phosphatidic acid A compound that has a glycerol molecule esterified by two fatty acid molecules and one phosphoric acid molecule (14.11).

Phosphatidyl ester An ester of a phosphatidic acid (14.11).

Phospholipid A glycerol ester that contains a phosphoric acid group (14.11).

Saponification Alkaline hydrolysis of an ester (especially a fat or oil) to produce an alcohol and a carboxylic salt (14.3).

Saponification number The number of milligrams of KOH required to saponify one gram of fat or oil. Saponification number provides a measure of the chain lengths of the fatty acid portions of oils and fats (14.13).

Soap The Na^+ or K^+ salt of a long-chain fatty acid (14.7).

Triacylglycerol An ester of an unbranched carboxylic acid having an even number of carbon atoms and the tryhydroxy alcohol, glycerol (14.1).

Waxes Esters of unbranched carboxylic acids and unbranched 1-alkanols, both of which contain chains of sixteen or more carbon atoms (14.6).

ADDITIONAL PROBLEMS

14.13 Write the structural formula of each of the following compounds.

[a] glyceryl tripalmitate [b] glyceryl tristearate
[c] glyceryl trimyristate (d) sodium oleate
(e) calcium laurate

14.14 Give an acceptable name for each of the following compounds.

[a]
$$\begin{array}{l} CH_2—O—\overset{\displaystyle O}{\overset{\|}{C}}\!+\!(CH_2)_2CH_3 \\ | \\ CH—O—\overset{\displaystyle O}{\overset{\|}{C}}\!+\!(CH_2)_2CH_3 \\ | \\ CH_2—O—\overset{\displaystyle O}{\overset{\|}{C}}\!+\!(CH_2)_2CH_3 \end{array}$$

[b] $CH_3(CH_2)_{12}CO_2H$

(c) $CH_3(CH_2)_7CH{=}CH(CH_2)_7CO_2H$

(d) $CH_3(CH_2)_{10}\overset{\overset{O}{\|}}{C}—O—(CH_2)_8CH_3$

[14.15] Explain how solid shortenings are manufactured from vegetable oils.

[14.16] Why is it not strictly correct to say that soap *dissolves* in water?

[14.17] How do synthetic detergents differ from soaps?

[14.18] In what way are synthetic detergents superior to soaps?

[14.19] Explain using equations why a soap might be ineffective if it is used in dishwater containing some citrus juice.

14.20 Aromatic sulfonic acids ($ArSO_3H$) are stronger acids than carboxylic acids. On the basis of this information, which do you expect to be more basic, a soap or an alkylbenzene sulfonate detergent?

14.21 Write equations for simple chemical tests that would make it possible to distinguish between the compounds in each of the following pairs.
[a] glyceryl tristearate and glyceryl trioleate
(b) glyceryl tristearate and the phosphatidic acid

$$\begin{array}{l} CH_2—O—\overset{\overset{O}{\|}}{C}—C_{17}H_{35} \\ | \\ CH—O—\overset{\overset{O}{\|}}{C}—C_{17}H_{35} \\ | \\ CH_2—O—\underset{\underset{O}{\|}}{\overset{\overset{O}{\|}}{P}}—OH \end{array}$$

[c] glyceryl trioleate and spermaceti

14.22 Write the chemical equation that describes the conversion of the major component of olive oil into [a] a soap, (b) shortening, [c] a brittle waxy substance.

[14.23] Explain, with structural formulas, the difference between a phosphatidic acid, a phosphatidyl ester, and a phospholipid.

[14.24] Which of the following detergents would be less of a pollutant? Explain.

$$CH_3\underset{\underset{CH_3}{|}}{C}HCH_2\underset{\underset{CH_3}{|}}{C}HCH_2\underset{\underset{CH_3}{|}}{C}HCH_2\underset{\underset{CH_3}{|}}{C}H—C_6H_4—SO_3Na \quad \text{or} \quad CH_3(CH_2)_{10}CH_2SO_3Na$$

[14.25] Write equations for the biochemical conversion of (a) hexanoic acid to octanoic acid (b) octanoic acid to hexanoic acid.

*14.26 Calculate the saponification number of [a] glyceryl trihexanoate, (b) glyceryl tripropanoate.

*[14.27] Calculate the molecular weight of a fat that has a saponification number of 132.7.

*14.28 Calculate the iodine number of [a] glyceryl lauropalmitooleate, (b) glyceryl trilinoleate.

*[14.29] A fat has a saponification number of 133.3 and an iodine number of 57.7. Propose a structure that is consistent with these data.

*[14.30] How would hydrogenation affect the saponification number and iodine number of the fat described in Problem 14.29?

---

* These problems refer to an elective topic (Sec. 14.13).

AMINES AND THEIR DERIVATIVES

Most organic bases are amines. Amines are derivatives of ammonia in which one, two, or all three of the hydrogens have been replaced by alkyl or aryl groups. Amines occur widely in plants and animals. In the form of amino acids, proteins, and nucleic acids (for example, DNA), amines play a dominant role in the chemistry of life. Amines also play a large role in industry. Many common products possess the amine functional group (medicines, drugs, and anesthetics) and many are manufactured from amines (nylon, and other fibers and plastics).

We shall see in this chapter that most of the chemical and physical properties of amines can be attributed to the unshared electron pair on the nitrogen atom.

15.1 CLASSIFICATION OF AMINES

We may replace one, two, or three of the hydrogens of ammonia to obtain **amines**.

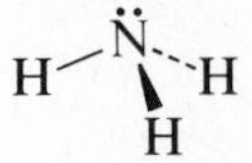

$$H-\ddot{N}(-H)-H$$

ammonia

$$R-\ddot{N}(-H)-H \qquad R-\ddot{N}(-R')-H \qquad R-\ddot{N}(-R')-R''$$

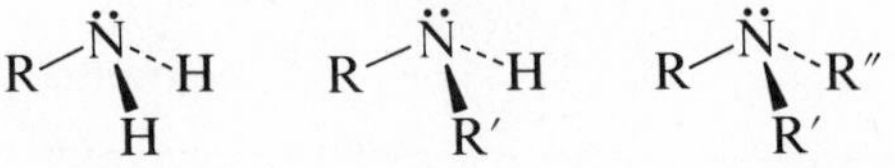

1° amine   2° amine   3° amine

In the above formulas, R, R′, and R″ may be alkyl or aryl groups. If one hydrogen has been replaced by a hydrocarbon group, the amine is called a *primary* (1°) *amine*. If two hydrogens have been replaced by hydrocarbon groups, the amine is called a *secondary* (2°) *amine*. If all three hydrogens have been replaced by hydrocarbon groups, the amine is called a *tertiary* (3°) *amine*. Notice that the terms primary, secondary, and tertiary have a slightly different meaning when applied to amines than they do when applied to alcohols (Sec. 10.2).

Problem 15.1

Classify each of the following amines as primary, secondary, or tertiary.

(a) $CH_3CH(CH_3)-NH_2$

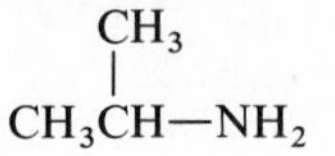

(b) CH_3-NH_2

(c) $CH_3CH_2-N(H)-CH_3$

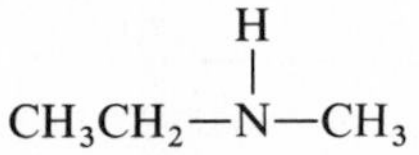

(d) $C_6H_5-NHCH_3$

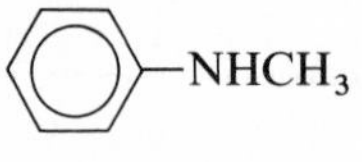

(e) $C_6H_5-NH-C_6H_5$

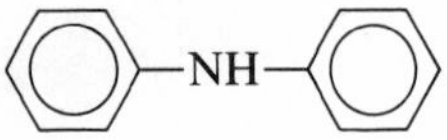

(f) piperidine (six-membered ring with N–H)

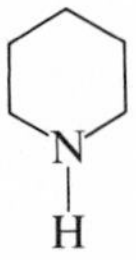

(g) N-methylpyrrolidine ($N-CH_3$ in a five-membered ring)

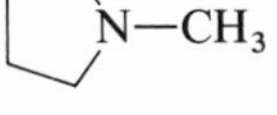

15.2 STRUCTURE OF AMINES

The unshared electron pair on the nitrogen atom prevents amines from being planar (Sec. 1.14).

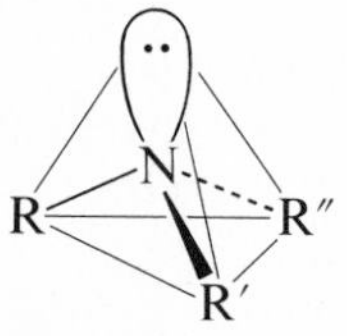

$$R-\ddot{N}(-R')-R''$$

The four orbitals of nitrogen are sp^3-hybridized, and they are therefore tetrahedrally oriented in space (actually, the R—N—R bond angles are 106–108°). In an amine in which all three groups are different, we should expect the possibility of enantiomers because such molecules are chiral (Chapter 8). Enantiomers have never been isolated, however, because the two enantiomers readily interconvert by "flipping" through a planar transition state:

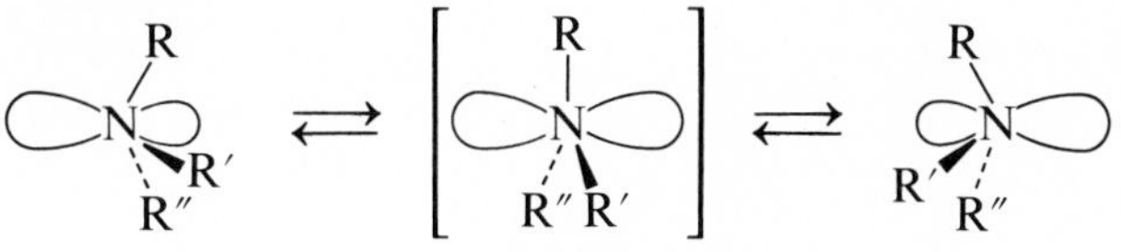

R, N, R′, R″ ⇌ [R, N, R″ R′] ⇌ R, N, R′, R″

planar
transition
state

Problem 15.2
The orbital hybridization of the nitrogen atom of amines is sp^3. What is the orbital hybridization of the nitrogen atom in the planar transition state in the above equation?

15.3 NOMENCLATURE

Generally we name amines by giving the names of the hydrocarbon groups, usually in order of increasing size, followed by the ending *-amine*. The entire name is one word. If the same hydrocarbon group occurs more than once, we use the prefixes *di-* or *tri-*. Some examples will serve as illustrations.

Primary Amines

CH_3NH_2	$CH_3CH_2NH_2$	$CH_3CH(CH_3)NH_2$	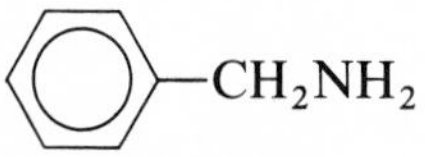 C_6H_5—CH_2NH_2
methylamine	ethylamine	isopropylamine	benzylamine

Secondary Amines

$CH_3NHCH_2CH_2CH_3$	$(CH_3CH_2)_2NH$	$(CH_3CH(CH_3)CH_2)_2NH$
N-methyl-*N*-propylamine	diethylamine	diisobutylamine

Tertiary Amines

$(CH_3)_3N$	$CH_3N(CH_2CH_2CH_3)CH_2CH_3$	$CH_2{=}CHCH_2N(CH_3)_2$
trimethylamine	*N*-methyl-*N*-ethyl-*N*-propylamine	*N*,*N*-dimethyl-*N*-allylamine

The *N*- designates that the group is bonded to nitrogen. The *N*- designation is used when it is necessary to avoid ambiguity.

When the hydrocarbon groups are too complex to name as above, we name the longest hydrocarbon chain that bears the amine group *alkanamine* and treat the other substituents as before. Some examples will illustrate this system.

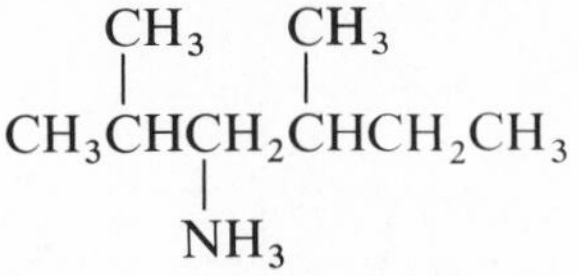

$$\begin{array}{c} CH_3 \quad\quad CH_3 \\ | \quad\quad\quad | \\ CH_3CHCH_2CHCH_2CH_3 \\ | \\ NH_3 \end{array} \qquad\qquad \begin{array}{c} CH_3 \\ | \\ ClCH_2CHCH_2NH_2 \end{array}$$

2,4-dimethyl-3-hexanamine 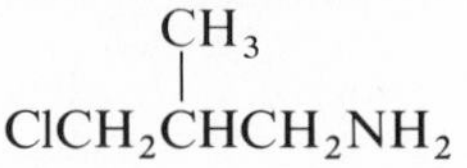   3-chloro-2-methyl-1-propanamine

Some amines have special names. Amines in which the nitrogen is attached to an aromatic ring are named as derivatives of *aniline*. Again, we use the designation *N*- to show that the substituent is attached to the nitrogen atom:

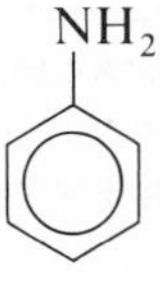

NH_2 on benzene ring	$NHCH_3$ on benzene ring	$N(CH_3)_2$ on benzene ring	NH_2 on benzene ring with CH_3 at 3 and 5
aniline	*N*-methylaniline	*N,N*-dimethylaniline	3,5-dimethylaniline

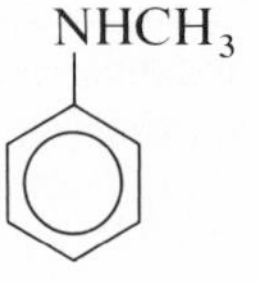

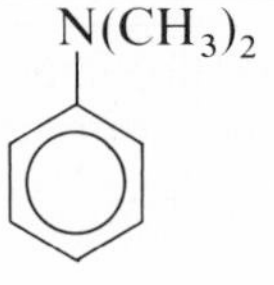

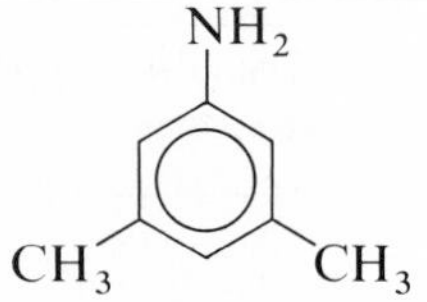

Other special names of aromatic amines will be noted when they arise.

Problem 15.3

Give the structural formulas of the following compounds.

(a) tripropylamine
(b) dipentylamine
(c) *N,N*-diethylaniline
(d) *N*-methyl-3,4-dichloroaniline

Problem 15.4

Name each of the following amines.

(a) 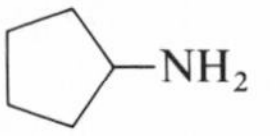 cyclopentyl–NH_2
(b) $CH_3CH_2CH_2NH_2$
(c)  benzene ring–$N(CH_2CH_2CH_3)_2$

15.4 PHYSICAL PROPERTIES OF AMINES: HYDROGEN BONDING

Amines are moderately polar but not as polar as alcohols of comparable molecular weights, and N—H · · · N hydrogen bonds are not as strong as O—H · · · O hydrogen bonds. For these reasons, amines have lower boiling points than alcohols of comparable molecular weights but higher boiling points than hydrocarbons.

Primary and secondary amine molecules can form moderately strong hydrogen bonds to each other; however, tertiary amines cannot. For this reason, primary and secondary amines usually have higher boiling points than tertiary amines of comparable molecular weights. All amines, including tertiary amines, can form hydrogen bonds with water, so they are moderately soluble in water. Table 15.1 lists the physical properties of a few typical amines.

Problem 15.5

Trimethylamine and propylamine have the same molecular weight. How do you explain the large difference in their boiling points? Is your explanation consistent with the fact that they have similar water solubilities? Explain.

Problem 15.6

Ethylamine has a higher boiling point than dimethylamine, and butylamine has a higher boiling point than diethylamine. Suggest an explanation for these differences.

Problem 15.7

Refer to Table 10.2 to find alcohols whose molecular weights correspond to those of methylamine, ethylamine, and propylamine. Compare the boiling point of each of these amines with that of the corresponding alcohol. How do you explain the difference in boiling points between amines and alcohols?

15.5 BASICITY OF AMINES

The most notable property of amines is their basicity (Table 15.1). Amines are bases because the nitrogen atom has an unshared electron pair that can accept a proton to form an ammonium ion:

$$\underset{\text{base}}{\mathrm{R'{-}\overset{\overset{\displaystyle R}{|}}{\underset{\underset{\displaystyle R''}{|}}{N}}{:}}} + \underset{\text{acid}}{\mathrm{HA}} \rightleftarrows \underset{\substack{\text{(ammonium ion)}\\ \text{conjugate acid}}}{\left[\mathrm{R'{-}\overset{\overset{\displaystyle R}{|}}{\underset{\underset{\displaystyle R''}{|}}{N}}{-}H}\right]^+} + \underset{\text{conjugate base}}{\mathrm{A^-}}$$

Water solutions of amines are basic because the amine molecule can accept a proton from a water molecule to produce a hydroxide ion:

$$\mathrm{R'{-}\overset{\overset{\displaystyle R}{|}}{\underset{\underset{\displaystyle R''}{|}}{N}}{:}} + \mathrm{H_2O} \rightleftarrows \left[\mathrm{R'{-}\overset{\overset{\displaystyle R}{|}}{\underset{\underset{\displaystyle R''}{|}}{N}}{-}H}\right]^+ + \mathrm{OH^-}$$

With most amines the equilibrium favors reactants; values of the base dissociation constant, K_b (Sec. 3.3), are less than 1 (Table 15.1).

Table 15.1 Physical Properties of Some Common Amines

Name	Formula	Boiling Point, °C	Water solubility,* g/100 ml H_2O	K_b, 25°C
Ammonia	NH_3	−33	52	1.8×10^{-5}
Primary Amines				
Methylamine	CH_3NH_2	−7	vs	4.4×10^{-4}
Ethylamine	$CH_3CH_2NH_2$	17	vs	4.7×10^{-4}
Propylamine	$CH_3CH_2CH_2NH_2$	49	vs	3.8×10^{-4}
Butylamine	$CH_3CH_2CH_2CH_2NH_2$	78	vs	4.1×10^{-4}
Aniline	C_6H_5—NH_2	184	3.4 (20°C)	3.8×10^{-10}
Secondary Amines				
Dimethylamine	$(CH_3)_2NH$	7	vs	5.0×10^{-4}
Diethylamine	$(CH_3CH_2)_2NH$	56	vs	9.6×10^{-4}
Dipropylamine	$(CH_3CH_2CH_2)_2NH$	111	s	9.5×10^{-4}
N-Methylaniline	C_6H_5—$NHCH_3$	196	vss	7.1×10^{-10}
Tertiary Amines				
Trimethylamine	$(CH_3)_3N$	3.5	vs	0.63×10^{-5}
Triethylamine	$(CH_3CH_2)_3N$	90	1.5 (20°C)	5.2×10^{-4}
Tripropylamine	$(CH_3CH_2CH_2)_3N$	156	vss	4.4×10^{-4}
N,N-Dimethylaniline	C_6H_5—$N(CH_3)_2$	194	vss	12×10^{-10}

* vs = very soluble, s = soluble, ss = slightly soluble, vss = very slightly soluble.

Problem 15.8
Refer to Table 15.1 and select the stronger base in each of the following pairs.
(a) ethylamine and diethylamine
(b) methylamine and aniline
(c) *N*-methylaniline and *N,N*-dimethylaniline

Problem 15.9
Write the equation for the equilibrium that exists when diethylamine dissolves in water.

Problem 15.10
Write the mathematical expression for the K_b of diethylamine.

15.6 EFFECTS OF STRUCTURE ON BASICITY

The basicities of the amines in Table 15.1 show two important trends:

1. Alkylamines are more basic than ammonia.
2. Aromatic amines are less basic than ammonia.

The basicity of an amine depends on the stability of its conjugate acid, the ammonium ion, compared with the stability of the amine. Any factor that stabilizes the ammonium ion results in greater basicity; any factor that stabilizes the amine (that is, its unshared electron pair) results in lower basicity.

Alkylamines are more basic than ammonia because the electron-releasing inductive effect (Sec. 13.4) of alkyl groups stabilizes the positive charge of the ammonium ion.

$$NH_3 + H_2O \rightleftarrows NH_4^+ + OH^-$$

$$R\text{—}NH_2 + H_2O \rightleftarrows R\text{—}NH_3^+ + OH^-$$

The inductive effect of R on the amine is less important than that on the ammonium ion, but probably destabilizes the electron pair on nitrogen by increasing the electron density at nitrogen. The overall effect of R is to stabilize the ammonium ion relative to the amine.

The aromatic ring has a significant effect on the unshared electron pair of an amine. We can draw several resonance structures of aniline in which the unshared electron pair is delocalized around the ring:

$$C_6H_5NH_2 \longleftrightarrow C_6H_5NH_2 \longleftrightarrow {}^-C_6H_5{=}NH_2^+ \longleftrightarrow {}^-C_6H_5{=}NH_2^+ \longleftrightarrow {}^-C_6H_5{=}NH_2^+$$

The structure of aniline is therefore a resonance hybrid of all these structures.

We cannot draw corresponding structures for the anilinium ion (that is, the ammonium ion of aniline).

$$C_6H_5NH_3^+ \longleftrightarrow C_6H_5NH_3^+$$

In summary, the aromatic ring stabilizes the amine relative to the ammonium ion. The result of this is that aromatic amines are less basic than ammonia.

Problem 15.11

Choose the stronger base in each of the following pairs. Reason by analogy to the amines listed in Table 15.1.

(a) isopropylamine and aniline
(b) isopropylamine and ammonia
(c) aniline and diphenylamine (If in doubt, draw resonance structures.)

Problem 15.12

p-Chloroaniline has $K_b = 1.2 \times 10^{-10}$; *p*-toluidine $\left(CH_3-C_6H_4-NH_2\right)$ has $K_b = 1.2 \times 10^{-9}$. Explain why *p*-chloroaniline is a weaker base than aniline, whereas *p*-toluidine is a stronger base than aniline.

15.7 PREPARATION OF AMINES

We shall examine two general methods of preparing amines—alkylation of ammonia, and reduction of nitro compounds and nitriles.

Alkylation of Ammonia

Ammonia reacts with alkyl halides by nucleophilic substitution (an S_N2 reaction; Sec. 9.6) to give alkylammonium halides.

$$\ddot{N}H_3 + R{-}Cl \longrightarrow R{-}\overset{+}{N}H_3Cl^-$$

The ammonium salts can be converted to the amine by adding a strong base. (Recall that an ammonium ion is an acid because it is a proton donor.)

$$R{-}\overset{+}{N}H_3 + OH^- \rightleftarrows R{-}\ddot{N}H_2 + H_2O$$

Because alkylamines and ammonia have similar base strengths, ammonia can act as a base in the second reaction above to produce the alkylamine:

$$R{-}\overset{+}{N}H_3 + :NH_3 \rightleftarrows R{-}NH_2 + NH_4^+$$

One disadvantage of preparing alkylamines by the alkylation of ammonia is that the alkylamine produced is capable of reacting further with alkyl halide to produce secondary and tertiary amines:

$$R{-}NH_2 + R{-}Cl \longrightarrow R_2\overset{+}{N}H_2\,Cl^-$$

$$R_2\overset{+}{N}H_2\,Cl^- + NH_3 \rightleftarrows R_2NH + NH_4^+Cl^-$$

$$R_2NH + R{-}Cl \longrightarrow R_3\overset{+}{N}H\,Cl^-$$

$$R_3\overset{+}{N}H\,Cl^- + NH_3 \rightleftarrows R_3N + NH_4^+\,Cl^-$$

In industrial processes, the primary, secondary, and tertiary amines are separated by fractional distillation. This reaction does not occur with aryl halides except at high pressure and high temperature because aryl halides are unreactive and not susceptible to nucleophilic substitution reactions (Sec. 9.8).

Quaternary ammonium salts. Tertiary amines can react further with alkyl halides to produce **quaternary ammonium salts**:

$$R_3N\!: + R{-}Cl \longrightarrow R{-}\overset{\displaystyle R}{\underset{\displaystyle R}{\overset{|}{\underset{|}{N^+}}}}{-}R \quad Cl^-$$

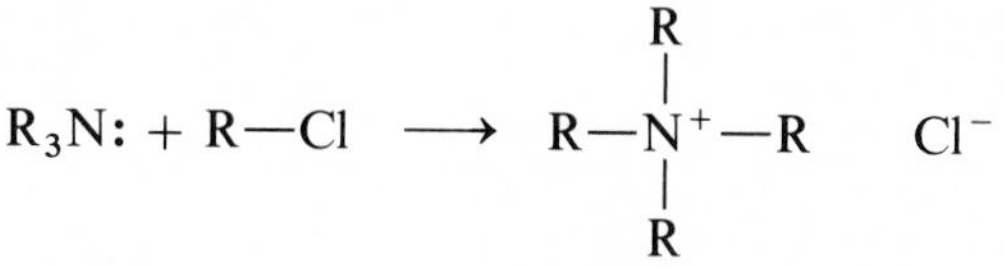

quaternary ammonium chloride

The quaternary ammonium ion has no proton to donate, so it is not acidic. Therefore, unlike other ammonium ions, it does not exist in equilibrium with a conjugate base.

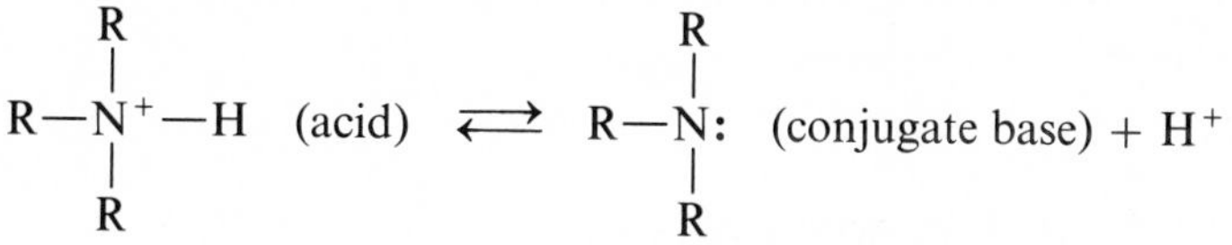

$$R{-}\overset{\displaystyle R}{\underset{\displaystyle R}{\overset{|}{\underset{|}{N^+}}}}{-}H \text{ (acid)} \rightleftarrows R{-}\overset{\displaystyle R}{\underset{\displaystyle R}{\overset{|}{\underset{|}{N}}}}\!: \text{ (conjugate base)} + H^+$$

tertiary ammonium ion

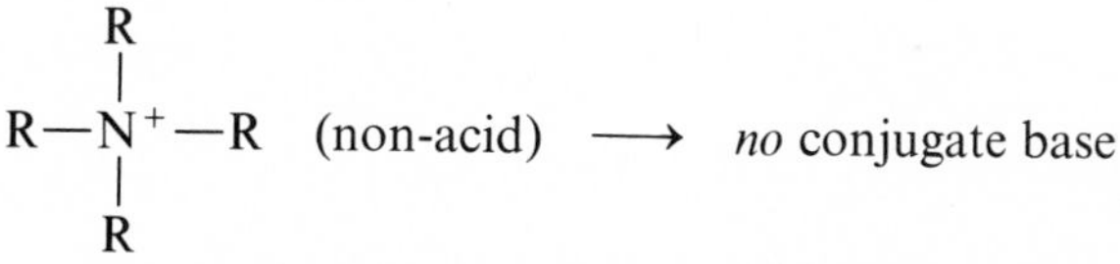

$$R{-}\overset{\displaystyle R}{\underset{\displaystyle R}{\overset{|}{\underset{|}{N^+}}}}{-}R \text{ (non-acid)} \longrightarrow \textit{no} \text{ conjugate base}$$

quaternary ammonium ion

For this reason, quaternary ammonium hydroxides are very strong bases—comparable to NaOH and KOH. Other ammonium hydroxides exist in equilibrium with their amines:

$$R_3\overset{+}{N}H + OH^- \rightleftarrows R_3N\!: + H_2O$$

Quaternary ammonium hydroxides are not capable of reaching this equilibrium and therefore exist completely in the ionic hydroxide form.

Choline is an important quaternary ammonium salt.

$$CH_3{-}\overset{\displaystyle CH_3}{\underset{\displaystyle CH_3}{\overset{|}{\underset{|}{N^+}}}}{-}CH_2CH_2OH \quad OH^-$$

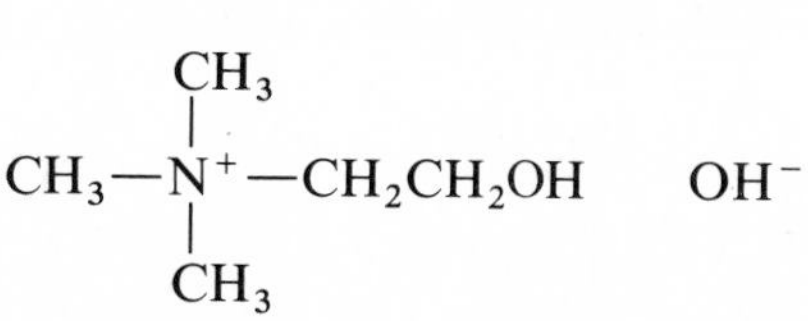

choline

It is involved in carbohydrate and protein metabolism and in reactions of fats. Acetylcholine, the acetic ester of choline, is involved in the transmission of nerve impulses.

$$CH_3-\overset{\overset{CH_3}{|}}{\underset{\underset{CH_3}{|}}{N^+}}-CH_2CH_2-O-\overset{\overset{O}{\|}}{C}-CH_3 \qquad OH^-$$

acetylcholine

Reduction of Nitro Compounds and Nitriles

Various unsaturated nitrogen compounds can be reduced to amines. The reductions are usually carried out by direct hydrogenation using a metal catalyst, or by a chemical reducing agent such as lithium aluminum hydride.

Nitriles can be reduced by several methods:

$$R-C\equiv N + H_2 \xrightarrow{Ni} RCH_2NH_2$$

$$R-C\equiv N + LiAlH_4 \longrightarrow RCH_2NH_2$$

Nitriles are prepared by an S_N2 reaction using an alkyl halide and NaCN (Sec. 9.4). Reduction of nitriles proves a convenient method of converting a primary alkyl halide into an amine that has one additional carbon atom:

$$RBr + NaCN \longrightarrow R-CN + NaBr$$

$$R-CN + H_2 \xrightarrow{Ni} RCH_2NH_2$$

Nitro compounds are also easily reduced:

$$R-NO_2 + H_2 \xrightarrow{Ni} R-NH_2 + H_2O$$

$$R-NO_2 + Fe \xrightarrow[-Fe^{3+}]{HCl\,(aq)} R-NH_3{}^+Cl^- \xrightarrow{OH^-} R-NH_2 + H_2O + Cl^-$$

This method is especially useful in the preparation of aromatic amines. Aniline, for example, is manufactured by nitrating benzene and reducing the nitrobenzene:

$$C_6H_6 + HNO_3 \xrightarrow{H_2SO_4} C_6H_5NO_2 + H_2O$$

$$C_6H_5NO_2 + H_2 \xrightarrow{Ni} C_6H_5NH_2$$

Problem 15.13

Write equations to show all the steps necessary to convert each of the following reactants into the amine shown. If more than one method is available, indicate which is best.

	Reactant	Product
(a)	methane	methylamine
(b)	benzene	aniline
(c)	methylamine	trimethylamine
(d)	1-chlorobutane	pentylamine
(e)	methane	tetramethylammonium bromide

15.8 REACTIONS OF AMINES

The reactions discussed in this section are typical of amines.

Salt Formation
Acylation
Sulfonation
Reaction with Nitrous Acid

Salt Formation

We have already seen that amines are basic. They react with aqueous mineral acids (HCl, H_2SO_4, and so forth) to form ammonium salts. Ammonium salts are water-soluble, so we can use this reaction as a test for amines that are insoluble in water:

$$\underset{\text{insoluble in water}}{R'\!-\!\underset{\substack{|\\R''}}{\overset{\substack{R\\|}}{N}}\!:} + H_3O^+X^- \longrightarrow \underset{\text{water-soluble}}{R'\!-\!\underset{\substack{|\\R''}}{\overset{\substack{R\\|}}{N^+}}\!-\!H} + X^- + H_2O$$

(R′ and R″ = alkyl, aryl, or H)

We can regenerate the original amine easily by adding a strong base, such as NaOH:

$$R'\!-\!\underset{\substack{|\\R''}}{\overset{\substack{R\\|}}{N^+}}\!-\!H + X^- + NaOH \longrightarrow R'\underset{\substack{|\\R''}}{\overset{\substack{R\\|}}{N}}\!: + NaX + H_2O$$

Acylation

Primary and secondary amines react with acyl halides or carboxylic anhydrides (Chapter 13) to form amides:

$$R{-}NH_2 + Cl{-}\overset{\overset{\displaystyle O}{\|}}{C}{-}R' \xrightarrow{\text{base}} R{-}\underset{\underset{\displaystyle H}{|}}{\ddot{N}}{-}\overset{\overset{\displaystyle O}{\|}}{C}{-}R' + Cl^- + H{-}base^+$$

(The base may be excess amine or a different base such as NaOH.)

$$R{-}NH_2 + R'{-}\overset{\overset{\displaystyle O}{\|}}{C}{-}O{-}\overset{\overset{\displaystyle O}{\|}}{C}{-}R' \longrightarrow R{-}\underset{\underset{\displaystyle H}{|}}{\ddot{N}}{-}\overset{\overset{\displaystyle O}{\|}}{C}{-}R' + R'{-}\overset{\overset{\displaystyle O}{\|}}{C}{-}OH$$

Problem 15.14

Write equations for the reaction of dimethylamine with $CH_3\overset{\overset{\displaystyle O}{\|}}{C}{-}Cl$ and with $CH_3\overset{\overset{\displaystyle O}{\|}}{C}{-}O{-}CH_3$.

Sulfonation: Hinsberg Separation

Amines also react with the acid chlorides of sulfonic acids. Benzenesulfonyl chloride is such an acid chloride. It is prepared by reacting the sulfonic acid with thionyl chloride, $SOCl_2$:

$$C_6H_5{-}\overset{\overset{\displaystyle O}{\|}}{\underset{\underset{\displaystyle O}{\|}}{S}}{-}OH + SOCl_2 \longrightarrow C_6H_5{-}\overset{\overset{\displaystyle O}{\|}}{\underset{\underset{\displaystyle O}{\|}}{S}}{-}Cl$$

benzenesulfonic acid   benzenesulfonyl chloride

Amines react with benzenesulfonyl chloride to form benzenesulfonamides according to the equation

$$R{-}NH_2 + C_6H_5{-}SO_2Cl \xrightarrow{\text{base}} R{-}\underset{\underset{\displaystyle H}{|}}{\ddot{N}}{-}SO_2{-}C_6H_5 + Cl^- + H\text{-}base^+$$

a benzenesulfonamide

This reaction is very useful for the separation of primary, secondary, and tertiary amines. The separation method, called the **Hinsberg separation**, depends on the differences in solubility of the products of the reaction. Primary amines yield a sulfonamide that has a hydrogen attached to nitrogen. The $—SO_2—$ group makes this hydrogen quite acidic, and the amide is thus soluble in aqueous NaOH solution, in which it forms the sulfonamide salt. Secondary amines yield amides that have no N—H. These amides are not acidic and do not dissolve in aqueous NaOH solution. Tertiary amines do not react with benzenesulfonyl chloride. The following equations summarize the reactions of the three classes of amines:

$$\underset{\text{1° amine}}{R—NH_2} + C_6H_5—SO_2Cl \longrightarrow \underset{\text{(acidic)}}{R—\ddot{N}H—SO_2—C_6H_5}$$

$$\xrightarrow[H_2O]{NaOH} \underset{\text{(water-soluble sulfonamide salt)}}{R—\ddot{N}^-—SO_2—C_6H_5 \; Na^+}$$

$$\underset{\text{2° amine}}{R_2NH} + C_6H_5—SO_2Cl \longrightarrow \underset{\text{(neutral)}}{R_2\ddot{N}—SO_2—C_6H_5} \xrightarrow[H_2O]{NaOH} \underset{\text{(amide is insoluble)}}{\text{no reaction}}$$

$$R_2\ddot{N}—SO_2—C_6H_5 \xrightarrow{H_3O^+} \underset{\text{(amide is insoluble)}}{\text{no reaction}}$$

$$\underset{\text{3° amine}}{R_3N} + C_6H_5—SO_2Cl \longrightarrow \underset{\text{(basic)}}{\text{no reaction}} \xrightarrow[H_2O]{NaOH} \underset{\text{(tertiary amine is insoluble)}}{\text{no reaction}}$$

$$\xrightarrow{H_3O^+} \underset{\text{(tertiary amine salt is soluble)}}{R_3NH^+}$$

In the laboratory we carry out the Hinsberg separation by shaking the amine mixture with a mixture of benzenesulfonyl chloride and aqueous NaOH. Primary amines end up in the aqueous layer as the sulfonamide salt. After separating the aqueous layer, we obtain the solid sulfonamide by acidifying the solution. Secondary amines end up as a water-insoluble sulfonamide that is also insoluble in acid. Tertiary amines also appear as an insoluble compound but differ from secondary sulfonamides in being acid-soluble.

Problem 15.15

Write equations showing how to separate a mixture of methylamine, dimethylamine, and trimethylamine by the Hinsberg separation method.

Reaction With Nitrous Acid

Nitrous acid ($HO—\ddot{N}=O$) is stable only at low temperatures. It is usually prepared in the presence of the amine by adding a solution of sodium nitrite to a solution of the amine in a mineral acid such as HCl or H_2SO_4. Nitrous acid is produced when sodium nitrite reacts with acid:

$$[Na^+NO_2^-] + [H_3O^+Cl^-] \xrightarrow{0\text{–}5^\circ} HO—\ddot{N}=O + Na^+Cl^- + H_2O$$

Amines react with nitrous acid to give a variety of products, depending on whether the amine is primary, secondary, or tertiary.

Primary amines react with nitrous acid to liberate N_2 gas:

$$R—NH_2 + HO—N=O \xrightarrow{H^+} N_2\uparrow + H_2O + R—OH + \text{alkenes} + \text{other products}$$

The reaction proceeds through an alkanediazonium ion intermediate.

$$R—NH_2 + HO—N=O + HCl \longrightarrow R—\overset{+}{N}\equiv N\colon + Cl^- + H_2O$$

an alkanediazonium chloride

$$R—\overset{+}{N}\equiv N\colon \longrightarrow R^+ + N_2\uparrow$$

The carbocation, R^+, can react in various ways to give a variety of products as shown in the following scheme:

$$R—\overset{+}{O}H_2 \xleftarrow{H_2O} R^+ \xrightarrow{X^-} R—X \text{ (an alkyl halide)}$$

$$R—\overset{+}{O}H_2 \xrightarrow{-H^+} R—OH \text{ (an alcohol)}$$

$$R^+ \xrightarrow{-H^+} \text{an alkene}$$

Primary **alkanediazonium salts** ($R—\overset{+}{N}\equiv N\ X^-$) are so unstable that they produce nitrogen gas immediately, even if the reaction mixture is maintained at 0°C. Primary **arenediazonium salts** are more stable, however, and may be kept for relatively long periods of time at 0–5°C. We shall examine the reactions of these arenediazonium salts in Section 15.9.

Secondary amines react with nitrous acid to produce *N*-nitrosoamines.

$$R—\underset{H}{\overset{R'}{N}}\colon + HO—\ddot{N}=O \xrightarrow{H^+} R—\underset{\cdot\cdot}{\overset{R'}{N}}—\ddot{N}=O + H_2O$$

an *N*-nitrosoamine

These compounds are relatively stable and usually separate from their reaction mixtures as yellow, oily liquids. Nitrosoamines are well-known carcinogens.

Tertiary amines dissolve in the acid solution to produce ammonium salts and *N*-nitrosoammonium salts. *N*-Nitrosoammonium salts of tertiary alkylamines usually decompose rapidly to give complex mixtures, and therefore they do not lead to synthetically useful products.

$$R_3N\!: + HO{-}N{=}O + HX \rightleftarrows \underset{\text{ammonium salt}}{R_3\overset{+}{N}HX^-} + \underset{N\text{-nitrosoammonium salt}}{R_3\overset{+}{N}{-}N{=}O\,X^-}$$

Tertiary arylamines, however, form *N*-nitrosoammonium salts that rearrange to give a nitrosoamine with the nitroso group in the para position if that position is available. The following reaction sequence shows what happens:

$$C_6H_5{-}NR_2 + HO{-}N{=}O \xrightarrow{HX} C_6H_5{-}\overset{+}{N}R_2{-}N{=}O \quad X^-$$

N-nitrosoammonium salt

$$\longrightarrow O{=}N{-}C_6H_4{-}\overset{+}{N}R_2{-}H \quad X^-$$

$$\xrightarrow{\text{base}} O{=}N{-}C_6H_4{-}NR_2$$

p-nitroso-*N*,*N*-dialkylaniline

The overall equation is

$$C_6H_5{-}NR_2 + HO{-}N{=}O \xrightarrow[(2)\ OH^-]{(1)\ HX} O{=}N{-}C_6H_4{-}NR_2 + H_2O$$

Problem 15.16

Give the products of the reaction of each of the following amines with nitrous acid at 0–5°C.

(a) $CH_3CH_2NH_2$ (b) $(CH_3CH_2)_2NH$ (c) $(CH_3CH_2)_3N$

(d) $C_6H_5{-}NH_2$ (e) $C_6H_5{-}NHCH_3$ (f) $C_6H_5{-}N(CH_3)_2$

15.9 DIAZONIUM SALTS: STABILITY

The reaction of nitrous acid with primary arylamines is so important in synthesis that we shall examine it separately.

Primary arylamines react with nitrous acid to produce arenediazonium salts that are stable at 0–5°C. The process is called **diazotization**.

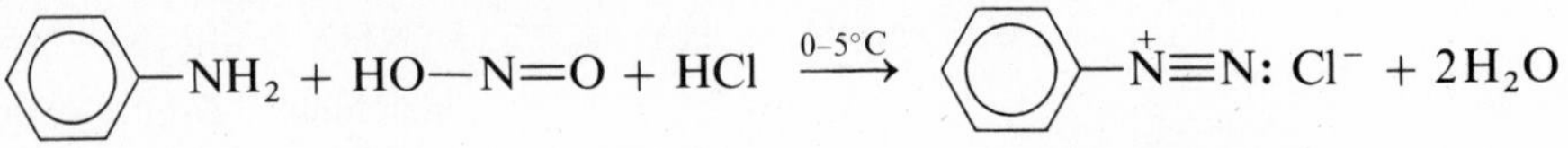

$$C_6H_5{-}NH_2 + HO{-}N{=}O + HCl \xrightarrow{0\text{–}5^\circ C} C_6H_5{-}\overset{+}{N}{\equiv}N{:}\ Cl^- + 2H_2O$$

benzenediazonium chloride

Arenediazonium ions are more stable than alkanediazonium ions due to the resonance structures III, IV, and V (see below), which involve electron delocalization over the ring. Alkanediazonium ions do not have resonance structures that correspond to structures III, IV, and V.

Arenediazonium ion

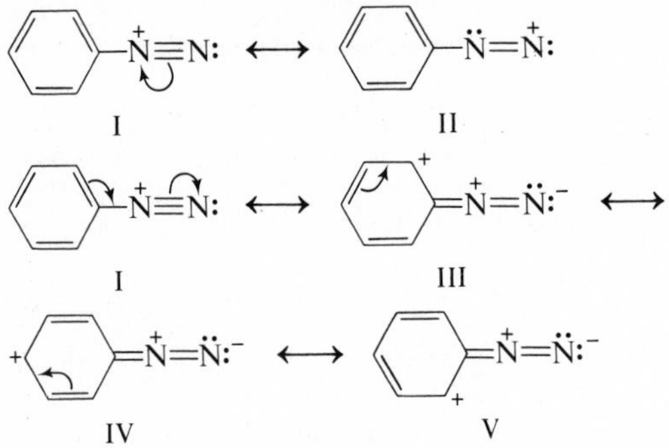

$$\underset{\text{I}}{C_6H_5{-}\overset{+}{N}{\equiv}N{:}} \longleftrightarrow \underset{\text{II}}{C_6H_5{-}\ddot{N}{=}\overset{+}{N}{:}}$$

$$\underset{\text{I}}{C_6H_5{-}\overset{+}{N}{\equiv}N{:}} \longleftrightarrow \underset{\text{III}}{\overset{+}{C_6H_5}{=}\overset{+}{N}{=}\ddot{N}{:}^-} \longleftrightarrow$$

$$\underset{\text{IV}}{{}^{+}C_6H_5{=}\overset{+}{N}{=}\ddot{N}{:}^-} \longleftrightarrow \underset{\text{V}}{C_6H_5^{+}{=}\overset{+}{N}{=}\ddot{N}{:}^-}$$

Alkanediazonium ion

$$\underset{\text{I}}{CH_3CH_2{-}\overset{+}{N}{\equiv}N{:}} \longleftrightarrow \underset{\text{II}}{CH_3CH_2{-}\ddot{N}{=}\overset{+}{N}{:}}$$

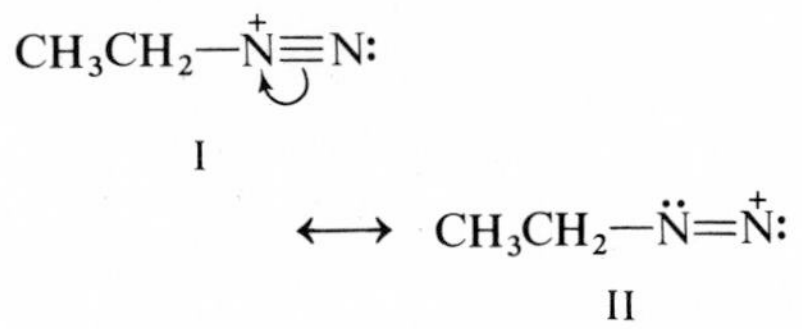

15.10 DIAZONIUM SALTS: SUBSTITUTION REACTIONS

Arenediazonium salts undergo a variety of substitution reactions in which the N_2 is replaced. Normally these reactions are carried out by adding the reagent to the cool diazotization mixture and then heating the mixture. The intermediate diazonium salts are not isolated because on warming they tend to explode. Some of the more useful substitution reactions are summarized in Figure 15.1.

Problem 15.17

Write equations showing how to convert 2,4,6-tribromoaniline into each of the following compounds. (a) 1,3,5-tribromobenzene, (b) 2,4,6-tribromophenol, (c) 2,4,6-tribromo-1-fluorobenzene, (d) 1,2,4,6-tetrabromobenzene, (e) 2,4,6-tribromo-1-chlorobenzene, (f) 2,4,6-tribromo-1-iodobenzene, (g) 2,4,6-tribromobenzoic acid. (Recall that nitriles can be hydrolyzed to carboxylic acids; Sec. 13.5.)

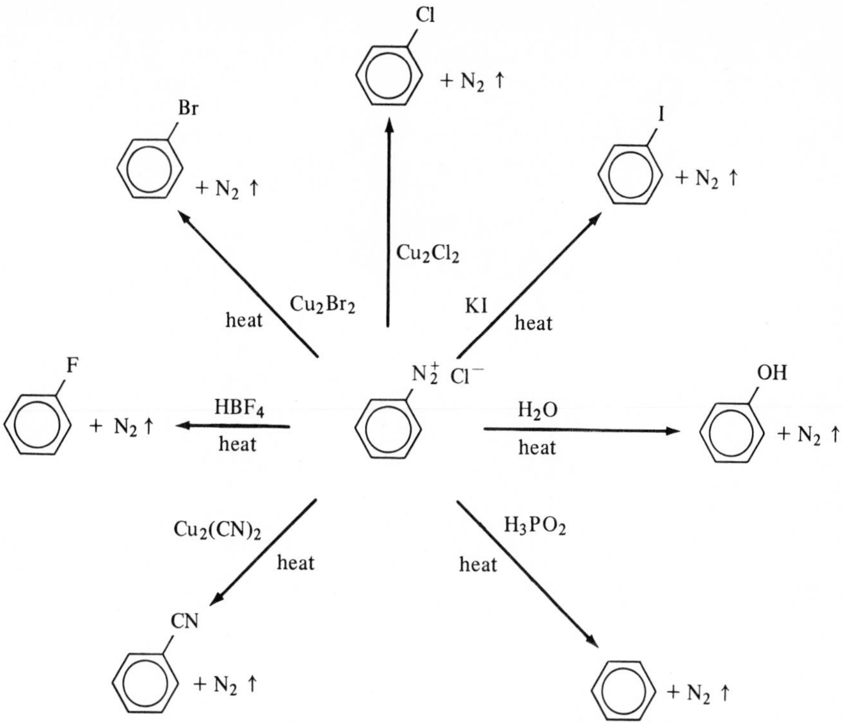

Figure 15.1. Substitution reactions of arenediazonium salts.

15.11 DIAZONIUM SALTS: COUPLING REACTIONS

Because they are positively charged, arenediazonium ions can act as electrophiles. They do so when they react with highly activated ring systems such as anilines or phenols. The reaction, called **coupling**, follows the mechanism described in Sections 7.9 and 7.10. Coupling gives **azo compounds**—compounds that contain the azo ($-\ddot{N}=\ddot{N}-$) functional group.

$$C_6H_5-\overset{+}{N}\equiv N: + C_6H_5-\ddot{N}H_2 \longrightarrow C_6H_5-\ddot{N}=\ddot{N}-C_6H_5(H)=\overset{+}{N}H_2 \xrightarrow{\text{base}} C_6H_5-\ddot{N}=\ddot{N}-C_6H_4-\ddot{N}H_2 + \text{H base}^+$$

p-aminoazobenzene (an azo compound)

$$C_6H_5-\overset{+}{N}\equiv N: + C_6H_5-OH \longrightarrow C_6H_5-\ddot{N}=\ddot{N}-C_6H_4-OH$$

p-hydroxyazobenzene

The general reaction is

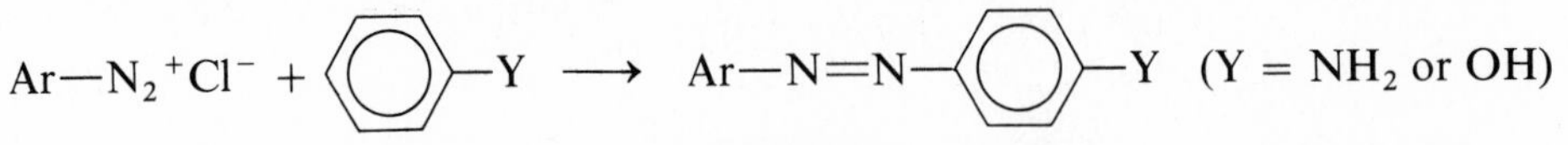

$$Ar{-}N_2{}^+Cl^- + \text{(ring)}{-}Y \longrightarrow Ar{-}N{=}N{-}\text{(ring)}{-}Y \quad (Y = NH_2 \text{ or } OH)$$

an azo compound

Problem 15.18
Outline a mechanism for the diazo coupling reaction of benzenediazonium ion with phenol shown above .

Their conjugated double bond systems make azo compounds colored. Many of them are used as dyes. Some important azo dyes are methyl orange, chrysamine G, and congo red. Their syntheses are outlined below.

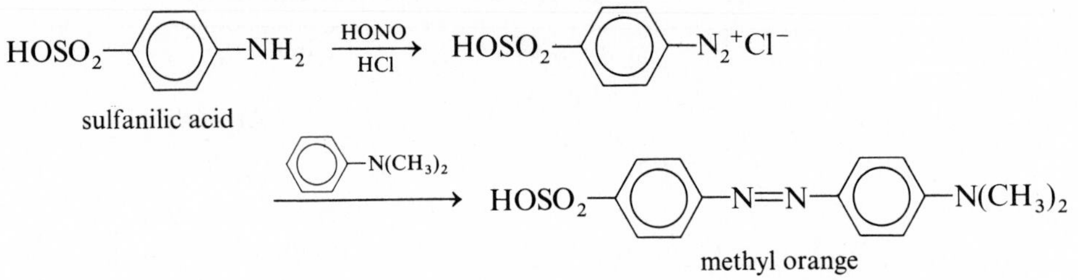

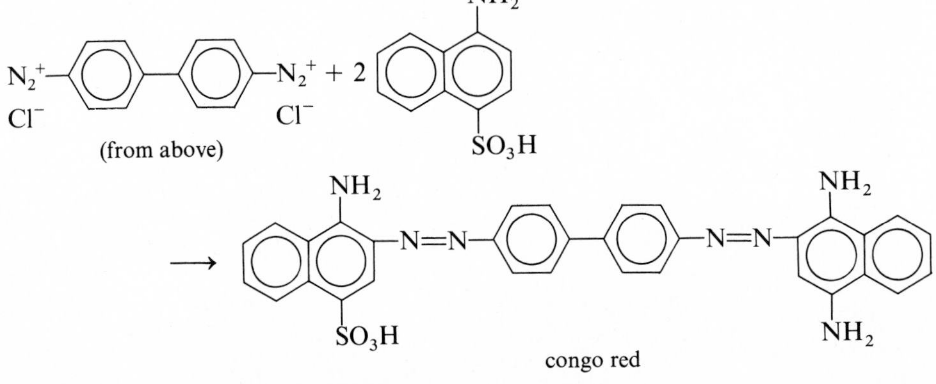

$HOSO_2$—NH$_2$ $\xrightarrow[HCl]{HONO}$ $HOSO_2$—$N_2{}^+Cl^-$

sulfanilic acid

—$N(CH_3)_2$ ⟶ $HOSO_2$—N=N—$N(CH_3)_2$

methyl orange

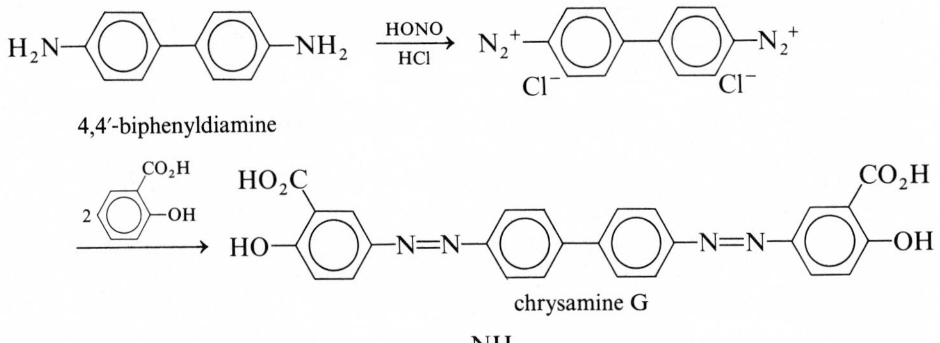

H_2N—NH$_2$ $\xrightarrow[HCl]{HONO}$ $N_2{}^+$—$N_2{}^+$ Cl^- Cl^-

4,4′-biphenyldiamine

2 (CO_2H, OH) ⟶ HO_2C, HO—N=N—N=N—OH, CO_2H

chrysamine G

$N_2{}^+$ Cl^-—$N_2{}^+$ Cl^- + 2 (NH_2, SO_3H)

(from above)

⟶ NH_2, SO_3H —N=N—N=N— NH_2, NH_2

congo red

All three of these dyes are useful because they have reactive functional groups through which they can become fixed to the fabric. As a result they bond strongly, and do not fade easily when the fabric is washed.

Problem 15.19

Write equations to describe the synthesis of the azo compound orange II:

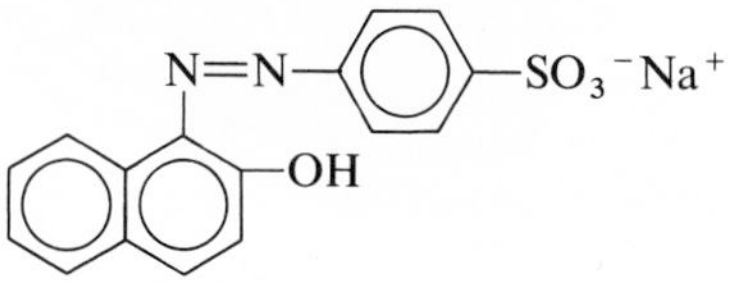

N=N—C_6H_4—$SO_3^-Na^+$ (on 2-naphthol, with —OH)

15.12 HETEROCYCLIC AMINES

Cyclic compounds that have atoms other than carbon as part of the ring are called **heterocyclic compounds**. We have already encountered a few heterocyclic compounds —furan, tetrahydrofuran, dioxane and ethylene oxide (Chapter 11). In these compounds the hetero atom is oxygen.

If a cyclic amine obeys Hückel's rule (Sec. 7.5)—that is, if it is conjugated around the ring, planar, and has $4\pi + 2$ pi electrons—it is aromatic.

Nonaromatic heterocyclic amines resemble ordinary alkylamines in their basicities (Table 15.2). Aromatic heterocyclic amines, on the other hand, are quite different. The nonaromatic amines pyrrolidine and piperidine, for example, have K_bs of 1.3×10^{-3} and 1.6×10^{-3}, respectively; diethylamine has $K_b = 0.6 \times 10^{-4}$ (Table 15.1). The remaining four amines in Table 15.2 are all aromatic and are much weaker bases.

We explain the lower basicity of aromatic amines in terms of the availability of the electron pair on nitrogen. In aromatic amines, the electron pair on nitrogen occupies an sp^2 orbital. In a nonaromatic amine, the electron pair occupies an sp^3 orbital. An sp^2 orbital has more *s* character and is therefore closer to the nucleus than an sp^3 orbital. Electrons in an sp^2 orbital are therefore held more tightly than sp^3 electrons and are not as readily available for bonding to a hydrogen ion as sp^3 electrons.

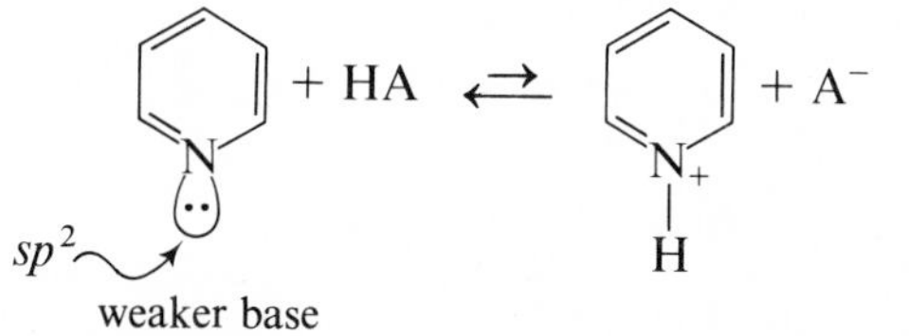

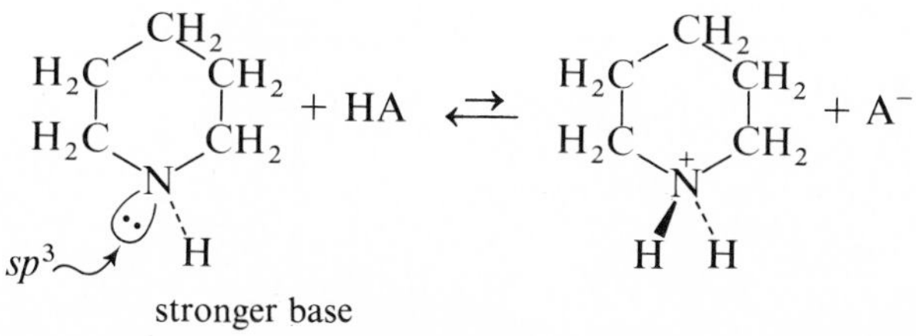

$$\text{pyridine} + HA \rightleftarrows \text{pyridinium (N}^+\text{—H)} + A^-$$

sp^2 (lone pair on N)

weaker base

$$\text{piperidine (}H_2C, CH_2, N\text{—H)} + HA \rightleftarrows \text{piperidinium (}N^+H_2\text{)} + A^-$$

sp^3 (lone pair on N)

stronger base

Table 15.2 Properties of Some Heterocyclic Amines

Name	Formula	Boiling Point, °C	Water Solubility*	K_b, 25°C
Pyrrolidine		88.5	∞	1.3×10^{-3}
Piperidine		106	∞	1.6×10^{-3}
Pyrrole		131	I	2.5×10^{-14}
Pyridine		115	∞	1.7×10^{-9}
Imidazole		256	vs	1.6×10^{-7}
Pyrimidine		124	s	5×10^{-12}

* ∞ = soluble in all proportions, s = soluble, I = insoluble, vs = very soluble.

Perhaps as interesting is the large difference in the K_bs of pyridine and pyrrole. Both are aromatic amines. Pyridine is a much stronger base ($K_b = 1.7 \times 10^{-9}$) than pyrrole ($K_b = 2.5 \times 10^{-14}$). When pyridine reacts with a hydrogen ion its aromatic system is not involved; that is, its six π electrons remain intact. Both pyridine and its conjugate acid obey Hückel's rule, and both are aromatic:

$$+ \text{HA} \rightleftarrows \quad + \text{A}^-$$

aromatic aromatic

In pyrrole the electrons needed to react with a hydrogen ion are part of the six π electron system:

$$\text{pyrrole (N–H)} + HA \rightleftarrows \text{pyrrolium } (NH_2^+) + A^-$$

aromatic   nonaromatic

The conjugate acid is not aromatic because it does not have six $(4n + 2)$ π electrons. Therefore, when a pyrrole molecule reacts with a hydrogen ion, it loses its resonance energy.

Problem 15.20

Which amine in each of the following pairs is the stonger base?

(a) N-methylpyrrolidine (N–CH_3) and pyrrolidine (N–H)   (b) N-methylpyrrole (N–CH_3) and N-methylpyrrolidine (N–CH_3)

Problem 15.21

When imidazole reacts with an acid, which nitrogen atom accepts the hydrogen ion? [Explain.]

15.13 HETEROCYCLIC AMINES IN NATURE: AN ELECTIVE TOPIC

elective topic

All the ring systems listed in Table 15.2 occur widely in both plants and animals. Many enzymes contain heterocyclic amines as part of their molecular structures. The amino acid histidine contains an imidazole ring. Nicotine contains a pyridine

histidine (imidazole ring, N–H, bearing CH_2CHCO_2H with NH_2)   nicotine (pyridine ring joined to pyrrolidine ring, N–CH_3)   nicotinic acid (pyridine ring bearing CO_2H)

histidine   nicotine   nicotinic acid

ring and a pyrrolidine ring. Nicotine occurs in tobacco and is toxic. It is used as an insecticide. Oxidation of nicotine yields nicotinic acid, which is a vitamin. It is part of the coenzyme NAD (nicotinamide adenine dinucleotide), an important oxidizing agent in biological systems. (It should be noted that nicotinic acid does not arise from oxidation of nicotine in the body.)

Two complex heterocyclic compounds that are required for normal body function but cannot be synthesized in the body are folic acid and riboflavin. These compounds must be ingested as preformed vitamins in foods.

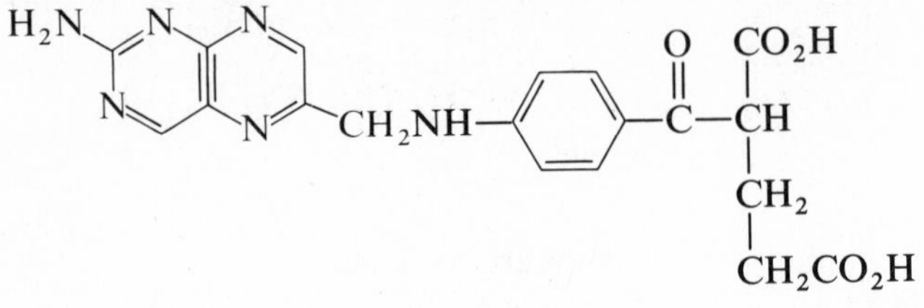

folic acid

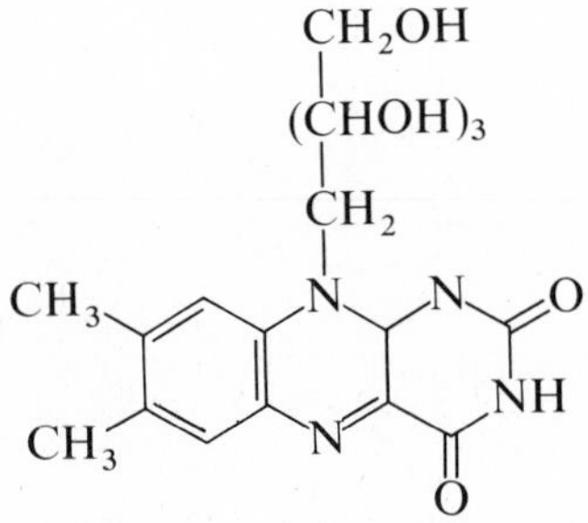

riboflavin

15.14 ALKALOIDS: AN ELECTIVE TOPIC

elective topic

Naturally occurring amino compounds can be found in various parts of many plants—leaves, bark, roots, and fruits. They are called **alkaloids** on account of their alkalinity. The alkaloids are well known and all exhibit pronounced and varied physiological effects. Nearly all of them are toxic.

Alkaloids have the ending *-ine* because they are all amines. Some important alkaloids are shown below.

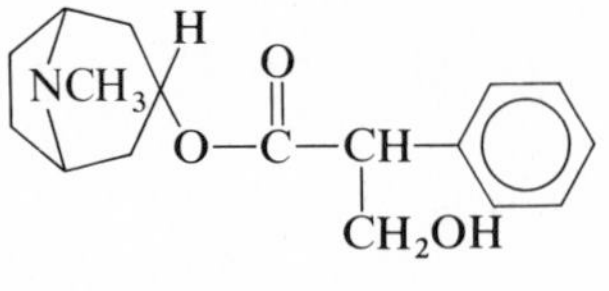

Atropine. Used to dilate the pupil of the eye during examinations.

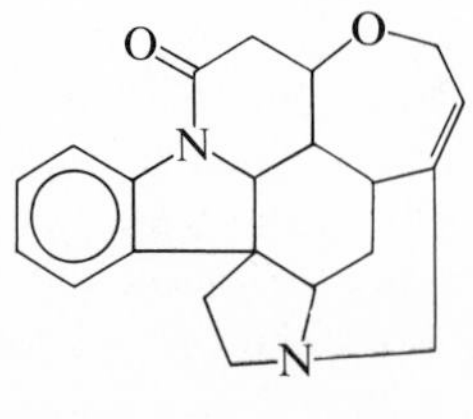

Strychnine. Used as a rodent poison. It is highly toxic and stimulates the central nervous system.

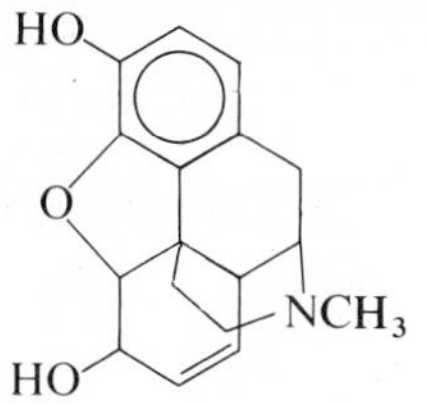

Morphine. An analgesic (pain reliever) usually used as the sulfate salt. It is addicting.

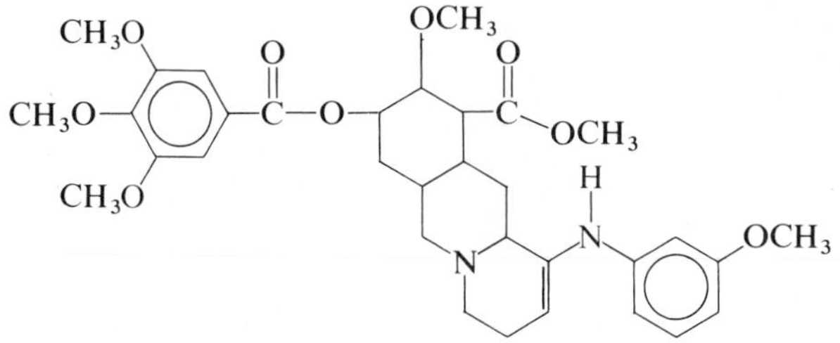

Reserpine. A tranquilizer used to lower the blood pressure.

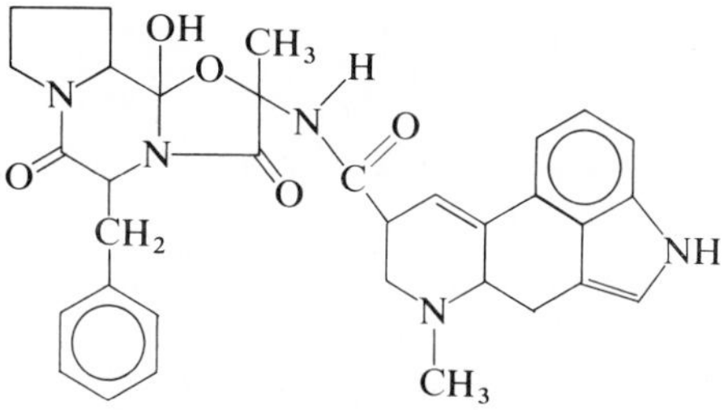

Ergotamine. An analgesic vasodilator used to treat migraine headaches.

15.15 NITROGEN FIXATION: AN ELECTIVE TOPIC

elective topic

Nitrogen compounds are essential to all life forms. The major supply of nitrogen on earth is N_2 in the atmosphere. The $N{\equiv}N$ bond in gaseous nitrogen is the strongest known covalent bond (225 kcal/mole). The conversion of atmospheric nitrogen into a form useful to life (NH_3, NO_2^-, NO_3^-) requires a very large amount of energy. Only a few organisms, primarily certain microorganisms (Rhizobium) that live in the roots of legumes such as peas or beans, are able to carry out this conversion, which is called **nitrogen fixation**.

$$N_2 + 6H \xrightarrow{\text{organism}} 2NH_3$$

Other plants and animals depend on these organisms to supply their nitrogenous foods.

Industrial nitrogen fixation involves the direct combination of nitrogen and hydrogen at high temperatures and pressures in the presence of a catalyst. The process was originally developed by the German chemist Fritz Haber, and bears his name.

The **Haber process:**

$$N_2 + 3H_2 \xrightarrow[\text{Fe-Mo-Al oxides}]{\text{300 atmospheres, 500°C}} 2\,NH_3$$

Ammonia produced by the Haber process is supplied as fertilizer to those plants that are unable to fix nitrogen and, through those plants, to animals (including ourselves).

Certain plants and microorganisms are able to convert NH_3, NO_3^-, and other simple nitrogenous substances into N_2. This process is called **denitrification** and is the reverse of nitrogen fixation. Denitrification serves to maintain the balance between atmospheric and chemically bound nitrogen.

15.16 ANALYSIS AND IDENTIFICATION OF AMINES

The most characteristic property of amines is their basicity. Most amines that are insoluble in water are soluble in aqueous mineral acid solutions. Therefore, a water-insoluble compound that dissolves in 10% aqueous HCl is probably an amine.

Amines also undergo *sodium fusion* as a qualitative test (Sec. 9.11). The Hinsberg separation is useful in the separation and identification of primary, secondary, and tertiary amines (Sec. 15.8).

NEW TERMS

Acylation Reaction of a 1° or 2° amine with an acyl halide or carboxylic anhydride to form an amide:

$$R_2NH + Cl{-}\overset{\overset{\displaystyle O}{\|}}{C}{-}R' \xrightarrow{\text{base}} R_2N{-}\overset{\overset{\displaystyle O}{\|}}{C}{-}R'$$

(R may be alkyl, aryl, or H; R′ may be alkyl or aryl) (15.8).

Alkaloid Any naturally occurring amino compound of plant origin (15.14).

Amine Derivative of ammonia in which one, two, or three of the hydrogen atoms have been replaced by alkyl or aryl groups (15.1).

Azo compound Compound that contains the $-\ddot{N}{=}\ddot{N}-$ functional group (15.11).

Classification of amines According to the number of alkyl or aryl groups attached to the nitrogen atom: *primary (1°) amines are* RNH_2; *secondary (2°) amines are* $RNHR'$; *tertiary (3°) amines* are $R\underset{\underset{\displaystyle R''}{|}}{N}R'$ (15.1).

Coupling of diazonium salts Formation of an azo compound by the reaction of an arenediazonium salt with a highly activated aromatic compound such as aniline or phenol (15.11).

Denitrification The reverse of nitrogen fixation; that is, conversion of NH_3, NO_2^-, NO_3^-, and so forth into N_2 (15.15).

Diazonium salts $R{-}\overset{+}{N}{\equiv}N{:}\ X^-$ (R may be alkyl or aryl; X is any anion, usually Cl^-) (15.8, 15.9).

Diazotization Process of forming a diazonium salt by the reaction of a 1° arylamine with nitrous acid (HONO) (15.9).

Haber process The catalytic reaction of N_2 and H_2 under high pressures and at high temperatures to produce ammonia (15.15).

Heterocyclic compound A cyclic compound that has one or more atoms other than carbon as part of the ring (15.12).

Hinsberg separation Separation and identification of 1°, 2°, and 3° amines on the basis of their different reactions with benzenesulfonic acid or other arenesulfonic acid (15.8).

Nitrile $R{-}C{\equiv}N$ (R may be alkyl or aryl) (15.7).

Nitrogen fixation Conversion of atmospheric nitrogen (N_2) into a form useful to life (NH_3, NO_2^-, NO_3^-) (15.15).

Quaternary ammonium salts

$$R_4N^+\ X^-$$

(X is any anion; R is alkyl, aryl, or any combination) (15.7).

Sulfonation Reaction of a 1° or 2° amine with the acid chloride of a sulfonic acid to form a sulfonamide:

$$R_2NH + Cl{-}SO_2{-}R' \xrightarrow{\text{base}} R_2N{-}SO_2{-}R'$$

(R may be alkyl, aryl, or H; R′ is usually aryl) (15.8).

ADDITIONAL PROBLEMS

15.22 Draw the structural formulas of the following amines.

[a] dipropylamine
(b) trimethylamine
[c] *N*-methylbenzylamine
(d) dibenzylamine
[e] *N,N*-dibenzylaniline
(f) *p*-bromoaniline
[g] cyclobutylamine
(h) 1,2-diaminocyclohexane

15.23 Name each of the following amines.

[a] $CH_3CH_2NHCH_2{-}C_6H_5$

[b] $CH_3CH_2CH_2CH_2CH_2CH_2CH_2NH_2$

[c] $C_6H_5{-}N(CH_2CH(CH_3)CH_3)_2$

[d] cyclohexane (chair) $-NHCH_3$

(e) 2,4,6-trichlorophenyl $-N(CH_3)_2$ (Cl at both ortho positions and para position)

(f) (g)

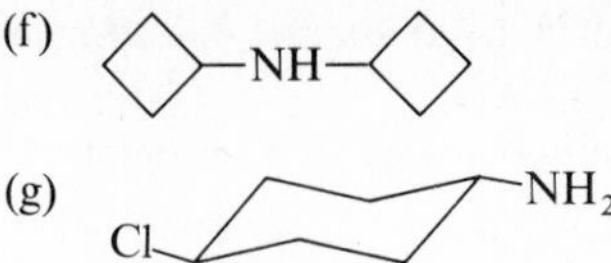

[15.24] Classify each amine in Problems 15.22 and 15.23 as primary, secondary, or tertiary.

15.25 Predict which compound in each pair below has the higher boiling point and the greater water solubility.
[a] 1-aminohexane and 1-hexanol (b) 1-Aminohexane and ethylamine
[c] propylamine and dipropylamine

[15.26] (a) Write the equation for the acid-base reaction that occurs when trimethylamine reacts with water.
(b) Label all the Brønsted–Lowry acids and bases.

15.27 Arrange the following compounds in order of increasing base strength: ammonia, aniline, *p*-nitroaniline, dimethylamine.

15.28 Select the stronger base in each of the following pairs.
[a] methylamine and aniline (b) ammonia and dimethylamine
[c] pentylamine and ammonia (d) benzylamine and aniline

15.29 Give the major organic product of each of the following reactions.

[a] C_6H_5—$CH_2Cl + CH_3NH_2 \longrightarrow$

[b] product of (a) $\xrightarrow{NaOH/H_2O}$

(c) C_6H_5—$N(CH_3)_2 + CH_3I \longrightarrow$

(d) product of (c) $\xrightarrow{NaOH/H_2O}$

[e] C_6H_5—$CH_2Cl \xrightarrow{NaCN}$

[f] product of (e) $\xrightarrow{LiAlH_4}$

(g) CH_3—C_6H_4—$NO_2 \xrightarrow{H_2/Ni}$

15.30 Use as organic starting compounds only alkyl halides, benzene, or toluene, and outline the synthesis of each of the following compounds. (You may use any necessary solvents and inorganic reagents.)
[a] cyclohexylamine [b] aniline
[c] CH_3—C_6H_4—NH_2 (d) C_6H_5—$CH_2CH_2NH_2$
(e) tetraethylammonium chloride

15.31 Give the major organic product, if any, of each of the following reactions.

[a] $C_6H_5-N(CH_3)_2 + HCl\ (aq) \longrightarrow$

(b) $C_6H_5-N(CH_3)_2 + CH_3\overset{O}{\overset{\|}{C}}-Cl \longrightarrow$

[c] $CH_3-C_6H_4-NH_2 + CH_3\overset{O}{\overset{\|}{C}}-Cl \longrightarrow$

(d) $CH_3CH_2CH_2NHCH_3 + C_6H_5-SO_2Cl \longrightarrow$

[e] (piperidine) $NH + C_6H_5-SO_2Cl \longrightarrow$

(f) (piperidine) $NH \xrightarrow{NaNO_2/HCl\ (ag)}$

[g] $C_6H_5-CH_2NH_2 \xrightarrow{NaNO_2/HCl\ (ag)}$

(h) (N-methylpiperidine) $N-CH_3 \xrightarrow{NaNO_2/HCl\ (ag)}$

[i] $Cl-C_6H_4-NH_2 \xrightarrow[0\text{-}5^\circ C]{NaNO_2/HCl\ (ag)}$

(j) product of (i) $\xrightarrow[heat]{H_3PO_2}$

[k] product of (i) $\xrightarrow[heat]{Cu_2(CN)_2}$

(l) product of (i) $\xrightarrow[heat]{H_2O}$

[m] product of (i) $\xrightarrow[heat]{Cu_2Cl_2}$

(n) product of (i) $\xrightarrow[heat]{KI}$

[o] product of (i) + (2,6-dimethylphenol: ring with CH_3, $-OH$, CH_3) $\longrightarrow$

(p) product of (i) + $C_6H_5-N(CH_3)_2 \longrightarrow$

15.32 Write equations for a laboratory synthesis of each of the following compounds starting with benzene, any alkyl halides, and any necessary solvents and inorganic reagents.

[a] $C_6H_5-N{=}N-C_6H_4-OH$

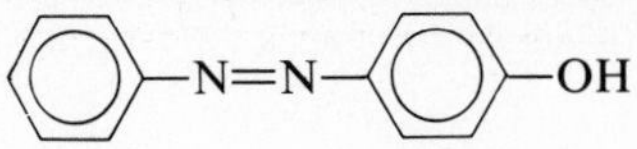

(b) $C_6H_5-N{=}N-C_6H_4-N(CH_2CH_3)_2$

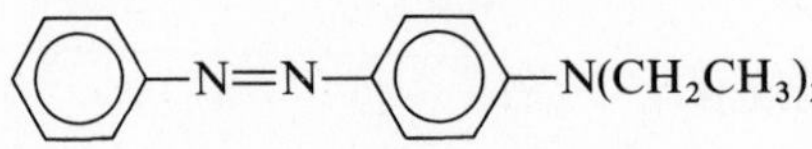

[15.33] A basic compound A (C_7H_9N) reacts with $NaNO_2$ in the presence of HCl at 0°C to give a rapid evolution of gas. Compound A reacts with benzenesulfonyl chloride to give a compound that is soluble in aqueous NaOH. Give a plausible structural formula for A, and write equations for all reactions.

15.34 Supply structural formulas for compounds B through D, which undergo the following sequence of reactions.

$$B\,(C_7H_8) \xrightarrow{Cl_2/\text{light}} C\,(C_7H_7Cl) + HCl$$

$$C \xrightarrow{NH_3} D\,(C_7H_{10}NCl) \xrightarrow{OH^-} E\,(C_7H_9N) + Cl^- + H_2O$$

[15.35] Compound F, on diazotization with $NaNO_2$ and aqueous HCl at 0–5°C followed by warming, gave 3,5-dimethylphenol. What is compound F? Write equations for all reactions.

15.36 The dye para red can be synthesized from *p*-nitroaniline and 2-naphthol. Write equations to show how this synthesis might be accomplished.

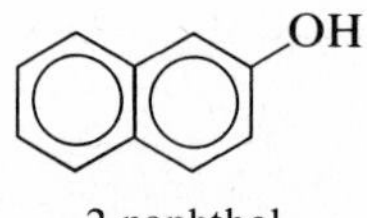

OH (2-naphthol structure)

2-naphthol

OH, $-N{=}N-C_6H_4-NO_2$ (para red structure)

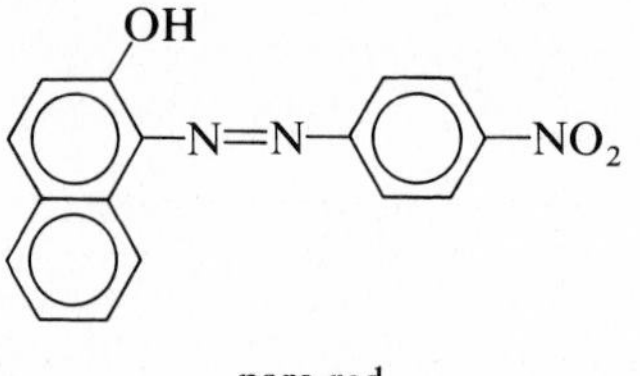

para red

15.37 Describe simple chemical tests that would make it possible to distinguish between the compounds in each of the following pairs.

[a] aniline and benzoic acid

[b] *p*-chloroaniline and *p*-xylene

(c) *N*-methylaniline and *p*-toluidine $\left(CH_3-C_6H_4-NH_2\right)$

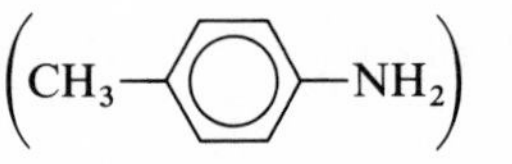

(d) triethylamine and ethylbutylamine

CARBOHYDRATES

The carbohydrates include sugars, starches, and cellulose. Sugars and starches are important sources of energy for plants and animals; cellulose is the structural material of plants. Carbohydrates are also the basic raw materials in the manufacture of paper and many textiles and plastics.

The term **carbohydrate** literally means hydrate of carbon. This name arose from the fact that many carbohydrates have the molecular formula $C_x(H_2O)_y$. Glucose, for example, has the formula $C_6H_{12}O_6$ [or $C_6(H_2O)_6$].

Carbohydrates possess two functional groups that we have already studied—the aldehyde or keto carbonyl group and the alcoholic hydroxyl group. As we shall see in this chapter, the chemistry of carbohydrates is essentially the chemistry of these two functional groups.

16.1 CLASSIFICATION

We may classify carbohydrates into three groups: monosaccharides, oligosaccharides, and polysaccharides.

Monosaccharides are the simple sugars; that is, they cannot be hydrolyzed into smaller, simpler carbohydrate units. Monosaccharides are polyhydroxyaldehydes or polyhydroxyketones that have only one aldehyde or keto group. Two common examples are glucose (also called dextrose) and fructose.

$$\begin{array}{cc} \text{H}\diagdown\;\diagup\!\!\!\diagup\text{O} & \\ \text{C} & \text{CH}_2\text{OH} \\ | & | \\ \text{CHOH} & \text{C=O} \\ | & | \\ \text{CHOH} & \text{CHOH} \\ | & | \\ \text{CHOH} & \text{CHOH} \\ | & | \\ \text{CHOH} & \text{CHOH} \\ | & | \\ \text{CH}_2\text{OH} & \text{CH}_2\text{OH} \\ \text{glucose} & \text{fructose} \end{array}$$

An **oligosaccharide** is a carbohydrate molecule that can be hydrolyzed to give several monosaccharide molecules (usually from two to ten). If hydrolysis of the oligosaccharide molecule gives two monosaccharide molecules, we call the oligosaccharide a *disaccharide*; if it gives three, we call it a *trisaccharide*; and so on.

Polysaccharides are carbohydrates that yield, on hydrolysis, more than ten monosaccharide molecules.

We can illustrate these terms with the following examples.

$$\underset{\text{a disaccharide}}{\text{Sucrose (one mole)}} \xrightarrow{\text{hydrolysis}} \underset{\text{monosaccharides}}{\text{glucose (one mole)} + \text{fructose (one mole)}}$$

$$\underset{\text{a polysaccharide}}{\text{Starch (one mole)}} \xrightarrow{\text{hydrolysis}} \underset{\text{a monosaccharide}}{\text{glucose} (>1000 \text{ moles})}$$

Problem 16.1

The carbohydrate raffinose is found in sugar beets. Hydrolysis of one mole of raffinose yields one mole each of fructose, glucose, and galactose. These three products do not undergo further hydrolysis. Classify raffinose, fructose, glucose, and galactose.

16.2 PROPERTIES OF CARBOHYDRATES

The simple monosaccharides are very soluble in water. This water solubility is not surprising if we consider the number of hydroxyl groups in each molecule and the large extent of potential hydrogen bonding. Hydrogen bonding also explains their physical state—the simple monosaccharides are white crystalline solids. Most monosaccharides have a sweet taste.

The oligosaccharides are also crystalline and water-soluble, although the larger ones are less soluble.

The polysaccharides, including starch and cellulose, are insoluble in water. Starch forms a suspension in water but is not soluble in the strict sense. The larger molecules such as starch and cellulose are not sweet to the taste.

Problem 16.2
A carbohydrate is very soluble in water and has a sweet taste. To what general class of carbohydrates does it probably belong? What experiment or test would you perform to confirm your answer?

16.3 CLASSIFICATION OF MONOSACCHARIDES

We classify a monosaccharide according to the number of carbon atoms in the molecule and whether it is an aldehyde or a ketone. We indicate the number of carbon atoms by the prefixes *tri-*, *tetr-*, *pent-*, *hex-*, and so forth. The ending *-ose* designates a carbohydrate.

Number of Carbon Atoms	General Name
3	triose
4	tetrose
5	pentose
6	hexose

To classify monosaccharides further we use the prefixes *aldo-* and *keto-* to indicate whether the molecule contains an aldehyde or keto group. Thus an aldehyde sugar that has five carbon atoms is an *aldopentose*; a ketone sugar that has six carbon atoms is a *ketohexose.*

Problem 16.3
Classify the following monosaccharides.

(a)
$$\begin{array}{c} CH_2OH \\ | \\ C{=}O \\ | \\ CHOH \\ | \\ CH_2OH \end{array}$$

(b)
$$\begin{array}{c} CHO \\ | \\ CHOH \\ | \\ CHOH \\ | \\ CH_2OH \end{array}$$

(c)
$$\begin{array}{c} CHO \\ | \\ CHOH \\ | \\ CHOH \\ | \\ CHOH \\ | \\ CHOH \\ | \\ CH_2OH \end{array}$$

(d)
$$\begin{array}{c} CH_2OH \\ | \\ C{=}O \\ | \\ CHOH \\ | \\ CHOH \\ | \\ CH_2OH \end{array}$$

16.4 STEREOCHEMISTRY OF MONOSACCHARIDES

If we examine any of the structural formulas in Problem 16.3, we see that they all contain chiral carbon atoms (atoms bonded to four different atoms or groups; Sec. 8.8). Each chiral atom is capable of two arrangements, so a molecule that has *n* chiral atoms is capable of existing in a maximum of 2^n stereoisomeric forms.

Problem 16.4
Designate the chiral atoms in each of the structural formulas of Problem 16.3. How many stereoisomers are possible for each?

16.5 FISCHER PROJECTION FORMULAS

The problem that arises immediately when discussing the stereochemistry of carbohydrates is how to draw stereoisomers simply. Probably the most common method involves **Fischer projection formulas**. In these formulas we draw the carbon chain in a vertical position and assume that the horizontal bonds all project toward the observer.

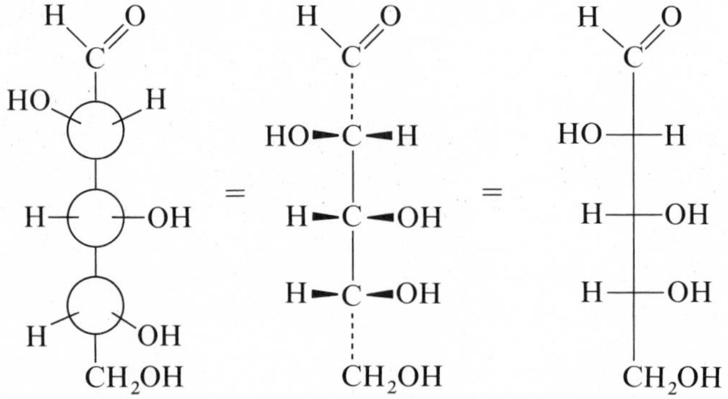

ball-and-stick drawing = line-dash-wedge formula = Fischer projection formula

---

Problem 16.5
Make a model of the aldopentose drawn above. Compare it with each of the drawings shown.

When we use Fischer projection formulas we must be careful not to lift the drawing (mentally) out of the plane of the paper, since this would change the meaning of the vertical and horizontal lines. Furthermore, we may not rotate the drawing by

90°, but we may rotate it by 180°. The following examples illustrate the latter restriction.

CHO / H—C—OH / CH_2OH = (Newman: CHO; H, OH; CH_2OH) = (Newman: CH_2OH; HO, H; CHO) = HO—C—H with CH_2OH on top and CHO at bottom

rotation by 180°

same structure

CHO / H—C—OH / CH_2OH ≠ $HOCH_2$—C—CHO with H on top and OH at bottom = (Newman: H; $HOCH_2$, CHO; OH)

rotation by 90° not allowed

= (Newman: CHO; HO, H; CH_2OH) = HO—C—H with CHO on top and CH_2OH at bottom

different structures

Problem 16.6
Draw the Fischer projection formula for each of the aldotetroses and each of the aldopentoses.

16.6 D AND L DESIGNATION OF MONOSACCHARIDES

The simplest optically active monosaccharide is the aldotriose glyceraldehyde. Glyceraldehyde has one chiral carbon atom, so it can exist in two enantiomeric forms:

CHO H—C—OH CH_2OH	CHO HO—C—H CH_2OH
(+)-glyceraldehyde	(−)-glyceraldehyde

The two enantiomers are known to have the configurations shown above. The (+) sign means dextrorotatory, and the (−) sign means levorotatory (Sec. 8.7). Before their absolute configurations were known, Emil Fischer (the great pioneer of carbohydrate chemistry) arbitrarily assigned the above structures to (+)- and (−)-glyceraldehyde. The later determination of their absolute configurations proved that Fischer's guess was correct.

Fischer designated the configuration of the dextrorotatory enantiomer as D-(+)-glyceraldehyde and that of the levorotatory enantiomer as L-(−)-glyceraldehyde. The prefixes D and L are also used with the names of higher monosaccharides to designate the configuration of the *highest numbered chiral carbon atom.* If that carbon atom has the same configuration as the chiral carbon atom of (+)-glyceraldehyde, we designate it D. If it has the same configuration as the chiral carbon atom of (−)-glyceraldehyde, we designate it L. As before, the (+) and (−) signs designate the direction of rotation of plane-polarized light. We should note that D and L prefixes do not indicate anything about the direction of rotation of plane-polarized light except for glyceraldehyde.

The following examples illustrate these designations.

^{1}CHO
H—^{2}—OH
HO—^{3}—H
H—^{4}—OH
H—^{5}—OH
$^{6}CH_2OH$

D-(+)-glucose
$[\alpha]_D^{20°} = +112°$

highest numbered chiral carbon

$^{1}CH_2OH$
^{2}=O
HO—^{3}—H
H—^{4}—OH
H—^{5}—OH
$^{6}CH_2OH$

D-(−)-fructose
$[\alpha]_D^{20°} = -132°$

Problem 16.7
Give the D and L designations of each of the following monosaccharides.

(a)
CHO
H—OH
H—OH
CH_2OH

(b)
CHO
HO—H
HO—H
CH_2OH

(c)
CHO
HO—H
H—OH
CH_2OH

(d)
CHO
H—OH
HO—H
CH_2OH

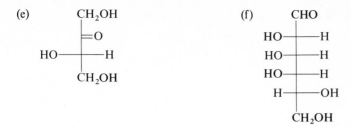

Problem 16.8
Give the *R–S* designations of the chiral carbons in each structure of Problem 16.7 (see Sec. 8.14).

Problem 16.9
Explain how the D–L and *R–S* designations differ.

16.7 MUTAROTATION OF GLUCOSE

Glucose undergoes an interesting series of changes in solution which suggest that its structure is not a simple open chain. Ordinary D-(+)-glucose (purified by allowing it to crystallize from its ethanol solution) has a melting point of 146°C and a specific rotation of +112°. If we crystallize the glucose from acetic acid, or if we allow it to crystallize from a water solution kept above 98°C, the glucose that precipitates has a melting point of 150°C and a specific rotation of +19°. The ordinary form is called the *α-form* and the form that has the higher melting point is called the *β-form.*

If we dissolve either form of D-(+)-glucose in water and measure its specific rotation, we find that the value changes gradually until it reaches a final value of +53°. *Both forms change to give the same final specific rotation.* This process is called **mutarotation**.

We can summarize the mutarotation of D-(+)-glucose by the following reaction sequence:

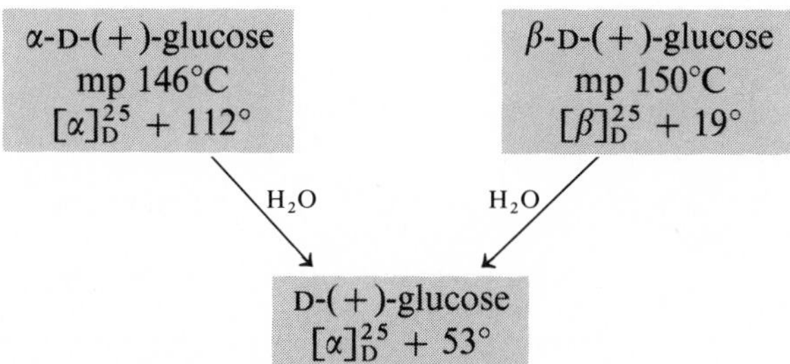

We can explain these striking changes by the existence of cyclic hemiacetal forms of glucose (Sec. 12.9). D-(+)-Glucose exists in two cyclic hemiacetal forms that differ in the orientation of the hemiacetal hydroxyl group. The α-form has the 1-OH group *trans* to the CH_2OH group; the β-form has the 1-OH group *cis* to the CH_2OH group. This means that the 1-OH is axial in the α-form and equatorial in the β-form. We can show the formation of the α- and β-forms of glucose by the following reaction schemes:

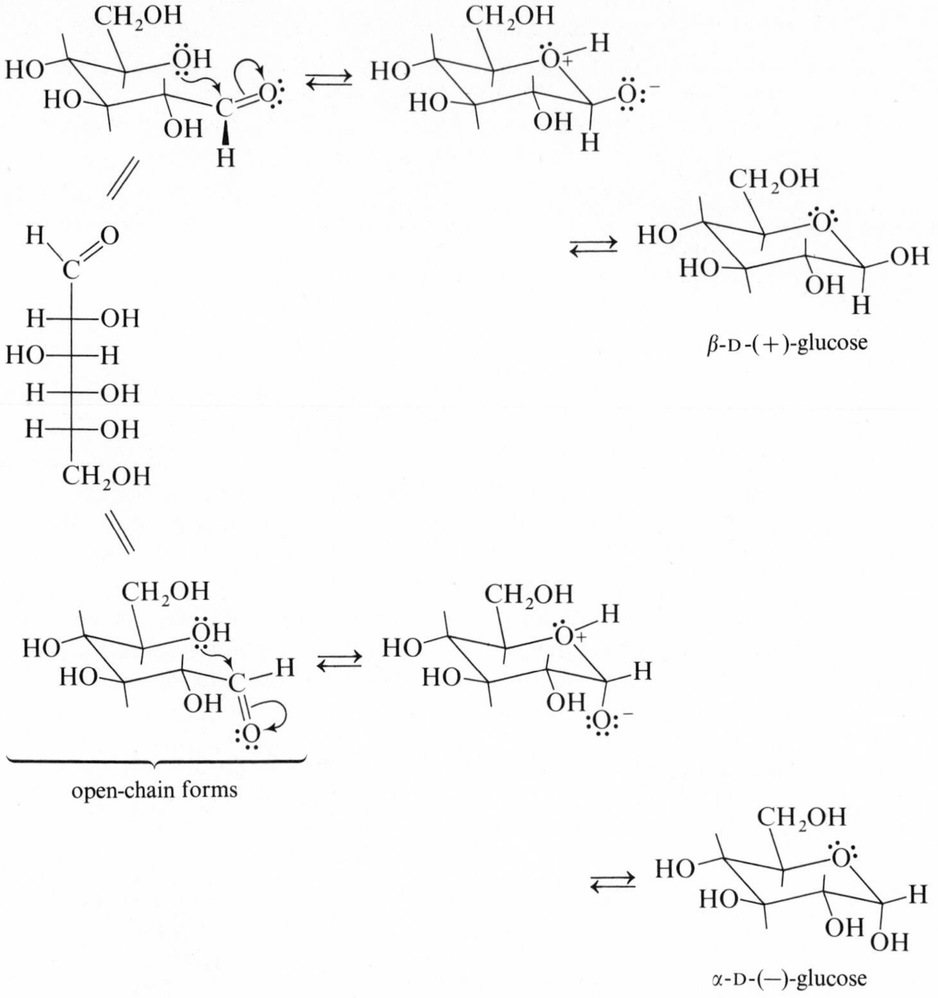

β-D-(+)-glucose

open-chain forms

α-D-(—)-glucose

Mutarotation occurs because the α-form and the β-form are in equilibrium with the open-chain form. The solution whose specific rotation is $+53°$ is therefore an equilibrium mixture of about one-third α-form and two-thirds β-form. The amount of open-chain form is very small at equilibrium.

Problem 16.10

Suggest a reason why the equilibrium mixture contains more of the β-form than the α-form.

16.8 FORMULAS OF CYCLIC HEMIACETAL STRUCTURES

In the previous section we represented the cyclic hemiacetal form of D-(+)-glucose as a line drawing (Fig. 4.4) of the chair conformation (Sec. 4.14). We expect glucose to exist in the chair conformation by analogy to cyclohexane. In the chair conformation of glucose all the hydroxyl groups (except the hemiacetal hydroxyl group) are in equatorial positions. This probably accounts for the great stability and preponderance of glucose in naturally occurring oligosaccharides and polysaccharides.

Another way to represent cyclic structures is by the **Haworth formula**. In these formulas we use a planar hexagon to represent the ring.

Haworth formula

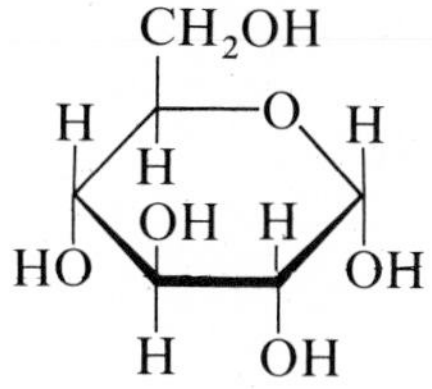

α-form

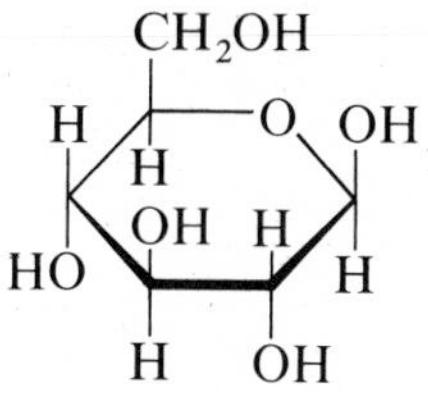

β-form

open-chain formula

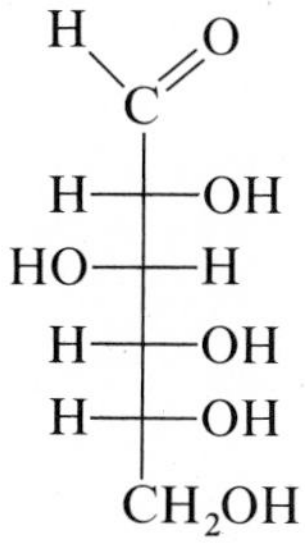

conformational formula

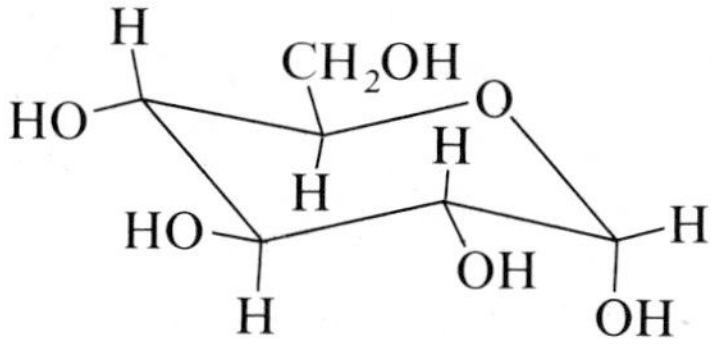

α-form

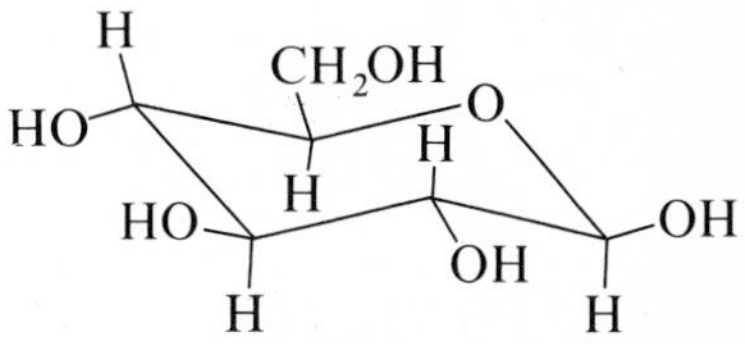

β-form

The Haworth formula does not give an accurate representation of the conformation of the molecule. It is useful, however, because it allows us to visualize more easily the relationship between the open-chain formula and the cyclic hemiacetal formula (Fig. 16.1).

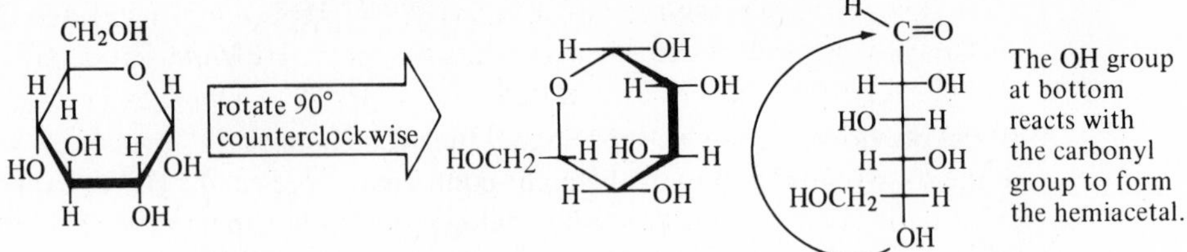

Figure 16.1. Relationship between Haworth formula and open-chain Fischer projection formula.

Problem 16.11
For each of the Fischer projection formulas given, draw the Haworth formula and the conformational formula of the corresponding six-membered hemiacetal form.

(a)

CHO
H—OH
H—OH
H—OH
H—OH
CH_2OH

(b)

CHO
HO—H
H—OH
HO—H
HO—H
CH_2OH

The α-β assignments were originally based on Fischer projection formulas. The α-form has the hemiacetal OH group on the right, and the β-form has it on the left:

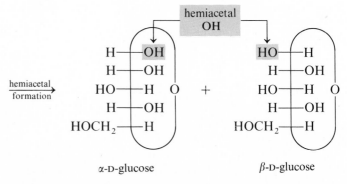

CHO
H—OH
HO—H
H—OH
H—OH
CH_2OH

=

CHO
H—OH
HO—H
H—OH
$HOCH_2$—H
OH

$\xrightarrow[\text{formation}]{\text{hemiacetal}}$

hemiacetal OH

H—OH
H—OH
HO—H O
H—OH
$HOCH_2$—H

α-D-glucose

+

HO—H
H—OH
HO—H O
H—OH
$HOCH_2$—H

β-D-glucose

These Fischer hemiacetal formulas do not represent the real molecules as well as the cyclic formulas do. We may translate the above definitions to cyclic formulas as follows: the α-form of any D-aldohexose is the one in which the hemiacetal hydroxyl group is *trans* to the CH_2OH group. In the β-form, the hemiacetal OH is *cis* to the CH_2OH group (see above paragraph).

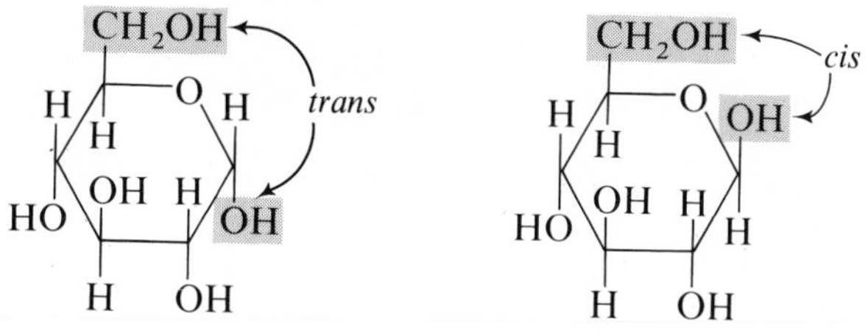

α-D-glucose

β-D-glucose

Problem 16.12
Draw the Haworth formulas of the α- and β-forms of the monosaccharides of Problem 16.11.

16.9 NOMENCLATURE OF MONOSACCHARIDES

Cyclic hemiacetals containing five-membered rings also occur, although they are not as common as the six-membered ring structures. The six-membered cyclic structures

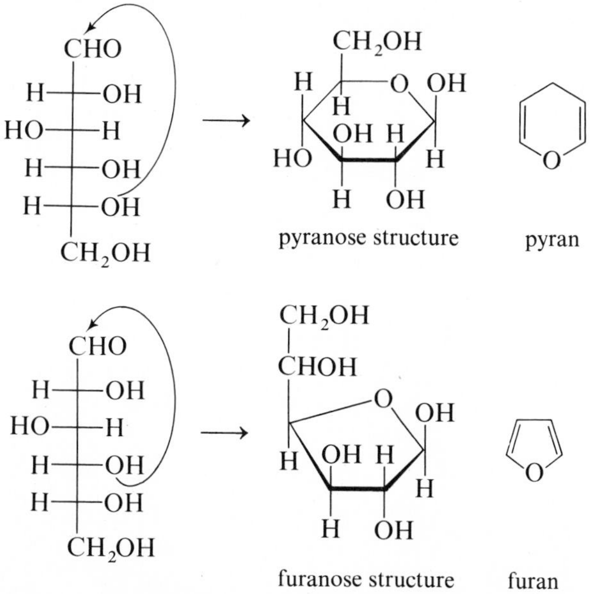

pyranose structure

pyran

furanose structure

furan

are called **pyranose** structures, and the five-membered cyclic structures are called **furanose** structures. These names are derived from the cyclic ethers *pyran* and *furan*.

We can now give descriptive names to the cyclic hemiacetals of the monosaccharides. We designate the configuration of the hemiacetal as α- or β-; we designate whether the structure is D or L; we may give the sign of optical rotation; and we give the prefix that identifies the monosaccharide, followed by the ending *-pyranose* or *-furanose* to designate the ring size. Some examples will illustrate.

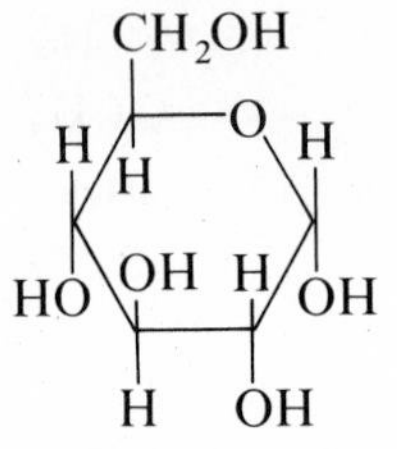

α-D-(+)-glucopyranose
(from D-(+)-glucose)

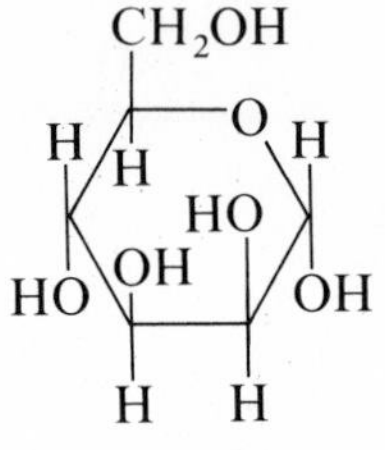

α-D-(+)-mannopyranose
(from D-(+)-mannose)

Problem 16.13

Draw the Haworth formula of each of the following compounds.

(a) β-D-glucopyranose
(b) β-D-mannopyranose
(c) α-D-arabinopyranose
(d) β-D-ribofuranose

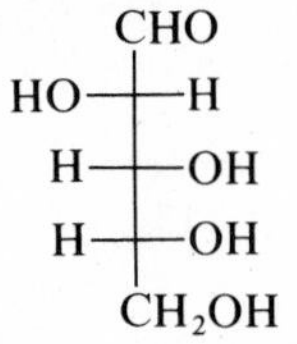

D-arabinose

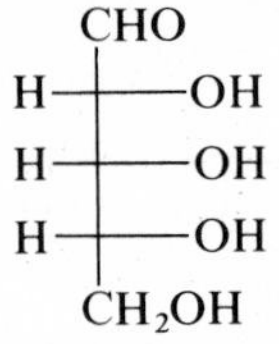

D-ribose

16.10 GLYCOSIDES

At one point in his research on glucose, Fischer attempted to prepare the dimethyl acetal:

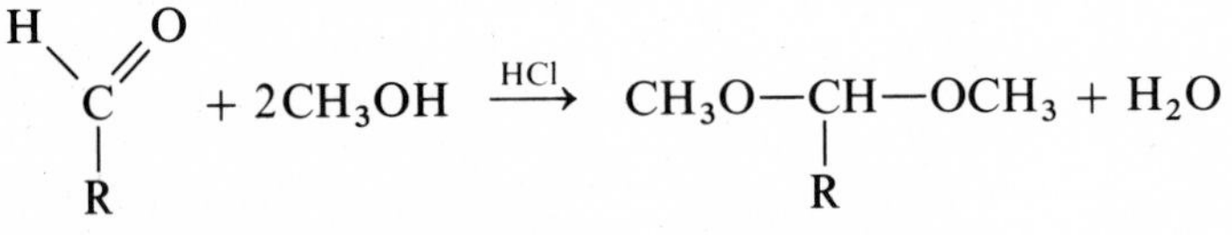

aldehyde — dimethyl acetal

Instead of the dimethyl acetal he obtained a product that had only one methoxy group. This product was the methyl ether of the α-form of D-glucopyranose. Later, the methyl ether of the β-form was also isolated:

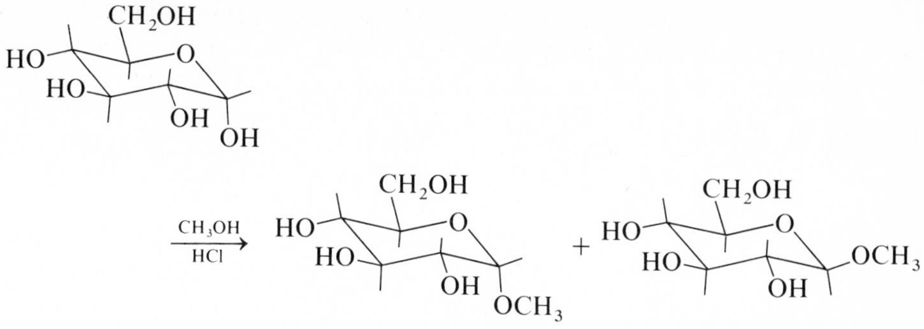

$$\xrightarrow[\text{HCl}]{\text{CH}_3\text{OH}}$$

methyl α-D-glucopyranoside + methyl β-D-glucopyranoside

These methyl ethers are actually acetals in which the two alkoxy groups, R and R′, are different:

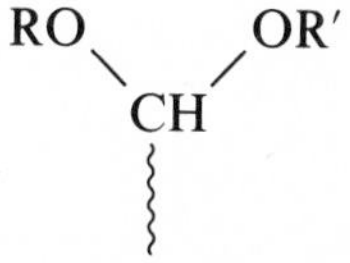

$$\text{RO}-\text{CH}(-\text{OR}')-$$

These acetals have the general name **glycoside**. The names of individual glycosides are based on the names of the monosaccharides from which they are formed—glucoside from glucose, mannoside from mannose, and so forth.

Unlike the hemiacetals, the glycosides are unreactive in neutral or basic solution. Thus glycosides are not in equilibrium with the open-chain sugars or with other glycosides.

Problem 16.14

Give the Haworth and conformational formulas of the methyl glycosides of α- and β-D-mannopyranose.

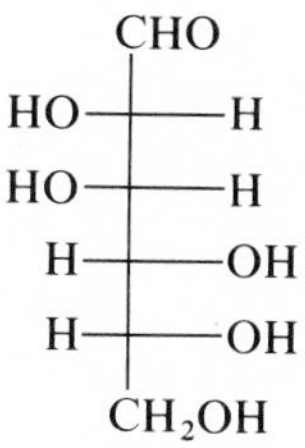

	CHO	
HO		H
HO		H
H		OH
H		OH
	CH_2OH	

D-mannose

16.11 OXIDATION OF MONOSACCHARIDES

Fehling's, Tollens's, and Benedict's solutions oxidize aldehydes and α-hydroxyketones. These oxidizing agents therefore oxidize both aldoses and ketoses. Even though the aldose or ketose exists almost entirely in the hemiacetal or hemiketal form, this form is in equilibrium with small amounts of the open-chain form:

α-form ⇌ open-chain form ⇌ β-form

α-form: CH_2OH; O; H, H, H; HO, OH, H, OH; H, OH

Open-chain form:

$$\begin{array}{c} CHO \\ H-C-OH \\ HO-C-H \\ H-C-OH \\ H-C-OH \\ CH_2OH \end{array}$$

β-form: CH_2OH; O, OH; H, H; HO, OH, H, H; H, OH

This open-chain form contains an aldehyde group that is readily oxidized.

$$\begin{array}{c} CHO \\ H-C-OH \\ HO-C-H \\ H-C-OH \\ H-C-OH \\ CH_2OH \end{array} \xrightarrow{Ag(NH_3)_2^+} \begin{array}{c} CO_2NH_4 \\ H-C-OH \\ HO-C-H \\ H-C-OH \\ H-C-OH \\ CH_2OH \end{array} + Ag\downarrow$$

Aldoses and ketoses are therefore referred to as **reducing sugars** because they reduce Fehling's, Tollens's, or Benedict's solutions. Glycosides do not react with these oxidizing agents, however, because glycosides do not have an oxidizable group and do not exist in equilibrium with the open-chain aldose structure.

Problem 16.15

Which of the following compounds react with Tollens's solution? Write equations for the reactions of those that do.

(a) CH_2OH; O, OH; H; HO, OH, H, H; H, OH

(b) CH_2OH — CHOH; O; H; H, OH, H, OH; H, OH

(c) CH_2OH; O, OCH_3; H, H; HO, OH, H, H; H, OH

(d)

$$\begin{array}{c} CH_2OH \\ C=O \\ HO-C-H \\ H-C-OH \\ H-C-OH \\ CH_2OH \end{array}$$

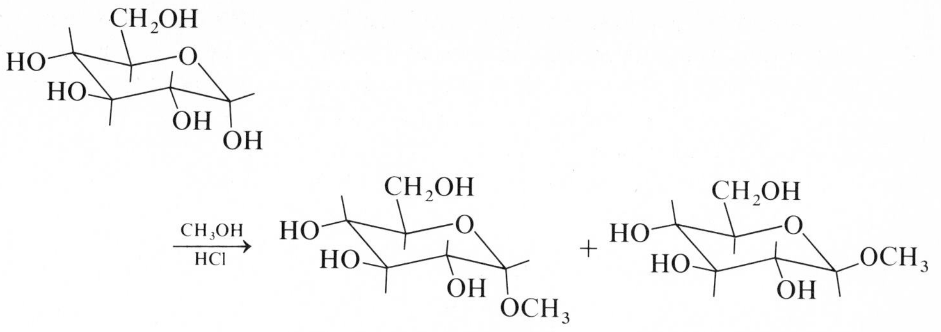

methyl α-D-glucopyranoside methyl β-D-glucopyranoside

These methyl ethers are actually acetals in which the two alkoxy groups, R and R′, are different:

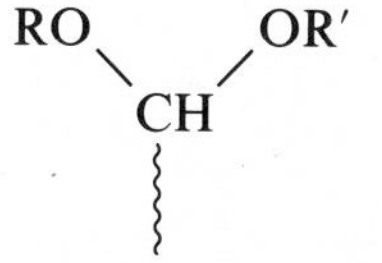

These acetals have the general name **glycoside**. The names of individual glycosides are based on the names of the monosaccharides from which they are formed—glucoside from glucose, mannoside from mannose, and so forth.

Unlike the hemiacetals, the glycosides are unreactive in neutral or basic solution. Thus glycosides are not in equilibrium with the open-chain sugars or with other glycosides.

Problem 16.14

Give the Haworth and conformational formulas of the methyl glycosides of α- and β-D-mannopyranose.

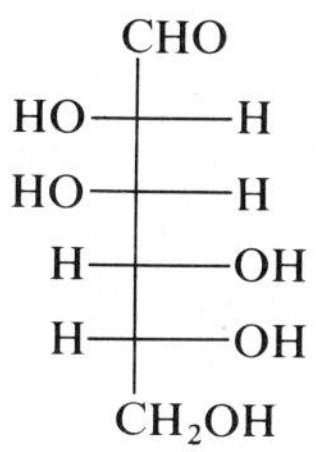

D-mannose

16.11 OXIDATION OF MONOSACCHARIDES

Fehling's, Tollens's, and Benedict's solutions oxidize aldehydes and α-hydroxyketones. These oxidizing agents therefore oxidize both aldoses and ketoses. Even though the aldose or ketose exists almost entirely in the hemiacetal or hemiketal form, this form is in equilibrium with small amounts of the open-chain form:

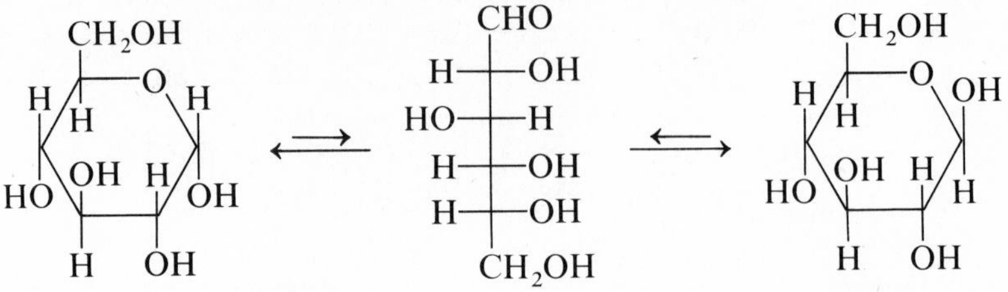

α-form ⇌ (CHO; H—OH; HO—H; H—OH; H—OH; CH_2OH) ⇌ β-form

This open-chain form contains an aldehyde group that is readily oxidized.

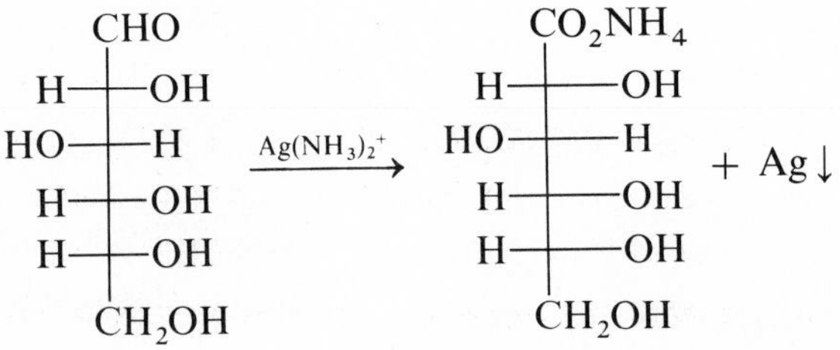

$$\text{CHO; H-OH; HO-H; H-OH; H-OH; CH}_2\text{OH} \xrightarrow{Ag(NH_3)_2^+} \text{CO}_2\text{NH}_4\text{; H-OH; HO-H; H-OH; H-OH; CH}_2\text{OH} + Ag\downarrow$$

Aldoses and ketoses are therefore referred to as **reducing sugars** because they reduce Fehling's, Tollens's, or Benedict's solutions. Glycosides do not react with these oxidizing agents, however, because glycosides do not have an oxidizable group and do not exist in equilibrium with the open-chain aldose structure.

Problem 16.15

Which of the following compounds react with Tollens's solution? Write equations for the reactions of those that do.

(a) (b) 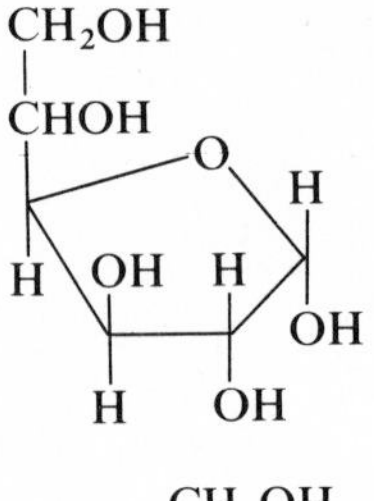 (c) 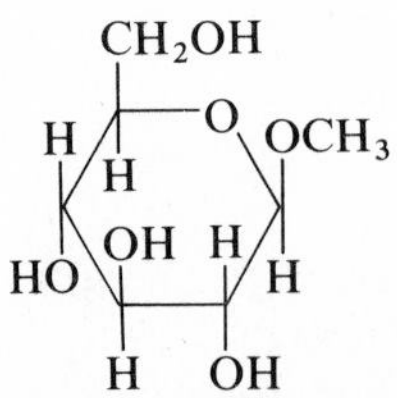 (d)

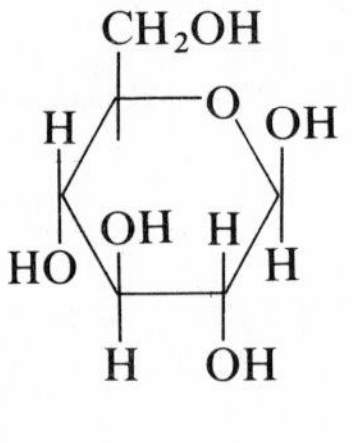

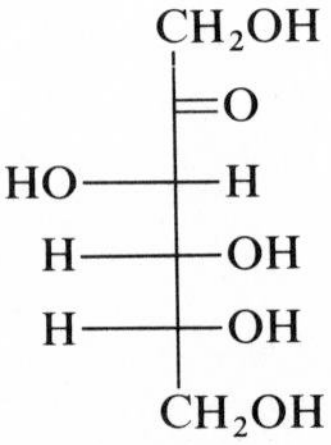

The hemiacetal hydroxyl group is very reactive; we have seen that it reacts readily with methanol in the presence of an acid to yield the glycosides. The remaining hydroxyl groups are much less reactive. To cause them to react we must use more highly reactive reagents.

When we react a monosaccharide with methyl iodide in the presence of silver oxide, all the available hydroxyl groups are methylated to the methyl ethers:

$$\text{D-glucopyranose} + 5\,CH_3I + 2\tfrac{1}{2}\,Ag_2O \longrightarrow \text{methyl-2,3,4,6-tetra-O-methyl-D-glucopyranoside} + 5\,AgI + 2\tfrac{1}{2}\,H_2O$$

D-glucopyranose

methyl-2,3,4,6-tetra-O-methyl-D-glucopyranoside

The silver oxide serves to drive the reaction to the right by removing iodide ions as the insoluble AgI.

The acetal methoxy group is easily hydrolyzed to the hemiacetal form, but the remaining methoxy groups are much less reactive.

$$\text{methyl-2,3,4,6-tetra-O-methyl-D-glucopyranoside} \xrightarrow{H_3O^+} \text{2,3,4,6-tetra-O-methyl-D-glucopyranose} + CH_3OH$$

Methylation of sugars is a useful reaction for structure determinations. We can see how it is used in the following exercise:

Exercise: A cyclic form of glucose was methylated with CH_3I and Ag_2O. After hydrolysis of the methyl glucose, the product had the open-chain structure shown below. Was the original cyclic form a pyranose or a furanose? Draw its Haworth formula.

CHO
H—C—OCH_3
CH_3O—C—H
H—C—OH
H—C—OCH_3
CH_2OCH_3

This OH is unmethylated.
It was involved in the pyranose ring.

Solution: The absence of a methoxy group at C_4 tells us that C_4 was involved in the methyl glucoside ring. The original compound was therefore a furanose.

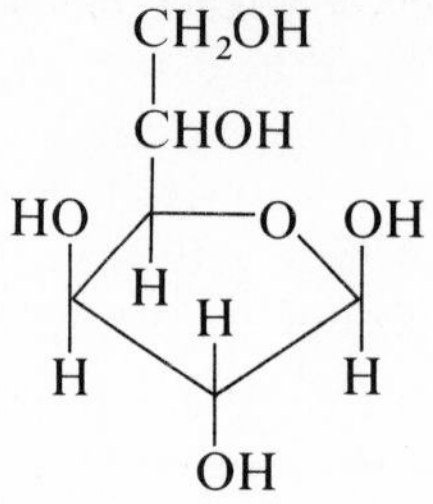

CH_2OH
CHOH
HO O OH
H H
H H
OH

original cyclic form

Problem 16.16

A cyclic form of glucose was methylated with CH_3I and Ag_2O. After hydrolysis of the methyl glucoside, the product had the following open-chain structure. Draw the Haworth formula of the original cyclic form.

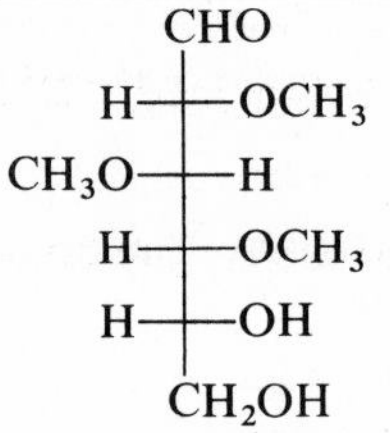

CHO
H—OCH_3
CH_3O—H
H—OCH_3
H—OH
CH_2OH

16.13 ACYLATION OF SUGARS

The free hydroxyl groups of carbohydrates can also undergo acylation to produce the ester. We accomplish this by reacting the carbohydrate with a carboxylic anhydride. If we use acetic anhydride, the reaction is called acetylation.

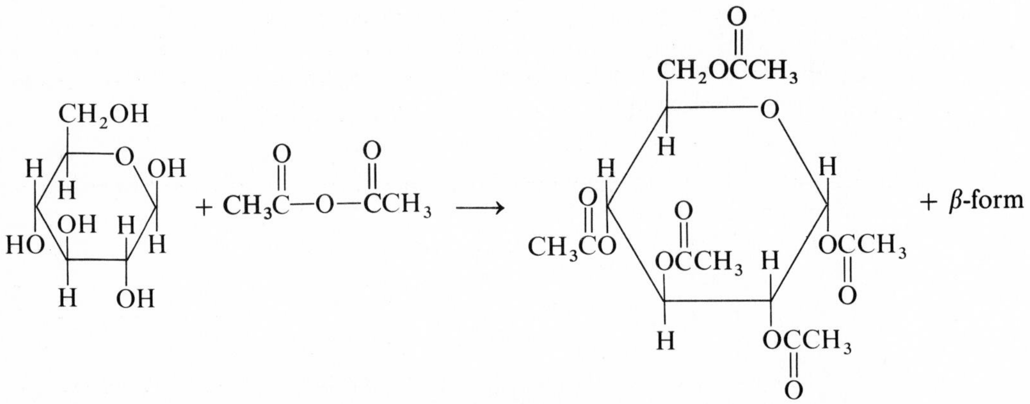

$$\text{glucose (}\alpha\text{-pyranose)} + CH_3\overset{O}{\overset{\|}{C}}—O—\overset{O}{\overset{\|}{C}}CH_3 \longrightarrow \text{penta-}O\text{-acetyl-}\alpha\text{-glucopyranose} + \beta\text{-form}$$

Acetylation is useful in determining the structure of carbohydrates.

Problem 16.17
Draw the Haworth formula of the product of acetylation of β-D-galactopyranose.

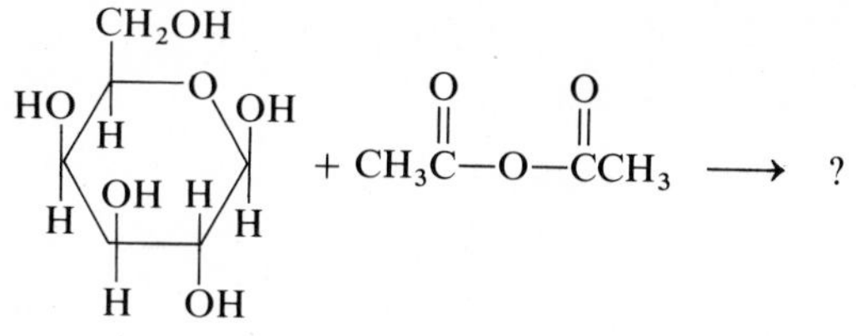

$$\text{β-D-galactopyranose} + CH_3\overset{O}{\overset{\|}{C}}-O-\overset{O}{\overset{\|}{C}}CH_3 \longrightarrow ?$$

β-D-galactopyranose

16.14 KILIANI–FISCHER SYNTHESIS

Aldoses react with hydrogen cyanide to produce the cyanohydrin (Sec. 12.9) in which a new chiral center is formed:

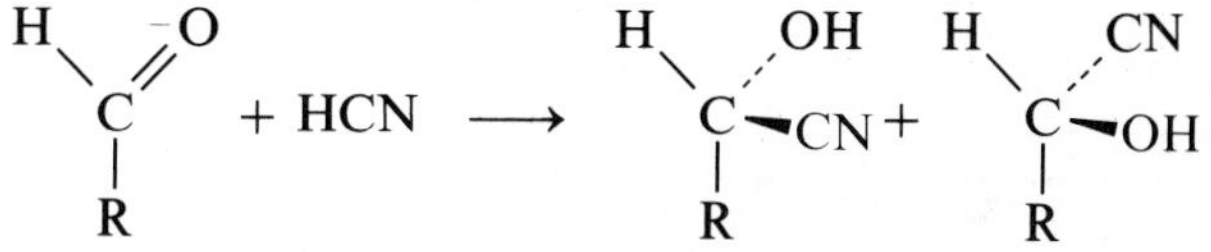

$$R-CHO + HCN \longrightarrow R-C(H)(OH)-CN + R-C(H)(CN)-OH$$

Because there are other chiral atoms in an aldose, these two cyanohydrins are diastereomers, so they have different properties and can be separated easily.

The **Kiliani–Fischer synthesis** employs this reaction to produce monosaccharides that have one more carbon atom than the starting monosaccharide. The synthesis involves forming the cyanohydrin as above, hydrolysis of the cyano group to the carboxylic acid, and finally reduction of the carboxyl group to the aldehyde. Figure 16.2 illustrates the overall method using D-glyceraldehyde.

Problem 16.18
Outline a Kiliani–Fischer synthesis as in Figure 16.2 to show the synthesis of two aldopentoses from D-erythrose and two aldopentoses from D-threose.

Through the use of the Kiliani–Fischer synthesis we can synthesize all the D-aldohexoses starting from D-glyceraldehyde. Figure 16.3 shows the family of D-aldohexoses and their relationships to each other and to D-glyceraldehyde.

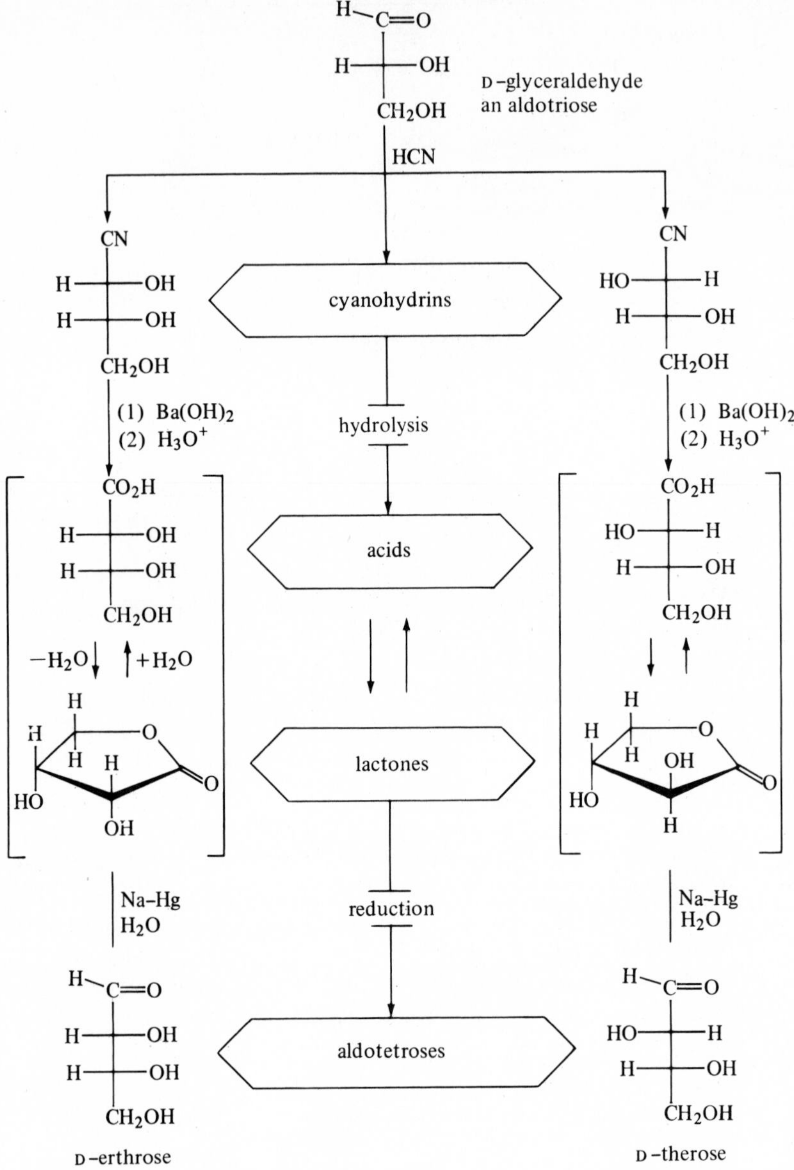

Figure 16.2. Kiliani–Fischer synthesis of two aldotetroses (D-erythrose and D-threose) from an aldotriose (D-glyceraldehyde).

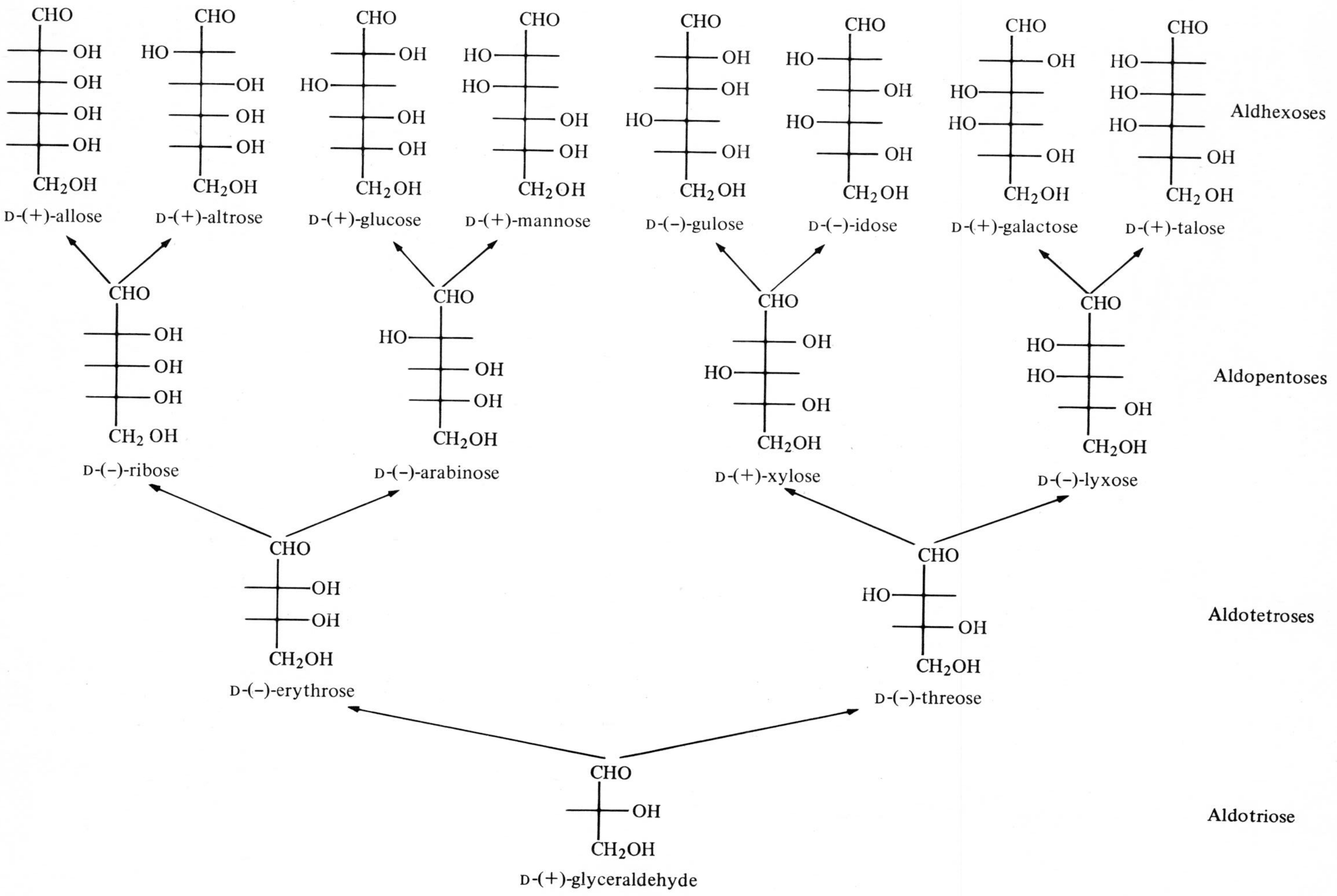

Figure 16.3. The family of D-aldohexoses and their relationships to each other and to D-glyceraldehyde.

Aldoses and ketoses, like other aldehydes and ketones, react with phenylhydrazine to form phenylhydrazones (Sec. 12.11). If an excess of phenylhydrazine is present, it serves as oxidizing agent and oxidizes the adjacent carbon atom to a carbonyl group. This new carbonyl group then reacts with phenylhydrazine to produce the **osazone**:

$$\begin{array}{l} H\diagdown \\ \quad C{=}O \\ \quad | \\ \quad CHOH \\ \quad | \\ \quad CHOH \\ \quad | \\ \quad CHOH \\ \quad | \\ \quad CHOH \\ \quad | \\ \quad CH_2OH \\ \text{a sugar} \end{array} + C_6H_5NHNH_2 \longrightarrow \begin{array}{l} H\diagdown \\ \quad C{=}NNH{-}C_6H_5 \\ \quad | \\ \quad CHOH \\ \quad | \\ \quad CHOH \\ \quad | \\ \quad CHOH \\ \quad | \\ \quad CHOH \\ \quad | \\ \quad CH_2OH \\ \text{a phenylhydrazone} \end{array}$$

$$\xrightarrow{C_6H_5NHNH_2} \begin{array}{l} H\diagdown \\ \quad C{=}NNH{-}C_6H_5 \\ \quad | \\ \quad C{=}O \\ \quad | \\ \quad CHOH \\ \quad | \\ \quad CHOH \\ \quad | \\ \quad CHOH \\ \quad | \\ \quad CH_2OH \end{array} + C_6H_5{-}NH_2 + NH_3 \xrightarrow{C_6H_5NHNH_2} \begin{array}{l} H\diagdown \\ \quad C{=}NNH{-}C_6H_5 \\ \quad | \\ \quad C{=}NNH{-}C_6H_5 \\ \quad | \\ \quad CHOH \\ \quad | \\ \quad CHOH \\ \quad | \\ \quad CHOH \\ \quad | \\ \quad CH_2OH \\ \text{an osazone} \end{array}$$

Fischer discovered this reaction. He found that the two aldohexoses D-glucose and D-mannose give identical osazones and concluded that these two aldoses differ only in the configuration of the alpha carbon (see Fig. 16-3).

Problem 16.19
Which other pairs of aldohexoses would give identical osazones? Draw their structures. How are each of these pairs of aldohexoses related to the aldopentoses?

16.16 SOME IMPORTANT MONOSACCHARIDES

By far the most commonly occurring monosaccharide is glucose. Other important aldohexoses are D-(+)-mannose and D-(+)-galactose. D-(+)-Mannose occurs in ivory nut and is obtained by hydrolysis of ivory nut shavings. D-(+)-Galactose is a constituent of many oligosaccharides and polysaccharides. It is usually prepared by hydrolysis of lactose (milk sugar).

Of the aldopentoses, D-(−)-ribose, L-(+)-arabinose, and D-(+)-xylose occur commonly. D-(−)-Ribose is a constituent of RNA (*r*ibose *n*ucleic *a*cid), a ribose phosphate polymer similar in structure to DNA. RNA is involved in protein synthesis in cells. L-(+)-Arabinose and D-(+)-xylose are found widely as polysaccharide constituents in plants, especially woods.

Problem 16.20
Note that the only L sugar mentioned in this section is L-(+)-arabinose. Draw the structural formula of L-(+)-arabinose. (Hint: it is the enantiomer of D-(−)-arabinose.)

D-(−)-Fructose, also called levulose or fruit sugar, is the most widely occurring ketohexose. It is found in a large number of fruits and in honey. Its cyclic form occurs mainly in the furanose structure. The hemiacetal is formed by reaction of the carbonyl group with the 5-hydroxyl oxygen atom:

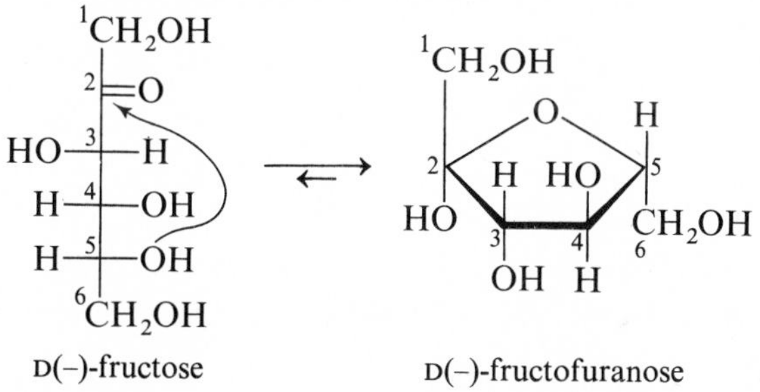

D(−)-fructose D(−)-fructofuranose

16.17 DISACCHARIDES

The most widely occurring disaccharides are sucrose, lactose, and maltose. The linkages that join monosaccharide units together to form di- and oligosaccharides are interesting, because they determine the properties of the carbohydrate.

Sucrose is ordinary table sugar obtained from cane and sugar beet. It is manufactured from the purified cane or beet juice by evaporating the water to leave pure, white crystals. Sucrose has the structural formula

conformational formula Haworth formula

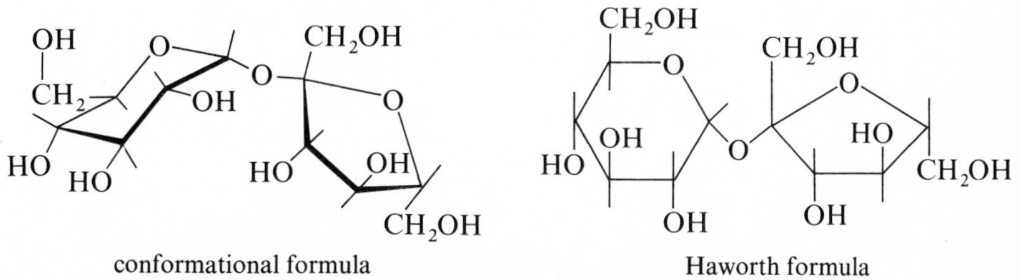

sucrose (α-D-glucopyranosyl-β-D-fructofuranoside)

The glucose unit is joined to the fructose unit through a glycoside linkage. We should note that neither monosaccharide unit exists as the hemiacetal. For this reason, sucrose is not in equilibrium with an open-chain carbonyl form. Therefore sucrose is not a reducing sugar; that is, it does not react with Fehling's, Tollens's, or Benedict's solutions.

Lactose occurs in the milk of humans and most other mammals. On hydrolysis it yields an equimolar mixture of D-glucose and D-galactose. It is a reducing sugar because it has a free hemiacetal group that is in equilibrium with the free aldehyde group:

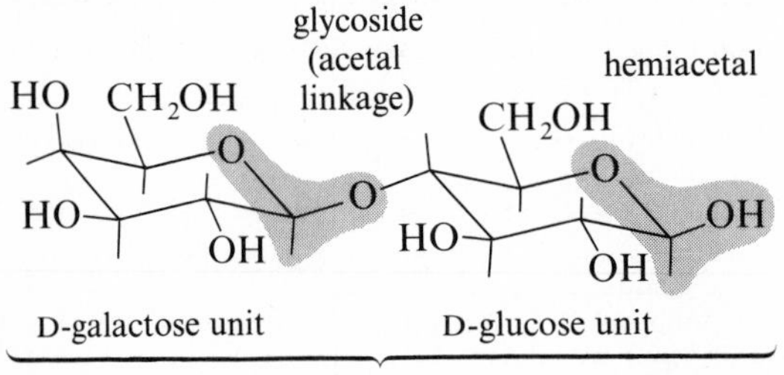

glycoside (acetal linkage) hemiacetal

D-galactose unit D-glucose unit

lactose

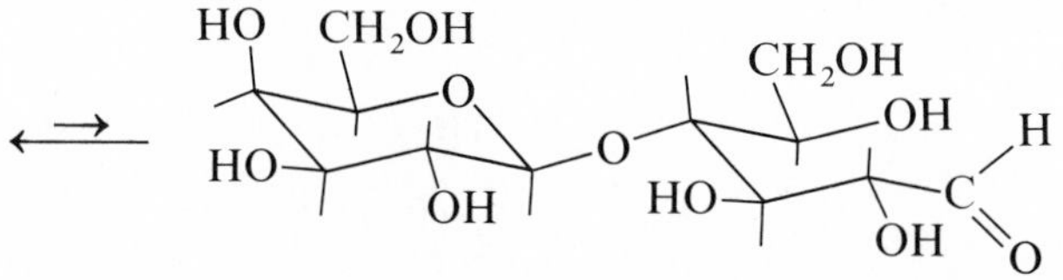

$\longleftrightarrow$

If we hydrolyze starch with the enzyme *diastase*, we obtain as one product the disaccharide *maltose*. Maltose contains two glucose units joined through a glycoside linkage similar to that of lactose. Maltose has a hemiacetal group and is therefore a reducing sugar. The glycoside linkage in maltose is an α- linkage.

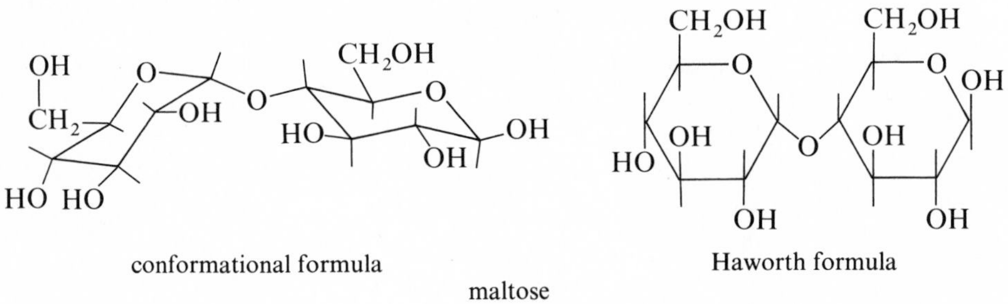

conformational formula Haworth formula

maltose

Problem 16.21
Cellobiose is obtained by the partial hydrolysis of cellulose. It is a stereoisomer of maltose; the only difference is that it has a β-glycoside linkage. Draw both the conformational formula and the Haworth formula of cellobiose.

Problem 16.22
Are the glycoside linkages in sucrose and lactose α- or β-?

Several trisaccharides occur naturally. *Raffinose* (fructose, glucose, galactose) occurs in sugar beets and other higher plants. *Melezitose* (glucose, fructose, glucose) occurs in the sap of coniferous trees.

Most of the carbohydrates found in nature exist as polysaccharides, and most polysaccharides consist entirely of D-glucose units. Polysaccharides differ from one another in several ways: the nature of the monosaccharide units, the glycoside linkages that join them together (α- or β-), and the number of monosaccharide units they contain. We shall examine starch, glycogen, and cellulose, all three of which consist entirely of D-glucose units.

Starch occurs in two forms, α-amylose and amylopectin. α-Amylose consists of chains of up to 3000 D-glucose units joined together by α-glycoside linkages to the 4-hydroxyl group of the next glucose unit:

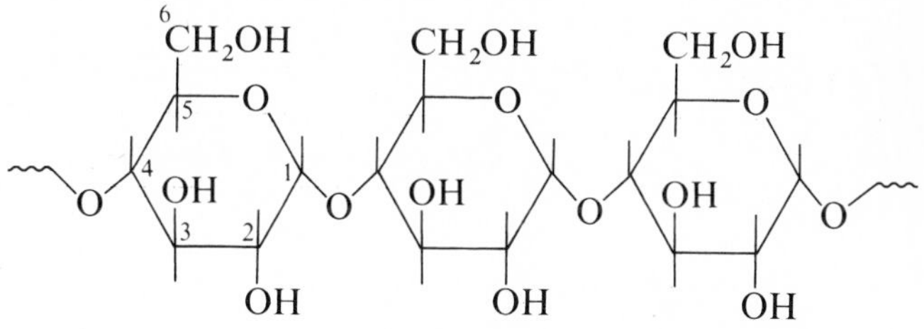

α-amylose

Amylopectin has the same type of linkages but contains branches as well. These branches are an average of 12 glucose units long, and they occur on approximately every twelfth glucose unit:

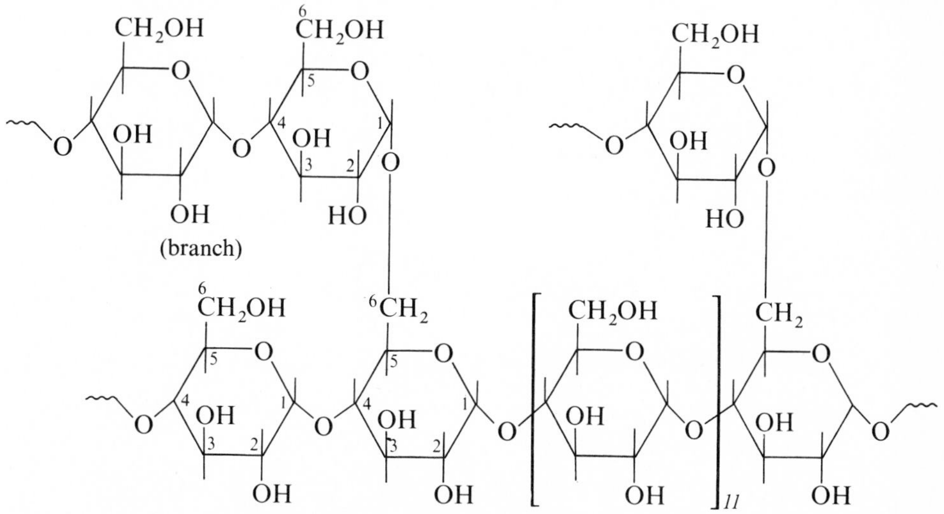

main chain amylopectin

These polysaccharides are hydrolyzed in the gastrointestinal tract by the enzyme *α-amylase,* which is present in pancreatic juice and saliva. In this hydrolysis the α-glycosidic linkages of α-amylose break to form a mixture of glucose, maltose, and polysaccharides of intermediate length called *dextrins.* To break the branching linkage of amylopectin (the α-glycoside linkage to the 6- carbon) another enzyme [an $\alpha(1 \rightarrow 6)$ glucosidase] is required. When both enzymes are present, amylopectin breaks down to glucose and maltose.

Glycogen is a polysaccharide similar to amylopectin except that it is more highly branched. It contains branches every eight to ten glucose units. Glycogen is found in animal tissues, especially liver and muscle. It is used to store food energy, as are fats. However, glycogen food energy is more quickly available than that of stored fats.

Cellulose is the most abundant organic compound found in nature. It comprises over 50% of all the organic carbon in the biosphere! It is the main structural material of plants. Wood, for example, is 50% cellulose.

Cellulose is a stereoisomer of α-amylose. It differs from α-amylose in being joined by β-glycoside linkages. Most mammals do not produce the enzymes necessary to hydrolyze these β-glycoside linkages, so mammals cannot digest cellulose. Cows, however, are able to utilize cellulose because bacteria present in their digestive tracts secrete *cellulases* which hydrolyze cellulose to glucose.

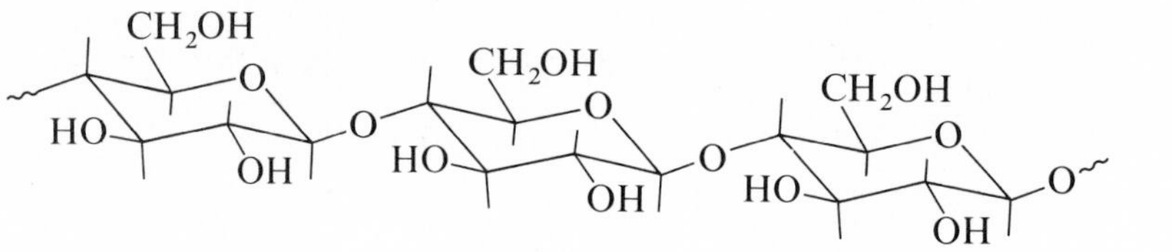

cellulose

Aside from the differences in their ease of hydrolysis, starch and cellulose exhibit striking physical differences. Starch forms globular molecules whereas cellulose forms long filaments of great strength. This difference is due to the glycoside linkage. Starch molecules tend to coil up on themselves to form clusters. Cellulose molecules are much more linear and tend to align themselves in a parallel manner. We shall examine these structural differences in Chapter 17.

Problem 16.23

You may have noticed that α-amylose and amylopectin structures were drawn with Haworth formulas, and cellulose was drawn with conformational formulas. Can you suggest a reason for this choice? [Draw α-amylose and amylopectin with conformational formulas and cellulose with Haworth formulas.]

Sugar Alcohols

The carbonyl groups of monosaccharides react readily with H_2 in the presence of a Ni, Pt, or Pd catalyst to yield sugar alcohols, referred to by the general name **alditols**. Glucose yields *glucitol* (also called sorbitol):

$$\begin{array}{c} \text{H}\diagdown\ \diagup\!\!\!\!\diagup\text{O} \\ \text{C} \\ \text{H}-\!\!\!+\!\!-\text{OH} \\ \text{HO}-\!\!\!+\!\!-\text{H} \\ \text{H}-\!\!\!+\!\!-\text{OH} \\ \text{H}-\!\!\!+\!\!-\text{OH} \\ \text{CH}_2\text{OH} \\ \text{glucose} \end{array} + H_2 \xrightarrow{\text{Ni}} \begin{array}{c} \text{CH}_2\text{OH} \\ \text{H}-\!\!\!+\!\!-\text{OH} \\ \text{HO}-\!\!\!+\!\!-\text{H} \\ \text{H}-\!\!\!+\!\!-\text{OH} \\ \text{H}-\!\!\!+\!\!-\text{OH} \\ \text{CH}_2\text{OH} \\ \text{glucitol} \end{array}$$

Mannose similarly yields *mannitol.* In the cell these reductions occur under the influence of enzymes.

A widely occurring sugar alcohol is glycerol. It occurs as carboxylic acid esters in fats (Chapter 14).

Problem 16.24
Write an equation for the reduction of mannose to mannitol.

Problem 16.25
What sugar gives glycerol on reduction?

Sugar Acids

Oxidation of the aldehyde group of monosaccharides yields **aldonic acids**. This oxidation is usually carried out with bromine and water:

$$\begin{array}{c} \text{H}\diagdown \\ \text{C}{=}\text{O} \\ \text{H}-\!\!\!+\!\!-\text{OH} \\ \text{HO}-\!\!\!+\!\!-\text{H} \\ \text{H}-\!\!\!+\!\!-\text{OH} \\ \text{H}-\!\!\!+\!\!-\text{OH} \\ \text{CH}_2\text{OH} \\ \text{glucose} \end{array} + Br_2 + H_2O \longrightarrow \begin{array}{c} \text{HO}\diagdown \\ \text{C}{=}\text{O} \\ \text{H}-\!\!\!+\!\!-\text{OH} \\ \text{HO}-\!\!\!+\!\!-\text{H} \\ \text{H}-\!\!\!+\!\!-\text{OH} \\ \text{H}-\!\!\!+\!\!-\text{OH} \\ \text{CH}_2\text{OH} \\ \text{gluconic acid} \end{array} + 2\,HBr$$

In living systems this oxidation is catalyzed by enzymes.

An important sugar acid is vitamin C, also called *ascorbic acid.* Ascorbic acid is the lactone (internal ester) of an unsaturated aldonic acid. Lack of ascorbic acid in the diet leads to scurvy, a deficiency disease. Scurvy became common in the fifteenth century during long ocean voyages of exploration. It was 1750 before the disease was

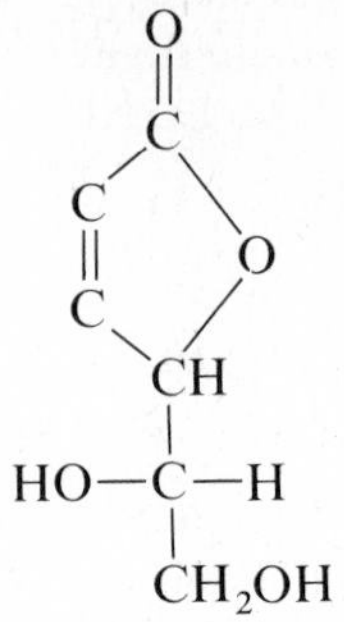

L-ascorbic acid

linked to the poor diet of the sailing men, who rarely obtained the fresh fruits and vegetables that are rich in vitamin C. Citrus fruits, especially limes, became a shipboard staple when their value in preventing scurvy became known. Citrus juices contain large amounts of vitamin C.

Phosphoric Acid Esters

Esters of phosphoric acids and monosaccharides are found in all living cells. They play an important role in metabolism. Some phosphoric acid esters are shown below.

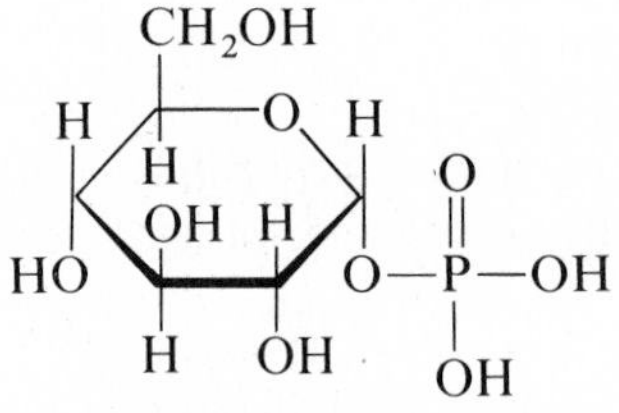

α-D-glucose 1-phosphoric acid

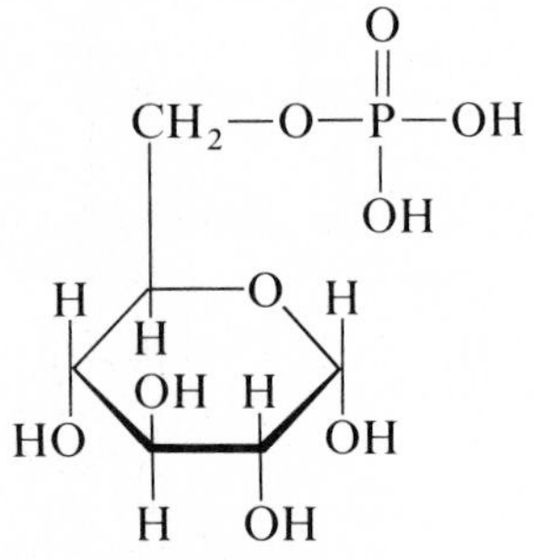

α-D-glucose 6-phosphoric acid

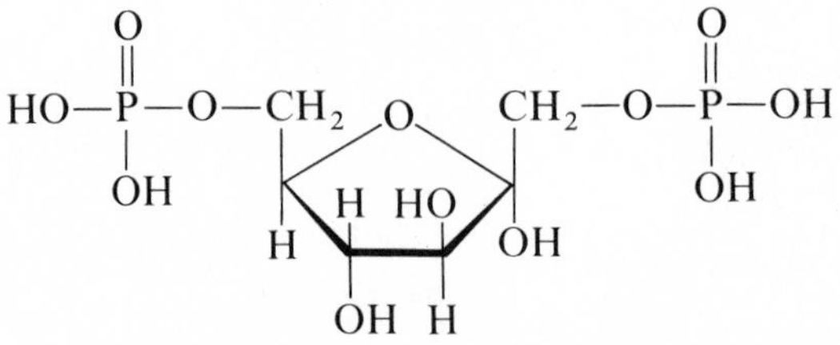

α-D-fructose 1,6-diphosphoric acid

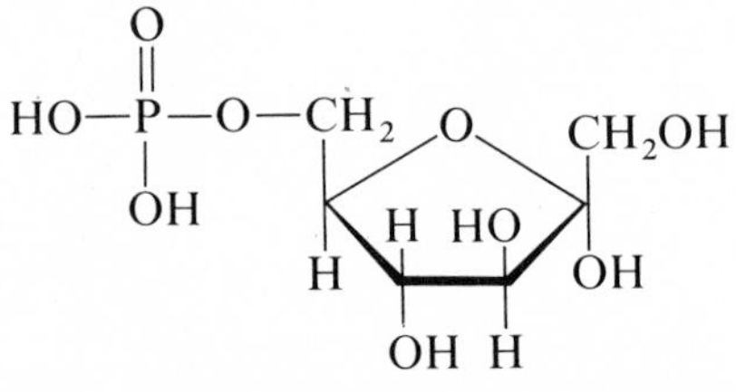

α-D-fructose 6-phosphoric acid

phosphoric acid esters

Amino Sugars

Some sugars contain an amino group in place of one of the hydroxyl groups. These amino sugars are found in structural tissues such as the exoskeletons of insects and crustaceans as well as in cartilage. Two of these amino sugars are D-glucosamine and D-galactosamine.

CHO
H—C—NH_2
HO—C—H
H—C—OH
H—C—OH
CH_2OH

D-glucosamine

CHO
H—C—NH_2
HO—C—H
HO—C—H
H—C—OH
CH_2OH

D-galactosamine

Deoxy Sugars

Reduction of a hydroxyl group produces a deoxy sugar. The most important deoxy sugar is 2-deoxy-D-ribose, which is the sugar component of DNA (*d*eoxyribo*n*ucleic *a*cid). L-Rhamnose (6-deoxy-L-mannose) and L-fucose (6-deoxy-L-galactose) are found in certain bacterial cell walls.

CHO
H—C—H
H—C—OH
H—C—OH
CH_2OH

2-deoxy-D-ribose

CHO
H—C—OH
H—C—OH
HO—C—H
HO—C—H
CH_3

L-rhamnose

CHO
HO—C—H
H—C—OH
H—C—OH
HO—C—H
CH_3

L-fucose

Problem 16.26

Draw the structural formula of an example of each of the following sugar derivatives. Do not repeat any of the compounds shown in this section.

(a) a phosphoric acid ester
(b) an alditol
(c) an aldonic acid
(d) an amino sugar
(e) a deoxy sugar

16.20 PHOTOSYNTHESIS: AN ELECTIVE TOPIC

elective topic

Green plants produce carbohydrates by a beautifully complex series of reactions that begins with carbon dioxide and water. The overall equation is

$$n\text{CO}_2 + n\text{H}_2\text{O} \xrightarrow{\text{energy}} (\text{CH}_2\text{O})_n + n\text{O}_2$$

The sun supplies the necessary energy, and the green pigment chlorophyll is the major catalyst. Other cellular components including several catalytic enzymes are also involved.

Most of the details of the photosynthetic process, complicated as they are, have been worked out by M. Calvin of the University of California. He studied the reaction by using carbon dioxide labeled with the radioactive isotope $^{14}_{6}\text{C}$ and tracing the fate of the radioactive carbon atoms.

Two important light-absorbing reactions are now known:

$$\text{ADP} + \text{HPO}_4{}^{2-} + \text{light energy} \xrightarrow{\text{enzyme}} \text{ATP} + \text{H}_2\text{O}$$

$$\text{NADP}^+ + \text{H}_2\text{O} + \text{light energy} \xrightarrow{\text{energy}} \text{NADPH} + \tfrac{1}{2}\text{O}_2 + \text{H}^+$$

Figures 16.4 and 16.5 show these reactions with the structural formulas of ADP, ATP, NADP^+, and NADPH.

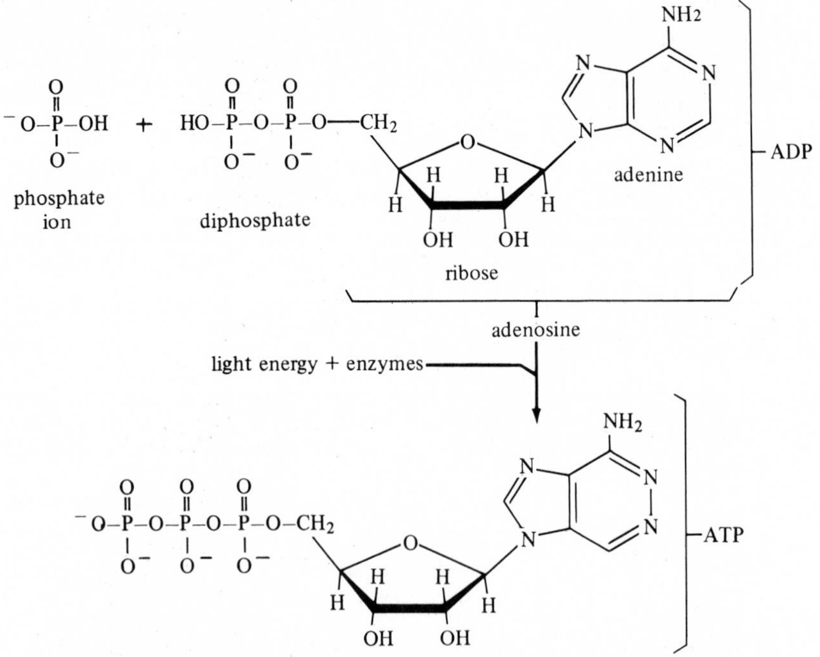

Figure 16.4. The light-induced conversion of ADP (adenosine *di*phosphate) into ATP (adenosine *tri*phosphate).

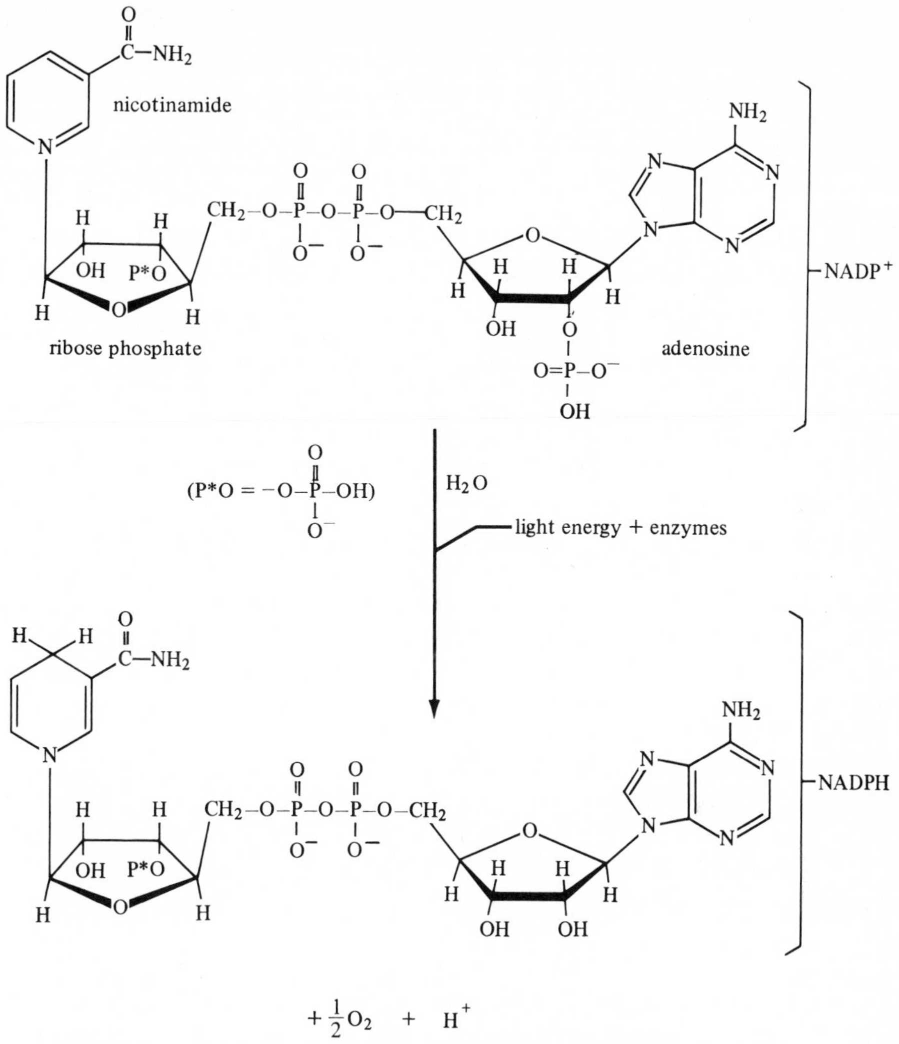

$$(P^*O = -O-\overset{O}{\underset{O^-}{P}}-OH)$$

$\xrightarrow[\text{light energy + enzymes}]{H_2O}$

NADP$^+$ → NADPH

$$+\tfrac{1}{2}O_2 + H^+$$

Figure 16.5. The light-induced conversion of NADP$^+$ (*n*icotinamide *a*denine *d*inucleotide *p*hosphate) into NADPH (*n*icotinamide *a*denine *d*inucleotide *p*hosphate-*H*), the reduced (or hydrogenated) form of NADP$^+$.

These reactions are reversible and serve to store light energy in the form of chemical bonds for use when needed in other reactions. We must recall at this point that reactions in biological systems must occur at a nearly constant temperature. Therefore most of the energy is transferred by the formation and breakage of chemical bonds rather than by the absorption or evolution of heat, which would change the temperature of the cell.

The incorporation of CO_2 into an organic molecule occurs without light. The required energy is supplied by ATP and NADPH. In this reaction, ribulose-1,5-diphosphate reacts with CO_2 to produce an unstable six-carbon intermediate which breaks down to two molecules of 3-phosphoglyceric acid.

$$\begin{array}{c} CH_2OPO_3^{2-} \\ | \\ C{=}O \\ | \\ H-C-OH \\ | \\ H-C-OH \\ | \\ CH_2OPO_3^{2-} \end{array} + CO_2 \xrightarrow[\text{enzymes}]{\text{ATP, NADPH}} \left[\begin{array}{c} CH_2OPO_3^{2-} \\ | \\ HO-C-CO_2H \\ | \\ C{=}O \\ | \\ H-C-OH \\ | \\ CH_2OH \end{array} \right] \xrightarrow{H_2O} \begin{array}{c} CH_2OPO_3^{2-} \\ | \\ HO-C-H \\ | \\ COO^- \\ + \\ CO_2^- \\ | \\ H-C-OH \\ | \\ CH_2OPO_3^{2-} \end{array}$$

ribulose-1,5-diphosphate — unstable intermediate — 3-phosphoglyceric acid (abbreviated 3-PGA)

3-Phosphoglyceric acid is then reduced to the aldehyde 3-phosphoglyceraldehyde, which is the key intermediate in the synthesis of pentoses, hexoses, glycerol, and various other compounds found in living systems.

Problem 16.27
Are the two molecules of 3-PGA shown in the above equation the same or are they enantiomers? [Explain.]

Problem 16.28
What kind of reactions are needed to convert 3-PGA into glycerol?

16.21 METABOLISM OF CARBOHYDRATES

Animals use carbohydrates as foods. The chemical equation is the reverse of the photosynthetic equation shown in Section 16.20:

$$(CH_2O)_n + nO_2 \longrightarrow CO_2 + H_2O + \text{energy}$$

This reaction is one of our major sources of energy.

The individual reactions that take place in the cell during the breakdown of carbohydrates to CO_2 and H_2O are extremely complex. In general, they involve enzyme-catalyzed reactions that are usually the reverse of those involved in the production of carbohydrates from CO_2 and H_2O, but that do not follow the same pathway as the forward reactions.

NEW TERMS

Acylation of sugars Esterification of the OH groups of carbohydrates by reaction with a carboxylic anhydride. If acetic anhydride is used, the process is called *acetylation.* (16.13).

Alditol A sugar derivative in which the carbonyl group has been reduced to a—CHOH group. Alditols are also called *sugar alcohols* (16.19).

Aldonic acid A sugar derivative in which the aldehyde group has been oxidized to a carboxyl group (16.19).

Alkylation of sugars Formation of the ethers of the OH groups of carbohydrates by reaction with CH_3I in the presence of Ag_2O (16.12).

α-β designation In the α-form of a D-aldohexose, the hemiacetal OH group is *trans* to the CH_2OH group; in the β-form, the hemiacetal OH group is *cis* to the CH_2OH group (16.7).

Amino sugar A sugar derivative that contains an amine group in place of one of the OH groups (16.19).

Carbohydrates Naturally occurring compounds that have the formula $C_x(H_2O)_y$ (Introduction).

Classification of monosaccharides The prefixes *aldo-* and *keto-* denote that the monosaccharide has an aldehyde group or a keto group, respectively. The suffix *-ose* denotes a carbohydrate. The designations *tri-*, *tetr-*, *pent-*, *hex-*, and so forth denote the number of carbon atoms. Thus an aldopentose is a five-carbon sugar that has an aldehyde group; a ketohexose is a six-carbon sugar that has a keto group (16.3).

Conformational formula Representation of cyclic structures of monosaccharides. β-D-glucose is represented as (16.8):

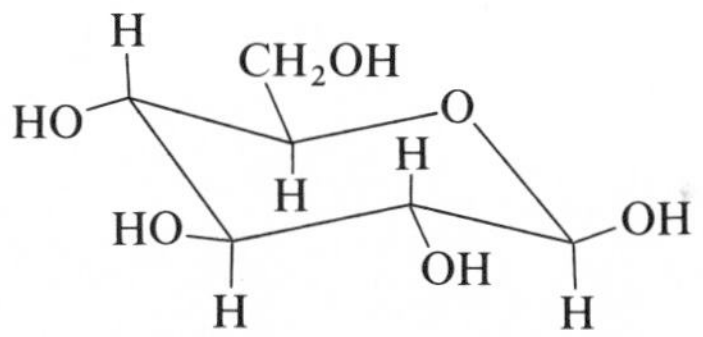

D-L designation Designation of the highest numbered chiral carbon atom. That carbon atom is designated D- if it has the same configuration as (+)-glyceraldehyde; it is designated L- if it has the same configuration as (–)-glyceraldehyde (16.6).

Deoxy sugar A sugar derivative in which one OH group has been reduced to H (16.19).

Fischer projection formula The two-dimensional projection of a monosaccharide molecule in which the carbon chain is shown as a vertical line and the H and OH groups extend horizontally. The horizontal lines are understood to project out above the page. D-Glucose has the Fischer projection formula (16.5).

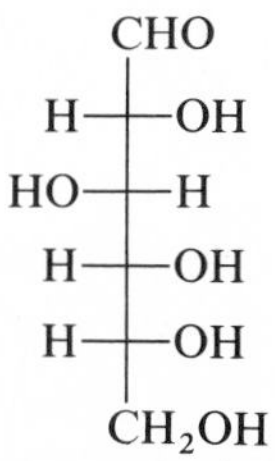

Furanose structure The five-membered cyclic structure of a monosaccharide (16.9).

Glycoside Cyclic acetal of a monosaccharide (16.10).

Haworth formula Representation of cyclic structures of monosaccharides using a hexagon. The Haworth formula of β-D-glucose is (16.8):

CH_2OH
H O OH
H
OH H
HO H
H OH

Kiliani–Fischer synthesis Synthesis of a monosaccharide with one more carbon atom than the starting monosaccharide (Fig. 16.2) (16.14).

Monosaccharide A simple sugar (carbohydrate) that cannot be hydrolyzed into simpler carbohydrate units (16.1).

Mutarotation Process in which either the pure α- or the pure β-hemiacetal form of a monosaccharide changes to give the same equilibrium mixture of α- and β-forms when dissolved in water (16.7).

Oligosaccharide A carbohydrate molecule that can be hydrolyzed to give from one to ten monosaccharide molecules. Individual oligosaccharides are called *disaccharides*, *trisaccharides*, and so forth depending on the number of monosaccharide molecules produced on hydrolysis (16.1).

Osazone A sugar derivative prepared by reacting a monosaccharide with phenylhydrazine (16.15).

Phosphoric acid ester An ester produced by esterifying one OH group of a monosaccharide with phosphoric acid (16.19).

Photosynthesis Biosynthesis of carbohydrates from CO_2 and H_2O in green plants (16.20).

Polysaccharide A carbohydrate molecule that yields more than ten monosaccharide molecules on hydrolysis (16.1).

Pyranose structure The six-membered cyclic structure of a monosaccharide (16.9).

Reducing sugar Any sugar that gives a positive test with Tollens's, Fehling's, or Benedict's solutions. Reducing sugars all have the hemiacetal or open-chain structures. Glycosides (acetals) are not reducing sugars (16.11).

ADDITIONAL PROBLEMS

[16.29] Give an example of each of the following.

(a) monosaccharide
(b) disaccharide
(c) aldopentose
(d) ketopentose
(e) aldohexose in the furanose form
(f) ketohexose in the pyranose form
(g) glycoside
(h) a D-ketopentose
(i) an L-ketohexose

[16.30] Describe the term mutarotation with equations and a specific example.

16.31 Draw Fischer projection formulas for [a] all the stereoisomeric ketopentoses, (b) all the stereoisomeric ketohexoses.

16.32 Draw the Haworth and conformational formulas of the α- and β-pyranose forms of each of the following sugars.

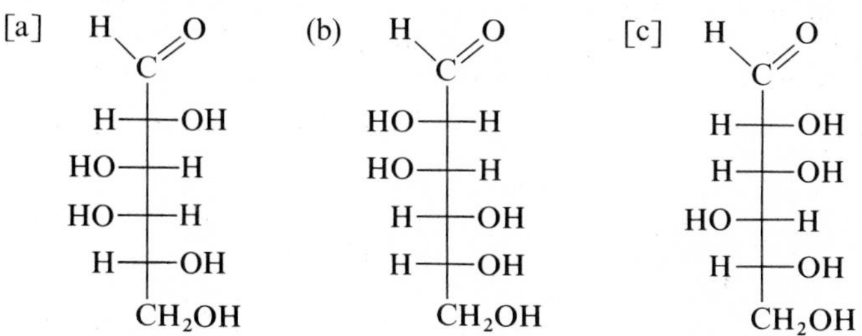

16.33 Draw the Haworth formula of each of the following compounds.
[a] β-D-glucofuranose [b] β-D-arabinofuranose
[c] α-D-altropyranose (d) α-D-galactopyranose
(e) methyl β-D-idopyranoside

[16.34] Which of the compounds in Problem 16.33 are reducing sugars? Write equations for their reactions with Tollens's reagent.

[16.35] Draw the structural formula of the product of the reaction of sucrose (Sec. 16.17) with CH_3I and Ag_2O. When this compound (an octamethyl sucrose) is treated with H_3O^+ it yields two sugar derivatives. Draw their Haworth formulas.

16.36 Draw the structural formula of the acetylation product of [a] lactose, (b) maltose.

16.37 Draw the Fischer projection formulas of the products of the Kiliani–Fischer synthesis using D-glucose as starting material.

[16.38] Write an equation using Fischer projection formulas for the reaction of D-arabinose with excess phenylhydrazine. Which other aldopentose gives the same product?

[16.39] Describe the structural difference between starch and cellulose.

16.40 Which D-aldohexoses undergo reduction to produce achiral alditols (sugar alcohols)?

[16.41] Give a mechanism for the formation of D-glucopyranose (see Sec. 12.9).

17 BEHAVIOR OF VERY LARGE MOLECULES: BIOPOLYMERS

Up to this point we have studied mainly relatively small molecules—molecules having molecular weights up to about 400.* There is also a class of molecules that are distinguished by their large size. These **macromolecules** have very high molecular weights. Many even exceed a million, and their properties are unlike those of small molecules. They include some of the most important substances in human biochemistry—proteins including muscle, tendons, hair, and enzymes; nucleic acids, which have a role in heredity; cellulose, the structural substance of plants; starch, a major component of our diet; and rubber. In addition, macromolecules dominate the chemical industry in products such as plastics and fibers. A majority of all practicing chemists work on some aspect of the chemistry of macromolecules.

We may immediately ask several questions about macromolecules: How can we understand the structures of such large molecules? How do their properties differ

* Polysaccharides such as starch and cellulose (Sec. 16.18) are macromolecules.

from those of small molecules? How does nature make such molecules, and how can synthetic ones be tailor-made to serve specific functions?

In the pages that follow we shall attempt to find answers to the first three of these questions. In Chapter 18 we shall attempt to answer the fourth question by examining the chemistry of synthetic macromolecules.

17.1 POLYMERS

It would be virtually impossible to understand the chemistry of macromolecules if it were not for one very important structural feature: Macromolecules are composed of repeating units. Polystyrene, for example, consists of long molecular chains of the same unit,

$$-CH_2-\underset{\displaystyle C_6H_5}{\underset{|}{CH}}-,$$

which repeats many times. We therefore write the structural formula of polystyrene as

$$\left(-CH_2-\underset{\displaystyle C_6H_5}{\underset{|}{CH}}-\right)_n$$

where n is the average number of repeating units in the molecule. Because macromolecules consist of repeating units we call them **polymers** (*poly* = many, *mer* = unit). We arbitrarily define polymers as having values of n greater than 10. Smaller molecules are called *dimers, trimers, tetramers*, and so forth. The original molecule, styrene, is called a **monomer**

$$\underset{\displaystyle C_6H_5}{\underset{|}{CH}}{=}CH_2$$

styrene

Problem 17.1
Explain the difference, if any, between the terms *macromolecule* and *polymer*.

17.2 MOLECULAR WEIGHTS OF MACROMOLECULES

In general, all molecules in a sample of polymer are different. A given sample of polymer will nearly always contain a range of molecular sizes, so when we speak of the molecular weight of a polymer we always mean its *average molecular weight*. Some biopolymers such as enzymes and nucleic acids have uniform molecular sizes. These are special cases, however. Most natural polymers—starch, cellulose, fibrous proteins such as wool and silk, and rubber—and all synthetic polymers contain a range of molecular sizes. This nonuniformity of molecular size confers unusual physical properties on polymeric substances. Its major effect is that all samples of polymers are mixtures.

17.3 PHYSICAL PROPERTIES OF POLYMERS

Physical State

Substances composed of small molecules exist as gases, liquids, and crystalline solids. Polymeric substances, on the other hand, range from liquids, to plastics, to brittle solids. Most solid polymers are not crystalline, but instead are glasses or supercooled liquids.

Melting Point

Most pure, low molecular weight compounds have sharp melting points; usually they change from the solid to the liquid phase within a temperature range of a degree or two. In fact, one way to know if a substance is pure is to determine its melting point range. If the substance contains impurities it is a mixture, and will therefore have a lower melting point. Mixtures usually melt over a wide temperature range—from several degrees to as much as 50° or more—depending on the amount of impurity.

Because all polymers are mixtures they melt over wide temperature ranges, often as great as 100°. Most polymers do not pass abruptly from a definite solid phase to a definite liquid phase. Instead, they soften gradually, and the exact point at which they become liquids is usually difficult to determine.

Viscosity

One of the most notable properties of polymers is the viscosity of their solutions. A 10% water solution of sugar or another low molecular weight substance is not noticeably different from pure water. A 10% solution of a high polymer, on the other hand, may be quite viscous. An example is commercial Jello, which is a dilute solution of gelatin (a protein) in water. In the laboratory we can tell when we have prepared a polymer by the viscosity of the product.

Problem 17.2
List and describe all the ways in which polymers differ from substances composed of small molecules.

Problem 17.3
We have just carried out a laboratory synthesis and have found that our product has a melting range of 120–150°C.

(a) What can we conclude about the nature of the product?
(b) Is it definitely polymeric?
(c) If not, what additional test would be necessary to establish its identity as a polymer?

In Section 16.18 we examined the structures of starch and cellulose. Let us now re-examine these two substances in the context of their polymeric structures.

Cellulose differs from starch in two structurally important ways: Cellulose is a linear, unbranched polymer of glucose units, whereas starch (mostly amylopectin) contains branches; and in cellulose the glucose units are joined together by β-glycosidic linkages, whereas in starch (both α-amylose and amylopectin) the glucose units are joined together by α-glycosidic linkages.

Cotton fibers are nearly pure cellulose. We know that cellulose molecules are insoluble in water and are capable of aligning themselves into strong, fiber-forming filaments. Starch, on the other hand, is granular, does not form fibers, and disperses in water to form a pasty mass. We can explain the properties of these two substances in terms of their molecular structures as follows.

Problem 17.4
On the basis of your own experience, compare the physical properties of starch and cellulose.

Problem 17.5
What two structural features distinguish starch from cellulose?

Cellulose

Because of their β-glycosidic linkages and lack of branching, molecules of cellulose form linear, extended chains that align themselves in a parallel fashion. The chains are held in place by hydrogen bonding between chains (Fig. 17.1). The result is a fiber with a high tensile strength. Cellulose chains have an average molecular weight of 600,000 (3000–4000 glucose units).

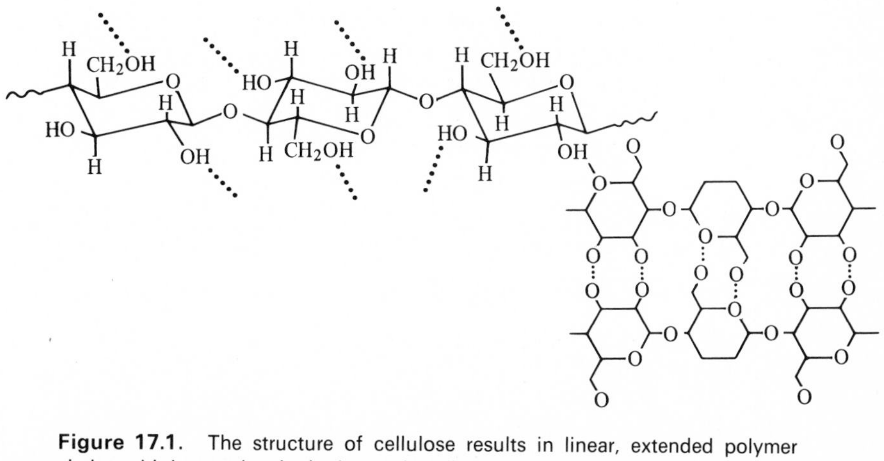

Figure 17.1. The structure of cellulose results in linear, extended polymer chains with intermolecular hydrogen bonding.

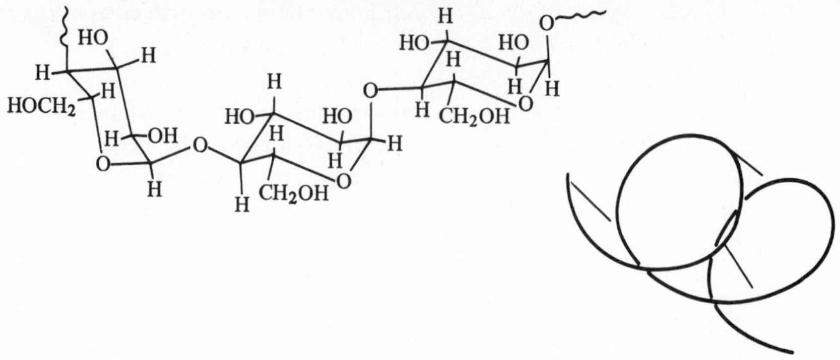

Figure 17.2. The structure of amylose (starch) results in coiled polymer chains with intramolecular hydrogen bonding.

Starch

Amylopectin is a mixture of branched molecules whose glucose units are joined by α-glycosidic linkages. The α-glycosidic linkages in α-amylose result in a coiled or spiral chain as shown in Figure 17.2. The result of these two factors (chain branching and the α-glycosidic linkages) is that starch molecules tend to become tangled rather than to align themselves as cellulose molecules do. Hydrogen bonding occurs primarily between parts of the same molecule and therefore contributes little to the tensile strength of the polymer.

Problem 17.6
Explain how chain branching in starch inhibits the formation of linear, extended chains.

17.5 PROTEINS

Proteins are one of the most important classes of biopolymers. They have molecular weights as low as 5000 and as high as millions. They occur in plants and animals as hair, wool, horn, organ tissues, silk, and many other structures.

Simple proteins hydrolyze to give only amino acids. **Conjugated proteins** are more complex substances that contain, in addition to amino acids, one or more nonprotein substances such as carbohydrate or lipid groups or metal ions.

The acid-catalyzed **hydrolysis** of any sample of protein regardless of its source always yields combinations of the same twenty or so amino acids. The hydrolysis is shown by the equation:

$$\text{Protein} + n\text{H}_2\text{O} \xrightarrow[\text{then neutralize}]{\text{acid, heat}} n\text{H}-\underset{\displaystyle \text{H}}{\underset{|}{\text{N}}}-\overset{\displaystyle \text{R}}{\overset{|}{\text{CH}}}-\overset{\displaystyle \text{O}}{\overset{\|}{\text{C}}}-\text{OH}$$

in which R may vary. Simple proteins are thus polyamides of α-amino acids.

The reverse of the above equation shows the formation of a diamide (dipeptide) as the first step in the formation of the polyamide:

$$\mathrm{H{-}\underset{\displaystyle H}{\underset{|}{N}}{-}\overset{\displaystyle R}{\overset{|}{C}}H{-}\overset{\displaystyle O}{\overset{\|}{C}}{-}OH + H{-}\underset{\displaystyle H}{\underset{|}{N}}{-}\overset{\displaystyle R}{\overset{|}{C}}H{-}\overset{\displaystyle O}{\overset{\|}{C}}{-}OH}$$

$$\longrightarrow \mathrm{H{-}\underset{\displaystyle H}{\underset{|}{N}}{-}\overset{\displaystyle R}{\overset{|}{C}}H{-}\overset{\displaystyle O}{\overset{\|}{C}}{-}\underset{\displaystyle H}{\underset{|}{N}}{-}\overset{\displaystyle R}{\overset{|}{C}}H{-}\overset{\displaystyle O}{\overset{\|}{C}}{-}OH + H_2O}$$

a dipeptide

(Note that by convention, polypeptides are always drawn with the free amino group on the left end and the free carboxyl group on the right end.) The dipeptide can react further to form tripeptides, tetrapeptides, and so forth. The new bond that forms is called a **peptide bond**.

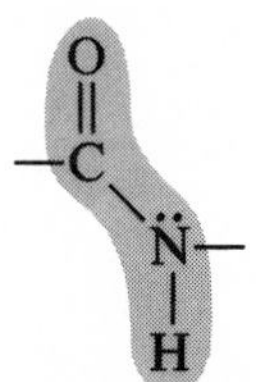

the peptide bond

Problem 17.7
Draw the structural formula of each of the following polypeptides.
(a) polyalanine (b) polyglycine (see Table 17.1)

Problem 17.8
(a) Complete and balance the following equation.

$$\mathrm{H_2NCH_2\overset{\displaystyle O}{\overset{\|}{C}}{-}NHCH_2\overset{\displaystyle O}{\overset{\|}{C}}{-}NHCH_2CO_2H + H_2O \xrightarrow[\text{then neutralize}]{\text{acid, heat}}}$$

(b) How would you describe the above reaction?
(c) How would you describe the starting compound in the above reaction?
(d) Calculate the molecular weight of the starting compound.

17.6 AMINO ACIDS

Only twenty or so different amino acids make up all of the known proteins. All of these amino acids are **α-amino acids**; that is, the amino group is on the carbon atom adjacent

Table 17.1 Natural Amino Acids

Name	Abbreviation	Formula
Alanine	Ala	$CH_3CH(NH_2)CO_2H$
Arginine	Arg	$H_2NC(=NH)NHCH_2CH_2CH_2CH(NH_2)CO_2H$
Asparagine	Asn	$H_2NC(=O)CH_2CH(NH_2)CO_2H$
Aspartic acid	Asp	$HO_2CCH_2CH(NH_2)CO_2H$
Cysteine	Cys	$HSCH_2CH(NH_2)CO_2H$
Cystine	Cys–Cys	$HO_2CCH(NH_2)CH_2SSCH_2CH(NH_2)CO_2H$
Glutamic acid	Glu	$HO_2CCH_2CH_2CH(NH_2)CO_2H$
Glutamine	Gln	$H_2NCOCH_2CH_2CH(NH_2)CO_2H$
Glycine	Gly	$CH_2(NH_2)CO_2H$
Histidine	His	(imidazol-4-yl)$-CH_2CH(NH_2)CO_2H$
Hydroxyproline	Hyp	(4-hydroxypyrrolidine ring, N–H)$-CO_2H$
*Isoleucine	Ile	$CH_3CH_2CH(CH_3)CH(NH_2)CO_2H$

Table 17.1 (*continued*)

Name	Abbreviation	Formula
*Leucine	Leu	$(CH_3)_2CHCH_2CH(NH_2)CO_2H$
*Lysine	Lys	$H_2NCH_2CH_2CH_2CH_2CH(NH_2)CO_2H$
*Methionine	Met	$CH_3SCH_2CH_2CH(NH_2)CO_2H$
*Phenylalanine	Phe	C_6H_5—$CH_2CH(NH_2)CO_2H$
Proline	Pro	pyrrolidine ring (N—H)—CO_2H
Serine	Ser	$HOCH_2CH(NH_2)CO_2H$
*Threonine	Thr	$CH_3CH(HO)CH(NH_2)CO_2H$
*Tryptophan	Trp	indole ring (N—H)—$CH_2CH(NH_2)CO_2H$
Tyrosine	Tyr	HO—C_6H_4—$CH_2CH(NH_2)CO_2H$
*Valine	Val	$(CH_3)_2CHCH(NH_2)CO_2H$

The α-amino acid unit —$CHCO_2H$ (with —NH below) is shown shaded. The eight essential amino acids are preceded by an asterisk.

to the carboxyl group. All these naturally occurring amino acids have at least one amino group and one carboxyl group. They have the general structure

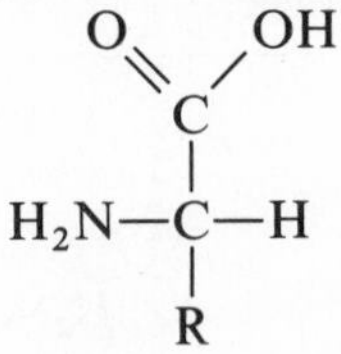

$$\begin{array}{c} O \quad OH \\ \diagdown\!\!\diagdown \;\; / \\ C \\ | \\ H_2N—C—H \\ | \\ R \end{array}$$

in which R may be aliphatic or aromatic and may contain additional functional groups such as amino, carboxyl, hydroxyl, and sulfhydryl (—SH).

All the twenty-odd amino acids that occur naturally in proteins are L-amino acids, defined as in Section 16.6 except that NH_2 replaces OH. The Fischer projection formulas are

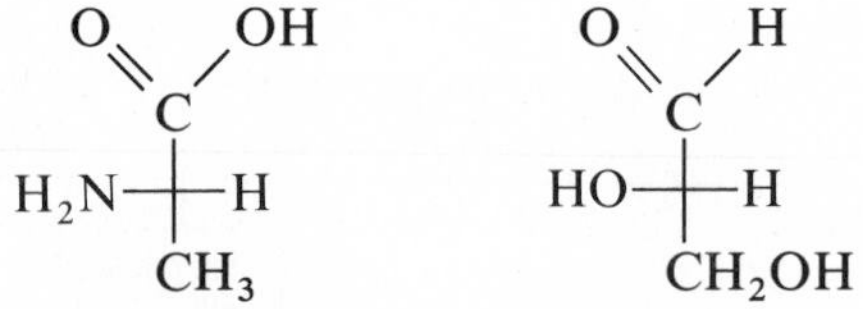

$$\begin{array}{c} O \quad OH \\ \diagdown\!\!\diagdown \;\; / \\ C \\ H_2N \!-\!\!\!+\!\!\!-\! H \\ CH_3 \end{array} \qquad\qquad \begin{array}{c} O \quad H \\ \diagdown\!\!\diagdown \;\; / \\ C \\ HO \!-\!\!\!+\!\!\!-\! H \\ CH_2OH \end{array}$$

L-alanine L-glyceraldehyde

The only exception is glycine, which is achiral. Table 17.1 lists the important naturally occurring amino acids.

All living organisms can synthesize amino acids. Many higher organisms, however, are unable to synthesize all the amino acids necessary to their existence. Such organisms must receive these amino acids in their diet. Humans require eight amino acids in this category; these are called **essential amino acids**, and are designated in Table 17.1.

Problem 17.9
Draw the Fischer projection formula of (a) D-alanine, (b) glycine, (c) L-leucine, (d) L-serine.

Problem 17.10
Draw the structural formula of each of the following dipeptides. (a) Phe·Ala, (b) Gly·Val (*Note*: In drawing these structures, show the amino group on the left and the carboxyl group on the right: Gly = $H_2N—CH_2—CO_2H$.)

17.7 ACID-BASE PROPERTIES OF AMINO ACIDS

Because amino acids possess both a basic and an acidic group, they occur primarily in the salt form called the **zwitterion**, which is both an anion and a cation. (*Zwitter* in German means hybrid.)

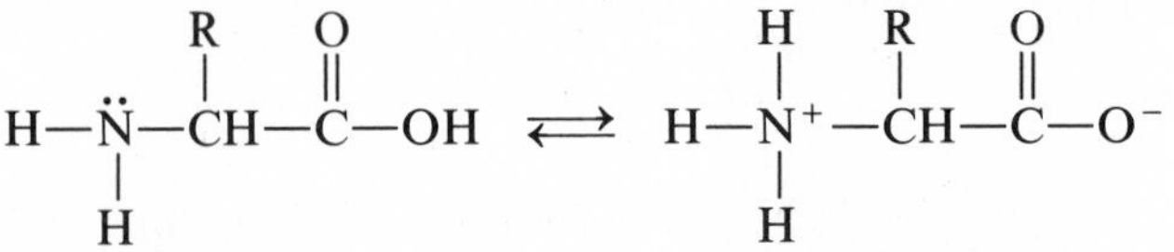

$$H—\ddot{N}(H)—CH(R)—C(=O)—OH \rightleftarrows H—N^{+}(H)(H)—CH(R)—C(=O)—O^{-}$$

zwitterion

Table 17.2 Isoelectric Points of Some Amino Acids

Amino Acid	R Group	Isoelectric Point (pH)
Glycine	H—	5.97
Alanine	CH_3—	6.02
Leucine	$CH_3CH(CH_3)CH_2$—	5.98
Lysine	$H_2NCH_2CH_2CH_2CH_2$—	9.7
Aspartic acid	$HO-C(=O)-CH_2$—	2.7

The physical properties of amino acids are consistent with an ionic structure: Amino acids have high melting points, usually above 200°C; they are very soluble in water and insoluble in nonpolar solvents such as benzene or ether.

Amino acids can react with both acids and bases; they are therefore amphoteric.

$$\underset{\text{anionic form}}{H_2N-CH(R)-C(=O)-O^-} \underset{H_3O^+}{\overset{OH^-}{\rightleftharpoons}} \underset{\text{zwitterion form}}{H_3\overset{+}{N}-CH(R)-C(=O)-O^-} \underset{OH^-}{\overset{H_3O^+}{\rightleftharpoons}} \underset{\text{cationic form}}{H_3\overset{+}{N}-CH(R)-C(=O)-OH}$$

The pH of the solution determines which species predominates. The pH at which the zwitterion concentration is at its maximum is called the **isoelectric point**. The nature of R determines the isoelectric point. R groups that contain an additional amino group have isoelectric points well above pH 7. Those that have an additional carboxyl group have isoelectric points well below pH 7. Groups that have neutral R groups have isoelectric points slightly below pH 7 because the carboxyl group "acid" is slightly stronger than the amino group "base." Table 17.2 lists the isoelectric points of a few amino acids.

Problem 17.11

Which of the amino acids in Table 17.1 have isoelectric points greater than 7? Which have isoelectric points lower than 7? [Explain.]

17.8 POLARITY OF AMINO ACID R GROUPS

In the polypeptide chain, the amino and carboxyl groups are involved in peptide bonds, so they are no longer acidic or basic. For this reason, the properties of a protein chain depend mainly on the nature of the R groups along the chain.

If R contains a polar functional group that can form hydrogen bonds with water, we call that R group **hydrophilic** (water-loving). If R is a hydrocarbon, we call it **hydrophobic** (water-hating) because it does not interact with water. Hydrophobic groups tend to cluster together within the polypeptide chain and squeeze out the water molecules. Hydrophilic groups, on the other hand, seek out water molecules and therefore tend to stretch out the polypeptide chain within the water medium. The structure of a protein depends ultimately on the number and locations of hydrophobic and hydrophilic groups along the chain. We shall examine this aspect of protein structure in Section 17.12.

Problem 17.12

Classify the R groups of the following amino acids as hydrophilic or hydrophobic.

(a) valine (b) phenylalanine (c) serine
(d) glutamic acid (e) hydroxyproline

17.9 ANALYSIS OF PROTEINS

The determination of which amino acids make up a protein requires acid-catalyzed hydrolysis and identification of the amino acids present in the hydrolysis product mixture.

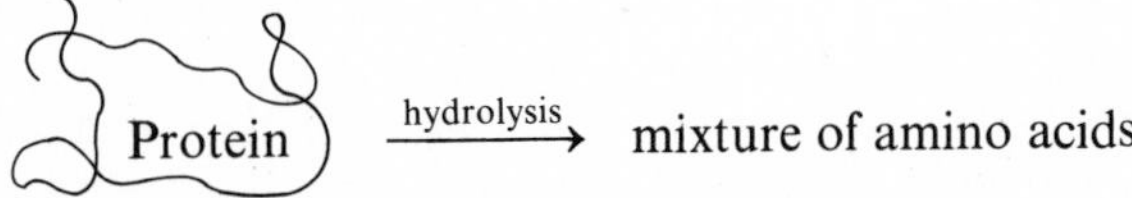

The next step in our analysis of a protein is the determination of its molecular weight. From these two pieces of information we know how many of each kind of amino acid unit (called amino acid *residues*) are present in the protein. We still have no clue, however, about the *sequence* of the amino acid residues within the protein.

17.10 PROTEIN PRIMARY STRUCTURE

When we consider that some proteins have molecular weights in the millions we can appreciate that the combinations of twenty-odd amino acids are almost limitless. For example, a tripeptide made up of histidine (H), proline (P), and serine (S) can occur in six different combinations: HPS, HSP, PSH, PHS, SHP, SPH. With five amino acids there are 120 different pentapeptides. The number of combinations is $n!$ (n factorial); $3! = 3 \times 2 \times 1 = 6$; $5! = 5 \times 4 \times 3 \times 2 \times 1 = 120$. A nonapeptide with equal molar amounts of nine amino acids can have 362,880 structural isomers! We may add to this complexity the fact that all the amino acids except glycine have at least one chiral center.

The exact arrangement of amino acids in a particular protein is fixed, however. In silk fibroin, for example, every other amino acid unit is glycine, every fourth one is alanine, and every sixteenth one is tyrosine. Other amino acids are also present in small amounts.

The exact linear sequence of amino acids determines a protein's **primary structure**. The primary structure is not a haphazard arrangement, but has a definite sequential order.

Problem 17.13

Draw the structural formulas of (a) all the dipeptides of alanine and glycine, (b) all the tripeptides of alanine, glycine, and tyrosine.

17.11 DETERMINATION OF PROTEIN PRIMARY STRUCTURE

The amino acid residue containing the free α-amino group at the end of the polypeptide chain is called the **N-terminal amino acid residue**; the amino acid residue con-

$$\underset{\text{N-terminal amino acid residue}}{H_2N\overset{R}{\overset{|}{C}}H\overset{O}{\overset{\|}{C}}}-NH\overset{R}{\overset{|}{C}}H\overset{O}{\overset{\|}{C}}-NH\overset{R}{\overset{|}{C}}H\overset{O}{\overset{\|}{C}}-NH\overset{R}{\overset{|}{C}}H\overset{O}{\overset{\|}{C}}-NH\overset{R}{\overset{|}{C}}H\overset{O}{\overset{\|}{C}}-\underset{\text{C-terminal amino acid residue}}{NH\overset{R}{\overset{|}{C}}H\overset{O}{\overset{\|}{C}}-OH}$$

taining the free carboxyl group at the other end of the chain is called the **C-terminal amino acid residue**. We can determine the N-terminal and C-terminal amino acid residues in a chain of any length.

To determine the amino acid sequence (primary structure) of a polypeptide, we must break down the chain into smaller polypeptide units by partial hydrolysis and then determine the N-terminal and C-terminal amino acid residues of each of these small polypeptides. By noting the amino acid residues that overlap, we can reconstruct the primary structure. Let us examine the methods of terminal residue analysis.

N-Terminal Residue Analysis

Of the methods now available, that of P. Edman, University of Lund, Sweden, is the most versatile. In the Edman method we label the terminal α-amino group by reacting the polypeptide with phenylisothiocyanate. When the labeled polypeptide is hydrolyzed with acid, the labeled amino acid comes off as a phenylthiohydantoin, which is then identified by comparison with thiohydantoins prepared from known amino acids. The remainder of the chain is left intact, and the procedure can be repeated. The reactions are as follows.

Edman Degradation Method

$$C_6H_5{-}N{=}C{=}S + H_2N{-}\underset{R}{CH}{-}\overset{O}{\overset{\|}{C}}{-}NH{-}\underset{R'}{CH}{-}\overset{O}{\overset{\|}{C}}\sim\sim$$

phenylisothiocyanate N-terminal amino acid residue

$$\Big\downarrow OH^-$$

$$C_6H_5{-}NH{-}\overset{S}{\overset{\|}{C}}{-}NH{-}\underset{R}{CH}{-}\overset{O}{\overset{\|}{C}}{-}NH{-}\underset{R'}{CH}{-}\overset{O}{\overset{\|}{C}}\sim\sim$$

labeled polypeptide

$$\Big\downarrow H_2O^+$$

$$\text{phenylthiohydantoin (ring: } C_6H_5{-}N{-}C({=}S){-}NH{-}CH(R){-}C({=}O){-}) + H_2N{-}\underset{R'}{CH}{-}\overset{O}{\overset{\|}{C}}\sim\sim$$

phenylthiohydantoin polypeptide less one amino acid

The remaining polypeptide can be treated by the Edman method to determine the next amino acid residue, and so forth.

Problem 17.14
Write equations for the Edman degradation of the tripeptide Phe·Ala·Gly.

C-Terminal Residue Analysis

We determine the C-terminal amino acid residue by using digestive enzymes called *carboxypeptidases.* These enzymes attack only the peptide bond of amino acids that have a free carboxyl group. The problem with this method is that the enzyme continues to work on the remaining polypeptide until it is completely hydrolyzed. During the reaction the mixture is analyzed periodically to determine which amino acid was produced first in the largest amount; that amino acid is the C-terminal amino acid. The amino acid produced in the second greatest is the second amino acid from the carboxyl end, and so forth.

Problem 17.15
Write the equation for the carboxypeptidase-catalyzed hydrolysis of Phe·Ala·Gly.

Partial Hydrolysis

Polypeptide chains can be partially hydrolyzed into smaller polypeptide units. Such smaller polypeptides are obviously easier to analyze by terminal residue analysis than larger ones. We can partially hydrolyze a polypeptide by heating it for a brief time in acid solution, or we can use enzymes. Some enzymes offer the advantage that they attack only specific peptide linkages, so we can determine how the fragments were joined together.

Let us now examine a simple case: Analysis of a polypeptide followed by molecular weight determination showed that it was a hexapeptide with the amino acids Ala_2, Glu, Gly, Phe, Ser. (Commas mean that the sequence is not specified.) Reaction of the hexapeptide with phenylisothiocyanate produced the phenylthiohydantoin of phenylalanine. Carboxypeptidase hydrolysis of the hexapeptide yielded glycine.

From the above we know the N-terminal and C-terminal residues:

$$\text{Phe}(\text{Ala}_2, \text{Glu}, \text{Ser})\text{Gly}$$

Partial hydrolysis of the hexapeptide gave fragments whose sequences were shown by terminal residue analysis to be Phe·Ala·Glu, Glu·Ser, and Ser·Ala·Gly. (The dots indicate the actual sequence.) Combining all the facts above and overlapping the fragments, we obtain

Phe·Ala·Glu
Glu·Ser
Ser·Ala·Gly
Phe·Ala·Glu·Ser·Ala·Gly = original hexapeptide

Problem 17.16

A polypeptide was found to have the composition Tyr, Cys, Arg_2, Gly, Ala, Lys. Reaction with phenylisothiocyanate gave the phenylthiohydantoin of tyrosine. Carboxypeptidase hydrolysis gave glycine. Partial hydrolysis gave the following fragments: Lys·Ala, Arg·Gly, Ala·Arg·Arg, and Tyr·Cys·Lys. What is the sequence of the original polypeptide?

17.12 PROTEIN SECONDARY STRUCTURE

Protein molecules orient themselves in definite patterns just as cellulose and starch do. Silk, for example, forms a highly oriented fiber that is slightly elastic. Wool fibers, on the other hand, consist of bundles of regularly coiled molecules and are much more elastic than silk.

We can explain these differences between silk and wool in terms of their **secondary structures**; that is, the orientation of polymer molecules that results from hydrogen bonding between parts of a given chain or between two chains. Silk has the **β-keratin** or stretched molecular form, in which hydrogen bonding occurs between different chains. Wool has the **α-keratin** or unstretched structure, in which hydrogen

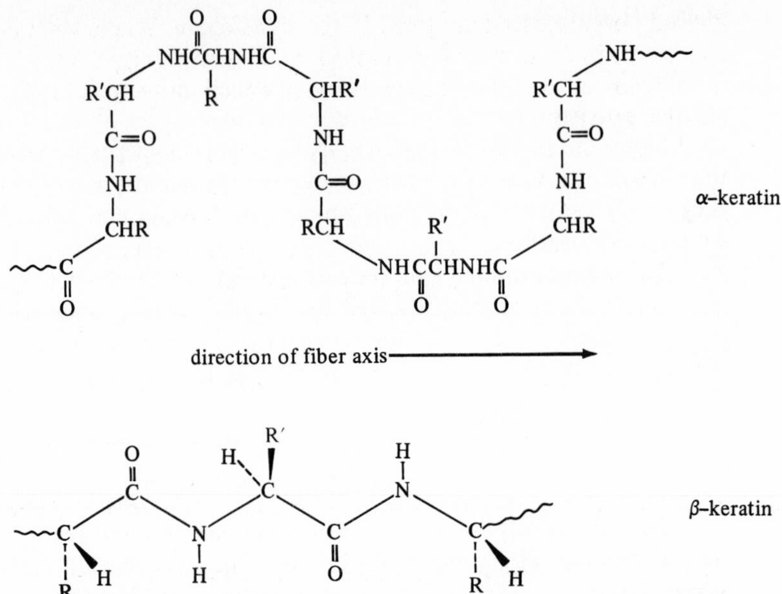

Figure 17.3. α-Keratin and β-keratin forms of protein secondary structure.

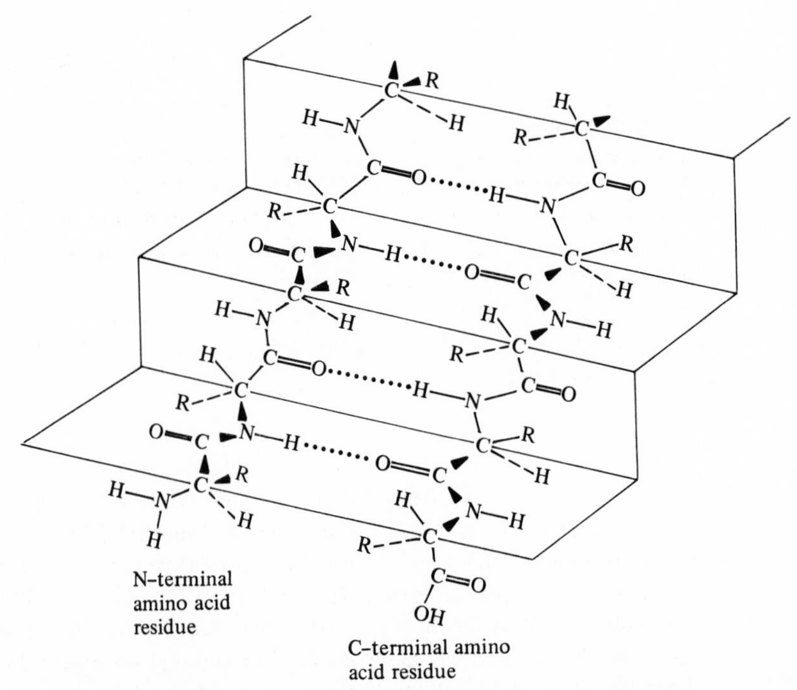

Figure 17.4. The pleated sheet structure of β-keratin.

bonding occurs between atoms in the same chain. When we stretch wool fibers we convert them from the α-keratin into the β-keratin form (Fig. 17.3). In the β-keratin form, hydrogen bonding occurs between chains (as shown in Fig. 17.4) to form the pleated sheet structure.

The α-keratin form is an example of a helical structure. Most proteins contain at least some α-helical structure in which the molecule coils on itself in a right-handed helix. The α-helix has about 3.6 amino acid units per coil, and the coil is held together by hydrogen bonding. These structures were determined by Linus Pauling and Robert Corey using X-ray diffraction analysis.

Problem 17.17

Assuming that there are 3.6 amino acid units per coil of the α-helix, show where hydrogen bonding occurs. Use the α-keratin formula of Fig. 17.3.

17.13 PROTEIN DENATURATION

We can disrupt the delicate secondary structure of many proteins by a process called **denaturation**. We denature a protein by heating it or by treating it with acid, base, or organic solvents at room temperature. This process results in the breakdown of the hydrogen bonding that holds the three-dimensional structure together. Examples of denaturation are the coagulation of milk by acidification, and the hardening of an egg by boiling.

17.14 BIOLOGICAL FUNCTIONS OF PROTEINS

Proteins serve numerous functions in living systems. We may classify them into two broad classes—structural proteins and regulatory proteins.

Structural proteins form the structural components of animals in the form of α-keratin: skin, hair, nails, feathers, connective tissues, bone, cartilage, tendons, muscle, and insect exoskeletons.

Regulatory proteins serve more strictly chemical functions such as catalysis (enzymes), storage (milk, eggs), hormonal control (insulin), transport (hemoglobin), and protection (antibodies).

17.15 NUCLEIC ACIDS

Deoxyribonucleic acid (DNA) is found in the nuclei of all cells and is the major component of the chromosomes. DNA carries the genetic information characteristic of the individual, and thus it plays a central role in the transmission of traits from one generation to the next.

Until 1953 the mechanism of DNA replication was unknown. It was known, however, that DNA consists of very long chains made up of alternating phosphate and sugar groups (deoxyribose) with one nitrogenous base attached to each sugar

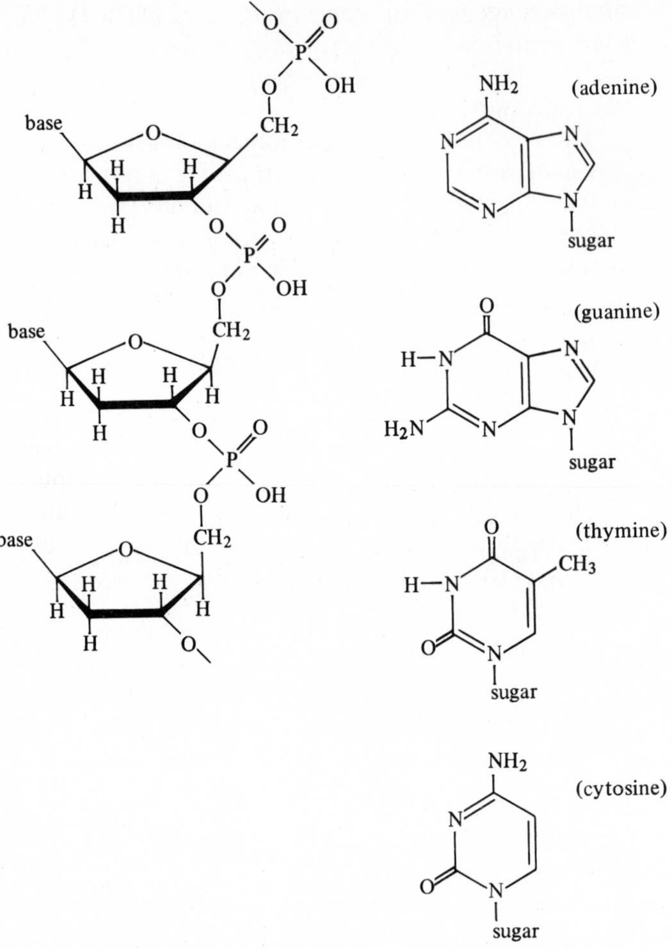

Figure 17.5. Schematic structure of a single DNA strand. The four bases are shown at the right with their points of attachment to the deoxyribose units.

(Fig. 17.5). Four nitrogenous bases are found in the DNA of all organisms. The sequence in which they occur in the chain allows the infinite variety possible in genetic material. The four bases are adenine, guanine, thymine, and cytosine. The first two are called *purine bases*; the latter two are called *pyrimidine bases*.

adenine guanine thymine cytosine

purine bases pyrimidine bases

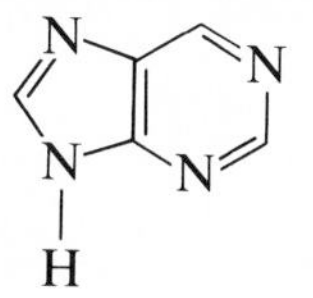

purine 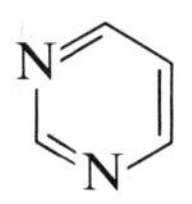 pyrimidine 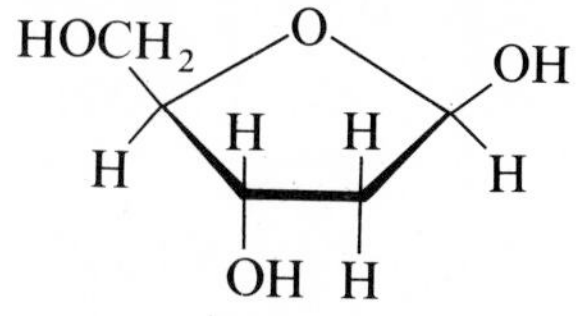 2-deoxyribose

The adenine/thymine ratio and the guanine/cytosine ratio are always very close to one regardless of the source of DNA, as explained below.

In 1953, J. D. Watson and F. H. C. Crick proposed a molecular structure that accounted for the known features of DNA. Their structure also suggested a mechanism for the replication of genetic material.

Watson and Crick proposed that DNA consists of two intertwined chains with the phosphate groups on the outside of the double helix and the bases on the inside. Hydrogen bonding between opposite bases on the two chains holds them together. They used models to show that a smooth, regular double helix is possible only if hydrogen bonding occurs between adenine and thymine and between guanine and cytosine. Other combinations of bases opposite each other on the chains result in unacceptable bond lengths and an irregular double helix. In other words, if one chain has an adenine group in a given position, the other chain must have a thymine group at the corresponding position opposite the adenine. Figure 17.6 shows these two base pairings, and Figure 17.7 diagrams the overall double helix structure.

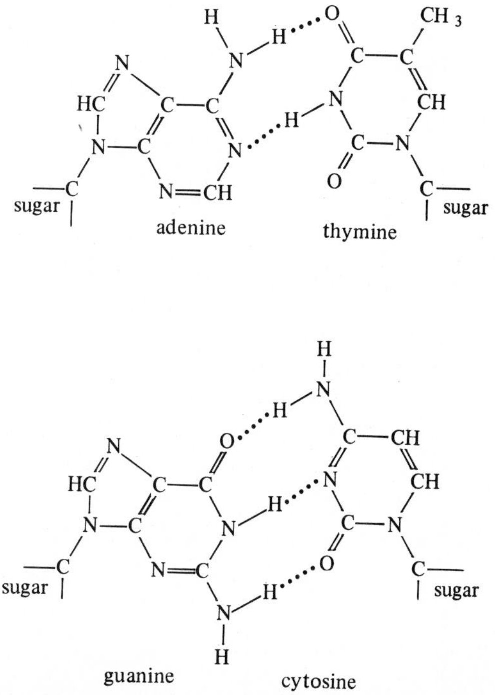

sugar adenine thymine sugar

sugar guanine cytosine sugar

Figure 17.6. Pairing of adenine-thymine and guanine-cytosine.

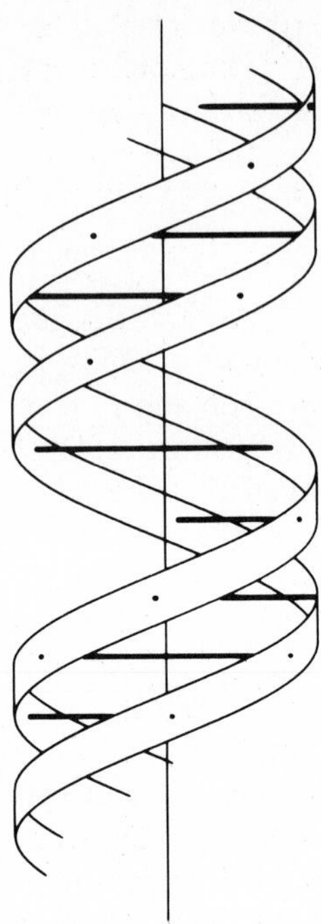

Figure 17.7. Schematic representation of the DNA double helix. The two ribbons represent the two phosphate-sugar chains and the horizontal rods represent the base pairs that hold the chains together. The vertical line represents the fiber axis.

During replication, then, the hydrogen bonds break and the single chains separate. The single strands are complements of each other, and each can act as template for the enzyme-catalyzed formation of a new double helix which is the exact replicate of the original.

RNA (*ribonucleic acid*) is another polymeric compound that occurs in the cell and is involved with DNA in the transmission of genetic information. RNA differs from DNA in that ribose replaces deoxyribose, and uracil replaces thymine.

uracil ribose

Unlike DNA, RNA molecules exist as single strands. There are three general types of RNA. *Ribosomal RNA* has a structural role and constitutes the major amount of RNA in a cell. *Messenger RNA* functions as a template for protein synthesis in association with the ribosomes. Soluble RNA or *transfer RNA* is a smaller molecule, also involved in protein synthesis. Its role is to bind to individual amino acid molecules and insert them in the proper place in the growing protein chain.

Problem 17.18

Using the symbols A = adenine, G = guanine, C = cytosine, and T = thymine, construct the complementary DNA strand from the partial single DNA strand given.

```
├A
├A
├T
├G
├C
├T
├A
├T
├C
├G
├G
├T
```

NEW TERMS

α-Amino acid Compound in which an amino group is bonded to the α-carbon atom of a carboxylic acid (17.6):

$$\mathrm{R{-}\underset{\displaystyle NH_2}{\underset{|}{C}H}{-}CO_2H}$$

Amphoteric A compound that can react both as an acid and as a base (17.7).

Cellulose A linear, unbranched polymer of β-D-glucose units (17.4).

Denaturation Disruption of protein secondary structure by heating or by treating the protein with acid, base, or organic solvents at room temperature (17.13).

Essential amino acids The eight amino acids that humans require but cannot synthesize. The essential amino acids must be available in the diet (17.6).

Hydrolysis Disruption of protein primary structure into its constituent amino acids by heating with aqueous acid (17.5).

Hydrophilic group An R group of an amino acid that can form hydrogen bonds with water. Hydrophilic means *water-loving* (17.8).

Hydrophobic group A hydrocarbon R group of an amino acid. Hydrophobic means *water-hating* (17.8).

Isoelectric point The pH at which the zwitterion concentration (of an amino acid) is at its maximum (17.7).

α-Keratin structure Unstretched secondary structure in which hydrogen bonding occurs between atoms in the same chain (17.12).

β-Keratin structure Stretched secondary structure in which hydrogen bonding occurs between atoms in different chains (17.12).

Macromolecule A molecule of very high molecular weight, also called *polymer* (Introduction).

Monomer The starting molecule in the synthesis of a polymer (17.1).

Nucleic acids DNA and similar compounds (the RNAs) that are the essential components in the transmission of genetic information (17.15).

Peptide bond Amide bond that joins amino acid units together in peptides and proteins (17.5):

$$-\overset{\overset{\displaystyle O}{\|}}{C}-\underset{\underset{\displaystyle H}{|}}{\ddot{N}}-$$

Polymer A macromolecule composed of a large number of small, repeating molecular units. Small polymers are represented as dimers, trimers, tetramers, and so forth, depending on the number of repeating units (17.1).

Protein A **simple protein** hydrolyzes to give only amino acids. A **conjugated protein** hydrolyzes to give amino acids and more complex substances such as carbohydrate or lipid groups or metal ions (17.5).

Protein primary structure The exact linear sequence of amino acid units in a protein (17.10).

Protein secondary structure The orientation of a protein molecule that results from hydrogen bonding between parts of a given chain or between two chains (17.12).

Starch A branched polymer of α-D-glucose units (17.4).

C-terminal amino acid residue The amino acid unit on the end of a polypeptide that has the free $-CO_2H$ group (17.11).

N-terminal amino acid residue The amino acid unit on the end of a polypeptide that has the free $-NH_2$ group (17.11).

Viscosity Resistance of a fluid to flow (17.3).

Zwitterion A species that is both an anion and a cation. For an amino acid, the zwitterion is (17.7):

$$R-\underset{\underset{\displaystyle {}^{+}NH_3}{|}}{CH}-CO_2^-$$

ADDITIONAL PROBLEMS

[17.19] Define each of the following terms.
(a) macromolecule (b) polymer (c) monomer
(d) dimer

[17.20] Is it strictly correct to speak of the molecular weight of a polymer? Explain.

17.21 What physical properties would you examine if you wanted to determine whether a substance was polymeric? What would you expect to observe for each property?

[17.22] What properties would you expect for a polyglucose that has α- and β-glycosidic linkages randomly distributed?

17.23 Draw the structural formula of [a] poly(aspartic acid) and (b) poly(hydroxyproline).

17.24 Write equations for the acid-catalyzed hydrolysis of the polypeptides of Problem 17.23.

17.25 Draw the Fischer projection formulas of [a] L-lysine, (b) D-phenylalanine.

[17.26] Draw the structural formula and outline the peptide bonds in the tripeptide Ala·Ala·Ala.

17.27 Which of the amino acids of Table 17.1 have isoelectric points near pH 7?

[17.28] Draw the structural formula of the hexapeptide Phe·Ala·Val·Ser·Asp·Phe. Label all the hydrophobic and hydrophilic groups.

17.29 How many isomeric tetrapeptides are possible if they are composed of four different amino acids in equal amounts?

[17.30] Explain the difference between the primary and secondary structure of a protein.

[17.31] (a) Write equations for the N-terminal residue analysis of Leu·Phe·Ala by the Edman method. (b) Write the equation for the carboxypeptidase-catalyzed C-terminal residue analysis of the same tripeptide.

17.32 A polypeptide has the composition His, Arg, Ala_2, Ile, Lys, Phe, Glu. Terminal residue analysis showed that it has alanine at the N-terminal position and Ile at the C-terminal position. Partial hydrolysis of the original polypeptide gave fragments that have the compositions Ala·His, Lys·Phe·Ile, Ala·Ala, Glu·Lys, and His·Arg·Glu. What is the primary structure of the polypeptide?

[17.33] Draw the β-keratin structure of polyglycine and show how hydrogen bonding occurs.

17.34 What chemical reaction occurs during protein hydrolysis? Show with an equation.

[17.35] Describe what occurs during protein denaturation.

[17.36] How does the Watson–Crick model of DNA explain the adenine/thymine and guanine/cytosine ratio obtained in all DNA?

[17.37] How does the Watson–Crick model of DNA explain the replication of genetic material?

---

18 BEHAVIOR OF VERY LARGE MOLECULES–SYNTHETIC POLYMERS: AN ELECTIVE CHAPTER

We saw in Chapter 17 that polymers have a wide range of physical properties from liquids to plastics, to rubbery solids, to brittle solids. The goal of polymer chemists is to understand and synthesize materials that can be used for desired, specific purposes. Over the past fifty years a massive polymer industry has arisen. The results are known and used by us all: fibers of all types, such as the acrylics and polyesters; plastics used in a variety of forms as utensils, tools, bearings, and shields; cements and finishes such as the well-known epoxy products.

The job of the synthetic polymer chemist is to tailor-make new polymers with specific properties. In this chapter, we shall explore how this task is carried out.

We name a polymer in terms of the monomer from which it was made or might have been made. Polystyrene is prepared by the reaction of styrene with a suitable catalyst.

$$n\,CH_2{=}CH(C_6H_5) \xrightarrow{\text{catalyst}} \left(\!-CH_2-CH(C_6H_5)-\right)_n$$

styrene — polystyrene

We form the name of the polymer by adding the prefix *poly-* to the name of the monomer. If the monomer name is complex or consists of two or more words, we must enclose the monomer name in parentheses. The following examples will illustrate how to name polymers.

Monomer	Polymer
Vinyl chloride, $CH_2{=}CHCl$	Poly(vinyl chloride), $\left(-CH_2-CH(Cl)-\right)_n$
α-Methylstyrene, $CH_2{=}C(CH_3)(C_6H_5)$	Poly(α-methylstyrene), $\left(-CH_2-C(CH_3)(C_6H_5)-\right)_n$
Ethylene glycol, $HOCH_2CH_2OH$ Terephthalic acid, $HO_2C-C_6H_4-CO_2H$	Poly(ethylene terephthalate), $\left(-OCH_2CH_2O-\overset{O}{\overset{\Vert}{C}}-C_6H_4-\overset{O}{\overset{\Vert}{C}}-\right)_n$

Problem 18.1

Give the name of each of the following polymers.

(a) $\left(-CH_2-CH(CN)-\right)_n$, prepared from acrylonitrile, $CH_2{=}CHCN$

(b) $\left(-CF_2-CF_2-\right)_n$, prepared from tetrafluoroethylene, $CF_2{=}CF_2$

(c) $\left(-CH_2-C(CH_3)(CO_2CH_3)-\right)_n$, prepared from methyl methacrylate $CH_2{=}C(CH_3)CO_2CH_3$

18.2 CLASSIFICATION OF POLYMERS BY STRUCTURE

We may classify polymers according to their molecular structure in a way that allows us to explain their physical properties. **Linear polymers** consist of chains which are independent of one another, so they move relatively freely. Linear polymers melt when they are heated and resolidify when they are cooled. Because of this property, linear polymers are called **thermoplastic polymers**.

In other polymers the chains are joined to each other by **cross-links**. Cross-links are chemical bonds that link one polymer chain to another. The resulting molecules form a three-dimensional network and are called **network polymers** or **cross-linked polymers**. Cross-linked polymers do not melt when they are heated because the molecules are locked into their relative positions by the cross-links. Once the cross-links are formed in a sample of polymer, the polymer is infusible. For this reason, cross-linked polymers are called **thermosetting polymers**.

Linear and cross-linked polymers also differ in their solubility properties. Linear polymers may dissolve in certain solvents. Cross-linked polymers are insoluble in all solvents because their molecules are inseparable.

Figure 18.1 summarizes the structure and properties of these two groups of polymers.

Problem 18.2

We have just isolated a substance from a growing plant. It has a wide melting point range and its 10% solution in acetone has the consistency of molasses. What conclusions can you draw about its chemical structure? (See Sec. 17.3.)

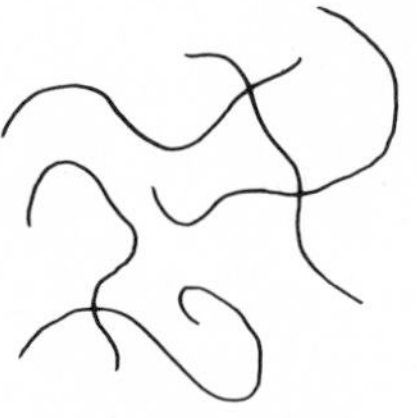

Linear polymer
Thermoplastic
Fusible, soluble

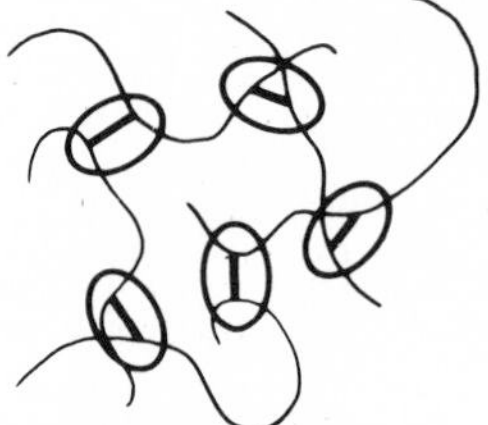

Cross-linked polymer
(network polymer)
Thermosetting polymer
Infusible, insoluble

Figure 18.1. Contrast between linear and cross-linked polymers. Cross-links are shown circled.

18.3 CLASSIFICATION OF POLYMERS BY REACTION TYPE

Synthetic polymers are generally prepared by one of two general reaction types—addition and condensation. Polymers such as polystyrene are called **addition polymers** because the monomer molecules add to each other without losing any atoms:

$$M^+ + CH_2{=}CH(C_6H_5) + CH_2{=}CH(C_6H_5) + CH_2{=}CH(C_6H_5) + \cdots$$

$$\longrightarrow M{-}CH_2{-}CH(C_6H_5){-}CH_2{-}CH(C_6H_5){-}CH_2{-}CH(C_6H_5){\sim}, \text{ written as } \left(CH_2CH(C_6H_5)\right)_n$$

Most addition polymers are prepared from alkenes by reactions analogous to the above reaction of styrene. The general equation is

$$n\,CH_2{=}\underset{\displaystyle X}{\underset{|}{C}}H \longrightarrow \left(CH_2{-}\underset{\displaystyle X}{\underset{|}{C}}H\right)_n \qquad \text{(X varies widely)}$$

Condensation polymers are produced by condensation reactions; that is, a reaction in which a small molecule (usually water) is lost when two larger molecules combine. An example is the synthesis of poly(ethylene terephthalate) by esterification:

$$n\,HO{-}CH_2CH_2{-}OH + n\,HO{-}\overset{O}{\overset{\|}{C}}{-}C_6H_4{-}\overset{O}{\overset{\|}{C}}{-}OH$$

ethylene glycol　　terephthalic acid

$$\longrightarrow \left(O{-}CH_2CH_2{-}O{-}\overset{O}{\overset{\|}{C}}{-}C_6H_4{-}\overset{O}{\overset{\|}{C}}\right)_n + n\,H_2O$$

poly(ethylene terephthalate)

Polypeptides (Sec. 17.5) are also condensation polymers. In general,

$$n\,X{-}R{-}X + n\,Y{-}R'{-}Y \longrightarrow (R{-}R')_n + 2n\,XY$$

We shall examine addition and condensation polymerization in the next two sections.

Problem 18.3

Classify each reaction below as an addition or a condensation.

(a) $n\,CH_2{=}\underset{\underset{CN}{|}}{CH} \longrightarrow \left(\!-CH_2-\underset{\underset{CN}{|}}{CH}-\!\right)_n$

(b) $n\,H\overset{\overset{O}{\|}}{C}H \longrightarrow \left(\!-CH_2-O-\!\right)_n$

(c) $n\,O{=}C{=}N\!-\!(CH_2)_6\!-\!N{=}C{=}O + n\,HO\!-\!(CH_2)_6OH$

$$\longrightarrow \left(\!-O(CH_2)_6O-\overset{\overset{O}{\|}}{C}-NH-(CH_2)_6-NH-\overset{\overset{O}{\|}}{C}-\!\right)_n$$

(d) n (phenol, C_6H_5OH) $+\; n\,H\overset{\overset{O}{\|}}{C}H \longrightarrow \left(\!-C_6H_3(OH)-CH_2-\!\right)_n + H_2O$

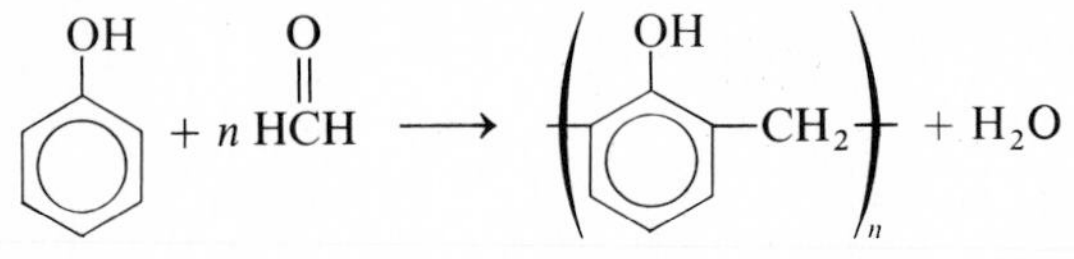

18.4 ADDITION POLYMERIZATION

In designating the structural formulas of addition polymers we have been noncommittal about the ends of the chains. We may ignore the chain ends because in a polymer of large n, the ends are relatively rare and therefore have little or no effect on the properties of the polymer. The groups attached to the ends may be the **initiator** that was used to bring about polymerization. We do not call such an initiator a catalyst because a catalyst by definition remains unchanged at the end of the reaction. In polymerizations the initiator molecules remain as part of the polymer chains.

Three general types of initiators are used in addition polymerization of alkenes (Sec. 5.6): free radicals, cations, and anions. We shall examine each of these briefly. Heterogeneous catalytic systems, called Ziegler–Natta catalysts, are also used. (We shall examine Ziegler–Natta catalysts in Sec. 18.7.)

Free Radical Initiation

Alkene monomers undergo addition polymerization when they are mixed with small amounts of compounds that decompose to produce free radicals. Peroxides are the most commonly used free radical generators because they decompose readily when warmed. Peroxides have the $-\ddot{\underset{\cdot\cdot}{O}}-\ddot{\underset{\cdot\cdot}{O}}-$ functional group. Benzoyl peroxide is an example:

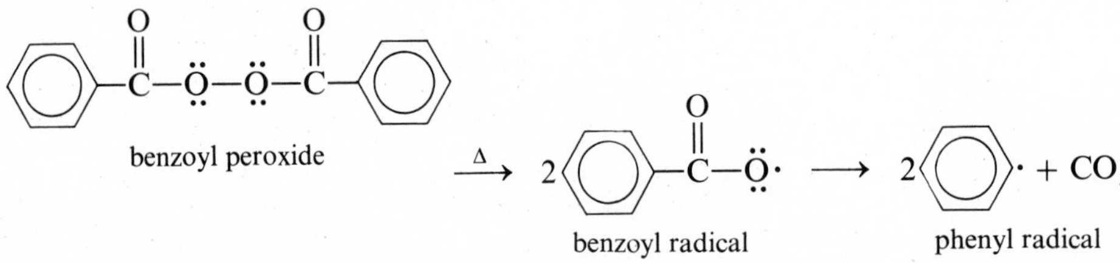

$$C_6H_5-\overset{\overset{O}{\|}}{C}-\ddot{\underset{\cdot\cdot}{O}}-\ddot{\underset{\cdot\cdot}{O}}-\overset{\overset{O}{\|}}{C}-C_6H_5 \xrightarrow{\Delta} 2\,C_6H_5-\overset{\overset{O}{\|}}{C}-\ddot{\underset{\cdot\cdot}{O}}\cdot \longrightarrow 2\,C_6H_5\cdot + CO_2$$

benzoyl peroxide — benzoyl radical — phenyl radical

The active initiator is believed to be the phenyl radical. Oxygen is a diradical and may also be used to initiate polymerization.

The mechanism of the benzoyl peroxide-catalyzed polymerization of ethylene is as follows.

$$C_6H_5\cdot + CH_2{=}CH_2 \longrightarrow C_6H_5{-}CH_2{-}\dot{C}H_2$$

$$C_6H_5{-}CH_2{-}\dot{C}H_2 + CH_2{=}CH_2 \longrightarrow C_6H_5{-}CH_2{-}CH_2{-}CH_2{-}\dot{C}H_2$$

$$\longrightarrow \longrightarrow \longrightarrow \text{(after } n \text{ steps)} \longrightarrow C_6H_5{+}CH_2CH_2{+}_nCH_2\dot{C}H_2$$

The chain may terminate by the combination of any two radicals:

$$2\,C_6H_5{+}CH_2CH_2{+}_nCH_2\dot{C}H_2 \longrightarrow C_6H_5{+}CH_2CH_2{+}_{2n+2}C_6H_5$$

or

$$C_6H_5{+}CH_2CH_2{+}_nCH_2\dot{C}H_2 + \cdot C_6H_5 \longrightarrow C_6H_5{+}CH_2CH_2{+}_{n+1}C_6H_5$$

Cationic Initiation

Lewis acids (electron pair acceptors) may also initiate the addition polymerization of alkenes. Using a proton donor such as gaseous HCl as initiator, we may describe the mechanism as follows:

$$:\ddot{\underset{..}{Cl}}:H + CH_2{=}CH(C_6H_5) \longrightarrow :\ddot{\underset{..}{Cl}}:^- + H{-}CH_2{-}\overset{+}{C}H(C_6H_5)$$

$$CH_3{-}\overset{+}{C}H(C_6H_5) + CH_2{=}CH(C_6H_5) \longrightarrow CH_3{-}CH(C_6H_5){-}CH_2{-}\overset{+}{C}H(C_6H_5) \longrightarrow \longrightarrow \longrightarrow$$

$$\longrightarrow \text{(after } n \text{ steps)} \longrightarrow H{+}CH_2{-}CH(C_6H_5){+}_n CH_2{-}\overset{+}{C}H(C_6H_5)$$

Chain termination may occur by elimination of a proton:

$$:\ddot{\underset{..}{Cl}}:^- + H\left(CH_2\underset{\underset{C_6H_5}{|}}{CH}\right)_n\underset{\underset{C_6H_5}{|}}{CH}{-}\overset{+}{CH} \longrightarrow H\left(CH_2\underset{\underset{C_6H_5}{|}}{CH}\right)_n\underset{\underset{C_6H_5}{|}}{CH}{=}CH + HCl$$

or by combination of the carbocation with the anion, Cl^-:

$$:\ddot{\underset{..}{Cl}}:^- + H\left(CH_2\underset{\underset{C_6H_5}{|}}{CH}\right)_n CH_2{-}\underset{\underset{C_6H_5}{|}}{\overset{+}{CH}} \longrightarrow H\left(CH_2\underset{\underset{C_6H_5}{|}}{CH}\right)_n CH_2{-}\underset{\underset{C_6H_5}{|}}{CHCl}$$

Anionic Initiation

When X in the alkene $CH_2{=}CHX$ is an electron-withdrawing group such as $-CN$, $-\overset{O}{\overset{\|}{C}}-R$, or $-C_6H_5$, addition polymerization may be initiated by an anionic initiator. An example of an anion that may be used as initiator is the R group of a Grignard reagent, RMgCl:

$$\overset{-}{R}{:}\overset{+}{M}gCl + CH_2{=}\underset{\underset{X}{|}}{CH} \longrightarrow R{-}CH_2{-}\underset{\underset{X}{|}}{\overset{\bar{..}}{C}H} + \overset{+}{M}gCl$$

$$R{-}CH_2{-}\underset{\underset{X}{|}}{\overset{\bar{..}}{C}H} + CH_2{=}\underset{\underset{X}{|}}{CH} \xrightarrow{\text{etc.}} R\left(CH_2\underset{\underset{X}{|}}{CH}\right)_n CH_2{-}\underset{\underset{X}{|}}{\overset{\bar{..}}{C}H}$$

Carbanions cannot lose the very basic $H{:}^-$ ion, and combination with the cation $\overset{+}{M}gCl$ leads to a very reactive end group (a Grignard reagent). Termination must therefore be brought about by quenching the anionic ends with an acid:

$$R\left(CH_2\underset{\underset{X}{|}}{CH}\right)_n CH_2\underset{\underset{X}{|}}{\overset{\bar{..}}{C}H} + H{:}A \longrightarrow R\left(CH_2\underset{\underset{X}{|}}{CH}\right)_n CH_2{-}\underset{\underset{X}{|}}{CH_2} + {:}A^-$$

Problem 18.4

In each of the mechanisms given in Section 18.4 the initiator was shown attacking the $CH_2{=}$ carbon rather than the $=CHX$ carbon. Explain why this is reasonable; that is, explain why the intermediate $I{-}\overset{*}{C}H_2CHX$ is more likely to occur than $\overset{*}{C}H_2{-}\underset{\underset{X}{|}}{\overset{*}{C}H}{-}I$. (I represents the initiator and * represents an unpaired electron, a positive charge, or a pair of electrons.)

The condensation reactions that are most useful for the general synthesis of condensation polymers are esterification, amidation, condensation of phenols with formaldehyde, and condensation of urea with formaldehyde. We shall examine each of these briefly.

Polyesters

Several polyesters are commercially important. The best known fiber-forming polyester is derived from ethylene glycol and terephthalic acid. The fiber is known as Dacron; when fabricated as a film it is called Mylar. In practice the product is manufactured by the **trans-esterification** process using ethylene glycol and dimethyl terephthalate. In this process the glycol exchanges with the methyl alcohol part of dimethyl terephthalate:

$$HO{-}CH_2CH_2{-}OH + CH_3O{-}\overset{\overset{\displaystyle O}{\|}}{C}{-}C_6H_4{-}\overset{\overset{\displaystyle O}{\|}}{C}{-}OCH_3$$

$$\rightleftarrows \left(OCH_2CH_2{-}O{-}\overset{\overset{\displaystyle O}{\|}}{C}{-}C_6H_4{-}\overset{\overset{\displaystyle O}{\|}}{C} \right)_n + 2CH_3OH$$

The equilibrium is continually upset by removing the methanol by distillation as it forms. This allows the attainment of close to 100% yields.

Polyamides

Diamines and diacids react to form polyammonium salts which yield polyamides when heated. Nylon 66 is made commercially by the reaction of hexamethylenediamine with adipic acid:

$$\underset{\text{hexamethylenediamine}}{n\,H_2N(CH_2)_6NH_2} + \underset{\text{adipic acid}}{HO_2C(CH_2)_4CO_2H}$$

$$\longrightarrow \underset{\text{poly(hexamethylenediammonium adipate)}}{[(H_3\overset{+}{N}(CH_2)_6\overset{+}{N}H_3)(\overset{-}{O}_2C(CH_2)_4CO_2^{\,-})]_n}$$

$$\xrightarrow{\text{heat}} \underset{\text{Nylon 66}}{\left(\overset{\overset{\displaystyle H}{|}}{N}(CH_2)_6\overset{\overset{\displaystyle H}{|}}{N}{-}\overset{\overset{\displaystyle O}{\|}}{C}(CH_2)_4\overset{\overset{\displaystyle O}{\|}}{C} \right)_n} + 2n\,H_2O$$

The name Nylon 66 derives from the fact that both starting compounds contain six carbon atoms.

Phenol-formaldehyde Polymers

Phenol reacts with formaldehyde in the presence of either an acid or a base to produce polymers known as *Bakelite plastics.* The reaction leads to network polymers because the phenol molecule can react at both the ortho and para positions.

$$C_6H_5OH + HCHO \xrightarrow{H^+} o\text{-}HOC_6H_4CH_2OH$$

$$n\,(o\text{-}HOC_6H_4CH_2OH) + \xrightarrow{H^+} \left(C_6H_3(OH)\text{—}CH_2 \right)_n + n\,H_2O$$

$$\downarrow HCHO/H^+$$

$$\sim CH_2\text{—}C_6H_3(OH)\text{—}CH_2\text{—}C_6H_2(OH)(CH_2\text{—}C_6H_2(OH)\text{—}CH_2\text{—}\wr)\text{—}CH_2 \sim$$

a cross-linked polymer

Urea-formaldehyde Polymers

Urea and formaldehyde undergo a condensation reaction much like the one between phenol and formaldehyde. The first step is catalyzed by base and the second by acid. Polymerization occurs in the second step by acidifying the mixture produced in the first step:

$$\underset{\text{urea}}{(NH_2)_2C{=}O} + 2\,HCHO \xrightarrow[\text{(first step)}]{OH^-} (H\text{—}N\text{—}CH_2OH)_2C{=}O \xrightarrow[\text{(second step)}]{H^+} \underset{\text{cross-linked polymer}}{\left(N\text{—}CH_2\right)_n\text{—}C{=}O\text{—}\left(N\text{—}CH_2\right)_m} + (n+m)H_2O$$

These polymers also form cross-links and are used in plastic kitchenware, utensils, and other molded articles.

Problem 18.5

Give the structural formula of the polymeric product of each of the following condensation reactions.

(a) $n\ HOCH_2CH_2CO_2H \longrightarrow$ a polyester

(b) $n\ H_2NCH_2CH_2CO_2H \longrightarrow$ a polyamide

(c) OH / benzene ring / CH_3 (p-cresol) + $\overset{O}{\overset{\|}{HCH}} \longrightarrow$ a phenol-formaldehyde polymer

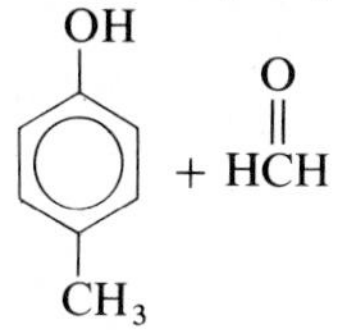

Does cross-linking occur in any of the above polymers?

18.6 NATURAL RUBBER

We saw in Section 6.15 that natural rubber occurs as two stereoisomeric forms of polyisoprene, the all *cis*, called **hevea**, and the all *trans*, called **gutta-percha**. Hevea is a soft, elastic substance; gutta-percha is a hard, brittle solid.

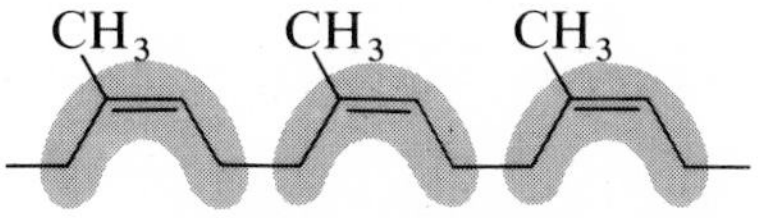

CH_3 CH_3 CH_3

hevea (all *cis*)

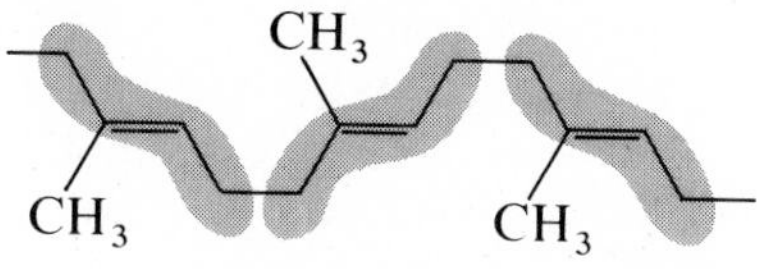

CH_3

CH_3 CH_3

gutta-percha (all *trans*)

We can explain the different properties of the two isomers of polyisoprene by the way in which the polymer molecules align themselves. In the all *trans* isomer, the methyl groups lie on alternating sides of the molecule. The all *trans* molecules thus assume a relatively linear conformation, and they form highly aligned and rigid crystal networks. The all *cis* isomer, as shown above, has all the methyl groups facing the same direction. Mutual repulsion between neighboring methyl groups hinders the linear conformation, and the molecules adopt a curved conformation. The resulting tangle caused by random coiling of the molecules prevents the mass from achieving a highly aligned network. The result is a soft, pliable, elastic polymer.

Problem 18.6

Is natural rubber a 1,2- or a 1,4- addition polymer of isoprene? (Isoprene is

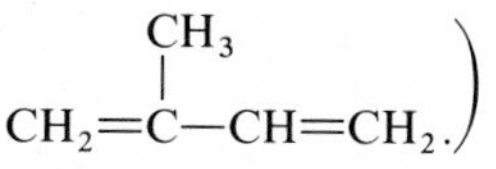

$$CH_2{=}\overset{\overset{\displaystyle CH_3}{|}}{C}{-}CH{=}CH_2.)$$

Write an equation for the polymerization of isoprene.

18.7 STEREOREGULAR ADDITION POLYMERIZATION: ZIEGLER–NATTA CATALYSTS

Attempts to prepare synthetic products comparable to natural rubber (hevea or gutta-percha) failed when the usual free radical or ionic initiation methods were used. Recall that free radicals and carbocations are planar. Carbanions behave as though they are planar. Thus it is usually impossible to arrive at one stereochemical product by the free radical- or ionic-initiated polymerization of isoprene:

$$\mathrm{I^*} + \left[\begin{array}{c} \mathrm{CH_3(CH_2{=})C{-}C(H){=}CH_2} \\ \downarrow\uparrow \\ \mathrm{CH_3(CH_2{=})C{-}C(H){=}CH_2} \end{array}\right] \longrightarrow \left[\begin{array}{c} \mathrm{(CH_3)(I{-}CH_2)C{=}C(H)(\overset{*}{C}H_2{\sim})} \\ + \\ \mathrm{(CH_3)(I{-}CH_2)C{=}C(\overset{*}{C}H_2{\sim})(H)} \end{array}\right] \longrightarrow \begin{array}{l}\text{random}\\ \text{arrangement}\\ \text{of } \textit{cis-trans}\\ \text{units}\end{array}$$

(I* = initiator; * = ·, +, or −)

A mixture of *cis* and *trans* isomers would occur at each step of the polymerization process.

In the early 1950s, K. Ziegler, in Germany, and G. Natta, in Italy, independently discovered a catalyst system that produces stereochemically controlled polymers from alkenes and dienes. These catalysts consist of heterogeneous mixtures of an alkyl aluminum and a transition metal halide; for example, $Al(C_2H_5)_3$ and $TiCl_4$. Many other catalyst combinations are also known. The use of **Ziegler–Natta catalysts** allowed for the first time the synthesis of **stereoregular** polymers such as hevea or gutta-percha. In addition, these catalysts allow the stereoregular polymerization of alkenes having the general formula $CH_2{=}CHX$. Three types of polymers are possible from the polymerization of $CH_2{=}CHX$:

1. **Atactic polymers**—X groups are arranged randomly.

$$\sim\mathrm{CH_2{-}C(H)(X){-}CH_2{-}C(X)(H){-}CH_2{-}C(X)(H){-}CH_2{-}C(H)(X){-}CH_2{-}C(X)(H){-}CH_2{-}C(H)(X){-}CH_2{-}C(H)(X)}\sim$$

2. **Isotactic polymers**—X groups all occur on the same side of the chain.

$$\sim\mathrm{CH_2{-}C(H)(X){-}CH_2{-}C(H)(X){-}CH_2{-}C(H)(X){-}CH_2{-}C(H)(X){-}CH_2{-}C(H)(X){-}CH_2{-}C(H)(X){-}CH_2{-}C(H)(X)}\sim$$

3. **Syndiotactic polymers**—X groups occur on alternating sides of the chain.

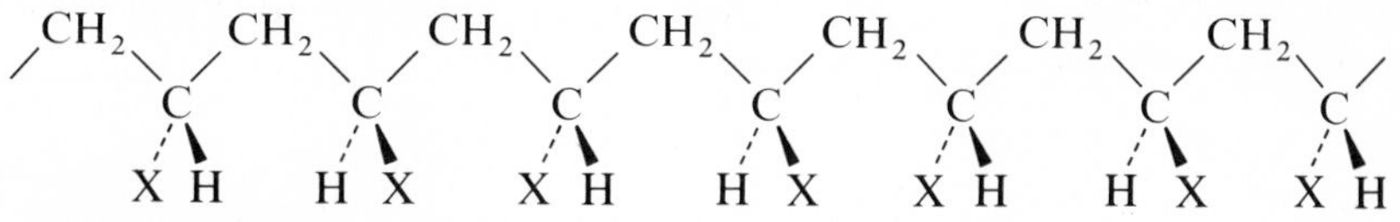

```
 CH2     CH2      CH2      CH2      CH2      CH2      CH2
/   \   /   \    /   \    /   \    /   \    /   \    /   \   /
     C        C         C         C         C         C        C
    /\       /\        /\        /\        /\        /\       /\
   X  H     H  X      X  H      H  X      X  H      H  X     X  H
```

By making the proper choice of a catalyst system it is now possible to prepare isotactic or syndiotactic polymers at will. Before the 1950s, only atactic polymers could be prepared synthetically.

Problem 18.7
Draw the structural formula of (a) isotactic poly(vinyl chloride), (b) syndiotactic poly(vinyl alcohol).

NEW TERMS

Addition polymer Polymer formed by the addition of monomer molecules without the loss of any atoms (18.3).

Condensation polymer Polymer produced by a condensation reaction (18.3).

Cross-link A bond that joins two separate polymer chains together (18.2).

Initiator The reagent used to initiate the reaction that leads to the formation of a polymer (18.4).

Linear polymer A polymer that consists of simple chains and melts when heated; also called a **thermoplastic polymer** (18.2).

Network polymer A polymer in which individual chains are bonded together by *cross-links* to form a three-dimensional network. Network polymers do not melt; they are called **thermosetting polymers** (18.2).

Stereoregular polymer A polymer in which the stereoisomeric configuration is perfectly regular along the chain. **Isotactic** polymers have the same configuration throughout the chain. **Syndiotactic** polymers have an alternating configuration throughout the chain. **Atactic** polymers have a random configuration throughout the chain (18.7).

Trans-esterification The reaction of an ester with an alcohol in which the alcohol groups exchange (18.5):

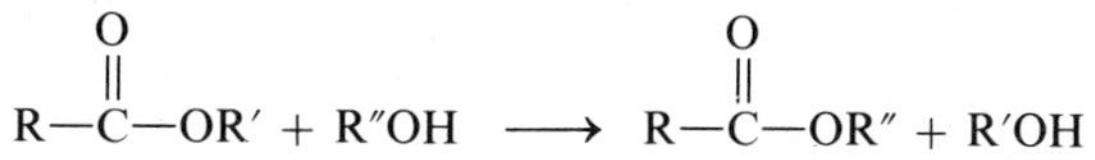

$$R{-}\overset{\overset{O}{\|}}{C}{-}OR' + R''OH \longrightarrow R{-}\overset{\overset{O}{\|}}{C}{-}OR'' + R'OH$$

Ziegler–Natta catalysts Heterogeneous mixture of an alkyl aluminum and a transition metal halide; for example, $Al(CH_2CH_3)_3$ and $TiCl_4$; used to promote stereoregular polymerization (18.7).

ADDITIONAL PROBLEMS

[18.8] Define each of the following terms.

(a) thermoplastic
(b) thermosetting
(c) cross-link
(d) linear polymer
(e) network polymer
(f) addition polymer
(g) condensation polymer

18.9 Draw the structural formula of the products of each of the following polymerization reactions. Tell whether each is an addition polymer or a condensation polymer.

[a] $HOCH_2CH_2OH +$ (benzene ring with adjacent HO_2C and CO_2H groups) $\longrightarrow$

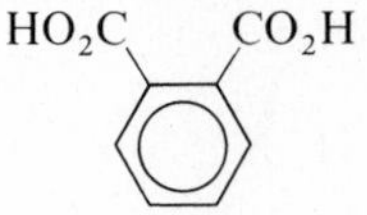

(phthalic acid)

(b) $CH_2{=}CH{-}O{-}\overset{O}{\overset{\|}{C}}CH_3 \xrightarrow{\text{catalyst}}$

[c] $CH_2{=}CCl_2 \xrightarrow{\text{catalyst}}$

(vinylidine chloride)

[18.10] Give an acceptable name for each of the polymers of Problem 18.9.

[18.11] Which of the following monomers or monomer combinations would you expect to lead to cross-linking? Write structural formulas to demonstrate your answers.

(a) $CH_2{=}CHCH{=}CH_2$

(b) 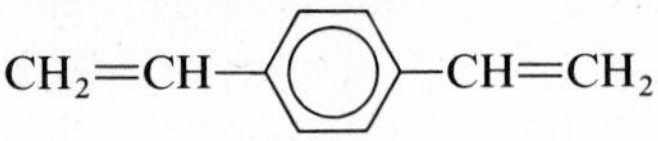 $CH_2{=}CH{-}C_6H_4{-}CH{=}CH_2$ (para-substituted benzene ring)

(c) $C_6H_5{-}CH{=}CH_2$

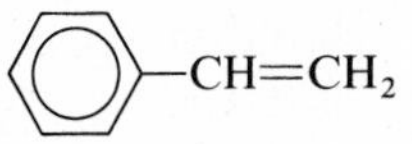

(d) $HOCH_2\overset{OH}{\overset{|}{C}}HCH_2OH + HO_2C{-}C_6H_4{-}CO_2H$ (para-substituted benzene ring)

(e) $HOCH_2CH_2OH +$ (benzene ring with three CO_2H groups at the 1,3,5 positions: CO_2H, HO_2C, CO_2H)

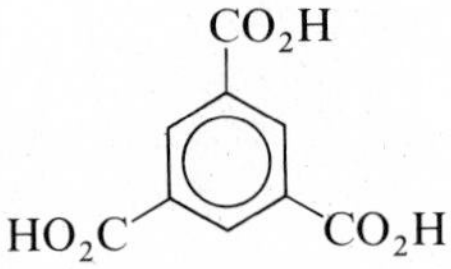

18.12 Classify each of the reactions of Problem 18.11 as addition or condensation.

[18.13] Write equations for the mechanism of the free radical-initiated polymerization of styrene $\left(C_6H_5{-}CH{=}CH_2\right)$.

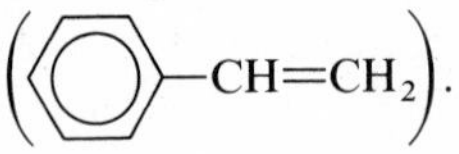

18.14 Write equations for the mechanism of the acid-initiated polymerization of styrene.

[18.15] Draw a segment of the structural formula of the polymer obtained by the free radical-initiated 1,4- addition polymerization of isoprene [$CH_2{=}C(CH_3)CH{=}CH_2$]. Show the random *cis–trans* orientation.

18.16 Draw a segment of the structural formulas of each of the following polymers.

[a] isotactic polystyrene (b) syndiotactic polystyrene

THE USE OF SPECTROSCOPY IN ORGANIC CHEMISTRY: AN ELECTIVE CHAPTER

Molecules absorb energy in discrete quantities called **photons** or **quanta**. We therefore say that molecular energies are **quantized**. By this we mean that a molecule can possess only certain fixed energies, and, like atoms, molecules exist in discrete energy levels. The quantization of molecular energy has given chemists their most powerful analytical tool: absorption spectroscopy.

Chemists have divided the electromagnetic spectrum into several regions based on the different methods used to produce and detect the energy in each region. We shall first examine the ultraviolet and infrared regions of the electromagnetic spectrum and then a slightly different kind of spectroscopy: nuclear magnetic resonance spectroscopy.

The most familiar form of **electromagnetic radiation** is light. A beam of any such electromagnetic radiation consists of an oscillating electric and magnetic field that travels through a vacuum at a velocity of 3×10^{10} cm/second (186,000 miles/second). The frequency of oscillation can vary enormously. The entire spectrum of electromagnetic radiation is shown in Figure 19.1. On the low frequency end of the spectrum are the radio waves and on the high frequency end are the gamma rays. Somewhere intermediate in frequency is the region visible to our eyes. This region lies roughly between 400 and 700 nm of wavelength. A **nanometer** (nm) is 10^{-9} meter.

The energy of electromagnetic radiation is related to its frequency by the equation

$$E = h\nu \qquad \text{or} \qquad E = \frac{hc}{\lambda}$$

where E is the energy of a photon whose frequency is ν, and h is a universal constant called Planck's constant ($h = 6.26 \times 10^{-27}$ erg sec). E is also given by the second equation if we recall that frequency is related to wavelength, λ, by the equation

$$\nu = \frac{c}{\lambda}$$

where c is the velocity of light ($c = 3.0 \times 10^{10}$ cm/sec).

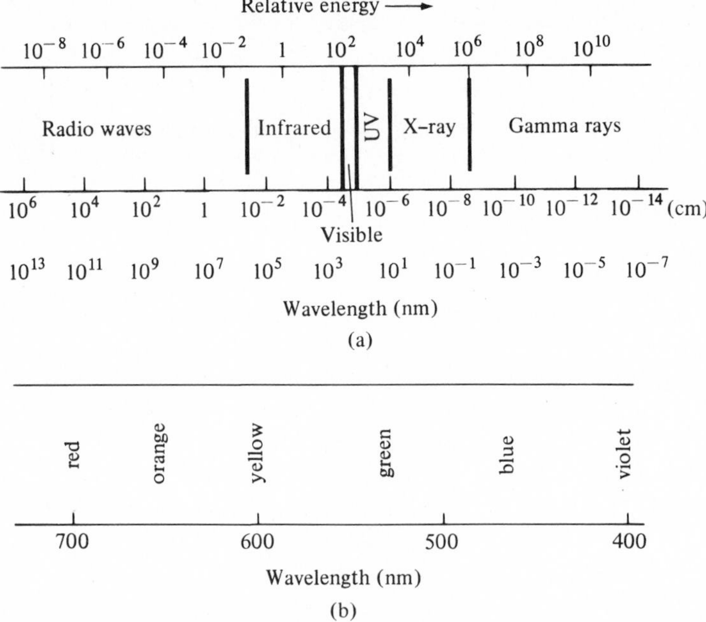

Figure 19.1. (a) The electromagnetic spectrum showing the various spectral regions. Wavelengths are given in units of centimeters (cm) and in nanometers (nm) (1 nm = 10^{-9} meters). (b) The visible region of the electromagnetic spectrum.

Problem 19.1

Calculate the energy in ergs of a photon whose wavelength is 250 nm. (Recall that a nanometer $= 10^{-9}$ meter $= 10^{-7}$ cm.)

Problem 19.2

Verify that the velocity of light is 186,000 miles/sec using the value of 3.0×10^{10} cm/sec given above.

19.2 ENERGY LEVELS IN ATOMS AND MOLECULES

Electrons in atoms are fixed in discrete energy levels. An electron may "jump" from one energy level to another if it absorbs a photon of exactly the required amount of energy. A photon of another energy could not be absorbed because the electron would then have an amount of energy that does not correspond to any of its fixed energy levels. The same behavior occurs with molecules as with atoms. Thus, if a sample of a compound is exposed to radiation containing a continuous range of frequencies, only those photons are absorbed that have the correct energies to excite the electrons or bonds within the molecules. The emerging radiation therefore contains gaps. Analysis of the emergent radiation leads to an **absorption spectrum** of the sample, as shown in Figure 19.2. This absorption spectrum is a characteristic property of the substance and may be used along with other properties (such as melting and boiling points) for purposes of identification. In addition, such spectra afford the trained scientist enormous amounts of information about the intimate detailed structure of the substance. In the spectrum shown in Figure 19.2, the information includes the number of and energy differences between electronic energy levels in acetone.

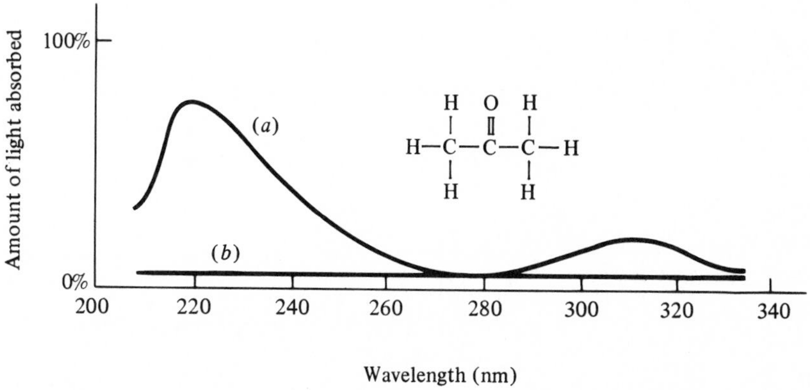

Figure 19.2. Ultraviolet absorption spectrum of acetone dissolved in isooctane (a) and pure isooctane (b). Acetone is seen to absorb light of wavelengths near 220 and 315 nm.

19.3 ANALYSIS OF ELECTROMAGNETIC RADIATION

Sources of electromagnetic radiation usually emit a range of frequencies rather than one single frequency. The sun, for example, radiates nearly the entire electromagnetic spectrum. Analysis of such a spectrum involves dispersing the radiation into its component frequencies or "colors" and measuring the intensity of each. Dispersion of visible light as well as of much of the ultraviolet and infrared regions is readily accomplished by use of a prism (Fig. 19.3). Another dispersion device is the diffraction

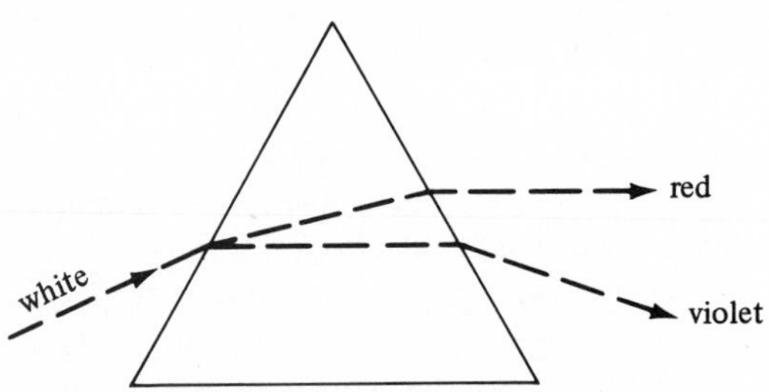

Figure 19.3. Dispersion of light by a prism.

grating. The intensity of each color is determined in many ways. Typically, a sensing device is used that is sensitive to electromagnetic radiation. The detector transforms the intensity of the energy into variations in electrical conductance. In this way, the radiation striking the detector causes an electrical impulse that activates a visual device such as a dial or a pen on a chart.

The basic components of a typical spectrophotometer are shown in Figure 19.4. The source emits mixed radiation of given intensity, I_o, which passes through the sample and emerges with intensity I_s. The emergent beam then passes into the disperser, which can be rotated so that a beam of any desired frequency can be directed at the detector. The electrical impulse of the detector can be recorded in any of several ways. Typically, it is used to move a pen that traces a line on a moving paper. The spectrum is thus drawn out automatically.

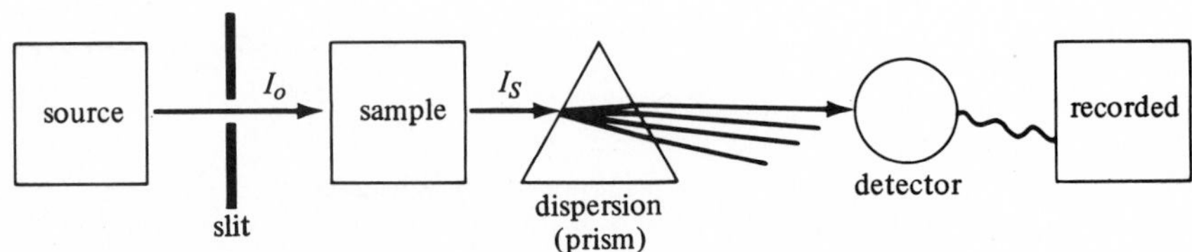

Figure 19.4. Schematic drawing of the basic components of a typical spectrophotometer.

19.4 ELECTRONIC SPECTRA: VISIBLE AND ULTRAVIOLET SPECTROSCOPY

Electromagnetic energy in the visible and ultraviolet (UV) range usually gives rise to electronic transitions in molecules. Spectra in this range are consequently called **electronic spectra**.

Electronic spectra of molecules typically involve energy level transitions of electrons involved in double and triple bonds. Electrons involved in single covalent bonds are more tightly bound and require energies that are beyond (or higher than) the usual UV range. Since electronic energy levels are relatively far apart compared with vibrational energy levels (see Sec. 19.5), UV spectra of most molecules are relatively simple. Typically, a given functional group will exhibit its transition relatively independent of other functional groups in the molecule. Thus electronic spectra are useful in determining the functional groups present in a molecule, provided that they contain multiple bonds.

Electronic spectra usually have broad peaks. For this reason they are less useful than infrared spectra (Sec. 19.5) as diagnostic tests for functional groups. Table 19.1 lists some electronic spectral data for several functional groups.

Table 19.1 Electronic Spectral Data of Some Functional Groups

Functional Group	Example	Formula	Wavelength, nm
C=C	Ethene	$CH_2{=}CH_2$	162
	1-Hexene	$CH_2{=}CH(CH_2)_3CH_2$	180
	Cyclohexene		183.5
Polyenes	1,3-Butadiene	$CH_2{=}CH{-}CH{=}CH_2$	210
	1,5-Hexadiene	$CH_2{=}CHCH_2CH_2CH{=}CH_2$	178
	1,3,5-Hexatriene	$CH_2{=}CHCH{=}CHCH{=}CH_2$	268
	1,3,5,7-Octatetraene	$CH_2{=}CHCH{=}CHCH{=}CHCH{=}CH_2$	304
C≡C	1-Butyne	$HC{\equiv}CCH_2CH_3$	172
Aromatic	Benzene		254, 203.5
C=O	Acetaldehyde	$CH_3\overset{\overset{O}{\|}}{C}H$	289, 182
	Acetone	$CH_3\overset{\overset{O}{\|}}{C}CH_3$	275, 190
	Cyclohexanone	=O	291
CO_2H	Acetic acid	CH_3CO_2H	204
CO_2R	Ethyl acetate	$CH_3CO_2CH_2CH_3$	204
$CONH_2$	Acetamide	$CH_3\overset{\overset{O}{\|}}{C}{-}NH_2$	205
NO_2	Nitromethane	CH_3NO_2	279, 202

An interesting use for electronic spectra is in determining whether polyenes (Sec. 6.2) are conjugated or isolated. Conjugation increases the wavelength of the absorption peak. Isolated double bonds are independent of one another and their absorption peaks are similar to those of alkenes.

Problem 19.3

An unknown compound A has the molecular formula C_4H_8O. It has an absorption peak at 278 nm. What structures are possible for A?

Problem 19.4

An unknown compound B has the molecular formula C_8H_{12}. Hydrogenation gives C_8H_{18}. Compound B has an absorption peak at 182 nm. What structures are possible for B?

Problem 19.5

An unknown compound C has the molecular formula $C_4H_8O_3$. It has an absorption peak at 204 nm. Give a possible structural formula for C.

19.5 VIBRATIONAL SPECTRA: INFRARED SPECTROSCOPY

Molecules, even in the solid phase, are in a state of perpetual random motion, including various modes of vibration of the atoms within the molecules. Like electronic energy, molecular vibrational energy is also quantized; that is, molecules possess discrete vibrational energy levels. A particular vibration of an atomic grouping, then, may have only certain fixed frequencies. Figure 19.5 shows some of these vibrations for the water molecule along with the frequency of the infrared radiation required to excite each to the next higher vibrational energy level.

Vibrational (infrared) spectra are most often given in terms of wave number. The **wave number** is a frequency unit and equals the number of cycles of the wave in each centimeter (wave number $= 1/\lambda$). The unit of wave number is cm^{-1}. Infrared spectra are also given in wavelength units. The **micron** (μ) ($1\mu = 10^{-6}$ meter $= 10^{-4}$ cm) is the wavelength unit used most often in reporting infrared spectra.

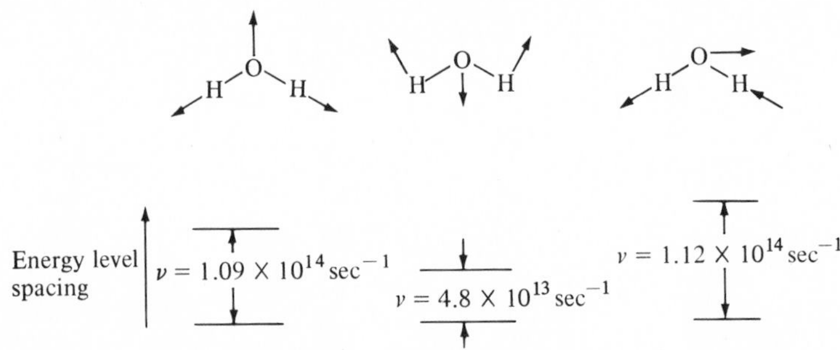

Figure 19.5. Some vibrations of the water molecule.

Problem 19.6
Calculate the frequency in cycles per second (sec^{-1}) and in wave numbers (cm^{-1}) for radiation whose wavelength is 5 μ.

Compared with electronic energy levels, vibrational energy levels are quite close together. Typical separations between electronic energy levels are about twenty times as great as those between vibrational levels.

Vibrational transitions such as the ones mentioned here give rise to infrared (IR) spectra. Because the transitions are so plentiful, the resulting IR spectra are much more complex than UV spectra. Although this makes an exact analysis more difficult, this greater complexity makes IR spectra more useful as "fingerprints" of molecules.

Rather complete analyses of IR spectra have been accomplished for many compounds, and they have resulted in tremendous advances in the determination of structures of complex molecules. A simple example of the use of IR spectra in structure determination might be the following. Two isomers of C_2H_6O are possible:

```
    H  H                 H       H
    |  |                 |       |
H—C—C—O—H          H—C—O—C—H
    |  |                 |       |
    H  H                 H       H
```

The first structure corresponds to ethyl alcohol and has, in addition to the various C—H and C—O vibrations, the O—H vibration. This grouping has an unmistakable characteristic IR absorption at 3400 cm^{-1} (3.0 μ) that can easily be used as a positive identification of the alcohol (Fig. 19.6).

Most atomic groupings can be identified by IR spectroscopy, and it is often possible to distinguish between different spatial arrangements of the same groupings in a molecule. For example, the two butenes

```
H          H          H          CH3
  \       /             \       /
   C=C                   C=C
  /       \             /       \
CH3        CH3        CH3          H
```

can be distinguished in the IR region because the *cis* form

```
  ↖H          H↗
     \       /
      C=C
     /  ↓  \
  CH3        CH3
```

has a symmetrical vibration that does not occur in the *trans* form. Other vibrations will differ in similar ways, and the resulting spectra are quite distinct.

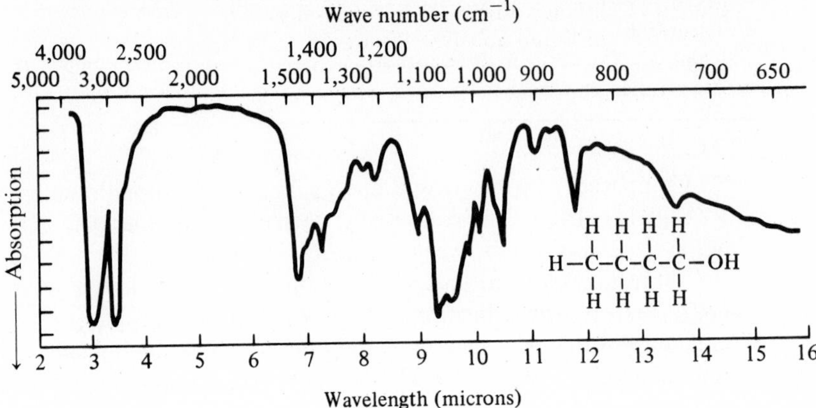

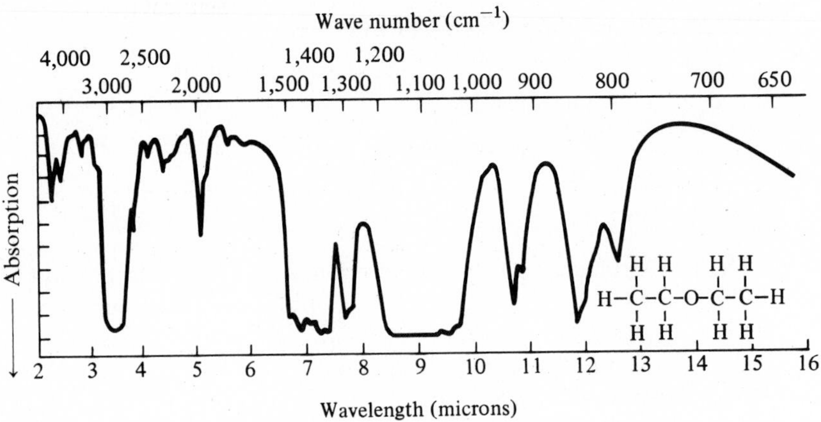

Figure 19.6. Infrared spectra of butyl alcohol and diethyl ether.

19.6 CHARACTERISTIC FREQUENCIES OF FUNCTIONAL GROUPS

As in the case of water (Fig. 19.5), each covalent bond in a molecule has its characteristic vibration modes and therefore its infrared absorption peaks. But what is even more important is that the absorption peaks for a given covalent bond are quite independent of the rest of the molecule (Table 19.2). As an example, the OH absorption of an alcohol occurs near 3300 cm^{-1} regardless of the nature of the alkyl group. If this were not true, infrared spectra would be much less useful in identifying functional groups in molecules.

Table 19.3 lists the characteristic infrared absorption frequencies for several commonly occurring functional groups.

Table 19.2 Infrared Absorption Frequencies of Some Covalent Bonds

Bond	Frequency Range, cm^{-1}
C—C, C—N, C—O	800–1300
C=C, C=N, C=O	1500–1900
C≡C, C≡N	2000–2300
C—H, N—H, O—H	2850–3650

As a diagnostic test for functional groups, the infrared spectrum is most useful in the region 4000–1500 cm^{-1}. At lower wave numbers the peaks overlap and are often difficult to identify. That lower wave number region, however, is quite useful as a "fingerprint" of the molecule. Because there are so many peaks in that region, no two compounds will have exactly the same spectrum. We may therefore assume that two substances are identical if they have identical infrared spectra.

Table 19.3 Infrared Absorption Frequencies of Some Functional Groups

Functional Group	Frequency Range, cm^{-1}
A. Hydrocarbons	
Alkane C—H (stretching)	2853–2962
Alkene C—H (stretching)	3010–3095
C=C (stretching)	1620–1680
Alkyne ≡C—H (stretching)	~3300
C≡C (stretching)	2100–2260
Aromatic Ar–H (stretching)	~3030
Aromatic substitution patterns (C—H out-of-plane bending)	
Monosubstituted	690–710, 730-770
o-Disubstituted	735–770
m-Disubstituted	680–725, 750–810
p-Disubstituted	790–840
B. Hydroxyl Groups	
OH (alcohols, phenols) (stretching)	3200–3600
OH (carboxylic acids) (stretching)	2500–3000
C. Carbonyl Group C=O (aldehydes, ketones, esters, carboxylic acids) (stretching)	1690–1750
D. Amines N—H (stretching)	3300–3500
E. Nitriles C≡N (stretching)	2220–2260
F. Alkyl Halides R—F (stretching)	1000–1350
R—Cl (stretching)	750–850
R—Br (stretching)	500–680
R—I (stretching)	200–500

Problem 19.7

Using only the IR frequency given for each compound, suggest which covalent bonds might be present in each of the unknown compounds below. Refer only to Table 19.2.

(a) 2250 cm^{-1} (b) 3400 cm^{-1} (c) 3025 cm^{-1}

Problem 19.8

On the basis of the following additional data for each of the compounds in Problem 19.7, what can you now conclude about the identity of each compound?

(a) The compound has the molecular formula C_2H_3N and is neutral to litmus.
(b) The compound is basic to litmus.
(c) The compound is neutral to litmus; it does not decolorize Br_2/CCl_4 solution in the dark, and it is insoluble in water.

Problem 19.9

A compound whose molecular formula is C_7H_7Br has (among others) the following IR absorption peaks: 3025 cm^{-1}, 830 cm^{-1}, and 600 cm^{-1}. Suggest a probable structure for the compound.

19.7 NUCLEAR MAGNETIC RESONANCE SPECTROSCOPY

Nuclear magnetic resonance is a phenomenon that differs from other energy-matter interactions. A sample placed in an apparatus with radiation in the frequency range of radio waves shows no observable interaction between the radio waves and the sample. If, however, the interaction is allowed to occur in the presence of a very strong magnetic field, then certain frequencies of the radio waves are observed to interact with the substance in the sample, and the interaction is called **magnetic resonance**. At different magnetic field strengths, the particular radio frequencies affected are different.

The absorption of ultraviolet light was shown to be associated with changes in electronic energy levels. Similarly, the absorption of infrared radiation gives rise to jumps in vibrational and rotational energy levels associated with parts of molecules. The magnetic resonance phenomenon is associated with changes in the energy levels of spinning nuclei.

The nuclei of certain atoms possess mechanical spin. Of these atoms, hydrogen is the most important because it is present in essentially all organic compounds. Since atomic nuclei are electrically charged, their spins result in the creation of a magnetic field. If placed in an external magnetic field such as that of a strong electromagnet, these spinning nuclei can orient themselves only in discrete ways. For the hydrogen nucleus, two orientations are possible—parallel to the external magnetic field (low energy state) and opposed to the external field (high energy state). Thus, for the hydrogen nucleus in an external magnetic field, two energy levels exist. Under the influence of the external magnet, nuclei tend to achieve one of these orientations. Since the nucleus is spinning, its rotational axis tends to precess (Fig. 19.7). This type of gyroscopic motion can be seen in a common top, which precesses under the influence of the earth's gravitational field.

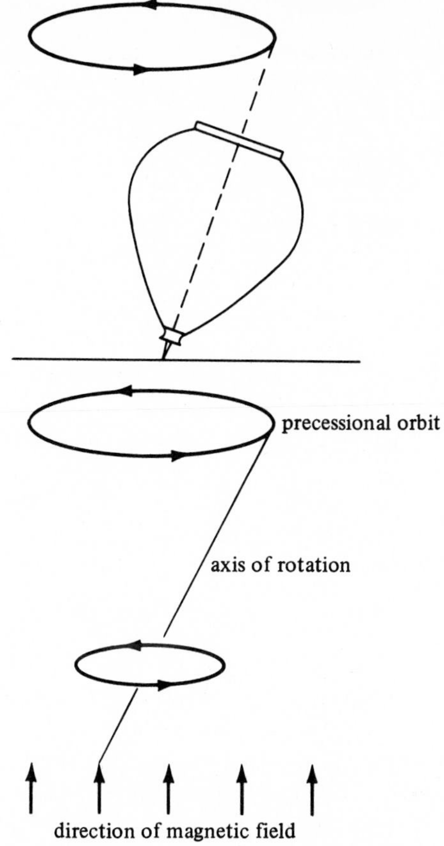

Figure 19.7. Precession of a nucleus under the influence of a magnetic field compared to that of top under the influence of gravity.

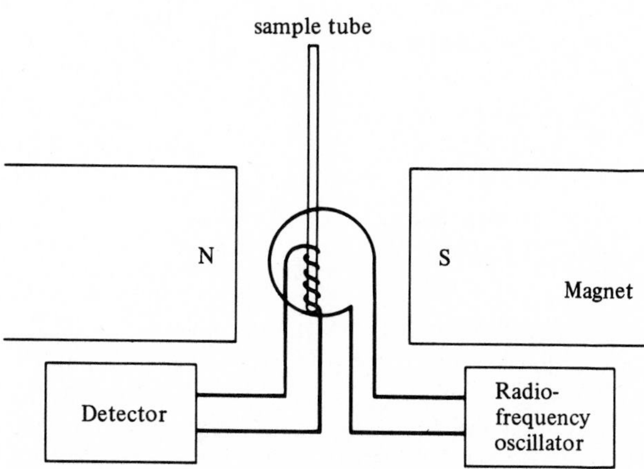

Figure 19.8. Schematic drawing of the basic components of a nuclear magnetic resonance spectrometer.

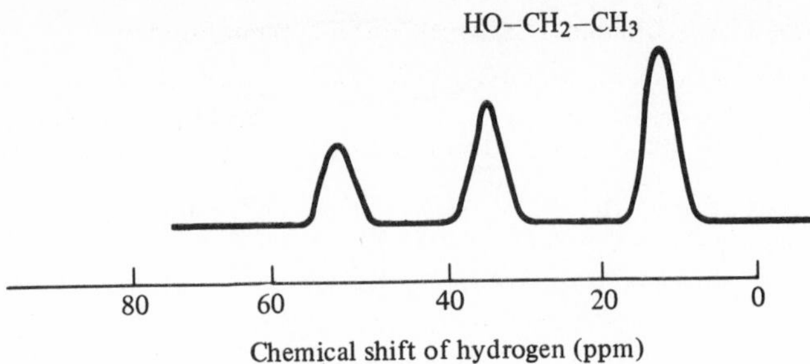

Figure 19.9. Nuclear magnetic resonance spectrum of ethyl alcohol.

The transition of a hydrogen nucleus from one orientation to the other can be brought about by the absorption or emission of exactly the required amount of energy, $E = h\nu$, where ν is the frequency of the electromagnetic radiation used. For magnetic fields in the vicinity of 14,000 gauss, the frequency of such energies lies in the radio frequency region of the spectrum—about 60 million hertz (commonly abbreviated as 60 MHz) [one hertz (Hz) = 1 sec^{-1}].

In actual practice, we observe the phenomenon by placing the sample in a radio beam of fixed frequency (in the most common instruments, 60 MHz) and in a variable magnetic field (Fig. 19.8). The field is varied continuously, and a response is noted when the correct combination of field and radio frequency occurs and the nuclei absorb the energy. This is the **nuclear magnetic resonance** (NMR) phenomenon.

But of what use is this phenomenon in the solution of chemical problems? Since hydrogen atoms are always bonded to other atoms by covalent bonds, the electron density in the vicinity of the hydrogen nucleus will vary with the nature of the atom to which the hydrogen is bonded. This electron density produces a shielding of the hydrogen nucleus from the externally applied field, so that the energy required to cause a transition in spin orientation depends on the environment of the nucleus. Fortunately, very minor structural changes near a given hydrogen atom produce rather drastic results in the energy required for nuclear resonance. An example of the selectivity of the NMR spectrum of a compound is shown in Figure 19.9. Moreover, the area under each peak is proportional to the number of hydrogen atoms involved. Thus the NMR spectrum gives us an accurate estimate of the number of each kind of hydrogen atom in the molecule.

19.8 THE CHEMICAL SHIFT

The resonance position of a given hydrogen atom (proton) relative to a standard proton is called the **chemical shift**. The most commonly used standard is tetramethylsilane [$(CH_3)_4Si$, called TMS]. TMS has sixteen equivalent protons and exhibits a single NMR resonance. The chemical shift is the displacement from the TMS resonance and is usually expressed in δ (delta) units,

$$\delta = \frac{\Delta\nu \text{ in Hz}}{\text{operating frequency of the instrument in Hz}}$$

Table 19.4 Chemical Shifts of Some Typical Protons

Type of Proton	Chemical Shift, ppm
Primary alkyl $RC\underline{H}_3$	0.9
Secondary alkyl $RC\underline{H}_2R$	1.3
Tertiary alkyl $R_3C\underline{H}$	1.5
Vinylic $\mathrm{>C{=}C<}$ (with $\underline{H}$ on C)	4.6–5.9
Acetylenic $-C{\equiv}C-\underline{H}$	2–3
Aromatic $Ar-\underline{H}$	6–8.5
Allylic $-\overset{\vert}{C}{=}\overset{\vert}{C}-C\underline{H}_3$	1.7–1.8
Benzylic $Ar-C\underline{H}_3$	2–3
Aldehyde $R-\overset{O}{\overset{\Vert}{C}}-\underline{H}$	9–10
Ketone (α-H) $R-\overset{O}{\overset{\Vert}{C}}-\overset{\vert}{\underset{\vert}{C}}-\underline{H}$	2–2.7
Ester $R-\overset{O}{\overset{\Vert}{C}}-O-\overset{\vert}{\underset{\vert}{C}}-\underline{H}$	2–2.2
Ether $R-O-\overset{\vert}{\underset{\vert}{C}}-\underline{H}$	3.3–4
Alcohol $HO-\overset{\vert}{\underset{\vert}{C}}-\underline{H}$	3.4–4
Alkyl halide $F-\overset{\vert}{\underset{\vert}{C}}-\underline{H}$	4–4.5
$Cl-\overset{\vert}{\underset{\vert}{C}}-\underline{H}$	3–4
$Br-\overset{\vert}{\underset{\vert}{C}}-\underline{H}$	2.7–4
$I-\overset{\vert}{\underset{\vert}{C}}-\underline{H}$	2–4
Hydroxyl $R-O\underline{H}$	1–6[a]
$Ar-O\underline{H}$	4–12[a]
$R-CO_2\underline{H}$	10–12[a]
$R-N\underline{H}_2$	1–5[a]

[a] The chemical shift varies with the solvent used and with concentration and temperature.

in which $\Delta\nu$ is the difference between the absorption frequency of the compound and that of TMS in Hz. ($\Delta\nu$ is the observed shift in radio frequency from TMS.) The δ unit for measuring chemical shifts is independent of the magnetic field because both the numerator and the denominator in the definition of δ are in frequency units (Hz). The chemical shift is therefore a fraction of the total magnetic field strength. Chemical shifts are typically very small (usually less than 500 Hz) compared with total field strengths of 60 or 100 *million* Hz. Chemical shift δ units are therefore usually expressed as *parts per million* (ppm).

Table 19.4 lists the chemical shifts for some hydrogen-containing compounds. We may generalize some of the data given in Table 19.4 as follows:

1. Aromatic protons occur at higher δ values than alkyl protons.
2. Protons attached to saturated carbon occur at lower δ values than protons attached to unsaturated carbon.
3. The chemical shift of a proton increases as the electronegativity of the atom to which it is bonded increases. (For example, see the alkyl halides in Table 19.4.)

Problem 19.10

The spectrum of *p*-xylene has two peaks. One is at 7.0 ppm and the other is at 2.3 ppm. Identify each of these peaks (that is, tell which hydrogens give rise to each peak).

19.9 SPIN–SPIN SPLITTING

Because a spinning nucleus exhibits a small magnetic field of its own, the presence of such a nucleus can influence other nearby nuclei. Thus a proton in an NMR spectrometer experiences not only the field of the instrument's magnet, but also the fields of the protons on adjacent carbon atoms. We can show this effect as follows, where the small arrows symbolize the magnetic field, *h*, produced by the CH_2 (methylene) protons, and the large arrow symbolizes the magnetic field, H_o, of the instrument.

```
      ↑H  H
       |  |
  H—O—C—C—H
       |  |
      ↑H  H
       ↑
       |
       |
       Ho
```

In the example shown, two methylene protons are on the carbon adjacent to the protons undergoing resonance; that is, the methyl protons. The two methylene protons can exert their individual magnetic fields either parallel to or opposed to the external field H_o. The methyl protons therefore experience three different total magnetic fields depending on the spin orientation of the adjacent methylene protons:

$$\begin{array}{ccc} H_o & h & h \\ \uparrow & + \uparrow & + \uparrow \end{array} = H_o + 2h = \Uparrow$$

$$\left.\begin{array}{l} \uparrow + \uparrow + \downarrow \\ \uparrow + \downarrow + \uparrow \end{array}\right\} = H_o = \uparrow$$

$$\uparrow + \downarrow + \downarrow = H_o - 2h = \downarrow$$

The result is that the resonance of the methyl protons is split into three peaks. The methyl protons also split the methylene protons. In this case the three methyl protons may align themselves into *four* different combinations:

$$\uparrow\ \uparrow\ \uparrow \quad = H_o + 3h$$

$$\left.\begin{array}{l} \uparrow\ \uparrow\ \downarrow \\ \uparrow\ \downarrow\ \uparrow \\ \downarrow\ \uparrow\ \uparrow \end{array}\right\} = H_o + h$$

$$\left.\begin{array}{l} \uparrow\ \downarrow\ \downarrow \\ \downarrow\ \uparrow\ \downarrow \\ \downarrow\ \downarrow\ \uparrow \end{array}\right\} = H_o - h$$

$$\downarrow\ \downarrow\ \downarrow \quad = H_o - 3h$$

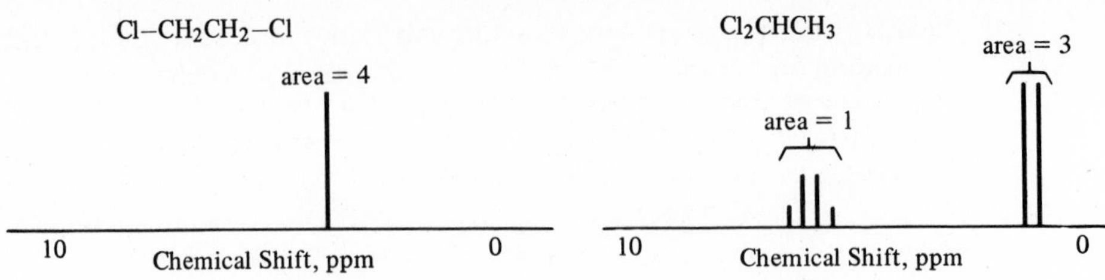

Figure 19.10. Predicted spectra of Cl—CH_2CH_2—Cl and Cl_2CHCH_3.

We can predict the splitting pattern by a simple formula: n equivalent protons will split a proton on an adjacent carbon into $n + 1$ peaks. We must also point out that equivalent protons, for example the three protons on the methyl group, do not split each other.

The separation of split peaks is called the coupling constant, J. One way to determine whether several peaks are split or whether they are simply different protons with different chemical shifts is by using the coupling constant. If two groups of peaks represent protons that are splitting each other, both sets of peaks will have the same coupling constant (Fig. 19.10).

Protons bonded to oxygen or nitrogen do not normally split protons on adjacent carbon atoms. Hydroxyl and amino groups exchange their protons through the formation of ions (proton transfer). The result is that their resonances usually show broad, unsplit peaks.

Let us now look at an example of the power of NMR spectroscopy in solving structure problems. Two compounds have the molecular formula $C_2H_4Cl_2$:

```
    H  H                  H  H
    |  |                  |  |
Cl—C—C—Cl    and    Cl—C—C—H
    |  |                  |  |
    H  H                  Cl H
1,2-dichloroethane    1,1-dichloroethane
```

We can predict the NMR spectrum of each of these compounds very easily using the knowledge that we have at hand. The spectrum of the 1,2-dichloro isomer should show a single unsplit resonance peak because all four protons are equivalent and equivalent protons do not split each other. The 1,1-dichloro isomer has two different types of protons—a $CHCl_2$ and a CH_3. The methyl protons should occur at a lower value and they should be split into a doublet ($n + 1 = 2$) because there is one proton on the adjacent carbon. The $CHCl_2$ proton should occur at a higher value because it has two chlorine atoms bonded to the same carbon. This proton should be split into a quartet ($n + 1 = 4$). The predicted spectra of the two compounds are shown in Figure 19.10.

Problem 19.11
Predict the NMR spectra of the isomers of C_2H_4BrCl.

NEW TERMS

Absorption spectrum Analysis of the radiation that emerges unabsorbed by a sample; usually a plot of the amount of light absorbed versus the wavelength (19.2).

Chemical shift The resonance position in the NMR spectrum of a given hydrogen atom relative to a standard hydrogen atom [usually tetramethylsilane, $(CH_3)_4Si$] (19.8).

Electromagnetic radiation Radiation (light, ultraviolet, infrared, radiowaves, and so forth) that travels at a velocity of 3×10^{10} cm/sec (see Fig. 19.1) (19.1).

Electronic spectrum Spectrum in the visible and ultraviolet region produced by electronic transitions in an atom or molecule (19.4).

Hertz (Hz) Frequency unit; 1 hertz = 1 cycle per sec = 1 sec^{-1} (19.7).

Micron (μ) Wavelength unit used in infrared spectra; one micron = 10^{-6} meter (19.5).

Nanometer (nm) Unit of length equal to 10^{-9} meter (19.1).

Nuclear magnetic resonance (NMR) Nuclear transition that occurs when atomic nuclei in a magnetic field absorb electromagnetic radiation in the radio frequency region of the spectrum (19.7).

Photon Discrete quantity of energy; also called a **quantum** (Introduction).

Quantization Existence of only discrete energy states for an atom or molecule (Introduction).

Spin–spin splitting Splitting of resonances due to the influence of nearby nuclei which, because of their spins, produce local magnetic fields that add to or subtract from the magnetic field of the spectrometer (19.9).

Vibrational (infrared) spectrum Spectrum in the infrared region produced by vibrational transitions in molecules (19.5).

Wave number (cm^{-1}) Frequency unit used in infrared spectra; wave number = 1 ÷ wavelength (in cm) (19.5).

ADDITIONAL PROBLEMS

19.12 What is the frequency of radiation whose wavelength is 720 nm?

[19.13] An unknown compound C has the molecular formula $C_5H_{10}O$. C has a UV absorption peak at 280 nm. Hydrogenation converts C into D, whose molecular formula is $C_5H_{12}O$. Give likely structures for C and D.

19.14 A hydrocarbon E has the molecular formula C_4H_6. It has a UV absorption peak at 172 nm. Give all the possible structures for E.

[19.15] The IR spectrum of compound E (Problem 19.14) shows no peaks above 3200 cm^{-1}. Can you now decide which of your possible structures is the correct one for E?

19.16 Compound F has the molecular formula C_6H_8. Hydrogenation of F gives G (C_6H_{12}). Heating F with a Ni catalyst converts it into benzene. The UV spectrum of F shows a peak at 180 nm. What is the structure of F?

[19.17] Predict the NMR spectra of the following compounds.

(a) CH_3CH_2Br (b) $CH_3CHClCH_3$ (c) C_6H_5—CH_3 (benzene ring)—CH_3

19.18 Compound H ($C_4H_{10}O$) has no UV absorption above 200 nm. It has an IR peak at 3300 cm^{-1}. The NMR spectrum consists of two single peaks: one at 1.0 ppm (area = 9) and one at 4 ppm (area = 1). Give the structure of H.

[19.19] Compound I (C_8H_{10}) has IR absorptions at 3030 cm^{-1}, 2960 cm^{-1}, and 830 cm^{-1}. Its NMR spectrum consists of two singlets: at 7.0 ppm (area = 4) and at 2.2 ppm (area = 6). Give the structure of I.

19.20 Compound J (C_7H_8O) has IR absorptions at 3028 cm^{-1}, 700 cm^{-1}, and 760 cm^{-1}. There is no absorption above 3028 cm^{-1}. The NMR spectrum of J consists of a multiplet (ill-defined, area = 5) at 7 ppm and a single peak at 3.8 ppm (area = 3). Give the structure of J.

[19.21] Compound K, an isomer of J (Problem 19.20), has nearly the same IR peaks as J but also has a peak at 3300 cm^{-1}. The NMR spectrum has three single peaks at 7.0 ppm (area = 5), at 3.5 ppm (area = 2), and at 2.8 ppm (area = 1). Give the structure of K.

19.22 Compound L has the molecular formula C_4H_6O. It has, among others, the following spectral features: IR: 1650, 1740 cm^{-1}. NMR: 1.0 ppm (area = 3), 5.0 ppm (area = 1), 5.3 ppm (area = 1), and 9.5 ppm (area = 1). Propose a structure for L.

ANSWERS TO SELECTED PROBLEMS

CHAPTER 1
PRINCIPLES OF BONDING

1.1. A functional group occurs in many molecules and exhibits properties that are relatively independent of the rest of the molecule.

1.2. (a) 1, (b) 3, (c) 4, (d) 6, (e) 5, (f) 2, (g) 1, (h) 7

1.3. (a) $\mathrm{Li}\ 21 + \mathrm{Cl}\ 287 \longrightarrow [\mathrm{Li}\ 2]^{+} + [288]^{-}$

(b) $\mathrm{Mg}\ 282 + 2\mathrm{Br}\ 28187 \longrightarrow [\mathrm{Mg}\ 28]^{2+} + 2[\mathrm{Br}\ 28188]^{-}$

1.4. (a) Na, (b) N, (c) N, (d) F, (e) Cl, (f) H

1.5. (a), (c), (d), (e)

1.6. (a) $\mathrm{H{:}\underset{\cdot\cdot}{\ddot{Cl}}{:}}$, (b) H:H, (c) $\mathrm{H{:}\underset{\cdot\cdot}{\ddot{S}}{:}}$, (d) $\mathrm{{:}\underset{\cdot\cdot}{\ddot{F}}{:}\underset{\cdot\cdot}{\ddot{F}}{:}}$

1.7. (a) $\mathrm{{:}\dot{N}{::}\ddot{O}{:}}$ + others, (b) H:C⋮⋮C:H

1.8. H H H
H:C:Ö:C:H, coordinate covalent bond
H + H

1.9.

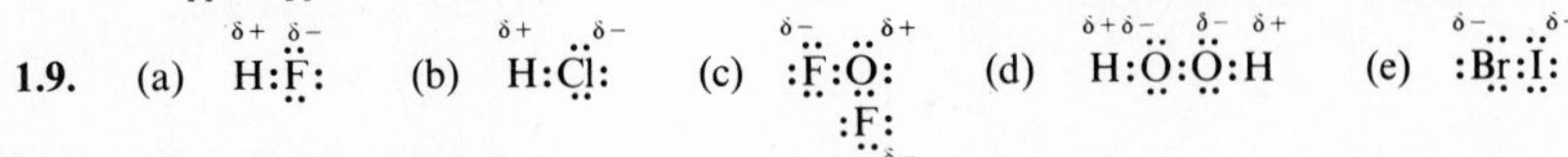

1.10. Ethyl alcohol molecules can form hydrogen bonds to each other. Methyl ether molecules cannot because they have no O—H hydrogens.

1.11. (a) H_2O, (b) HF, (c) HF

1.12. (a) H:Ö:N::Ö: has no formal charges, (b) H:Ö:N⁺::Ö:, :Ö:⁻ (c) :C⁻::O⁺:,

(d) :N::Ö: has no formal charges, (e) ⁺:N:Ö:⁻

1.13. (a) four sp^3 hybrid orbitals, (b) three sp^2 hybrid orbitals, (c) two sp hybrid orbitals

1.14. (a) tetrahedral, (b) planar trigonal (three orbitals in a plane and at 120° angles),

(c) linear (two orbitals at 180° to one another)

1.15.

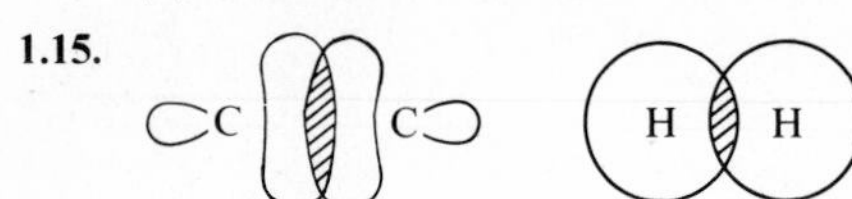

1.16. (a) H
H:C:H; C is sp^2 hybridized; planar, 120° bond angles
+

(b) H
H:C:⁻; C is sp^3 hybridized; tetrahedral, ~109.5° bond angles
H

1.17. With no hybridization: H—O—H angle = 90°. With sp^3 hybridization: H—O—H ≅ 109.5° (actually 105°).

1.18. (e) ionic, (f) covalent, (g) covalent, (h) ionic

1.19. (e) Na^+ :Br:⁻ (f) ⁻:Ö:S⁺::Ö:, (g) H:Cl:, (h) $2K^+$:Ö:$^{2-}$

1.20. (h) covalent, 2H· + :S: ⟶ H:S:H, (i) ionic, Na· + :Cl: ⟶ Na^+ + :Cl:⁻

(j) covalent, H· + :I: ⟶ H:I:, (k) ionic, ·Ca· + 2:Cl: ⟶ Ca^{2+} + 2:Cl:⁻

(l) covalent, 2H· + :Ö: ⟶ H:Ö:H, (m) covalent, 2:Cl: + :S: ⟶ :Cl:S:Cl:

(n) covalent, :Cl: + :I: ⟶ :Cl:I:, (o) ionic, 2Li· + :Ö: ⟶ $2Li^+$ + :Ö:$^{2-}$

1.21. (e) C_4H_{10} has the higher molecular weight

(f) H H H
H:C:C:C:ÖH
H H H

(g) H
H:C:ÖH
H

(h) H
H:C:Ö:H
H

(f), (g), (h) can form intermolecular hydrogen bonds

1.23. (e) :Ö:⁻
:Ö::S::Ö:
:Ö:⁻

(f) :Ö:⁻
⁻:Ö:S⁺:Ö:⁻

(g) ⁻:Ö:C::C:C::O:⁺ (h) ⁻:Ö:N::Ö:

1.24. (d) δ+ δ− H:S: H δ+ (e) δ+ δ− H:I: (f) δ− :F: δ+ ; δ− :F:C:Cl: δ− ; :Cl: δ−

1.25. (i)

1.27. (f) sp^3, (g) sp^3, (h) sp^3, (i) sp^3, (j) sp^2

CHAPTER 2
PRINCIPLES OF STRUCTURE

2.1. (a)

```
   H             H
   |             |
H—C—H,           C---H
   |            / \
   H           H   H
```

(b)

```
   H             H
   |             |
H—C—Cl,          C---H
   |            / \
   H           H   Cl
```

(c)

```
   H  H  H
   |  |  |
H—C—C—C—H,
   |  |  |
   H  |  H
      |
    H—C—H
      |
      H
```

```
          H
    H    /
     \  C—H
        |
  H     C---H
   \   / \
    C     C—H
   / \   / |
  H   H H  H
```

(d)

```
   H             H
   |             |
H—C—Cl,          C---Cl
   |            / \
   Cl          H   Cl
```

(e)

```
   Cl            Cl
   |             |
H—C—Cl,          C---Cl
   |            / \
   Cl          H   Cl
```

2.3. They are not isomers because they are identical. The difference suggested by the graphic formulas is not real.

2.4. (a)

```
   H  H  H  H  H
   |  |  |  |  |
H—C—C—C—C—C—H,
   |  |  |  |  |
   H  H  H  H  H
```

$CH_3CH_2CH_2CH_2CH_3$;

```
   H  H  H  H
   |  |  |  |
H—C—C—C—C—H,
   |  |  |  |
   H  |  H  H
      |
    H—C—H
      |
      H
```

$CH_3CHCH_2CH_3$ (with CH_3 on the second carbon); CH_3CCH_3 (with CH_3 above and below the central carbon),

```
         H
         |
       H—C—H
         |
   H     |    H
   |     |    |
H—C — C — C—H;
   |     |    |
   H     |    H
         |
       H—C—H
         |
         H
```

(b)

```
   H  H  H
   |  |  |
H—C—C—C—Cl,
   |  |  |
   H  H  H
```

$CH_3CH_2CH_2Cl$;

```
   H  H  H
   |  |  |
H—C—C—C—H,
   |  |  |
   H  Cl H
```

CH_3CHCH_3 (with Cl on the central carbon)

(c) $H{-}CH_2{-}CH_2{-}CHCl{-}Cl$ (H–C(H)(H)–C(H)(H)–C(H)(Cl)–Cl), $CH_3CH_2CHCl_2$; H–C(H)(H)–C(H)(Cl)–C(H)(H)–Cl, CH_3CHCH_2Cl (Cl on C-2);

Cl–C(H)(H)–C(H)(H)–C(H)(H)–Cl, $ClCH_2CH_2CH_2Cl$; H–C(H)(H)–C(Cl)(Cl)–C(H)(H)–H, CH_3CCH_3 (two Cl on C-2)

2.6. (a) None (b) $\ce{Cl(H)C=C(Cl)CH_2CH_3}$ [Cl, H on one carbon; Cl, CH_2CH_3 on the other] and [Cl, H on one carbon; CH_2CH_3, Cl on the other]

(c) Cl, Br / C=C / Cl, Br [Cl cis to Cl] and Cl, Br / C=C / Br, Cl [Cl trans to Cl] (d) Cl, Br / C=C / H, CH_3 and Cl, Br / C=C / CH_3, H

2.8. (a) alcohol hydroxyl, aldehyde; (b) two alcohol hydroxyls, ketone; (c) aryl, ester

2.10. (f) and (g) are constitutional isomers; (e) are stereoisomers; and (h) are not isomers.

2.11. (d) $CH_2{=}CHCCl_3$, $CH_2{=}C(Cl)CHCl_2$, $CH(Cl){=}CHCHCl_2$ (*cis* and *trans*),

$CH(Cl){=}C(Cl)CH_2Cl$ (*cis* and *trans*), $Cl_2C{=}C(Cl){-}CH_3$, $Cl_2C{=}CHCH_2Cl$, cyclopropane CH_2–CCl_2–$CH(Cl)$ (ring: CH_2, CH–Cl, C bearing two Cl)

cyclopropane $CH(Cl)$–$CH(Cl)$–$CHCl$ (*cis* and *trans*)

(e) $CH_3CH_2C{\equiv}CH$, $CH_3C{\equiv}CCH_3$, $CH_2{=}CHCH{=}CH_2$, $CH_2{=}C{=}CHCH_3$,

cyclobutene (CH_2–CH=CH–CH_2), methylenecyclopropane (CH_2, CH_2, $C{=}CH_2$), 3-methylcyclopropene ($CH{=}CH$, $CHCH_3$), bicyclobutane (CH, CH_2, CH, CH_2), 1-methylcyclopropene (CH_2, $CH{=}CCH_3$)

2.12. (f) aryl, carboxyl; (g) alcohol hydroxyl; (h) ester; (i) aldehyde; (j) alcohol hydroxyl, keto, carboxyl

2.13. (a)

$$CH_3CH_2CH_2C({=}O){-}OH \quad \text{and} \quad CH_3CH(CH_3)C({=}O){-}OH$$

(b)

$$HOCH_2CH_2CH_2C({=}O){-}H \quad \text{and} \quad CH_3C(OH)(CH_3){-}C({=}O){-}H$$

2.16. (e) and (f) are alkenes; (g) is an alkyne; (h) is an aromatic hydrocarbon. All except (h) are aliphatic.

2.17. (f) is an alkane; (h) is either an alkene or a cycloalkane; (g) is an alkyne (C_nH_{2n-2}); (i) is either an alkyne or a diene; (j) is aromatic.

CHAPTER 3
PRINCIPLES OF REACTIVITY

3.1. HCl, $CH_3\overset{\overset{O}{\|}}{C}{-}OH$, H_2SO_4, HSO_4^-, H_2O, NH_3, $C_6H_5{-}OH$

3.2. (a) NH_2^-, (b) O^{2-}, (c) CH_3^-
3.3. (a) $H_3SO_4^+$, (b) $H_2NO_3^+$
3.4. (a) acids: $HCl > H_3O^+$; bases: $H_2O > Cl^-$
(b) acids: $H_3O^+ > CH_3CO_2H$; bases: $CH_3CO_2^- > H_2O$
(c) acids: $NH_4^+ > H_2O$; bases: $OH^- > NH_3$
3.5. (a) 0.1 molar, (b) 0.008 molar
3.6. (a) 1.7×10^{-12}, (b) 1.4×10^{-5}, (c) 2.2×10^{-8}, (d) 2.1×10^{-4}

3.7. (a)

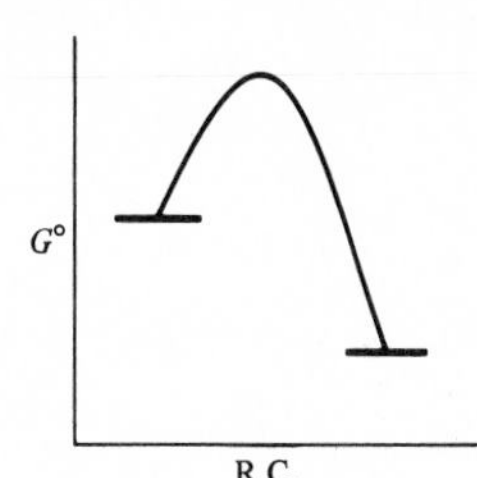

(b)

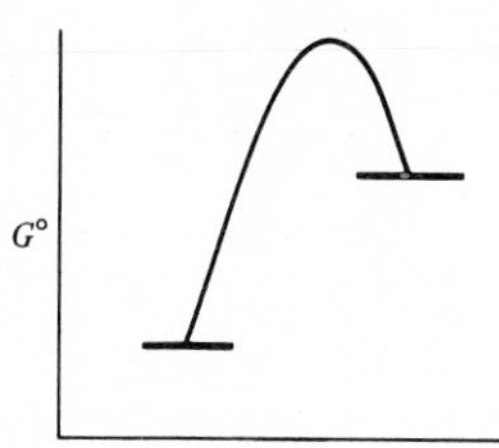

(c)

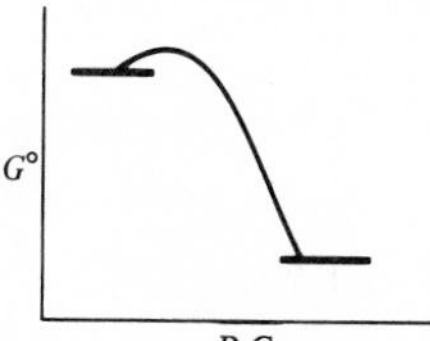

(d)

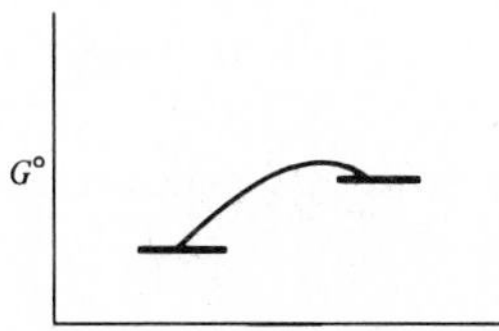

3.8.

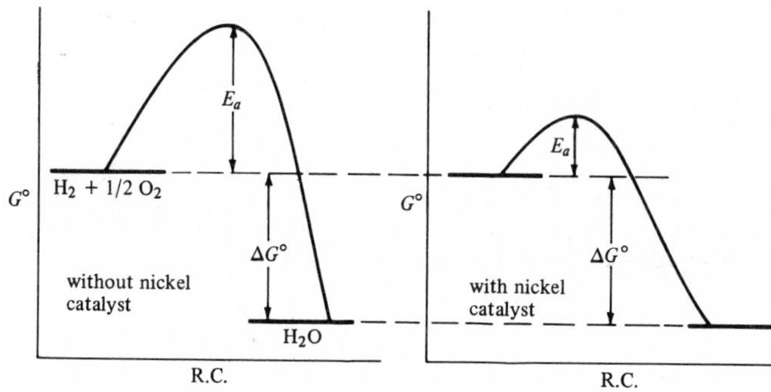

$\Delta G°$ remains the same, E_a decreases with Ni catalyst.

3.9. (a) $H^\cdot + H^\cdot$, (b) $:\ddot{\underset{\cdot\cdot}{Cl}}\cdot + \cdot\ddot{\underset{\cdot\cdot}{Cl}}:$, (c) $C_6H_5\cdot + \cdot CH_3$

3.10. (a) $H:^- + H^+$, (b) $:\ddot{\underset{\cdot\cdot}{Cl}}:^- + :\ddot{\underset{\cdot\cdot}{Cl}}^+$, (c) $C_6H_5:^- + {}^+CH_3$ or $C_6H_5^+ + :CH_3^-$

3.11. (a) 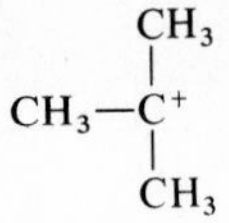 $CH_3-\overset{CH_3}{\underset{CH_3}{C^+}}$ (b) 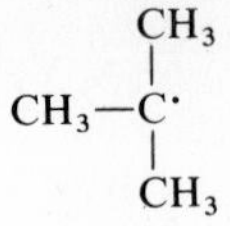 $CH_3-\overset{CH_3}{\underset{CH_3}{C\cdot}}$ (c) $CH_3CH_2CH_2\ddot{C}H_2^-$ (d)

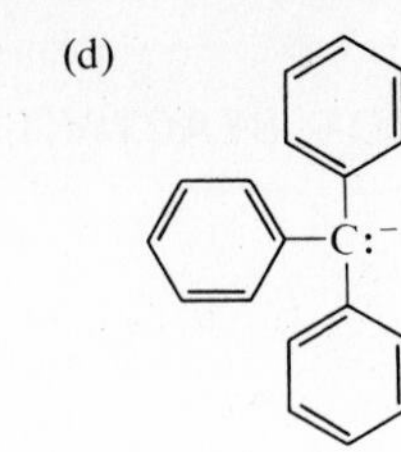

(e) 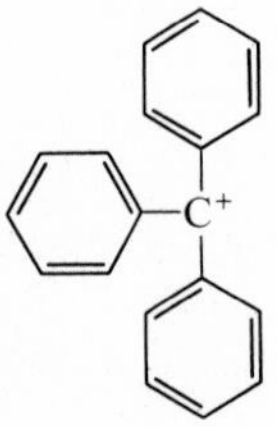 $(C_6H_5)_3C^+$ (f) 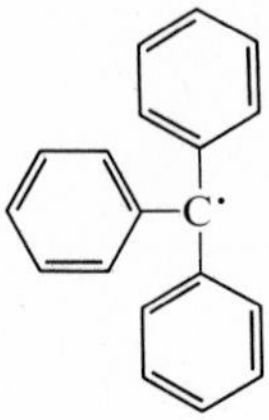 $(C_6H_5)_3C\cdot$

3.12. $R_3C^+ + :NH_3 \longrightarrow R_3C-N^+H_3$ (C⁺ has an empty orbital and is an electron pair acceptor.)

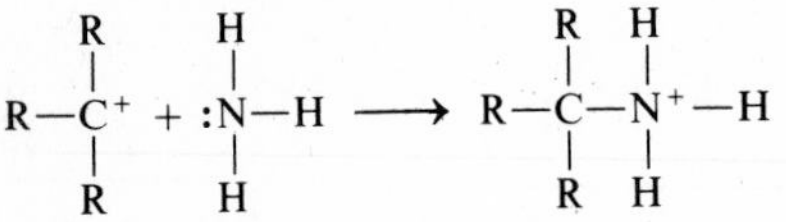

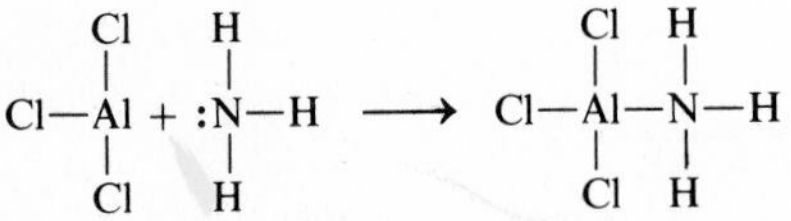

$Cl_3Al + :NH_3 \longrightarrow Cl_3Al-NH_3$ (Al has an empty orbital and is an electron pair acceptor.)

3.13. (a) any H^+ donor; (b) BF_3, $AlCl_3$, $ZnCl_2$, $FeCl_3$ and other Lewis acids that do not donate H^+.

3.14. Acids: (f) H_2S (SH^-), (h) HI (I^-), (i) HSO_3^- (SO_3^{2-}), (j) SH^- (S^{2-}), (k) $H_2PO_4^-$ (HPO_4^{2-}) (Conjugate bases are in parentheses.)
Bases: (i) HSO_3^- (H_2SO_3), (j) SH^- (H_2S), (k) $H_2PO_4^-$ (H_3PO_4). (Conjugate acids are in parentheses.)
(g) BF_3 is neither a Brønsted–Lowry acid nor a base.
The following are both Brønsted–Lowry acids and Lewis acids: H_2S, HI, HSO_3^-, SH^-, $H_2PO_4^-$.
All Brønsted–Lowry bases are also Lewis bases.

3.17. (d) 1.8×10^{-16} (e) ~ 0

3.18. 4.5×10^{-8}

3.20. (b)

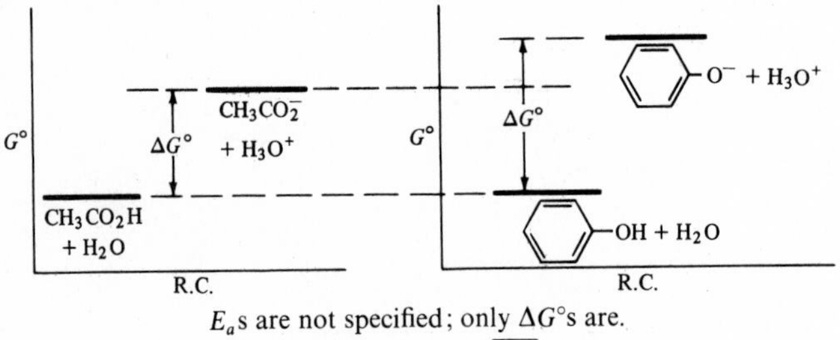

Left diagram: $G°$ vs R.C.; $CH_3CO_2H + H_2O$ (lower), $CH_3CO_2^- + H_3O^+$ (upper), separated by $\Delta G°$.

Right diagram: $G°$ vs R.C.; $C_6H_5OH + H_2O$ (lower), $C_6H_5O^- + H_3O^+$ (upper), separated by $\Delta G°$.

E_as are not specified; only $\Delta G°$s are.

$|\Delta G°| = CH_3CO_2H < C_6H_5OH$

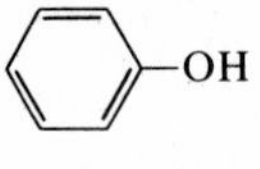

3.21. (f) $\cdot\ddot{O}:H$, (g) $H:\overset{H}{\underset{H}{C^+}}$, (h) $H:\overset{H\,H}{\underset{H\,H}{C:C:^-}}$ (i) $2:\ddot{Cl}\cdot$, (j) $:\ddot{Cl}:^- + {}^+\ddot{Cl}:$

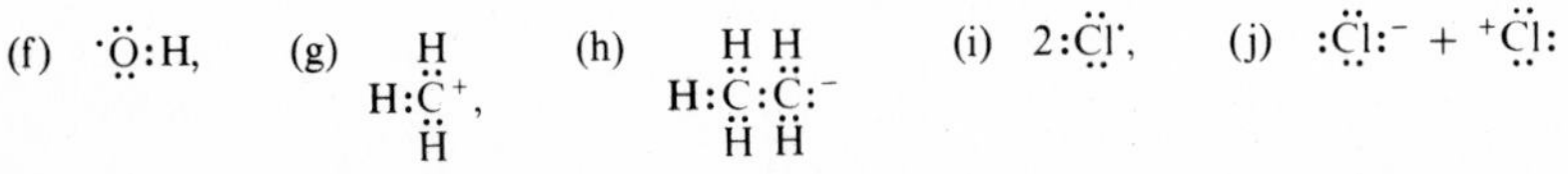

3.24. (f) $H_3O^+ + I^-$, (g) BCl_4^-, (h) $FeBr_4^-$, (i) 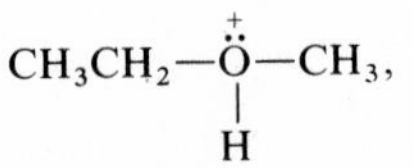 $CH_3CH_2-\overset{+}{\underset{H}{\ddot{O}}}-CH_3$,

(j) $CH_3CH_3 + {}^-OCH_3$

3.26. All of them.

CHAPTER 4
ALKANES

4.1.

C_5H_{12}: $CH_3CH_2CH_2CH_2CH_3$, $CH_3CH_2\underset{\large CH_3}{\underset{|}{C}}HCH_3$, $CH_3\overset{CH_3}{\overset{|}{\underset{\underset{\large CH_3}{|}}{C}}}CH_3$

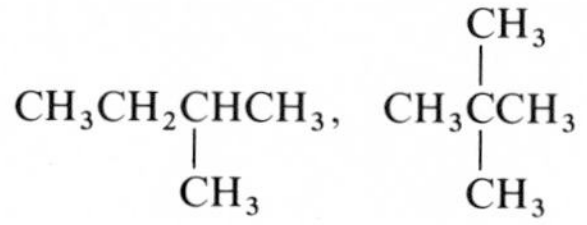

C_6H_{14}: $CH_3CH_2CH_2CH_2CH_2CH_3$, $CH_3CH_2CH_2\underset{\large CH_3}{\underset{|}{C}}HCH_3$, $CH_3CH_2\underset{\large CH_3}{\underset{|}{C}}HCH_2CH_3$,

$CH_3CH_2\overset{CH_3}{\overset{|}{\underset{\underset{\large CH_3}{|}}{C}}}CH_3$, $CH_3\overset{CH_3}{\overset{|}{C}}H\underset{\large CH_3}{\underset{|}{C}}HCH_3$

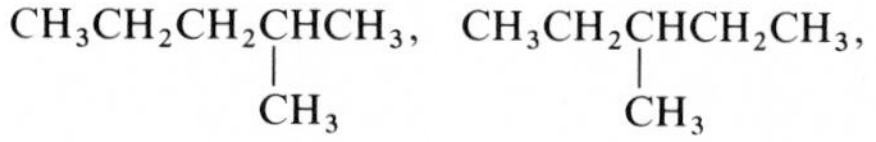

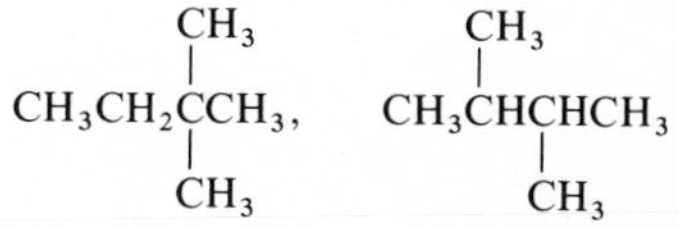

4.2.

CH₃ / H, CH₃ / H, H / H — H₃C H / H, H / H, CH₃ — CH₃ / H, H / H, H / CH₃

H₃C H / CH₃, H / H, H — CH₃ / CH₃, H / H, H / H — H₃C CH₃ / H, H / H, H

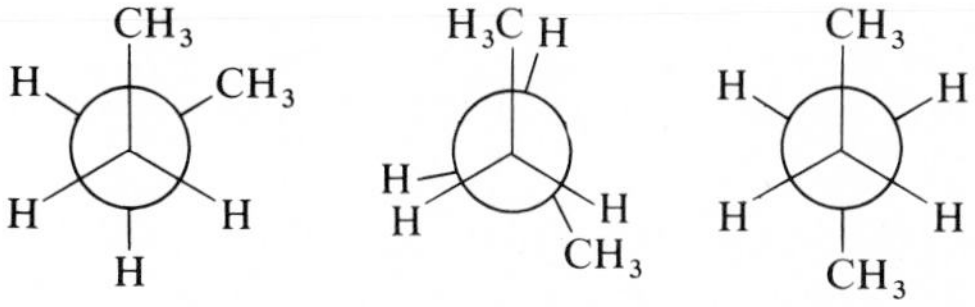

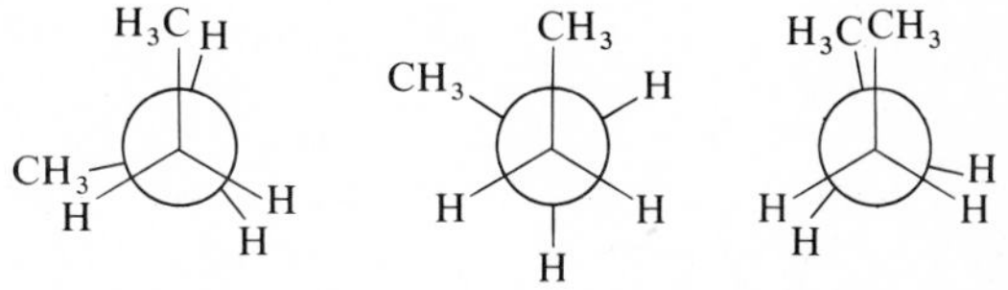

4.3. (a) pentyl, (b) hexyl, (c) octyl

4.4. $CH_3CH_2CH_2—$ (propyl), $CH_3\underset{|}{C}HCH_3$ (isopropyl)

4.5. (a) both 1°; (b) CH_3s are 1°, $—\overset{|}{\underset{|}{C}}—$ is 4°; (c) all 2°; (d) 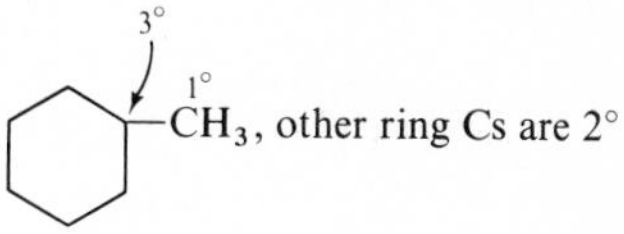 (cyclohexyl ring, 3° carbon indicated) $—CH_3$ (1°), other ring Cs are 2°

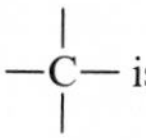

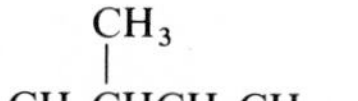

4.6. $CH_3\overset{CH_3}{\overset{|}{C}}HCH_2CH_2—$ (isopentyl), $CH_3\overset{CH_3}{\overset{|}{\underset{\underset{\large CH_3}{|}}{C}}}CH_2—$ (neopentyl),

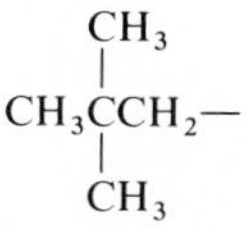

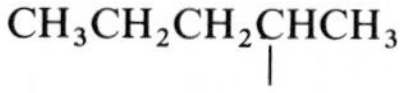

$CH_3CH_2CH_2\underset{|}{C}HCH_3$ (*sec*-pentyl, *ambiguous*), $CH_3CH_2\overset{CH_3}{\overset{|}{\underset{\underset{\large CH_3}{|}}{C}}}—$ (*tert*-pentyl)

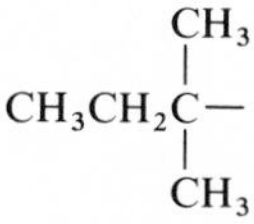

4.7. (a) butane, (b) 2-methylbutane (the 2- is unnecessary in this case because the name is unambiguous without it), (c) dimethylpropane, (d) 3,3-dimethylpentane

4.8. (a) 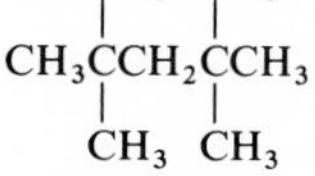 $CH_3\overset{CH_3}{\overset{|}{\underset{\underset{\large CH_3}{|}}{C}}}CH_2\overset{CH_3}{\overset{|}{\underset{\underset{\large CH_3}{|}}{C}}}CH_3$ (b) $CH_3CH_2\overset{CH_2CH_3}{\overset{|}{C}}HCH_2\underset{\large CH_2CH_3}{\underset{|}{C}}HCH_2CH_3$

(c) $CH_3\overset{CH_3}{\overset{|}{C}}H—\overset{CH_3}{\overset{|}{\underset{\underset{\large CH_3CH_2}{|}}{C}}}—\underset{\large CH_2CH_2CH_3}{\underset{|}{C}}HCH_2CH_2CH_2CH_2CH_3$ (d) 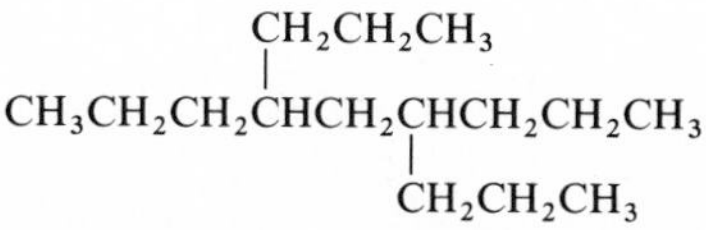 $CH_3CH_2CH_2\overset{CH_2CH_2CH_3}{\overset{|}{C}}HCH_2\underset{\large CH_2CH_2CH_3}{\underset{|}{C}}HCH_2CH_2CH_3$

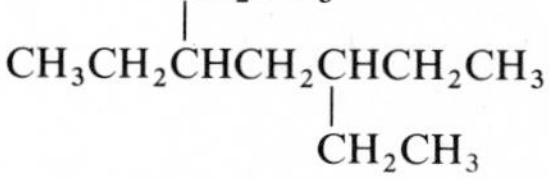

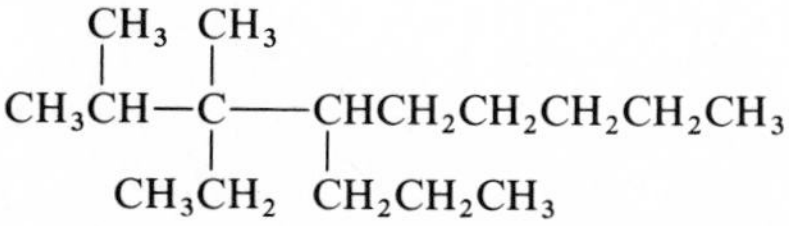

4.9. (a) 2-methylpentane, (b) 2,2,6-trimethylheptane, (c) 2,5-dimethylhexane, (d) 2,3-dimethylhexane

4.10. (a) 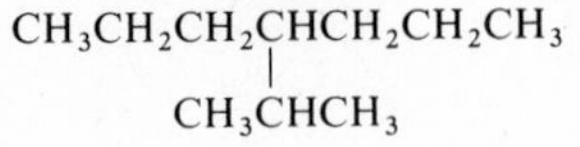
$$\begin{array}{l} CH_3CH_2CH_2\underset{\displaystyle |}{C}HCH_2CH_2CH_3 \\ \qquad\quad\;\; CH_3CHCH_3 \end{array}$$

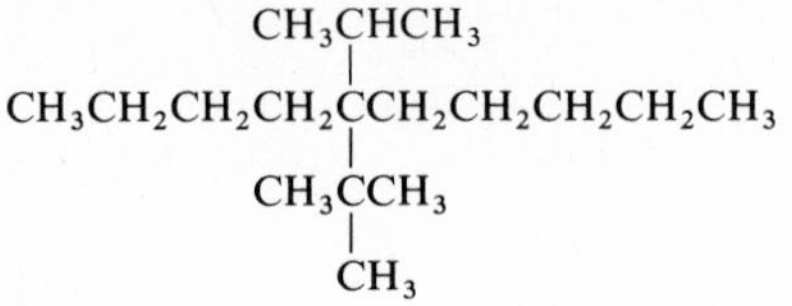

(b)
$$\begin{array}{c} CH_3CHCH_3 \\ | \\ CH_3CH_2CH_2CH_2CCH_2CH_2CH_2CH_2CH_3 \\ | \\ CH_3CCH_3 \\ | \\ CH_3 \end{array}$$

(c)
$$\begin{array}{l} CH_3CH_2CHCH_3 \\ \qquad\quad\;\; | \\ \qquad\quad\; OH \end{array}$$

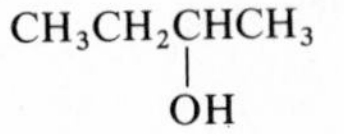

(d) 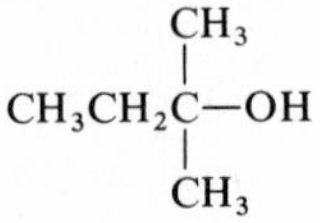
$$\begin{array}{r} CH_3 \\ | \;\;\;\;\;\;\; \\ CH_3CH_2C{-}OH \\ | \;\;\;\;\;\;\; \\ CH_3 \end{array}$$

4.12.
$$\begin{array}{l} (CH_3CH_2CH)_2CuLi + CH_3CH_2Cl \\ \qquad\qquad | \\ \qquad\quad\; CH_3 \end{array}$$

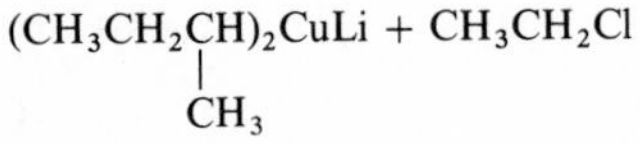

4.13. The Corey–House synthesis permits the synthesis of unsymmetrical alkanes R—R′.

4.14. Reduction of alkyl halides yields alkanes containing the same number of carbon atoms as the starting alkyl halide. In the Wurtz and Corey–House methods, the product alkane is larger than the starting alkyl halides.

4.15. (a) $2CH_3CH_2CH_2Br + 2Na \xrightarrow{\text{boil}} CH_3CH_2CH_2CH_2CH_2CH_3 + 2NaBr$

(b) 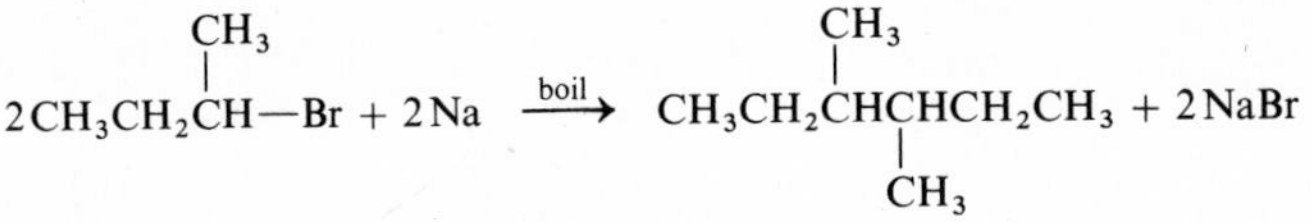
$$\begin{array}{c} CH_3 \\ | \\ 2CH_3CH_2CH{-}Br + 2Na \end{array} \xrightarrow{\text{boil}} \begin{array}{c} CH_3 \\ | \\ CH_3CH_2CHCHCH_2CH_3 \\ | \\ CH_3 \end{array} + 2NaBr$$

(c) $CH_3CH_2CH_2Cl + Zn + CH_3CO_2H \longrightarrow CH_3CH_2CH_3 + Zn^{2+} + Cl^- + CH_3CO_2{}^-$

4.16. Only (a).

(a) $CH_3CH_2CH_2Br + 2Li \longrightarrow CH_3CH_2CH_2Li + LiBr$
$2CH_3CH_2CH_2Li + CuI \longrightarrow (CH_3CH_2CH_2)_2CuLi + LiI$
$(CH_3CH_2CH_2)_2CuLi + BrCH_2CH_2CH_3 \longrightarrow CH_3CH_2CH_2CH_2CH_2CH_3 + CH_3CH_2CH_2Cu + LiBr$

(c) $(CH_3CH_2)_2CuLi + BrCH_3 \longrightarrow CH_3CH_2CH_3 + CH_3CH_2Cu + LiBr$

4.17. $CH_3CH_3 \longrightarrow CH_3CH_2\cdot + H\cdot$
$\qquad\qquad \searrow 2CH_3\cdot$

Recombination:

$CH_3CH_2\cdot + CH_3\cdot \longrightarrow CH_3CH_2CH_3$
$2CH_3CH_2\cdot \longrightarrow CH_3CH_2CH_2CH_3$
$CH_3CH_2\cdot + H\cdot \longrightarrow CH_3CH_3$
$CH_3\cdot + H\cdot \longrightarrow CH_4$
$2H\cdot \longrightarrow H_2$

4.18. (a) $C_5H_{12} + 8O_2 \longrightarrow 5CO_2 + 6H_2O$
(b) $C_2H_6 + Cl_2 \longrightarrow C_2H_5Cl + HCl$
(c) $C_2H_6 + 6Cl_2 \longrightarrow C_2Cl_6 + 6HCl$
(d) $C_2H_6 + HONO_2 \longrightarrow C_2H_5NO_2 + H_2O$

4.19. (a) A large excess of CH_4, (b) A large excess of Cl_2

4.20. (a) 1,2-dimethylcyclobutane, (b) 1-cyclopropyl-2,4-dimethylpentane

4.21. (a) 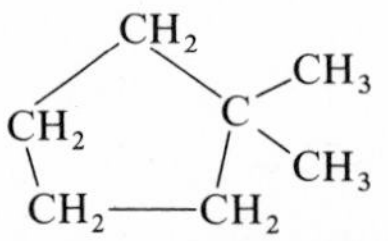
$$\begin{array}{ccc} & CH_2 & \\ \diagup & & \diagdown \;\; CH_3 \\ CH_2 & & C \\ \backslash & & / \;\; \diagdown CH_3 \\ CH_2 & {-\!\!-} & CH_2 \end{array}$$

(b)
$$\begin{array}{l} CH_3CH_2CH{-}CHCH_2CH_2CH_3 \\ \qquad\quad\;\;\; | \qquad | \\ CH_3CH_2CH_2CH{-}CHCH_2CH_3 \end{array}$$

(c)
$$\begin{array}{ccc} & CH_2{-}CH_2 & \\ \diagup & & \diagdown \\ CH_2 & & CH_2 \\ | & & \diagdown \\ | & & CHCH_3 \\ | & & \diagup \\ CH_2 & & CH_2 \\ \diagdown & & \diagup \\ & CH_2{-}CH_2 & \end{array}$$

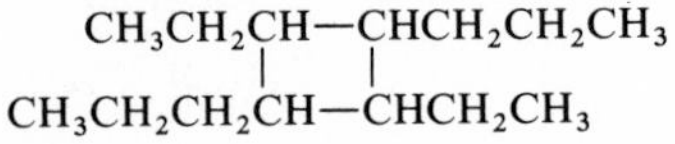

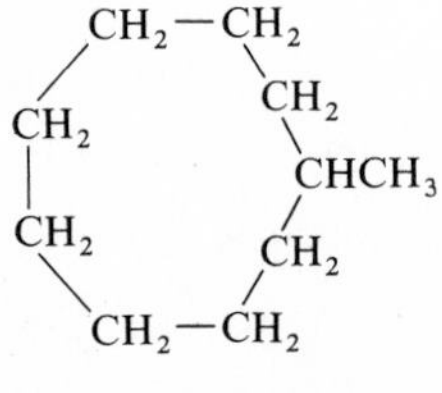

4.22. (a) 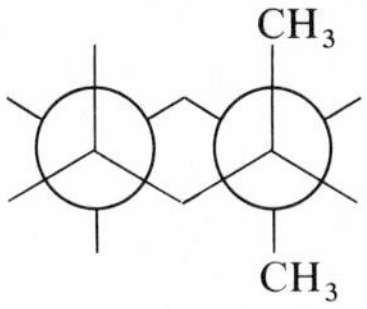 CH3 … CH3 (b) 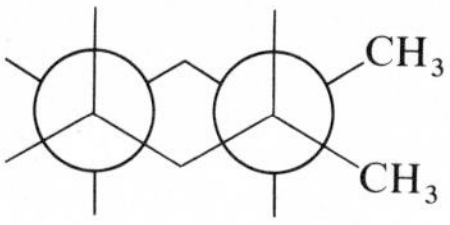 CH3, CH3 (c) 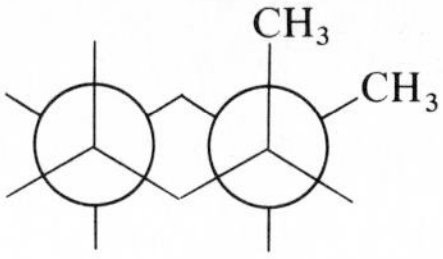 CH3, CH3

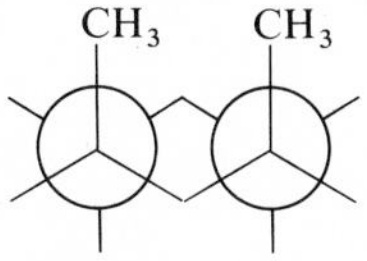

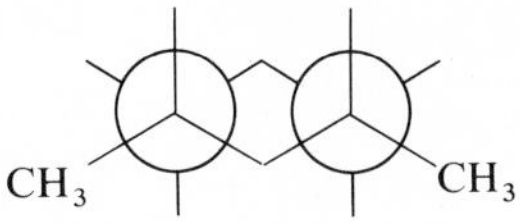

4.23. 1,3-: (a) CH3 CH3 (b) CH3 CH3 (c) CH3 CH3

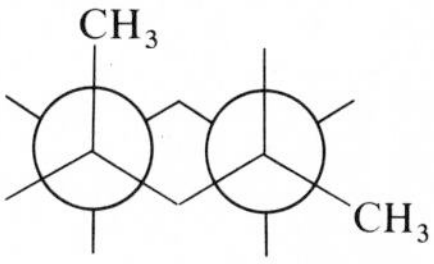

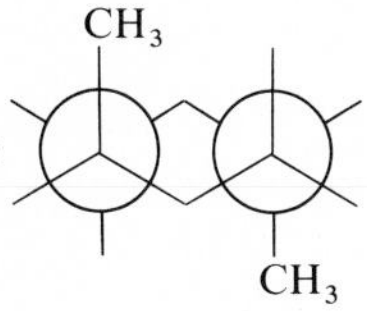

1,4-: (a) CH3 CH3 (b) 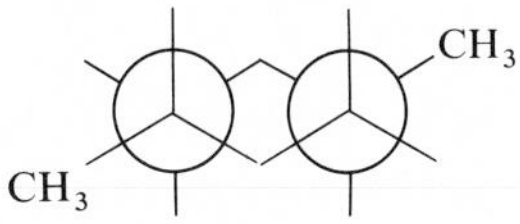 CH3 CH3 (c) 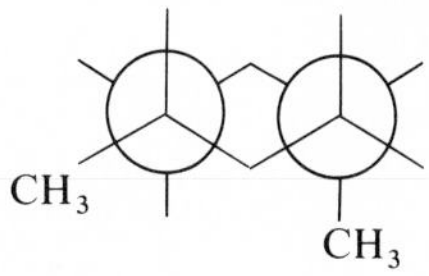 CH3 CH3

4.24. (a) 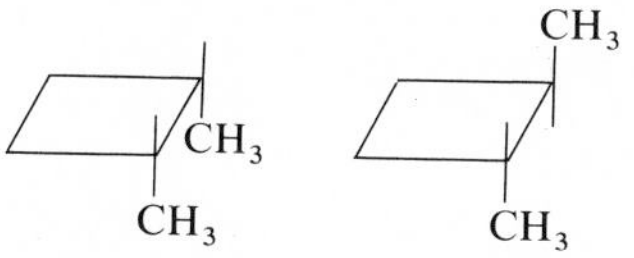 CH3 CH3 CH3 CH3 (b) 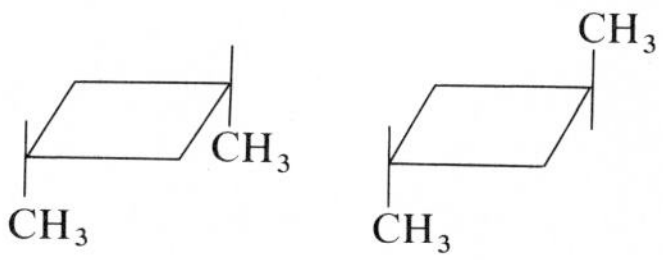 CH3 CH3 CH3 CH3

(c) 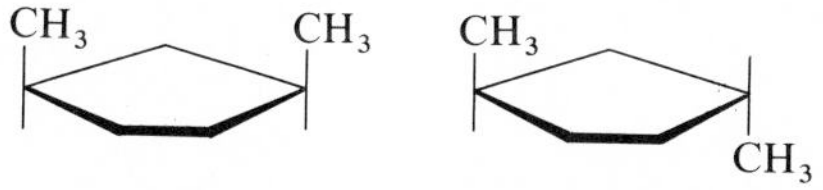 CH3 CH3 CH3 CH3

4.25. (a) e,e; (b) e,a (the two are equivalent); (c) *trans* (e,e)

4.26.

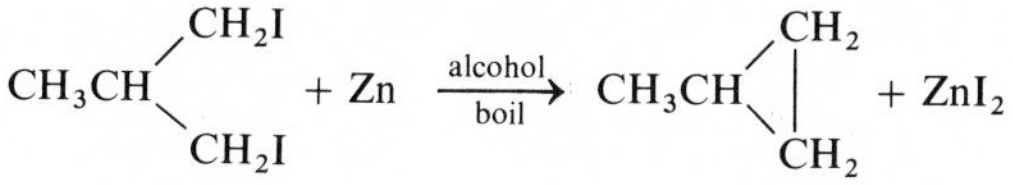

$$CH_3CH(CH_2I)_2 + Zn \xrightarrow[\text{boil}]{\text{alcohol}} CH_3CH\langle(CH_2)_2\rangle + ZnI_2$$

4.27. (a) $CH_3CH_2CH_2Cl$, (b) $ClCH_2CH_2CH_2Cl$, (c) $DCH_2CH_2CH_2D$

4.28. H H H H H H / H C C C H / C C C / H H H H H H

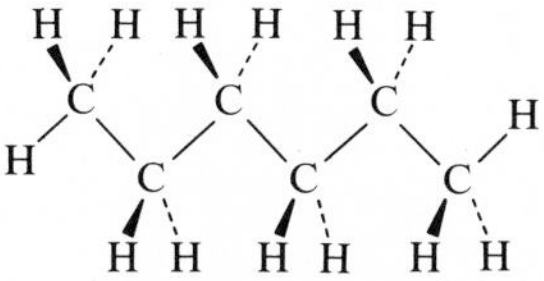

4.30. (d) 2,3-dimethylpentane, (e) 3,3-dimethylhexane, (f) 3-ethylpentane

4.31. (e) 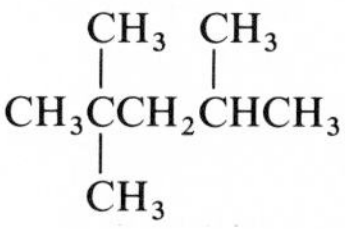 $CH_3C(CH_3)_2CH_2CH(CH_3)CH_3$ (f) 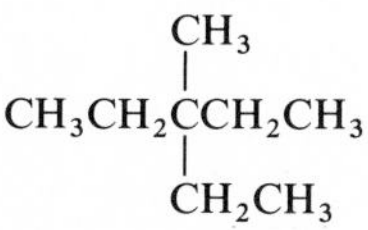 $CH_3CH_2C(CH_3)(CH_2CH_3)CH_2CH_3$

(g) $CH_3CH_2CH_2CH_2CH_2CH_2CH_2CH_3$ (h) 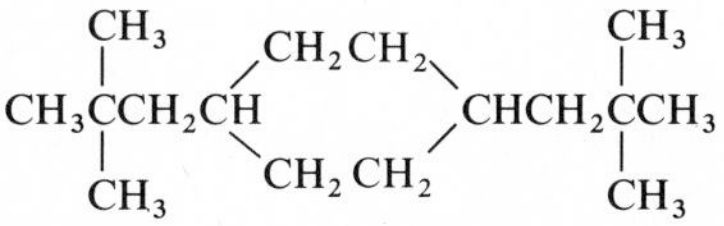 $CH_3C(CH_3)_2CH_2CH\langle CH_2CH_2, CH_2CH_2\rangle CHCH_2C(CH_3)_2CH_3$

4.33. (b) $2CH_3CH_2CH_2CH_2Br + 2Na \xrightarrow{\text{boil}}$ *n*-octane

$CH_3CH_2CH_2CH_2CH_2CH_2CH_2CH_2Br + Zn + CH_3CO_2H \xrightarrow{\text{boil}}$ *n*-octane + Zn^{2+} + Br^- + $CH_3CO_2^-$

4.34. (d) $CH_2(CH_2CH_2)_2CH_2 + Cl_2 \xrightarrow[\text{or } \Delta]{\text{light}} CH_2(CH_2CH_2)_2CH{-}Cl + HCl$

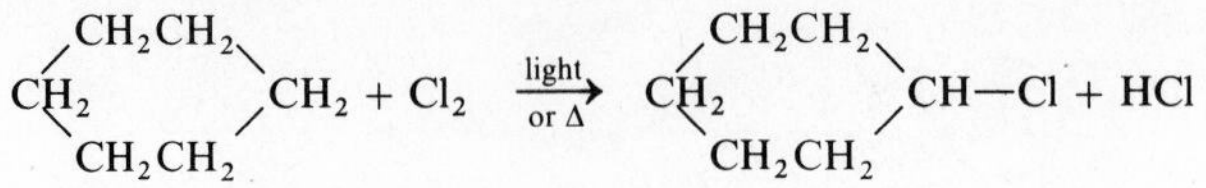

(e) $CH_3CH_3 + HONO_2 \longrightarrow CH_3CH_2NO_2 + H_2O$

(f) $C_9H_{20} + 14O_2 \longrightarrow 9CO_2 + 10H_2O$

4.36. (c) 1,2-dipropylcyclobutane, (d) 1,3-diethyl-1,3-dimethylcyclopentane

4.37. (d)

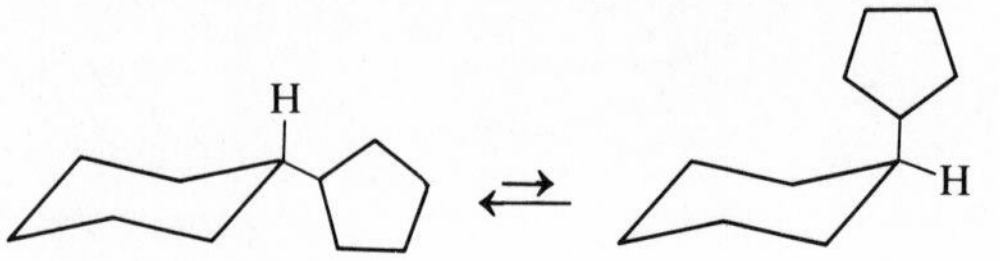

H

⇄

H

(e) CH_2CH_3

CH_2CHCH_3
|
CH_3

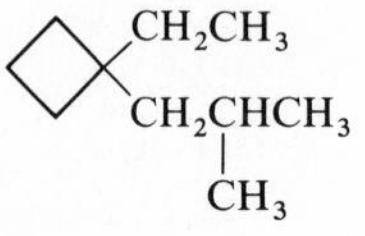

(f)

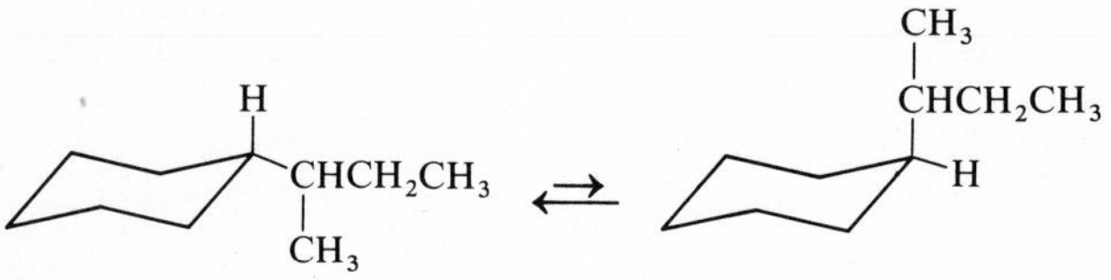

H

$CHCH_2CH_3$
|
CH_3

⇄

CH_3
|
$CHCH_2CH_3$

H

4.40. (b) ▷$-CH_2CH_2CH_3$

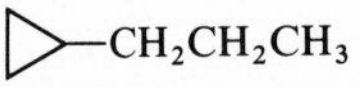

4.41. (d) ▷ $+ HI \longrightarrow CH_3CH_2CH_2I$

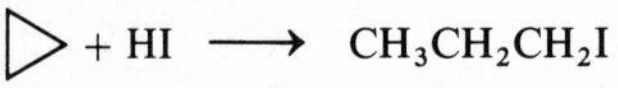

(e) ▷ $+ Br_2 \xrightarrow{\text{dark}} BrCH_2CH_2CH_2Br$

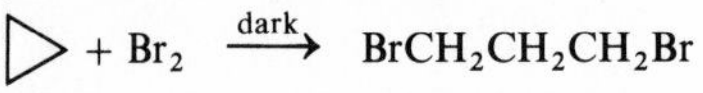

(f) ▷ $+ HD \longrightarrow CH_3CH_2CH_2D$

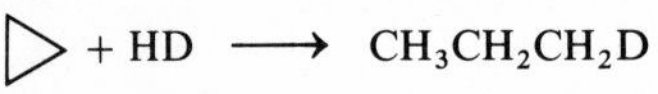

4.42. (c) $(CH_3CH(CH_3)CH_2)_2CuLi + BrCH_2CH(CH_3)CH_3 \longrightarrow CH_3CH(CH_3)CH_2CH_2CH(CH_3)CH_3 + CH_3CH(CH_3)CH_2Cu + LiBr$

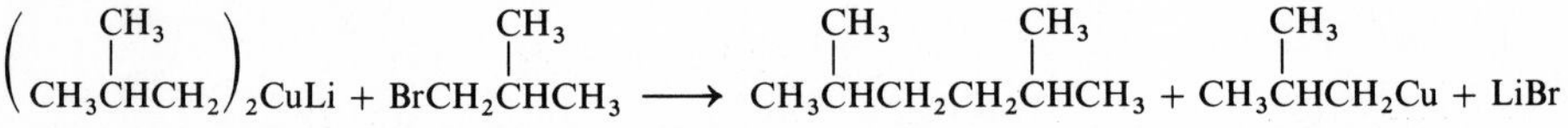

(d) $(CH_3CH(CH_3))_2CuLi + BrCH_3 \longrightarrow CH_3CH(CH_3)CH_3 + CH_3CH(CH_3)Cu + LiBr$

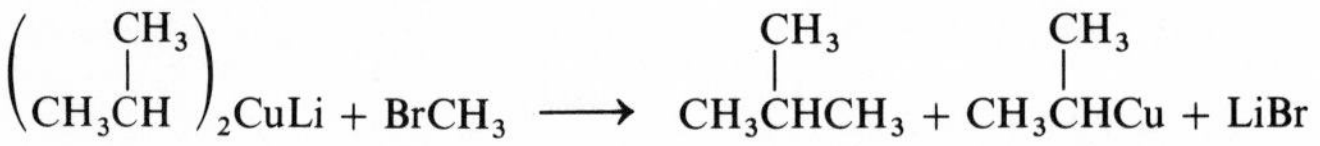

CHAPTER 5
UNSATURATED HYDROCARBONS: ALKENES

5.1. (a) $CH_3CH{=}CHCH_2CH_3$ (b) ▷ (c) 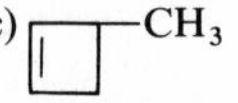 $-CH_3$ (d)

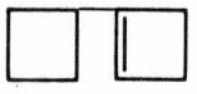

5.2. (a) 3-hexene, (b) 4-methyl-1-heptene, (c) 3-allylcyclohexene,
(d) 4-methyl-2-pentene, (e) 3,5-divinylcyclopentene

5.3. The correct names are (a) 2-pentene, (b) 3-ethyl-1-pentene, (c) 1-methylcyclobutene,
(d) 3-methyl-1-butene.

5.4. In a sigma bond, orbital overlap is along the direction of the orbital and therefore more effective.

In a pi bond, orbital overlap is perpendicular to the direction of the orbital and therefore less effective.

5.5.

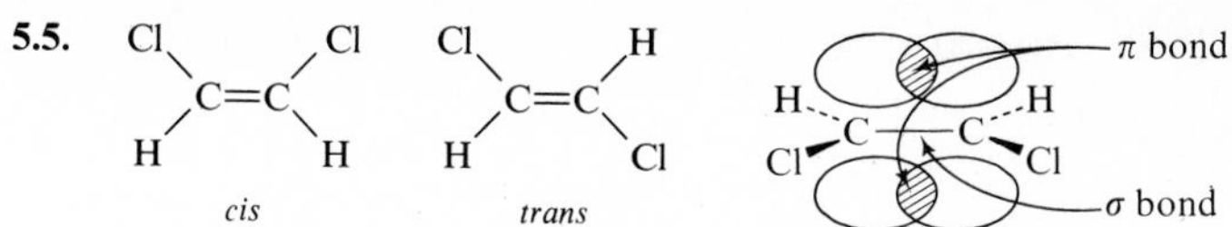

5.6. (a) and (c) do not exist as *cis-trans* isomers;

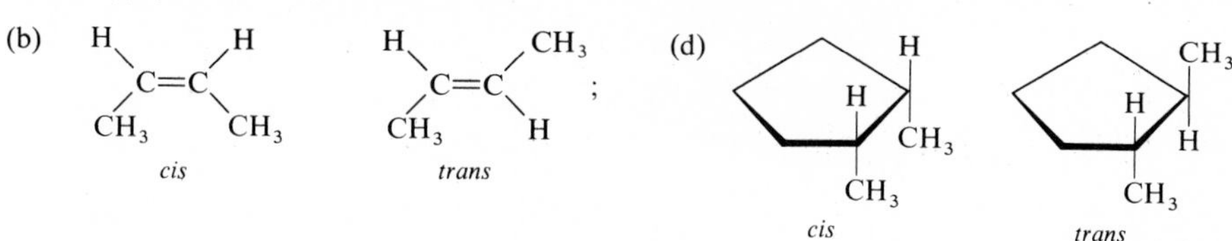

5.7. (a) H—Cl (b) CH_3Cl (c) (d) (e) (f)

(e) (no overall polarity) (f) (no overall polarity)

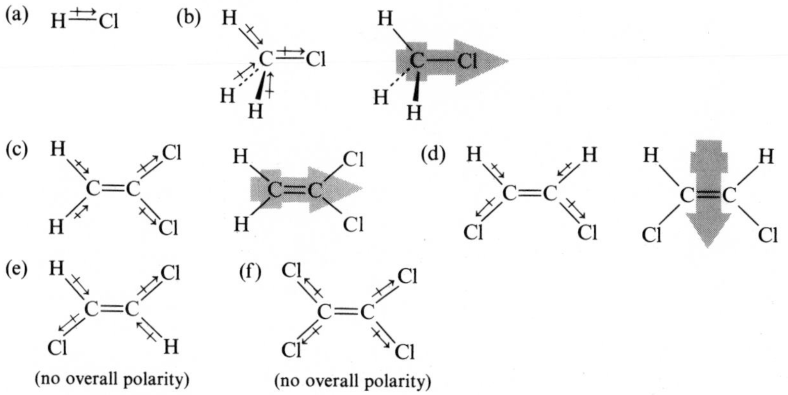

5.8. Alkanes and alkenes have low polarities and therefore do not pack together tightly.

5.9. (a) $CH_3CH_2CH_2OH \xrightarrow[\Delta]{H_2SO_4} CH_3CH{=}CH_2 + H_2O$

(b) $CH_3CH_2CH_2Cl + KOH \xrightarrow[\Delta]{\text{ethanol}} CH_3CH{=}CH_2 + KCl + H_2O$

5.10. (a) $CH_3CH_2CH{=}CH_2$

(b) $CH_3CH_2CH{=}CH_2$

(c) $CH_3C(CH_3){=}CH_2$

(d) $CH_3C(CH_3){=}CH_2$

(e) $CH_3CH_2CH{=}CHCH_3 + CH_3CH_2CH_2CH{=}CH_2$
(major product)

(f) $CH_3CH{=}C(CH_3)CH_2CH_2CH_3 + CH_3CH_2C(CH_3){=}CHCH_2CH_3$ (major products) $+ CH_3CH_2C({=}CH_2)CH_2CH_2CH_3$

5.11. $CH_3C(CH_3)(OH)CH_2CH_3$

5.12.

Formation of a new C—C sigma bond:	−84 kcal/mole
Dissociation of a C—C pi bond:	+62 kcal/mole
Total ΔH°:	−22 kcal/mole

5.13. Teflon is made from $CF_2{=}CF_2$; poly(methyl methacrylate) from

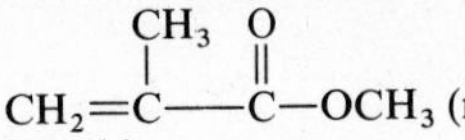

$$CH_2{=}\overset{\displaystyle CH_3}{\overset{|}{C}}{-}\overset{\displaystyle O}{\overset{\|}{C}}{-}OCH_3$$ (methyl methacrylate).

5.14. Possible structures are

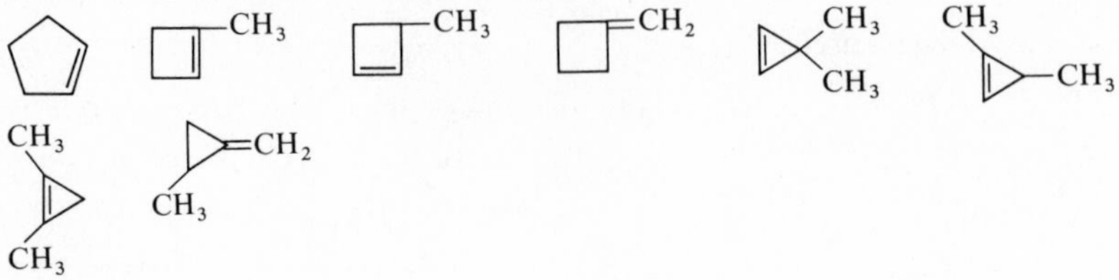

cyclopentene; 1-methylcyclobutene (CH_3); 3-methylcyclobutene (CH_3); methylenecyclobutane ($=CH_2$); 1,1-dimethylcyclopropane (CH_3, CH_3); 1,3-dimethylcyclopropene (CH_3, CH_3); 1,2-dimethylcyclopropene (CH_3, CH_3); 2-methylmethylenecyclopropane ($=CH_2$, CH_3)

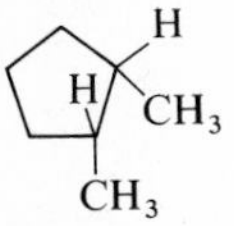

5.15. 1,2-dimethylcyclopentane (H, H, CH_3, CH_3)

5.16. $(CH_3)_2C{=}C(CH_3)_2$

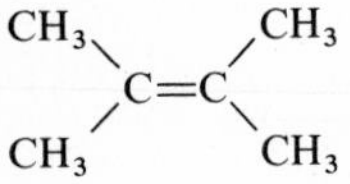

5.17.

cyclopropane $+ Br_2 \xrightarrow{CCl_4} BrCH_2CH_2CH_2Br$

cyclopropane $+ H_2SO_4 \longrightarrow CH_3CH_2CH_2OSO_3H$

$>C{=}C< + Br_2 \xrightarrow{CCl_4} -\underset{Br}{\underset{|}{\overset{|}{C}}}-\underset{Br}{\underset{|}{\overset{|}{C}}}-$

$>C{=}C< + MnO_4^- \xrightarrow{H_2O} -\underset{OH}{\underset{|}{\overset{|}{C}}}-\underset{OH}{\underset{|}{\overset{|}{C}}}- + MnO_2$

$>C{=}C< + H_2SO_4 \longrightarrow -\underset{H}{\underset{|}{\overset{|}{C}}}-\underset{OSO_3H}{\underset{|}{\overset{|}{C}}}$

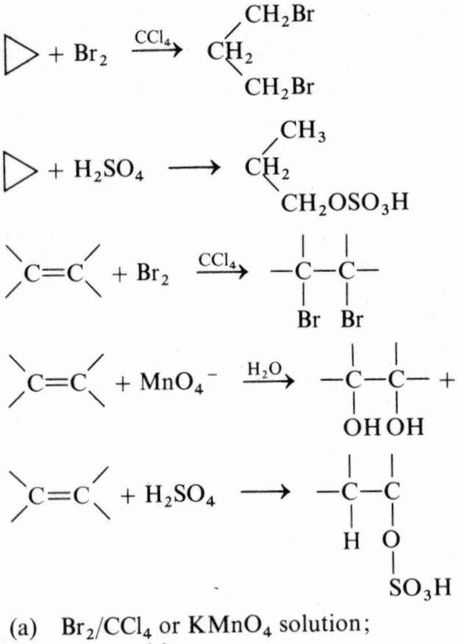

5.18. (a) Br_2/CCl_4 or $KMnO_4$ solution;
(b) same as (a)

5.19. C_5H_{10}: $CH_2{=}CHCH_2CH_2CH_3$ 1-pentene; $CH_3CH{=}CHCH_2CH_3$ 2-pentene (*cis* and *trans*); $CH_2{=}CHCH(CH_3)CH_3$ 3-methyl-1-butene

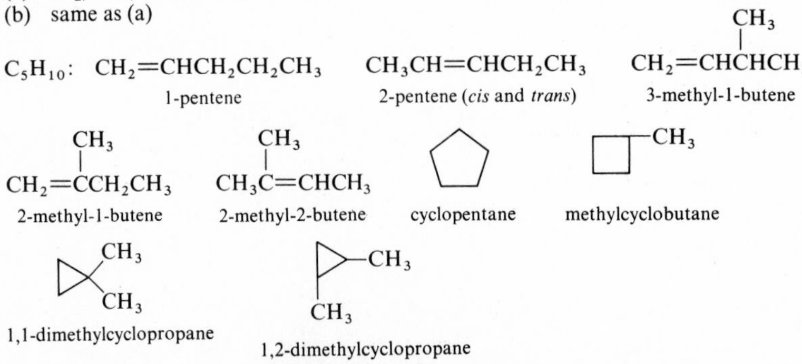

$CH_2{=}C(CH_3)CH_2CH_3$ 2-methyl-1-butene; $CH_3C(CH_3){=}CHCH_3$ 2-methyl-2-butene; cyclopentane; methylcyclobutane (CH_3)

1,1-dimethylcyclopropane (CH_3, CH_3); 1,2-dimethylcyclopropane (*cis* and *trans*) (CH_3, CH_3)

5.20. (d)

$$\begin{array}{c} \quad\quad\;\; CH_3\;\; CH_3 \\ \quad\quad\;\; |\quad\;\; | \\ CH_3CH_2C{=}CCH_2CH_3 \end{array}$$

(e)

$$\begin{array}{ccc} H & & H \\ & C{=}C & \\ CH_3 & & CH_2CH_3 \end{array}$$

(f) (cyclopentene ring with H and CH_3 on each of two ring carbons)

H, H, CH_3, CH_3

5.23. (c)

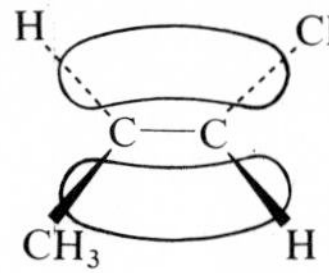

(d)

H H

5.24. (e)

$$\begin{array}{ccc} H & & H \\ & C{=}C & \\ CH_3 & & CH_3 \end{array} \qquad \begin{array}{ccc} H & & CH_3 \\ & C{=}C & \\ CH_3 & & H \end{array}$$

cis *trans*

(d) and (f) do not exhibit *cis-trans* isomerism

5.25. (h)

$$\begin{array}{ccc} Br & & F \\ & C{=}C & \\ F & & Br \end{array}$$

5.27. (b)

$$\begin{array}{c} CH_3\;\; CH_3 \\ CH_3C{-}CHCH_3 \\ OH \end{array} \xrightarrow[\Delta]{H_2SO_4} \begin{array}{c} CH_3\;\; CH_3 \\ CH_3C{=}CCH_3 \end{array} + H_2O$$

$$\begin{array}{c} CH_3\;\; CH_3 \\ CH_3C{-}CHCH_3 \\ Br \end{array} + KOH \xrightarrow[\Delta]{\text{ethanol}} \begin{array}{c} CH_3\;\; CH_3 \\ CH_3C{=}CCH_3 \end{array} + KBr + H_2O$$

5.29. Products are:

(a) $CH_3CH_2CH_2CH_3$, (b) $\begin{array}{c} Br \\ CH_3CHCHCH_3, \\ Br \end{array}$ (c) $\begin{array}{c} Cl \\ CH_3CHCH_2CH_3, \end{array}$

(d) $\begin{array}{c} OH \\ CH_3CHCHCH_3, \\ OH \end{array}$ (e) $\begin{array}{c} O \\ \| \\ 2CH_3CH \end{array}$

5.32. The original compound is cyclohexene:

cyclohexene $+ H_2 \xrightarrow{Ni}$ cyclohexane (C_6H_{12})

cyclohexene $\xrightarrow[\text{reduction}]{\text{ozonolysis, then}}$ $\begin{array}{c} CH\;\; CH \\ \|\quad\;\; \| \\ O\quad\;\; O \end{array}$ (joined by a four-carbon chain)

5.34. (c) (cyclopentane ring bearing I and CH_3 on the same carbon)

I, CH_3

CHAPTER 6
UNSATURATED HYDROCARBONS: DIENES AND ALKYNES

6.1. (a) 1-methyl-1,3-cyclopentadiene, (b) 2-methyl-1,3-cyclopentadiene, (c) 5-methyl-1,3-cyclopentadiene, (d) 4,6-dimethyl-2,4-heptadiene

6.2.

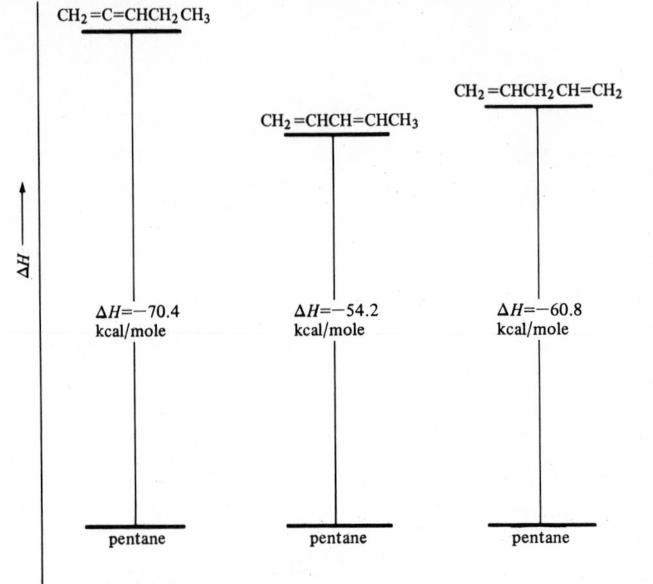

6.3. (a)

$$\underset{\text{OH}}{\text{CH}_2}\overset{\text{CH}_3}{\text{CH}}-\overset{\text{CH}_3}{\text{CH}}\underset{\text{OH}}{\text{CH}_2} \xrightarrow[\Delta]{H_2SO_4} CH_2{=}\overset{\text{CH}_3}{\text{C}}-\overset{\text{CH}_3}{\text{C}}{=}CH_2 + 2H_2O$$

(b)

$$\underset{\text{Br}}{\text{CH}_2}\overset{\text{CH}_3}{\text{CH}}-\overset{\text{CH}_3}{\text{CH}}\underset{\text{Br}}{\text{CH}_2} + KOH \xrightarrow[\Delta]{\text{ethanol}} CH_2{=}\overset{\text{CH}_3}{\text{C}}-\overset{\text{CH}_3}{\text{C}}{=}CH_2 + 2KBr + 2H_2O$$

6.4. (a) cyclopentadiene $+ Cl_2 \longrightarrow$ 3,4-dichlorocyclopentene (Cl, Cl) (1,2-) $+$ 3,5-dichlorocyclopentene (Cl, Cl) (1,4-)

(b) cyclopentadiene $+ H_2O \xrightarrow{H_2SO_4}$ cyclopentenol (HO) (1,2-) $+$ cyclopentenol (OH) (1,4-)

In (b), 1,2- addition and 1,4- addition give the same product.

6.5.

butadiene + maleic anhydride (O, O, O): diene, dienophile,

H_3C–C(=CH_2)–C(=CH_2)–CH_3 + CH_2=CH–CH=O: diene, dienophile,

cyclopentadiene + maleic anhydride (O, O, O): diene, dienophile

cyclopentadiene + cyclopentadiene: diene, dienophile

6.6. (a) diene + dienophile → (methylcyclohexene product, CH_3)

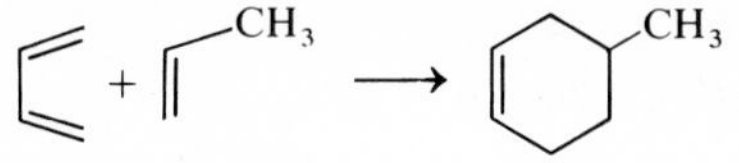

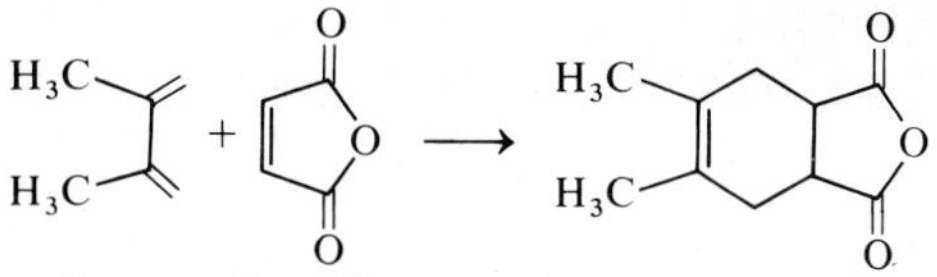

(b) diene (H_3C, H_3C) + dienophile (maleic anhydride) → (product with H_3C, H_3C)

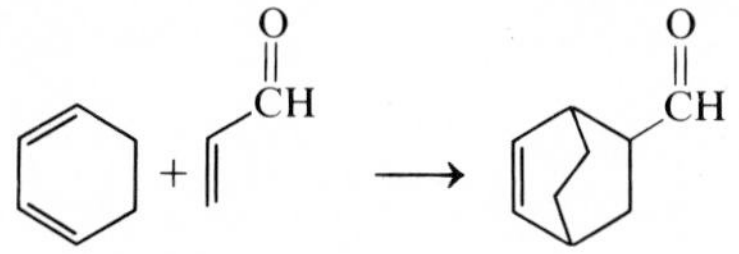

(c) diene + dienophile ($CH{=}O$) → (bicyclic product with $\overset{O}{\overset{\|}{C}}H$)

6.7. (a) 1-butyne, (b) 3-methyl-1-pentyne

6.8. (a) $CH_3CH_2CBr_2CH_3$ (b) $CH_3CCl_2{-}CCl_2CH_3$ (c) $CH_3\overset{O}{\overset{\|}{C}}CH_2CH_3$

6.9. (a) $CH_3C{\equiv}CCH_3 + NaNH_2 \xrightarrow{NH_3}$ no reaction

$CH_3CH_2C{\equiv}CH + NaNH_2 \xrightarrow{NH_3} CH_3CH_2C{\equiv}CNa + NH_3$

(b) $CH_3C{\equiv}CCH_3 + Ag(NH_3)_2OH \xrightarrow{H_2O}$ no reaction

$CH_3CH_2C{\equiv}CH + Ag(NH_3)_2OH \xrightarrow{H_2O} CH_3CH_2C{\equiv}CAg\downarrow + 2NH_3 + H_2O$

(c) $CH_3C{\equiv}CCH_3 + Cu(NH_3)_2OH \xrightarrow{H_2O}$ no reaction

$CH_3CH_2C{\equiv}CH + Cu(NH_3)_2OH \xrightarrow{H_2O} CH_3CH_2C{\equiv}CCu\downarrow + 2NH_3 + H_2O$

6.10. (a) 1,3-Butadiene would decolorize twice as much Br_2/CCl_4 solution as an equal amount of 1-butene.
(b) 1-Butyne gives a precipitate with $Ag(NH_3)_2OH$: 2-butyne does not.

6.11. hevea:

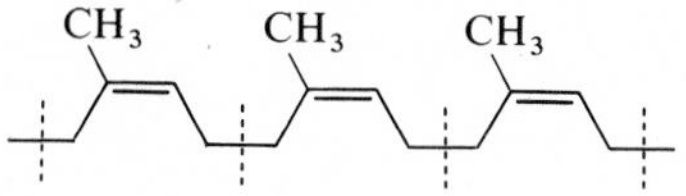

CH_3 CH_3 CH_3

gutta-percha:

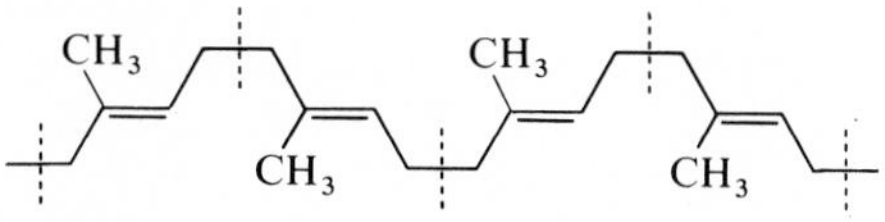

CH_3 CH_3 CH_3 CH_3

retinal:

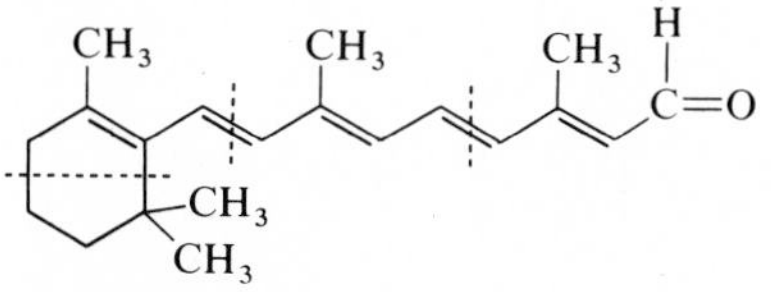

CH_3 CH_3 CH_3 H $C{=}O$ CH_3 CH_3

neoretinal b:

CH_3 CH_3 CH_3 CH_3 CH_3 $C{=}O$ H

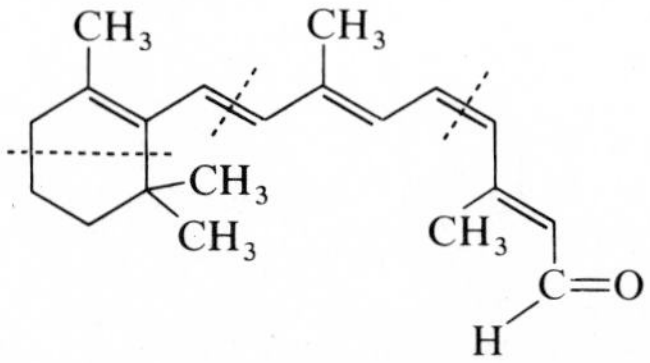

vitamin A_1:

CH_3 CH_3 CH_3 CH_2OH CH_3 CH_3

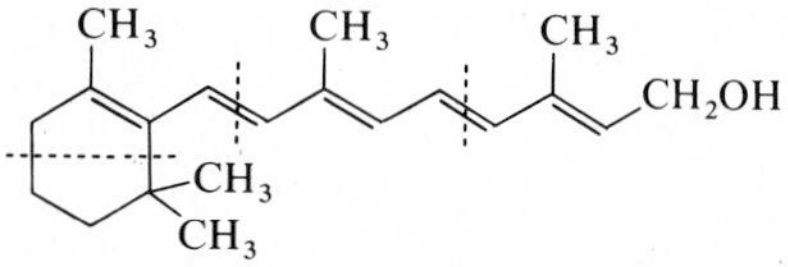

6.12. (d) 2,3-pentadiene, (e) 1,3-cyclohexadiene, (f) dicyclohexylethyne

6.14. (a)

1,3-cyclohexadiene + HCl ⟶ 3-chlorocyclohexene + 3-chlorocyclohexene (identical molecules)

(1,2- addition) (1,4- addition)

(b)

1,3-cyclohexadiene + Br_2 ⟶ 3,4-dibromocyclohexene + 3,6-dibromocyclohexene

(1,2- addition) (1,4- addition)

(c)

1,3-cyclohexadiene + H_2O $\xrightarrow{H_3O^+}$ 2-cyclohexen-1-ol (HO) + 2-cyclohexen-1-ol (OH) (identical molecules)

(1,2- addition) (1,4- addition)

6.15. (b)

cyclopentadiene + $CH_2=CH-\overset{O}{\overset{\|}{C}}-H$

6.17. (a)

$$HC\equiv CCH_2CH_2CH_2CH_3 + 2HCl \longrightarrow CH_3\overset{Cl}{\underset{Cl}{C}}CH_2CH_2CH_2CH_3$$

(b)

$$HC\equiv CCH_2CH_2CH_2CH_3 + Br_2 \longrightarrow H\overset{Br}{C}=\underset{Br}{C}CH_2CH_2CH_2CH_3$$

(c)

$$HC\equiv CCH_2CH_2CH_2CH_3 + 2Br_2 \longrightarrow H\overset{Br}{\underset{Br}{C}}-\overset{Br}{\underset{Br}{C}}CH_2CH_2CH_2CH_3$$

(d)

$$HC\equiv CCH_2CH_2CH_2CH_3 + H_2O \xrightarrow[HgSO_4]{H_2SO_4} CH_3\overset{O}{\overset{\|}{C}}CH_2CH_2CH_2CH_3$$

(e) $HC\equiv CCH_2CH_2CH_2CH_3 + NaNH_2 \xrightarrow{NH_3} NaC\equiv CCH_2CH_2CH_2CH_3 + NH_3$

6.19. (d) 2-Pentyne absorbs two moles of H_2 per mole; cyclopentene absorbs one mole of H_2 per mole.

(e) 1-Hexyne forms a precipitate with $Ag(NH_3)_2OH$; 2-hexyne does not.

6.20. (b)

β-pinene (CH_2; CH_3, CH_3) + O_3 ⟶ ozonide (CH_2, O, O, O; CH_3, CH_3) $\xrightarrow{Zn/H_2O}$ $H\overset{O}{\overset{\|}{C}}H$ + ketone (O; CH_3, CH_3)

(e) CH_3–C(=O)–CH_2CH_2–CH=O + $2CH_3\overset{O}{\overset{\|}{C}}CH_3$ + $6H\overset{O}{\overset{\|}{C}}-\overset{O}{\overset{\|}{C}}H$ + $4H\overset{O}{\overset{\|}{C}}-\underset{CH_3}{\underset{|}{C}}{=}O$

(f) 2 CH_3–C(=O)–$CH_2CH_2CH_2$–C(CH_3)(CH_3)–C(=O)–CH=O + $4H\overset{O}{\overset{\|}{C}}-\overset{O}{\overset{\|}{\underset{CH_3}{\underset{|}{C}}}}$ + $4H\overset{O}{\overset{\|}{C}}-\overset{O}{\overset{\|}{C}}H$

6.21. (b) H—C——C—CH_2CH_3

6.22. (e) non-zero, (f) zero: $CH_3CH_2—C\equiv C—CH_2CH_3$,

(g) zero: $Cl—C\equiv C—Cl$, (h) non-zero, (i)

6.25. B: $HC\equiv CCH_2CH_3$

CHAPTER 7
AROMATIC HYDROCARBONS

7.1. (a) $C_6H_6Cl_2$, (b) $C_6H_5Cl + HCl$

7.2.

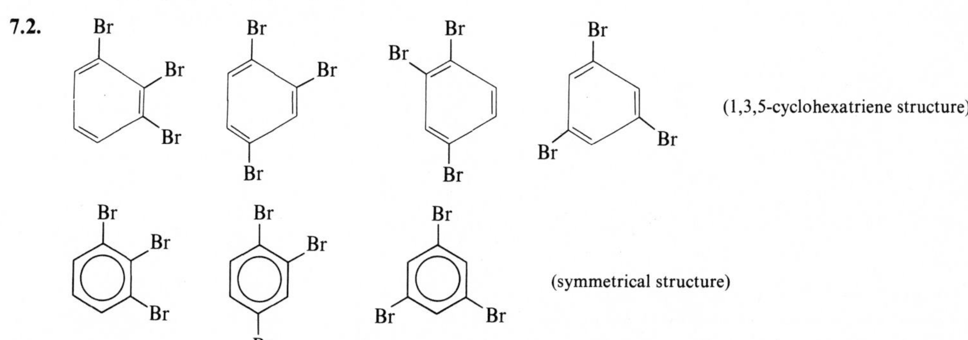

Yes.

7.3. – 1.8 kcal/mole

7.4. (c) and (d) only

7.5.

7.6. (b) and (d)

7.7. 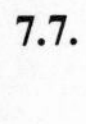 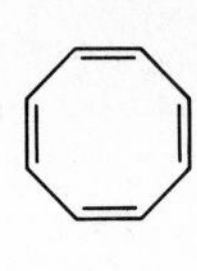 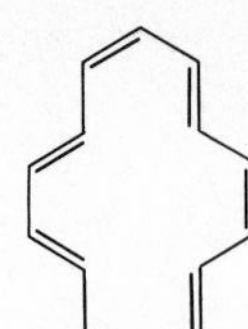 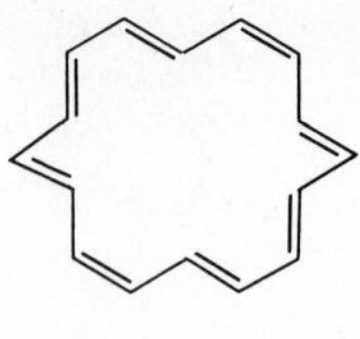 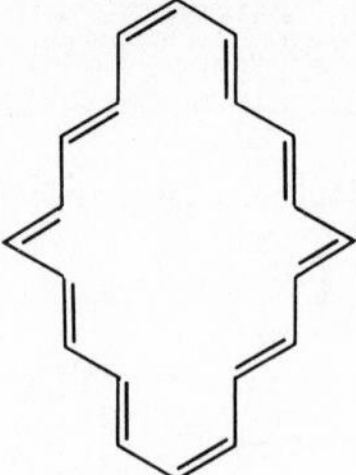

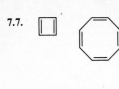

do not obey Hückel's rule

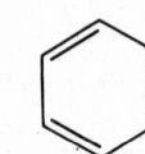
($n = 1$)
($4n + 2 = 6$)

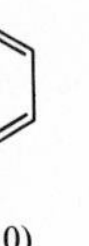
($n = 2$)
($4n + 2 = 10$)

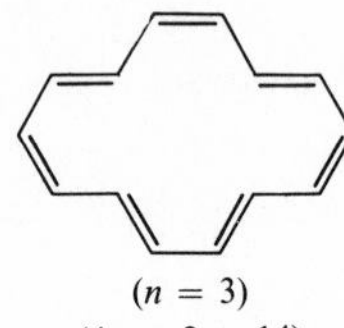
($n = 3$)
($4n + 2 = 14$)

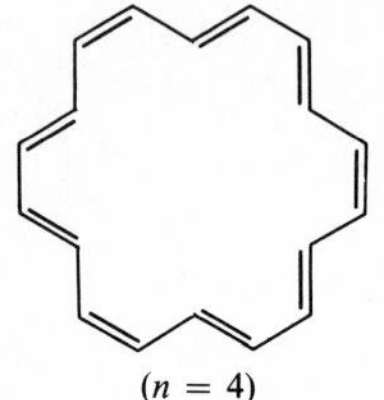
($n = 4$)
($4n + 2 = 18$)

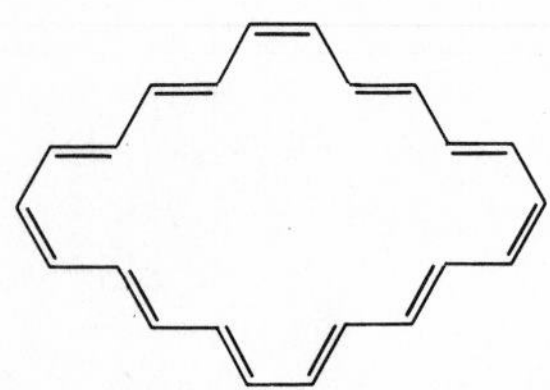
($n = 5$) ($4n + 2 = 22$)

obey Hückel's rule

7.8. (a) 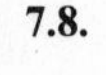 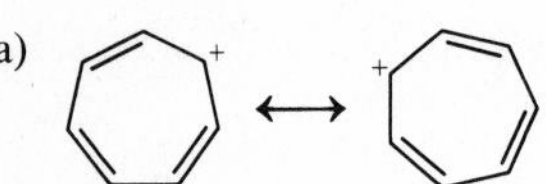 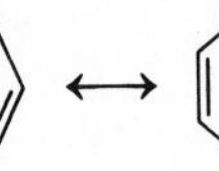 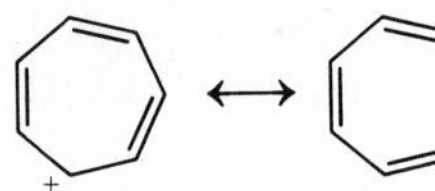

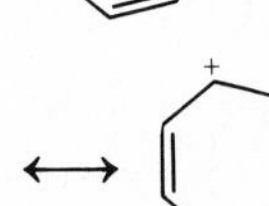 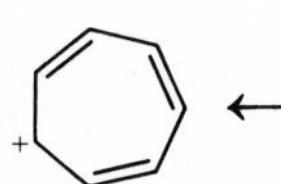 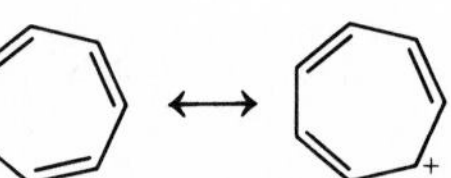

(b) 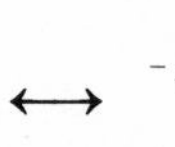

(c)

7.9. It does not satisfy Hückel's rule.

7.10. (a) 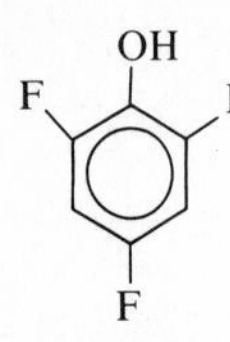

(b) CH_3, CH_3

(c) CH_3, CH_3

(d) CH_3, NO_2, NO_2

(e) CH_2CH_3, NO_2

(f) CH_2Cl, Cl

(g) CH_2

7.11. The pi electrons are delocalized over only four carbon atoms and not over the entire ring.

7.12. Linear. $R-\overset{+}{C}=\ddot{O}: \longleftrightarrow R-C\equiv\overset{+}{O}:$

7.13.

$C_6H_5-C(CH_3)_2-CH(CH_3)CH_3$

7.14. (a), (b) are electron-releasing; (c), (d), (e) are electron-attracting.

7.15. (a) p-$NO_2C_6H_4CH(CH_3)_2$ (b) o-ClC_6H_4OH + p-ClC_6H_4OH (c) m-BrC_6H_4CN (d) m-$NO_2C_6H_4C(=O)-CH_3$

(e) m-$CCl_3C_6H_4SO_3H$

7.16.

A: $C_6H_5-CH_2CH_3$, B: m-$CH_3C_6H_4CH_3$

7.17. (a) $C_6H_5-CH(Br)CH_2CH_3$ + HBr

(b) $Br-C_6H_4-CH_2CH_2CH_3$ + HBr + ortho isomer

(c) $C_6H_5-CH=C(CH_3)CH_3$ + KBr + H_2O

7.19. (a) Br_2/CCl_4 indicates $CH_2=CHC\equiv CCH=CH_2$.
(b) $Ag(NH_3)_2OH$ indicates $HC\equiv CCH=CHCH=CH_2$.
(c) Oxidation with $KMnO_4$ yields benzoic acid with propylbenzene and phthalic acid with 2-ethyltoluene.

7.20.

$$C_6H_{12} \text{ (cyclohexane, H, H)} + Br_2 \xrightarrow[\text{or heat}]{\text{light}} C_6H_{11}Br \text{ (Br, H)} + HBr$$

$$C_6H_{10} \text{ (cyclohexene, H, H)} + Br_2 \xrightarrow[\text{temperature, dark}]{\text{room}} C_6H_{10}Br_2 \text{ (Br, H, H, Br)}$$

$$C_6H_6 + Br_2 \xrightarrow[\text{dark}]{\text{Fe}} C_6H_5Br + HBr$$

7.23.

	Number of pi electrons	*Obeys Huckel's rule* $(4n + 2)$
Benzene	6	yes, $n = 1$
Naphthalene	10	yes, $n = 2$
Anthracene	14	yes, $n = 3$
Phenanthrene	14	yes, $n = 3$
Pyrene	16	no
1,2-Benzopyrene	20	no
10-Methyl-1,2-benzanthracene	18	yes, $n = 4$

7.24. Cyclopentadienyl anion has 6 π electrons ($4n + 2 = 6$, where $n = 1$); the cation has 4 π electrons, and does not obey Hückel's rule.

7.25. (e) 4-nitrotoluene, (f) 2,4-dinitrotoluene, (g) 3,5-dimethylnitrobenzene, (h) 3,4-difluoroethylbenzene

7.26. H_2SO_4

7.27. (d)

$$C_6H_6 + HONO_2 \xrightarrow{H_2SO_4} C_6H_5NO_2 + H_2O$$

(e)

$$C_6H_6 + ClCH_2CH_3 \xrightarrow{AlCl_3} C_6H_5CH_2CH_3$$

(f)

$$C_6H_6 + Cl{-}\overset{O}{\overset{\|}{C}}CH_3 \xrightarrow{AlCl_3} C_6H_5\overset{O}{\overset{\|}{C}}CH_3$$

7.28. (c)

$$C_6H_6 + HONO_2 \xrightarrow{H_2SO_4} C_6H_5NO_2 + H_2O$$

$$C_6H_5NO_2 + Cl_2 \xrightarrow{Fe} \text{3-ClC}_6H_4NO_2$$

(d)

$$CH_3\overset{O}{\overset{\|}{C}}{-}C_6H_5\ [7.27(f)] + Cl_2 \xrightarrow{Fe} CH_3\overset{O}{\overset{\|}{C}}{-}C_6H_4Cl\ (meta) + HCl$$

7.29.

$$C_6H_5{-}CH_2CH_2CH_3 \quad \text{or} \quad C_6H_5{-}\underset{}{\overset{CH_3}{\overset{|}{C}}}HCH_3$$

7.31. (c) Br_2/CCl_4 solution; (d) $KMnO_4/OH^-$/heat

CHAPTER 8
MOLECULAR SHAPES: CHIRALITY AND OPTICAL ACTIVITY

8.1. (a) conformers, (b) and (c) constitutional isomers, (d) stereoisomers
8.3. (a), (b), (c), (d), (f), (g), (h) are achiral; (e) is chiral
8.4. All except (e) have a plane of symmetry.
8.5. None is chiral.

8.6. (c) and (d)
8.7. $+148°$
8.8. $-0.24°$

8.9. (a) CH_3—(CH)—CH_2CH_3, OH (b) CH_3—(CH)—CCH_3, O, CH_3 (c) CH_3

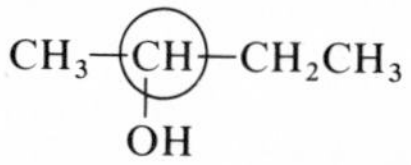

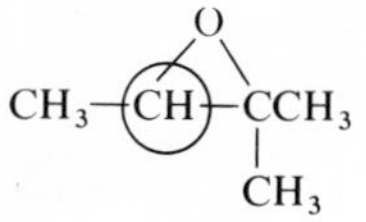

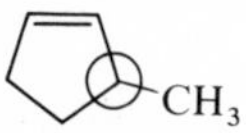

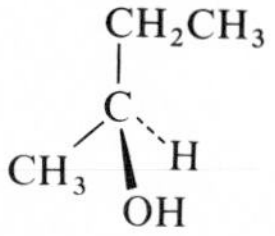

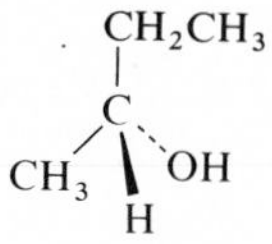

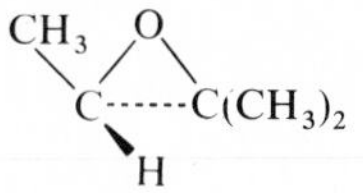

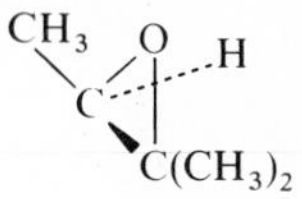

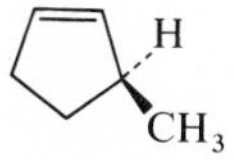

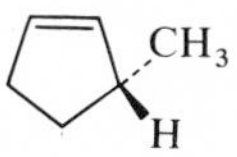

(d) C—H, Cl, NO_2

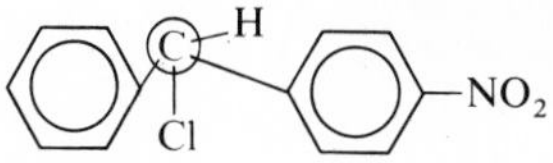

8.10. (a) CH_2CH_3, C, H, CH_3, OH; CH_2CH_3, C, OH, CH_3, H (b) CH_3, O, C, $C(CH_3)_2$, H; CH_3, O, H, C, $C(CH_3)_2$

(c) H, CH_3; CH_3, H (d) NO_2, C, H, Cl; NO_2, C, Cl, H

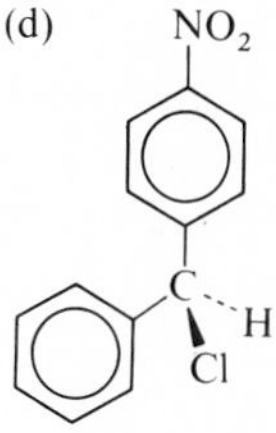

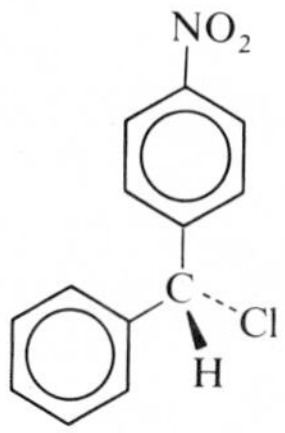

8.11. (a)

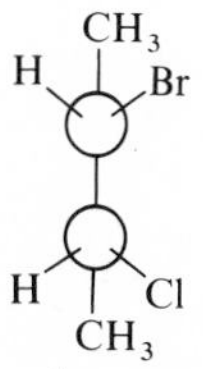

A, 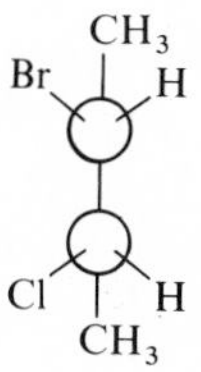 **B**, 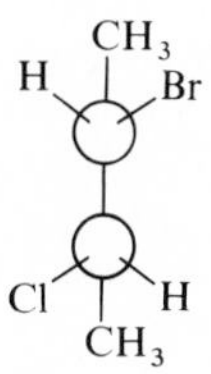 **C**, 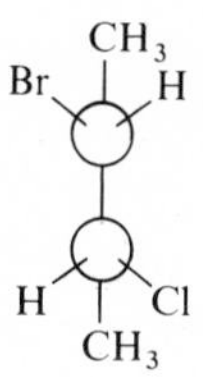 **D**

AB and **CD** are pairs of enantiomers; **AC**, **AD**, **BC**, and **BD** are diastereomers. There are no meso compounds.

(b)

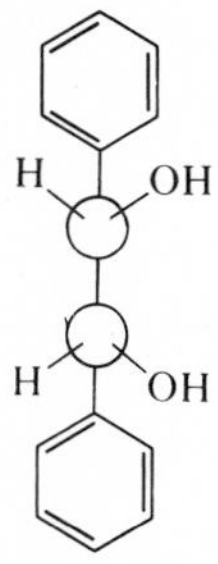

A, 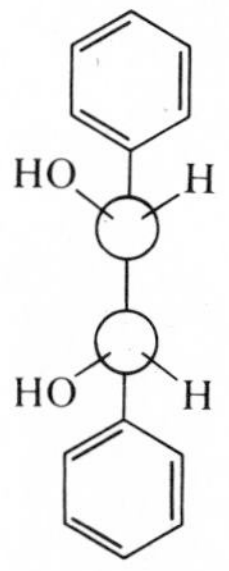 **B**, 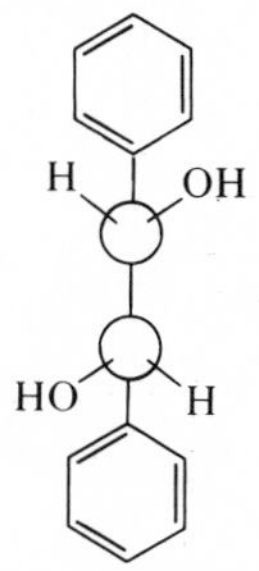 **C**, 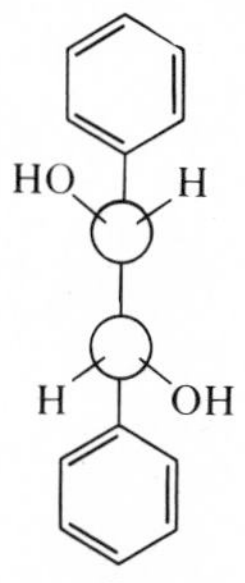 **D**

A and **B** are identical (meso compound); **CD** are enantiomers; **AC** and **AD** are diastereomers.

8.12. (a) identical, (b) identical, (c) enantiomers, (d) enantiomers, (e) identical, (f) enantiomers

8.13.

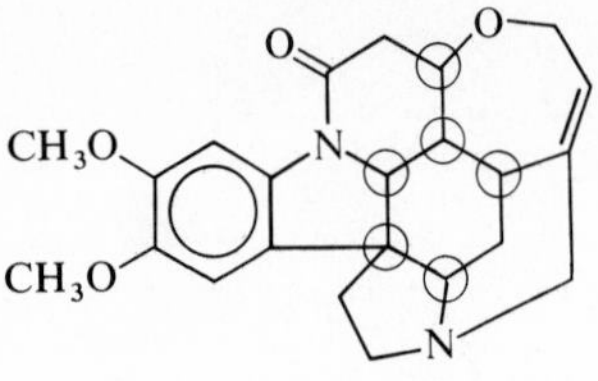

CH_3O N O N CH_3O

The chiral atoms are circled

8.14.

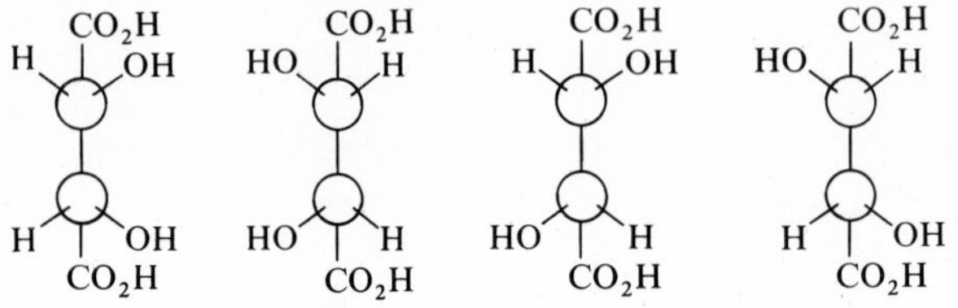

CO_2H	CO_2H	CO_2H	CO_2H
H, OH	HO, H	H, OH	HO, H
H, OH	HO, H	HO, H	H, OH
CO_2H	CO_2H	CO_2H	CO_2H
A	**B**	**C**	**D**

A and **B** are identical (meso compound); **CD** are enantiomers; **AC** and **AD** are diastereomers.

8.15. (a) $CH_3CH(Cl)CH_3$ (achiral) (b) $CH_3CH(Cl)CH_2CH_3$ (a racemic mixture)

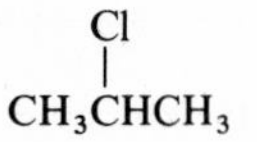

(c) $CH_3CH_2C(CH_3)(H)CH_2O{-}C(=O)CH_3$ (a single chiral product)

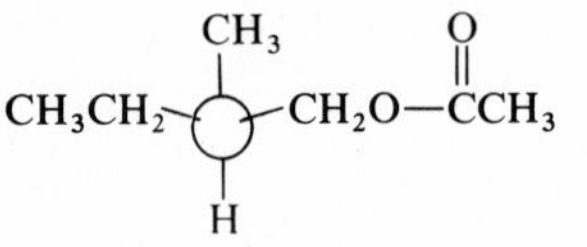

(d) $CH_3CH_2C(CH_3)(H)CH_2O{-}C(=O)C(H)(F)CH_3 + CH_3CH_2C(CH_3)(H)CH_2O{-}C(=O)C(F)(H)CH_3$ (a mixture of unequal amounts of diastereomers)

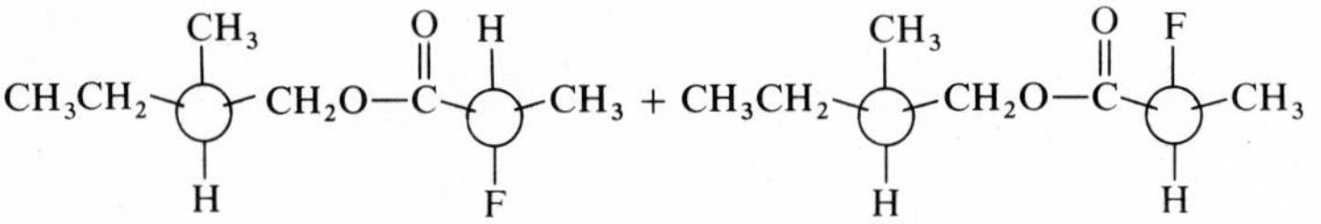

(e) $CH_3CH_2C(CH_3)(H)CH_2O{-}C(=O)C(H)(F)CH_3 + CH_3CH_2C(H)(CH_3)CH_2O{-}C(=O)C(F)(H)CH_3$ (enantiomeric pair, A, in equal amounts)

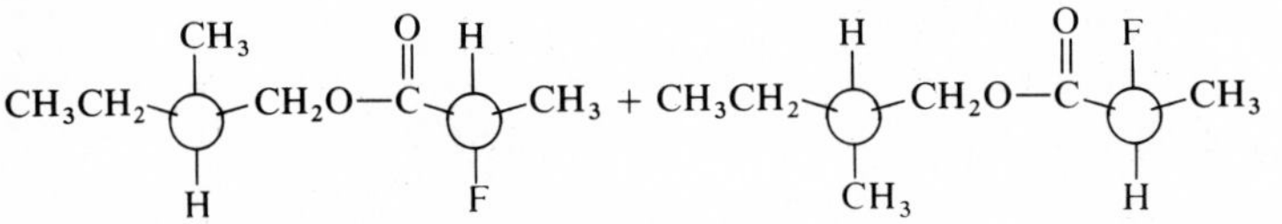

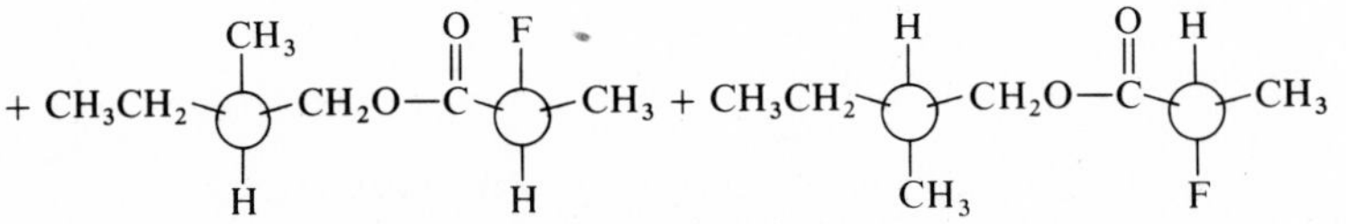

$+ CH_3CH_2C(CH_3)(H)CH_2O{-}C(=O)C(F)(H)CH_3 + CH_3CH_2C(H)(CH_3)CH_2O{-}C(=O)C(H)(F)CH_3$ (enantiomeric pair, B, in equal amounts)

The amount of pair A does not equal the amount of pair B.

8.16. (a) *S*, (b) *R*, (c) *R*

8.17. No.

8.18.

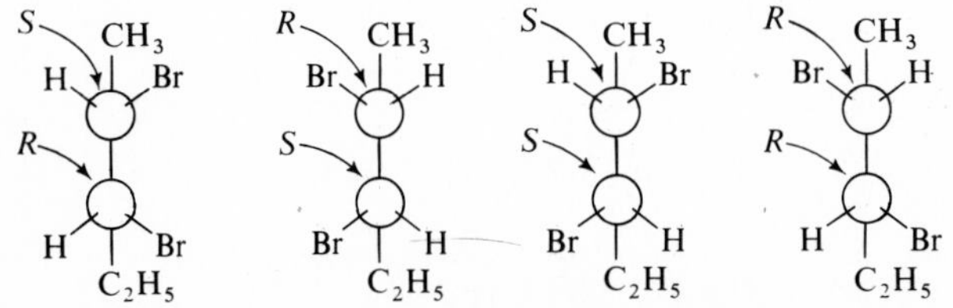

S CH_3	*R* CH_3	*S* CH_3	*R* CH_3
H, Br	Br, H	H, Br	Br, H
R	*S*	*S*	*R*
H, Br	Br, H	Br, H	H, Br
C_2H_5	C_2H_5	C_2H_5	C_2H_5
A	B	C	D

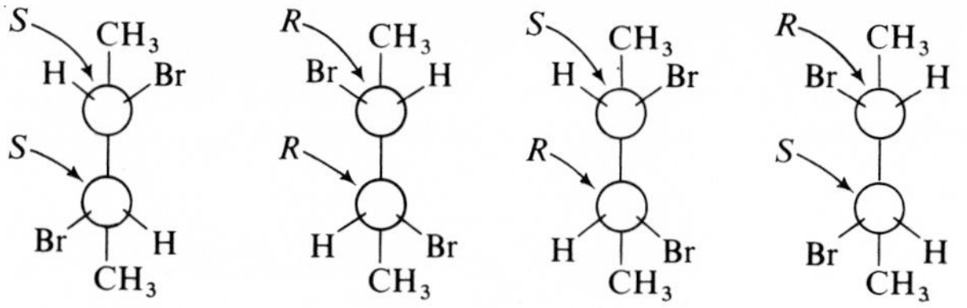

A: upper carbon (S): CH_3, H, Br; lower carbon (S): Br, H, CH_3
B: upper carbon (R): CH_3, Br, H; lower carbon (R): H, Br, CH_3
C: upper carbon (S): CH_3, H, Br; lower carbon (R): H, Br, CH_3
D: upper carbon (R): CH_3, Br, H; lower carbon (S): Br, H, CH_3

8.19. (a) $\mathbf{A} = \begin{Bmatrix} S \\ R \end{Bmatrix}$, $\mathbf{B} = \begin{Bmatrix} R \\ S \end{Bmatrix}$, $\mathbf{C} = \begin{Bmatrix} S \\ S \end{Bmatrix}$, $\mathbf{D} = \begin{Bmatrix} R \\ R \end{Bmatrix}$

(b) $\mathbf{A} = \begin{Bmatrix} S \\ R \end{Bmatrix}$, $\mathbf{B} = \begin{Bmatrix} R \\ S \end{Bmatrix}$, $\mathbf{C} = \begin{Bmatrix} S \\ S \end{Bmatrix}$, $\mathbf{D} = \begin{Bmatrix} R \\ R \end{Bmatrix}$

8.21. No.

8.22. (a) *E*-2-butene, (b) $(CH_3)(H)C{=}C(CH_2CH_3)(CH_3)$

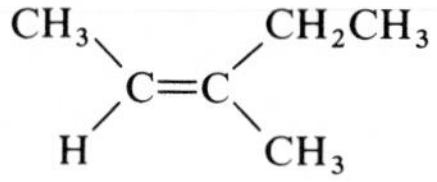

8.23. (d), (e), and (f)
8.25. (b) enantiomers, (d) same, (f) same, (h) same, (j) same

8.27. CH_3–C(H)(OH)–CH_2CH_3 and its mirror image CH_3–C(H)(HO)–CH_2CH_3

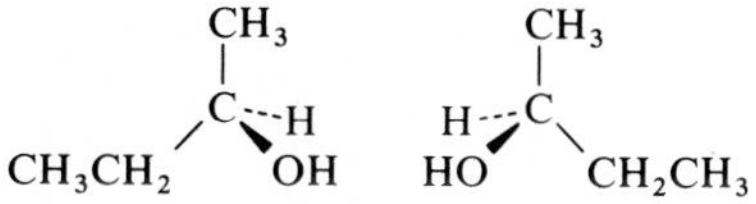

8.29. $+0.0425°$
8.30. (b) *S*, (d) *S*, *S*
8.32. (c) *Z*

8.35. (a) CH_3–C(H)(OH)–CH_2CH_3 + enantiomer (c) CH_3–C(CH_2CH_3)(C_6H_5)–O–C(=O)–CH_3

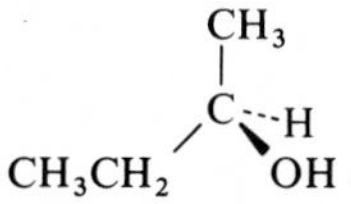

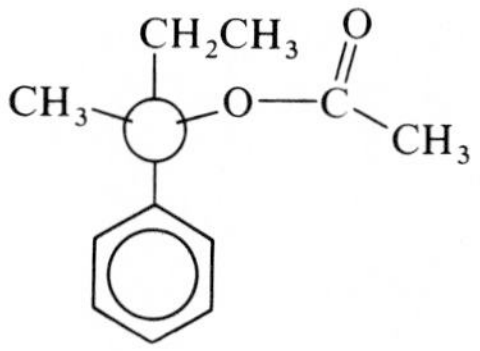

8.36. (d) and (f) are chiral and optically active; (e) is chiral but not optically active.

CHAPTER 9
ORGANIC HALOGEN COMPOUNDS

9.1. (a) 1-fluoropropane, (b) *cis*-1,2-dichlorocyclohexane, (c) 2,4-dichloro-1-butene, (d) *cis*-1-bromo-4-fluoro-2-butene, (e) bromocyclopentane
9.2. (a) $CH_2{=}CHCH_2I$, 3-iodo-1-propene; (b) $CH_2{=}CHI$, iodoethene;

(c) Cl, Cl-benzene (meta), same or 1,3-dichlorobenzene; (d) CH_3, Cl-benzene (meta), same or 3-chlorotoluene;

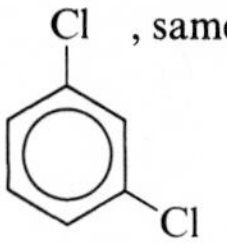

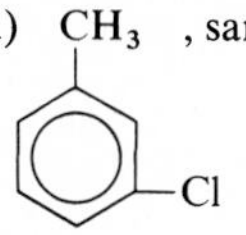

(e) C_6Cl_6 (hexachlorobenzene), same

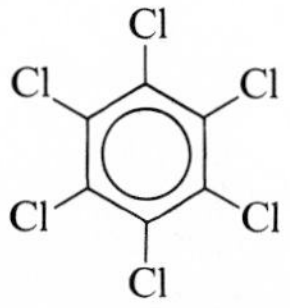

9.3. (a) $2CH_3CH_2CH_2Br + HgF_2 \longrightarrow 2CH_3CH_2CH_2F + HgBr_2$

(b) $CH_3CH_2CH_2Br + NaI \longrightarrow CH_3CH_2CH_2I + NaBr$

(c) $CH_3CH(OH)CH_3 + SOCl_2 \longrightarrow CH_3CH(Cl)CH_3 + SO_2 + HCl$

(d) $C_6H_5{-}CH_3 + Br_2 \xrightarrow{\text{light}} C_6H_5{-}CH_2Br + HBr$

(e) $C_6H_6 + Br_2 \xrightarrow{\text{Fe}} C_6H_5{-}Br + HBr$

9.4. (a) $CH_3CH_2CH_2Br + NaSH \longrightarrow CH_3CH_2CH_2SH + NaBr$

(b) $C_6H_5{-}CH_2Br + NaNH_2 \longrightarrow C_6H_5{-}CH_2NH_2 + NaBr$

(c) $C_6H_5{-}CH_2Br + NaC{\equiv}CCH_3 \longrightarrow C_6H_5{-}CH_2C{\equiv}CCH_3 + NaBr$

(d) $CH_3CH_2CH_2CH_2Br + NaOH \longrightarrow CH_3CH_2CH_2CH_2OH + NaBr$

(e) $C_6H_5{-}CH_2Br + NaI \longrightarrow C_6H_5{-}CH_2I + NaBr$

9.5. (a) $CH_2{=}C(CH_3)_2$ (b) $CH_3CH_2CH_2CH_2NH_2$ (major) + $CH_3CH_2CH{=}CH_2$ (minor)

(c) $C_6H_5{-}CH_2{-}O{-}C_2H_5$ (d) $CH_2{=}CHCH_2SH$

9.6. A tertiary halide, a polar solvent such as water, and a weak nucleophile.

9.7. S_N1

9.8. (a) $C_2H_5OCH_2CH_2CH_2CH_2CH_3$, (b) $CH_2{=}C(CH_3)_2$, (c) $(CH_3)_3C{-}OH$

9.9. reactant is (*R*), product is (*S*)

9.10. $C_2H_5O{-}C(H)(CH_3)(C_2H_5)$

9.11. (a) $HO{-}C(H)(CH_3)(C_2H_5)$ (b) $C_6H_5(CH_3)(C_2H_5)C{-}OH + HO{-}C(CH_3)(C_6H_5)(C_2H_5)$

(c) $CH_3(H)C{=}C(H)CH_3$ (major product) (d) $C_6H_5(CH_3)C{=}C(H)CH_3$ (major product)

9.12. $:\ddot{Cl}^+$ $\longleftrightarrow$ $:\ddot{Cl}^+$ $\longleftrightarrow$ $:\ddot{Cl}^+$

9.13. Grignard reagents react rapidly with water.
9.14. RMgBr is the strongest base.

9.15. (a) $CH_4 \text{ (excess)} + Br_2 \xrightarrow{\text{light}} CH_3Br + HBr$

$CH_3Br + NaI \longrightarrow CH_3I + NaBr$

$CH_3I + Mg \xrightarrow[\text{ether}]{\text{dry}} CH_3MgI$

(b) $C_6H_6 + Br_2 \xrightarrow{\text{Fe}} C_6H_5\text{—}Br$

$C_6H_5\text{—}Br + Mg \xrightarrow[\text{ether}]{\text{dry}} C_6H_5\text{—}MgBr$

9.16. (a) $CH_3CH{=}CHCH_2Cl$ gives AgCl immediately with alcoholic $AgNO_3$; $CH_2{=}CHCH_2CH_2Cl$ reacts more slowly.
(b) $CH_3CH{=}CHCH_2Cl$ gives AgCl immediately with alcoholic $AgNO_3$; $CH_3CH_2CH{=}CHCl$ does not react.
(c) $CH_3CH_2CH_2CH_2Cl$ gives a sodium fusion test; pentane does not.
(d) CH_3CH_2SH has a very nauseating odor. It gives, on sodium fusion, H_2S, which can be identified by its odor of rotten eggs.

9.18. (e) H(Cl)C=C(Cl)(CH_3) — H and Cl on one carbon, Cl and CH_3 on the other
(f) H($ClCH_2$)C=C(CH_2Cl)(H)
(g)

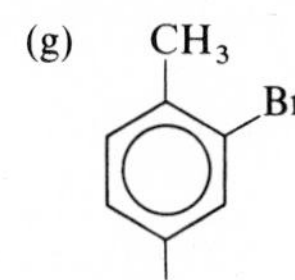

(h) I—(cyclopentene)—I

9.19. (e) 1-chloro-4-bromobenzene, (f) 1,2-dichloroethane,
(g) 1,1,1,3,3,3-hexachloropropane, (h) 4-bromocyclopentene

9.20. (f) $ClCH_2\text{—}CH_2Cl \xrightarrow{NaNH_2} HC{\equiv}CH \xrightarrow{2HBr} CH_3CHBr_2$
(from (c))

(g) $CH_3CH{=}CH_2$ (from (a)) + HBr $\longrightarrow$ $CH_3CH(Br)CH_3$

(h) $CH_3CH_2CH_2CH_3 \text{ (excess)} + Cl_2 \xrightarrow{\text{light}} CH_3CH_2CH(Cl)CH_3 + CH_3CH_2CH_2CH_2Cl$

$CH_3CH_2CH(Cl)CH_3 \xrightarrow{KOH/C_2H_5OH} CH_3CH{=}CHCH_3$

$CH_3CH_2CH_2CH_2Cl \xrightarrow{KOH/C_2H_5OH} CH_3CH_2CH{=}CH_2 \xrightarrow{HBr} CH_3CH_2CH(Br)CH_3 \xrightarrow[C_2H_5OH]{KOH} CH_3CH{=}CHCH_3$

$CH_3CH{=}CHCH_3 \xrightarrow{Cl_2} CH_3CH(Cl)CH(Cl)CH_3$

(i) $C_6H_5CH_3 + Br_2 \xrightarrow{\text{Fe}}$ p-$BrC_6H_4CH_3$ + o-$BrC_6H_4CH_3$ + HBr

(j) $CH_4 \text{ (excess)} + Br_2 \xrightarrow{\text{light}} CH_3Br + HBr$

9.21. (d) $2\,C_6H_5CH_2Br + HgF_2 \longrightarrow 2\,C_6H_5CH_2F + HgBr_2$

(e) $C_6H_5CH(CH_3)Br + NaI \longrightarrow C_6H_5CH(CH_3)I + NaBr$

(f) $C_6H_5CH_2CH_2OH + SOCl_2 \longrightarrow C_6H_5CH_2CH_2Cl + SO_2 + HCl$

9.22. (d) $C_6H_5CH_2CN + NaCl$

(e) $C_6H_5CH_2OH + NaI$

(f) $CH_2{=}CHCH_2OC_2H_5 + NaBr$

9.23. (c) $C_6H_5CH_2CH_2CH_2OH$ (d) $C_6H_5CH{=}C(CH_3)CH_2CH_3$

9.24. (d) and (e) go through the S_N2 mechanism; (f) does not react.
9.26. Elimination occurs only with (f); (d) and (e) do not undergo elimination.
9.29. (f) is chiral; (e), (g), and (h) are achiral.
9.34. (d) Na fusion is positive with vinyl chloride.

(e) $AgNO_3/C_2H_5OH$ gives an immediate AgCl precipitate with $C_6H_5CHClC_6H_5$

(f) $AgNO_3/C_2H_5OH$ gives an immediate AgBr precipitate with *p*-chlorobenzyl bromide.

CHAPTER 10
ALCOHOLS AND PHENOLS

10.1. (a) 1°, (b) 1°, (c) 1°, (d) 3°, (e) 2°
10.2. Methanol is neither; it is in a class by itself. 2-Propenol-1, ethanol, benzyl alcohol, 1,2-ethanediol are 1°. 1,2,3-Propanetriol has two 1° and one 2° OH groups.

10.3. (a) $CH_3C(CH_3)_2CH_2CH_2CH_2OH$ (b) $CH_3C(CH_3)(OH)CH(CH_3)CH_3$ (c) cyclohexane-1,3,5-triol (OH, HO, OH) (d) cyclobutane with H, OH on one carbon and CH_3, H on the opposite carbon

10.4. (a) 2-methyl-1-butanol, (b) 2-ethyl-1,3-propanediol, (c) *trans*-1,3-cyclobutanediol, (d) 1-phenylethanol

10.5. (a) phenol with OH, CH_3 (ortho) and CH_3 (para) (b) phenol with OH and OH (meta) (c) phenol with OH and CH_3 (meta) (d) phenol with OH and Cl (para)

10.6. (a) catechol, (b) hydroquinone, (c) 2,3,5-trimethylphenol, (d) 3-methyl-4,5-dinitrophenol

10.8.

$$NH_2^- > RO^- > OH^- > C_6H_5\text{—}O^-$$

10.9.

$$C_6H_5\text{—}OH + OH^- \longrightarrow C_6H_5\text{—}O^- + H_2O$$

10.10.

$$H_2O + H_2O \rightleftarrows H_3O^+ + OH^-$$

$$CH_3CH_2OH + H_2O \rightleftarrows H_3O^+ + CH_3CH_2O^-$$

$$C_6H_5\text{—}OH + H_2O \rightleftarrows H_3O^+ + C_6H_5\text{—}O^-$$

$$NH_3 + H_2O \rightleftarrows H_3O^+ + NH_2^-$$

10.11.

$$C_6H_5\text{—}OH > H_2O > CH_3CH_2OH > NH_3$$

10.13. *p*-Cresol is more soluble in NaOH solution than it is in water.

10.14.

$$C_6H_5OH + H_2SO_4 \longrightarrow C_6H_5\overset{+}{O}H_2 + HSO_4^-$$

10.15. Both methods involve high temperatures and high pressures.

10.16. Wild yeast present on the grapes or in the air causes the fermentation.

10.17. $CaO + H_2O \longrightarrow Ca(OH)_2$

10.18. 104 proof

10.19. (a) $CH_3CHOHCH_3$, (b) $CH_3CH_2CH_2OH$, (c) $CH_3CH{=}CH_2$, (d) $CH_3CH_2CH_2OH$, (e) $(CH_3)_2C{=}CH_2$, (f) $(CH_3)_3COH$

10.20. (b), (c) react immediately; (d) does not react; (a) shows little or no reaction after 30 minutes.

10.21. Five-membered ring containing O, with base attached at the top carbon, $HOCH_2$ and OH substituents (base; O; $HOCH_2$; OH)

10.22. (a) $CH_3CH_2CH_2Br$, (b) $CH_3CH_2\overset{O}{\overset{\|}{C}}H$, (c) $CH_3CH_2CH_2\text{—}O\text{—}SO_3H$,

(d) $CH_3CH_2CH_2ONa + H_2$, (e) no reaction, (f) $CH_3CH_2CH_2O\text{—}\overset{O}{\overset{\|}{C}}CH_3$

10.23. (a) $CH_3CHBrCH_3$, (b) $CH_3\overset{O}{\overset{\|}{C}}CH_3$, (c) $CH_3\overset{OSO_3H}{\overset{|}{C}H}\text{—}CH_3$,

(d) $CH_3\overset{ONa}{\overset{|}{C}H}CH_3 + H_2$, (e) no reaction, (f) $(CH_3)_2CH\text{—}O\text{—}\overset{O}{\overset{\|}{C}}CH_3$

10.24.

$$C_6H_5\text{—}OH + NaOH \longrightarrow C_6H_5\text{—}ONa + H_2O$$

$$2\,C_6H_5\text{—}OH + 2\,Na \longrightarrow 2\,C_6H_5\text{—}ONa + H_2$$

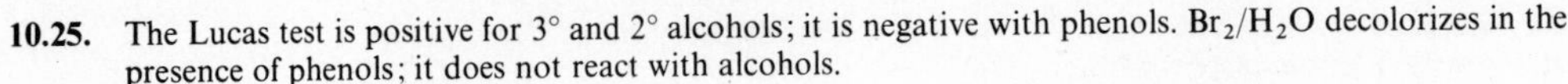

10.25. The Lucas test is positive for 3° and 2° alcohols; it is negative with phenols. Br_2/H_2O decolorizes in the presence of phenols; it does not react with alcohols.

10.26. $CH_3CH_2ONa + CH_3CH_2SH \rightleftarrows CH_3CH_2OH + CH_3CH_2SNa$. Products.

10.27. (b) cyclopentane ring with H, H; CH_3, OH   (d) ring with H, H; OH, OH   (f) $CH_3\overset{CH_3}{\underset{CH_3}{C}}—ONa$ (h) cyclohexene ring with OH

10.29. (d) 3-bromophenol, (e) 2-bromo-4-ethylphenol, (f) 2-cyclopentenol

10.31.

	Higher Boiling Point	*Higher H_2O Solubility*
(c)	glycerin (intermolecular H bonding)	same (H bonding with H_2O)
(d)	glycerin (more intermolecular H bonding and higher molecular weight)	same (more H bonding with H_2O)

10.32. (b) $C_6H_5—SH + OH^- \rightleftarrows C_6H_5—S^- + H_2O$

(d) $CH_3OH + OH^- \leftrightarrows CH_3O^- + H_2O$

(f) $C_6H_5—OH + CH_3CH_2O^- \rightleftarrows C_6H_5—O^- + CH_3CH_2OH$

10.34. (b) $6CH_3CH_2CH{=}CH_2 + (BH_3)_2 \longrightarrow 2(CH_3CH_2CH_2CH_2)_3B$

$(CH_3CH_3CH_2CH_2)_3B + 3H_2O_2 \xrightarrow{OH^-} 3CH_3CH_2CH_2CH_2OH + B(OH)_3$

(d) $CH_3CH_2CH{=}CH_2 + HCl \longrightarrow CH_3CH_2CHClCH_3$

(f) $3CH_3CH_2CH_2CH_2OH + 2H_2CrO_4 + 6H_3O^+ \longrightarrow 3CH_3CH_2CH_2\overset{O}{\overset{\|}{C}}H + 2Cr^{3+} + 14H_2O$

10.35. (e) $3\,C_6H_5—CH_2CH_2OH + PBr_3 \longrightarrow 3\,C_6H_5—CH_2CH_2Br + P(OH)_3$

(from (b))

(f) $C_6H_5—CH{=}CH_2 + HCl \longrightarrow C_6H_5—CHClCH_3$

(g) $C_6H_5—CH_2CH_2OH + CH_3CO_2H \xrightarrow{H^+} C_6H_5—CH_2CH_2—O—\overset{O}{\overset{\|}{C}}—CH_3 + H_2O$

10.36. (b) $Br—C_6H_4—OH + NaOH \longrightarrow Br—C_6H_4—ONa + H_2O$

(d) $3CH_3CH_2CH_2OH + PBr_3 \longrightarrow 3CH_3CH_2CH_2Br + P(OH)_3$
$CH_3CH_2CH_2Br + NaSH \longrightarrow CH_3CH_2CH_2SH + NaBr$

10.37. (c) Phenol dissolves in aqueous NaOH solution; chlorobenzene does not.
(e) Br_2/CCl_4 solution loses its color with allyl alcohol. No reaction with *tert*-butyl alcohol.
(g) Same as (c).

10.39. B is a cresol.

CHAPTER 11
ETHERS AND EPOXIDES

11.1. $CH_3CH_2CH_2CH_2OH$, $\quad CH_3CH_2\overset{\overset{OH}{|}}{C}HCH_3$, $\quad CH_3\overset{\overset{CH_3}{|}}{C}HCH_2OH$, and $\quad CH_3\underset{\underset{OH}{|}}{\overset{\overset{CH_3}{|}}{C}}CH_3$ are alcohols.

$CH_3{-}O{-}CH_2CH_2CH_3$, $\quad CH_3CH_2{-}O{-}CH_2CH_3$, and $\quad CH_3{-}O{-}\overset{\overset{CH_3}{|}}{C}HCH_3$ are ethers.

11.2. (a) $CH_3CH_2CH_2{-}O{-}C_6H_5$ $\quad$ (b) $CH_3{-}O{-}CH_2CH_2{-}O{-}CH_3$

(c) cyclobutane ring bearing OCH_2CH_3 and H on one carbon, CH_3CH_2O and H on the opposite carbon (trans)

(d) $CH_3\overset{\overset{CH_3}{|}}{C}HCH_2{-}O{-}CH_2\overset{\overset{CH_3}{|}}{C}HCH_3$

11.3. (a) methyl cyclopentyl ether, methoxycyclopentane;
(b) ethyl isopropyl ether, 2-ethoxypropane

11.5. diethyl ether.

11.6. Dehydration of alcohols gives alkenes as well as ethers; and only symmetrical ethers can be prepared without considerable waste.

11.7. (a) $2CH_3OH \xrightarrow[\text{heat}]{H_2SO_4} CH_3{-}O{-}CH_3 + H_2O$

(b) $C_6H_5{-}OH + ClCH_2{-}C_6H_5 + NaOH \xrightarrow{H_2O} C_6H_5{-}O{-}CH_2{-}C_6H_5 + NaCl + H_2O$

(c) $CH_3CH_2CH_2Br + HO{-}C_6H_4{-}Cl + NaOH$

$\xrightarrow{H_2O} CH_3CH_2CH_2{-}O{-}C_6H_4{-}Cl + NaBr + H_2O$

(a) can be prepared by both methods; (b) and (c) can be prepared only by the Williamson synthesis.

11.8. (a) $CH_3CH_2I + CH_3\overset{\overset{CH_3}{|}}{C}HI + H_2O$

(b) $C_6H_5{-}OH + CH_3I$

(c) no cleavage.

11.9. The $R{-}O^-$ group is highly basic and is therefore a poor leaving group.

11.10. (a) $C_6H_5{-}\overset{O}{\widehat{CH{-}CH_2}}$ $\quad$ (b) $CH_3{-}\overset{O}{\widehat{CH{-}CH}}{-}CH_3$

11.11. (a) $CH_3CH{=}CHCH_3 + HOCl \longrightarrow CH_3\overset{\overset{OH}{|}}{C}H{-}\overset{\overset{Cl}{|}}{C}HCH_3$

$CH_3\overset{\overset{OH}{|}}{C}H{-}\overset{\overset{Cl}{|}}{C}HCH_3 + NaOH \longrightarrow CH_3\overset{O}{\widehat{CH{-}CH}}CH_3 + NaCl + H_2O$

(b)

$$\text{C}_6\text{H}_5\text{-CH=CH-C}_6\text{H}_5 + \text{HOCl} \longrightarrow \text{C}_6\text{H}_5\text{-}\underset{}{\overset{\text{OH}}{\text{CH}}}\text{-}\overset{\text{Cl}}{\text{CH}}\text{-C}_6\text{H}_5$$

$$\text{C}_6\text{H}_5\text{-}\overset{\text{OH}}{\text{CH}}\text{-}\overset{\text{Cl}}{\text{CH}}\text{-C}_6\text{H}_5 + \text{NaOH} \longrightarrow \text{C}_6\text{H}_5\text{-}\overset{\text{O}}{\text{CH-CH}}\text{-C}_6\text{H}_5 + \text{NaCl} + \text{H}_2\text{O}$$

11.12. (a)

$$CH_3\overset{O}{CH-CH}CH_3 + H_2O \longrightarrow CH_3\overset{OH}{CH}-\overset{OH}{CH}CH_3$$

(b)

$$CH_3\overset{O}{CH-CH}CH_3 + HCl \longrightarrow CH_3\overset{OH}{CH}-\overset{Cl}{CH}CH_3$$

(c)

$$CH_3\overset{O}{CH-CH}CH_3 + NH_3 \longrightarrow CH_3\overset{OH}{CH}-\overset{NH_2}{CH}CH_3$$

(d)

$$CH_3\overset{O}{CH-CH}CH_3 + CH_3CH_2OH \longrightarrow CH_3\overset{OH}{CH}-\overset{OCH_2CH_3}{CH}CH_3$$

(e)

$$CH_3\overset{O}{CH-CH}CH_3 + NaOH \xrightarrow{H_2O} CH_3\overset{OH}{CH}-\overset{OH}{CH}CH_3 + NaOH$$

The reactions of $\text{C}_6\text{H}_5\text{-}\overset{\text{O}}{\text{CH-CH}}\text{-C}_6\text{H}_5$ are identical to the reactions of $CH_3\overset{O}{CH-CH}CH_3$.

11.13. The base-catalyzed ring opening reaction is helped by bond angle strain.

11.15. (d) cyclopentenyl$-\dot{O}CH_2CH_3$ (e) $CH_3-O-CH_2-O-CH_3$

11.16. (d) 2-ethoxypropane, (e) diisobutyl ether

11.19. (d)

$$\text{C}_6\text{H}_5\text{-OH} + CH_3CH_2CH_2Br + NaOH \longrightarrow \text{C}_6\text{H}_5\text{-O-}CH_2CH_2CH_3 + NaBr + H_2O$$

(e)

$$\text{C}_6\text{H}_5\text{-}CH_2OH \xrightarrow[\text{heat}]{H_2SO_4} \text{C}_6\text{H}_5\text{-}CH_2-O-CH_2\text{-C}_6\text{H}_5 + H_2O$$

11.21. (c)

$$\text{C}_6\text{H}_5\text{-O-}CH_2\text{-C}_6\text{H}_5 + HI \longrightarrow \text{C}_6\text{H}_5\text{-OH} + ICH_2\text{-C}_6\text{H}_5$$

11.22. (b)

$$\text{cyclohexene} + HOCl \longrightarrow \text{2-chlorocyclohexanol (OH, Cl)}$$

$$\text{2-chlorocyclohexanol (OH, Cl)} + NaOH \longrightarrow \text{cyclohexene oxide (O)} + NaCl + H_2O$$

11.24.

$$\underset{\text{C}}{\text{C}_6\text{H}_5\text{-O-}CH_2CH_3} + HI \longrightarrow \underset{\text{D }(C_6H_6O)}{\text{C}_6\text{H}_5\text{-OH}} + CH_3CH_2I$$

11.26. Some possible structures are:

$CH_3CH_2\overset{O}{CH-CH}CH_2CH_3$, $CH_3\overset{O}{CH-CH}CH_2CH_2CH_3$, $\overset{O}{CH_2-CH}CH_2CH_2CH_2CH_3$,
and so forth.

CHAPTER 12
ALDEHYDES AND KETONES

12.1. (a) ketone, (b) neither (ester), (c) ketone, (d) aldehyde

12.2. (a) $H:\overset{+\ddot{O}:H}{\ddot{C}}:H \longleftrightarrow H:\overset{:\ddot{O}:H}{\underset{+}{C}}:H$ (b) $H:\overset{:\ddot{O}:^-}{\underset{:\ddot{O}:H}{C}}:H$

12.3. (a), (b), (c), and (d) The alcohol has the higher bp; they have similar water solubilities.
(e) The ether and the ketone have similar molecular weights; they should have similar bp and water solubilities.

12.4. (a) $CH_3CH_2CH_2CH_2CH_2CH_2CH_2\overset{O}{\overset{\|}{C}}H$ (b) $CH_3CH_2\overset{Cl}{\overset{|}{C}}H\overset{O}{\overset{\|}{C}}H$

(c) $CH_3\overset{CH_3}{\overset{|}{C}}H-\overset{O}{\overset{\|}{C}}-\overset{CH_3}{\overset{|}{C}}HCH_3$ (d) $CH_3CH_2\overset{CH_3}{\overset{|}{C}}H-\overset{O}{\overset{\|}{C}}-CH_2\overset{CH_3}{\overset{|}{C}}HCH_3$

(e) $CH_3CH_2\underset{C_6H_5}{\underset{|}{C}}H\overset{O}{\overset{\|}{C}}CH_2CH_2CH_2CH_3$

12.5. (a) cyclobutanone, (b) 2-methylpropanal or isobutyraldehyde,
(c) 1,3-diphenylpropanone or dibenzyl ketone

12.6.

	Compound	*Number of bonds to O or X*
(b) $CH_3\overset{O}{\overset{\|}{C}}-OH$, (d) $H\overset{O}{\overset{\|}{C}}-OCH_2CH_3$, (f) $CH_3\overset{O}{\overset{\|}{C}}-Cl$		3 (highest oxidation state)
(c) $CH_3\overset{O}{\overset{\|}{C}}CH_3$, (e) cyclopentanone, (g) $CH_3-\overset{Cl}{\underset{Cl}{C}}H$		2
(a) CH_3CH_3, (h) $CH_3CH_2CH_3$, (i) cyclohexane		0 (lowest oxidation state)

12.7. (a) oxidizing agent, (b) neither, (c) neither, (d) neither, (e) reducing agent,
(f) neither

12.8. (a)

$$3\,C_6H_5CH(OH)CH_3 + H_2CrO_4 \longrightarrow 3\,C_6H_5COCH_3 + Cr^{3+} + 4H_2O$$

$$C_6H_5CCl_2CH_3 + 2NaOH \longrightarrow C_6H_5COCH_3 + 2NaCl + H_2O$$

(b)

$$3CH_3CH(OH)CH_2CH_3 + H_2CrO_4 \longrightarrow 3CH_3COCH_2CH_3 + Cr^{3+} + 4H_2O$$

(c)

$$3\,C_6H_5CH_2OH + H_2CrO_4 \longrightarrow 3\,C_6H_5CHO + Cr^{3+} + 4H_2O$$

$$C_6H_5CHCl_2 + 2NaOH \longrightarrow C_6H_5C(=O){-}H + 2NaCl + H_2O$$

(d)

$$3\,\text{cyclopentanol} + H_2CrO_4 \longrightarrow 3\,\text{cyclopentanone} + Cr^{3+} + 4H_2O$$

12.9. The base is a nucleophile and could react with the carbonyl carbon. An acid catalyst could react with the nucleophile (a base) to form its conjugate acid, which would be less nucleophilic.

12.10. (a)

$$C_6H_5CHO + CH_3CH_2OH \rightleftarrows C_6H_5CH(OH)OCH_2CH_3$$

$$C_6H_5CH(OH)OCH_2CH_3 + CH_3CH_2OH \overset{HCl}{\rightleftarrows} C_6H_5CH(OCH_2CH_3)_2 + H_2O$$

(b)

$$CH_3COCH_3 + CH_3CH_2OH \rightleftarrows CH_3C(OH)(OCH_2CH_3)CH_3$$

$$CH_3C(OH)(OCH_2CH_3)CH_3 + CH_3CH_2OH \overset{HCl}{\rightleftarrows} CH_3C(OCH_2CH_3)_2CH_3 + H_2O$$

12.11. (a) $CH_3C(OMgBr)(CH_3)CH_3$, $CH_3C(OH)(CH_3)CH_3 + MgBrOH$

(b) 1-phenylcyclohexyl–OMgBr, 1-phenylcyclohexanol (OH)

(c) CH_3CH_2OMgBr, $CH_3CH_2OH + MgBrOH$

(d) $CH_3CH(CH_3){-}CH(OMgBr)CH_3$, $CH_3CH(CH_3){-}CH(OH)CH_3 + MgBrOH$

12.12. (a)

$$CH_3\overset{\overset{O}{\|}}{C}H + HCN \longrightarrow CH_3\overset{\overset{OH}{|}}{C}H{-}CN$$

(b)

$$H\overset{\overset{O}{\|}}{C}H + HCN \longrightarrow H{-}\underset{\underset{CN}{|}}{\overset{\overset{OH}{|}}{C}}{-}H$$

(c)

$$CH_3\overset{\overset{O}{\|}}{C}CH_3 + HCN \longrightarrow CH_3\underset{\underset{CN}{|}}{\overset{\overset{OH}{|}}{C}}CH_3$$

12.13. (a)

$$CH_3\overset{\overset{NOH}{\|}}{C}H + H_2O$$

(b)

$$C_6H_5{-}\overset{\overset{CH_3}{|}}{C}{=}N{-}NH{-}C_6H_5 + H_2O$$

12.14. (a)

$$C_6H_5{-}\overset{\overset{O}{\|}}{C}{-}C_6H_5 + H_2 \xrightarrow{Ni} C_6H_5{-}\overset{\overset{OH}{|}}{C}H{-}C_6H_5$$

$$C_6H_5{-}\overset{\overset{O}{\|}}{C}{-}C_6H_5 + LiAlH_4 \longrightarrow C_6H_5{-}\overset{\overset{OH}{|}}{C}H{-}C_6H_5$$

(b)

$$C_6H_5{-}\overset{\overset{O}{\|}}{C}{-}C_6H_5 \xrightarrow[HCl]{Zn{-}Hg} C_6H_5{-}CH_2{-}C_6H_5$$

12.15. Alcohols ($K_a \sim 10^{-18}$) most closely resemble aldehydes and ketones ($K_a \sim 10^{-19}$–10^{-20}) in acidity.

12.16. Keto and enol forms differ in the positions of one of their hydrogen atoms.

12.17. (a)

$$\text{cyclohexanone } (C_6H_{10}{=}O) \rightleftharpoons \text{cyclohexen-1-ol } (C_6H_9{-}OH)$$

(b)

$$CH_3\overset{\overset{O}{\|}}{C}H \rightleftharpoons CH_2{=}\overset{\overset{OH}{|}}{C}H$$

(c)

$$C_6H_5{-}\overset{\overset{O}{\|}}{C}CH_3 \rightleftharpoons C_6H_5{-}\overset{\overset{OH}{|}}{C}{=}CH_2$$

12.18. (a)

$$CH_3CH_2CH_2\overset{\overset{OH}{|}}{C}H\underset{\underset{CH_2CH_3}{|}}{C}H\overset{\overset{O}{\|}}{C}H$$

(b)

$$CH_3\underset{\underset{CH_3}{|}}{\overset{\overset{OH}{|}}{C}}{-}CH_2\overset{\overset{O}{\|}}{C}CH_3$$

(c)

$$C_6H_5{-}CH_2\overset{\overset{OH}{|}}{C}H{-}\underset{\underset{C_6H_5}{|}}{C}H{-}\overset{\overset{O}{\|}}{C}H$$

12.19.

$$C_6H_5{-}\overset{\overset{OH}{|}}{C}H\underset{\underset{CH_3}{|}}{C}H\overset{\overset{O}{\|}}{C}H$$

The other product is

$$CH_3CH_2\overset{\overset{OH}{|}}{C}H\underset{\underset{CH_3}{|}}{C}HCHO.$$

12.20. [structure: C_6H_5–C(=O)–H] α-carbon has no H

12.21. (b) and (e)

12.23. (e) [cyclohexanone] (f) [dicyclopropyl ketone, C_3H_5–C(=O)–C_3H_5] (g) [C_6H_5–C(=O)H] (h) [3,5-dinitrophenyl–C(=O)CH_3; NO_2, NO_2]

12.24. (e) 2-cyclopentenone (f) 2-methylpropanal, (g) dibenzyl ketone, (h) hydroxyethanal

12.25.

	More H_2O-Soluble	*Higher Boiling Point*
(c)	similar	alcohol
(d)	$CH_3\overset{O}{\overset{\|\|}{C}}CH_3$	$CH_3\overset{O}{\overset{\|\|}{C}}CH_3$

12.26. (e) $CH_3\overset{O}{\overset{\|\|}{C}}$–$C_6H_5$ because carbon has two bonds to 0; (f) $CH_3CCl_2CH_2CH_3$, because one carbon has two bonds to Cl;

(g) same, because two bonds to O are equivalent in oxidation state to two bonds to Cl.

12.27. (e) H_2CrO_4, (f) H_2CrO_4, (g) H_2O/H_3O^+

12.28. (b) C_6H_5–$CH_2CH(OCH_2CH_3)_2$ + H_2O

(d) C_6H_5–C(OH)(CH_3)–C_6H_5 + MgBrOH

(f) C_6H_5–$CH_2CH(OH)CN$

(h) C_6H_5–$CH_2CH(OH)CO_2H$

(j) C_6H_5–$CH_2CH{=}NH$

(l) C_6H_5–$CH_2CH{=}N{-}NH$–C_6H_5

(n) C_6H_5–C(CH_3)=NOH

(p) C_6H_5–CH(OH)CH$_3$

(r) C_6H_5–CH_2CH_3

(t) C_6H_5–C(OH)(CH$_3$)–CH$_2$–C(=O)–C_6H_5

(v) C_6H_5–CO_2^- + CHI_3

12.29. (c) cyclopentanone (=O) $\rightleftarrows$ cyclopentenol (–OH)

12.30. (b)

$$HC{\equiv}CH + H_2O \xrightarrow[HgSO_4]{H_2SO_4} CH_3\overset{O}{\overset{\|}{C}}H$$

$$C_6H_5\text{–MgBr (Part a)} + CH_3CHO \longrightarrow \xrightarrow{H_3O^+} C_6H_5\text{–}\overset{OH}{\overset{|}{C}}HCH_3$$

$$C_6H_5\text{–}\overset{OH}{\overset{|}{C}}HCH_3 \xrightarrow{H_2CrO_4} C_6H_5\text{–}\overset{O}{\overset{\|}{C}}CH_3$$

12.32. (b) $CH_3\overset{O}{\overset{\|}{C}}CH_2CH_2CH_3$ gives a positive iodoform test.

(d) $CH_3\overset{O}{\overset{\|}{C}}$–$C_6H_5$ gives a positive iodoform test.

(f) The aldehyde gives a positive Fehling's test.

12.34. B is one of several alcohols:

$C_{10}H_7$–$\overset{OH}{\overset{|}{C}}HCH{=}CH_2$, C_6H_5–$\overset{OH}{\overset{|}{C}}H$–$C_6H_5$, $C_{10}H_7$–$CH{=}CHCH_2OH$,

$C_{10}H_6$(CH_2OH)($CH{=}CH_2$), and so forth. It does not have the —$\overset{OH}{\overset{|}{C}}HCH_3$ group

12.35.

C: CH_3–C_6H_4–CHO

CHAPTER 13
CARBOXYLIC ACIDS AND THEIR DERIVATIVES

13.1. (a) $CH_3CH_2CH_2CH_2CH_2CH_2C(=O)-OH$, (b) $CH_3CH_2CH_2CH_2CH_2CH(CH_3)CH(CH_3)C(=O)-OH$,

(c) $CH_3CH(CH_3)C(=O)-OH$, (d) $CH_3CH_2CH_2CH_2C(=O)-ONa$

13.2. (a) 2,4-dichlorobenzoic acid, (b) sodium 2,4-dichlorobenzoate, (c) 3,3-dimethylbutanoic acid, (d) 3-methyl-2-butenoic acid

13.3. The carboxylic acid can form intermolecular hydrogen bonds; therefore its boiling point is higher than the boiling point of the aldehyde. The water solubility is higher than that of the aldehyde because it has two oxygen atoms that can form hydrogen bonds to water; the aldehyde has only one.

13.4. The carboxylic acid has two oxygen atoms that can enter into hydrogen bonding.

13.5. (a) benzoic acid, (b) *p*-nitrobenzoic acid, (c) *m*-chlorobenzoic acid

13.6. (a) benzoate ion, (b) *p*-nitrobenzoate ion, (c) *m*-chlorobenzoate ion

13.7. (a) difluoroacetic acid, (b) *p*-ethylbenzoic acid, (c) 2-chlorobutanoic acid

13.8. (a) $CH_3CH_2CH_2OH \xrightarrow{H_2CrO_4} CH_3CH_2CO_2H$

(b) $C_6H_5-CH_2Cl + NaCN \longrightarrow C_6H_5-CH_2CN + NaCl$

$C_6H_5-CH_2CN \xrightarrow{H_3O^+} C_6H_5-CH_2CO_2H$

$C_6H_5-CH_2Cl + Mg \xrightarrow[\text{ether}]{\text{dry}} C_6H_5-CH_2MgCl \xrightarrow{CO_2} C_6H_5-CH_2C(=O)-OMgCl \xrightarrow{H_3O^+} C_6H_5-CH_2CO_2H$

(c) $C_6H_5-CH_2Cl \xrightarrow[OH^-,\ \text{heat}]{KMnO_4} C_6H_5-CO_2^- \xrightarrow{H_3O^+} C_6H_5-CO_2H$

(d) $C_6H_5-CH_2C(=O)H \xrightarrow{Ag(NH_3)_2} C_6H_5-CH_2CO_2^- \xrightarrow{H_3O^+} C_6H_5-CH_2CO_2H$

(e) $C_6H_5-CH_2C(=O)CH_3 \xrightarrow{Cl_2,\ OH^-} C_6H_5-CH_2CO_2^- + CHCl_3$

$C_6H_5-CH_2CO_2^- \xrightarrow{H_3O^+} C_6H_5-CH_2CO_2H$

(f) $C_6H_5-CH_2C(=O)CH_3 \xrightarrow{KMnO_4,\ OH^-} C_6H_5-CO_2^- \xrightarrow{H_3O^+} C_6H_5-CO_2H$

(g) $C_6H_5-CH_2C(=O)H + HCN \longrightarrow C_6H_5-CH_2CH(OH)CN \xrightarrow{H_3O^+} C_6H_5-CH_2CH(OH)CO_2H$

(h) $CH_3C(CH_3)_2-Cl + Mg \xrightarrow[\text{ether}]{\text{dry}} CH_3C(CH_3)_2MgCl \xrightarrow{CO_2} CH_3C(CH_3)_2-CO_2MgCl \xrightarrow{H_3O^+} CH_3C(CH_3)_2-CO_2H$

13.9. only $R{-}\overset{O}{\overset{\|}{C}}{-}H \longrightarrow R{-}\overset{O}{\overset{\|}{C}}{-}OH$

13.10. (a) 3-methylbutanoyl chloride, (b) *p*-nitrobenzoyl bromide

13.11. (a)

$$3\,CH_3CH_2\overset{O}{\overset{\|}{C}}{-}OH + PBr_3 \longrightarrow CH_3CH_2\overset{O}{\overset{\|}{C}}{-}Br + P(OH)_3$$

(b)

$$CH_3\overset{O}{\overset{\|}{C}}{-}OH + SOCl_2 \longrightarrow CH_3\overset{O}{\overset{\|}{C}}{-}Cl + SO_2 + HCl$$

(c)

$$C_6H_5{-}\overset{O}{\overset{\|}{C}}{-}OH + SOCl_2 \longrightarrow C_6H_5{-}\overset{O}{\overset{\|}{C}}{-}Cl + SO_2 + HCl$$

13.12. (a) *p*-chlorobenzoic anhydride, (b) 3,3-dimethylbutanoic anhydride

13.13. (a)

$$Cl{-}C_6H_4{-}\overset{O}{\overset{\|}{C}}{-}OH + NaOH \xrightarrow{H_2O} Cl{-}C_6H_4{-}\overset{O}{\overset{\|}{C}}{-}ONa + H_2O$$

$$Cl{-}C_6H_4{-}\overset{O}{\overset{\|}{C}}{-}OH + SOCl_2 \longrightarrow Cl{-}C_6H_4{-}\overset{O}{\overset{\|}{C}}{-}Cl + SO_2 + HCl$$

$$Cl{-}C_6H_4{-}\overset{O}{\overset{\|}{C}}{-}ONa + Cl{-}\overset{O}{\overset{\|}{C}}{-}C_6H_4{-}Cl \longrightarrow Cl{-}C_6H_4{-}\overset{O}{\overset{\|}{C}}{-}O{-}\overset{O}{\overset{\|}{C}}{-}C_6H_4{-}Cl + NaCl$$

(b)

$$CH_3C(CH_3)_2{-}CO_2H + NaOH \xrightarrow{H_2O} CH_3C(CH_3)_2{-}CO_2Na + H_2O$$

$$CH_3C(CH_3)_2{-}CO_2H + SOCl_2 \longrightarrow CH_3C(CH_3)_2{-}\overset{O}{\overset{\|}{C}}{-}Cl + SO_2 + HCl$$

$$CH_3C(CH_3)_2{-}\text{[illegible]}\ \longrightarrow CH_3C(CH_3)_2{-}\overset{O}{\overset{\|}{C}}{-}O{-}\overset{O}{\overset{\|}{C}}{-}C(CH_3)_2CH_3 + NaCl$$

13.14. (a) cyclohexyl acetate, [illegible]

13.15. (a)

$$CH_3\overset{O}{\overset{\|}{C}}{-}Cl + HO{-}C_6H_{11} \ \text{[illegible]}\ {-}C_6H_{11} + H_2O + Cl^-$$

(b)

$$H\overset{O}{\overset{\|}{C}}{-}OH + HO{-}C_6H_5 \underset{}{\overset{H^+}{\rightleftarrows}} H\overset{O}{\overset{\|}{C}}{-}O{-}C_6H_5 + H_2O$$

13.16. (a) propanamide, (b) 3-chlorobenzamide

13.17. (a)

$$CH_3CH_2\overset{O}{\overset{\|}{C}}OH + SOCl_2 \longrightarrow CH_3CH_2\overset{O}{\overset{\|}{C}}{-}Cl + SO_2 + HCl$$

$$CH_3CH_2\overset{O}{\overset{\|}{C}}{-}Cl + 2NH_3 \longrightarrow CH_3CH_2\overset{O}{\overset{\|}{C}}{-}NH_2 + NH_4Cl$$

(b)

$$m\text{-}ClC_6H_4{-}\overset{O}{\overset{\|}{C}}{-}OH + SOCl_2 \longrightarrow m\text{-}ClC_6H_4{-}\overset{O}{\overset{\|}{C}}{-}Cl \xrightarrow{NH_3} m\text{-}ClC_6H_4{-}\overset{O}{\overset{\|}{C}}{-}NH_2 + NH_4Cl$$

13.18.

$$CH_3\overset{\ddot{O}:}{\overset{\|}{C}}{-}\ddot{O}CH_2CH_3 + :NH_3 \rightleftarrows CH_3{-}\overset{:\ddot{O}:^-}{\overset{|}{\underset{^+NH_3}{\underset{|}{C}}}}{-}\ddot{O}CH_2CH_3$$

$$\rightleftarrows CH_3{-}\overset{:\ddot{O}:^-}{\overset{|}{\underset{:NH_2}{\underset{|}{C}}}}{-}\overset{H}{\underset{+}{\ddot{O}}}CH_2CH_3 \longrightarrow CH_3{-}\overset{\ddot{O}:}{\overset{\|}{C}}{-}NH_2 + HOCH_2CH_3$$

13.19. (a) $FeCl_3$ gives a colored solution with phenol but not with butanol.
(b) Butanoic acid turns blue litmus paper red; butanol does not.
(c) Butanoic acid reacts with $NaHCO_3$ to give CO_2 bubbles; phenol does not.

13.20. (b)

$$HO{-}\overset{O}{\overset{\|}{C}}CH_2CH_2CH_2CH_2\overset{O}{\overset{\|}{C}}{-}OH$$

(d)

$$CH_3CH_2CH_2CH_2\overset{O}{\overset{\|}{C}}{-}OH$$

(f) phthalic anhydride structure (benzene ring fused to a five-membered ring C(=O)–O–C(=O))

(h)

$$CH_3CH_2CH_2CH_2CH_2\overset{O}{\overset{\|}{C}}{-}NH_2$$

(j)

$$CH_3\overset{O}{\overset{\|}{C}}{-}NH_2$$

(l)

$$CH_3CH_2CH_2CH_2\overset{O}{\overset{\|}{C}}{-}O{-}C_6H_{11}\ \text{(cyclohexyl)}$$

(n)

$$CH_3\overset{O}{\overset{\|}{C}}{-}N(CH_2CH_3)_2$$

(p)

$$CH_3{-}C_6H_4{-}\overset{O}{\overset{\|}{C}}{-}O{-}\overset{O}{\overset{\|}{C}}{-}C_6H_4{-}CH_3$$

(r)

$$O{=}C{-}O{-}C_6H_5\ \text{attached to 1-naphthyl}$$

13.21. (b) 2,4-dichloropentanoic acid, (d) pentanedioic anhydride,
(f) succinic acid, (h) dicyclohexyl oxalate, (j) sodium propanoate

13.22. (b) propanoic acid

13.24. (b) benzoic acid, (d) 2-nitrobutanoic acid

13.26. (b)

$$CH_3CH_2CH_2\overset{\overset{\displaystyle O}{\|}}{C}{-}OH + NH_3 \longrightarrow CH_3CH_2CH_2\overset{\overset{\displaystyle O}{\|}}{C}{-}\bar{O}NH_4^+$$

$$\xrightarrow{\Delta} CH_3CH_2CH_2\overset{\overset{\displaystyle O}{\|}}{C}{-}NH_2 + H_2O$$

(d)

$$3CH_3CH_2CH_2\overset{\overset{\displaystyle O}{\|}}{C}{-}OH + PBr_3 \longrightarrow 3CH_3CH_2CH_2\overset{\overset{\displaystyle O}{\|}}{C}{-}Br + P(OH)_3$$

(f) no reaction

13.27. (e)

$$CH_3CH_2\overset{\overset{\displaystyle O}{\|}}{C}CH_3 \xrightarrow{Br_2,\ OH^-} CH_3CH_2\overset{\overset{\displaystyle O}{\|}}{C}{-}O^- + CHBr_3$$

$$\xrightarrow{H_3O^+} CH_3CH_2CO_2H$$

(f) $CH_3CH_2CH_2Cl + Mg \xrightarrow{\text{dry ether}} CH_3CH_2CH_2MgCl \xrightarrow{CO_2}$

$$CH_3CH_2CH_2\overset{\overset{\displaystyle O}{\|}}{C}{-}OMgCl \xrightarrow{H_3O^+} CH_3CH_2CH_2CO_2H$$

(g) $CH_3CH_2CH_2CH_2OH \xrightarrow{H_2CrO_4} CH_3CH_2CH_2CO_2H$

13.28. (c)

$$o\text{-}CH_3C_6H_4\overset{\overset{\displaystyle O}{\|}}{C}{-}OH + SOCl_2 \longrightarrow o\text{-}CH_3C_6H_4\overset{\overset{\displaystyle O}{\|}}{C}{-}Cl + SO_2 + HCl$$

(d)

$$o\text{-}CH_3C_6H_4\overset{\overset{\displaystyle O}{\|}}{C}{-}Cl\ [\text{from (c)}] + 2NH_3 \longrightarrow o\text{-}CH_3C_6H_4\overset{\overset{\displaystyle O}{\|}}{C}{-}NH_2 + NH_4Cl$$

13.32. E: glutaric anhydride (six-membered cyclic anhydride, ring O with two C=O) F: $HO_2C(CH_2)_3CO_2H$ (chain with two CO_2H groups)

CHAPTER 14
FATS, OILS, WAXES, AND THEIR DERIVATIVES

14.1. (a)

$$\begin{array}{l} CH_2{-}O{-}\overset{\overset{O}{\|}}{C}(CH_2)_7CH{=}CH(CH_2)_7CH_3 \\ | \\ CH{-}O{-}\overset{\overset{O}{\|}}{C}(CH_2)_7CH{=}CH(CH_2)_7CH_3 \\ | \\ CH_2{-}O{-}\overset{\overset{O}{\|}}{C}(CH_2)_7CH{=}CH(CH_2)_7CH_3 \end{array}$$

(b)

$$\begin{array}{l} CH_2{-}O{-}\overset{\overset{O}{\|}}{C}(CH_2)_7CH{=}CH(CH_2)_7CH_3 \\ | \\ CH{-}O{-}\overset{\overset{O}{\|}}{C}(CH_2)_{10}CH_3 \\ | \\ CH_2{-}O{-}\overset{\overset{O}{\|}}{C}(CH_2)_{14}CH_3 \end{array}$$

14.2.

$$\begin{array}{l} CH_2-O-\overset{O}{\overset{\|}{C}}(CH_2)_{12}CH_3 \\ | \\ CH-O-\overset{O}{\overset{\|}{C}}(CH_2)_{12}CH_3 \\ | \\ CH_2-O-\overset{O}{\overset{\|}{C}}(CH_2)_{12}CH_3 \end{array} + 3NaOH \longrightarrow \begin{array}{l} CH_2OH \\ | \\ CHOH \\ | \\ CH_2OH \end{array} + 3CH_3(CH_2)_{12}\overset{O}{\overset{\|}{C}}-ONa$$

14.3. (a) glyceryl trilaurate, (b) sodium palmitate

14.4. Waxes are esters of 1-alkanols; fats and oils are esters of glycerol.

14.5. No.

14.6.

$$2CH_3(CH_2)_{12}\overset{O}{\overset{\|}{C}}-ONa + Ca^{2+} \longrightarrow \left(CH_3(CH_2)_{12}\overset{O}{\overset{\|}{C}}-O\right)_2Ca + 2Na^+$$

14.7. (a)

$$CH_3(CH_2)_7CH{=}CH(CH_2)_7\overset{O}{\overset{\|}{C}}-O^- + H_2O \rightleftharpoons CH_3(CH_2)_7CH{=}CH(CH_2)_7\overset{O}{\overset{\|}{C}}-OH + OH^-$$

(b)

$$CH_3(CH_2)_7CH{=}CH(CH_2)_7\overset{O}{\overset{\|}{C}}-O^- + H_3O^+ \longrightarrow CH_3(CH_2)_7CH{=}CH(CH_2)_7\overset{O}{\overset{\|}{C}}-OH + H_2O$$

The above are the net ionic equations.

14.8. The molecule has two carbon–carbon double bonds.

14.9. *Phospholipid* is a general term that describes any glycerol ester which contains phosphoric acid groups. *Phosphatidic acids* are glycerol derivatives in which two of the glycerol OH groups have been esterified by fatty acids, and the remaining glycerol OH group has been esterified by a phosphoric acid molecule. *Phosphatidyl esters* are phosphatidic acids in which the free phosphoric acid group has been esterified by an alcohol group.

14.10.

$$CH_3CH_2CH_2\overset{O}{\overset{\|}{C}}-OH + CoA-SH \underset{}{\overset{\text{(enzyme)}}{\rightleftharpoons}} CH_3CH_2CH_2\overset{O}{\overset{\|}{C}}-S-CoA + H_2O$$

$$CH_3\overset{O}{\overset{\|}{C}}-S-CoA + CO_2 \overset{\text{(enzyme)}}{\rightleftharpoons} HO-\overset{O}{\overset{\|}{C}}-CH_2-\overset{O}{\overset{\|}{C}}-S-CoA$$

$$HO-\overset{O}{\overset{\|}{C}}-CH_2-\overset{O}{\overset{\|}{C}}-S-CoA + CH_3CH_2CH_2\overset{O}{\overset{\|}{C}}-S-CoA \text{ (from above)} \overset{\text{(enzyme)}}{\rightleftharpoons} HO-\overset{O}{\overset{\|}{C}}-\underset{\underset{CH_2CH_2CH_3}{|}}{\underset{C=O}{\underset{|}{CH}}}-\overset{O}{\overset{\|}{C}}-S-CoA$$

$$HO-\overset{O}{\overset{\|}{C}}-\underset{\underset{CH_2CH_2CH_3}{|}}{\underset{C=O}{\underset{|}{CH}}}-\overset{O}{\overset{\|}{C}}-S-CoA \overset{\text{(enzyme)}}{\underset{-CO_2}{\rightleftharpoons}} CH_3CH_2CH_2\overset{O}{\overset{\|}{C}}CH_2\overset{O}{\overset{\|}{C}}-S-CoA$$

$$CH_3CH_2CH_2\overset{O}{\overset{\|}{C}}CH_2\overset{O}{\overset{\|}{C}}-S-CoA \overset{\text{(enzyme)}}{\underset{+2H}{\rightleftharpoons}} CH_3CH_2CH_2\overset{OH}{\overset{|}{C}}HCH_2\overset{O}{\overset{\|}{C}}-S-CoA$$

$$CH_3CH_2CH_2\overset{OH}{\overset{|}{C}}HCH_2\overset{O}{\overset{\|}{C}}-S-CoA \overset{\text{(enzyme)}}{\underset{-H_2O}{\rightleftharpoons}} CH_3CH_2CH_2CH{=}CH\overset{O}{\overset{\|}{C}}-S-CoA$$

$$CH_3CH_2CH_2CH{=}CH\overset{O}{\overset{\|}{C}}-S-CoA \overset{+2H}{\underset{\text{(enzyme)}}{\rightleftharpoons}} CH_3CH_2CH_2CH_2CH_2\overset{O}{\overset{\|}{C}}-S-CoA$$

$$CoA-SH + CH_3CH_2CH_2CH_2CH_2\overset{O}{\overset{\|}{C}}-OH \overset{H_2O}{\underset{\text{(enzyme)}}{\rightleftharpoons}} CH_3CH_2CH_2CH_2CH_2\overset{O}{\overset{\|}{C}}-S-CoA$$

14.11.

$$CH_3CH_2CH_2CH_2CH_2\overset{O}{\overset{\|}{C}}-OH + CoA-SH \underset{}{\overset{\text{(enzyme)}}{\rightleftharpoons}} CH_3CH_2CH_2CH_2CH\ \overset{O}{\overset{\|}{C}}-S-CoA + H_2O$$

$$\Big\updownarrow \begin{matrix}-2H \\ \text{(enzyme)}\end{matrix}$$

$$CH_3CH_2CH_2\overset{OH}{\overset{|}{C}}HCH_2\overset{O}{\overset{\|}{C}}-S-CoA \overset{\begin{matrix}H_2O\\ \text{(enzyme)}\end{matrix}}{\rightleftharpoons} CH_3CH_2CH_2CH{=}CH\overset{O}{\overset{\|}{C}}-S-CoA$$

$$\Big\updownarrow \begin{matrix}-2H \\ \text{(enzyme)}\end{matrix}$$

$$CH_3CH_2CH_2\overset{O}{\overset{\|}{C}}CH_2\overset{O}{\overset{\|}{C}}-S-CoA \overset{\begin{matrix}\text{CoA-SH}\\ \text{(enzyme)}\end{matrix}}{\rightleftharpoons} CH_3CH_2CH_2\overset{O}{\overset{\|}{C}}-S-CoA + CH_3\overset{O}{\overset{\|}{C}}-S-CoA$$

$$\Big\updownarrow \begin{matrix}H_2O \\ \text{(enzyme)}\end{matrix}$$

$$CH_3CH_2CH_2\overset{O}{\overset{\|}{C}}-OH + CoA-SH$$

14.12. (a) olive oil = corn oil > peanut oil > beef tallow = cottonseed oil
(b) corn oil > cottonseed oil > peanut oil > olive oil > beef tallow

14.13. (d)

$$CH_3(CH_2)_7CH{=}CH(CH_2)_7\overset{O}{\overset{\|}{C}}-ONa$$

(e)

$$(CH_3(CH_2)_{10}\overset{O}{\overset{\|}{C}}-O)_2Ca$$

14.14. (c) oleic acid, (d) 1-nonyldodecanoate
14.20. Soaps.
14.21. (b) Phosphatidic acid is acidic and would redden blue litmus paper.

14.22. (b) Glyceryltrioleate + H_2 $\xrightarrow{Ni}$

$$\begin{array}{l} CH_2-O-\overset{O}{\overset{\|}{C}}(CH_2)_7CH{=}CH(CH_2)_7CH_3 \\ | \\ CH-O-\overset{O}{\overset{\|}{C}}(CH_2)_7CH{=}CH(CH_2)_7CH_3 \\ | \\ CH_2-O-\overset{O}{\overset{\|}{C}}(CH_2)_{16}CH_3 \end{array}$$

(partially hydrogenated)

14.26. (b) S.N. = 648
14.28. (b) I.N. = 173.4

CHAPTER 15
AMINES AND THEIR DERIVATIVES

15.1. 1° amines: (a), (b)
2° amines: (c), (d), (e), (f)
3° amines: (g)
15.2. The bonding orbitals are sp^2.

15.3. (a) $CH_3CH_2CH_2{-}N{-}CH_2CH_2CH_3$ with $CH_2CH_2CH_3$ on N

(b) $CH_3CH_2CH_2CH_2CH_2{-}N(H){-}CH_2CH_2CH_2CH_2CH_3$

(c) 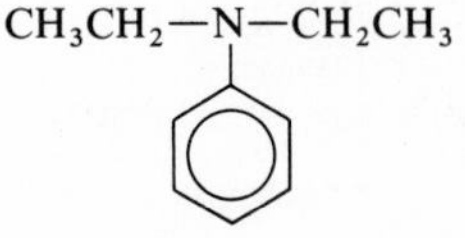 $CH_3CH_2{-}N{-}CH_2CH_3$ with a benzene ring on N

(d) 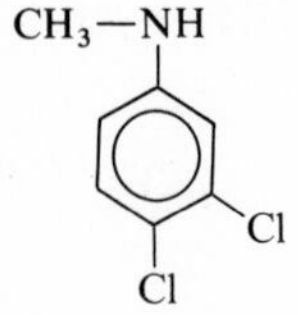 $CH_3{-}NH$ on a benzene ring bearing Cl, Cl

15.4. (a) cyclopentylamine, (b) propylamine, (c) *N,N*-dipropylaniline

15.5. Trimethylamine cannot form intermolecular hydrogen bonds. They have similar water solubilities because both can form hydrogen bonds with water.

15.6. Primary amines have two N—H hydrogens. Secondary amines have only one N—H hydrogen.

15.7. Hydrogen bonds between oxygen atoms (—O—H·····O—) are stronger than hydrogen bonds between nitrogen atoms (—N—H·····N—).

15.8. (a) diethylamine, (b) methylamine, (c) *N,N*-dimethylaniline

15.9. $(CH_3CH_2)_2NH + H_2O \rightleftarrows (CH_3CH_2)_2NH_2^+ + OH^-$

15.10. $K_b = \dfrac{[(CH_3CH_2)_2NH_2^+][OH^-]}{[(CH_3CH_2)_2NH]}$

15.11. (a) isopropylamine, (b) isopropylamine, (c) aniline

15.12. The inductive effect of the chlorine (withdrawing electron density) is greater than its resonance effect (supplying electron density).

15.13. (a) $CH_4 + Cl_2 \xrightarrow{\text{light}} CH_3Cl + HCl$

$CH_3Cl + 2NH_3 \longrightarrow CH_3NH_2 + NH_4Cl$

(b) 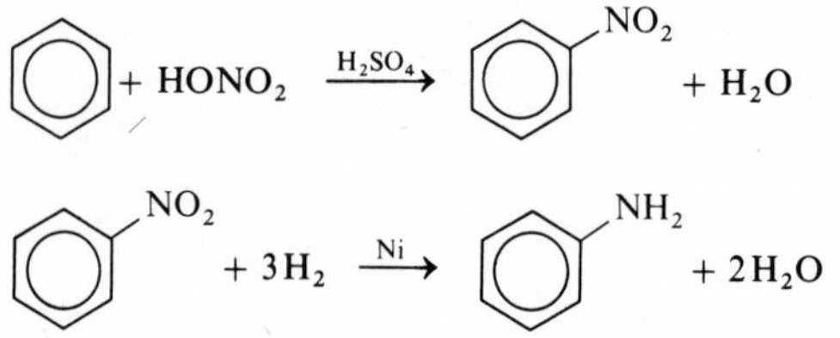 $C_6H_6 + HONO_2 \xrightarrow{H_2SO_4} C_6H_5NO_2 + H_2O$

$C_6H_5NO_2 + 3H_2 \xrightarrow{Ni} C_6H_5NH_2 + 2H_2O$

(c) $CH_3NH_2 + CH_3Cl \longrightarrow (CH_3)_2NH_2{}^+Cl^- \xrightarrow{OH^-} (CH_3)_2NH + H_2O + Cl^-$

$(CH_3)_2NH + CH_3Cl \longrightarrow (CH_3)_3NH^+Cl^- \xrightarrow{OH^-} (CH_3)_3N + H_2O + Cl$

(d) $CH_3CH_2CH_2CH_2Cl + KCN \longrightarrow CH_3CH_2CH_2CH_2CN + KCl$

$\downarrow H_2/Ni$

$CH_3CH_2CH_2CH_2CH_2NH_2$

(e) $(CH_3)_3N$ [from (c)] $+ CH_3Br \longrightarrow (CH_3)_4\overset{+}{N}Br^-$

15.14.

$CH_3C(=O){-}Cl + (CH_3)_2NH \xrightarrow{\text{base*}} CH_3C(=O){-}N(CH_3)_2 + (CH_3)_2\overset{+}{N}H_2Cl^-$

(*the base may be excess dimethylamine)

$CH_3C(=O){-}O{-}CH_3 + (CH_3)_2NH \longrightarrow CH_3C(=O){-}N(CH_3)_2 + CH_3OH$

15.15.

$$\left.\begin{array}{c} CH_3NH_2 \\ + \\ (CH_3)_2NH \\ + \\ (CH_3)_3N \end{array}\right\} + C_6H_5\text{—}SO_2Cl \longrightarrow \left\{\begin{array}{c} CH_3NHSO_2\text{—}C_6H_5 \\ + \\ (CH_3)_2N\text{—}SO_2\text{—}C_6H_5 \\ + \\ (CH_3)_3N \text{ (no reaction)} \end{array}\right\} \xrightarrow[H_2O]{NaOH}$$

$$\left.\begin{array}{l} CH_3\ddot{\bar{N}}\text{—}SO_2\text{—}C_6H_5\ Na^+ \text{ (soluble)} \\ + \\ (CH_3)_2N\text{—}SO_2\text{—}C_6H_5 \text{ (no reaction, insoluble)} \\ + \\ (CH_3)_3N \text{ (no reaction, insoluble)} \end{array}\right\} \xrightarrow{H_3O^+} \begin{array}{c} (CH_3)_2N\text{—}SO_2\text{—}C_6H_5 \text{ (insoluble)} \\ + \\ (CH_3)_3NH^+ \text{ (soluble)} \end{array}$$

15.16. (a) $N_2 + H_2O + CH_2{=}CH_2 + CH_3CH_2OH$

(b) $(CH_3CH_2)_2N\text{—}N{=}O$ (*N*-nitrosodimethylamine) + H_2O

(c) $(CH_3CH_2)_3NH^+ X^- + (CH_3CH_2)_3\overset{+}{N}\text{—}N{=}O\ X^-$

(d) $C_6H_5\text{—}\overset{+}{N}{\equiv}N\!:\ X^- + 2H_2O$

(e) $C_6H_5\text{—}N(CH_3)\text{—}N{=}O + H_2O$

(f) $O{=}N\text{—}C_6H_4\text{—}N(CH_3)_2$ (para)

15.17.

Let Ar=Br—(benzene ring, with Br on each adjacent ring position, 2,6-dibromo relative to the open bond)—

(a) $Ar\text{—}NH_2 + HONO + HCl \xrightarrow{0\text{–}5°C} Ar\text{—}N_2{}^+Cl^- + H_2O$

$$\downarrow H_3PO_2/\text{heat}$$

$$ArH$$

(b) $Ar\text{—}N_2{}^+Cl^- + H_2O \xrightarrow{\text{heat}} Ar\text{—}OH$

(c) $Ar\text{—}N_2{}^+Cl^- + HBF_4 \xrightarrow{\text{heat}} Ar\text{—}F$

(d) $Ar\text{—}N_2{}^+Cl^- + Cu_2Br_2 \xrightarrow{\text{heat}} Ar\text{—}Br$

(e) $Ar\text{—}N_2{}^+Cl^- + Cu_2Cl_2 \xrightarrow{\text{heat}} Ar\text{—}Cl$

(f) $Ar\text{—}N_2{}^+Cl^- + KI \xrightarrow{\text{heat}} Ar\text{—}I$

(g) $Ar\text{—}N_2{}^+Cl^- + Cu_2CN_2 \xrightarrow{\text{heat}} Ar\text{—}CN$

$$\downarrow H_2O/H^+$$

$$Ar\text{—}CO_2H$$

15.18. (Reaction mechanism: benzenediazonium ion + phenol → azo intermediate → (B:) p-hydroxyazobenzene + HB^+)

15.19. $HOSO_2-C_6H_4-NH_2 + HONO + HCl \xrightarrow{0\text{–}5°C} HOSO_2-C_6H_4-\overset{+}{N}_2Cl^-$

2-naphthol + $^-Cl\overset{+}{N}_2-C_6H_4-SO_2OH \longrightarrow$ 1-($N{=}N-C_6H_4-SO_3H$)-2-naphthol $\xrightarrow{OH^-}$ 1-($N{=}N-C_6H_4-SO_3Na$)-2-naphthol

15.20. (a) azetidine $\ddot{N}-CH_3$ (b) pyrrolidine $\ddot{N}-CH_3$

15.21. The 3° nitrogen.

15.22. (b) $(CH_3)_3N$ (d) $C_6H_5-CH_2NHCH_2-C_6H_5$

(f) $Br-C_6H_4-NH_2$ (h) 1,2-diaminocyclohexane (NH_2, NH_2)

15.23. (e) *N,N*-dimethyl-2,4,6-trichloroaniline, (f) dicyclobutylamine, (g) *trans*-1-chloro-4-aminocyclohexane

15.25. (b) 1-aminohexane (higher bp), ethylamine (higher H_2O solubility)

15.27. *p*-nitroaniline < aniline < ammonia < dimethylamine

15.28. (b) dimethylamine, (d) benzylamine

15.29. (c) $C_6H_5-N^+(CH_3)_3\ I^-$

(d) no reaction (g) $CH_3-C_6H_4-NH_2$

15.30. (d) $C_6H_5-CH_3 + Cl_2 \xrightarrow{light} C_6H_5-CH_2Cl \xrightarrow{NaCN} C_6H_5-CH_2CN \xrightarrow{LiAlH_4 \text{ or } H_2/Ni} C_6H_5-CH_2CH_2NH_2$

(e) $CH_3CH_2Cl + NH_3 \longrightarrow CH_3CH_2NH_3{}^+Cl \xrightarrow{OH^-} CH_3CH_2NH_2$

$CH_3CH_2NH_2 + CH_3CH_2Cl \longrightarrow (CH_3CH_2)_2NH_2{}^+Cl^- \xrightarrow{OH^-} (CH_3CH_2)_2NH$

$(CH_3CH_2)_2NH + CH_3CH_2Cl \longrightarrow (CH_3CH_2)_3NH^+Cl^- \xrightarrow{OH^-} (CH_3CH_2)_3N$

$(CH_3CH_2)_3N + CH_3CH_2Cl \longrightarrow (CH_3CH_2)_4N^+Cl^-$

15.31. (b) no reaction

(d) $CH_3CH_2CH_2—N(CH_3)—SO_2—C_6H_5$ (f) piperidine N—N=O

(h) N-methylpiperidinium (N–H) Cl^- + N-methyl-N-nitrosopiperidinium Cl^- (j) $Cl—C_6H_5$

(l) $Cl—C_6H_4—OH$ (n) $Cl—C_6H_4—I$ (p) $Cl—C_6H_4—N{=}N—C_6H_4—N(CH_3)_2$

15.32. (b) $C_6H_5—NH_2$ (from above) + $CH_3CH_2Cl \longrightarrow C_6H_5—\overset{+}{N}H_2CH_2CH_3\ Cl^-$ $\xrightarrow{OH^-}$ $C_6H_5—NHCH_2CH_3$ $\xrightarrow{CH_3CH_2Cl}$ $C_6H_5—\overset{+}{N}H(CH_2CH_3)_2\ Cl^-$ $\xrightarrow{OH^-}$ $C_6H_5—N(CH_2CH_3)_2$

$C_6H_5—N(CH_2CH_3)_2 \xrightarrow[\text{(from above)}]{C_6H_5—N_2{}^+Cl^-} C_6H_5—N{=}N—C_6H_4—N(CH_2CH_3)_2$

15.34.
B: $C_6H_5—CH_3$ C: $C_6H_5—CH_2Cl$ D: $C_6H_5—CH_2\overset{+}{N}H_3\ Cl^-$ E: $C_6H_5—CH_2NH_2$

15.36.
$O_2N—C_6H_4—NH_2 + HONO + HCl \xrightarrow{0–5°C} O_2N—C_6H_4—N_2{}^+Cl^-$

$\xrightarrow{\text{2-naphthol (OH)}}$ 1-($N{=}N—C_6H_4—NO_2$)-2-naphthol (naphthalene ring bearing $N{=}N—C_6H_4—NO_2$ and —OH)

15.37. (c) Hinsberg test, (d) Hinsberg test

CHAPTER 16
CARBOHYDRATES

16.1. Raffinose is a trisaccharide; fructose, glucose, and galactose are monosaccharides.
16.2. It is probably a monosaccharide or a small oligosaccharide. Hydrolysis.
16.3. (a) ketotetrose, (b) aldotetrose, (c) aldohexose, (d) ketopentose

16.4. The chiral atoms are designated by (*):

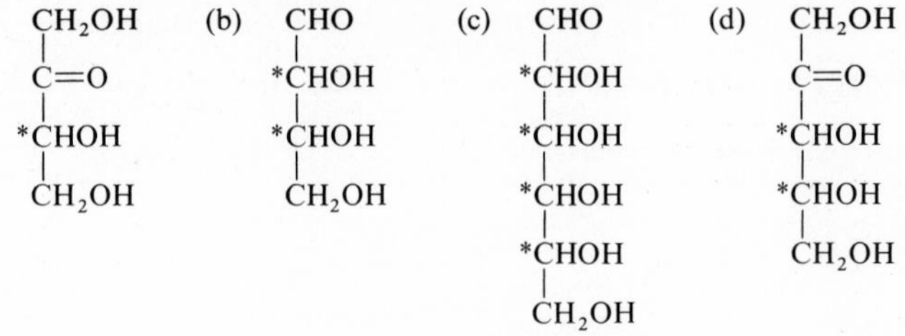

(a) $2^1 = 2$ stereoisomers; (b) $2^2 = 4$ stereoisomers; (c) $2^4 = 16$ stereoisomers; (d) $2^2 = 4$ stereoisomers

16.6.

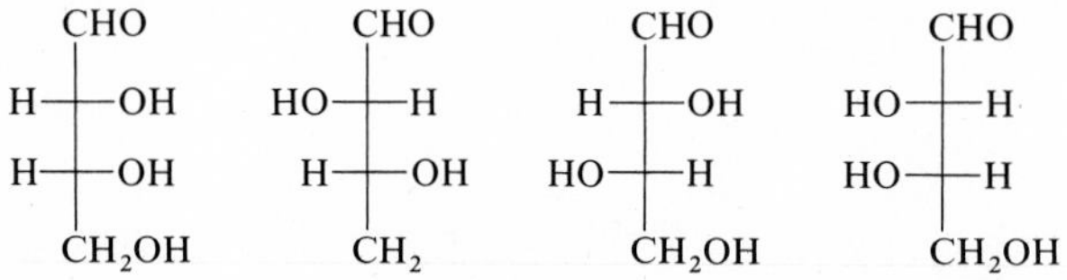

aldotetroses

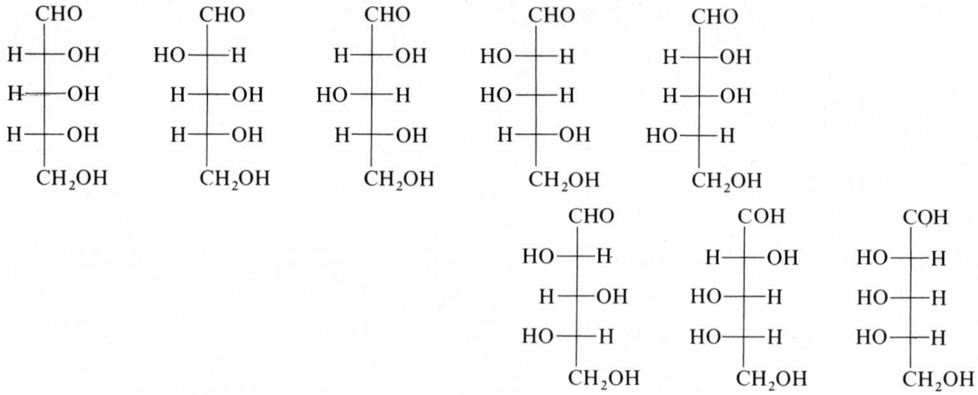

aldopentoses

16.7. (a) D, (b) L, (c) D, (d) L, (e) L, (f) D

16.8. (a) *R*, | *R* (b) *S*, | *S* (c) *S*, | *R* (d) *R*, | *S* (e) *S*, (f) *S* | *S* | *S* | *R*

16.9. D-L refers only to the highest numbered chiral carbon; *R-S* designation may be applied to every chiral atom in the molecule.

16.10. In the β-form, the hemiacetal hydroxyl group is in the less hindered equatorial conformation; in the α-form it is axial.

16.11. (a)

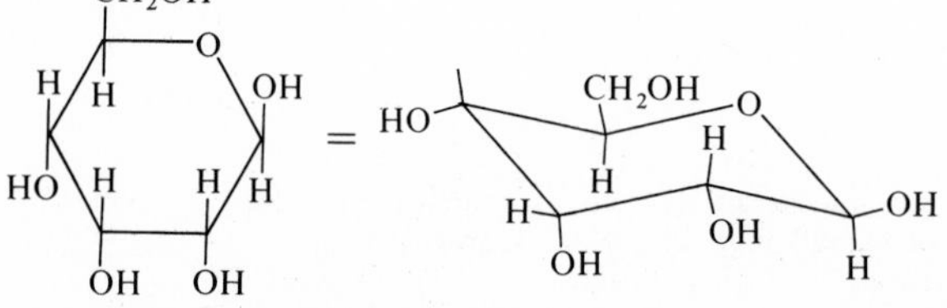

(β-form shown)

(b) [structure: Haworth and chair conformations]

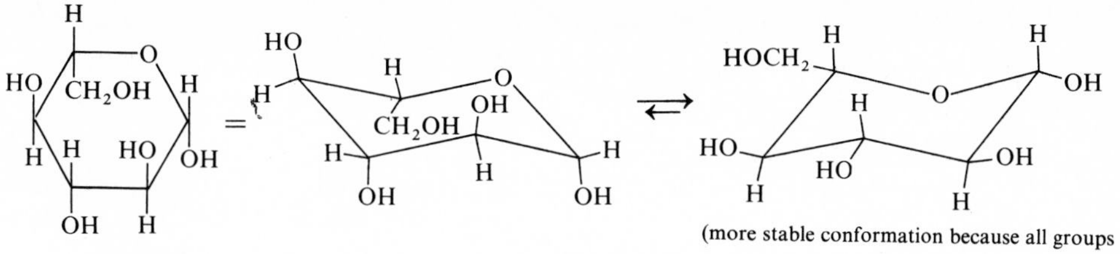

(more stable conformation because all groups are equatorial)

16.12. (a) [structure] is α; the β-form is shown in 16.11 (a).

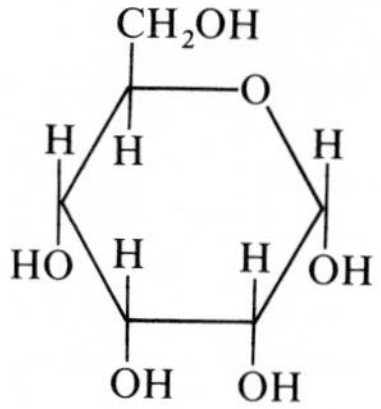

(b) [structure] is α; the β-form is shown in 16.11 (b).

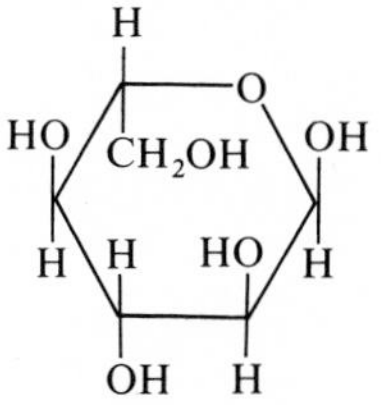

16.13. (a) [structure] (b) [structure] (c) [structure]

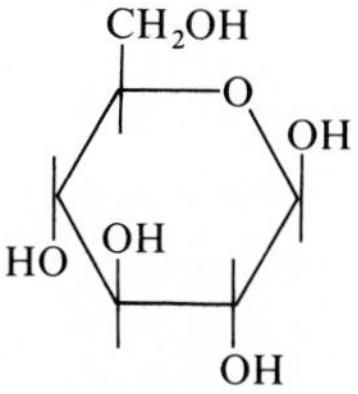

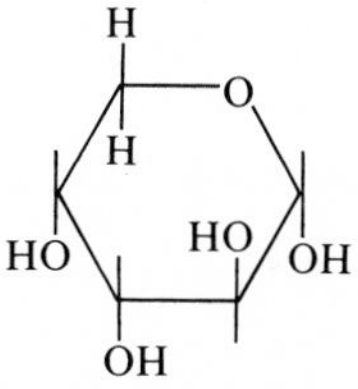

(d) [structure]

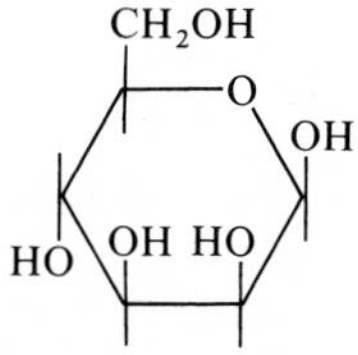

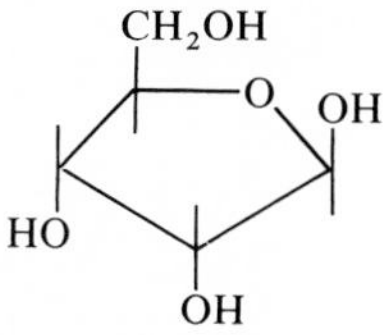

16.14. [structures]

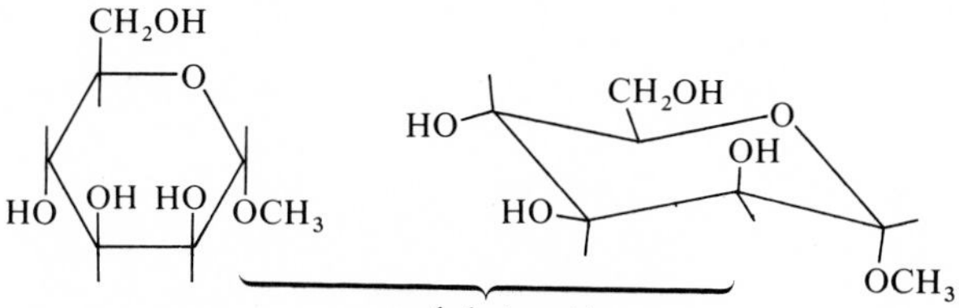

α-methyl glycoside

[structures]

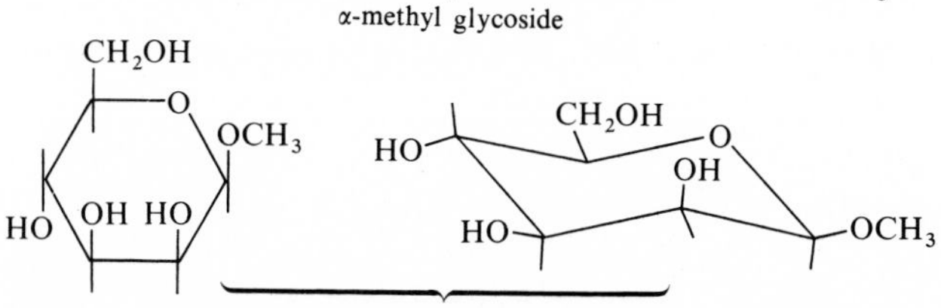

β-methyl glycoside

16.15. (c) does not.

(a) [Structure: pyranose ring with CH_2OH, O, OH, HO, OH, OH] $\xrightarrow{Ag(NH_3)_2^+}$ [Fischer projection: $C(=O)ONH_4$; –OH; HO–; –OH; –OH; CH_2OH] + Ag ↓

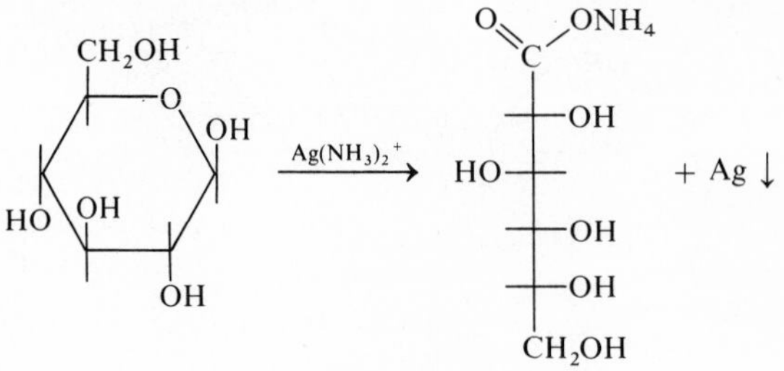

(b) [Structure: furanose ring with CH_2OH, CHOH, HO, O, OH, OH] $\xrightarrow{Ag(NH_3)_2^+}$ [Fischer projection: $C(=O)ONH_4$; –OH; HO–; –OH; –OH; CH_2OH] + Ag ↓

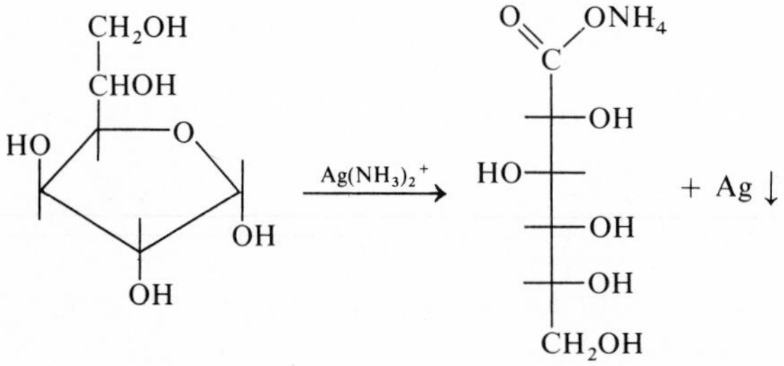

(d) [Fischer projection: CH_2OH; =O; HO–; –OH; –OH; CH_2OH] $\xrightarrow{Ag(NH_3)_2^+}$ [Fischer projection: $C(=O)ONH_4$; =O; HO–; –OH; –OH; CH_2OH] + other oxidation products + Ag ↓

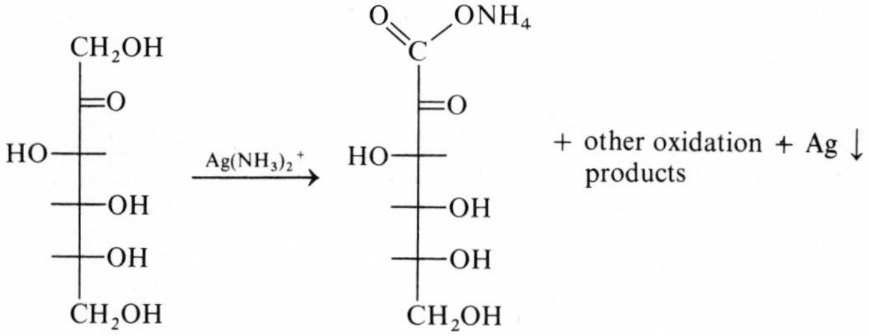

16.16. [Structure: pyranose ring with CH_2, O, OH, HO, OH, OH]

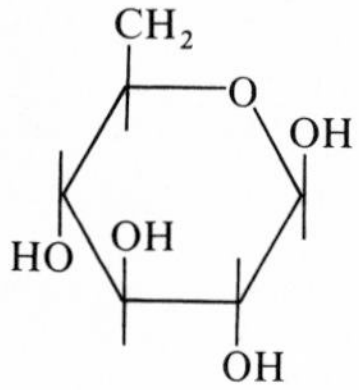

16.17. [Structure: pyranose ring with $CH_2O{-}\overset{O}{\overset{\|}{C}}CH_3$, O, $CH_3\overset{O}{\overset{\|}{C}}O$, $O\overset{O}{\overset{\|}{C}}CH_3$, $O\overset{O}{\overset{\|}{C}}CH_3$, $O\underset{\|}{C}CH_3$ (=O)]

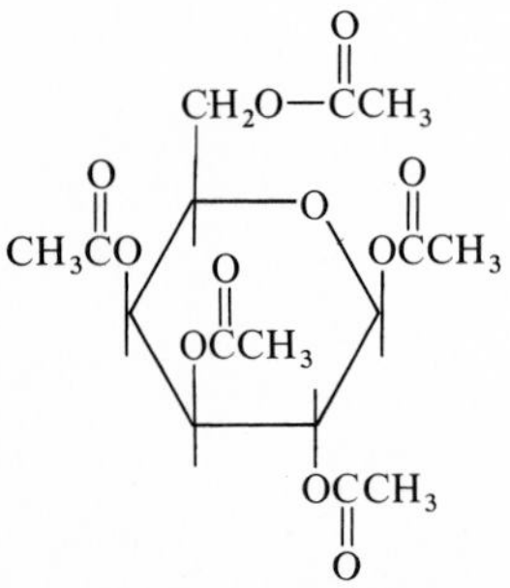

16.18.

CHO
H—OH
H—OH
CH_2OH
D-erythrose

CHO
HO—H
H—OH
CH_2OH
D-threose

HCN (each sugar gives two cyanohydrins)

From D-erythrose:

CN
—OH
—OH
—OH
CH_2OH

(1) $Ba(OH)_2$
(2) H_3O^+

CO_2H
H—OH
H—OH
H—OH
CH_2OH

⇌

CH_2OH
O
O
HO
OH

Na-Hg
H_2O

CHO
H—OH
H—OH
H—OH
CH_2OH

CN
HO—H
H—OH
H—OH
CH_2OH

(1) $Ba(OH)_2$
(2)H_3O^+

CO_2H
HO—H
H—OH
H—OH
CH_2OH

⇌

CH_2
O
OH
O
HO

Na-Hg
H_2O

CHO
HO—H
H—OH
H—OH
CH_2OH

From D-threose:

CN
H—OH
HO—H
H—OH
CH_2OH

(1) $Ba(OH)_2$
(2) H_3O^+

CO_2H
H—OH
HO—H
H—OH
CH_2OH

⇌

CH_2OH
HO
O
O
OH

Na-Hg
H_2O

CHO
H—OH
HO—H
H—OH
CH_2OH

CN
HO—H
HO—H
H—OH
CH_2OH

(1) $Ba(OH)_2$
(2) H_3O^+

CO_2H
HO—H
HO—H
H—OH
CH_2OH

⇌

CH_2OH
HO
O
OH
O

Na-Hg
H_2O

CHO
HO—H
HO—H
H—OH
CH_2OH

16.19. D-(+)-allose and D-(+)-altrose
D-(−)-gulose and D-(−)-idose
D-(+)-galactose and D-(+)-talose
} Each of these pairs gives the same osazone.

Two aldohexoses that give the same osazone must be derived from the same aldopentose.

16.20.

CHO
H—OH
HO—H
HO—H
CH_2OH

16.21.

CH_2OH HO HO O OH O CH_2OH O HO OH OH = CH_2OH O HO OH HO O CH_2OH O OH OH OH

16.22. Sucrose is α-; lactose is β-.
16.23. It is easier to draw them as shown.

16.24.

CHO
HO—H
HO—H
H—OH
H—OH
CH_2OH

$+ H_2 \xrightarrow{Ni}$

CH_2OH
HO—H
HO—H
H—OH
H—OH
CH_2OH

16.25. either glyceraldehyde or CH_2OH—C=O—CH_2OH

16.26. (a)

CH_2OH HO OH HO O—P(=O)(OH)—OH

α-D-mannose 1-phosphoric acid

(b)

CH_2OH
HO—H
HO—H
H—OH
H—OH
CH_2OH

mannitol

(c)

CO_2H
HO—H
HO—H
H—OH
H—OH
CH_2OH

mannonic acid

(d)

CHO
H_2N—H
HO—H
H—OH
H—OH
CH_2OH

D-mannosamine

(e)

CHO
H—H
HO—H
H—OH
H—OH
CH_2OH

2-deoxy-D-glucose

16.27. They are the same.
16.28. Reduction of the carboxyl group to a CH_2OH and hydrolysis of the phosphate ester to the CH_2OH.

16.31. (b) ketohexoses

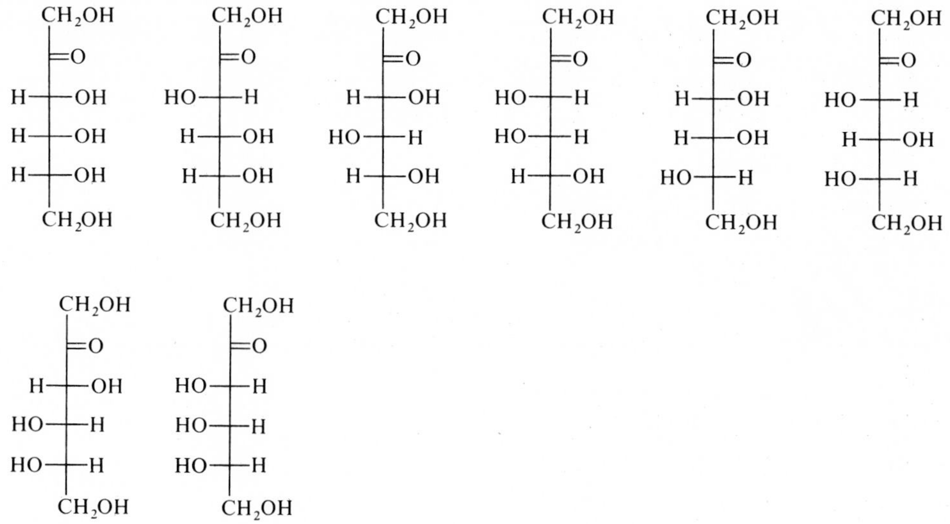

16.32. (b)

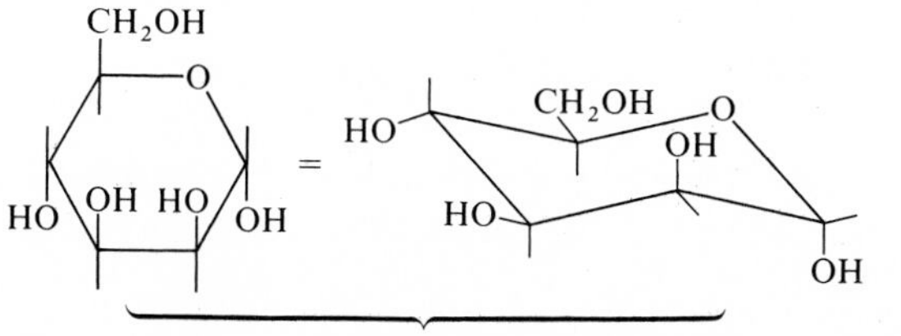

α

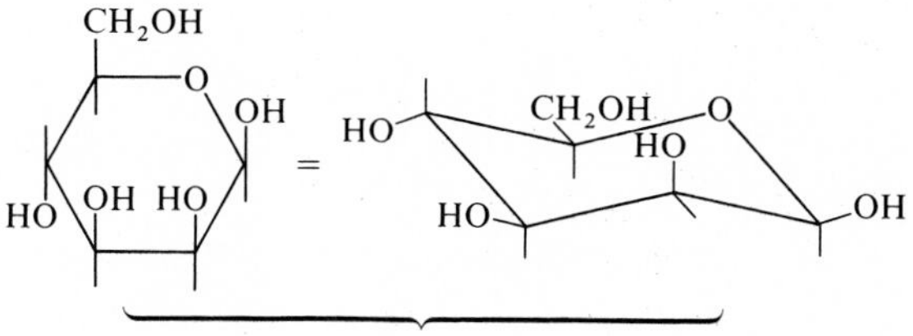

β

16.33. (d) (e)

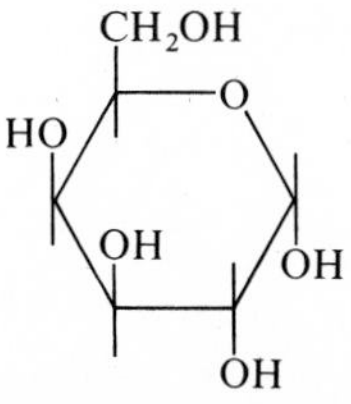

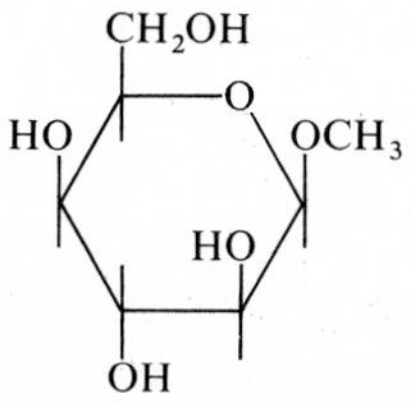

16.36. (b)

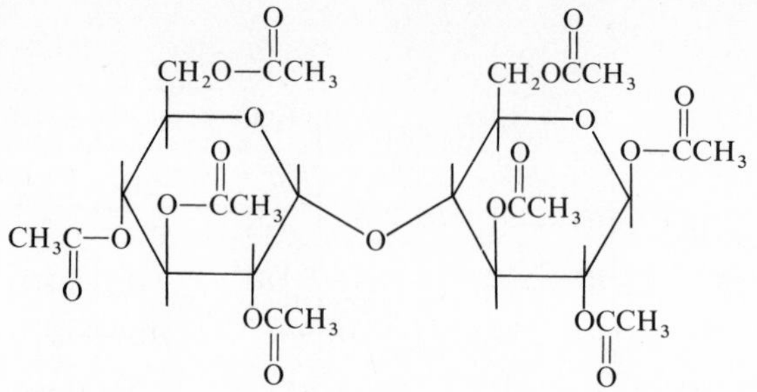

maltose octaacetate

16.37.

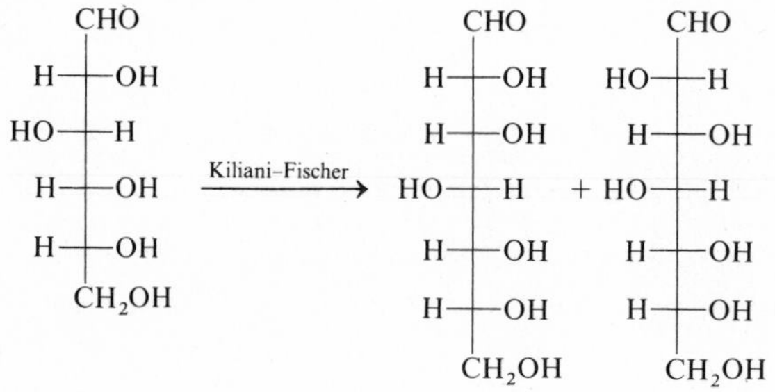

$$\begin{array}{c} CHO \\ H-C-OH \\ HO-C-H \\ H-C-OH \\ H-C-OH \\ CH_2OH \end{array} \xrightarrow{\text{Kiliani–Fischer}} \begin{array}{c} CHO \\ H-C-OH \\ H-C-OH \\ HO-C-H \\ H-C-OH \\ H-C-OH \\ CH_2OH \end{array} + \begin{array}{c} CHO \\ HO-C-H \\ H-C-OH \\ HO-C-H \\ H-C-OH \\ H-C-OH \\ CH_2OH \end{array}$$

16.40. D-allose, D-galactose

CHAPTER 17
BEHAVIOR OF VERY LARGE MOLECULES: BIOPOLYMERS

17.1. Macromolecule means a large molecule. Polymer means a molecule that is composed of many (poly) repeating units (mers). They are used synonymously.

17.2. Polymers have average molecular weights rather than exact molecular weights. Polymer solutions are viscous; polymers have wide melting ranges; polymers are amorphous rather than crystalline.

17.3. (a) It is impure.
(b) Not necessarily.
(c) Determine the viscosity of its solution.

17.5. Cellulose is a linear, unbranched polymer; starch is a branched polymer. Cellulose consists of β-glucoside linkages; starch consists of α-glucoside linkages.

17.6. The branches interfere with packing into tight, linear chains.

17.7. (a) $\left(\begin{array}{c} H \\ | \\ -N- \end{array} \begin{array}{c} \\ \\ C \\ | \\ CH_3 \end{array} - \begin{array}{c} O \\ \| \\ C- \end{array} \right)_n$ (b) $\left(\begin{array}{c} H \\ | \\ -N- \end{array} CH_2 - \begin{array}{c} O \\ \| \\ C- \end{array} \right)_n$

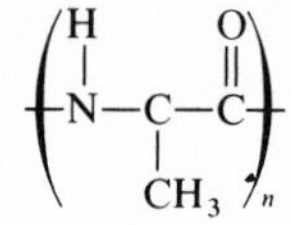

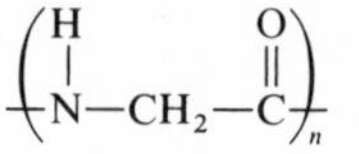

17.8. (a) $H_2NCH_2\overset{O}{\overset{\|}{C}}-NHCH_2\overset{O}{\overset{\|}{C}}-NHCH_2CO_2H + 2H_2O \longrightarrow 3H_2NCH_2\overset{O}{\overset{\|}{C}}-OH$
(b) hydrolysis (c) a tripeptide (d) 189

17.9. (a) Fischer projection: CO_2H (top), H (left), NH_2 (right), CH_3 (bottom)

(b) Fischer projection: CO_2H (top), H (left), NH_2 (right), H (bottom)

(c) Fischer projection: CO_2H (top), H_2N (left), H (right), $CH_2CH(CH_3)_2$ (bottom)

(d) Fischer projection: CO_2H (top), H_2N (left), H (right), CH_2OH (bottom)

17.10. (a) $H_2NCH(CH_2C_6H_5)C(=O)$—$NHCH(CH_3)CO_2H$

(b) $H_2NCH_2C(=O)$—$NHCH(CH(CH_3)_2)CO_2H$

17.11. Arg, His, Lys, and Trp have high isoelectric points. Asp, Glu, and Tyr have low isoelectric points.

17.12. (a), (b) are hydrophobic; (c), (d), (e) are hydrophilic.

17.13. (a) $H_2NCH_2C(=O)$—$NHCH(CH_3)CO_2H$

(Gly·Ala)

$H_2NCH(CH_3)C(=O)$—$NHCH_2CO_2H$

(Ala·Gly)

(b) $H_2NCH_2C(=O)$—$NHCH(CH_3)C(=O)$—$NHCH(CH_2C_6H_4OH)CO_2H$

(Gly·Ala·Tyr)

$H_2NCH_2C(=O)$—$NHCH(C_6H_4OH)C(=O)$—$NHCH(CH_3)CO_2H$

(Gly·Tyr·Ala)

$H_2NCH(CH_3)C(=O)$—$NHCH_2C(=O)$—$NHCH(C_6H_4OH)CO_2H$

(Ala·Gly·Tyr)

$H_2NCH(CH_3)C(=O)$—$NHCH(C_6H_4OH)C(=O)$—$NHCH_2CO_2H$

(Ala·Tyr·Gly)

$H_2NCH(C_6H_4OH)C(=O)$—$NHCH_2C(=O)$—$NHCH(CH_3)CO_2H$

(Tyr·Gly·Ala)

H_2N—$CH(C_6H_4OH)C(=O)$—$NHCH(CH_3)C(=O)$—$NHCH_2CO_2H$

(Tyr·Ala·Gly)

17.14. *First Degradation*:

$$C_6H_5\text{-N=C=S} + H_2N\underset{\underset{C_6H_5}{|}}{\underset{|}{C}H}(CH_2)\overset{O}{\overset{\|}{C}}\text{-NH}\underset{|}{C}H(CH_3)\overset{O}{\overset{\|}{C}}\text{-NHCH}_2CO_2H \xrightarrow{OH^-}$$

$$C_6H_5\text{-NH-}\overset{S}{\overset{\|}{C}}\text{-NHCH}(CH_2C_6H_5)\overset{O}{\overset{\|}{C}}\text{-NHCH}(CH_3)\overset{O}{\overset{\|}{C}}\text{-NHCH}_2CO_2H$$

$$\xrightarrow{H_3O^+} \text{(phenylthiohydantoin: } C_6H_5\text{-N, C=S, NH, C=O, CH-}C_6H_5\text{)} + H_2NCH(CH_3)\overset{O}{\overset{\|}{C}}\text{-NHCH}_2CO_2H$$

Second Degradation:

$$C_6H_5\text{-N=C=S} + H_2NCH(CH_3)\overset{O}{\overset{\|}{C}}\text{-NHCH}_2CO_2H$$

$$\xrightarrow{OH^-} C_6H_5\text{-NH-}\overset{S}{\overset{\|}{C}}\text{-NHCH}(CH_3)\overset{O}{\overset{\|}{C}}\text{-NHCH}_2CO_2H \xrightarrow{H_3O^+} \text{(phenylthiohydantoin: } C_6H_5\text{-N, C=S, NH, C=O, CHCH}_3\text{)} + H_2NCH_2CO_2H$$

17.15.

$$H_2NCH(CH_2C_6H_5)\overset{O}{\overset{\|}{C}}\text{-NHCH}(CH_3)\overset{O}{\overset{\|}{C}}\text{-NHCH}_2CO_2H \xrightarrow{\text{carboxypeptidase}} H_2NCH(CH_2C_6H_5)\overset{O}{\overset{\|}{C}}\text{-NHCH}(CH_3)\overset{O}{\overset{\|}{C}}\text{-OH} + H_2NCH_2CO_2H$$

$$\downarrow \text{carboxypeptidase}$$

$$H_2NCH(CH_2C_6H_5)\overset{O}{\overset{\|}{C}}\text{-OH} + H_2NCH(CH_3)CO_2H$$

17.16. Tyr·Cys·Lys·Ala·Arg·Arg·Gly

17.17.

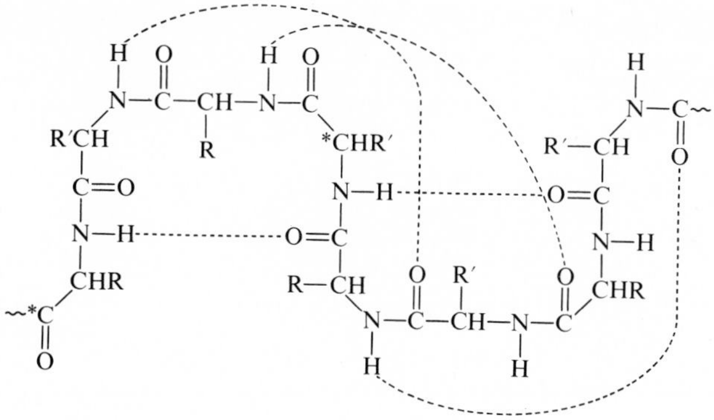

H O H O H
N—C—CH—N—C N—C~
R′CH R *CHR′ R′—CH O
C=O N—H O=C
N—H O=C N—H
CHR R—CH O R′ O CHR
~*C N—C—CH—N—C
O H H

17.18.

T—
T—
A—
C—
G—
A—
T—
A—
G—
C—
C—
A—

17.21. Melting point should have a wide range. Viscosity of its solution should be high.

17.23. (b)

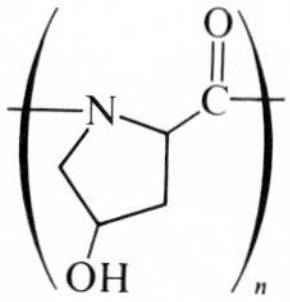

17.24. (b) + $n H_2O$ $\xrightarrow{H^+}$ H—N—C(=O)—OH

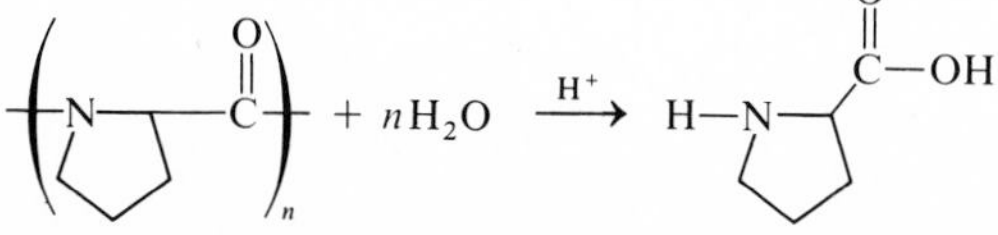

17.25. (b)

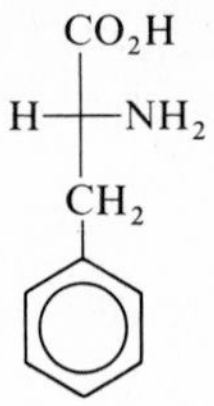

$$\begin{array}{c} CO_2H \\ H-\!\!\!+\!\!\!-NH_2 \\ CH_2 \\ | \\ C_6H_5 \end{array}$$

17.27. Ala, Asn, Cys, Cys-Cys, Gln, Gly, Hyp, Ile, Leu, Met, Phe, Pro, Ser, Thr, Val
17.29. 24
17.32. Ala·Ala·His·Arg·Glu·Lys·Phe·Ile
17.34. Hydrolysis of the peptide bond:

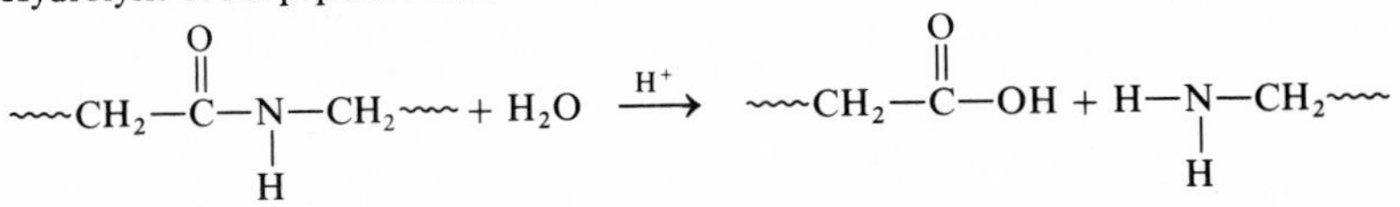

$$\sim CH_2-\overset{O}{\overset{\|}{C}}-\underset{H}{\underset{|}{N}}-CH_2\sim + H_2O \xrightarrow{H^+} \sim CH_2-\overset{O}{\overset{\|}{C}}-OH + H-\underset{H}{\underset{|}{N}}-CH_2\sim$$

CHAPTER 18
BEHAVIOR OF VERY LARGE MOLECULES: SYNTHETIC POLYMERS: AN ELECTIVE CHAPTER

18.1. (a) polyacrylonitrile, (b) poly(tetrafluoroethylene), (c) poly(methyl methacrylate)
18.2. It is macromolecular.
18.3. (a), (b), (c) additions; (d) condensation
18.4. The initiator attacks the monomer to form the more stable radical, cation, or anion.

18.5. (a) $\left(-O-CH_2CH_2\overset{O}{\overset{\|}{C}}-\right)_n$ (b) $\left(-\overset{H}{\overset{|}{N}}-CH_2CH_2\overset{O}{\overset{\|}{C}}-\right)_n$ (c) $\left(-C_6H_2(OH)(CH_3)-CH_2-\right)_n$ (OH and CH_3 para on the ring)

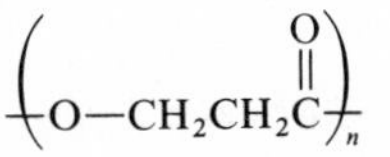

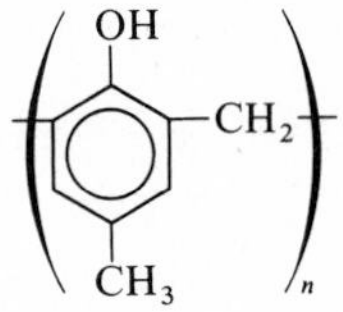

No.

18.6. Natural rubber is a 1,4- addition polymer:

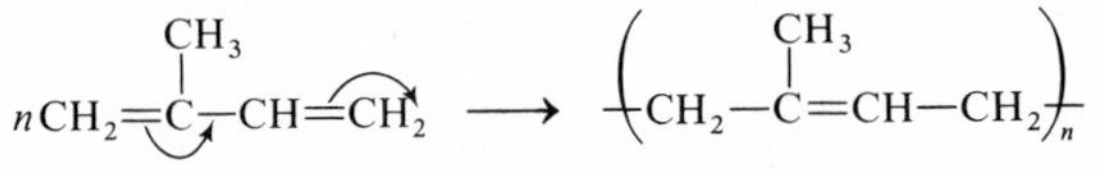

$$n\,CH_2=\overset{CH_3}{\overset{|}{C}}-CH=CH_2 \longrightarrow \left(-CH_2-\overset{CH_3}{\overset{|}{C}}=CH-CH_2-\right)_n$$

18.7. (a) $\sim CH_2-C(H)(Cl)-CH_2-C(H)(Cl)-CH_2-C(H)(Cl)-CH_2-C(H)(Cl)\sim$ (all Cl on the same side)

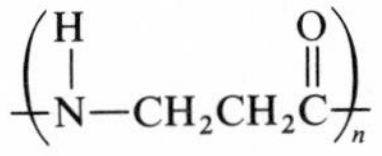

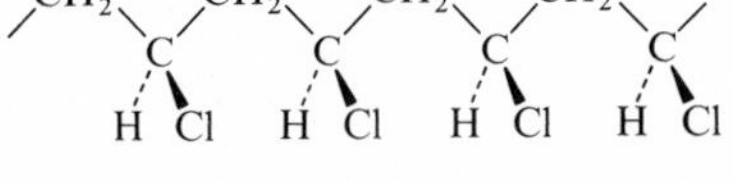

(b) $\sim CH_2-C(H)(OH)-CH_2-C(HO)(H)-CH_2-C(H)(OH)-CH_2-C(HO)(H)\sim$ (H OH HO H H OH HO H)

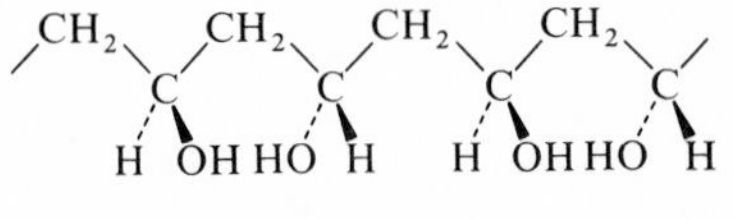

18.9. (b) $\left(-CH_2\underset{\underset{CH_3-C=O}{|}}{\underset{O}{\underset{|}{CH}}}-\right)_n$ (addition)

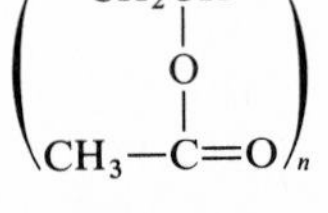

18.12. (a), (b), (c) addition; (d), (e) condensation

18.14. $Cl{-}H + CH_2{=}CH(C_6H_5) \longrightarrow Cl^- + CH_3{-}CH^+(C_6H_5)$

$CH_3{-}CH^+(C_6H_5) + CH_2{=}CH(C_6H_5) \longrightarrow CH_3{-}CH(C_6H_5)CH_2CH^+(C_6H_5) \longrightarrow \longrightarrow$ polymer

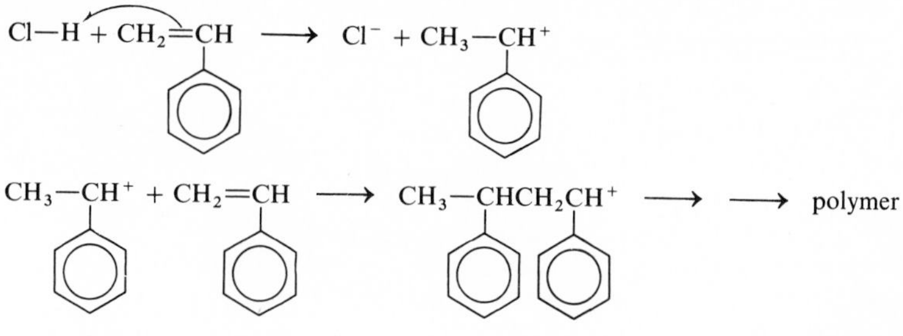

18.16. (b) 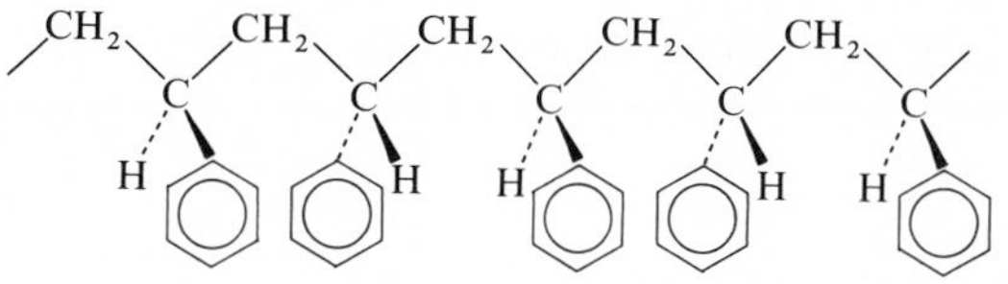 [structure: chain of CH_2 and C units, with H and phenyl substituents alternating on the C atoms]

CHAPTER 19
THE USE OF SPECTROSCOPY IN ORGANIC CHEMISTRY: AN ELECTIVE CHAPTER

19.1. 7.5×10^{-12} erg

19.2. $$\frac{(3.0 \times 10^{10}\ \text{cm/sec})}{(2.54\ \text{cm/in})(12\ \text{in/ft})(5280\ \text{ft/mi})} = 186{,}000\ \text{miles/sec}$$

19.3. The peak at 278 nm suggests a carbonyl group:

$CH_3CH_2CH_2\overset{O}{\overset{\|}{C}}H$, $CH_3\overset{H_3C}{\overset{|}{C}}H\overset{O}{\overset{\|}{C}}H$, $CH_3CH_2\overset{O}{\overset{\|}{C}}CH_3$

19.4. $CH_2{=}CHCH_2CH{=}CHCH_2CH{=}CH_2$

19.5. $HOCH_2CH_2CH_2CO_2H$, $CH_3\underset{OH}{\underset{|}{C}}HCH_2CO_2H$, $CH_3CH_2\underset{OH}{\underset{|}{C}}HCO_2H$, $CH_3\overset{CH_3}{\overset{|}{\underset{OH}{\underset{|}{C}}}}{-}CO_2H$,

$HO{-}\overset{CH_3}{\overset{|}{C}}HCH_2CO_2H$

19.6. frequency $= 6 \times 10^{13}$ cycles/sec
wave number $= 1/\lambda = 2000\ \text{cm}^{-1}$

19.7. (a) $C{\equiv}C$ or $C{\equiv}N$; (b) C—H, N—H, or O—H; (c) same as (b)

19.8. (a) $CH_3C{\equiv}N$; (b) it is a primary or secondary amine; (c) aromatic

19.9. $Br{-}C_6H_4{-}CH_3$ (para)

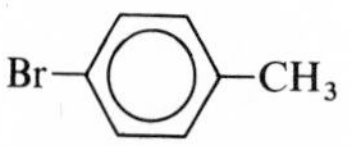

19.10. aromatic H atoms (7.0 ppm); methyl H atoms (2.3 ppm)

19.11. $Cl-\overset{H}{\underset{H}{C}}-\overset{H}{\underset{H}{C}}-Br$ with the lower H atoms labelled (1) (2)

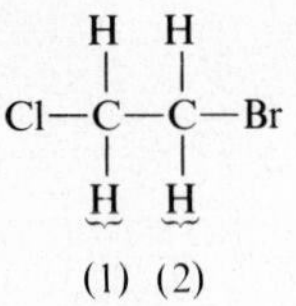

H(1) and H(2) split each other into two triplets. The triplet due to H(1) would occur at the higher chemical shift because Cl is more electronegative than Br.

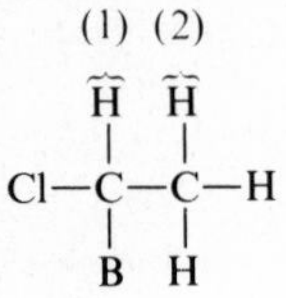

$Cl-\overset{H}{\underset{B}{C}}-\overset{H}{\underset{H}{C}}-H$ with the upper H atoms labelled (1) (2)

H(1) at higher chemical shift is a quartet; H(2) is a doublet.

19.12. $4.2 \times 10^{14}\ sec^{-1}$

19.14. $CH_3CH_2C{\equiv}CH$ or $CH_3C{\equiv}CCH_3$

19.16.

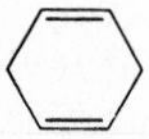

19.18. $CH_3\overset{CH_3}{\underset{CH_3}{C}}-OH$

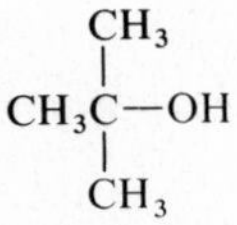

19.20. $-OCH_3$

19.22. $CH_3\underset{H}{C}=\underset{H}{C}\overset{O}{\overset{\|}{C}}-H$

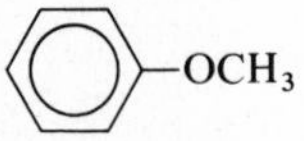

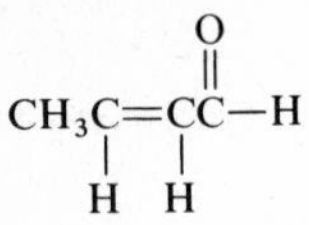

INDEX

A

B

C

D

E

F

G

H

I

K

L

M

N

O

S

T

U

V

W

X

Z